U0921012

中国高等植物

彩色图鉴

Higher Plants of China in Colour

《中国高等植物彩色图鉴》编委会 主编

Edited by
Editorial Committee of
Higher Plants of China in Colour

科学出版社
北京

内 容 简 介

本套图鉴精选中国境内野生高等植物和重要栽培植物1万余种，配以图片近2万张，每一物种以中英文形式简要介绍植物的中文名、拉丁学名、形态特征、花果期、生境和分布。图鉴共分为9卷，收载苔藓植物100科、蕨类植物40科、裸子植物11科、被子植物232科，共计383科，且除苔藓植物之外，已收全所有科。本套图鉴是继《中国高等植物图鉴》、《中国植物志》、Flora of China之后，又一部大型植物分类学巨著。本卷为第7卷。

本书适合植物学领域的科研人员、管理人员及爱好植物学的普通大众阅读和收藏。

This set of pictorial books contains nearly 20 thousand photographs, presenting the cream of wild higher plants and important cultivated plants in China, the species of which number more than 10 thousand. Each of the species is concisely introduced in both Chinese and English from such aspects as Chinese name, Latin name, morphological features, flowering and fruiting season, habitat and distribution. Divided into nine volumes, this work includes 100 bryophyte families, 40 pteridophyte families, 11 gymnosperm families and 232 angiosperm families, 383 families altogether; the inclusion of all the said families is complete except for the bryophytes. The set of pictorial books is another monumental work on plant taxonomy, after *Iconographia Cormophytorum Sinicorum*, *Flora Reipublicae Popularis Sinicae*, and *Flora of China*. This is volume Ⅶ of the series.

This work is intended for scientific researchers and administrators in the field of botany and also for botany enthusiasts. As well as for reading, the work can be a classic collection.

图书在版编目(CIP)数据

中国高等植物彩色图鉴＝Higher Plants of China in Colour. 第7卷，被子植物. 玄参科—菊科：汉英 /《中国高等植物彩色图鉴》编委会主编；陈又生分册主编.—北京：科学出版社，2016.1

ISBN 978-7-03-047068-3

Ⅰ. ①中… Ⅱ. ①中… ②陈… Ⅲ. ①高等植物-中国-图集 ②玄参科-中国-图集 ③菊科-中国-图集 Ⅳ. ①Q949.4-64

中国版本图书馆CIP数据核字（2016）第013511号

责任编辑：王 静 付 聪 马 俊 / 责任校对：张怡君 赵桂芬

责任印制：肖 兴 / 书籍设计：北京美光设计制版有限公司

科学出版社 出版

北京东黄城根北街16号

邮政编码：100717

http://www.sciencep.com

北京汇瑞嘉合文化发展有限公司 印刷

科学出版社发行 各地新华书店经销

*

2016年1月第 一 版 开本：787×1092 1/16

2016年1月第一次印刷 印张：43 3/4

字数：2 002 000

定价：648.00元

（如有印装质量问题，我社负责调换）

中国高等植物
彩色图鉴
Higher Plants of China in Colour
《中国高等植物彩色图鉴》编委会 主编
Edited by Editorial Committee of Higher Plants of China in Colour
第7卷 被子植物 玄参科－菊科
Volume Ⅶ Angiosperms Scrophulariaceae—Compositae
卷编辑 陈又生
Edited by Yousheng CHEN

《中国高等植物彩色图鉴》编委会

Editorial Committee of Higher Plants of China in Colour

Chairman of the Editorial Committee

Wentsai WANG

Vice Chairmen of the Editorial Committee

Shenghua WU Zhenyu LI

Members of the Editorial Committee (in the order of Chinese pinyin)

Bin CHEN Yousheng CHEN Xiao CHENG Yong FEI Xiaohua JIN
Xiwen LI Zhenyu LI Qinwen LIN Bing LIU Bo LIU
Yitao LIU Gang LU Ching-I PENG Haining QIN Guoda TAO
Wentsai WANG Shenghua WU Nianhe XIA Shengxiang YU Li ZHANG
Shuren ZHANG Xianchun ZHANG Qin ZUO

Chief Planner

Shenghua WU

Co-Planners

Yong FEI Guoda TAO Xiao CHENG Haining QIN Zhenyu LI Jing WANG
Yitao LIU Mu ZANG

Authors

Yong FEI Yitao LIU Bing LIU Guoda TAO Xiao CHENG
Bin CHEN Yousheng CHEN Liguo FU Guosheng HE Xiaohua JIN
Zhenyu LI Qinwen LIN Bo LIU Wentsai WANG Shenghua WU
Songzhi XU Yechun XU Shengxiang YU Xunlin YU Li ZHANG
Daigui ZHANG Shuren ZHANG Xianchun ZHANG Xinxin ZHU Qin ZUO

Secretaries to the Editorial Committee

Xiaohong ZHONG Songzhi XU Jiuqiong SUN

Responsible Editors

Jing WANG Cong FU Jun MA

Collaborators

Institute of Botany, Chinese Academy of Sciences
Kunming Institute of Botany, Chinese Academy of Sciences
South China Botanical Garden, Chinese Academy of Sciences
Fairylake Botanical Garden, Shenzhen & Chinese Academy of Sciences

丛书图片主要拍摄者

(按姓氏汉语拼音排序)

阿不都拉·阿巴斯 白鹭 白重炎 毕延超 邴艳红 曹同 车晋滇 陈彬 陈高
陈丽 陈庆 陈鑫 陈炳华 陈世品 陈贤兴 陈又生 陈志雄 成晓 程文达
迟敏杰 戴攀峰 邓涛 邓云飞 丁炳扬 丁学欣 董仕勇 杜诚 杜巍 杜玉芬
段长虹 段士民 方振东 方振兴 费勇 冯君茹 傅连中 甘啟良 高贤明 高信芬
高云东 葛斌杰 耿玉英 古训铭 顾余兴 管开云 郭世伟 韩国营 郝加琛 郝云庆
何海 何理 何春梅 何国生 和兆荣 侯元同 胡光万 胡国雄 华国军 黄健
黄江华 黄圣卓 黄向旭 黄俞淞 惠肇祥 季定乾 贾渝 姜林 蒋宏 蒋蕾
蒋日红 金伟涛 金孝锋 金效华 康世昌 赖阳均 郎楷永 黎斌 黎兴江 李东
李恒 李凯 李敏 李攀 李不言 李策宏 李东辉 李家湘 李建民 李剑武
李良千 李文军 李先源 李晓东 李小杰 李新华 李新伟 李学东 李泽贤 李振宇
李志奇 李智选 李中阳 梁同军 廖明林 廖云标 林敏 林祁 林维 林广旋
林建勇 林俊杰 林茂祥 林秦文 林哲丽 刘冰 刘博 刘静 刘军 刘坤
刘夙 刘翔 刘鑫 刘演 刘莹 刘大伟 刘光裕 刘海桑 刘红梅 刘伦辉
刘全儒 刘晟源 刘怡涛 刘正宇 刘宗才 柳永红 卢刚 卢元 罗柳青 骆适
吕碧凤 吕志学 马林 马炜梁 马文章 马欣堂 莫水松 牟善杰 沐先运 慕泽泾
南程慧 倪静波 倪素碧 农东新 潘勃 潘建斌 彭博 彭镜毅 彭日成 乔明明
秦卫华 秦祥堃 覃海宁 邱志敬 仁琛 任飞 任丽华 任明波 任昭杰 尚策
邵剑文 沈阳肇 施忠辉 石硕 寿海洋 税玉民 宋纬文 宋柱秋 买买提明·苏来曼
苏丽飞 苏享修 孙航 孙苗 孙观灵 孙明洲 孙卫邦 孙小美 谭运洪 陶国达
田乾福 田新民 童毅 童毅华 汪远 王辰 王东 王泓 王晖 王健
王进 王强 王颖 王耘 王喆 王长荣 王钧杰 王慷林 王清隆 王秋美
王文卿 王雅琼 王亚玲 王英伟 王玉兵 王正元 王祝年 韦宏金 韦毅刚 韦玉梅
卫然 魏来 温韩东 温九良 翁茂伦 吴丰 吴磊 吴双 吴棣飞 吴凤琴
吴光弟 吴国晞 吴林芳 吴声华 吴望辉 吴问舫 吴永红 吴增源 伍凯 武全安
武素功 武玉东 夏念和 肖翠 肖亮 肖艳 肖红菊 谢磊 辛夷 辛晓伟
辛益群 辛宇明 熊源新 徐徭 徐锦泉 徐克学 徐连升 徐申健 徐文斌 徐晔春
徐永福 许为斌 寻路路 严新富 严岳鸿 阳文静 杨浩 杨永 杨成梓 杨建昆
杨金财 杨科明 杨青山 杨世雄 杨奕绯 杨增宏 姚永飚 叶德平 叶建飞 叶喜阳
叶幸儿 易思荣 殷建涛 尹志坚 于胜祥 郁文彬 喻勋林 袁彩霞 曾孝濂 曾云保
张力 张良 张强 张伟 张莹 张勇 张彩飞 张重岭 张代贵 张凤秋
张海华 张宏伟 张金龙 张金政 张守君 张淑梅 张树仁 张维柱 张宪春 张霄林
张志翔 赵宏 赵伟 赵大昌 郑宝江 郑希龙 郑小明 钟智明 周繇 周重建
周海成 周家宝 周兰平 周喜乐 周小林 周浙昆 朱弘 朱大海 朱仁斌 朱淑霞
朱维明 朱鑫鑫 David E. Boufford Dmitry Sokoloff Jan Thomas Johansson
Jozef Lemmens Kirill Tkachenko Pavel Novák Ralf Knapp Richard Ree
Susan Kelley

香港植物标本室(免费提供)

Major Photographers of the Series

(in the order of Chinese pinyin)

Abdulla ABASI	Lu BAI	Chongyan BAI	Yanchao BI	Yanhong BING	Tong CAO
Jindian CHE	Bin CHEN	Gao CHEN	Li CHEN	Qing CHEN	Xin CHEN
Binghua CHEN	Shipin CHEN	Xianxing CHEN	Yousheng CHEN	Chihhsiung CHEN	Xiao CHENG
Wenda CHENG	Minjie CHI	Panfeng DAI	Tao DENG	Yunfei DENG	Bingyang DING
Xuexin DING	Shiyong DONG	Cheng DU	Wei DU	Yufen DU	Changhong DUAN
Shimin DUAN	Zhendong FANG	Zhenxing FANG	Yong FEI	Junru FENG	Lianzhong FU
Qiliang GAN	Xianming GAO	Xinfen GAO	Yundong GAO	Binjie GE	Yuying GENG
Xunming GU	Yuxing GU	Kaiyun GUAN	Shiwei GUO	Guoying HAN	Jiachen HAO
Yunqing HAO	Hai HE	Li HE	Chunmei HE	Guosheng HE	Zhaorong HE
Yuantong HOU	Guangwan HU	Guoxiong HU	Guojun HUA	Jian HUANG	Jianghua HUANG
Shengzhuo HUANG	Xiangxu HUANG	Yusong HUANG	Zhaoxiang HUI	Dingqian JI	Yu JIA
Lin JIANG	Hong JIANG	Lei JIANG	Rihong JIANG	Weitao JIN	Xiaofeng JIN
Xiaohua JIN	Shihchang KANG	Yangjun LAI	Kaiyong LANG	Bin LI	Xingjiang LI
Dong LI	Heng LI	Kai LI	Min LI	Pan LI	Buyan LI
Cehong LI	Donghui LI	Jiaxiang LI	Jianmin LI	Jianwu LI	Liangqian LI
Wenjun LI	Xianyuan LI	Xiaodong LI	Xiaojie LI	Xinhua LI	Xinwei LI
Xuedong LI	Zexian LI	Zhenyu LI	Zhiqi LI	Zhixuan LI	Zhongyang LI
Tongjun LIANG	Minglin LIAO	Yunbiao LIAO	Min LIN	Qi LIN	Wei LIN
Guangxuan LIN	Jianyong LIN	Junjie LIN	Maoxiang LIN	Qinwen LIN	Zheli LIN
Bing LIU	Bo LIU	Jing LIU	Jun LIU	Kun LIU	Su LIU
Xiang LIU	Xin LIU	Yan LIU	Ying LIU	Dawei LIU	Guangyu LIU
Haisang LIU	Hongmei LIU	Lunhui LIU	Quanru LIU	Shengyuan LIU	Yitao LIU
Zhengyu LIU	Zongcai LIU	Yonghong LIU	Gang LU	Yuan LU	Liuqing LUO
Shi LUO	Pifong LU	Zhixue LÜ	Lin MA	Weiliang MA	Wenzhang MA
Xintang MA	Shuisong MO	Shannjye MOORE	Xianyun MU	Zejing MU	Chenghui NAN
Jingbo NI	Subi NI	Dongxin NONG	Bo PAN	Jianbin PAN	Bo PENG
Ching-I PENG	Richeng PENG	Mingming QIAO	Weihua QIN	Xiangkun QIN	Haining QIN
Zhijing QIU	Chen REN	Fei REN	Lihua REN	Mingbo REN	Zhaojie REN
Ce SHANG	Jianwen SHAO	Yangzhao SHEN	Zhonghui SHI	Shuo SHI	Haiyang SHOU

Yumin SHUI	Weiwen SONG	Zhuqiu SONG	Mamtimin SULAYMAN	Lifei SU	Xiangxiu SU
Hang SUN	Miao SUN	Guanling SUN	Mingzhou SUN	Weibang SUN	Xiaomei SUN
Yunhong TAN	Guoda TAO	Qianfu TIAN	Xinmin TIAN	Yi TONG	Yihua TONG
Yuan WANG	Chen WANG	Dong WANG	Hong WANG	Hui WANG	Jian WANG
Jin WANG	Qiang WANG	Ying WANG	Yun WANG	Zhe WANG	Changrong WANG
Junjie WANG	Kanglin WANG	Qinglong WANG	Chiumei WANG	Wenqing WANG	Yaqiong WANG
Yaling WANG	Yingwei WANG	Yubing WANG	Zhengyuan WANG	Zhunian WANG	Hongjin WEI
Yigang WEI	Yumei WEI	Ran WEI	Lai WEI	Handong WEN	Jiuliang WEN
Maolun WENG	Feng WU	Lei WU	Shuang WU	Difei WU	Fengqin WU
Guangdi WU	Guoxi WU	Linfang WU	Shenghua WU	Wanghui WU	Wenfang WU
Yonghong WU	Zengyuan WU	Kai WU	Quan'an WU	Sugong WU	Yudong WU
Nianhe XIA	Cui XIAO	Liang XIAO	Yan XIAO	Hongju XIAO	Lei XIE
Yi XIN	Xiaowei XIN	Yiqun XIN	Yuming XIN	Yuanxin XIONG	Yao XU
Jinquan XU	Kexue XU	Liansheng XU	Shenjian XU	Wenbin XU	Yechun XU
Yongfu XU	Weibin XU	Lulu XUN	Hsinfu YEN	Yuehong YAN	Wenjing YANG
Hao YANG	Yong YANG	Chengzi YANG	Jiankun YANG	Jincai YANG	Keming YANG
Qingshan YANG	Shixiong YANG	Yifei YANG	Zenghong YANG	Yongbiao YAO	Deping YE
Jianfei YE	Xiyang YE	Xing'er YE	Sirong YI	Jiantao YIN	Zhijian YIN
Shengxiang YU	Wenbin YU	Xunlin YU	Caixia YUAN	Xiaolian ZENG	Yunbao ZENG
Li ZHANG	Liang ZHANG	Qiang ZHANG	Wei ZHANG	Ying ZHANG	Yong ZHANG
Caifei ZHANG	Chongling ZHANG	Daigui ZHANG	Fengqiu ZHANG	Haihua ZHANG	Hongwei ZHANG
Jinlong ZHANG	Jinzheng ZHANG	Shoujun ZHANG	Shumei ZHANG	Shuren ZHANG	Weizhu ZHANG
Xianchun ZHANG	Xiaolin ZHANG	Zhixiang ZHANG	Hong ZHAO	Wei ZHAO	Dachang ZHAO
Baojiang ZHENG	Xilong ZHENG	Xiaoming ZHENG	Zhiming ZHONG	You ZHOU	Chongjian ZHOU
Haicheng ZHOU	Jiabao ZHOU	Lanping ZHOU	Xile ZHOU	Xiaolin ZHOU	Zhekun ZHOU
Hong ZHU	Dahai ZHU	Renbin ZHU	Shuxia ZHU	Weiming ZHU	Xinxin ZHU
David E. Boufford	Dmitry Sokoloff	Jan Thomas Johansson		Jozef Lemmens	Kirill Tkachenko
Pavel Novák	Ralf Knapp	Richard Ree	Susan Kelley		

Hong Kong Herbarium (free of charge)

丛书文字主要编写者

(按姓氏汉语拼音排序)

陈世龙　陈又生　陈之端　成　晓　崔逸群　邓云飞　杜　宁　段士民　樊　杰　方瑞征
费　勇　傅立国　高　凡　高　乞　谷粹芝　郭　慧　韩　宇　郝加琛　何　理　侯学良
侯元同　胡光万　黄向旭　黄俞淞　蒋　宏　金孝锋　金效华　赖阳均　雷立公　黎　斌
李　恒　李　嵘　李秉滔　李宏哲　李剑武　李梦华　李文军　李锡文　李晓贤　李新华
李新伟　李章海　李振宇　李中阳　廖文波　林秦文　刘　冰　刘　博　刘　演　刘大伟
刘海桑　刘全儒　刘衍男　卢金梅　马欣堂　潘　勃　覃海宁　邱志敬　萨　仁　尚　策
税玉民　孙　苗　孙久琼　万　涛　王　东　王　晖　王　健　王　强　王德艺　王文采
王英伟　卫　然　魏　来　吴　磊　吴鹏程　吴声华　向巧萍　向小果　谢　磊　徐松芝
徐晓婷　薛大伟　闫瑞亚　严岳鸿　杨世雄　叶建飞　游旨价　于胜祥　袁　慷　张　力
张　梅　张　强　张重岭　张钢民　张红瑞　张树仁　张宪春　张志翔　赵　宏　赵存峰
周兰平　左　勤

Major Textwriters of the Series

(in the order of Chinese pinyin)

Shilong CHEN	Yousheng CHEN	Zhiduan CHEN	Xiao CHENG	Yiqun CUI
Yunfei DENG	Ning DU	Shimin DUAN	Jie FAN	Ruizheng FANG
Yong FEI	Liguo FU	Fan GAO	Qi GAO	Cuizhi GU
Hui GUO	Yu HAN	Jiachen HAO	Li HE	Xueliang HOU
Yuantong HOU	Guangwan HU	Xiangxu HUANG	Yusong HUANG	Hong JIANG
Xiaofeng JIN	Xiaohua JIN	Yangjun LAI	Ligong LEI	Bin LI
Heng LI	Rong LI	Bingtao LI	Hongzhe LI	Jianwu LI
Menghua LI	Wenjun LI	Xiwen LI	Xiaoxian LI	Xinhua LI
Xinwei LI	Zhanghai LI	Zhenyu LI	Zhongyang LI	Wenbo LIAO
Qinwen LIN	Bing LIU	Bo LIU	Yan LIU	Dawei LIU
Haisang LIU	Quanru LIU	Yannan LIU	Jinmei LU	Xintang MA
Bo PAN	Haining QIN	Zhijing QIU	Ren SA	Ce SHANG
Yumin SHUI	Miao SUN	Jiuqiong SUN	Tao WAN	Dong WANG
Hui WANG	Jian WANG	Qiang WANG	Deyi WANG	Wentsai WANG
Yingwei WANG	Ran WEI	Lai WEI	Lei WU	Pengcheng WU
Shenghua WU	Qiaoping XIANG	Xiaoguo XIANG	Lei XIE	Songzhi XU
Xiaoting XU	Dawei Xue	Ruiya YAN	Yuehong YAN	Shixiong YANG
Jianfei YE	Zhijia YOU	Shengxiang YU	Qian YUAN	Li ZHANG
Mei ZHANG	Qiang ZHANG	Chongling ZHANG	Gangmin ZHANG	Hongrui ZHANG
Shuren ZHANG	Xianchun ZHANG	Zhixiang ZHANG	Hong ZHAO	Cunfeng ZHAO
Lanping ZHOU	Qin ZUO			

丛书前言

中国是世界上植物最丰富的国家之一，已知有三万五千多种野生和重要栽培的高等植物，其中特有种达一万五千多种，形成复杂而独具特色的植物区系。中国的先人们创造了古老而辉煌的农业文明，选育出水稻、大豆、茶、枣、桃、柿等重要作物，其中水稻的栽培历史可追溯到约七千年前新石器时期的河姆渡文化，如今稻米已成为世界上近一半人口的粮食。丰富的植物资源和灿烂的历史文化，使中国成为“花园之母”和世界农作物七大起源中心之一。

中国植物学家为了系统地展示中国植物的多样性，历经艰辛，相继编研了《中国高等植物图鉴》和《中国植物志》，并与外国专家合作出版Flora of China等大型志书，这些著作在国内外应用广泛、影响巨大，客观地展现了不同时期的植物分类学研究和植物资源调查的成果，成为植物分类学领域最重要的大型经典著作。但是，它们都有一个共同的缺憾，即仅有黑白线条图，难以充分表达植物各器官的质地和颜色等自然状态下的外貌特征，其效果难以满足部分读者鉴赏植物的需要。

大多数发达国家都有自己的植物彩色图鉴，这些图鉴不仅展示了本国的生物多样性，还兼备工具书功能和富有感染力的艺术效果，具有很高的应用和收藏价值。迄今为止，国内出版的植物彩色图书多为地区性的，或局限于某一类植物的，如观赏植物、栽培作物和药用植物。作为世界生物多样性大国，中国应当拥有一套全面体现本国野生植物多样性的的大型鉴赏类彩色图册。

将灿烂的瞬间变为永恒是广大植物爱好者和摄影爱好者的追求。为了填补上述空白，台湾吴声华研究员策划并启动了这项工作。在海峡两岸学者的共同努力下，本书的规模在不断扩大，从最初的云南植物写真集扩展到全国性大型彩色植物图鉴。中国科学院植物研究所王文采院士出任丛书编委会主任，吴声华研究员和中国科学院植物研究所李振宇研究员任副主任。编委会遴选国内从事植物分类学研究的专家担任各卷卷编辑，邀请中国大陆、台湾和香港近200位植物学家承担各科的编写和审稿工作，卷编辑在专家审稿的基础上，再次对本卷内容进行核查。近400位摄影作者提供了大量精美的植物彩色照片。丛书还采用了著名的动植物科学画大师曾孝濂先生绘制的20余幅优雅而灵动的彩色图片。

本丛书划分为九卷，共收录中国高等植物1万余种，种类以野生植物为主，同时收载重要的栽培植物，精选图片近2万张。本丛书中科的系统排列如下：苔藓植物主要参考《中国苔藓志》中的系统；蕨类植物按张宪春2015年在《石松类和蕨类名词及名称》提出的系统；裸子植物和被子植物的系统排列按第尔斯(L. Diels, 1936)于A. Engler's Syllabus der Pflanzenfamilien中采用的系统。仅第三卷将毛茛科分为星叶草科、毛茛科和芍药科，将木兰科分为木兰科、八角科、五味子科和水青树科。全书收载中国高等植物383科，其中苔藓植物100科，占全国苔藓科总数的大多数；其余是蕨类植物40科，裸子植物11科，被子植物232科，分别代表了国产三大门类所有的科。本丛书收载的植物中有一些是Flora of China出版后发表的国产新种，如香港鹅耳枥(*Carpinus insularis*)、球柱楼梯草(*Elatostema globosostigmatum*)和西藏小囊兰(*Micropera tibetica*)，以及中国分布新记录，如轮叶三棱栎(*Trigonobalanus verticillata*)和格力兜兰(*Paphiopedilum gratrixianum*)。

为了方便更多的读者阅读，本丛书的文字采用中英文，简要介绍各种植物的中文名、拉丁学名、形态特征、花果期、生境和分布。

本书在编写过程中，承中国科学院植物研究所中国植物图像库和中国自然标本馆提供了许多方便和帮助，在此向他们表示衷心的感谢。

感谢国家出版基金和科学出版社对本丛书出版的大力支持。

由于编著者的业务水平有限、错漏之处，欢迎批评指正。

《中国高等植物彩色图鉴》编委会

2015年10月31日

Preface to the Series

As one of the countries with the richest diversity of plant species in the world, China has more than 35 000 known species of wild and important cultivated higher plants, among which there are over 15 000 endemic species, forming a complex and unique flora. The ancestors of the Chinese people created an ancient and splendid agricultural civilization. They selected and cultivated significant crops like rice, soya bean, tea, jujube, peach, and persimmon. Among these crops, the cultivation history of rice can be traced back to the Hemudu culture of the Neolithic Period around 7000 years ago. Nowadays, rice has become the staple food for nearly half of the world's population. With abundant plant resources and a long history and great culture, China is renowned as 'the mother of gardens' and is one of the seven important centers of origin for crops in the world.

In order to present the diversity of China's plants systematically, botanists from China have made pain-taking efforts to compile a series of large-volume floras including *Iconographia Cormophytorum Sinicorum*, *Flora Reipublicae Popularis Sinicae* (Chinese version) by themselves, and *Flora of China* (English version) with the collaboration of international specialists. These books are well known both in China and abroad and have been used extensively for studying Chinese plants and plants from adjacent areas, these works present the results of plant taxonomic study and study of plant resources in China at different periods, and constitute some of the most important large classic volumes in the field of plant taxonomy. However, in all these works the plants are only partly illustrated by black and white line-drawings, unable to present the texture and colour of flowers and leaves fully in their natural state, and they hardly reveal the spectacular beauty and fascination of the wealth of plant species.

Most developed countries have colour pictorial books of their plants, which form greatly desirable works, because they are not only a presentation of the plant diversity of the countries, but are also an attractive record of the beauty of the nature. So far, most of the Chinese colour pictorial books of plants are regional treatments, or concentrate on particular groups, such as ornamental plants, cultivated crops and medicinal plants or certain taxonomic groups. As a country with a high level of biodiversity, China merits a large-scale colour pictorial book with high appreciation value featuring the wild plants that occur within its territory.

It is the goal of every lover of plants and plant photography to capture the essence of plant beauty and make it permanent. In order to fill the above-mentioned gap, Professor Shenghua WU from Taiwan, planned and launched the present project. With the joint effort of specialists from all over China, the scale of the book has expanded from the initial pictorial book of plants of Yunnan to a many-volume colour pictorial book of plants of the whole country. Academician Wentsai WANG of the Institute of Botany, Chinese Academy of Sciences, took up the post of the chairman of the editorial committee, and the positions of vice chairmen of the editorial committee were assumed by Prof. Shenghua WU, Taiwan, and Prof. Zhenyu LI of the Institute of Botany, Chinese Academy of Sciences. The editorial committee then selected experienced plant taxonomists as volume editors for each volume, and invited nearly 200 botanists from mainland China, Taiwan and Hong Kong to undertake the compilation and reviewing work for each plant family by volumes. Nearly 400 photographers provided numerous beautiful full colour plant photos. The well known zoological and botanical artist, Xiaolian ZENG, kindly allowed the use of more than twenty of his elegant and vivid plant portraits in this series.

This whole work is divided into nine volumes, depicting more than 10 thousand species of higher plants from China, dealing mainly with wild plants, but also including some important cultivated plants, and has involved the careful selection of nearly 20 thousand photographs. The system arrangement for plant families are as follows: bryophytes are mainly arranged according to the system used in *Flora Bryophytorum Sinicorum*; pteridophytes are arranged according to the system proposed by Professor Xianchun ZHANG in *A Glossary of Terms and Names of Lycopods and Ferns* (2015); gymnosperms and angiosperms are arranged according to the system used in *A. Engler's Syllabus der Pflanzenfamilien* (L. Diels, 1936), with the difference that in Volume III, Ranunculaceae is divided into Circaesteraceae, Ranunculaceae, and Paeoniaceae, and Magnoliaceae is divided into Magnoliaceae, Illiciaceae, Schisandraceae, and Tetracentraceae. The higher plants of China included in this work comprise 383 families; with 100 families of bryophytes, which represent the majority of the bryophyte families in China; the others are 40 pteridophyte families, 11 gymnosperm families and 232 angiosperm families, which represent all the families distributed in China respectively. The work includes some new additions of species published since *Flora of China*, such as *Carpinus insularis*, *Elatostema globosostigmatum*, *Micropera tibetica*, and new distribution records for China, such as *Trigonobalanus verticillata* and *Paphiopedilum gratrixianum*.

关于本图鉴

1995年夏天，我参加由中国科学院昆明植物研究所臧穆教授带领的云南野外工作，同行的还有国际真菌学会理事长德籍的Franz Oberwinkler教授与法国的学者。臧教授爽朗好客，外国人都喜欢他的热情。那年去丽江，再去南部的西双版纳。西双版纳热带植物园的陶国达先生带领我们的野外工作，他是当地植物鉴定首席专家，知道好的树林在何处。一天，在傣族传统农家的木架房子吃中饭。臧教授建议陶先生既然喜欢摄影，何不出一本版纳植物图鉴，问我能不能帮忙在台湾找出版。我答应回去问问。

先问自然科学博物馆的李家维馆长，他对植物研究及保育充满热诚，对这项工作有兴趣。但未久他感觉这项工作所需时间过久，博物馆经费也不足以出版。我又问其他出版公司，没有得到响应。我想应该先有成果再问出版吧，就请陶先生持续植物拍摄。臧教授和夫人黎兴江教授推荐了费勇帮忙这项工作。1997年夏天，我在昆明机场与臧教授和费勇会合，一同飞去版纳。费勇年纪与我相当，长得瘦黑，话不太多。陶先生带领我们野外工作。回程时费勇说他想找几个同事一起负责滇西北的植物拍摄工作，与陶先生滇南的工作结合成为云南植物图鉴。回台后看陶先生给我的幻灯片，感觉质量不是太好，问他才知道所用的相机是正牌，镜头却是小厂牌。我汇钱请他购买一套相机，以利拍摄质量。

1998年，我到昆明植物所，和陶国达、费勇及孙航，讨论植物图鉴工作。费勇对此工作充满兴趣，人缘也好，决定由他征集昆明植物所人员拍摄的植物照片，并且中、英文字也由他撰写。翌年臧穆夫妇介绍昆明植物所的著名画家曾孝濂先生。曾先生长期考察云南山野的植物与动物，画作结合了科学性与艺术美感，是中国写实花鸟画得最好的。

2000年，费勇在日本富山县中央植物园半年，其间拍摄植物园栽培的中国植物。那年秋天费勇带我去大理点苍山和楚雄紫溪山。一天，我们在大理古城一间白族旧庭院吃风味晚餐，他兴致好，畅所欲言。费勇起初给我的印象是有些木讷，几次往来后就把我当熟人。几次的讨论，感觉他满心想做好这件事，并不在意条件。大理巷弄中有摊贩卖当地特产乳扇，他说闺女爱吃，买了两大张带走。

2001年年初一个早上，臧教授发来邮件，通知我费勇前一日在丽江不幸去世。一个年轻健康生命的突然离去，令人难以承受。出席完上午的会议后即打电话到昆明。黎教授说费勇到丽江出差，半夜室友听到声响，见他口吐白沫急送医院。地方医院初以为是癫痫，到清晨就不治了。

几个月后我有事联络曾孝濂先生，他告诉我费勇太太想与我联系。费勇太太姓向，我们称小向。她电话中希望植物图鉴工作能够继续，而且费勇的几个同事愿意帮忙。当年夏天在昆明的一个晚上，小向同昆明植物所的成晓、孙航、周浙昆一起和我见面，商讨后续的工作。成晓说他与费勇是同学，同时毕业，同时上班，他一定会帮忙。他确实尽力后续工作的联系与推动。2002年在昆明，几个朋友见面，小向带初中的女儿同来。女儿乖巧懂事，我说长得像费勇，她眼眶微微红了。成晓研究蕨类，他的岳父武素功先生及岳母方瑞征女士也是昆明植物所学者，两位在图片提供及文稿修订均提供协助。昆明植物所李锡文教授对植物分类的造诣比较全面，负责图片和文稿审查。昆明植物所还有多位专家对本书工作做出贡献，不在此逐一罗列。

2001年，曾孝濂先生介绍昆明植物所的画家刘怡涛先生。刘先生在版纳热带植物园待过，建立独特的版纳风光绘画风格，也喜好摄影，带过我几次野外工作。他建议我把植物图鉴工作扩大到全中国。艺术家天生具有美感，曾、刘两位画家拍摄的植物图，构图与取景皆有独到之处。2003年我到河北与吉林进行野外工作，2004年到新疆与吉林时决定把植物图鉴范围扩大到全中国。我和小向说明书的分量和质量要到位，才能彰显费勇的努力精神。费勇原本即有中国植物图鉴的梦想，干脆一次到位。

2004年，中国科学院植物研究所覃海宁博士来台，我们是1994年在英国邱园认识的。中国植物图像库在海宁领导下建立得有声有色。海宁总是满脸笑容，热诚谦虚，听我说植物图鉴的事，立即寻思找人帮忙。他人面广，介绍不少人，拍摄较好的有福建的何国生、四川的吴光弟、广西的刘演、广东的李泽贤。刘演的图片色彩饱满令人赞叹。我去爱丁堡皇家植物园时知道David Chamberlain博士是杜鹃花科专家，他同意审查杜鹃花科及小檗科图片。彭镜毅介绍哈佛大学David E. Boufford博士，他的图片是从中国西南的横断山脉植物调查工作所拍摄。David又推荐Susan Kelley及Richard Ree提供植物图片。

中科院植物所吴鹏程教授是苔藓专家，1990年我在芬兰赫尔辛基大学即将取得博士学位时他在赫大待了几个月。吴教授介绍几位中科院植物所的专家帮忙图片审查及文字撰写。台湾真菌学前辈吕理燊博士介绍昆明市农业局副局长惠肇祥先生提供杂草图片，惠先生又介绍北京的车晋滇先生提供华北的杂草图片。台湾赖明洲教授介绍上海自然博物馆的秦祥堃先生提供华东植物图片，又介绍中国科学院沈阳应用生态研究所的赵大昌先生提供长白山植物图片。我2004年到乌鲁木齐开会，组织会议的新疆大学阿不都拉教授拍了不少新疆植物图片，也提供给我。

大学同学康世昌是植物及计算机高手，拍摄的植物图片也提供给我。他早预想到网络世界的影响力，不推荐大部头实体书的出版构想。多年前他写个网址要我去看，那是我不知道的“Google”，可以查询信息。网络上图片的数量越来越多，趋势是如此。我在芬兰的指导教授Tuomo Niemelä出版过大型真菌的小书，亲自编排，图片与文字搭配得美感十足，我每翻阅总是心情愉悦。我向Tuomo请教对这套植物实体书的意见。他说网络的数据有时会消失，且许多没经过审查。我想这套书终要完成，无法顾及趋势与新世代人类的想法。

2007年年初，我在网络发现中科院植物所的中国植物图像库有影像部分。负责的是李敏，我问他图片提供者，他推荐几位拍摄较好的。多数是中科院植物所的年轻人，有刘冰、林秦文、于胜祥、李敏、高贤明、郈艳红，还有陕西的王䓨。我当时已收集中国植物5000种的图片，工作超过10年理应收尾。然不加入这批有许多北方植物的图片实在不舍。刘冰是植物分类奇葩，这么年轻就拍到数量惊人的植物图片。刘冰和刘怡涛是给这项工作提供图片最多的两位。刘博帮忙不少文字撰写及图片审查，工作积极。当年年底，图片收集到6500种以上，接着准备文字、图片审查等出书的各项工作。

2010年在台湾“中央研究院”召开一项研讨会，覃海宁和李敏也来了，他们的报告显示中国植物图像库已收到数十万张图片。我如果再搜寻一次图片，能收到更好及更多的图片，但面对许多人殷盼这套书问世，时间的延长，压力更大。终究，我相信费勇会支持这最后一批图片的征求。湖南喻勋林及张代贵两位教授寄来许多华中植物图片，浙江张宏伟先生及安徽施忠辉先生也送来图片。吉林通化的周繇教授寄来他辛苦拍摄的长白山植物图片。近三年送来较多图片的还有朱鑫鑫、陈又生、徐晔春、陈彬、陈世品、何海、周喜乐，以及蕨类的张宪春和兰科的金效华。好友张力负责苔藓部分。

“中央研究院”彭镜毅博士提供了许多秋海棠科图片，也修订这科的文稿。牟善杰是台湾的蕨类学家，提供一些蕨类图片给本书，也审查过蕨类图片及文字。我在台湾大学念博士时，善杰是大学生，见他圆圆的笑脸，成天在标本馆研究。2010年11月，44岁的他突然中风走了，令人感慨！吕碧凤小姐是台湾优秀的业余蕨类专家，提供一些好的蕨类图片。还要感谢提供及审查图片的几位同事：王秋美、陈志雄、胡维新、黄俊霖、严新富和邱少婷。

早期收到的是幻灯片及少数印好的照片，2005年以后送来的是数码影像。数码图片干净，缺点是饱和度、清晰度和锐利度表现稍差，绿色部分有时偏黄。图效调整可改善这些问题。幻灯片的影像则会受到底片、冲洗、保存、扫描等质量的影响，好质量的并不多。图片须裁切出重点部位，再调整影像效果，这些工作大多是我处理。商请到一批人分别撰写文字。虽然有范本给撰写人参考，但各人的写法与仔细程度难免不一，有疏漏或小错误的情形普遍存在。起初我自己参考文献逐一查核，修订了约两千种的文字，但工作量太大，无法继续亲为。文字工作贡献较多的有费勇、刘博、萨仁、谷粹芝、成晓、杜宁、李锡文、徐晓婷和方瑞征等。

吴鹏程教授与科学出版社生物分社社长王静女士提及这项工作，王静有兴趣了解出版的可能，我们2010年在北京见面。自己过于深入这项工作，甚至如排版形式、字形等都亲自研究。像是自己养大的小孩，不放心交给他人处理，而且书稿已经在台湾找设计公司开始排版了。王静有毅力，持续两年逐渐消减我的疑虑。二十年来两岸的社会经济形势改变，使得这套书在大陆出版成为自然。德高望重的王文采院士及植物分类权威李振宇教授鼎力相助、组织动员，国家出版基金给予资助，促使整体工作能顺利完成。

吴声华

2015年10月20日

About the Pictorial Series

In the summer of 1995, I took part in the Yunnan fieldwork led by Prof. Mu ZANG from Kunming Institute of Botany, Chinese Academy of Sciences. Joining us were German professor Franz Oberwinkler, director general of International Mycological Association, and some French scholars. Prof. ZANG was candid, cordialand hospitable, which impressed everyone, especially the foreign guests. We first went to Lijiang and then Sipsongpanna in the south. During this fieldwork, we were guided by Mr. Guoda TAO from Xishuangbanna Tropical Botanical Garden, Chinese Academy of Sciences. He was the chief expert of plant identification in the area, knowing which areas of the woods were worth this field inspection of ours. One day, when we were having lunch together in a traditional wooden house of an ethnic Dai family, Prof. ZANG proposed to Mr. TAO: "Since you are so fond of photography, why not compile a pictorial book of Banna's plants?" Prof. ZANG then turned to me, asking whether I could give help in getting the book published in Taiwan, and I promised to give it a try after returning to Taiwan.

I first contacted Dr. Chiawei LI, director of Museum of Natural Science, who was passionate about plant research and conservation and interested in the project. But before long, his passion faded due to his sense that the project was likely to take too long a time, and the Museum did not have sufficient fund to support the publishing. I then inquired of other publishing companies, but none of them gave a positive response. These setbacks sent me thinking that perhaps we should make some tangible achievements first before our work could be accepted for publication. So I asked Mr. TAO to proceed with shooting plants. Prof. ZANG and his wife Prof. Xingjiang LI recommended Yong FEI to provide assistance to the work. In the summer of 1997, I met Prof. ZANG and Yong FEI at Kunming Airport, and we flew to Sipsongpanna together for the fieldwork led by Mr. TAO. Of the same age as mine, Yong FEI was a thin and swarthy man, not very talkative. On our way back, Yong FEI said he was considering asking several of his colleagues to join him in shooting plants of the northwest of Yunnan so that the pictures taken in the two areas (the SouthYunnan and the Northwest Yunnan) could be combined to make a single pictorial book that could be called "Plants of Yunnan". After returning to Taiwan, I browsed the slides given by Mr. TAO, feeling that their quality was not ideal. Having asked Mr. TAO about this, I learned that it had been caused by his camera whose main body was of good brand and quality but whose lens was made by a mediocre producer. I remitted money to him for purchasing a new camera set, hoping that the quality of photos could be ensured by a high-quality camera.

In 1998, I visited Kunming Institute of Botany to discuss with Guoda TAO, Yong FEI and Hang SUN about the work of pictorial book for plants. Yong FEI was full of enthusiasm on the work, and had good relations with people, so we decided to commission him to collect plant photos taken by staff from the Kunming Institute, and to compose text both in English and Chinese. The next year, Prof. Mu ZANG and Prof. LI introduced to me Mr. Xiaolian ZENG, who had been engaging in the investigation of plants and animals in the wilds of Yunnan Province for a long time and was also a famous painter from the Kunming Institute of Botany. His paintings are the best realistic bird-and-flower works in China, blending scientificity with artistic beauty.

In 2000, Yong FEI spent half a year in Botanic Gardens of Toyama (Japan), taking photos of Chinese plants grown in the Gardens. In autumn of the same year, with Yong FEI as my guide, we went to Diancang Mountain in Dali and Zixi Mountain in Chuxiong. One day, when we were having local delicacies for super in an old courtyard of Bai nationality, Yong FEI got into high spirit and chatted with me without restraint. My first impression of Yong FEI was that he was a bit unapproachable, but after several rounds of conversations, he regarded me as his close friend. After several discussions with him, I felt that he very much concentrated on doing the work well, paying no attention to remuneration. In a lane of Dali, we found a vendor selling milk fan cake, a kind of local specialty, and he bought two big pieces, saying that they were for his daughter who liked such food.

One early morning in the early 2001, Prof. ZANG sent me an email, saying sadly that Yong FEI passed away in Lijiang the day before. It was really unbearable to hear of the sudden passing of such a young life. As soon as the meeting in that morning ended, I called to Kunming. The call was answered by Prof. LI who said that Yong FEI had been on a working trip at the time. At midnight, his roommates were awakened by some noises and found him foaming at the mouth. He was rushed to a local hospital and initially diagnosed as only having a fit of epilepsy, but no amount of treatment took effect on him; he passed away just as dawn came.

Several months later, when contacting Mr. Xiaolian ZENG, I was told that Yong FEI's wife was looking for me. The family name of Yong FEI's wife was XIANG, so we called her Little XIANG, a traditional way of Chinese people addressing their acquaintances who were younger than themselves. Little XIANG expressed her wish over phone that the project of the pictorial book should go on as usual and she also said that several of Yong FEI's colleagues were willing to help. One summer evening of the same year in Kunming, Little XIANG, together with Xiao CHENG, Hang SUN and Zhekun ZHOU all from the Kunming Institute of Botany, had a meeting with me to discuss about the remaining work of the project. Xiao CHENG said he and Yong FEI were classmates, graduating and first getting employed at the same time, so he would definitely offer his help. And in fact he did try his best to facilitate the progress of the work through networking. In 2002, we had a gathering in Kunming. Little

XIANG brought her daughter there, who was then a junior-secondary-school student. The girl was both clever and well-behaved, and when I said to her that she looked like her father, her eyes moistened slightly. Xiao CHENG was a fern researcher. His father-in-law Mr. Sugong WU and mother-in-law Mrs. Ruizheng FANG were also scholars of the Kunming Institute of Botany, both of whom offered their assistance in providing photos and editing texts. Prof. Xiwen LI also from the Kunming Institute of Botany, who had comprehensive attainments in plant taxonomy, was responsible for examining photos and texts. There were many other experts from the Kunming Institute of Botany who made contributions to this book, but due to space constraint, their names are not listed here one by one.

In 2001, Mr. Xiaolian ZENG introduced to me Mr. Yitao LIU, another painter from Kunming Institute of Botany. Mr. LIU used to stay in Xishuangbanna Tropical Botanical Garden, where he developed his distinctive painting style with which to depict typical Sipsongpanna's landscape. He was also a lover of photography, and used to be my fieldwork guide for several times. Mr. LIU suggested that I expand the pictorial plant book project to cover the whole territory of China. Due to the innate aesthetic sense of artist, the plant photos taken by the two painters - Mr. ZENG and Mr. LIU - had unique characteristics both in picture composing and view finding. My fieldwork in Hebei and Jilin in 2003 and then my travelling in Xinjiang and Jilin in 2004 prompted my final decision to expand the pictorial plant book project to the whole country. I explained to Little XIANG that only when the book was comprehensive enough and of high quality, could Yong FEI's hardworking spirit and aspiration in this regard be fully manifested. And only in this way could Yong FEI's cherished dream of compiling a pictorial book on plants of China be realized without unnecessary pre-steps.

2004 saw Dr. Haining QIN's visit to Taiwan. Dr. QIN was from Institute of Botany, Chinese Academy of Sciences, and we got to know each other at British Kew Gardens in 1994. Under the leadership of Haining, the construction of Plant Photo Bank of China was making marvelous progress. Haining was a cordial and modest man, with his face always shining with smile. Upon knowing that I was conducting the project of pictorial plant book, he offered to give help. Taking advantage of his wide network, he brought in many talents, among whom Guosheng HE from Fujian, Guangdi WU from Sichuan, Yan LIU from Guangxi, and Zexian LI from Guangdong were all good at photography. In terms of color, Yan LIU's photos were particularly good, which was admirable. In addition, Dr. David Chamberlain, an expert in Ericacea, whom I got to know when I visited Royal Botanic Garden Edinburgh, agreed to review the photos of Ericaceae and Berberidaceae. Besides, Dr. Ching-I PENG introduced Dr. David E. Boufford from Harvard University who provided photos taken when he was investigating the plants of the Hengduan Mountains in the southwest of China. And David also recommended Susan Kelley and Richard Ree who both offered their plant photos.

Prof. Pengcheng WU from Institute of Botany, Chinese Academy of Sciences. was an expert in bryophytes. In 1990, he stayed in University of Helsinki, Finland for a few months when I was about to obtain my doctorate awarded by the University. Prof. WU introduced several experts from Institute of Botany to help review photos and write text. Dr. Liisin LEU, a Taiwan veteran in mycology, recommended Mr. Zhaoxiang HUI, deputy director of Kunming Municipal Bureau of Agriculture, to provide photos of weeds. And Mr. HUI invited Mr. Jindian CHE from Beijing to provide photos of weeds in Northern China. Prof. Mingjou LAI from Taiwan involved Mr. Xiangkun QIN from Shanghai Natural History Museum in contributing photos of plants in Eastern China, and then recommended Mr. Dachang ZHAO from Shenyang Institute of Applied Ecology, Chinese Academy of Sciences to offer photos of plants in Changbai Mountains. In 2004, I went to Urumqi to attend a conference whose organizer, Prof. Abdulla from Xinjiang University, gave me many photos of Xinjiang plants taken by himself.

My college classmate Shihchang KANG, an expert in plants and computer, also sent me plant photos taken by himself. Having long foreseen the power of internet, he did not quite agree with the idea of publishing a bulky physical book. Years ago, he wrote down a website address and asked me to visit it. The website, which I had never heard of before, was 'Google', a 'search engine' enabling us to search for information easily. And it turned out that this became a strong upward trend, with more and more photos being uploaded onto internet for people to view or download. However, Prof. Tuomo Niemelä, my Finnish adviser, had a different view on this phenomenon. He had published a handbook about large fungi, whose formatting was done by himself. The photos and text were arranged so well that a full sense of beauty permeated the entire book, and this always made me in a good mood each time I read it. When being consulted about the idea of publishing a physical plant book like this one, he encouraged me to continue doing so, saying that sometimes online data and materials would vanish for no reason and many online materials could not be said to be authentic because they had not undergone necessary review and approval. With this encouragement, I decided to carry out this project through to the end, paying no attention to the trends and fashionable ideas of new generations.

At the beginning of 2007, I found on internet that the Plant Photo Bank of China owned by Institute of Botany contained image data being managed by Min LI. So I asked him for sources of these photos.

Min LI recommended several persons whose photos in the Database were regarded as excellent. Most of these photo-takers were young people from the Institute of Botany. They were Bing LIU, Qinwen LIN, Shengxiang YU, Min LI, Xianming GAO, Yanhong BING. Besides, Yun WANG from Shaanxi was also added to the list of recommendation. By that time, I had already collected photos of 5000 species of plants in China through over 10 years of my hardwork which could have very well wound up. However, it would have been regrettable if I had not added so many fine photos of plants in Northern China to this important book. Bing LIU was a wonder in plant taxonomy - so young as he was, he had taken astonishingly large number of plant photos. It was Bing LIU and Yitao LIU who provided the largest number of photos for this work. Bo LIU, who was very active in work, helped a lot in writing text and reviewing photos. By the end of the same year, we had collected photos of more than 6500 species, paving the way for doing other publication-related preparatory work such as text writing and photo reviewing.

In 2010, "Academia Sinica" held a seminar in Taiwan, at which Haining QIN and Min LI delivered their reports which revealed that the Plant Photo Bank of China had collected hundreds of thousands of photos. In this circumstance, one more round of photo searching and collecting would certainly make more and better photos available for this upcoming book. Only, it would take more time. With so many people looking forward to the publication of the book, the longer time we took in publishing, the heavier pressure we would face. But in final analysis, I believed that Yong FEI, if he were still alive, would support this last round of photo searching and collecting. Prof. Xunlin YU and Prof. Daigui ZHANG from Hunan sent me many photos of plants of Central China. Mr. Hongwei ZHANG from Zhejiang and Mr. Zhonghui SHI from Anhui also sent photos to me. Prof. You ZHOU from Tonghua of Jilin contributed the photos of plants of Changbai Mountain that he took with great efforts. In the recent three years, a lot of photos were also provided by Xinxin ZHU, Yousheng CHEN, Yechun XU, Bin CHEN, Shipin CHEN, Hai HE, and Xile ZHOU. Xianchun ZHANG offered many photos of ferns. Xiaohua JIN submitted many photos of Orchidaceae plants. My good friend Li ZHANG was responsible for bryophytes.

Dr. Ching-I PENG from "Academia Sinica" provided a lot of pictures of Begoniaceae and edited the draft for this family. Shannjye MOORE, an expert of ferns from Taiwan, contributed some pictures of ferns to this book, and reviewed the pictures and text for the fern part. When I studied for doctorate in Taiwan University, Shannjye was still an undergraduate of the University. With a lovely round face often with smile, he was always seen studying in herbarium. In November 2010, however, he suddenly died of a stroke at the age of 44, making us very sad and regretful. Miss Pifong LU, an excellent amateur expert of ferns from Taiwan, contributed some good pictures of ferns. I also would like to express my thanks to the following colleagues who provided and reviewed pictures for me: Chiumei WANG, Chihhsiung CHEN, Weihsin HU, Chunlin HUANG, Hsinfu YEN and Shauting CHIU.

What we received in earlier stages were slides and a small number of prints,and after 2005, all contributions were in the form of digital image. Digital photos are clean, but their saturation, definition and sharpness are not very ideal, with green parts tending to turn slightly yellowish. Fortunately, these problems could be solved through photo-effect modification. As to the slides, high-quality ones were not many, as the quality of such images hinged on such factors as: quality of the film, developing process, storage condition and scanning, etc. The photos first needed some trimming so as to highlight their essential parts and then required modification to the image effect; most of the work was done by myself. In the meantime, we engaged a group of people to do text writing. Although templates were provided to text writers, inconsistency still appeared in some places due to different writing styles of different writers. There were also not a few oversights or slips caused by some writers who were not conscientious enough. At first, I myself did the correction and revision one by one against reference literature, finishing the work on about two thousand species, but as the amount of this kind of work was so big that I could not continue to do it all by myself. Here, I would like to list those who made greater contributions to the text. They are: Yong FEI, Bo LIU, Ren SA, Cuizhi GU, Xiao CHENG, Ning DU, Xiwen LI, Xiaoting XU and Ruizheng FANG, etc.

Prof. Pengcheng WU mentioned this work to Ms. Jing WANG, director of Biological Division of Science Press, who was interested in exploring the possibility of publishing the work, so we met each other in Beijing in 2010. Before this, I had devoted myself to the work so deeply that even small details like typesetting and font were studied and arranged by myself. Therefore, the work was like a child brought up by myself, so I would feel uneasy if I put it in the care of someone else, and moreover, we had already commissioned a design company in Taiwan to start typesetting the draft. However, my concern and worry were gradually dispelled by Jing WANG's sincerity and her perseverance in persuasion and explanation over two successive years. And the changes in social and economic situations across the Straits also made it natural for the book to be published in the Mainland. Also worthy of mentioning are: the generous support, organization and mobilization given or conducted by both Wentsai WANG, a renowned academician and Prof. Zhenyu LI, an expert on plant taxonomy, as well as the funding by National Publication Foundation. All this facilitated the smooth completion of the entire work.

Shenghua WU

20 October, 2015

To facilitate and attract readers from both China and abroad, the text of this book series is bilingual in Chinese and English, providing the Chinese name, Latin name, morphological features, flowering and fruiting season, habitat and distribution.

In the process of compiling this work, Plant Photo Bank of China (PPBC) and Chinese Field Herbarium, both of which are under the Institute of Botany of Chinese Academy of Sciences, provided great help with the selection of photographs, for which we express our gratitude.

We also thank National Publication Foundation and Science Press, Beijing, for their great support for the publication of this book series.

It will be appreciated if mistakes and omissions are brought to our attention.

Editorial Committee of *Higher Plants of China in Colour*
31 October, 2015

第7卷编审者分工

玄参科	刘　博　杨福生　向小果　郁文彬
紫葳科	刘　博　陈又生　李锡文
胡麻科－角胡麻科	陈又生
列当科	刘　博　陈又生　张志耘
苦苣苔科	邱志敬　刘　博
狸藻科	李振宇
爵床科	邓云飞　李锡文
苦槛蓝科－透骨草科	刘　博　陈又生
车前科	李振宇　刘　博
茜草科	陈又生　林秦文
忍冬科	林秦文　刘　博
五福花科	陈又生
败酱科－川续断科	陈又生　彭　华
葫芦科	金效华　许再文
桔梗科	刘　博　陈又生
草海桐科	陈又生
花柱草科	陈　彬
菊科	陈又生　高天刚　袁　慊　李锡文

Authors and Reviewers of Volume Ⅶ

Scrophulariaceae	Bo LIU	Fusheng YANG	Xiaoguo XIANG	Wenbin YU
Bignoniaceae	Bo LIU	Yousheng CHEN	Xiwen LI	
Pedaliaceae—Martyniaceae	Yousheng CHEN			
Orobanchaceae	Bo LIU	Yousheng CHEN	Zhiyun ZHANG	
Gesneriaceae	Zhijing QIU	Bo LIU		
Lentibulariaceae	Zhenyu LI			
Acanthaceae	Yunfei DENG	Xiwen LI		
Myoporaceae—Phrymaceae	Bo LIU	Yousheng CHEN		
Plantaginaceae	Zhenyu LI	Bo LIU		
Rubiaceae	Yousheng CHEN	Qinwen LIN		
Caprifoliaceae	Qinwen LIN	Bo LIU		
Adoxaceae	Yousheng CHEN			
Valerianaceae—Dipsacaceae	Yousheng CHEN	Hua PENG		
Cucurbitaceae	Xiaohua JIN	Zaiwen XU		
Campanulaceae	Bo LIU	Yousheng CHEN		
Goodeniaceae	Yousheng CHEN			
Stylidiaceae	Bin CHEN			
Compositae	Yousheng CHEN	Tiangang GAO	Qian YUAN	Xiwen LI

目录 | Contents

第7卷
Volume Ⅶ

被子植物

玄参科－菊科

Angiosperms

Scrophulariaceae—Compositae

玄参科
Scrophulariaceae

毛瓣毛蕊花 *Verbascum blattaria*

大穗花

Pseudolysimachion dauricum (Steven) Holub

多年生草本。茎直立，每个茎节上有环连接叶柄基部。叶对生，具腺状柔毛。花序1个，从不为大型圆锥状，具腺毛；花冠粉色或粉色；冠筒长为花冠的约1/3。蒴果与花萼等长；宿存花柱长约1厘米。花期7-8月。生海拔1300米以下的干草地、沙丘或疏林。产华北和东北。俄罗斯(远东地区和东西伯利亚)、蒙古和朝鲜半岛亦有。

Perennial herbs. Stems erect, with a ring at each node connecting petioles base. Leaves opposite, glandular pubescent. Inflorescences 1, never in a large panicle, glandular hairy; corolla white or pink; tubes ca. 1/3 of corolla length. Capsules as long as calyx; persistent styles ca. 1 cm long. Fl. Jul-Aug. Steppes, dunes or sparse forests below 1300 m. Distributed in N and NE China. Also in Russia (Far East and E Siberia), Mongolia and Korean Peninsula.

穗花

Pseudolysimachion spicatum (L.) Opiz

多年生草本。茎直立或斜升，密被柔毛。叶对生，常于茎基部更密；叶长圆形至披针形，被黏腺毛。花序密被黏腺毛；花无柄；花冠紫色或蓝色；冠筒长为花冠的约1/3。幼蒴果长圆状球形，具多室腺毛。花期7-9月。生海拔2500米以下的草地或针叶林中。产新疆西北部。哈萨克斯坦、吉尔吉斯斯坦、俄罗斯、蒙古和欧洲亦有。

Perennial herbs. Stems erect or ascending, densely villous hairy. Leaves opposite, often more crowded at stem base. Leaves oblong to lanceolate, with viscid glandular hairs. Inflorescences densely with viscid glandular hairs; flowers subsessile; corolla purple or blue; tubes ca. 1/3 of corolla length. Young capsules oblong-globose, with multicellular glandular hairs. Fl. Jul-Sep. Meadows or coniferous forests below 2500 m. Distributed in NW Xinjiang. Also in Kazakhstan, Kyrgyzstan, Russia, Mongolia and Europe.

毛瓣毛蕊花

Verbascum blattaria L.

一年生或二年生草本植株具腺毛。茎不分枝。基生叶矩圆形，边具钝锯齿或基部羽状浅裂；茎生叶边具不规则浅尖齿。花序总状，花单生。花萼裂片矩圆状披针形；花冠黄色，后方三裂片的基部有绵毛；雄蕊5，花丝有紫色绵毛。蒴果卵状球形，长于宿存花萼，上部疏生微腺毛。花期5-6月，果期7-8月。生河滩、沼泽、芨芨草滩、路旁。产新疆北部。俄罗斯和欧洲亦有。

Annuals or biennials, pubescent. Stems unbranched. Basal leaf blade oblong, margin obtusely serrate to basally pinnately lobed; stem leaf blade margin irregularly and shallowly toothed. Raceme, solitary. Calyx lobes lanceolate; corolla yellow; upper lobes 3, woolly at base; stamens 5, filaments purple woolly. Capsules ovoid, longer than persistent calyx, apically sparsely glandular pilose. Fl. May-Jun. Fr. Jul-Aug. Grassland along rivers, trailsides. Distributed in N Xinjiang. Also in Russia and Europe.

大穗花
Pseudolysimachion dauricum

穗花
Pseudolysimachion spicatum

毛蕊花
Verbascum thapsus L.

二年生草本，全体密被灰黄色星状毛。叶倒披针状长圆形。花常少量簇生；花冠黄色；前方2个雄蕊花丝无毛，后方3个具柔毛。蒴果卵球形，与宿存花萼等长。花期6-8月，果期7-10月。生海拔1400-3200米的山坡、草地或河边。产中国西南、华东和西北。北半球其他地方亦有。

Biennials, densely with grayish yellow stellate hairs. Leaves oblanceolate-oblong. Flowers usually few fascicled; corolla yellow; filaments of anterior 2 stamens glabrous and of posterior 3 pubescent. Capsules ovoid, as long as persistent calyx. Fl. Jun-Aug. Fr. Jul-Oct. Slopes, grasslands or river banks at 1400-3200 m. Distributed in SW, E and NW China. Also in other places of Northern Hemisphere.

准噶尔毛蕊花
Verbascum songoricum
Schrenk ex Fisch. et Mey.

多年生草本，全株密被灰白色星状毛。基生叶长圆形至倒披针形，柄长10厘米，边缘具浅圆齿；茎生叶披针状长圆形至长圆形，无柄。圆锥花序长达40厘米；花2-7朵簇生；花冠黄色；花丝具白色绵毛，花药肾形。蒴果约与宿存花萼等长。花期6月，果期8月。生海拔400-600米的草滩或田边湿处。产新疆。俄罗斯、中亚和西南亚亦有。

准噶尔毛蕊花 *Verbascum songoricum*

Perennial herbs, densely with grayish stellate hairs. Basal leaves oblong to oblanceolate, petioles of basal leaves to 10 cm, margin shallowly crenate; cauline leaves lanceolate-oblong to oblong, sessile. Panicles to 40 cm long; flowers 2-7 fascicled; corolla yellow; filaments white woolly hairy, anthers reniform. Capsules nearly as long as persistent calyx.

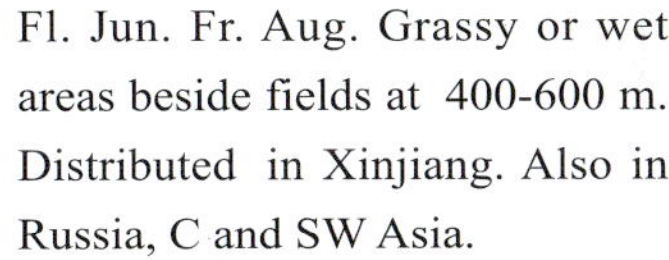

Fl. Jun. Fr. Aug. Grassy or wet areas beside fields at 400-600 m. Distributed in Xinjiang. Also in Russia, C and SW Asia.

毛蕊花 *Verbascum thapsus*

来江藤 *Brandisia hancei*

来江藤

Brandisia hancei Hook.

灌木，密被锈黄色星状绒毛。枝条渐无毛；叶卵状披针形。花单生叶腋；花萼宽钟形，10脉；花冠橙红色，外部具星状绒毛，下唇裂片舌状，上唇裂片三角形。蒴果卵球形。花期11月至翌年2月，果期翌年3-4月。生海拔500-2600米的林中或林缘。产中国西南、华中和华西。

Shrubs densely rust-yellow stellate-tomentose. Branches glabrescent, leaves ovate-lanceolate. Flowers solitary, axillary, calyx broadly campanulate, 10-veined, corolla orange-red, outside stellate tomentose, lower lips lobes ligulate, upper lips lobes triangular. Capsules ovoid. Fl. Nov to next Feb. Fr. next Mar-Apr. Forests or forest edges at 500-2600 m. Distributed in SW, C and W China.

岭南来江藤

Brandisia swinglei Merr.

灌木，密被灰褐色星状绒毛。叶卵形，密被灰褐色星状绒毛。花单生或有时成对；花萼钟形，长1.5厘米，外部具10脉；裂片5；花冠黄色，长约2.5厘米。蒴果短于花萼。花期6-11月，果期12月至翌年1月。生海拔500-1000米的山坡。产广西、广东和湖南。

Shrubs, densely gray-brown stellate-tomentose. Leaves ovate, densely gray-brown stellate-tomentose. Flowers solitary or sometimes paired; calyx campanulate, 1.5 cm long, outside 10-veined; lobes 5; corolla yellow, ca. 2.5 cm long. Capsules shorter than calyx. Fl. Jun-Nov. Fr. Dec to next Jan. Mountain slopes at 500-1000 m. Distributed in Guangxi, Guangdong and Hunan.

广西来江藤

Brandisia kwangsiensis H. L. Li

攀援灌木，高约1米，全株有褐黄色的星状毛，枝和叶上面毛渐脱落。叶卵状长圆形。花单生或成对；花冠紫红色。蒴果卵球形，内藏于花萼，密具星状绒毛。花期7-11月。生海拔900-2700米的灌丛或林下。产云南、贵州和广西。

Scandent shrubs, ca. 1 m tall, whole plant tawny stellate, branches and leaves adaxially glabrescent. Leaves ovate-oblong. Flowers solitary or paired; corolla purple-red. Capsules ovoid, included in calyx, densely stellate tomentose. Fl. Jul-Nov. Thickets or forests at 900-2700 m. Distributed in Yunnan, Guizhou and Guangxi.

茎花来江藤

Brandisia cauliflora Tsoong et Lu

藤状灌木。叶狭披针形，长5-10厘米，先端长渐尖，基部楔形，全缘；叶柄长5-8毫米，无毛。总状花序直接生于主茎上，长5-6厘米；苞片叶状，较小；花成对生于苞腋中；小苞片具柄，披针形，生于花梗顶端；花萼钟形，5浅裂，萼齿三角形；花冠鲜红色。蒴果卵球形，无毛，有锐尖头。花期6-7

岭南来江藤 *Brandisia swinglei*

广西来江藤 *Brandisia kwangsiensis*

茎花来江藤 *Brandisia cauliflora*

月，果期秋季。生矮山林中。产广西西南部。

Scandent shrubs. Leaves narrowly lanceolate, 5-10 cm long, apex narrowly acuminate, base cuneate, margin entire; petioles 5-8 mm long, glabrous. Racemes borne on main stem, 5-6 cm long; bracts petiolate, small; flowers paired at nodes; bracteoles petiolate, lanceolate, inserted at pedicel apex; calyx campanulate, 5-lobed, lobes triangular; corolla bright red. Capsules ovoid, glabrous, apex acute. Fl. Jun-Jul. Fr. autumn. Forests in low mountains. Distributed in SW Guangxi.

毛泡桐

Paulownia tomentosa (Thunb.) Steud.

乔木，高达20米。树冠宽，伞形。叶心形，下面密被毛。聚伞圆锥花序塔形至狭圆锥状；聚伞花序具3或4花；花冠紫色，漏斗状钟形。蒴果卵球形。花期4-5月，果期8-9月。通常栽培于海拔1800米以下。产中国东北、南达长江以南。朝鲜半岛、日本、欧洲和北美洲亦有引种栽培。

Trees, to 20 m tall. Crown broad, umbelliform. Leaves cordate, abaxially densely to sparsely hairy. Thyrses pyramidal to narrowly conical; cymes 3- or 4-flowered; corolla purple, funnelform-campanulate. Capsules ovoid. Fl. Apr-May. Fr. Aug-Sep. Usually cultivated below 1800 m. Distributed in NE China, southward to the South of Yangtz River. Also introduced in Korean Peninsula, Japan, Europe and North America.

楸叶泡桐

Paulownia catalpifolia T. Gong ex Hong

乔木。树冠高大，树干通直。叶常狭卵状心形，长为宽的约2倍，下面密被星状绒毛，上面无毛。花冠淡紫色。蒴果椭圆体形，长4.5-5.5厘米，幼时被星状绒毛；果皮厚达3毫米。花期4月，果期7-8月。生低海拔处。产山东(邹县)。

Trees. Crown large and high, trunk straight. Leaves often narrowly ovate- cordate, ca. 2 × as long as wide, abaxially densely stellate tomentose, adaxially glabrous. Corolla light purple. Capsules ellipsoid, 4.5-5.5 cm long, stellate hairy when young; pericarps to 3 mm thick. Fl. Apr. Fr. Jul-Aug. At low altitude. Distributed in Shandong (Zou County).

毛泡桐 *Paulownia tomentosa*

楸叶泡桐 *Paulownia catalpifolia*

白花泡桐 *Paulownia fortunei*

白花泡桐
Paulownia fortunei (Seem.) Hemsl.

乔木，高达30米。树冠圆锥形。幼枝、花序和果被黄褐色星状绒毛。叶狭卵状心形，下面具星状毛或腺点。花冠白色、紫色或淡紫色，管状漏斗形，长8-12厘米。蒴果长圆形至长圆状椭圆体形。花期3-4月，果期7-8月。野生或栽培于海拔2000米以下的山坡、林下、山谷或荒地。产中国西南、华南、华中和华东。老挝和越南亦有。

Trees, to 30 m tall. Crown conical. Young shoots, inflorescences and fruits yellowish brown stellate tomentose. Leaves narrowly ovate-cordate, abaxially stellate hairy or glandular. Corolla white, purple or light purple, tubular-funnelform, 8-12 cm long. Capsules oblong to oblong-ellipsoid. Fl. Mar-Apr. Fr. Jul-Aug. Wild or cultivated on mountain slopes, forests, mountain valleys or wastelands below 2000 m. Distributed in SW, S, C and E China. Also in Laos and Vietnam.

台湾泡桐
Paulownia kawakamii T. Itô

乔木。树冠伞形。叶心形，具黏腺毛，全缘或3-5裂或有角。聚伞圆锥花序宽圆锥状；聚伞花序通常具3花；花萼明显具棱，具绒毛；花冠浅紫色至蓝紫色，近钟形。蒴果卵球形。花期4-5月，果期8-9月。生海拔200-1500米的灌丛中、林中或荒地。产中国西南、华南、华中和华东。

Trees. Crown umbellate. Leaves cordate, viscid-glandular hairy, margin entire or 3-5-lobed or angled. Thyrses broadly conical; cymes often 3-flowered; calyx conspicuously ridged, tomentose; corolla pale violet to blue-purple, subcampanulate. Capsules ovoid. Fl. Apr-May. Fr. Aug-Sep. Bushes, forests or wastelands at 200-1500 m. Distributed in SW, S, C and E China.

美丽桐
Wightia speciosissima (D. Don) Merr.

乔木或半附生假藤本。叶常长圆形至椭圆形。聚伞圆锥花序长30厘米以上，疏具锈色毛；聚伞花序常具3花；花冠粉红色，长达3.5厘米。种子具狭翅。花期9-10月。生海拔2500米以下的林中或田边。产云南中部和南部。印度、尼泊尔、不丹、缅甸和越南亦有。

Trees or semiepiphytic pseudovines. Leaves often oblong to elliptic. Thyrses more than 30 cm long, sparsely rusty stellate hairy; cymes often 3-flowered; corolla pink, to 3.5 cm long. Seeds narrowly winged. Fl. Sep-Oct. Forests or field sides below 2500 m.

台湾泡桐 *Paulownia kawakamii*

美丽桐 *Wightia speciosissima*

Distributed in C and S Yunnan. Also in India, Nepal, Bhutan, Myanmar and Vietnam.

玄参
Scrophularia ningpoensis Hemsl.

草本，高达1.5米。侧根少，纺锤形至圆锥形。茎四方。叶对生。聚伞圆锥花序长达50厘米，大且疏松；聚伞花序顶生及腋生，常具2-4花；花冠褐紫色。蒴果卵球形。花期6-10月，果期9-11月。生海拔1700米以下的竹林、溪边、灌丛或高草丛。产中国西南、华南、华中、华东、华北和西北。

Herbs, to 1.5 m tall. Lateral roots few, fusiform to conical. Stems quadrangular. Leaves opposite. Thyrses to 50 cm long, largely lax; cymes terminal and axillary, often 2-4-flowered; corolla brown-purple. Capsules ovoid. Fl. Jun-Oct. Fr. Sep-Nov. Bamboo forests, along streams, thickets or tall grasses below 1700 m. Distributed in SW, S, C, E, N and NW China.

玄参 *Scrophularia ningpoensis*

高玄参
Scrophularia elatior Benth.

草本。具纤维状根。茎四方形，明显具翅，髓白色。叶柄长达10厘米，明显具翅；叶卵形至披针形，边缘具锯齿至重锯齿。聚伞圆锥花序顶生，长30厘米；聚伞花序具5-8花；花冠绿色。蒴果球状卵球形。花期7-9月，果期9-11月。生海拔2000-3000米的林中或湿草甸。产云南和新疆南部。印度北部和尼泊尔亦有。

Herbs. Rhizomes with fibrous roots. Stems quadrangular, conspicuously winged, pith white. Petioles to 10 cm long, conspicuously winged; leaves ovate to lanceolate, margin serrate to double serrate. Thyrses terminal, to 30 cm long; cymes 5-8-flowered; corolla green. Capsules globose-ovoid. Fl. Jul-Sep. Fr. Sep-Nov. Forests or wet grasslands at 2000-3000 m. Distributed in Yunnan and S Xinjiang. Also in N India and Nepal.

高玄参 *Scrophularia elatior*

长柱玄参 *Scrophularia stylosa*

藏玄参 *Oreosolen wattii*

长柱玄参

Scrophularia stylosa Tsoong

多年生草本。茎不分枝或上部具短分枝，中空。叶对生，下面两对极小；柄有狭翅；叶片长达9厘米，边缘有大尖齿。聚伞花序具1-3花，腋生。花萼具毛，裂片顶端尖；花冠淡黄色，长15-18毫米，裂片近圆形，边缘相互重叠；退化雄蕊倒心形；子房长约3毫米，花柱约8毫米。蒴果尖卵形。花期6月，果期7-9月。生海拔2000-3000米石崖上。特产陕西(太白山)。

Perennial herbs. Stems simple or apically short branched, hollow. Leaves opposite, lower 2 pairs smaller than others; petioles narrowly winged; leaf blade to 9 cm long, margin toothed. Cymes axillary, 1-3-flowered. Calyx glandular pubescent, lobes apex acute; corolla light yellow, 15-18 mm long, lobes rotund, with overlapping margins; staminode obcordate; ovary ca. 3 mm; style ca. 8 mm. Capsules narrowly ovoid. Fl. Jun. Fr. Jul-Sep. Rocks at 2000-3000 m. Endemic to Shaanxi (Taibai Mountain).

大花玄参 *Scrophularia delavayi*

大花玄参

Scrophularia delavayi Franch.

多年生草本。茎簇生，中空。叶卵形至卵状菱形。聚伞圆锥花序近头状或近穗状，1-3轮；聚伞花序具1-3花；花冠黄色，长0.9-1.5厘米，上唇和冠筒部分具柔毛，冠筒钟形。花期5-7月，果期8月。生海拔3100-3800米的草坡、灌丛湿处或岩石缝。产云南北部和四川西南部。

Perennial herbs. Stems fascicled, hollow. Leaves ovate to ovate-rhomboid. Thyrses subcapitate or subspicate, 1-3-whorled; cymes 1-3-flowered; corolla yellow, 0.9-1.5 cm long, pilose on upper lips and part of tube, tubes campanulate. Fl. May-Jul. Fr. Aug. Grassy slopes, moist areas in scrubs or rocky crevices at 3100-3800 m. Distributed in N Yunnan and SW Sichuan.

藏玄参

Oreosolen wattii Hook. f.

多年生草本，全株被腺毛。根粗大。下部叶鳞片状，上部叶簇生且莲座状，心形至卵形。花序低于5厘米；花萼裂片条状披针形；花冠黄色，长1.5-2.5厘米。蒴果卵球形。花期6月，果期8月。生海拔3000-5100米的高山草甸。产西藏和青海南部。印度、尼泊尔和不丹亦有。

Perennial herbs, plants entirely with glandular hairs. Roots thick. Lower leaves scalelike, uppers ones crowded and rosulate, cordate to ovate. Inflorescences less than 5 cm tall; calyx lobes linear-lanceolate; corolla yellow, 1.5-2.5 cm long. Capsules ovoid. Fl. Jun. Fr. Aug. Alpine meadows at 3000-5100 m. Distributed in Xizang and S Qinghai. Also in India, Nepal and Bhutan.

野甘草

Scoparia dulcis L.

草本或亚灌木。叶具柄，菱状卵形至菱状披针形。花常腋生，每节间(1或)2朵；花冠白色，喉部有密毛，筒极短，瓣片4，上方1枚稍大，钝头，缘有啮蚀状锯齿。生海拔1400米以下的荒地或路边，偶在山坡上。产云南、广西、广东、台湾和福建。热带和亚热带地区广布。

野甘草 *Scoparia dulcis*

Herbs or subshrubs. Leaves petiolate, rhomboid-ovate to rhomboid-lanceolate. Flowers usually axillary, (1 or)2 per node; corolla white, throat densely hairy, tubes very short, lobes 4, uppermost one slightly larger, obtuse, margin erose-serrulate. Wastelands or roadsides, occasionally on mountain slopes below 1400 m. Distributed in Yunnan, Guangxi, Guangdong, Taiwan and Fujian. Also throughout tropics and subtropics.

假马齿苋

Bacopa monnieri (L.) Wettst.

匍匐草本，节上生根。叶无柄，对生，长圆状倒披针形，先端钝。花单生叶腋，花梗长0.5-3.5厘米，顶部具2小苞片；花冠不明显二唇形，蓝色、紫色或白色；雄蕊4；柱头头状。花期5-10月。生海拔1100米以下的水边、湿地或沙滩。产云南中部和东部、华南和华东。世界热带地区亦有。

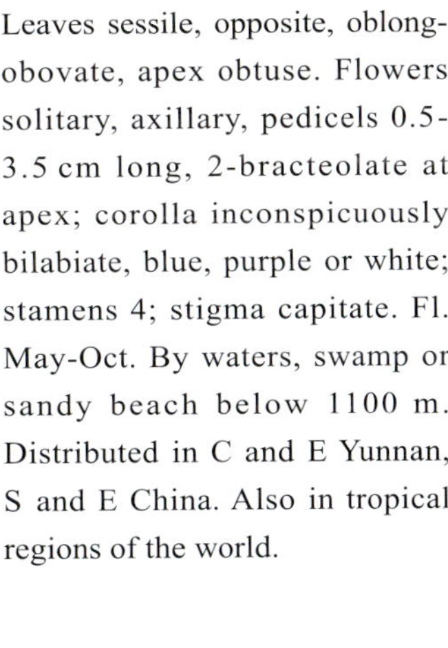

Herbs prostrate, rooting at nodes. Leaves sessile, opposite, oblong-obovate, apex obtuse. Flowers solitary, axillary, pedicels 0.5-3.5 cm long, 2-bracteolate at apex; corolla inconspicuously bilabiate, blue, purple or white; stamens 4; stigma capitate. Fl. May-Oct. By waters, swamp or sandy beach below 1100 m. Distributed in C and E Yunnan, S and E China. Also in tropical regions of the world.

假马齿苋 *Bacopa monnieri*

虻眼

Dopatrium junceum (Roxb.) Buch.-Ham. ex Benth.

一年生直立草本。茎稍肉质。叶有时鳞片状，向上渐小，无柄，披针形至近匙状披针形。花单生叶腋；花冠白色、玫瑰色或淡紫色，比萼长2倍，二唇形；雄蕊4，后方2枚能育。蒴果球形。花期8-11月。生海拔1800米以下的稻田或潮湿处。产中国西南、华南、华中、华北和华东。南亚、东南亚、日本和大洋洲亦有。

Annual herbs, erect. Stems slightly succulent. Leaves sometimes scalelike, gradually decreasing in size upward, sessile, lanceolate to subspatulate-lanceolate. Flowers solitary, axillary, corolla white, rose or pale purple, 2 × as long as calyx, bilabiate; stamens 4, posterior 2 fertile. Capsules globose. Fl. Aug-Nov. Rice paddy or wet places below 1800 m. Distributed in SW, S, C, N and E China. Also in S and SE Asia, Japan and Oceania.

虻眼 *Dopatrium junceum*

野地钟萼草 *Lindenbergia muraria*

野地钟萼草

Lindenbergia muraria (Roxb. ex D. Don) Bruhl

一年生草本。叶卵形，近膜质。花单生于叶腋；冠筒膜质，稍白色；裂片等大，长圆状卵形；花冠黄色；冠檐近无毛；下唇明显具褶；上唇截形，顶端微凹。蒴果卵球形，密被毛。种子黄色。花期7-9月，果期10月。生海拔800-2500米的河边或干旱山坡上。产中国西南、华南和华中。南亚、东南亚和西南亚亦有。

Annual herbs. Leaves ovate, submembranous. Flowers solitary in leaf axils; tubes membranous, somewhat white; lobes equal, oblong-ovate; corolla yellow; limbs subglabrous; lower lips conspicuously plicate; upper lips truncate, emarginate. Capsules ovoid, densely hairy. Seeds yellow. Fl. Jul-Sep. Fr. Oct. Along rivers or dry mountain slopes at 800-2500 m. Distributed in SW, S and C China. Also in S, SE and SW Asia.

钟萼草

Lindenbergia philippensis (Cham. et Schlecht.) Benth.

多年生粗壮直立坚挺草本。叶卵形至卵状披针形，纸质。花序顶生，穗状总状，密集；花萼具明显5脉，裂片近钻状三角形；花冠黄色带紫色斑点。蒴果狭卵球形，密被棕色硬毛。花果期11月至翌年3月。生海拔1200-2600米的干燥山坡或岩石缝中。产中国西南、华南和华中。南亚和东南亚亦有。

Perennial herbs, erect, stout. Leaves ovate to ovate-lanceolate, papery. Inflorescences terminal, spicate-racemose, dense; calyx conspicuously 5-veined, lobes subulate-triangular; corolla yellow with purple spots. Capsules narrowly ovoid, densely brown hirsute. Fl. and fr. Nov to next Mar. Dry slopes or rock crevices at 1200-2600 m. Distributed in SW, S and C China. Also in S and SE Asia.

毛麝香

Adenosma glutinosum (L.) Druce

一年生草本。茎密被具腺及无腺的绒毛。叶披针状卵形或宽卵形，下面被绒毛和黄色腺点，上面被绒毛。花单生于叶腋或呈总状花序生于茎顶或枝顶；花冠紫红色至紫色；下唇3(或4)裂。蒴果卵球形，具2纵沟。花果期7-10月。生海拔300-2000米的山坡或疏林下湿地。产云南、广西、海南、福建和江西。南亚、东南亚和大洋洲亦有。

Annual herbs. Stems densely villous with eglandular and glandular hairs. Leaves lanceolate-ovate to broadly ovate, abaxially villous and densely yellow punctate glandular, adaxially villous. Flowers axillary and solitary or in dense racemes apically on stems and branches; corolla purple-red to violet; lower lips 3(or 4)-lobed. Capsules ovoid, 2-grooved. Fl. and fr. Jul-Oct. Mountain slopes or wet places under sparse forests at 300-2000 m. Distributed in Yunnan, Guangxi, Hainan, Fujian and Jiangxi. Also in S and SE Asia, and Oceania.

钟萼草 *Lindenbergia philippensis*

毛麝香 *Adenosma glutinosum*

石龙尾

Limnophila sessiliflora (Vahl) Blume

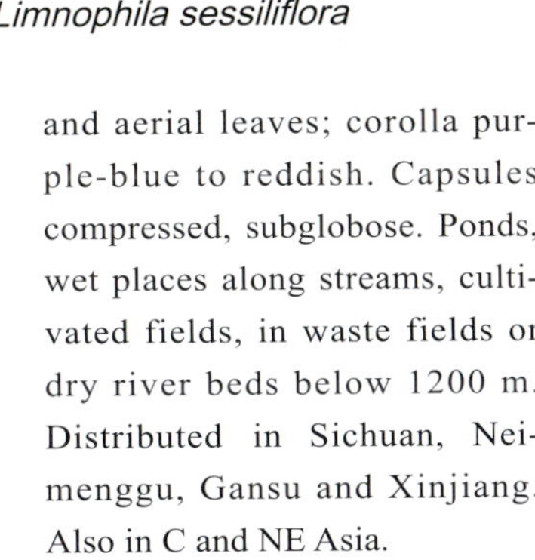

石龙尾 *Limnophila sessiliflora*

多年生两栖植物。水下叶多裂，裂片扁平或毛发状；气生叶轮生，椭圆状披针形，密被腺点，边缘具圆齿至开裂；脉1-3。花无柄，单生于沉水和气生叶叶腋；花冠蓝紫色至浅红色。蒴果扁平，近球形。生海拔1200米以下的水塘、溪边湿地、农耕地、弃荒地或干旱河床。产四川、内蒙古、甘肃和新疆。中亚和东北亚亦有。

Perennial herbs, amphibious. Submerged leaves multiparted, segments flattened or capillary; aerial leaves whorled, elliptic-lanceolate, densely glandular punctate, margin crenate to lobed; veins 1-3. Flowers sessile, solitary in axils of submerged and aerial leaves; corolla purple-blue to reddish. Capsules compressed, subglobose. Ponds, wet places along streams, cultivated fields, in waste fields or dry river beds below 1200 m. Distributed in Sichuan, Neimenggu, Gansu and Xinjiang. Also in C and NE Asia.

大叶石龙尾

Limnophila rugosa (Roth) Merr.

多年生草本。叶卵形，菱状椭圆形或椭圆形，3-9 × 1-5厘米，叶背沿脉被短硬毛，羽状脉。花萼果熟时平滑或仅具5条凸起的纵脉。蒴果浅棕色，卵球形。花果期8-11月。生海拔900米以下的水边、山谷或草地。产中国西南、华南、华中和华东。南亚、东南亚和日本亦有。

Perennial herbs. Leaves ovate, rhomboid-elliptic or elliptic, 3-9 × 1-5 cm, abaxially hirtellous along veins, pinni-nerved. Calyx smooth or longitudinally 5-nerved only when in fruits matured. Capsules pale brown, ovoid. Fl. and fr. Aug-Nov. Adjacent to water, valleys or grasslands below 900 m. Distributed in SW, S, C and E China. Also in S and SE Asia, and Japan.

大叶石龙尾 *Limnophila rugosa*

紫苏草

Limnophila aromatica (Lam.) Merr.

一年生或多年生草本。叶对生或近轮生，基部半抱茎。花单生于叶脉或顶生或腋生总状；花冠白色、蓝紫色或粉红色，1-1.3厘米，疏被细腺体，里面具白色长柔毛。花果期3-9月。生低海拔的水库边或其他湿处。产华南、台湾、福建和江西。南亚、东南亚和澳大利亚亦有。

Annual or perennial herbs. Leaves opposite or in whorls, base semiamplexicaul. Flowers solitary in leaf axils or in terminal or axillary racemes; corolla white, blue-purple or pink, 1-1.3 cm long, sparsely and finely glandular, inside white villous. Fl. and fr. Mar-Sep. Around reservoirs or other moist places at low elevations. Distributed in S China, Taiwan, Fujian and Jiangxi. Also in S and SE Asia, and Australia.

紫苏草 *Limnophila aromatica*

肉果草 *Lancea tibetica*

肉果草

Lancea tibetica Hook. f. et Thoms.

多年生小草本。叶片6-10，莲座状。花3-5成一束或呈总状；花冠深蓝色或紫色，喉部带黄色或紫色斑点；下唇中部裂片全缘，上唇直立，深2裂。果红色至深紫色，肉质，包于宿存花萼内。花期5-7月，果期7-9月。生海拔2000-4500米的草地、疏林或溪边。产中国西南和西北。印度、不丹和蒙古亦有。

Perennial herbs. Leaves 6-10, rosulate. Flowers in fascicles of 3-5 or in a raceme; corolla dark blue or purple, throat yellow or purple spotted; lower lips middle lobes entire, upper lips erect, deeply 2-lobed. Fruits red or dark purple, succulent, enveloped by persistent calyx. Fl. May-Jul. Fr. Jul-Sep. Grasslands, sparse forests or along streams at 2000-4500 m. Distributed in SW and NW China. Also in India, Bhutan and Mongolia.

苦玄参

Picria felterrae Lour.

草本，长达1米，基部节上生根。叶卵形至有时近圆形。总状花序4-8花；花冠白色或红褐色，约1.2厘米长。蒴果卵球形。生海拔700-1400米的疏林或田间。产云南南部、贵州、广西和广东。南亚和东南亚亦有。

Herbs, to 1 m tall, rooting from lower nodes. Leaves ovate to sometimes suborbicular. Raceme 4-8-flowered; corolla white or red-brown, ca. 1.2 cm long. Capsules ovoid. Sparse forests or fields at 700-1400 m. Distributed in S Yunnan, Guizhou, Guangxi and Guangdong. Also in S and SE Asia.

苦玄参 *Picria felterrae*

宽叶母草

Lindernia nummulariifolia (D. Don) Wettst.

一年生草本。叶宽卵形至圆卵形。花序顶生或腋生的近伞形，具少量花；花冠紫色，少有蓝色或白色；雄蕊4，全育。果比宿萼长约2倍。花期7-9月，果期8-11月。生海拔1800米以下的田边、溪边或湿地。产中国西南、华南、华中、华东和西北。南亚和东南亚亦有。

Annual herbs. Leaves broadly ovate to orbicular-ovate. Inflorescences terminal or axillary subumbels, few-flowered; corolla purple, rarely blue or white; stamens 4, all fertile. Fruits almost ca. 2 × longer than persistent calyx. Fl. Jul-Sep. Fr. Aug-Nov. By farmlands, along streams or wet places below 1800 m. Distributed in SW, S, C, E and NW China. Also in S and SE Asia.

宽叶母草 *Lindernia nummulariifolia*

母草 *Lindernia crustacea*

母草
Lindernia crustacea (L.) F. Muell.

一年生草本。叶三角状卵形至宽卵形。单花腋生或形成顶生的短总状；花萼坛状，外有疏粗毛；花冠紫色；雄蕊4，全育。果与宿萼近等长。花果期全年。生海拔1300米以下的湿地、田地、草地或路边。产中国西南、华南、华中和华东。热带和亚热带地区亦有。

Annual herbs. Leaves triangular-ovate to broadly ovate. Flowers axillary and solitary or in short apical racemes; calyx urceolate, sparsely setose outside; corolla purple; stamens 4, all fertile. Fruits almost as long as persistent calyx. Fl. and fr. all year. Moist areas, farmlands, grasslands or roadsides below 1300 m. Distributed in SW, S, C and E China. Also in tropical and subtropical areas.

陌上菜
Lindernia procumbens (Krock.) Philcox

直立丛生草本，高5-20厘米，基部多分枝，无毛。根细密。叶椭圆形至长圆形，稍菱形。单花腋生；花冠粉红色至紫色。蒴果球形至卵球形，与宿存花萼等长或稍长。花期7-10月，果期9-11月。生海拔1200米以下的水边或潮湿处。产中国西南、华南、华中、华东和东北。南亚、东南亚、中亚、东北亚和欧洲南部亦有。

Erect herbs, caespitose, 5-20 cm tall, basally much branched, glabrous. Roots slender. Leaves elliptic to oblong, somewhat rhomboid. Flower axillary, solitary; corolla pink to purple. Capsules globose to ovoid, as long as or slightly longer than persistent calyx. Fl. Jul-Oct. Fr. Sep-Nov. Next to water or wet areas below 1200 m. Distributed in SW, S, C, E and NE China. Also in S, SE, C and NE Asia, and S Europe.

陌上菜 *Lindernia procumbens*

细茎母草 *Lindernia pusilla*

细茎母草
Lindernia pusilla (Willd.) Bold.

一年生铺散草本，有时具长根状茎。茎近直立。叶卵形或心形。花序顶生，亚伞形总状花序，具花3-5朵；花萼深裂，外面具粗毛；花冠紫色。蒴果卵球形，与宿存花萼几等长。花期5-9月，果期9-11月。生海拔800-1600米的水边、潮湿地、稻田或林中。产云南、广西、海南和台湾。南亚、东南亚和新几内亚岛亦有。

Annual herbs, diffuse, sometimes long stoloniferous. Stems subercect. Leaves ovate or cordate. Inflorescences terminal, subumbellate-racemose, 3-5-flowered; calyx deeply lobed, coarsely hairy outside; corolla purple. Capsules ovoid, almost as long as persistent calyx. Fl. May-Sep. Fr. Sep-Nov. By water, wet places, rice fields or forests at 800-1600 m. Distributed in Yunnan, Guangxi, Hainan and Taiwan. Also in S and SE Asia, and New Guinea.

长蒴母草 *Lindernia anagallis*

长蒴母草
Lindernia anagallis (Burm.) Pennell

一年生草本。根纤维状。叶三角状卵形、卵形或长圆形。花单生叶腋；花萼基部贴生，裂片狭披针形，无毛；花冠白色或淡紫色。蒴果条状卵球形，大概为宿存花萼的2倍。花期4-9月，果期6-11月。生海拔约1500米的林缘、溪边、田边或潮湿地。产中国西南、华南、华中和华东。南亚、东南亚和澳大利亚亦有。

Annual herbs. Roots fibrous. Leaves triangular-ovate, ovate or oblong. Flowers axillary, solitary; calyx basally connate, lobes narrowly lanceolate, glabrous; corolla white or light purple. Capsules linear-ovoid, ca. 2 × as long as persistent calyx. Fl. Apr-Sep. Fr. Jun-Nov. Forest edges, along streams, fields or wet places at ca. 1500 m. Distributed in SW, S, C and E China. Also in S and SE Asia, and Australia.

泥花母草
Lindernia antipoda (L.) Alston

一年生草本。叶基部楔形且下延。总状花序顶生，具2-20花；花冠紫色、紫白色或白色；后对雄蕊能育，前对雄蕊败育。蒴果圆柱状，长约为宿存花萼的2倍。花果期春季至秋季。生海拔1700米以下的水田或潮湿的草地中。产中国西南、华南、华中和华东。南亚、东南亚、琉球群岛、太平洋岛屿和澳大利亚亦有。

Annual herbs. Leaves base cuneate and decurrent. Racemes terminal, 2-20-flowered; corolla purple, purple-white or white; posterior 2 stamens fertile, anterior 2 stamens abortive. Capsules cylindric, ca. 2 × or more as long as persistent calyx. Fl. and fr. spring to autumn. Rice fields or wet grasslands below 1700 m. Distributed

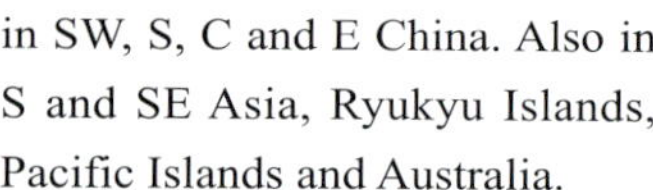
in SW, S, C and E China. Also in S and SE Asia, Ryukyu Islands, Pacific Islands and Australia.

泥花母草 *Lindernia antipoda*

旱田草
Lindernia ruellioides (Colsm.) Pennell

一年生草本。叶长圆形、椭圆形、卵状长圆形或圆形，具柄，边缘具锐锯齿。总状花序顶生，2-10花；花冠紫红色；后对雄蕊能育，前对雄蕊败育。果比宿萼长约2倍。花期6-9月，果期7-11月。生海拔1500米以下的草地、平原、山谷或林中。产中国西南、华南、华中和华东。南亚、东南亚、琉球群岛和新几内亚岛亦有。

旱田草 *Lindernia ruellioides*

紫萼蝴蝶草 *Torenia violacea*

单色蝴蝶草 *Torenia concolor*

Annual herbs. Leaves oblong, elliptic, ovate-oblong or orbicular, petiolate, margin acute-serrate. Racemes terminal, 2-10-flowered; corolla purple-red; posterior 2 stamens fertile, anterior 2 stamens abortive. Fruits almost ca. 2 × longer than persistent calyx. Fl. Jun-Sep. Fr. Jul-Nov. Grasslands, plains, mountain valleys or forests below 1500 m. Distributed in SW, S, C and E China. Also in S and SE Asia, Ryukyu Islands and New Guinea.

紫萼蝴蝶草

Torenia violacea (Azaola ex Blanco) Pennell

直立或铺散草本。叶卵形至狭卵形。花顶生成簇状或单生于叶腋；花萼紫红色；花冠淡黄色或白色，下唇3裂片各有一蓝色斑块，中裂片中央有一黄色斑块；花丝无附属物。花果期8-11月。生海拔200-2000米的山坡草地、林下或路边。产中国西南、华南、华中和华东。南亚和东南亚亦有。

Herbs erect or spreading outward. Leaves ovate to narrowly ovate. Flowers in terminal fascicles or solitary in leaf axils; calyx purple-red; corolla pale yellow or white, 3 lobes of lower lips each with a blue spot, middle lobes with a yellow spot at center; stamens unappendaged. Fl. and fr. Aug-Nov. Grasslands on slopes, forests or roadsides at 200-2000 m. Distributed in SW, S, C and E China. Also in S and SE Asia.

单色蝴蝶草

Torenia concolor Lindl.

匍匐草本。茎四方，节间生根。叶三角状卵形或狭卵形。花腋生或顶生，单生或成束；花萼具5翅，翅宽略超过1毫米，基部下延，齿2，长三角形，果时裂成5小齿；花冠蓝色或蓝紫色；花丝具附属物。花果期5-11月。生海拔1500(-2500)米以下的林下、山谷或路旁。产中国西南、华南和华东。老挝、越南和琉球群岛亦有。

Herbs prostrate. Stems quadrangular, rooting from nodes. Leaves triangular-ovate or narrowly ovate. Flowers axillary or terminal, solitary or in fascicles; calyx 5-winged, wings slightly more than 1 mm width, decurrent at base, calyx lobes 2, narrowly triangular, but becoming 5 teeth when in fruit; corolla blue or blue-purple; filaments with appendages. Fl. and fr. May-Nov. Forests, valleys or roadsides below 1500(-2500) m. Distributed in SW, S and E China. Also in Laos, Vietnam and Ryukyu Islands.

长叶蝴蝶草

Torenia asiatica L.

匍匐至近直立草本。叶三角状卵形、狭卵形或卵状圆形，无毛或疏具绒毛。单花腋生或簇；花萼二唇形，具5翅；花冠紫红色至蓝红色；外部雄蕊附属物条形。花果期6-9月。生海拔1700米以下的山坡、小径旁或阴处。产中国西南、华南、华中和华东。越南和日本亦有。

Herbs, creeping to suberect. Leaves triangular-ovate, narrowly ovate or ovate-orbicular, glabrous or sparsely villous. Flowers axillary and solitary or in fascicles; calyx 2-lipped, 5-winged; corolla purple-red to blue-red; anterior stamens appendages linear. Fl. and fr. Jun-Sep. Mountain slopes, trailsides or shady places below 1700 m. Distributed in SW, S, C and E China. Also in Vietnam and Japan.

长叶蝴蝶草 *Torenia asiatica*

尼泊尔沟酸浆 *Mimulus tenellus* var. *nepalensis*

尼泊尔沟酸浆

Mimulus tenellus Bunge var. **nepalensis** (Benth.) Tsoong ex H. P. Yang

多年生草本。茎近直立。花梗几乎和叶等长；单花腋生；花冠黄色，喉部具红点；花冠管漏斗形；花萼圆柱状，约1厘米或更长，先端截形。蒴果椭圆体形。花果期6-9月。生海拔500-3000米的水边或潮湿处。产中国西南、华中、华东和西北。印度、尼泊尔、越南和日本亦有。

Perennial herbs. Stems suberect. Pedicels nearly as long as whole leaf; flowers axillary, solitary; corolla yellow, throat with red spots; tubes funnelform; calyx cylindric, ca. 1 cm long or longer, apex truncate. Capsules ellipsoid. Fl. and fr. Jun-Sep. Water's edge or wet places at 500-3000 m. Distributed in SW, C, E and NW China. Also in India, Nepal, Vietnam and Japan.

南红藤 *Mimulus tenellus* var. *platyphyllus*

南红藤

Mimulus tenellus Bunge var. **platyphyllus** (Franch.) Tsoong ex H. P. Yang

多年生草本。茎棱角具狭翅。叶大、厚，掌状脉，基部宽楔形，全缘或中部以上具糙锯齿。花萼宽钟形，先端近倾斜；花冠漏斗状，黄色，喉部有红斑点。花果期6-9月。生海拔1900-2200米的林下或路旁。产云南和四川。

Perennial herbs. Stem angles narrowly winged. Leaves large, thick, palmately veined, base broadly cuneate, margin entire or coarsely serrate above middle. Calyx broadly campanulate, apex suboblique; corolla funnelform, yellow, throat red spotted. Fl. and fr. Jun-Sep. Forests or roadsides at 1900-2200 m. Distributed in Yunnan and Sichuan.

弹刀子菜 *Mazus stachydifolius*

弹刀子菜

Mazus stachydifolius (Turcz.) Maxim.

多年生草本，高10-50厘米，全体被白色长柔毛。茎生叶无柄；叶匙形。总状花序顶生，疏松；花冠蓝紫色，长1.5-2厘米。蒴果卵球形，2-3.5毫米，扁平。花期4-6月，果期7-9月。生海拔1500米以下的小径旁、草地或林缘湿处。产中国西南、华南、华中、华东、华北、西北和东北。俄罗斯、蒙古和朝鲜半岛亦有。

Perennial herbs, 10-50 cm tall, white villous. Stem leaves sessile; leaves spatulate. Racemes terminal, lax; corolla blue-purple, 1.5-2 cm long. Capsules ovoid, 2-3.5 mm long, flattened. Fl. Apr-Jun. Fr. Jul-Sep. Trailsides, grassland or wet places at edge of forests below 1500 m. Distributed in SW, S, C, E, N, NW and NE China. Also in Russia, Mongolia and Korean Peninsula.

早落通泉草

Mazus caducifer Hance

多年生草本，具白色柔毛。基部生多为莲座状，常早落；茎生叶对生，纸质。总状花序顶生，疏松，长35厘米；花萼钟形，具10脉；花冠浅蓝紫色；下唇中裂片反折，小于侧裂片；上唇裂片顶部锐尖。蒴果球形。花期4-5月，果期6-8月。生海拔约1300米的铁轨边、林中湿地或草地。产浙江、安徽和江西。

早落通泉草 *Mazus caducifer*

Perennial herbs, white villous. Basal leaves mostly in a rosette, often deciduous; stem leaves opposite, papery. Racemes terminal, lax, to 35cm long; calyx funnelform, veins 10; corolla light blue-purple; lower lips middle lobe exserted and smaller than lateral lobes; upper lips lobes apically acute. Capsules globose. Fl. Apr-May. Fr. Jun-Aug. Trailsides, wet places in forests or grasslands at ca. 1300 m. Distributed in Zhejiang, Anhui and Jiangxi.

岩白翠

Mazus omeiensis H. L. Li

多年生草本，植株无毛或疏具长柔毛。叶全基生，莲座状；叶倒卵状匙形至匙形，厚纸质至亚革质；花葶1(-4)。总状花序花稀疏；花冠淡蓝紫色。蒴果卵球形。花期4-7月，果期7-9月。生海拔500-2000米的岩缝中。产四川西南部和贵州西北部。

Perennial herbs, plants glabrous or sparsely villous. Leaves all basal, rosulate; leaves obovate-spatulate to spatulate, thick papery to subleathery; scapes 1 (-4). Racemes lax flowered; corolla light blue-purple. Capsules ovoid. Fl. Apr-Jul. Fr. Jul-Sep. Moist crevices at 500-2000 m. Distributed in SW Sichuan and NW Guizhou.

岩白翠 *Mazus omeiensis*

美丽通泉草 *Mazus pulchellus*

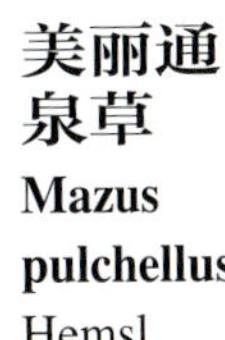

美丽通泉草

Mazus pulchellus Hemsl.

多年生草本，密被白色或锈色绒毛，渐无毛。叶全基生莲座状；花葶1-5。总状花序疏松，多花；花萼钟形，脉不明显；花冠红色、紫色或深紫红色；下唇裂片疏被流苏状齿；上唇裂片顶部截形。蒴果卵球形。花果期3-6月。生海拔约1600米的石缝或林中潮湿处。产云南东南部、四川东南部和湖北西部。

Perennial herbs, densely white or rusty pubescent, glabrescent. Leaves all basal, rosulate; scapes 1-5. Racemes lax, numerous-flowered; calyx campanulate, veins inconspicuous; corolla red, purple, or dark violet; lower lips lobes sparsely fimbriate-toothed; upper lips apex truncate. Capsules ovoid. Fl. and fr. Mar-Jun. Moist places in rocky crevices or forests at ca. 1600 m. Distributed in SE Yunnan, SE Sichuan and W Hubei.

低矮通泉草 *Mazus humilis*

低矮通泉草

Mazus humilis Hand.-Mazz.

多年生草本，高在10厘米以下，被白色长柔毛。叶倒卵状匙形至长圆状倒卵形，纸质。花萼漏斗形，具腺毛；花冠白色或白色具紫斑，长约1厘米。蒴果球形。花果期6月。生海拔2500-3500米的湿草地。产四川西南部、云南西北部和广西。

Perennial herbs, less than 10 cm tall, white villous. Leaves obovate-spatulate to oblong-obovate, papery. Calyx funnelform, glandular hairy; corolla white or white with purple spots, ca. 1 cm long. Capsules globose. Fl. and fr. Jun. Wet grassland at 2500-3500 m. Distributed in SW Sichuan, NW Yunnan and Guangxi.

通泉草

Mazus pumilus (Burm. f.) Steenis

一年生草本。叶倒卵状匙形至卵状倒披针形。总状花序生茎枝顶端，常在基部即生花，伸长，其多至20花；花萼钟形，5裂至近中部；花冠白色、紫色或蓝色。蒴果球形。花果期4-10月。生海拔1200-3800米的湿润草坡、沟边、路旁、荒废田地或林缘。除宁夏、青海和新疆外，中国各处均有。南亚、东南亚和东北亚亦有。

Annual herbs. Leaves obovate-spatulate to ovate-oblanceolate. Racemes terminal on tops of stems and branches, always with flowers at base, elongate, up to 20-flowered; calyx campanulate, 5-parted near to middle; corolla white, purple or blue. Capsules globose. Fl. and fr. Apr-Oct. Moist grassy slopes, by ditches and roads, waste fields or forest edges at 1200-3800 m. Distributed throughout China, except Ningxia, Qinghai and Xinjiang. Also in S, SE and NE Asia.

通泉草 *Mazus pumilus*

匍茎通泉草 *Mazus miquelii*

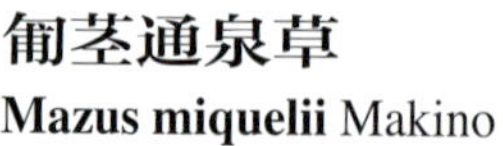

匍茎通泉草

Mazus miquelii Makino

多年生草本。茎直立至斜展，高10-15厘米。基部叶多数，莲座状，具柄；叶倒卵状匙形。总状花序顶生，伸长，疏松；花冠紫色或白色具紫斑；下唇中部倒卵形，短于侧裂片。蒴果球形。花果期2-8月。生海拔300米以下的路边湿地或疏林。产中国西南、东南、华中和华东。日本亦有。

Perennial herbs. Stems erect to obliquely ascending, 10-15 cm tall. Basal leaves numerous, rosulate, petiolate; leaves obovate-spatulate. Racemes terminal, elongated, lax; corolla purple or white with purplish spots; lower lip

野胡麻 *Dodartia orientalis*

middle lobe obovate, shorter than lateral lobes. Capsules globose. Fl. and fr. Feb-Aug. Roadsides in wet places or sparse forests below 300 m. Distributed in SW, SE, C and E China. Also in Japan.

野胡麻

Dodartia orientalis L.

草本。茎基部具黄褐色鳞片；枝细，之字形。叶基部对生，顶部常互生；叶鳞片状至宽条形，全缘或疏具齿。总状花序伸长，具3-7花；花冠紫色至深紫红色。蒴果褐色至深褐色，顶部细尖。种子黑色，卵球形。花果期5-9月。生海拔1200米以下的沟边、农耕地、废弃田地、沙漠或干旱河床。产四川、内蒙古、甘肃和新疆。中亚和东北亚亦有。

Herbs. Stems with brown-yellow scales near base; branches, slender, angled. Leaves basally opposite, apically often alternate; leaves scalelike to broadly linear, margin entire or sparsely toothed. Racemes elongated, 3-7-flowered; corolla purple to dark purple-red. Capsules brown to dark brown, apex apiculate. Seeds black, ovoid. Fl. and fr. May-Sep. Ditch banks, cultivated fields, waste fields, desert or dry river beds below 1200 m. Distributed in Sichuan, Neimenggu, Gansu and Xinjiang. Also in C and NE Asia.

宽叶柳穿鱼

Linaria thibetica Franch.

多年生草本，高达1米。叶互生，无柄，长椭圆形至卵状椭圆形。穗状花序顶生，具多数簇生的花；花萼于两面或仅内侧具多室毛；花冠淡紫色或黄色，距很短。蒴果球形。花期7-9月。生海拔2500-3800米的山坡草地、林缘或疏灌丛中。产云南西北部、四川西部和西藏东南部。

Perennial herbs, to 1 m tall. Leaves alternate, sessile, narrowly elliptic to ovate-elliptic. Spikes terminal, with numerous crowded flowers; calyx with multicellular glandular hairs on both surfaces or only on inside; corolla pale purple or yellow, with very short spur. Capsules globose. Fl. Jul-Sep. Grassy slopes, forest edges or sparse shrubs at 2500-3800 m. Distributed in NW Yunnan, W Sichuan and SE Xizang.

宽叶柳穿鱼 *Linaria thibetica*

新疆柳穿鱼 *Linaria vulgaris* subsp. *acutiloba*

柳穿鱼 *Linaria vulgaris*

柳穿鱼

Linaria vulgaris Mill.

多年生草本。叶常多数且互生；叶条形。花序总状，花密集；花冠黄色；花冠筒管状，基部有长距，距长1-1.5厘米，中间裂片舌状，上唇长于下唇。蒴果卵球形。花期6-9月，果期10月。生海拔2200米以下的山坡、林中、小径旁、草甸或多沙石的草原。产华中、华东、华北和西北。朝鲜半岛亦有。

Perennial herbs. Leaves usually numerous and alternate; leaves linear. Inflorescences racemose, flowers crowded; corolla yellow; corolla tubes tubular, spur 1-1.5 cm long at tubes base, middle lobes ligular; upper lips longer than lower lips. Capsules ovoid. Fl. Jun-Sep. Fr. Oct. Mountain slopes, forests, trailsides, meadows or gravelly steppes below 2200 m. Distributed in C, E, N and NW China. Also in Korean Peninsula.

新疆柳穿鱼

Linaria vulgaris Mill. subsp. **acutiloba** (Fisch. ex Rchb.) Hong

多年生草本，除花序外无毛。茎常在上部分枝。叶互生，条形，具3脉，长3-8厘米，宽5-15毫米。花序轴及花梗无毛；花萼裂片披针形至卵状披针形，宽超过1.5毫米，内面近无毛；花冠黄色，距长10-15毫米。蒴果卵球状。种子盘状，边缘有宽翅，成熟时中央常有瘤状突起。花期6-8月。生海拔1000-2200米的山谷草地或林下。产新疆。蒙古和俄罗斯亦有。

Perennial herbs, glabrous except for inflorescences. Stems often apically branched. Leaves alternate, linear, 3-veined, 3-8 × 0.5-1.5 cm. Axis and pedicels glabrous; calyx lobes lanceolate to ovate-lanceolate, more than 1.5 mm wide, inside subglabrous; corolla yellow, spur 10-15 mm long. Capsules ovoid. Seeds disc-like, margin broadly winged, center verrucose when mature. Fl. Jun-Aug. Ravine meadows or forests at 1000-2200 m. Distributed in Xinjiang. Also in Mongolia and Russia.

毛地黄

Digitalis purpurea L.

一年生或多年生草本，除花冠外全体被灰白色短柔毛和腺毛，有时茎几无毛。叶互生。花冠上唇甚短于下唇，紫色或白色，内面具斑点；2强雄蕊。蒴果长约1.5厘米。种子棍棒状，

毛地黄 *Digitalis purpurea*

具柔毛。花期5-6月。产四川、浙江、江苏和江西。原产欧洲。

Annual or perennial herbs, whole gray-pubescent and glandular hairy except corolla, sometimes glabrescent on stem. Leaves alternate. Upper lips of corolla extremely shorter than lower ones, purple or white, dotted inside; stamens didynamous. Capsules ca. 1.5 cm long. Seeds clavate, puberulent. Fl. May-Jun. Distributed in Sichuan, Zhejiang, Jiangsu and Jiangxi. Native to Europe.

地黄 *Rehmannia glutinosa*

天目地黄 *Rehmannia chingii*

地黄

Rehmannia glutinosa (Gaert.) Libosch. ex Fisch. et Mey.

草本，密具绒毛和腺毛或无腺毛。根茎肉质，粗达5.5厘米。茎紫红色。叶卵形至狭椭圆形。花冠管狭；裂片外部紫红色，内部黄紫色。蒴果卵球形至狭卵球形。花果期4-7月。生海拔1100米以下的山坡、山路旁或沙土上。产华中、华东、华北、西北和东北。

Herbs, densely villous with glandular or eglandular hairs. Rhizomes to 5.5 cm thick, fleshy. Stems purple-red. Leaves ovate to narrowly elliptic. Corolla tubes narrow; lobes outside purple-red, inside yellow-purple. Capsules ovoid to narrowly ovoid. Fl. and fr. Apr-Jul. Mountain slopes, trailsides or sandbanks below 1100 m. Distributed in C, E, N, NW and NE China.

天目地黄

Rehmannia chingii H. L. Li

草本，被绒毛。基生叶近莲座状；叶椭圆形，疏被白色绒毛，茎生叶向上渐小。花超出苞片；花冠紫红色；冠筒膨大；雄蕊4；花丝基部具短腺毛。蒴果卵球形。种子卵球形至狭卵球形。花期4-5月，果期5-6月。生海拔100-500米的山坡草甸。产浙江、安徽和江西。

Herbs, villous. Basal leaves subrosulate. Leaves elliptic, sparsely white villous, stem leaves gradually decreasing in size upward. Flowers exceeding bracts; corolla purple-red; tubes inflated; stamens 4; filaments basally short glandular hairy. Capsules ovoid. Seeds ovoid to narrowly ovoid. Fl. Apr-May. Fr. May-Jun. Grasslands on mountain slopes at 100-500 m. Distributed in Zhejiang, Anhui and Jiangxi.

裂叶地黄 *Rehmannia piasezkii*

裂叶地黄
Rehmannia piasezkii Maxim.

草本。叶狭椭圆形，长15厘米，羽状开裂，裂片近三角形，边缘具三角状的尖齿，两面均被白色柔毛；叶柄长约4厘米，带翅，向上叶柄逐渐缩小。小苞片2，与叶同形，不具柄，着生于花梗基部；花萼不等5裂；花冠紫红色，筒长3.5-4厘米，多少囊状，内面褶襞上被长腺毛。花期5-9月。生拔海800-1500米的山坡。产陕西和湖北。

Herbs. Leaves narrowly elliptic, 15 cm long, pinnately parted and lobes subtriangular, lobe margin triangularly toothed, white villous on both surfaces; petioles ca. 4 cm long, decurrent and gradually shorter upward. Bracteoles 2, similar to leaves in shape, sessile, inserted near base of pedicels; calyx unequally 5-lobed; corolla purple-red, tube 3.5-4 cm long, somewhat saccate, inside long glandular villous on plaits. Fl. May-Sep. Mountain slopes at 800-1500 m. Distributed in Shaanxi and Hubei.

呆白菜
Triaenophora rupestris (Hemsl.) Soler.

草本。茎、花梗、叶柄和花萼均被白色绵毛；茎单生或基部分枝，木质化；基部叶柄长3-6厘米。叶卵状长圆形至狭椭圆形，质厚，近革质。花冠紫红色；冠筒狭管状。种子小，长圆形。花期7-9月。生海拔200-1200米的崖面。产湖北。

Herbs. Stems, pedicels, petioles and calyces white woolly; stems simple or branched at base, woody; basal leaves petioles 3-6 cm long. Leaves ovate-oblong to narrowly elliptic, thick, subleathery. Corolla purple-red; tubes narrowly tubular. Seeds minute, oblong. Fl. Jul-Sep. Cliff faces at 200-1200 m. Distributed in Hubei.

呆白菜 *Triaenophora rupestris*

鞭打绣球
Hemiphragma heterophyllum Wall.

多年生铺散匍匐草本，全体被短柔毛。叶二型，主茎上为圆形、心形或肾形，枝上为簇生，针状。花单生叶腋，花冠白色至玫瑰色。果实卵球形，红色，浆果状，肉质。花期4-6月，果期6-8月。生海拔2000-4100米的高山草地或岩石缝。产中国西南、华中、华东和西北。南亚和东南亚亦有。

Perennial herbs, prostrate and diffuse, all pubescent. Leaves dimorphic, on main stems orbicular, cordate or reniform, on branches crowded, needlelike. Flowers solitary, axillary, corolla white to rose. Fruits ovoid, red, berrylike, fleshy. Fl. Apr-Jun. Fr. Jun-Aug. Alpine grasslands or rock crevices at 2000-4100 m. Distributed in SW, C, E and NW China. Also in S and SE Asia.

幌菊
Ellisiophyllum pinnatum (Wall. ex Benth.) Mak.

多年生匍匐草本，除花冠外全体密被短柔毛。叶卵形至长圆状卵形，纸质，疏具柔毛，中部以下浅裂，以上裂锐齿裂，裂片5-9；花冠白色，漏斗状，7-12毫米。种子少量，近球形。花果期7-9月。生海拔

鞭打绣球 *Hemiphragma heterophyllum*

幌菊 *Ellisiophyllum pinnatum*

1500-2500米的草地、溪边或疏林。产云南、四川、贵州、广西、河北、甘肃和江西。印度、不丹、菲律宾、日本和新几内亚岛亦有。

Perennial herbs, creeping, densely pubescent except for corolla. Leaves ovate to oblong-ovate, papery, sparsely villous, margin lobed below middle and acutely crenate above middle, segments 5-9; corolla white, funnelform, 7-12 mm long. Seeds few, subglobose. Fl. and fr. Jul-Sep. Grassland, along streams or sparse forests at 1500-2500 m. Distributed in Yunnan, Sichuan, Guizhou, Guangxi, Hebei, Gansu and Jiangxi. Also in India, Bhutan, the Philippines, Japan and New Guinea.

腹水草

Veronicastrum stenostachyum (Hemsl.) T. Yamaz.

多年生草本。根状茎短，横生；茎长于1米，多于顶端拱曲并生根。叶狭卵形至披针形。花序腋生，有时顶生于侧枝上；花冠白色、紫色或紫红色。蒴果卵球形。花期7-9月，果期10月。生海拔1300米以下的灌丛、林下或林缘。产四川、贵州北部、湖北西部、湖南西北部和陕西南部。

Perennial herbs. Rhizomes short, horizontal. Stems more than 1 m tall, mostly arching and rooting apically. Leaves narrowly ovate to lanceolate. Inflorescences axillary, sometimes terminal on leafy branches; corolla white, purple or purple-red. Capsules ovoid. Fl. Jul-Sep. Fr. Oct. Thickets, forests or forest edges below 1300 m. Distributed in Sichuan, N Guizhou, W Hubei, NW Hunan and S Shaanxi.

腹水草 *Veronicastrum stenostachyum*

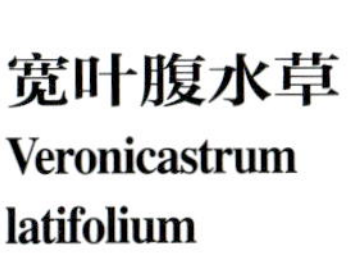

宽叶腹水草

Veronicastrum latifolium (Hemsl.) T. Yamaz.

多年生草本。茎常长于1米，顶部拱形且生根，常具黄色短曲毛。叶互生，卵状圆形至圆形，常疏具硬毛。花序腋生，稀顶生；花冠浅紫色或白色。蒴果卵球形。花期8-9月，果期10月。生海拔300-500米的林中、灌丛或有时倒挂于岩石。产四川中部至东部、贵州、湖北和湖南。

Perennial herbs. Stems usually more than 1 m tall, apically arching and rooting, usually with yellow, short curly hairs. Leaves alternate, ovate-orbicular to orbicular, usually sparsely hispidulous. Inflorescences axillary, rarely also terminal; corolla pale purple or white. Capsules ovoid. Fl. Aug-Sep. Fr. Oct. Forests, thickets or sometimes hanging from rocks at 300-500 m. Distributed in C to E Sichuan, Guizhou, Hubei and Hunan.

宽叶腹水草 *Veronicastrum latifolium*

爬岩红 *Veronicastrum axillare*

爬岩红
Veronicastrum axillare
(Siebold et Zucc.) T. Yamaz.

多年生草本。根状茎短，横走。叶互生，卵形至卵状披针形，纸质，无毛。花序腋生，稀顶生；花冠紫色至紫红色，冠檐辐射状，裂片狭三角形。蒴果卵球形。花期7-9月。生低海拔的林中、林缘草地或沟谷阴处。产华南和华东。日本亦有。

Perennial herbs. Rhizomes short, horizontal. Leaves alternate, ovate to ovate-lanceolate, papery, glabrous. Inflorescences axillary, rarely terminal; corolla purple to purple-red, limbs actinomorphic, lobes narrowly triangular. Capsules ovoid. Fl. Jul-Sep. Forests, grasslands at forest edges or shaded places in valleys at low elevations. Distributed in S and E China. Also in Japan.

大叶腹水草 *Veronicastrum robustum* subsp. *grandifolium*

大叶腹水草
Veronicastrum robustum
(Diels) D. Y. Hong subsp. **grandifolium** T. L. Chin et D. Y. Hong

草本。茎粗壮无毛，直立，每节处折曲。叶互生，无毛，卵状圆形，15-27 × 7-11厘米。单花序腋生，有时2或3个生于叶腋；花冠紫色或白色；冠檐辐射状。种子长圆形。花期6-7月。生疏林或灌丛中。产广西和湖南。

Herbs. Stems erect, twisted apically at each node, stout, terete, glabrous. Leaves alternate, glabrous, ovate-orbicular, 15-27 × 7-11 cm. Single inflorescences axillary, sometimes 2 or 3 in leaf axil; corolla purple or white; limbs actinomorphic. Seeds oblong. Fl. Jun-Jul. Sparse forests or thickets. Distributed in Guangxi and Hunan.

四方麻
Veronicastrum caulopterum
(Hance) Yamazaki

草本。茎直立，有分枝，具翅，无毛。叶互生，长圆形、卵形或披针形。花序顶生于主茎或叶状枝上；苞片长尾尖状；花冠红色、紫红色或深紫色；冠筒与冠檐等长；冠檐放射状。蒴果卵球形。花期8-11月。生海拔2000米以下的沟谷草地或疏林。产中国西南、华南、华中和华东。

Herbs. Stems erect, branched, with wings, glabrous. Leaves alternate, oblong, ovate or lanceolate. Inflorescences terminal on main stems and on leafy branches; bracts long caudate; corolla red, purple-red or dark purple; tubes as long as limb; limbs actinomorphic. Capsules ovoid. Fl. Aug-Nov. Valley meadows or sparse forests below 2000 m. Distributed in SW, S, C and E China.

四方麻 *Veronicastrum caulopterum*

草本威灵仙 *Veronicastrum sibiricum*

细叶穗花 *Pseudolysimachion linariifolium*

草本威灵仙

Veronicastrum sibiricum (L.) Pennell

草本。根状茎横生，节间短。茎直立、单生，圆柱形。叶4-6枚轮生，无柄，长圆形至宽条形。花序顶生；苞片具长尾尖；花冠紫红色、紫色或浅紫色。蒴果卵球形。花期7-9月，果期10月。生海拔2500米以下的山路旁、草坡或灌丛。产山东、华北、西北和东北。俄罗斯、蒙古、朝鲜半岛和日本亦有。

Herbs. Rhizomes horizontal, internodes short. Stems erect, simple, terete. Leaves in whorls of 4-6, sessile, oblong to broadly linear. Inflorescences terminal; bracts long caudate; corolla purple-red, purple or pale purple. Capsules ovoid. Fl. Jul-Sep. Fr. Oct. Trailsides, grassy slopes or thickets below 2500 m. Distributed in Shandong, N, NW and NE China. Also in Russia, Mongolia, Korean Peninsula and Japan.

细叶穗花

Pseudolysimachion linariifolium (Pall. ex Link) Holub

多年生草本。根状茎短。茎直立，单生，有毛。叶全部互生或下部的对生。总状花序单支或数支复出，长穗状；花梗长2-4毫米，被柔毛。花冠蓝色、紫色，筒部长约2毫米，后方裂片卵圆形，其余3枚卵形；花丝无毛，伸出花冠。蒴果长2-3.5毫米，宽2-3.5毫米。花期6-10月。生海拔200-2100米的山坡草地、灌丛间或疏林中。产中国东北、河北、内蒙古、陕西和山西等地。

Perennial herbs. Rhizomes short. Stems erect, often simple, usually with white curly hairs. Leaves alternate at least apically on stem. Inflorescences 1 or several, spikes; pedicel 2-4 mm, pubescent. Corolla blue, purple, tube ca. 2 mm, upper lobe ovate-orbicular, other 3 lobes ovate, filaments exserted, glabrous. Capsules 2-3.5 × 2-3.5 mm. Fl. Jun-Oct. Mea-dows, thickets, sparse forests at 200-2100 m. Distributed in NE China, Hebei, Neimenggu, Shaanxi and Shanxi.

水蔓菁

Pseudolysimachion linariifolium (Pallas ex Link) Holub subsp. **dilatatum** (Nakai et Kitag.) Hong

多年生草本。茎直立，常单生，常具白色弯曲毛。叶片宽线性至卵状圆形，宽0.5-2厘米，边缘齿状，至少在下部节上对生。花冠蓝色、紫色或稀白色。花期7-10月。生海拔200-2100米的草甸或灌丛。产中国西南、华南、华中、华东、华北和西北。

Perennial herbs. Stems erect, often simple, usually with white curly hairs. Leaves broadly linear to ovate-orbicular, 0.5-2 cm wide, margin always dentate, opposite at least on lower nodes. Corolla blue, purple or rarely white. Fl. Jul-Oct. Meadows or thickets at 200-2100 m. Distributed in SW, S, C, E, N and NW China.

水蔓菁 *Pseudolysimachion linariifolium* subsp. *dilatatum*

兔儿尾苗 *Pseudolysimachion longifolium*

兔儿尾苗

Pseudolysimachion longifolium (L.) Opiz

多年生草本。茎直立，常单生，叶腋具不育枝，每个节间有一环连接叶柄基部。叶对生，披针形。花序1至数个，从不为大型圆锥花序，具白色短曲毛；花冠紫色至蓝色；冠筒长为花冠的2/5-1/2。蒴果无毛；宿存花柱长约7毫米。花期6-8月。生海拔1500米以下的典型草原、山坡草地、林缘草甸或桦木林中。产内蒙古、新疆北部、黑龙江和吉林(汪清县)。西南亚、哈萨克斯坦、俄罗斯、蒙古、朝鲜半岛北部和欧洲亦有。

Perennial herbs. Stems erect, usually simple, with sterile branches in leaf axils, with a ring at each node connecting petioles bases. Leaves opposite, lanceolate. Inflorescences 1 or several, never in a large panicle, with white short curly hairs; corolla purple to blue; tubes 2/5-1/2 of corolla length. Capsules glabrous; persistent styles ca. 7 mm long. Fl. Jun-Aug. Steppes, grassy mountain slopes, meadows at forest edges or *Betula* forests below 1500 m. Distributed in Neimenggu, N Xinjiang, Heilongjiang and Jilin (Wangqing County). Also in SW Asia, Kazakhstan, Russia, Mongolia, N Korean Peninsula and Europe.

东北穗花 *Pseudolysimachion rotundum* subsp. *subintegrum*

东北穗花

Pseudolysimachion rotundum subsp. **subintegrum** (Nakai) D. Y. Hong

多年生草本。茎单生。叶对生，中下部的叶无柄，半抱茎，基部楔形，边缘具三角状锯齿。总状花序多单生，长穗状，花序轴密被白色短曲毛；花梗密被多细胞腺毛。花冠蓝色或蓝紫色，筒部短，不足全长1/3，后方1枚裂片卵圆形，其余3枚长卵形；花丝伸出花冠。蒴果长3-5毫米。花期6-8月。生海拔1600米以下的草甸、林缘草地及林中。产中国东北。

Perennial herbs. Stems simple. Leaves opposite, amplexicaul; no leaf petioles from middle to lower; leaf blade base cuneate, margin deltoid dentate. Inflorescences mostly 1, axis densely with white short curly hairs ; pedicel densely with multicellular glandular hairs. Corolla blue or blue-purple, tube short, less than 1/3 of corolla length; lobes patent, upper lobe ovate-orbicular, other 3 lobes narrowly ovate; filaments exserted outside corolla. Capsules 3-5 mm long. Fl. Jun-Aug. Meadows, forest margins, forests, grassy mountain slopes below 1600 m. Distributed in NE China.

密花婆婆纳

Veronica densiflora Ledeb.

多年生丛生草本。叶无柄，下部叶鳞片状；叶卵状圆形。花序顶生，总状或头状；花冠深蓝色；冠筒长1.5-2毫米；雄蕊反折；子房顶部被毛。蒴果倒

密花婆婆纳 *Veronica densiflora*

蚊母草 *Veronica peregrina*

婆婆纳 *Veronica polita*

卵球形，稍压扁。花期5-6月。生海拔3400米以下的多石山坡或林中或高山地区。产新疆(伊宁县)。哈萨克斯坦、俄罗斯和蒙古亦有。

Perennial herbs, caespitose. Leaves sessile, lower ones scale-like; leaves ovate-orbicular. Inflorescences terminal, racemose or capitate; corolla dark blue; tubes 1.5-2 mm long; stamens exserted; ovary apically hairy. Capsules obovoid, slightly compressed. Fl. May-Jun. Stony slopes , forests or alpine zone below 3400 m. Distributed in Xinjiang (Yining County). Also in Kazakhstan, Russia and Mongolia.

蚊母草
Veronica peregrina L.

一年生草本。主茎直立；枝平铺。叶无柄。总状花序顶生及腋生，疏松，果期长达20厘米；花萼4裂；花冠白色或浅蓝色，旋转。蒴果倒心形，强烈压扁。种子每室约40粒，两面凹入。花期5-6月。生海拔3000米以下的湿润弃荒地或路边。产中国西南、东南、华中、华北、华东和东北。俄罗斯、蒙古、朝鲜半岛、日本和欧洲亦有归化。原产北美洲。

Annual herbs. Main stems erect; branches diffuse. Leaves sessile. Racemes terminal and axillary, lax, to 20 cm long in fruits; calyx 4-lobed; corolla white or pale blue, rotate. Capsules obcordate, strongly compressed. Seeds ca. 40 per capsules, convex on both sides. Fl. May-Jun. Moist waste lands or roadsides below 3000 m. Distributed in SW, SE, C, N, E and NE China. Naturalized in Russia, Mongolia, Korean Peninsula, Japan and Europe. Native to North America.

婆婆纳
Veronica polita Fries

一年生草本。茎铺散，疏具柔毛。叶对生，卵形至近圆形。总状花序顶生，疏松，极长；花萼4裂，果期稍膨大；花冠通常蓝色，稀白色，辐状。蒴果肾形，稍微压扁。花期3-10月，果期11月。归化于海拔2200米以下的荒地。产中国西南、华中、华东、西北和北京。原产西南亚，归化于世界大部分地区。

Annual herbs. Stems diffuse, sparsely pubescent. Leaves opposite, ovate to suborbicular. Racemes terminal, lax, very long; calyx 4-lobed, slightly dilated in fruit; corolla usually blue, rarely white, rotate. Capsules reniform, very slightly compressed. Fl. Mar-Oct. Fr. Nov. Naturalized in waste fields below 2200 m. Distributed in SW, C, E and NW China, and Beijing. Native to SW Asia, naturalized in most parts of the world.

阿拉伯婆婆纳
Veronica persica Poir.

一年生或二年生草本。茎铺散，密具多室柔毛。叶卵状披针形至近圆形。总状花序疏松，很长；花萼4裂；花冠通常蓝色，喉部具疏毛。蒴果倒心形，强烈压扁。花期3-5月。生海拔1700米以下的路边或荒野。归化于中国西南、华中、华东和新疆。原产西南亚，自19世纪扩散至世界大部分地区。

Annual, sometimes biennials. Stems diffuse, densely pubescent with multicellular hairs. Leaves ovate-lanceolate to suborbicular. Racemes terminal, lax, very long; calyx 4-lobed; corolla usually blue, throat sparsely hairy. Capsules obcordate, strongly compressed. Fl. Mar-May. Beside roads or in wilds below 1700 m. Naturalized in SW, C and E China, and Xinjiang. Native to SW Asia, and since the 19th century spread over most parts of the world.

阿拉伯婆婆纳 *Veronica persica*

卷毛婆婆纳 *Veronica teucrium* subsp. *altaica*

卷毛婆婆纳

Veronica teucrium L. subsp. **altaica** Watzl

多年生草本。单茎直立或斜升，密被短的向上的曲毛。叶疏被柔毛。总状花序生上部叶叶腋，果期长12厘米；花冠亮蓝色、粉色或白色，旋转。蒴果倒心状卵球形，稍压扁，无毛。种子每室2-10粒，卵球形。花期5-6月。生海拔2000米以下的疏林或草地。产内蒙古北部、新疆北部和黑龙江西北部。哈萨克斯坦和俄罗斯亦有。

Perennial herbs. Stems erect or ascending, simple, densely with short and upward curly hairs. Leaves sparsely puberulent. Racemes axillary from upper leaves, to 12 cm long in fruits; corolla bright blue, pink or white, rotate. Capsules obcordate-ovoid, slightly compressed, glabrous. Seeds 2-10 per locule, ovoid. Fl. May-Jun. Sparse forests or grasslands below 2000 m. Distributed in N Neimenggu, N Xinjiang and NW Heilongjiang. Also in Kazakhstan and Russia.

石蚕叶婆婆纳

Veronica chamaedrys L.

多年生草本。叶卵形至卵状圆形，边缘具深刻钝齿。总状花序成对生于上部叶叶腋；花萼裂片4；花冠蓝色，旋转，内部近无毛；雄蕊短于花冠。蒴果倒心形，强烈压扁。种子扁平。花期5月。产辽宁(凤城)。哈萨克斯坦、俄罗斯和欧洲亦有。

Perennial herbs. Leaves ovate to ovate-orbicular, margin with deeply incised obtuse teeth. Racemes paired, axillary from upper leaves; calyx lobes 4; corolla blue, rotate, subglabrous inside; stamens shorter than corolla. Capsules obcordate, strongly compressed. Seeds flattened. Fl. May. Distributed in Liaoning (Fengcheng). Also in Kazakhstan, Russia and Europe.

四川婆婆纳

Veronica szechuanica Batal.

多年生草本。茎单一或有几分枝，高(5-)15-35厘米，有2列短柔毛。叶卵形，两面或仅上面

石蚕叶婆婆纳 *Veronica chamaedrys*

四川婆婆纳 *Veronica szechuanica*

察隅婆婆纳 *Veronica chayuensis*

具柔毛混生多室毛。总状花序2-6个，生于上部叶叶腋，3-5花形成伞状；花冠白色或稀淡紫色。花期7月。生海拔1600-3500米的山谷、草坡、林下或林缘。产四川东部、湖北西部、陕西南部、甘肃东南部和青海东部。

Perennial herbs. Stems simple or few branched, (5-) 15-35 cm tall, pubescent along 2 lines. Leaves ovate, both surfaces or adaxially hirsute with multicellular hairs. Racemes 2-6, axillary from upper leaves, 3-5-flowered, in corymbs; corolla white or rarely pale purple. Fl. Jul. Valleys, grassy slopes, forests or forest edges at 1600-3500 m. Distributed in E Sichuan, W Hubei, S Shaanxi, SE Gansu and E Qinghai.

察隅婆婆纳

Veronica chayuensis Hong

多年生草本。茎单生。叶对生。花1-3朵簇生上部叶腋；苞片宽条形；花梗仅1-1.5毫米长。花萼裂片条状椭圆形；花冠白色，筒内部无毛，前方裂片倒卵状椭圆形，侧2裂片倒卵形，后方裂片横地矩圆形；雄蕊短于花冠。蒴果扁平，倒心状肾形。种子多粒。花期8月。生海拔3500-4200米的山坡水边碎砾石堆、草丛中及林下。产西藏南部和云南。

Perennial herbs. Stems solitary. Leaves opposite. Flowers 1-3, fascicled in axils of upper leaves; bracts broadly linear. Pedicel 1-1.5 mm long. Calyx 4-lobed; corolla white, tube glabrous inside, lower lobe obovate-elliptic, lateral 2 lobes obovate, upper lobe apex subtruncate; stamens shorter than corolla. Capsules compressed, reniform. Seeds numerous. Fl. Aug. Gravelly slopes by water, meadows, forests at 3500-4200 m. Distributed in S Xizang and Yunnan.

华中婆婆纳

Veronica henryi Yamazadi

多年生草本。茎直立，常紫红色。叶4-6对；叶卵形至狭卵形。总状花序1-4对，生于上部叶叶腋，疏松，具2-7花；花萼4裂；花冠白色或浅红色，具紫色条纹，旋转；喉部具毛。蒴果折扇状菱形，强烈压扁。种子每室2-10粒，扁平。花期4-5月。生海拔500-2300米的阴湿处。产中国西南、华中和华东。

Perennial herbs. Stems erect, often purple-red. Leaves 4-6 pairs; leaves ovate to narrowly ovate. Racemes 1-4 pairs, axillary from upper leaves, lax, 2-7-flowered; calyx 4-lobed; corolla white or pale red, with purple striate, rotate; throat hairy. Capsules pliciform-rhomboid, strongly compressed. Seeds 2-10 per capsules, flattened. Fl. Apr-May. Shaded moist places at 500-2300 m. Distributed in SW, C and E China.

华中婆婆纳 *Veronica henryi*

鹿蹄草婆婆纳

Veronica piroliformis Franch.

多年生草本。茎短，长1-5厘米，节间多而短。叶簇生，常莲座状，多为匙形。花冠紫色、蓝色或白色。蒴果折扇状菱形，基部近于平截或大于120°的角，两侧角急尖或稍钝。花期6-7月。生海拔2600-4000米的山坡草甸、林下或岩石缝中。产云南西北部和四川西南部。

Perennial herbs. Stems short, 1-5 cm long, nodes many and short. Leaves crowded, often rosulate, mostly spatulate. Corolla purple, blue or white. Capsules plicate-rhombic, base subtruncate or at an angle above 120°, two lateral angles acute or obtuse. Fl. Jun-Jul. Slope meadows, forests or crevices of rocks at 2600-4000 m. Distributed in NW Yunnan and SW Sichuan.

鹿蹄草婆婆纳
Veronica piroliformis

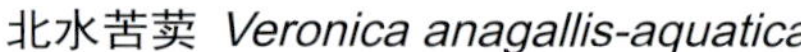
北水苦荬 *Veronica anagallis-aquatica*

北水苦荬

Veronica anagallis-aquatica L.

多年生草本，全株无毛。叶椭圆形至卵形，有时卵状长圆形。总状花序腋生，多花，长于叶；花萼4裂；花冠浅蓝色、浅紫色或白色。蒴果近球形。花期4-9月。生海拔可达4000米的水边或沼地。产中国西南和长江以北。亚洲温带地区和欧洲亦有，归化于北美洲。

Perennial herbs, plants glabrous. Leaves elliptic to ovate, sometime ovate-oblong. Racemes axillary, many-flowered, longer than leaves; calyx 4-lobed; corolla pale blue, pale purple or white. Capsules subglobose. Fl. Apr-Sep. By waters or marsh below 4000 m. Distributed in SW China and the north of Yangtze River. Also in temperate Asia and Europe, naturalized in North America.

水苦荬

Veronica undulata Wall. ex Jack

多年生草本。茎直立或基部匍匐，肉质。叶上部半抱茎，多为椭圆形至卵形。总状花序腋生，宽1-1.5厘米，多花；花梗直而横叉开，与花序轴几成直角；花柱长1-1.5毫米；花冠浅蓝色、浅紫色或白色；花序轴、花萼和蒴果上多少有腺毛。花期4-9月。生海拔2800米以下的水边或沼地。中国除西北和内蒙古以外各省区都有分布。南亚、东南亚、中亚和东亚亦有。

Perennial herbs. Stems erect or procumbent at base, succulent. Leaves amplexicaul upward, mostly elliptic to ovate. Racemes axillary, 1-1.5 cm wide, many-flowered; pedicels erect and divaricate, almost at a right angle to inflorescences rachis; styles 1-1.5 mm long; corolla pale blue, pale purple or white; inflorescences rachis, calyx and capsules glandular hairy. Fl. Apr-Sep. By waters or marsh below 2800 m. Throughout China except NW China and Neimenggu. Also in S, SE, C and E Asia.

水苦荬 *Veronica undulata*

长果水苦荬

Veronica anagalloides Guss.

一年生草本。叶无柄，半抱茎，披针形至条状披针形，长2-5厘米，宽0.5-1厘米。花梗长3-7毫米；花萼裂片椭圆形，果期直立，紧贴蒴果；花冠蓝色或淡紫色；花柱长约1.5毫米。蒴果椭圆形或宽椭圆形，超出花萼。花期6-7月，果期7-9月。生海拔300-2900米的水沟、河边及湿地。产青海、甘肃、陕西、山西、内蒙古、黑龙江和西藏。欧洲和亚洲亦有。

Annual herbs. Leaves sessile, semiamplexicaul, lanceolate to linear-lanceolate, 2-5 × 0.5-1 cm. Pedicel 3-7 mm long; calyx lobes

长果水苦荬 *Veronica anagalloides*

短穗兔耳草 *Lagotis brachystachya*

紫叶兔耳草 *Lagotis praecox*

elliptic, erect and appressed to capsules in fruit; corolla blue or pale purple; style ca. 1.5 mm long. Capsules ellipsoid to broadly so, longer than calyx. Fl. Jun-Jul. Fr. Jul-Sep. Ditches, river sides or wet places at 300-2900 m. Distributed in Qinghai, Gansu, Shaanxi, Shanxi, Neimenggu, Heilongjiang and Xizang. Also in Europe and Asia.

短穗兔耳草
Lagotis brachystachya Maxim.

草本，高4-8厘米。叶基生，排成莲座状；叶宽条形至披针形；花葶少量，细弱，不超出叶。穗状花序紧密；花冠白色、淡红色或紫色。果实红色，卵球形。花果期5-8月。生海拔3200-4500米的高山草地、河岸或湖岸沙质草地。产四川、甘肃、青海和新疆。

Herbs, 4-8 cm tall. Leaves basal, rosulate. Leaves broadly linear to lanceolate; scapes few, slender, not exceeding leaves. Spikes dense; corolla white, reddish or purple. Fruits red, ovoid. Fl. and fr. May-Aug. Alpine grassland, sandy grassland on riverbanks or lake shores at 3200-4500 m. Distributed in Sichuan, Gansu, Qinghai and Xinjiang.

紫叶兔耳草
Lagotis praecox W. W. Sm.

多年生草本。叶全部基生，肾形、圆形或卵形，背面和叶柄紫色。穗状花序卵球形；萼裂成2枚，裂片披针状长圆形，边缘具细流苏状；花冠蓝色，冠筒伸直。果实椭圆状长圆形。花果期7-8月。生海拔4500-5200米的高山草地或沙质及砾质地上。产云南西北部和四川西部。

Perennial herbs. Leaves all basal, reniform, orbicular or ovate, purple abaxially and on petioles. Spikes ovoid; calyx divided into 2-lobes, lobes lanceolate-oblong, minutely fimbriate on margin; corolla blue, tubes straight. Fruits elliptic-oblong. Fl. and fr. Jul-Aug. Alpine grasslands or sandy and gravelly areas at 4500-5200 m. Distributed in NW Yunnan and W Sichuan.

倾卧兔耳草
Lagotis decumbens Rupr.

草本。根状茎短；茎匍匐或斜升。叶卵状长圆形至卵状椭圆形；叶3或4。花萼2裂至近基部；花冠浅蓝色，冠筒直伸，下唇3或4裂，平展，上唇近圆形。花期6-7月。生海拔4800-5500米的冰川、石质山坡或溪边。产西藏和新疆。吉尔吉斯斯坦和塔吉克斯坦亦有。

Herbs. Rhizomes short; stems prostrate to ascending. Leaves ovate-oblong to ovate-elliptic; leaves 3 or 4. Calyx 2-lobed nearly to base; corolla pale blue, tubes straight; lower lips 3- or 4-parted, spreading flat, upper lips suborbicular. Fl. Jun-Jul. Moraines, stony slopes or along streams at 4800-5500 m. Distributed in Xizang and Xinjiang. Also in Kyrgyzstan and Tajikistan.

倾卧兔耳草 *Lagotis decumbens*

云南兔耳草 *Lagotis yunnanensis*

云南兔耳草
Lagotis yunnanensis W. W. Sm.

多年生草本。茎生叶2-6枚，无柄；叶卵形至长圆形，纸质。花萼佛焰苞状，淡黄绿色，边缘常为细流苏状；花冠白色，少紫色，冠筒伸直；下唇裂片2-4，上唇不裂。花期6-8月。生海拔3300-4700米的高山草地。产中国西南。

Perennial herbs. Stem leaves 2-6, sessile; leaves ovate to oblong, papery. Calyx spathelike, pale yellow-green, usually minutely fimbriate on margin; corolla white, rarely purple, tubes straight; lower lips lobes 2-4, upper lips unlobed. Fl. Jun-Aug. Alpine grasslands at 3300-4700 m. Distributed in SW China.

亚中兔耳草
Lagotis integrifolia (Willd.) Schischk.

草本。茎粗壮。穗状花序密，果期伸长；花冠暗白色、浅蓝色或紫色；花冠管长于冠檐，基部向前弯曲；下唇裂片2或3，常于果期反卷；上唇不裂，全缘或具短2或3齿。果矩圆形。花果期6-8月。生海拔2400-3100米的灌丛中或沙砾质坡地。产山西、内蒙古和新疆。哈萨克斯坦、吉尔吉斯斯坦、俄罗斯和蒙古亦有。

Herbs. Stems stout. Spikes dense, elongated in fruit; corolla dull white, pale blue or purple; tubes longer than limb, basally curved forward; lower lips lobes 2 or 3, usually revolute in fruit; upper lips unlobed, margin entire or shortly 2- or 3-toothed. Fruits oblong. Fl. and fr. Jun-Aug. Among thickets or gravel slopes at 2400-3100 m. Distributed in Shanxi, Neimenggu and Xinjiang. Also in Kazakhstan, Kyrgyzstan, Russia and Mongolia.

全缘兔耳草
Lagotis integra W. W. Sm.

草本。茎生叶3-11，近无柄，全缘或具不明显锯齿。穗状花序长5-15厘米；花冠浅黄色、浅绿色或稀紫色；下唇2裂；上唇不裂。果实黑色，圆锥形。花果期6-8月。生海拔3200-4800米的高山草地或针叶林中。产云南西北部、西藏东部、四川西部和青海南部。

Herbs. Stem leaves 3-11, subsessile, margin entire or obscurely incised. Spikes 5-15 cm long; corolla pale yellow, light green or rarely purple; lower lips 2-lobed; upper lips unlobed. Fruits black, conical. Fl. and fr. Jun-Aug. Alpine grassland or conifer forests at 3200-4800 m. Distributed in NW Yunnan, E Xizang, W Sichuan and S Qinghai.

独脚金
Striga asiatica (L.) O. Kuntze

一年生半寄生草本，株高10-20(-30)厘米，被刚毛。茎直立。叶条形，基部狭披针形。花单朵腋生或顶生穗状花序。花萼长4-8毫米，有棱10，5裂；花冠筒顶端急剧弯曲，筒长0.8-1.5厘米；顶端弯曲；唇短2裂。蒴果卵状，包于宿存的萼内。花期秋天。生海拔800米以下的庄稼地和荒草地。产云南、贵州、广西、广东、湖南、江西、福建和台湾。

Annual herbs, hemiparasitic, 10-20(-30) cm tall, entirely hirsute. Stems erect. Leaf blade linear, base narrowly lanceolate. Flowers axillary, solitary or in a spike upward. Calyx 4-8 mm, 10-ribbed; lobes 5, subulate; apically curved; upper lip 2-lobed. Capsules ovoid, enveloped in per-

全缘兔耳草 *Lagotis integra*

亚中兔耳草 *Lagotis integrifolia*

独脚金 *Striga asiatica*

山罗花 *Melampyrum roseum*

滇川山罗花
Melampyrum klebelsbergianum

sistent calyx. Fl. autumn. Crop fields and waste grasslands below 800 m. Distributed in Yunnan, Guizhou, Guangxi, Guangdong, Hunan, Jiangxi, Fujian and Taiwan.

山罗花

Melampyrum roseum Maxim.

一年生草本。茎直立，近四方形。叶条形、条状披针形、椭圆形、卵状椭圆形或狭卵形。苞片基部具尖齿或整个边缘具刚毛状齿；花冠紫色、紫红色或红色，具白边；上唇盔状，内密被须毛。蒴果卵球形。种子黑色。花期夏季至秋季。生海拔1500米以下的山坡灌丛或高草丛中。产中国西南、华南、东南、华中、华北、华西、华东和东北。俄罗斯(远东地区)、朝鲜半岛和日本亦有。

Annual herbs. Stems erect, subquadrangular. Leaves linear, linear-lanceolate, elliptic, ovate-elliptic or narrowly ovate. Bracts with pointed teeth at base or whole margin with setoselike teeth; corolla purple, purple-red or red, with white on sides; galea densely barbate inside. Capsules ovoid. Seeds black. Fl. summer to autumn. Thickets on slopes or among high grasses below 1500 m. Distributed in SW, S, SE, C, N, W, E and NE China. Also in Russia (Far East), Korean Peninsula and Japan.

滇川山罗花

Melampyrum klebelsbergianum Soó

一年生草本。茎直立，四方形。叶披针形，粗糙。花自第5至第9个茎节向上开放；花冠紫红色至红色，具白边；冠筒为冠檐的约2倍；上唇盔状，内密被须毛。蒴果卵状圆锥形，粗糙。种子黑色。花期6-8月。生海拔1200-3400米的草坡或林下。产云南西北部、四川和贵州(贵定县)。

Annual herbs. Stems erect, quadrangular. Leaves lanceolate, scabrous. Flowers starting from 5th to 9th node of stems; corolla purple-red to red, with white sides; tubes ca. 2 × as long as limbs; galea densely barbate inside. Capsules ovoid-conical, scabrous. Seeds black. Fl. Jun-Aug. Grassy slopes or forests at 1200-3400 m. Distributed in NW Yunnan, Sichuan and Guizhou (Guiding County).

松蒿

Phtheirospermum japonicum (Thunb.) Kanitz

一年生草本，被腺毛。叶对生，狭三角状卵形，1回羽状全裂，小裂片多为卵形，边缘具重锯齿或羽状深裂。花冠淡紫红色或红色，外面具柔毛，下裂片钝，上裂片三角状卵形。花果期6-10月。生海拔100-1900米的山坡灌丛湿地中。产中国除新疆以外的地区。俄罗斯、朝鲜半岛和日本亦有。

Annual herbs, with glandular hairs. Leaves opposite, narrowly triangular-ovate, pinnately cleft, segments mostly ovate, margin double serrate or pinnatisect. Corolla pale purple-red or red, outside villous, lower lobes obtuse, upper lobes triangular-ovate. Fl. and fr. Jun-Oct. Wet places in shrubbery on slopes at 100-1900 m. Distributed throughout China, except Xinjiang. Also in Russia, Korean Peninsula and Japan.

松蒿 *Phtheirospermum japonicum*

细裂叶松蒿 *Phtheirospermum tenuisectum*

小米草 *Euphrasia pectinata*

细裂叶松蒿

Phtheirospermum tenuisectum Bureau et Franch.

多年生草本。茎多数，丛生。叶三角状卵形，2-3回羽状全裂，小裂片条形，具多室腺柔毛。花单生；花冠黄色至橘黄色，外面具腺毛或无腺绒毛，喉部具毛，下裂片倒卵形。花果期5-10月。生海拔1900-4100米的草坡、灌丛或林下。产中国西南和西北。不丹亦有。

Perennial herbs. Stems numerous, caespitose. Leaves triangular-ovate, 2-3-pinnately cleft, segments linear, with villous multicellular glandular hairs. Flowers solitary; corolla usually yellow to orange-yellow, outside glandular and eglandular villous, throat hairy, lower lobes obovate. Fl. and fr. May-Oct. Grassy slopes, thickets or under woods at 1900-4100 m. Distributed in SW and NW China. Also in Bhutan.

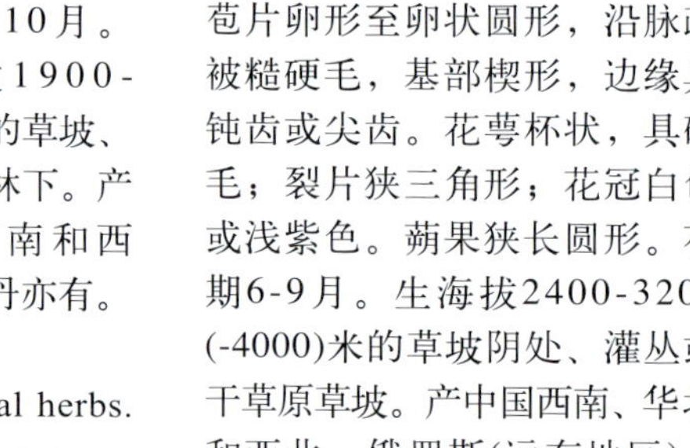

小米草

Euphrasia pectinata Tenore

一年生草本。茎单一或自基部分枝，高10-30(-45)厘米。叶和苞片卵形至卵状圆形，沿脉疏被糙硬毛，基部楔形，边缘具钝齿或尖齿。花萼杯状，具硬毛；裂片狭三角形；花冠白色或浅紫色。蒴果狭长圆形。花期6-9月。生海拔2400-3200(-4000)米的草坡阴处、灌丛或干草原草坡。产中国西南、华北和西北。俄罗斯(远东地区)、蒙古和欧洲亦有。

Annual herbs. Stems 10-30(-45) cm tall, simple or branched from base. Leaves and bracts ovate to ovate-orbicular, sparsely hispid along veins, base cuneate, margin with several obtuse to acute teeth. Calyx tubular, hispid; lobes narrowly triangular; corolla white or pale purple. Capsules narrowly oblong. Fl. Jun-Sep. Shaded grassy slopes, thickets or slopes in steppes at 2400-3200(-4000) m. Distributed in SW, N and NW China. Also in Russia (Far East), Mongolia and Europe.

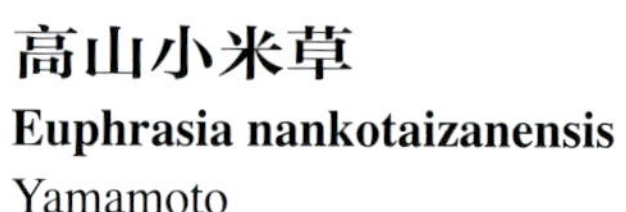

高山小米草

Euphrasia nankotaizanensis Yamamoto

多年生草本。叶卵圆形至狭卵形，长5-12毫米，先端圆，每边有3-4(-6)枚尖齿，两面被刚毛；无柄。穗状花序顶生；花萼果期长达7毫米，密被头状腺点，裂片卵状披针形，长为萼筒的2倍，先端钝；花冠黄色，

高山小米草 *Euphrasia nankotaizanensis*

被头状腺点，筒部长达1厘米，下唇中裂片宽1.5毫米。蒴果疏被刚毛。花期7-10月。生海拔2800-3600米的多石山坡。产台湾。

Perennial herbs. Stems ascending or erect, often basally prostrate, simple or branched, densely white pubescent. Leaves ovate to narrowly ovate, 5-12 mm long, apex rounded, margin acutely 3-4(-6)-toothed on each side, hispid. Spikes terminal; calyx to 7 mm in fruit, densely capitate glandular, lobes ovate-lanceolate, ca. 2 × as long as tube, apex obtuse; corolla yellow, capitate glandular; tube ca. 1 cm long; middle lobe of lower lip ca. 1.5 mm broad. Capsules narrowly ovoid, nearly as long as calyx, sparsely hirsute. Fl. Jul-Oct. Gravelly slopes at 2800-3600 m. Distributed in Taiwan.

疗齿草 *Odontites vulgaris*

白氏马先蒿 *Pedicularis paiana*

疗齿草
Odontites vulgaris Moench

一年生草本，高20-60厘米，全体被贴伏的白色毛。叶披针形至条状披针形。穗状花序顶生；下部的苞片叶状；花冠紫色、紫红色或淡红色，长8-10毫米，外被白色长柔毛。花期7-8月。生海拔2000米以下的草甸。产华北、西北和东北。哈萨克斯坦、乌兹别克斯坦、吉尔吉斯斯坦、塔吉克斯坦、俄罗斯、蒙古和欧洲亦有。

Annual herbs, 20-60 cm tall, entirely with appressed white hairs. Leaves lanceolate to linear-lanceolate. Spikes terminal; lower bracts leaflike; corolla purple, purple-red or pale red, 8-10 mm long, white villous outside. Fl. Jul-Aug. Meadows below 2000 m. Distributed in N, NW and NE China. Also in Kazakhstan, Uzbekistan, Kyrgyzstan, Tajikistan, Russia, Mongolia and Europe.

硕大马先蒿 *Pedicularis ingens*

白氏马先蒿
Pedicularis paiana Li

多年生草本，高35厘米。茎单出，有毛。叶多茎生，羽裂，裂片每边10-15。苞片叶状，羽裂。萼外有毛，管达1.5厘米，齿5枚；花冠长40-45毫米，盔宽6-8毫米，前端下缘终于1个不显著的小凸尖，上半沿下缘有密须毛，下唇裂片3枚亚相等；雄蕊花丝两对均有疏毛。花期7-8月，果期8-9月。生海拔2800-3000米的高山荒草坡中。产甘肃和四川西部。

Perennial herbs, 35 cm tall. Stems solitary, pubescent. Leaves mostly on stem, pinnatifid; segments 10-15 pairs. Bracts leaflike, pinnate. Calyx pubescent, tube to 1.5 cm, lobes 5; corolla 40-45 mm long, galea 6-8 mm wide, as long as tube, apex acute with 1 protrude, margin densely ciliate; lower lip 3-lobes, lobes ± equal; filaments sparsely pubescent. Fl. Jul-Aug. Fr. Aug-Sep. Alpine meadows at 2800-3000 m. Distributed in Gansu and W Sichuan.

硕大马先蒿
Pedicularis ingens Maxim.

多年生草本，干时变黑色。茎高于60厘米，直立，中空，被短柔毛，基部具长圆形鳞片。茎生叶抱茎，边缘具锐重锯齿。花序长达20厘米；花冠黄色，约2.5厘米。花期7-9月，果期9月。生海拔3000-4200米的高杂草坡或矮灌草坡。产四川北部、甘肃和青海东部。

Perennial herbs, drying black. Stems more than 60 cm tall, erect, hollow, pubescent, with oblong scales at base. Stem leaves clasping, margin incised-double dentate. Inflorescences to 20 cm long; corolla yellow, ca. 2.5 cm long. Fl. Jul-Sep. Fr. Sep. High weedy slopes or grassy and scrubby slopes at 3000-4200 m. Distributed in N Sichuan, Gansu and E Qinghai.

克氏马先蒿
Pedicularis clarkei Hook. f.

克氏马先蒿 *Pedicularis clarkei*

多年生草本，高升，干时变黑。茎不分枝，中空。叶基部耳形抱茎，轴有翅，裂片15-25对。花序稠密；苞片线状长圆形。萼钟形，5齿；花冠下唇3裂，在中裂与侧裂组成的2个缺的底部均有1齿，盔不作舟形，前方渐细而端有2浅裂的喙；花丝无毛。蒴果卵形尖头。种子具深网纹。花期7-9月，果期8-9月。生海拔3800-4500米的灌丛中或陡岩下。产西藏南部。

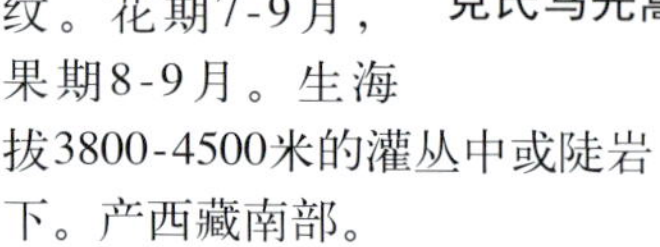

Perennial herbs, hirtellous, black when dry. Stems hollow, unbranched. Leaves base auriform, rachis winged, segments 15-25 pairs. Inflorescences dense; bracts leaflike, oblong. Calyx bell-shaped, 5-lobed; corolla lower lip 3-lobes, 1 tooth at the base of middle and lateral lobes; galea not navicular, beak apex 2-lobed; filaments glabrous. Capsules ovoid, apiculate. Seed with deep reticulate. Fl. Jul-Sep. Fr. Aug-Sep. Among dwarf scrubs, steep banks below cliffs at 3700-4500 m. Distributed in S Xizang.

毛盔马先蒿
Pedicularis trichoglossa Hook. f.

毛盔马先蒿 *Pedicularis trichoglossa*

多年生草本。叶长而狭，条状披针形，每侧有裂片20-25。花序总状；花萼密具黑紫色柔毛，5裂；花冠黑紫红色，冠筒在基处强烈向前弓曲，喙极长。蒴果阔卵球形。花期7-8月，果期8-9月。生海拔3500-5000米的林下多石开阔草甸或山麓砾石中。产云南西北部、四川西部、西藏南部和东南部、青海。印度、尼泊尔、不丹和缅甸亦有。

Perennial herbs. Leaves long and narrow, linear-lanceolate, with 20-25 segments in each side. Inflorescences racemose; calyx densely blackish purple villous, 5-lobed; corolla dark purple-red, tubes bending forward at base, beaks extremely long. Capsules broadly ovoid. Fl. Jul-Aug. Fr. Aug-Sep. Open stony meadows in forests or amidst boulder screes at 3500-5000 m. Distributed in NW Yunnan, W Sichuan, S and SE Xizang, and Qinghai. Also in India, Nepal, Bhutan and Myanmar.

红毛马先蒿
Pedicularis rhodotricha Maxim.

红毛马先蒿 *Pedicularis rhodotricha*

多年生草本，高8-35厘米。茎有2列毛。叶条状披针形，羽状深裂至全裂；裂片8-12对。花序头状至总状，常紧密；苞片叶状；花萼与花冠紫红色；盔具浅红色毛。花期6-8月，果期8-9月。生海拔2600-4000米的多石高山草甸或岩屑堆。产云南西北部和四川西部。

Perennial herbs, 8-35 cm tall. Stems with 2 lines of hairs. Leaves linear-lanceolate, pinnatipartite to pinnatisect; segments 8-12 pairs. Inflorescences capitate to racemose, usually dense; bracts leaflike; calyx and corolla purple-red; galea with pale red hairs. Fl. Jun-Aug. Fr. Aug-Sep. Stony alpine meadows or screes at 2600-4000 m. Distributed in NW Yunnan and W Sichuan.

维氏马先蒿
Pedicularis vialii Franch.

多年生草本，高达80厘米。叶有长柄；叶片披针状长圆形，中部者最大，上部者渐小，叶轴有翅，羽状全裂，裂片各边5-10。花序总状；苞片很不发达；萼长光滑，钟形，齿5枚；花冠连盔长仅1厘米，3裂，盔有长喙。果扁平，光滑。花期5-8月，果期7-9月。生海拔2700-4300米的针叶林下或草坡中。产四川西部、西藏东南部和云南西北部。

Perennial herbs, to 80 cm tall. Leaf with long petioles; leaf blade lanceolate ovate, middle leaves larger than upper leaves, rachis winged, pinnatisect; segments 5-10 pairs. Inflorescences raceme, bracts undeveloped; calyx glabrous, navicular, lobes 5; corolla with galea ca. 1 cm, 3-lobes; galea with beak. Capsules flat, glabrous. Fl. May-Aug. Fr. Jul-

维氏马先蒿 *Pedicularis vialii*

Sep. Coniferous forests or grassy slopes at 2700-4300 m. Distributed in W Sichuan, SE Xizang and NW Yunnan.

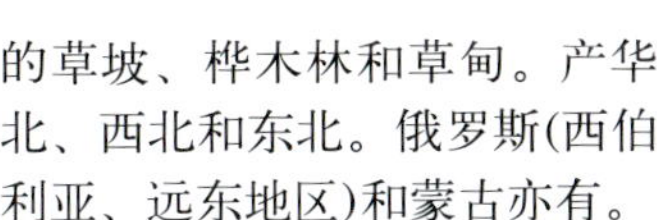

鼻喙马先蒿 *Pedicularis proboscidea*

鼻喙马先蒿

Pedicularis proboscidea Steven

一年生草本。根短，纤维状。茎直立，粗壮。基生叶具长柄，羽状全裂；裂片羽状半裂。花序长20厘米，密花；苞片条形；花萼卵形，先端深裂；裂片5，三角状披针形；花冠黄色；盔边缘具长须毛；喙直伸。蒴果斜卵球形。花期6-7月，果期7-8月。生高

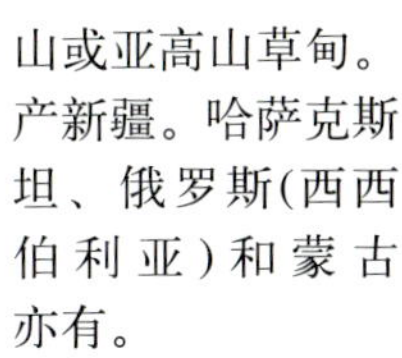

山或亚高山草甸。产新疆。哈萨克斯坦、俄罗斯(西西伯利亚)和蒙古亦有。

Perennial herbs. Roots short, fibrous. Stems erect, stout. Basal leaves long petiolate, pinnatisect; segments pinnatipartite. Inflorescences to 20 cm long, dense; bracts linear; calyx ovate, deeply cleft anteriorly; lobes 5, triangular-lanceolate; corolla yellow; galea margin long bearded; beaks straight. Capsules obliquely ovoid. Fl. Jun-Jul. Fr. Jul-Aug. Alpine or subalpine meadows. Distributed in Xinjiang. Also in Kazakhstan, Russia (W Siberia) and Mongolia.

红纹马先蒿

Pedicularis striata Pall.

多年生草本。叶披针形，羽状深裂至全裂。花序密穗状；花萼疏具柔毛，裂片5；花冠黄色，具紫红色条纹，长2.5-3.3厘米；盔在边缘一侧具1明显的齿。蒴果卵球形。花期6-7月，果期7-8月。生海拔300-2700米的草坡、桦木林和草甸。产华北、西北和东北。俄罗斯(西伯利亚、远东地区)和蒙古亦有。

Perennial herbs. Leaves lanceolate, pinnatipartite to pinnatisect. Inflorescences dense, spicate; calyx sparsely pubescent, lobes 5; corolla yellow with reddish purple stripes, 2.5-3.3 cm long; galea with a distinct tooth on one side of margin. Capsules ovoid. Fl. Jun-Jul. Fr. Jul-Aug. Grassy slopes, Betula forests and meadows at 300-2700 m. Distributed in N, NW and NE China. Also in Russia (Siberia, Far East) and Mongolia.

红纹马先蒿 *Pedicularis striata*

绒舌马先蒿 *Pedicularis lachnoglossa*

假山萝花马先蒿 *Pedicularis pseudomelampyriflora*

绒舌马先蒿

Pedicularis lachnoglossa Hook. f.

多年生草本。茎花葶状，不分枝。叶多基生，裂片细而密，茎生者不发达。花序伸长，尤其下部花多少疏远；花冠盔部有直而细的喙，喙端有刷毛，上面及缘有毛，下唇裂片锐头，管在萼内稍稍前俯；花丝无毛；花柱不伸出。蒴果黑色，长卵圆形。花期6-7月，果期8月。生海拔2500-5400米的高山草原与疏云杉林中多石之处。产中国西南。

Perennial herbs. Stems scape-like, unbranched. Base leaves numerous, segment clustered at base; stem leaves undeveloped. Inflorescences racemose, lower usually interrupted; corolla galea with erect and slender beak, densely red-brown pubescent abaxially and at margin; lower lip ciliate; tube forwar in calyx; filaments glabrous; styles not extended. Capsules black, long ovoid. Fl. Jun-Jul. Fr. Aug. Alpine meadows, among small shrubs on hillsides at 2500-5400 m. Distributed in SW China.

丹参花马先蒿

Pedicularis salviiflora Franch. ex Forbes et Hemsl.

多年生草本。茎下部常木质化，分枝对生。叶对生，卵形至长圆状披针形，密具柔毛，羽状半裂至羽状全裂。花序疏总状，长达25厘米；花冠玫红色至红色，大，有疏毛。蒴果卵球形。花期8-9月，果期10-11月。生海拔2000-3900米的草坡或林下。产云南西北部和四川西部。

Perennial herbs. Lower part of stems woody, branches opposite. Leaves opposite, ovate to oblong-lanceolate, densely pubescent, pinnatipartite to pinnatisect. Lax racemes, up to 25 cm long; corolla rose to red, large, sparsely hairy. Capsules ovoid. Fl. Aug-Sep. Fr. Oct-Nov. Grassy slopes or forests at 2000-3900 m. Distributed in NW Yunnan and W Sichuan.

假山萝花马先蒿

Pedicularis pseudomelampyriflora Botani

一年生草本。茎单出；枝3-4条轮生。叶3-6枚轮生，羽状深裂至全裂，裂片线形。花轮间断；苞片叶状。萼齿5枚，齿具白色胼胝；花冠长1.8-2.1厘米，管盔约相等，前者在中部以下作强烈的弓曲使花前俯，后者镰状弓曲，背部至额部有鸡冠状凸起，下唇长0.9-1.15厘米；花丝1对有毛。花期6-8月，果期9-10月。生海拔3000-3800米阳坡。特产四川西北部和云南西北部。

Annual herbs. Stems single; branches in whorls of 3-4. Leaves in whorls of 3-6; leaf blade pinnatipartite to pinnatisect; segments linear, serrate. Inflorescences with flowers in whorls in ± interrupted racemes; bracts leaflike. Calyx lobes 5, serrate with white callus; corolla 1.8-2.1 cm long, tube nearly as long as galea, strongly curved in calyx; galea falcate, crested; lower lip 0.9-1.15 cm; 2 filaments pubescent, 2 glabrous. Fl. Jun-Aug. Fr. Sep-Oct. Open moist areas, thicket margins at 3000-3800 m. Endemic to NW Sichuan and NW Yunnan.

山萝花马先蒿

Pedicularis melampyriflora Franch. ex Maxim.

一年生草本。老时茎基部木质化，多分枝。叶3-6轮排列，披

丹参花马先蒿 *Pedicularis salviiflora*

山萝花马先蒿 *Pedicularis melampyriflora*

长茎马先蒿 *Pedicularis longicaulis*

针状长圆形，羽状全裂。花序疏松而间断；花冠玫瑰色至紫色，在萼中强烈反折；盔直伸，不具鸡冠状凸起。花期7-8月，果期9-10月。生海拔2700-3600米的山坡或林中。产云南西北部和四川西南部。

Annual herbs. Stems woody when old, many-branched. Leaves in whorls of 3-6, lanceolate-oblong, pinnatisect. Inflorescences lax, interrupted; corolla rose to purple, strongly curved in calyx; galea straight, not crested. Fl. Jul-Aug. Fr. Sep-Oct. Mountain slopes or forests at 2700-3600 m. Distributed in NW Yunnan and SW Sichuan.

长茎马先蒿

Pedicularis longicaulis Franch. ex Maxim.

一年生或多年生草本，植株高达1米以上。茎中空。叶长圆状披针形，羽状半裂至全裂，裂片10-14对，具骨质齿。花序长总状，顶生，间断；花冠紫红色；盔先端强烈弯曲。花期8-9月，果期9-10月。生海拔3700-3900米的山坡草地。产云南北部。

Annual or perennial herbs, to more than 1 m tall. Stems hollow. Leaves oblong-lanceolate, pinnatipartite to pinnatisect; segments 10-14 pairs, teeth callose. Inflorescences long racemose, terminal, interrupted; corolla purple-red; galea strongly bent at apex. Fl. Aug-Sep. Fr. Sep-Oct. Grasslands on slopes at 3700-3900 m. Distributed in N Yunnan.

纤细马先蒿

Pedicularis gracilis Wall.

一年生草本。茎在4-6轮生枝。叶片卵状长圆形。总状花序；苞片叶状；花萼圆筒状，沿中脉被短柔毛，裂片5，全缘或有锯齿。花冠紫红色，1.2-1.5厘米，直管，7-8毫米，盔弯曲成直角的顶部，无冠，喙4-5.5毫米；下唇7-10毫米宽，无毛；花丝无毛。蒴果卵形。花期8-9月，果期9-10月。生海拔2000-4000米的高山草甸。产四川、西藏和云南。

Annual herbs. Stems branches in whorls of 4-6. Leaf blade ovate-oblong. Inflorescences racemose; bracts leaf-like; calyx cylindric, pubescent along mid-vein, lobes 5, entire or serrate; corolla purplish pink, 1.2-1.5 cm, tube straight, 7-8 mm, galea bent at a right angle apically, not crested, beak 4-5.5 mm, lower lip 7-10 mm wide, glabrous; filaments glabrous. Capsules ovoid. Fl. Aug-Sep. Fr. Sep-Oct. Alpine meadows on mountain slopes, grassy slopes at 2000-4000 m. Distributed in Sichuan, Xizang and Yunnan.

纤细马先蒿 *Pedicularis gracilis*

埃氏马先蒿 *Pedicularis artselaeri*

腋花马先蒿 *Pedicularis axillaris*

埃氏马先蒿

Pedicularis artselaeri Maxim.

多年生草本，干后稍黑色。根肉质。茎1至数条包于膜质鳞片和宿存的叶柄中。叶小裂片8-14对，羽状半裂。花萼裂片5，叶状；花冠紫色。蒴果完全包被于花萼中。花期5月，果期7-8月。生海拔1000-2800米的湿润处、多石山坡或林下。产四川东北部、湖北、河北、山西和陕西。

Perennial herbs, drying slightly black. Roots fleshy; stems 1 to several together, enveloped in membranous scales and marcescent petioles. Leaves segments 8-14 pairs, pinnatipartite. Calyx lobes 5, leaflike; corolla purple. Capsules completely enclosed by calyx. Fl. May. Fr. Jul-Aug. Moist places, rocky slopes or forests at 1000-2800 m. Distributed in NE Sichuan, Hubei, Hebei, Shanxi and Shaanxi.

腋花马先蒿

Pedicularis axillaris Franch. ex Maxim.

多年生草本。叶多对生，椭圆状披针形，羽状全裂；裂片5-12对，有锐锯齿；花冠紫色或浅绿色；花冠管直立，长为花萼的约2倍；盔以直角转折。蒴果扁，卵球形。花期6-8月，果期7-9月。生海拔2700-4000米的湿牧场、林下、灌丛阴湿处或开阔岩石缝。产云南西北部、四川西南部和西藏东南部。

Perennial herbs. Leaves mostly opposite, elliptic-lanceolate, pinnatisect; segments 5-12 pairs, incised-dentate; corolla purple or greenish white; tubes erect, ca. 2 × as long as calyx; galea bent at a right angle. Capsules compressed, ovoid. Fl. Jun-Aug. Fr. Jul-Sep. Moist pastures, shaded damp places in forests, thickets or open rock crevices at 2700-4000 m. Distributed in NW Yunnan, SW Sichuan and SE Xizang.

蔊菜叶马先蒿

Pedicularis nasturtiifolia Franch.

多年生草本，干时不变黑。叶卵形至长圆形，膜质，羽状全裂；裂片2-7对，边缘具重锯齿。花萼前方稍裂，叶状；花冠玫瑰色，下唇大，稍具缘毛；花丝2枚具短柔毛，2枚无毛。花期6-7月，果期7-8月。生海拔约2000米的潮湿处或林下。产四川东部、湖北西部和陕西。

蔊菜叶马先蒿 *Pedicularis nasturtiifolia*

巴塘马先蒿 *Pedicularis batangensis*

细管马先蒿 *Pedicularis gracilituba*

Perennial herbs, not drying black. Leaves ovate to oblong, membranous, pinnatisect; segments 2-7 pairs, margin double dentate. Calyx slightly cleft anteriorly, leaflike; corolla rose, lower lips large, slightly ciliate; filaments 2 pubescent, 2 glabrous. Fl. Jun-Jul. Fr. Jul-Aug. Moist places or forests at ca. 2000 m. Distributed in E Sichuan, W Hubei and Shaanxi.

巴塘马先蒿

Pedicularis batangensis Bureau et Franch.

多年生草本，高10-20厘米，干时黑色。叶长圆形至卵状长圆形，裂片条形至条状披针形。花散生；花冠粉红色至玫瑰色，直立，1.7-3厘米，密被短柔毛。蒴果卵球形，稍压扁，具细尖。花期6-8月，果期8-9月。生海拔2500-3100米的开阔石坡。产四川西北部和西部。

Perennial herbs, 10-20 cm tall, drying black. Leaves oblong to ovate-oblong, segments linear to linear-lanceolate. Flowers scattered; corolla pink to rose, erect, 1.7-3 cm long, densely pubescent. Capsules ovoid, slightly compressed, apiculate. Fl. Jun-Aug. Fr. Aug-Sep. Open rocky slopes at 2500-3100 m. Distributed in NW and W Sichuan.

花楸叶马先蒿

Pedicularis sorbifolia Tsoong

多年生草本，有疏毛。基出叶大，互生或对生；叶柄长达4.5厘米；叶披针状倒卵形，羽状全裂，裂片6-13对，卵形至长圆形，长达18毫米，边缘有锐重锯齿。花梗长3-8毫米；萼长几不达1厘米，前方开裂至中部，2-4浅裂；花冠管达9.5厘米，外面有毛；花丝无毛。花期8月，果期9月。生海拔约3300米的原始冷杉林下苔藓中。产四川西部。

Perennial herbs, slightly pubescent. Basal leaves large, alternate or opposite; petioles to 4.5 cm long; leaves lanceolate-obovate, pinnatisect, segments 6-13 pairs, ovate to oblong, to 18 mm long, margin incised-double dentate. Pedicel 3-8 mm long; calyx scarcely 1 cm long, cleft anteriorly to middle of tube, 2-4-lobed; corolla tube to 9.5 cm long, pubescent outside; filaments glabrous. Fl. Aug. Fr. Sep. Mossy places in old growth Abies forests at ca. 3300 m. Distributed in W Sichuan.

花楸叶马先蒿 *Pedicularis sorbifolia*

细管马先蒿

Pedicularis gracilituba Li

多年生草本。茎多数。叶互生，披针状长圆形，膜质，长2.5-3.5厘米，上面疏被短硬毛，羽状全裂，裂片4-6对，裂片卵形至卵状长圆形，边缘具深锐锯齿；叶柄长0.2-4厘米。花梗长可达1厘米；花萼先端尖裂，被短柔毛，裂片5；花冠紫色，外面有短柔毛；花丝无毛。花期6-7月，果期8月。生海拔3600-4000米的高山草地和林下。产四川西南部。

Perennial herbs. Stems many. Leaves alternate, lanceolate-oblong, membranous, 2.5-3.5 cm long, sparsely hispidulous adaxially, pinnatisect, segments 4-6 pairs, lobes ovate to ovate-oblong, margin deeply incised-dentate; petioles 0.2-4 cm long. Pedicel to 1 cm long; calyx slightly cleft anteriorly, pubescent, lobes 5; corolla purple, pubescent outside; filaments glabrous. Fl. Jun-Jul. Fr. Aug. Alpine meadows and forests at 3600-4000 m. Distributed in SW Sichuan.

大管马先蒿 *Pedicularis macrosiphon*

大管马先蒿
Pedicularis macrosiphon Franch.

多年生草本，通常密集丛生，干时变黑。叶卵状披针形至条状长圆形，羽状全裂；裂片7-12对，具重针状齿。花较宽大；花冠淡紫色至玫瑰色，4.5-6厘米或有时7-8厘米。蒴果完全包被于花萼内，长圆形至倒卵球形。花期5-8月，果期7-9月。生海拔1200-3500米的潮湿阴林或沟壑。产云南西北部和四川西北部。

Perennial herbs, usually densely tufted, drying black. Leaves ovate-lanceolate to linear-oblong, pinnatisect; segments 7-12 pairs, spinescent double dentate. Flowers widely spaced; corolla pale purple to rose, 4.5-6 cm long or sometimes to 7-8 cm long. Capsules completely enclosed by calyx, oblong to obovoid. Fl. May-Aug. Fr. Jul-Sep. Moist shaded forests or ravines at 1200-3500 m. Distributed in NW Yunnan and NW Sichuan.

藓生马先蒿
Pedicularis muscicola Maxim.

多年生草本。茎丛生。叶柄长达1.5厘米；叶片羽状全裂，每边4-9枚。花腋生；萼前方不裂，齿5；花冠管长4-7.5厘米，盔在基部即向左方扭折使其顶部向下，长喙卷曲或“S”字形，长达1厘米以上，下唇侧裂极大，中裂较狭，长圆形；花丝无毛。蒴果为宿萼所包。花期5-7月，果期8月。生海拔1700-2700米的杂林中。产中国西北和华北。

五角马先蒿 *Pedicularis pentagona*

Perennial herbs. Stems cespitose. Leaves petioles to 1.5 cm long; leaf blade pinnatisect; segments 4-9 pairs. Calyx cylindric, without cracking, lobes 5; corolla tube 4-7.5 cm; galea twisted; beak S-shaped, above 1 cm long; lateral lobe of lower lip larger than other lobe, middle lobe narrow, oblong; filaments glabrous. Capsules enclosed by calyx. Fl. May-Jul. Fr. Aug. Under shrubs, near water in valleys at 1700-2700 m. Distributed in NW and N China.

藓生马先蒿 *Pedicularis muscicola*

五角马先蒿
Pedicularis pentagona Li

多年生草本。根肉质。茎直立。叶对生具长柄，叶片披针形，羽状全裂，裂片4形。花序穗状；花成对着生；萼膨大，长卵形，具5条角，齿5；花冠达2.5厘米，向前弯曲，盔仅下唇的半长，下缘有一对细齿；雄蕊花丝近端处被毛；子房卵圆形，花柱细线形。蒴果卵圆形。花期7-8月，果期8-9月。生海拔2800-3300米干燥或阴湿的山坡。特产中国西南。

Perennial herbs. Roots fleshy. Stems erect. Leaves opposite, petioles long; leaf blade lanceolate, pinnatisect; segments 4 types. Inflorescences spicate; flowers opposite; calyx expanded, long ovate, ridge 5, lobes 5; corolla to 2.5 cm, bend forward, galea erect, 1/2 length of lower lip, beakless; margin of lower lip fimbriate; filaments pubescent; ovary oblong, styles linear. Capsules ovoid. Fl. Jul-Aug. Fr. Aug-Sep. Dry slopes or moist shaded banks in valleys at 2800-3300 m. Endemic to SW China.

二岐马先蒿
Pedicularis dichotoma Bonati

多年生草本，干后不为黑色。叶对生，羽状全裂，裂片5-7对。花序穗状，具2-18对花；花冠粉红色；盔顶端下弯成一直角。蒴果包被于增大的花萼中，卵球形。花期7-9月，果期8-9月。生海拔2700-4300米的开阔高山牧场或开阔林地。产云南西北部、四川西南部和西藏东部。

Perennial herbs, not drying black. Leaves opposite, pinnatisect, segments 5-7 pairs. Spikes with 2-18 pairs of flowers; corolla pink; galea ± bent at a right angle apically. Capsules enclosed by accrescent calyx, ovoid. Fl. Jul-Sep. Fr. Aug-Sep. Open alpine pastures or open forests at 2700-4300 m. Distributed in NW Yunnan, SW Sichuan and E Xizang.

大王马先蒿
Pedicularis rex C. B. Clarke ex Maxim.

多年生草本。叶轮生；上部叶柄多强烈膨大，在每一轮中互相结合成一斗状体，最下部者常不膨大而各自分离。花中等大小；花冠黄色。一多变异种。花期5-8月，果期8-9月。生海拔2500-4300米的山坡、平原或高山疏针叶林下。产云

二岐马先蒿 *Pedicularis dichotoma*

南、四川、西藏和贵州。印度北部和缅甸北部亦有。

Perennial herbs. Leaves whorled; upper petioles very dilated and connate each other into a funnel in a whorl, lowest ones not dilated and free each other in a whorl. Flowers medium-sized; corolla yellow. A variable species. Fl. May-Aug. Fr. Aug-Sep. Slopes, open pastures or alpine sparse needle-leaved forests at 2500-4300 m. Distributed in Yunnan, Sichuan, Xizang and Guizhou. Also in N India and N Myanmar.

大王马先蒿 *Pedicularis rex*

灌丛马先蒿

Pedicularis thamnophila (Hand.-Mazz.) H. L. Li

多年生草本，高20-60厘米。茎自下至上均具长分枝或仅于基部分枝。叶多数3枚一轮，羽状全裂；裂片9-12对。花序穗状；花冠黄色；花冠管长约为花萼的2倍；盔直立，额部弯曲。花期6-7月，果期8月。生海拔3200-4000米的云杉林、高山草甸或冷杉林林窗的草甸。产云南西北部、四川西南部和西藏东南部。

Perennial herbs, 20-60 cm tall. Stems long branched throughout entire length or near base only. Leaves mostly in whorls of 3, pinnatisect; segments 9-12 pairs. Inflorescences spicate; corolla yellow; tubes ca. 2 × as long as calyx; galea erect, bowed apically. Fl. Jun-Jul. Fr. Aug. *Picea* forests, alpine meadows or meadows of canopy gaps in *Abies* forests at 3200-4000 m. Distributed in NW Yunnan, SW Sichuan and SE Xizang.

灌丛马先蒿 *Pedicularis thamnophila*

华丽马先蒿 *Pedicularis superba*

斗叶马先蒿 *Pedicularis cyathophylla*

华丽马先蒿

Pedicularis superba Franch. ex Maxim.

多年生草本，高30-90厘米。叶每轮3或4，长椭圆形，羽状全裂；裂片12-15对。花序穗状；苞片叶状；花冠紫红色至红色，3.7-5厘米；盔直立。蒴果压扁，卵球形。花期6-8月，果期7-8月。生海拔2800-4000米的高山草甸、开阔多石牧场或林缘阴处。产云南西北部和四川西南部。

Perennial herbs, 30-90 cm tall. Leaves in whorls of 3 or 4, long elliptic, pinnatisect; segments 12-15 pairs. Inflorescences spicate; bracts leaflike; corolla purplish red to red, 3.7-5 cm long; galea erect. Capsules compressed, ovoid. Fl. Jun-Aug. Fr. Jul-Aug. Alpine meadows, open stony pastures or shaded places near forest edges at 2800-4000 m. Distributed in NW Yunnan and SW Sichuan.

斗叶马先蒿

Pedicularis cyathophylla Franch.

多年生草本。主根圆锥形。茎直立，不分枝，被毛。叶3-4枚轮生，基部结合成斗状体，叶片长椭圆形，羽状全裂，裂片有锯齿。花序穗状，苞片基部合生；萼长约15毫米，被毛，先端2齿；花冠紫红色，花管细，脉不扭转，盔前俯，喙7毫米；花丝被毛；柱头隐于盔中。花期5-7月，果期7-8月。生海拔4700米的高山草地上。特产四川西南部和云南西北部。

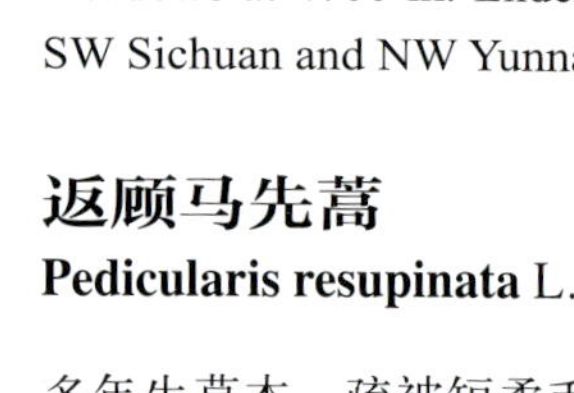

Perennial herbs. Roots conical. Stems erect, unbranched, pubescent. Leaves in whorls of 3 or 4, petioles and bract bases greatly dilated, connate, cupular; leaf blade long elliptic, pinnatisect, serrate. Inflorescences spicate; bracts base united; calyx ca. 15 mm, long pubescent, lobes 2; corolla purple-red, tube bent at a right angle apically, slender; galea strongly bent; beak ca. 7 mm; filaments pubescent; styles in galea. Fl. May-Jul. Fr. Jul-Aug. Alpine meadows at 4700 m. Endemic to SW Sichuan and NW Yunnan.

返顾马先蒿

Pedicularis resupinata L.

多年生草本，疏被短柔毛至无毛。叶卵形至长圆状披针形，膜质至纸质。花腋生；花冠粉红色至紫色。蒴果长圆状披针形，偏斜，稍长于花萼。花期6-8月，果期7-9月。生海拔300-2000米的草坡或开阔林中。产中国西南、华中、华东、华北、西北和东北。哈萨克斯坦、俄罗斯、蒙古、朝鲜半岛和日本亦有。

Perennial herbs, sparsely pubescent to glabrous. Leaves ovate to oblong-lanceolate, membra-

返顾马先蒿 *Pedicularis resupinata*

纤裂马先蒿 *Pedicularis tenuisecta*

nous to papery. Flowers axillary; corolla pink to purple. Capsules obliquely oblong-lanceolate, slightly longer than calyx. Fl. Jun-Aug. Fr. Jul-Sep. Grassy slopes or open forests at 300-2000 m. Distributed in SW, C, E, N, NW and NE China. Also in Kazakhstan, Russia, Mongolia, Korean Peninsula and Japan.

纤裂马先蒿

Pedicularis tenuisecta Franch. ex Maxim.

多年生草本，干后变黑。叶一至二回羽状全裂；裂片约10对。花序总状，多花；苞片叶状，长于花萼，短于花；花冠紫红色；盔长大，其端圆钝无明显的喙。蒴果斜披针状卵球形。花期8-9月，果期9-11月。生海拔1500-3700米的松柏林中的高山草甸。产云南西北部、四川西南部和贵州西部。老挝亦有。

Perennial herbs, drying black. Leaves 1- or 2-pinnatisect; segments ca. 10 pairs. Inflorescences racemose, many-flowered; bracts leaflike, longer than calyx, shorter than flowers; corolla purple-red; galea long and big, without distinct beaks at apex. Capsules obliquely lanceolate-ovoid. Fl. Aug-Sep. Fr. Sep-Nov. Alpine meadows in coniferous forests at 1500-3700 m. Distributed in NW Yunnan, SW Sichuan and W Guizhou. Also in Laos.

亨氏马先蒿

Pedicularis henryi Maxim.

多年生草本，干后黑色。叶纸质，一回羽状全裂；裂片6-12对，齿白色骨质。花生于叶腋而成长总状花序；花冠紫红色，冠筒直伸，盔顶端弯曲。蒴果斜披针状卵球形。花期5-9月，果期8-11月。生海拔400-1400米的空旷山坡、草甸或开阔林中。产中国西南、华南、华中和华东。老挝和越南亦有。

Perennial herbs, drying black. Leaves papery, 1-pinnatisect; segments 6-12 pairs, teeth white and callose. Inflorescences axillary, racemose; corolla purplish red, tubes straight, galea curved apically. Capsules obliquely lanceolate-ovoid. Fl. May-Sep. Fr. Aug-Nov. Open mountain slopes, meadows or open forests at 400-1400 m. Distributed in SW, S, C and E China. Also in Laos and Vietnam.

拉氏马先蒿

Pedicularis labordei Vaniot ex Bonati

多年生草本。叶长圆形，羽状半裂或有时1回羽状全裂；裂片5-8对。花序近头状，苞片叶状；花冠较短，紫红色，冠筒中部稍弯曲，盔无鸡冠状突起。蒴果斜卵球形。花期7-9月，果期8-10月。生海拔2800-3500米的高山草甸。产云南东部和西北部、四川西南部和贵州西北部。

Perennial herbs. Leaves oblong, pinnatipartite or sometimes 1-pinnatisect; segments 5-8 pairs. Inflorescences subcapitate; bracts leaflike; corolla short, purple-red, tubes slightly curved at middle, galea without crest. Capsules obliquely narrowly ovoid. Fl. Jul-Sep. Fr. Aug-Oct. Alpine meadows at 2800-3500 m. Distributed in E and NW Yunnan, SW Sichuan and NW Guizhou.

亨氏马先蒿 *Pedicularis henryi*

拉氏马先蒿 *Pedicularis labordei*

鹤首马先蒿 *Pedicularis gruina*

鹤首马先蒿
Pedicularis gruina Franch. ex Maxim.

多年生草本。茎常多条，常多分枝。叶卵状长圆形，长达2厘米，两面均有毛，羽状深裂或几全裂，裂片5-10对。花梗长5-16毫米，被有短柔毛；花萼密被锈色毛，前方开裂1/3-1/2；花冠红色至紫红色；花丝皆密被长柔毛。蒴果卵圆形，长1-1.2厘米。花期7-10月，果期9-10月。生海拔2600-3000米的高山草地中、杂木林下或沟边湿处。产云南西北部。

Perennial herbs. Stems often several, usually many branched. Leaves ovate-oblong, to 2 cm long, pubescent on both surfaces, pinnatipartite to pinnatifid; segments 5-11 pairs. Pedice 15-16 mm long, pubescent; calyx densely rust colored pubescent, 1/3-1/2 cleft anteriorly; corolla red to purplish red; filaments densely villous. Capsules ovoid, 1-1.2 cm long. Fl. Jul-Oct. Fr. Sep-Oct. Alpine meadows, mixed forests or damp soil by gully margins at 2600-3000 m. Distributed in NW Yunnan.

柳叶马先蒿
Pedicularis salicifolia Bonati

一年生草本。须状根。茎直立。叶无柄而对生，披针形至线形，全缘或有不显著的细齿。穗状花序顶生；苞片基部膨大。萼齿5枚，全缘；花冠深玫瑰色，无毛，管稍长于萼，盔端几方形，下唇3裂；雄蕊着生于管的中部，花丝有毛。蒴果包于萼内，具凸尖。种子长的1毫米。花果期7-9月。生海拔900-3500米的空旷多石的草滩中。特产云南西北部。

Annual herbs. Roots fibrous. Stems erect. Leaves sessile, opposite, lanceolate to linear, ± entire or serrate. Inflorescences spicate, terminal; bracts base expanded. Calyx lobes 5, entire; corolla dark rose, glabrous; tube a little longer than calys; galea apex truncate; lower lip 3-lobes; stamen at middle tube, filaments villous. Capsules enclosed by calyx, apex acuminate. Seed 1 mm long. Fl. and fr. Jul-Sep. Open stony pastures, forest margins at 900-3500 m. Endemic to NW Yunnan.

康泊东叶马先蒿
Pedicularis comptoniaefolia Franch.

多年生草本。茎上部常有分枝。叶革质，4枚轮生，线形，长达5厘米，羽状开裂，裂片圆形，边缘具重锯齿；叶柄长约3毫米。花序总状，花多而常有间断；苞片叶状，长过于萼；花萼钟形，裂片5，三角形，缘有长毛；花冠深红色；后方2枚花丝有疏毛。花期7-9月，果期9-12月。生海拔2400-3000米的干草坡与草滩中。产四川和云南。缅甸亦有。

Perennial herbs. Stems usually branched apically. Leaves leathery, in whorls of 4, linear, to 5 cm long, pinnatifid, segments rounded, margin double dentate; petioles ca. 3 mm long. Inflorescences racemose, many flowered, usually interrupted; bracts leaflike, longer than calyx; calyx campanulate, lobes 5, triangular, margin long ciliate; corolla dark red; posterior 2 filaments sparsely pubescent. Fl. Jul-Sep. Fr. Sep-Dec. Open dry pastures and meadows at 2400-3000 m. Distributed in Sichuan and Yunnan. Also in Myanmar.

皱褶马先蒿
Pedicularis plicata Maxim.

多年生草本。叶条状披针形，羽状深裂；裂片6-12对。花序长3-7厘米，基部间断；花冠黄色；花冠管在萼前裂口向前伸出；盔稍镰形，锯齿鸡冠状，具皱褶。花期7-8月。生海拔2900-5000米的高山区、石灰岩石上、湿山坡或湿石质草场。产西藏东南部、云南西北部、四川北部、甘肃和青海。

Perennial herbs. Leaves linear-lanceolate, pinnatipartite; segments 6-12 pairs. Inflorescences 3-7 cm long, interrupted basally; corolla yellow; tubes decurved through anterior slit of calyx; galea slightly falcate, serrate crested, plicate. Fl. Jul-Aug. Alpine regions, among limestone rocks, moist slopes or moist stony pastures at 2900-5000 m. Distributed in SE Xizang, NW Yunnan, N Sichuan, Gansu and Qinghai.

柳叶马先蒿 *Pedicularis salicifolia*

康泊东叶马先蒿 *Pedicularis comptoniaefolia*

皱褶马先蒿 *Pedicularis plicata*

岩居马先蒿 *Pedicularis rupicola*

岩居马先蒿

Pedicularis rupicola Franch. ex Maxim.

多年生草本。根粗壮，肉质。主茎直立。叶4枚成轮，羽状全裂，裂片6-9对。穗状花序顶生；苞片叶状；萼齿5，宽卵形，具小裂片和锯齿；花冠紫红色，中裂仅侧裂的半大，盔粗壮，额顶圆形；花丝无毛。蒴果长1.5-2.1厘米。种子有网纹。花期5-6月，果期7-8月。生海拔2700-4800米的高山草地中。特产云南西北部、四川西南部和西藏东南部。

Perennial herbs. Roots thick, fleshy. Central stem erect. Basal and stem leaves in whorls of 4; pinnatisect; segments 6-9 pairs. Inflorescences spicate; bracts leaflike. Calyx lobes 5, broadly ovate, distinctly lobulate and serrate; corolla purple-red, middle lobe ca. 1/2 as long as lateral lobes; galea thick, apex rounded; filaments glabrous. Capsules 1.5-2.1 cm. Seeds reticulation. Fl. May-Jun. Fr. Jul-Aug. Alpine meadows, rocky slopes at 2700-4800 m. Endemic to NW Yunnan, SW Sichuan and SE Xizang.

丽江马先蒿

Pedicularis likiangensis Franch. ex Maxim.

多年生草本，高(4-)9-18厘米。叶每轮4枚。短总状花序；花萼筒状；花冠红色或粉红色或淡紫红色，1.4-1.6厘米；盔3-4毫米长；下唇约8毫米长。蒴果卵球状披针形。花期6-8月，果期9月。生海拔3200-4600米的高山草甸或林缘。产云南西北部、四川西南部和西藏东部。

Perennial herbs, (4-)9-18 cm tall. Leaves in whorls of 4. Short racemes; calyx tubular; corolla red or pink or pale purple-red, 1.4-1.6 cm long; galea 3-4 mm long; lower lips ca. 8 mm long. Capsules ovoid-lanceolate. Fl. Jun-Aug. Fr. Sep. Alpine meadows or forest edges at 3200-4600 m. Distributed in NW Yunnan, SW Sichuan and E Xizang.

轮叶马先蒿

Pedicularis verticillata L.

多年生草本。叶常4枚一轮，长圆形至条状披针形。花序总状，常较密；花萼常红色，卵球形，膜质，密具柔毛；花冠紫色；盔稍镰形，前端圆。蒴果披针形。花期6-8月，果期7-9月。生海拔2100-4400米的湿处或高山牧场，在北极高地则生于苔原。产中国西南、华北、华西和东北。俄罗斯、日本、欧洲中部和南部、北美洲西北部和北极高地亦有。

Perennial herbs. Leaves usually in whorls of 4, oblong to linear-lanceolate. Inflorescences racemose, usually dense; calyx usually red, ovoid, membranous, densely villous; corolla purple; galea slightly falcate, rounded in front. Capsules lanceolate. Fl. Jun-Aug. Fr. Jul-Sep. Wet places or alpine pastures at 2100-4400 m. tundra in Arctic highlands Distributed in SW, N, W and NE China. Also in Russia, Japan, C and S Europe, NW North America and Arctic highlands.

丽江马先蒿 *Pedicularis likiangensis*

轮叶马先蒿 *Pedicularis verticillata*

甘肃马先蒿 *Pedicularis kansuensis*

退毛马先蒿 *Pedicularis glabrescens*

甘肃马先蒿

Pedicularis kansuensis Maxim.

一年生或二年生草本，全株具柔毛。叶长圆形，有时卵形，羽状全裂；裂片约10对，具骨质齿。花序具多层花；花冠紫粉色至紫红色，有时白色；花冠管近基部弧形下弯。花期6-8月，果期7-9月。生海拔1800-4600米的砾石地、亚高山草坡、田边湿草地、潮湿山坡或沟谷。产西藏东部和东北部、四川西部、甘肃南部与西南部和青海。

Annual or biennial herbs, pubescent throughout. Leaves oblong, sometimes ovate, pinnatisect; segments ca. 10 pairs, teeth callose. Inflorescences with many whorls; corolla purple-pink to purple-red, sometimes white; tubes decurved near base. Fl. Jun-Aug. Fr. Jul-Sep. Gravelly ground, grassy slopes in subalpine zone, damp grassy areas along field edges, damp slopes or valleys at 1800-4600 m. Distributed in E and NE Xizang, W Sichuan, S and SW Gansu and Qinghai.

退毛马先蒿

Pedicularis glabrescens Li

多年生草本。茎多条，不分枝。下部茎生叶对生或3枚轮生，上部者4枚轮生；叶长圆状披针形，长2-3厘米，羽状全裂，裂片6-8对，卵形至披针状长圆形。花序穗状；苞片条状披针形，叶状；花梗长2-5毫米，无毛；花萼斜宽卵形，膜质，无毛；花冠紫色；2枚花丝有毛。花期7月，果期8月。生海拔约3500米的湿润山坡。产云南西北部。

Perennial herbs. Stems several, unbranched. Proximal leaves opposite or in whorls of 3, distal ones in whorls of 4; leaves oblong-lanceolate, 2-3 cm long, pinnatisect, segments 6-8 pairs, ovate to lanceolate-oblong. Inflorescences spicate; bracts linear-lanceolate, leaflike; pedicel 2-5 mm long, glabrous; calyx obliquely broadly ovate, membranous, glabrous; corolla purple; 2 filaments pubescent. Fl. Jul. Fr. Aug. Damp slopes at ca. 3500 m. Distributed in NW Yunnan.

全萼马先蒿

Pedicularis holocalyx Hand.-Mazz.

一年生草本，干后不为黑色。茎单生，直立，常于顶部多分枝。叶3到4枚轮生；近轴叶柄长1厘米，具翅；叶羽状半裂；裂片6-9对。花序穗状；花小；花萼裂片合生成3个；花冠紫红色；冠筒在花萼中强烈向下弯曲；盔先端圆形。蒴果三角状披针形。花期6月，果期6-7月。生海拔约2000米的草坡。产四川东部和湖北西部。

Annual herbs, not drying black. Stems single, erect, often many

全萼马先蒿 *Pedicularis holocalyx*

穗花马先蒿 *Pedicularis spicata*

branched apically. Leaves in whorls of 3 or 4; proximal petioles to 1cm long, winged; leaves pinnatipartite; segments 6-9 pairs. Inflorescences spicate; flowers small; calyx lobes by fusion appearing 3; corolla purplish red; tubes strongly bent in calyx, deflexed; galea rounded in front. Capsules triangular-lanceolate. Fl. Jun. Fr. Jun-Jul. Grassy slopes at ca. 2000 m. Distributed in E Sichuan and W Hubei.

穗花马先蒿

Pedicularis spicata Pall.

一年生草本。远轴的苞片叶状或斜方状卵形；花丝2枚密被长柔毛，2枚无毛。花序穗状；花冠红色；花冠管基部弯成直角。蒴果狭卵球形，6-7毫米，先端急尖。花期6-9月，果期8-10月。生海拔1500-2600米的湿草甸、沼泽草甸或灌丛。产中国西南、华中、华北、西北和东北。俄罗斯(东西伯利亚和远东地区)、蒙古、朝鲜半岛北部和日本亦有。

Annual herbs. Distal bracts leaf-like or rhomboid-ovate; filaments 2 densely villous, 2 glabrous. Inflorescences spicate; corolla red; tubes bent at a right angle basally. Capsules narrowly ovoid, 6-7 mm long, apex acute. Fl. Jun-Sep. Fr. Aug-Oct. Wet or swampy meadows or thickets at 1500-2600 m. Distributed in SW, C, N, NW and NE China. Also in Russia (E Siberia and Far East), Mongolia, N Korean Peninsula and Japan.

四川马先蒿

Pedicularis szetschuanica Maxim.

一年生草本，高(10-)20(-30)厘米，有短柔毛。叶片长卵形至卵状长圆形或长圆状披针形。花序穗状，顶部带红色；花冠紫红色；花冠管长约为盔的2倍，额端强烈弧形下弯，顶部膨大。花期6-8月，果期8-9月。生海拔3400-4600米的高山草甸、草坡或沟壑。产四川北部和西部、西藏东部、甘肃西南部和青海东南部。

Annual herbs, (10-)20(-30)cm tall, pubescent. Leaves long ovate to ovate-oblong or oblong-lanceolate. Inflorescences spicate, tinged with red apically; corolla purple-red; tubes ca. 2 × as long as galea, strongly decurved basally, expanded apically. Fl. Jun-Aug. Fr. Aug-Sep. Alpine meadows, grassy slopes or ravines at 3400-4600 m. Distributed in N and W Sichuan, E Xizang, SW Gansu and SE Qinghai.

小唇马先蒿

Pedicularis microchilae Franch. ex Maxim.

一年生草本，高达40厘米，干时不变黑。叶裂片6-10对。花序穗状，间断；花冠在筒部和盔的交界处强烈上升，淡红色，盔紫色，长约2厘米；盔稍镰状，极长于下唇。花期6-8月，果期7-9月。生海拔2800-4000米的高山草甸或溪边灌丛旁。产云南西北部和四川西南部。

Annual herbs, to 40 cm tall, not drying black. Leaves segments 6-10 pairs. Inflorescences spicate, interrupted; corolla strongly ascending at junction of tubes and galea, pale red, with purple galea, ca. 2 cm long; galea slightly falcate, much longer than lower lips. Fl. Jun-Aug. Fr. Jul-Sep. Alpine meadows or thicket edges by streams at 2800-4000 m. Distributed in NW Yunnan and SW Sichuan.

四川马先蒿 *Pedicularis szetschuanica*

小唇马先蒿 *Pedicularis microchilae*

碎米蕨叶马先蒿 *Pedicularis cheilanthifolia*

球花马先蒿 *Pedicularis globifera*

碎米蕨叶马先蒿

Pedicularis cheilanthifolia Schrenk

多年生草本。根纺锤状。茎生叶4枚一轮；叶条状披针形，羽状全裂；裂片8-12对，羽状半裂，具重锯齿或锐锯齿。花序近头状或穗状；花冠紫红色至白色，有时黄色；盔镰形。蒴果披针状三角形。花期6-8月，果期7-9月。生海拔2100-5200米的石质或沙砾质山坡近顶峰处、草坡、河岸或溪边湿沙滩。产西藏北部、甘肃东部和西南部及西部、青海和新疆。印度、中亚和蒙古亦有。

Perennial herbs. Roots fusiform. Stem leaves in whorls of 4; leaves linear-lanceolate, pinnatisect; segments 8-12 pairs, pinnatifid, double dentate or incised-dentate. Inflorescences subcapitate or spicate; corolla purple-red to white, sometimes yellow; galea falcate. Capsules lanceolate-triangular. Fl. Jun-Aug. Fr. Jul-Sep. Stony or gravelly slopes near summits, grassy slopes, banks or damp sandy areas along streams at 2100-5200 m. Distributed in N Xizang, E, SW and W Gansu, Qinghai and Xinjiang. Also in India, C Asia and Mongolia.

球花马先蒿

Pedicularis globifera Hook. f.

多年生草本。茎生叶条状披针形，小于基生叶，羽状全裂，裂片10-18对，披针形至条形，浅裂或具齿，齿具胼胝。花序常密穗状；花萼长圆状钟形，前方开裂至1/3处，脉有密毛，5浅裂；花冠红色至白色；花丝无毛。蒴果卵状披针形，长约1.4厘米，锐头。花果期6-10月。生海拔3600-5400米的河谷水湿地或河滩莎草群落中。产西藏。尼泊尔和印度北部亦有。

Perennial herbs. Stem leaves linear-lanceolate, smaller than basal leaves, pinnatisect; segments 10-18 pairs, lanceolate to linear, lobed or dentate, teeth callose. Inflorescences often densely spicate; calyx oblong-campanulate, 1/3 cleft anteriorly, densely pilose along veins, 5-lobed; corolla red to white; filaments glabrous. Capsule ovoid-lanceolate, ca. 1.4 cm long, apiculate. Fl. and fr. Jun-Oct. Boggy places along rivers and streams or *Kobresia* meadows at 3600-5400 m. Distributed in Xizang. Also in Nepal and N India.

鸭首马先蒿

Pedicularis anas Maxim.

多年生草本。叶长圆状卵形至条状披针形，羽状半裂，具刺状齿。花序头状至穗状；花萼常具浅紫色斑点或带紫色；花冠紫色、黄色或紫色且下唇带黄色而盔暗紫红色。蒴果由花萼包被2/5。花期7-9月，果期8-10月。生海拔3000-4300米的高山草甸。产西藏、四川和甘肃南部。

Perennial herbs. Leaves oblong-ovate to linear-lanceolate, pinnatifid, spinescent-dentate. Inflorescences capitate to spicate; calyx often with purplish dots or tinged with purple; corolla purple, yellow or purple with pale yellow; lower lips and dark purplish red galea. Capsules 2/5 enclosed by calyx. Fl. Jul-Sep. Fr. Aug-Oct. Alpine meadows at 3000-4300 m. Distributed in Xizang, Sichuan and S Gansu.

鸭首马先蒿 *Pedicularis anas*

赛氏马先蒿

Pedicularis semenowii Regel

多年生草本。根肉质。茎不分枝。叶片线状披针形，羽状深裂，顶端有胼胝。花序头状；花梗长1.5厘米；苞片短于花；萼长钟状，膜质，齿5，基部三

赛氏马先蒿 *Pedicularis semenowii*

角形，缘有细锯齿；花冠长2.2-3厘米，盔前缘浅"S"字形弯曲，萼圆形，侧裂披针状卵形；雄蕊花丝1对有毛。蒴果斜卵圆形。花期5-6月，果期6-7月。生海拔2700-4500米的石山上。产新疆和西藏。

Perennial herbs. Roots fleshy. Stems unbranched. Leaves linear lanceolate; leaf blade pinnatisect, apex callus. Inflorescences capitate; pedicel ca. 1.5 cm; bracts shorter than flowers; calyx campanulate, membraceous, lobes 5, base triplicate, brim serrulate; corolla 2.2-3 cm, galea slightly S-shaped apically, apex round, lateral lobe lanceolate-oblong; 2 filaments pubescent, 2 glabrous. Capsules obliquely ovoid. Fl. May-Jun. Fr. Jun-Jul. Steppes, stony and rocky slopes at 2700-4500 m. Distributed in Xinjiang and Xizang.

柔毛马先蒿

Pedicularis mollis Wall. ex Benth.

一年生草本，高30-80厘米。茎直立，多叶。叶3-5枚轮生；羽状全裂，裂片10-15对。花序长穗状；萼钟形，多毛，5齿，缘有锯齿；花管短于萼，下唇有2折襞，3裂，盔伸直，较管宽一半，三角形尖端；花丝着生在子房中部，无毛；花柱不伸出。蒴果伸出萼外1倍。种子褐色。花果期7-9月。生海拔3000-4500米的河谷、沙滩。产西藏南部。

Annual herbs, 30-80 cm tall. Stems erect, leafy. Leaves in whorls of 3-5; leaf blade pinnatisect, segments 10-15 pairs. Inflorescences spike; calyx campanulate, tomentose, lobes 5, serrulate; corolla tube shorter than calyx, lower lip 2 plication, 3-lobes; galea erect, 0.5 × wider than tube, apex acute but beakless; filaments at middle ovary, glabrous; styles not extended. Capsules 2 × as long as calyx. Seed brownish. Fl. and fr. Jul-Sep. Sand dunes along beaches, field margins, dry ground at 3000-4500 m. Distributed in S Xizang.

密穗马先蒿

Pedicularis densispica Franch. ex Maxim.

一年生草本。叶长卵形至卵状长圆形，羽状半裂至全裂；裂片6-10对。密集穗状花序；苞片短于花；花冠玫瑰色至淡紫色；下唇等长于盔。蒴果卵球形，稍压扁。花期4-7月。生海拔1900-4400米的沼泽草甸或林中。产云南西北部、西藏东南部和四川西部。

Annual herbs. Leaves long ovate to ovate-oblong, pinnatipartite to pinnatisect; segments 6-10 pairs. Dense spikes; bracts shorter than flowers; corolla rose to pale purple; lower lips ca. as long as galea. Capsules ovoid, ± compressed. Fl. Apr-Jul. Swampy meadows or forests at 1900-4400 m. Distributed in NW Yunnan, SE Xizang and W Sichuan.

柔毛马先蒿 *Pedicularis mollis*

密穗马先蒿 *Pedicularis densispica*

万叶马先蒿 *Pedicularis myriophylla*

阿拉善马先蒿 *Pedicularis alaschanica*

万叶马先蒿

Pedicularis myriophylla Pall.

一年生草本。茎常单一，自基部分枝，细毛线4条。茎叶披针状长圆形，长约3厘米，两面均无毛，羽状全裂。花序穗状；花萼前方开裂至1/3；花冠玫瑰色，冠筒短，约与花萼等长或略长于花萼；仅前方1对花丝有长柔毛。蒴果披针状卵形，长约1.5厘米。花期6-7月，果期7-8月。生林中空地或草地。产新疆。蒙古和俄罗斯(西伯利亚)亦有。

Annual herbs. Stems usually single, branched throughout, with 4 lines of hairs. Stem leaves lanceolate-oblong, ca. 3 cm long, both surfaces glabrous, pinnatisect. Inflorescences racemose; calyx ca. 1/3 cleft anteriorly; corolla rose, tube short, ca. as long as or slightly exceeding calyx; only anterior filament pair villous. Capsules lanceolate-ovoid, ca. 1.5 cm long. Fl. Jun-Jul. Fr. Jul-Aug. Forest clearings or meadows. Distributed in Xinjiang. Also in Mongolia and Russia (Siberia).

阿拉善马先蒿

Pedicularis alaschanica Maxim.

多年生草本，高达35厘米，干时稍变黑色。叶披针状长圆形至卵状长圆形，羽状深裂；裂片7-9对。花序穗状，常于基部间断；花冠黄色；花冠管顶部稍弯曲；盔顶部稍弯曲，不明显鸡冠状。花期6-8月，果期9月。生海拔3900-5100米的河谷干燥石坡、谷床石堆、开阔山坡或灌丛。产西藏东北部和东南部、内蒙古、甘肃、宁夏和青海。

Perennial herbs, to 35 cm tall, drying slightly black. Leaves lanceolate-oblong to ovate-oblong, pinnatipartite; segments 7-9 pairs. Inflorescences spicate, usually interrupted basally; corolla yellow; tubes slightly bent apically; galea slightly bent apically, indistinctly crested. Fl. Jun-Aug. Fr. Sep. Dry rocky slopes in river valleys, among stones of valley beds, open hillsides or thickets at 3900-5100 m. Distributed in NE and SE Xizang, Neimenggu, Gansu, Ningxia and Qinghai.

塔氏马先蒿

Pedicularis tatarinowii Maxim.

一年生草本，高达50厘米。叶羽状半裂至深裂，卵状长圆形至长圆状披针形。花序总状；花冠紫红色；盔额强烈弯曲，

塔氏马先蒿 *Pedicularis tatarinowii*

鹬形马先蒿 *Pedicularis scolopax*

具冠马先蒿 *Pedicularis cristatella*

具不明显鸡冠状突起；下唇长于盔。蒴果斜卵球形。花期7-8月，果期8-9月。生海拔2000-2300米的高山草甸。产河北北部、山西北部和内蒙古南部。

Annual herbs, to 50 cm tall. Leaves pinnatifid to pinnatipartite, ovate-oblong to oblong-lanceolate. Inflorescences racemose; corolla purplish red; galea strongly bent apically, indistinctly crested; lower lips longer than galea. Capsules obliquely ovoid. Fl. Jul-Aug. Fr. Aug-Sep. Alpine meadows at 2000-2300 m. Distributed in N Hebei, N Shanxi and S Neimenggu.

鹬形马先蒿

Pedicularis scolopax Maxim.

多年生草本，高达30厘米有余，干时不变黑。根纺锤状，肉质。叶条状长圆形至长圆状披针形，羽状深裂；裂片6-8对。花序穗状，花4-6朵一轮，基部间断；花冠黄色，约1.5厘米。花期6-7月，果期8月。生海拔3500-4100米的高山灌丛草地。产甘肃北部和青海东北部。

Perennial herbs, to more than 30 cm tall, not drying black. Roots fusiform, fleshy. Leaves linear-oblong to oblong-lanceolate, pinnatipartite; segments 6-8 pairs. Inflorescences spicate, flowers in whorls of 4-6, interrupted basally; corolla yellow, ca. 1.5 cm long. Fl. Jun-Jul. Fr. Aug. Alpine shrubby grasslands at 3500-4100 m. Distributed in N Gansu and NE Qinghai.

具冠马先蒿

Pedicularis cristatella Pennell et H. L. Li

一年生草本。叶长圆状披针形至狭披针形；裂片6-12对，羽状半裂。花序长穗状，每轮具3或4花；花萼白色，膜质；花冠紫红色；盔先端弯曲，明显具冠。花期6-7月，果期7月。生海拔1900-3000米的崖壁、山谷草甸、开阔或灌丛草地。产四川北部和甘肃西南部。

Annual herbs. Leaves oblong-lanceolate to narrowly lanceolate; segments 6-12 pairs, pinnatifid. Inflorescences long spicate, with flowers in whorls of 3 or 4; calyx white, membranous; corolla reddish purple; galea bent apically, distinctly crested. Fl. Jun-Jul. Fr. Jul. Cliffs, meadows in valleys, open or shrubby grasslands at 1900-3000 m. Distributed in N Sichuan and SW Gansu.

半扭卷马先蒿

Pedicularis semitorta Maxim.

一年生草本，高达60厘米。叶羽状深裂；裂片8-10对。花序穗状，基部间断；花冠黄色，1.4-1.8厘米；花丝2枚具长柔毛，2枚无毛；喙常侧扭形成环状。花期6-7月，果期7-8月。生海拔2500-3900米的高山草甸。产四川北部、甘肃中部与西南部和青海东部。

Annual herbs, to 60 cm tall. Leaves pinnatisect; segments 8-10 pairs. Inflorescences spicate, interrupted basally; corolla yellow, 1.4-1.8 cm long; filaments 2 villous, 2 glabrous; beaks twisted laterally usually into a circle. Fl. Jun-Jul. Fr. Jul-Aug. Alpine meadows at 2500-3900 m. Distributed in N Sichuan, C and SW Gansu, and E Qinghai.

半扭卷马先蒿 *Pedicularis semitorta*

奥氏马先蒿 *Pcdicularis olivcriana*

奥氏马先蒿
Pcdicularis olivcriana Prain

多年生草本。根丛生，肉质。茎黑色。叶羽裂，轴有翅，裂片5-8对。花序长达20厘米；苞片与花等长。花连喙长14-16毫米；萼齿5；花管不弯曲；下唇楔形而前方宽，盔向右扭折，喙扭为S形；花丝生于花管顶端，被毛；花柱不伸出。蒴果长卵圆形。种子狭卵形。花期6-8月，果期7-9月。生海拔3400-4000米的林下及河岸柳林下。产东喜马拉雅。

Perennial herbs. Roots fascicled, fleshy. Stems dark. Leaves pinnatifid, incised-dentate; segments 5-8 pairs. Inflorescences to 20 cm; bracts as long as flowers. Flower 14-16 mm; cylax lobes 5; corolla tube unbending; lower lip wide, cuniform, galea marginally ciliate, beak S-shaped, slender; filaments at top tube, densely villous; styles non-extend. Capsules compressed, oblong. Seed ovate. Fl. Jun-Aug. Fr. Jul-Sep. Dry rocky places, sand dunes along rivers, open grassy meadows at 3400-4000 m. Distributed in E Himalaya.

准噶尔马先蒿
Pedicularis songarica Schrenk ex Fisch. et C. A. Mey.

多年生草本，干后黑色。根簇生，近纺锤形，肉质。茎常单生。叶多基生；叶柄长达4厘米；叶羽状全裂；裂片15-30对。花序短穗状，长6厘米，常密具花；花萼管状，裂片5；花冠浅黄色；盔稍镰状，顶部具短喙。蒴果披针状长圆形。花期6-7月，果期7-8月。产新疆。哈萨克斯坦亦有。

Perennial herbs, drying black. Roots fascicled, subfusiform, fleshy. Stems often single. Leaves mostly basal; petioles to 4 cm long; leaves pinnatisect; segments 15-30 pairs. Inflorescences short spicate, 6 cm long, often dense; calyx tubular, lobes 5; corolla pale yellow; galea slightly falcate, apex short beaked. Capsules lanceolate-oblong. Fl. Jun-Jul. Fr. Jul-Aug. Distributed in Xinjiang. Also in Kazakhstan.

水泽马先蒿
Pedicularis uliginosa Bunge

多年生草本。根簇生，稍加厚。茎单生，基部具膜质鳞片；基部叶柄稍短于叶长的一半。叶羽状全裂；裂片再羽状半裂。花最初为密总状，于果期伸长至17厘米；花冠紫红色；盔稍镰状，无喙。蒴果长圆状披针形。花期7-8月，果期8-9月。生林中阴处、溪边阴湿草甸或山顶。产新疆。西南亚、中亚和东北亚亦有。

准噶尔马先蒿 *Pedicularis songarica*

水泽马先蒿 *Pedicularis uliginosa*

尖果马先蒿 *Pedicularis oxycarpa*

Perennial herbs. Roots clustered, slightly thickened. Stems single, with membranous scales at base; basal leaf petioles slightly shorter than to 1/2 as long as leaves. Leaves pinnatisect; segments pinnatifid. Flowers initially in a dense racemes, later elongating to 17 cm long in fruits; corolla purple-red; galea ± falcate, beakless. Capsules oblong-lanceolate. Fl. Jul-Aug. Fr. Aug-Sep. Shaded glades in forests, shaded damp meadows along streams or summits of hills. Distributed in Xinjiang. Also in SW, C and NE Asia.

尖果马先蒿

Pedicularis oxycarpa Franch. ex Maxim.

多年生草本。叶互生，羽状深裂，裂片多达15对。总状花序；花冠白色，具浅紫色喙；花冠的盔在额部有高鸡冠状突起。蒴果披针状长圆形。花期5-8月，果期8-10月。生海拔2800-4400米的高山草地。产云南西北部和四川西南部。

Perennial herbs. Leaves alternate, pinnatipartite, with up to 15 pairs of segments. Racemes; corolla white, with purplish beak; galea with a high crest. Capsules lanceolate-oblong. Fl. May-Aug. Fr. Aug-Oct. Alpine grasslands at 2800-4400 m. Distributed in NW Yunnan and SW Sichuan.

扭旋马先蒿

Pedicularis torta Maxim.

多年生草本，高20-40(-70)厘米，干时不变黑。根多少肉质。叶羽状全裂；裂片9-16对。花序总状，多花；花冠黄色，盔紫色或紫红色；盔扭转；喙"S"字形，细弱。花期6-8月，果期8-9月。生海拔2500-4000米的高山草甸。产四川东部和北部、湖北西部、陕西和甘肃南部。

Perennial herbs, 20-40(-70) cm tall, not drying black. Roots ± fleshy. Leaves pinnatisect; segments 9-16 pairs. Inflorescences racemose, many-flowered; corolla yellow, with purple or purple-red galea; galea twisted; beak S-shaped, slender. Fl. Jun-Aug. Fr. Aug-Sep. Alpine meadows at 2500-4000 m. Distributed in E and N Sichuan, W Hubei, Shaanxi and S Gansu.

西藏马先蒿

Pedicularis tibetica Franch.

多年生草本。根肉质。叶长圆状倒披针形至条状长圆形，羽状深裂至全裂；裂片9-13对。花序总状；花瓣浅红色，下唇上有白色斑点；盔扭曲；喙半圆形；下唇不具缘毛。花期6-7月，果期8月。生海拔约4600米的高山草甸。产四川西部和西藏东部。

Perennial herbs. Roots fleshy. Leaves oblong-oblanceolate to linear-oblong, pinnatipartite to pinnatisect; segments 9-13 pairs. Inflorescences racemose; corolla reddish, with white spots on lower lip; galea twisted; beaks semicircular; lower lips not ciliate. Fl. Jun-Jul. Fr. Aug. Alpine meadows at ca. 4600 m. Distributed in W Sichuan and E Xizang.

扭旋马先蒿 *Pedicularis torta*

西藏马先蒿 *Pedicularis tibetica*

戴维氏马先蒿

Pedicularis davidii Franch.

多年生草本。叶卵状长圆形至披针状长圆形，羽状全裂；裂片9-14对，羽状半裂。花序总状，疏松；花冠紫色或红色，长为花萼的约2倍；盔扭曲。花期7-8月，果期8-9月。生海拔1400-4400米的草坡或平原、灌丛、林下或溪边。产四川、陕西西南部和甘肃西南部。

Perennial herbs. Leaves ovate-oblong to lanceolate-oblong, pinnatisect; segments 9-14 pairs, pinnatifid. Inflorescences racemose, lax; corolla purple or red, ca. 2 × as long as calyx; galea twisted. Fl. Jul-Aug. Fr. Aug-Sep. Grassy slopes or flats, thickets, woods or along streams at 1400-4400 m. Distributed in Sichuan, SW Shaanxi and SW Gansu.

戴维氏马先蒿 *Pedicularis davidii*

坛萼马先蒿 *Pedicularis urceolata*

拟鼻花马先蒿

Pedicularis rhinanthoides Schrenk

多年生草本。茎单出或多条，不分枝。基生叶线状长圆形，羽状全裂，裂片9-12对，卵形；叶柄长2-5厘米；茎叶少数，柄较短。花序总状；花萼前方开裂至一半，常具紫色斑点；花冠玫瑰色至紫色；前方1对花丝有毛。蒴果披针状卵形，长约1.9厘米。花期5-7月，果期7-9月。生海拔3000-5000米的潮湿草甸中。产新疆。蒙古、俄罗斯、印度和中亚亦有。

Perennial herbs. Stems single to numerous, unbranched. Basal leaves linear-oblong, pinnatisect, segments 9-12 pairs, ovate; petioles 2-5 cm long; stem leaves few, shorter petiolate. Inflorescences racemose; calyx 1/2 cleft anteriorly, often with purplish dots; corolla rose to violet-purple; anterior filament pair pubescent. Capsules lanceolate-ovoid, ca. 1.9 cm long. Fl. May-Jul. Fr. Jul-Sep. Moist meadows at 3000-5000 m. Distributed in Xinjiang. Also in Mongolia, Russia, India and C Asia.

拟鼻花马先蒿 *Pedicularis rhinanthoides*

坛萼马先蒿

Pedicularis urceolata Tsoong

一年生草本，高10-20厘米，老时基部多少木质，干时不变黑。茎生叶仅1对或缺无。花序具1-6对花；花冠玫瑰红色，筒部的额部和盔的基部黄色。花期7-8月，果期8-9月。生海拔约3800米的高山草甸。产四川西部。

Annual herbs, 10-20 cm tall, ± woody at base when old, not drying black. Stem leaves often only 1 pair or absent. Inflorescences with 1-6 pairs flower; corolla rose-red, with yellow at tubes apex and galea base. Fl. Jul-Aug. Fr. Aug-Sep. Alpine meadows at ca. 3800 m. Distributed in W Sichuan.

长舌马先蒿

Pedicularis dolichoglossa H. L. Li

多年生草本。茎单生，黑色。叶少量对生，椭圆状卵形，羽状半裂；裂片2-4对。花序近头状，多花；花冠黄色，具色泽暗淡的斑点；花冠管直立，超出花萼；盔顶端镰形，长达2厘米。产云南西北部。

Perennial herbs. Stems single, black. Leaves few, opposite, elliptic-ovate, pinnatipartite; segments 2-4 pairs. Inflorescences subcapitate, many-flowered; corolla yellow, with dull colored spots; tubes erect, exceeding calyx; galea falcate apically, to 2 cm long. Distributed in NW Yunnan.

舟形马先蒿

Pedicularis cymbalaria Botani

一年生或二年生草本，无毛。茎长15厘米，从根茎顶端发出多条，基部多分枝，均铺地上升。花成对生于枝端之叶腋中；苞片叶状；萼管状，被毛，上部膨大，3裂，萼齿5枚；花冠无斑

长舌马先蒿 *Pedicularis dolichoglossa*

三角叶马先蒿 *Pedicularis deltoidea*

点，盔上半部镰状弓曲，端尖削舟形，裂片相等而同形；雄蕊生于花管中部以下；子房具附属物1枚。蒴果长圆形。花期8月，果期9月。生海拔3400-4000米的高山草甸、石沙地或河岸上。特产云南西北部和四川西南部。

Annual or biennial herbs, glabrous. Stems 15 cm long, several, diffuse or procumbent, many branched basally. Flowers axillary; bract leafy; calyx tubular, pubescent, upper expansion, 3-dentate, lobes 5; corolla non-spot; galea falcate apically, apex navicular; middle lobe ca. as long as lateral lobes; stamens at lower tube; ovary with 1 appendix. Capsules lanceolate-oblong. Fl. Aug. Fr. Sep. Alpine meadows, rocky soils, shaded banks at 3400-4000 m. Endemic to NW Yunnan and SW Sichuan.

三角叶马先蒿
Pedicularis deltoidea Franch. ex Maxim.

一年生或二年生草本，被灰白色短柔毛。茎中空，开花时强烈木质化。叶片角状卵形，具短柄。总状花序，具苞片；花玫瑰红，盔的下缘多少向前膨臌，比上半部显然较宽，上部镰形弯曲，额圆凸，小喙形，下缘主齿极靠近前额基部；子房披针形。蒴果披针形。花期8-9月，果期9-10月。生海拔2600-3300米的草坡中。特产云南西北部和四川西南部。

Annual or biennial herbs, densely gray pubescent throughout. Stems hollow, flowering strongly lignification. Leaf blade triangular-ovate, short shank. Inflorescences racemose; bracts leaflike; corolla rose, galea lower ± expanded, larger than upper, apically falcate, front cycle, beaklike; lower lip shorter than galea, slightly praemorse, middle lobe larger than lateral pair; ovary lanceolate. Capsules lanceolate. Fl. Aug-Sep. Fr. Sep-Oct. Grassy slopes at 2600-3500 m. Endemic to NW Yunnan and SW Sichuan.

浅黄马先蒿
Pedicularis lutescens Franch.

多年生草本。茎多丛生。叶片纸质，羽裂。花序总状，紧密，花多向上生长；苞片比萼长，基部膨大；花梗短；萼圆筒形，萼齿5枚；花冠淡黄色，盔镰形弓曲，盔端尖削；雄蕊着生于花管下部1/4处；子房狭卵圆形；花柱伸出于盔外。蒴果披针形。花期7-8月，果期8-9月。生海拔3200-4000米的灌丛或高山草地上。特产中国西南。

Perennial herbs. Stems usually clustered. Leaf blade papery, pinnatifid. Inflorescences racemose, compact, ± upside; bracts longer than calyx; pedicel short; calyx cyclindric, lobes 5; corolla pale yellow; galea slightly falcate, apex acute; stamen at 1/4 of tube; ovary narrow oblong; styles longer than galea. Capsules lanceolate. Fl. Jul-Aug. Fr. Aug-Sep. Thickets, alpine meadows at 3000-4000 m. Endemic to SW China.

舟形马先蒿 *Pedicularis cymbalaria*

浅黄马先蒿 *Pedicularis lutescens*

团花马先蒿
Pedicularis sphaerantha Tsoong

团花马先蒿 *Pedicularis sphaerantha*

多年生草本，密生长毛。茎单出或数条。基出叶椭圆形至长圆形，长1-2厘米，羽状全裂，裂片5-7对，长圆形，羽状分裂，裂片有齿，叶柄长达1厘米，疏被短柔毛；茎生叶与基出叶相似，但叶柄较短。花序密集成团；花萼膜质；花冠红色而盔色较深；前方1对花丝有疏毛。花期7-8月，果期9月。生海拔3900-4800米的湿润草地或草坡。产西藏东部。

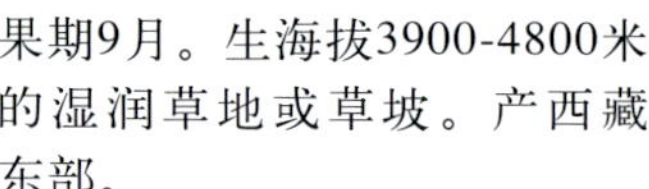

Perennial herbs, densely long pubescent. Stems 1 to several. Basal leaves elliptic to oblong, 1-2 cm long, pinnatisect, segments 5-7 pairs, oblong, pinnatifid, dentate, petioles to 1 cm long, sparsely pubescent; stem leaves similar to basal leaves but shorter petiolate. Inflorescences compact, globose; calyx membranous; corolla red, with dark red galea; anterior filament pair sparsely pubescent. Fl. Jul-Aug. Fr. Sep. Swampy meadows or grassy slopes at 3900-4800 m. Distributed in E Xizang.

全叶马先蒿
Pedicularis integrifolia Hook. f.

全叶马先蒿 *Pedicularis integrifolia*

多年生草本。根肉质纺锤形。叶披针形，均有波状圆齿。花1-3，簇生；苞片叶状；花冠深紫色，管长2厘米，伸直，下唇3裂，盔直立部分高4毫米，以直角转折为含有雄蕊的部分，长达6.5毫米，前方多少骤狭为S形弯曲的长喙，长1.5厘米。蒴果卵圆形。花期6-7月，果期7-9月。生海拔约4000米的高山草甸和云杉林中。产青海、四川、西藏和云南。

Perennial herbs. Roots fusiform, fleshy. Leaf blade lanceolate, serrate. Inflorescences 1-3-fascicled; bracts leaflike; corolla dark purple, tube erect, ca. 2 cm, slender, lower lip 3-lobes; galea erect part 4 mm tall, bent at a right angle apically, with stamen, to 6.5 mm long; beak S-shaped, ca. 1.5 cm, slender. Capsules ovoid. Fl. Jun-Jul. Fr. Jul-Sep. Alpine rocky meadows, *Picea* forests at ca. 4000 m. Distributed in Qinghai, Sichuan, Xizang and Yunnan.

谬氏马先蒿
Pedicularis mussotii Franch.

多年生草本。叶大多基生，羽状深裂至近羽状全裂；裂片6-13对。花萼裂片2或3，后裂片小或退化，侧裂片叶状，有齿；花腋生；花冠红色；盔弯成直角；喙半圆形。蒴果半球形。花期7-8月，果期8月。生海拔3600-4900米的高山草甸。产云南西北部和四川西部。

Perennial herbs. Leaves mostly basal, pinnatipartite to nearly pinnatisect; segments 6-13 pairs. Calyx lobes 2 or 3, posterior one small or absent, lateral pair leaflike, dentate; flowers axillary; corolla red; galea ± bent at a right angle; beaks semicircular. Capsules semi-globose. Fl. Jul-Aug. Fr. Aug. Alpine meadows at 3600-4900 m. Distributed in NW Yunnan and W Sichuan.

谬氏马先蒿 *Pedicularis mussotii*

哀氏马先蒿
Pedicularis elwesii Hook. f.

多年生草本。叶羽状半裂，裂片10-20(-30)对。短总状花序；花萼长圆状钟形，前方深裂至1/2；花冠紫色至紫红色，盔和啄偏扭，啄具长毛；下唇完全包被盔。花期6-8月，果期8-9月。生海拔3200-4600米的高山草甸。产云南西北部、西藏东部及南部和东南部。印度北部、尼泊尔、不丹和缅甸北部亦有。

Perennial herbs. Leaves pinnatipartite, segments 10-20(-30)

哀氏马先蒿 *Pedicularis elwesii*

pairs. Short racemes; calyx oblong-campanulate, parted to 1/2 in front; corolla purple to purple-red, galea and beaks curved, beaks with long hairs; lower lip completely enveloping galea. Fl. Jun-Aug. Fr. Aug-Sep. Alpine meadows at 3200-4600 m. Distributed in NW Yunnan, and E, S and SE Xizang. Also in N India, Nepal, Bhutan and N Myanmar.

皁莱氏马先蒿
Pedicularis fletcheri Tsoong

一年生草本，高达40厘米。叶长圆状披针形，羽状全裂；裂片约7对。花序总状，长达15厘米；花冠白色，下唇中央红色，约2.8厘米；喙向下弯曲，2裂。花期7月。生海拔3500-4200米的高山草甸。产西藏东南部。不丹亦有。

Annual herbs, to 40 cm tall. Leaves oblong-lanceolate, pinnatisect; segments ca. 7 pairs. Inflorescences racemose, to 15 cm long; corolla white, with red-tinged center to lower lip, ca. 2.8 cm long; beaks bent downward, 2-cleft. Fl. Jul. Alpine meadows at 3500-4200 m. Distributed in SE Xizang. Also in Bhutan.

魏氏马先蒿
Pedicularis wilsonii Botani

多年生草本。根茎短。茎几不发达，草质。基生叶少，与茎生叶集成疏丛，有长柄，背面网脉明显，网眼中叶面作碎冰纹凸起，有屑状物，深羽状开裂，茎叶小，常假对生。花单生叶腋。萼管钟形，有毛，具3齿，后方1齿线状全缘；花冠红色而大，盔前缘高约6毫米，镰状弓曲，具圆锥状喙，裂片3枚；雄蕊着生于花管中部，花丝有毛。特产四川。

Perennial herbs. Roots short. Stems undeveloped, herbaceous. Basal leaves few, clustered with stem leaves; basal leaves with long petioles, abaxiallary vein obvious, with protrude, crumb; pinnatipartite; stem leaves usually pseudo-opposite, smaller than basal leaves. Flowers axillary. Calyx navicular, pubescent, lobes 3, posterior one linear, entire; corolla red throughout; galea margin ca. 6 mm tall, falcate apically; beak bent downward, conical, 3-lobes; stamen at middle tube, filaments pubescence. Endemic to Sichuan.

皁莱氏马先蒿 *Pedicularis fletcheri*

魏氏马先蒿 *Pedicularis wilsonii*

独龙马先蒿 *Pedicularis dulongensis*

隐花马先蒿 *Pedicularis cryptantha*

独龙马先蒿

Pedicularis dulongensis H. P. Yang

多年生草本。叶片椭圆形，羽状全裂。花梗1-1.5厘米。花萼圆筒形钟状，密被白色长柔毛，5裂片，不等长；花冠红色，6-7厘米；管直立，4-5厘米，盔近镰刀形，喙弯曲，5-6毫米，下唇1.2-1.4 × 3-3.5厘米，中部裂片小；花丝被柔毛。蒴果卵球形长圆形，1-1.5厘米。花期7月，果期7-8月。生海拔3500-3600米的山坡湿润草甸上。产云南西北部。

Perennial herbs. Leaf blade elliptic, pinnatisect. Pedicel 1-1.5 cm. Calyx cylindric-campanulate, densely white villous with multicellular hairs; lobes 5, unequal; corolla red, 6-7 cm; tube erect, 4-5 cm; galea nearly falcate, very enlarged; beak incurved, 5-6 mm; lower lip 1.2-1.4 × 3-3.5 cm, middle lobe smaller than lateral pair; filaments pubescent. Capsules ovoid-oblong, 1-1.5 cm. Fl. Jul. Fr. Jul-Aug. Moist meadows on mountain slopes at 3500-3600 m. Distributed in NW Yunnan.

欧氏马先蒿

Pedicularis oederi Vahl

多年生草本。茎常鳞片状，具绵毛。叶多基生；叶条状披针形至条形，羽状全裂。花冠黄色，具紫色盔；盔前端圆形；柱头内藏或稍伸出。蒴果长卵球形至卵球状披针形。花期6-9月，果期7-10月。生海拔2600-5400米的高山草甸、牧场、湿润石灰岩、苔原或草坡。产中国西南、华北、华西和西北。南亚、西南亚、东北亚、欧洲中部和北部、北美洲北部亦有。

欧氏马先蒿 *Pedicularis oederi*

Perennial herbs. Stems usually scapelike, woolly. Leaves mostly basal; leaves linear-lanceolate to linear, pinnatisect. Corolla yellow, with purple galea; galea rounded in front; stigma included or slightly exserted. Capsules long ovoid to ovoid-lanceolate. Fl. Jun-Sep. Fr. Jul-Oct. Alpine meadows, pastures, damp limestone rocks, tundra or grassy slopes at 2600-5400 m. Distributed in SW, N, W and NW China. Also in S, SW and NE Asia, C and N Europe, and N North America.

隐花马先蒿

Pedicularis cryptantha Marq. et Shaw

多年生低矮草本。茎短缩，弯曲上升。羽状裂片每边8-12对。花腋生于基部，10-20枚，枝端亦有总状花序，离心开放；苞片叶状；花梗细长；萼管圆筒形，齿5枚；花冠黄色，盔镰状弓曲，前缘顶端三角形向前凸尖，下唇中裂圆形，侧裂肾脏形。花期5-8月，果期9-10月。生海拔2700-4700米的河岸湿处及阴湿林下。特产西藏东南部。

Perennial herbs. Stems procumbent to ascending. Leaves pinnatefid, segments 8-12 pairs. Flowers base axillary, 10-20-flowered, sometimes racemose apically, centrifugal; bracts leaflike; pedicel slender; calyx tube cylindric; lobes 5; corolla yellow; galea falcate, front rounded, apex slightly acute; lower lip with middle lobe rounded, lateral lobes

假多色马先蒿 *Pedicularis pseudoversicolor*

勒公氏马先蒿 *Pedicularis lecomtei*

kidney-shaped; filaments glabrous. Fl. May-Aug. Fr. Sep-Oct. Grassy stream banks and woods at 2700-4700 m. Endemic to SE Xizang.

假多色马先蒿

Pedicularis pseudoversicolor Hand.-Mazz.

多年生草本。叶互生，披针形，下面具白色簇毛，沿脉带紫色，羽状全裂。穗状花序密集多花；花冠黄色，仅在盔部紫红色；冠筒稍弯且顶端膨大；下唇远短于盔。花期6-8月，果期9月。生海拔(3600-)4300-4500米的高山草甸。产云南西北部和西藏南部。不丹亦有。

Perennial herbs. Leaves alternate, lanceolate, abaxially white scurfy, tinged with purple along veins, pinnatisect. Spikes densely flowered, corolla yellow, purple-red only on galea; tubes slightly bowed and expanded apically; lower lips much shorter than galea. Fl. Jun-Aug. Fr. Sep. Alpine meadows at (3600-)4300-4500 m. Distributed in NW Yunnan and S Xizang. Also in Bhutan.

勒公氏马先蒿

Pedicularis lecomtei Bonati

低矮多年生草本。根束生。茎不分枝，草质。叶基出，缘羽状全裂，裂片每边10-20枚。花序总状顶生；苞片叶状，被毛；花梗密被毛。萼圆筒形前方稍开裂，齿5枚；花冠黄色，喙紫色，盔镰状弓曲，具短喙，下唇开裂；雄蕊着生于花管近基处，前方1对花丝被毛；柱头微伸出。花期6-7月，果期7-8月。生海拔约3500米的多岩山坡上。特产云南西北部。

Perennial herbs. Roots fascicled. Stems unbranched, leathery. Leaves mostly basal; pinnatipartite; segments 10-20 pairs. Inflorescences racemose; bracts leaf-like, pubescent; pedicel, densely villous. Calyx cylinder, lobes 5; corolla yellow, with purple beak, galea ± falcate apically, with short beak; lower lip cracking; stamen at tube base, anterior filament pair densely villous. Fl. Jun-Jul. Fr. Jul-Aug. Rocky slopes at ca. 3500 m. Endemic to NW Yunnan.

普氏马先蒿

Pedicularis przewalskii Maxim.

多年生低矮草本。叶丛生，有柄；叶披针状线形，羽状浅裂成圆齿。开花次序系离心。萼管狭长，齿5枚；花冠紫红色，盔强壮，喙向前下方直伸，下唇3裂，中裂圆形有凹头，基部狭细为短柄；雄蕊生于管端，花丝有毛；花柱不伸出。蒴果长于萼1倍。花期6-7月，果期7-9月。生海拔4000-5300米的高山湿草地。产青藏高原。

Perennial herbs, low. Leaves clustered, with petioles; leaf blade lanceolate-linear, pinnatifid, round serrate. Inflorescences centrifugal. Calyx tube narrow long, lobes 5; corolla purple-red; galea bent at a right angle apically, stout; beak straight; lower lip deeply 3-lobed, middle lobe rounded to emarginated, base short petioles; stamen at base tube, filaments pubescent; styles non-extended. Capsules 2 × as long as calyx. Fl. Jun-Jul. Fr. Jul-Sep. Alpine meadows at 4000-5300 m. Distribtued in Tibetan Plateau.

普氏马先蒿 *Pedicularis przewalskii*

美丽马先蒿 *Pedicularis bella*

凸额马先蒿 *Pedicularis cranolopha*

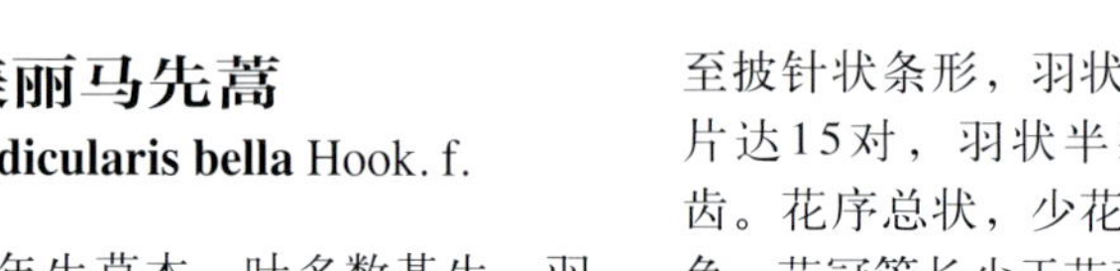

美丽马先蒿
Pedicularis bella Hook. f.

一年生草本。叶多数基生，羽状半裂。花腋生，具1-14花；整个花冠暗紫色或有些花冠的筒部淡黄色，盔紫色，下唇白色；盔镰形；喙“S”字形，先端不2裂。蒴果斜长圆形。花期6-7月，果期7-9月。生海拔3600-4900米的湿草甸、矮杜鹃花丛周围陡峭岩坡上或崖面上。产西藏南部。不丹和印度亦有。

Annual herbs. Leaves mostly basal, pinnatifid. Flowers axillary, 1-14-flowered; corolla dark purple throughout or some with pale yellow tube, purple galea, and white lower lip; galea falcate; beaks S-shaped, not 2-cleft at apex. Capsules obliquely oblong. Fl. Jun-Jul. Fr. Jul-Sep. Moist meadows, steep rocky slopes among dwarf *Rhododendron* or cliff faces at 3600-4900 m. Distributed in S Xizang. Also in Bhutan and India.

凸额马先蒿
Pedicularis cranolopha Maxim.

多年生草本。叶长圆状披针形至披针状条形，羽状全裂；裂片达15对，羽状半裂至重锯齿。花序总状，少花；花冠黄色，花冠管长少于花萼的3倍；盔顶部镰形。花期6-7月，果期8月。生海拔2600-4200米的高山草甸。产云南西北部、四川北部、甘肃西南部和青海东北部。

Perennial herbs. Leaves oblong-lanceolate to lanceolate-linear, pinnatisect; segments to 15 pairs, pinnatifid to double dentate. Inflorescences racemose, few-flowered; corolla yellow; tubes less than 3 × as long as calyx; galea falcate apically. Fl. Jun-Jul. Fr. Aug. Alpine meadows at 2600-4200 m. Distributed in NW Yunnan, N Sichuan, SW Gansu and NE Qinghai.

三色马先蒿
Pedicularis tricolor Hand.-Mazz.

一年生草本。叶披针形，羽状深裂。花序总状，达15花；萼齿常3枚，侧方的叶状，具白色长毛；花冠黄色，具红色盔，下唇边缘白色；花冠管长35-50毫米，盔额有鸡冠状突起。花期8-9月，果期9-10月。生海拔3000-3600米的高山草甸。产云南西北部。

Annual herbs. Leaves lanceolate, pinnatipartite. Inflorescences racemose, to 15-flowered; calyx 3-lobed, lateral ones leaflike, with long and white hairs; corolla yellow, with red galea, and white margin on lower lip; tubes 35-50 mm long, galea with crests. Fl. Aug-Sep. Fr. Sep-Oct. Alpine meadows at 3000-3600 m. Distributed in NW Yunnan.

中国马先蒿
Pedicularis chinensis Maxim.

一年生草本，高达30厘米。叶披针状长圆形至条状长圆形，羽状半裂；裂片7-13对，边缘具重锯齿。花序长总状；花萼管状，裂片2，叶状；花冠黄色；盔先端稍弯，形成近环状。蒴果长圆状披针形。花期7月，果期8月。生海拔1700-2900米的高山草甸。产河北、山西、内蒙古、陕西、甘肃中部和南部、青海东北部。

三色马先蒿 *Pedicularis tricolor*

中国马先蒿 *Pedicularis chinensis*

Annual herbs, to 30 cm tall. Leaves lanceolate-oblong to linear-oblong, pinnatifid; segments 7-13 pairs, margin double dentate. Inflorescences long racemose; calyx tubular, lobes 2, leaf-like; corolla yellow; galea slightly bent apically, forming nearly a circle. Capsules oblong-lanceolate. Fl. Jul. Fr. Aug. Alpine meadows at 1700-2900 m. Distributed in Hebei, Shanxi, Neimenggu, Shaanxi, C and S Gansu, and NE Qinghai.

管状长花马先蒿

Pedicularis longiflora Rudolph. var. **tubiformis** (Klotzsch) Tsoong

一年生草本。叶互生，披针形至狭长圆形，羽状浅裂至深裂；裂片5-9对。花萼筒状，2裂；花腋生；花冠黄色，较小；下唇具一条浅栗色的条纹。蒴果披针形。花果期5-10月。生海拔2700-5300米的高山草甸、泉边或溪边。产云南西北部、四川西部和西藏东南部。印度北部、尼泊尔和巴基斯坦亦有。

Annual herbs. Leaves alternate, lanceolate to narrowly oblong, pinnatifid to pinnatipartite; segments 5-9 pairs. Calyx tubular, lobes 2; flowers axillary; corolla yellow, small; lower lips with a narrow maroon stripe. Capsules lanceolate. Fl. and fr. May-Oct. Alpine meadows, seeps or along streams at 2700-5300 m. Distributed in NW Yunnan, W Sichuan and SE Xizang. Also in N India, Nepal and Pakistan.

二齿马先蒿 *Pedicularis bidentata*

二齿马先蒿

Pedicularis bidentata Maxim.

草本，高6-8厘米，全体被灰色短柔毛。叶基生，条状长圆形，波状裂。花腋生，具2-4花；花萼圆柱形，裂片2，椭圆形；花冠黄色；花冠管长于花萼的4倍；盔向下弯曲成马蹄形。产四川北部。

Herbs, 6-8 cm tall, gray pubescent throughout. Leaves basal, linear-oblong, undulate-lobed. Flowers axillary, 2-4-flowered; calyx cylindric, lobes 2, elliptic; corolla yellow; tubes more than 4 × as long as calyx; galea curving downward into a horse-shoe-shape. Distributed in N Sichuan.

管状长花马先蒿 *Pedicularis longiflora* var. *tubiformis*

刺齿马先蒿 *Pedicularis armata* var. *trimaculata*

粗管马先蒿 *Pedicularis latituba*

刺齿马先蒿

Pedicularis armata Maxim. var. **trimaculata** X. F. Lu

多年生草本。茎常簇生。叶条状长圆形，羽状半裂；裂片4-9对，边缘具重锯齿。多花腋生；花冠全黄色或下唇有3深红色或栗色的条形或狭椭圆形斑点。花期8-9月，果期9月。生海拔3000-4000米的潮湿地区高山草甸、阳坡或草坪。产甘肃和青海。

Perennial herbs. Stems usually tufted. Leaves linear-oblong, pinnatipartite; segments 4-9 pairs, margin double dentate. Flowers axillary, many-flowered; corolla yellow throughout or lower lips with 3 crimson or maroon spots. Fl. Aug-Sep. Fr. Sep. Alpine meadows in moist areas or sunny slopes or turf at 3000-4000 m. Distributed in Gansu and Qinghai.

粗管马先蒿

Pedicularis latituba Bonati

多年生草本，高仅达10厘米。叶披针状长圆形，羽状半裂至全裂；裂片5-11对。花少量腋生，较密；花冠紫红色；花冠筒3-4.5厘米长；具紫色柔毛；盔扭曲镰形，额部多少具鸡冠状突起。产四川西部和西藏东南部。

Perennial herbs, barely 10 cm tall. Leaves lanceolate-oblong, pinnatipartite to pinnatisect; segments 5-11 pairs. Flowers few, axillary, dense; corolla purple-red; tubes 3-4.5 cm long, purple pubescent; galea twisted falcate, ± crested in front. Distributed in W Sichuan and SE Xizang.

台氏管花马先蒿

Pedicularis siphonantha D. Don var. **delavayi** (Franch. et Maxim.) Tsoong

多年生草本。根常明显纺锤形。叶羽状全裂；裂片6-15对。花萼裂片常5；苞片叶状，花冠玫瑰红色，2裂，盔具不明显耳状突出或几无，顶端强烈扭曲，喙向外反卷。花期6-7月，果期7-8月。生海拔3000-4600米的高山湿草地或草甸中。产云南西北部和四川西部。

Perennial herbs. Roots usually strongly fusiform. Leaves pinnatisect; segments 6-15 pairs. Calyx lobes usually 5; bracts leaf like, corolla rose-red, 2-lipped, galea with inconspicuous auriculate protrusion or nearly none, apex strongly curved, beaks revolute; galea with inconspicuous auriculate protrusion or nearly none. Fl. Jun-Jul. Fr. Jul-Aug. Alpine wet grasslands or meadows at 3000-4600 m. Distributed in NW Yunnan and W Sichuan.

硕花马先蒿

Pedicularis megalantha Don

一年生草本，达45厘米。茎成丛或单条。叶片羽状深裂，裂片7-12对。花序显著离心，30厘米以上；苞片叶状；花梗长5-12毫米。萼长圆形，齿5枚；花冠管长3-6厘米，为萼2-4倍，盔被下唇包裹，前端渐细为环状的长喙；雄蕊生于花管的顶部，1对花丝有毛。蒴果为宿萼1倍；种子卵圆形。花期6-8月，果期7-9月。生海拔2300-4200米的溪流旁湿润处与林中。产西藏。

Annual herbs, to 45 cm tall. Stems cespitose or single. Leaf blade pinnatipartite; segments 7-12 pairs. Inflorescences centrifugal, more than 30 cm; bracts leaflike; pedicel 5-12 mm. Calyx oblong, lobes 5; corolla tube 3-6 cm, 2-4 × as long as calyx; galea bent at a right angle apically; beak circular; lower lip completely enveloping galea; stamen at tube base, anterior filament pair pubescent. Capsules 1 × than persistent cylax. Fl. Jun-Aug. Fr. Jul-Sep. Swampy places at forest margins, damp grassy slopes at 2300-4200 m. Distributed in Xizang.

台氏管花马先蒿
Pedicularis siphonantha var. *delavayi*

硕花马先蒿 *Pedicularis megalantha*

杜氏翅茎草 *Pterygiella duclouxii*

翅茎草
Pterygiella nigrescens Oliv.

草本，高25-35(-50)厘米，全体被具腺短柔毛。茎单生，四方形，棱上具翅。叶披针形至狭披针形，近抱茎。花冠黄色，1.6-1.8厘米，被短柔毛。蒴果卵球形。花期7-10月，果期9-10月。生海拔1700-2600米的灌丛。产云南。

Herbs, 25-35(-50) cm tall, glandular pubescent. Stems 1, quadrangular, winged along ribs. Leaves lanceolate to narrowly so, subamplexicaul. Corolla yellow, 1.6-1.8 cm long, pubescent. Capsules ovoid. Fl. Jul-Oct. Fr. Sep-Oct. Scrubs at 1700-2600 m. Distributed in Yunnan.

翅茎草 *Pterygiella nigrescens*

杜氏翅茎草
Pterygiella duclouxii Franch.

一年生草本，全体近于无毛或疏被毛。茎四方形，沿棱有狭翅。下部叶早落，无柄，条形。花冠黄色，外疏具柔毛；冠筒长8-10毫米；下唇裂片近圆形；上唇裂片具凹缺。蒴果卵球形。花期7-10月。生海拔1000-2800米的林缘、草坡或路旁。产云南、四川和广西。

Annual herbs, subglabrous to sparsely villous. Stems quadrangular, winged along ribs. Leaves early deciduous below, sessile, linear. Corolla yellow, outside sparsely pubescent; tubes 8-10 mm long; lower lips lobes suborbicular; upper lips emarginate. Capsules ovoid. Fl. Jul-Oct. Forest edges, grassy slopes or roadsides at 1000-2800 m. Distributed in Yunnan, Sichuan and Guangxi.

阴行草
Siphonostegia chinensis Benth.

一年生草本，全体密被锈色短毛。枝条具棱。叶宽卵形，密具柔毛，2回羽状全裂；羽片3对，最下1对2回羽状深裂。总状花序花少；下唇黄色；上唇紫红色。蒴果卵球状长圆形。花期6-8月。生海拔800-3400米的干山坡或草地上。产中国大部分地区。俄罗斯、朝鲜半岛和日本亦有。

Annual herbs, densely rubiginous hairy. Branches angled. Leaves broadly ovate, densely pubescent, 2-pinnatisect; pinnae 3 pairs, lowermost pair 2-pinnately parted. Racemes few-flowered; lower lips yellow; upper lips purple-red. Capsules ovoid-oblong. Fl. Jun-Aug. Dry mountain slopes or grasslands at 800-3400 m. Distributed in most parts of China. Also in Russia, Korean Peninsula and Japan.

阴行草 *Siphonostegia chinensis*

达乌里芯芭 *Cymbaria daurica*

达乌里芯芭
Cymbaria daurica L.

草本，密被白色绢毛。叶对生，条形至条状披针形，密被白色绢毛，常全缘，顶部渐尖和锐尖。单花腋生；花梗和花萼连接处有小苞片；花冠外面被白色毛，内部具腺点。种子卵球形。花期6-8月，果期7-9月。生海拔600-1100米的旱山坡或沙质草地。产河北、内蒙古、黑龙江和吉林。俄罗斯(东西伯利亚)和蒙古亦有。

Herbs, densely white sericeous. Leaves opposite, linear to linear-lanceolate, densely white sericeous, margin often entire, apex acuminate and apiculate. Flowers axillary, solitary; bracteoles inserted at calyx and pedicels junction; corolla outside white hairy, inside glandular. Seeds ovoid. Fl. Jun-Aug. Fr. Jul-Sep. Dry mountain slopes or sandy grasslands at 600-1100 m. Distributed in Hebei, Neimenggu, Heilongjiang and Jilin. Also in Russia (E Siberia) and Mongolia.

蒙古芯芭
Cymbaria mongolica Maxim.

草本，高5-20厘米，被柔毛。叶片在下部对生，向上逐渐变大且变窄，长圆状披针形。小苞片全缘或于花梗顶端具1或2个小齿；花冠2.5-3.5厘米。种子狭卵球形。花期4-8月。生海拔800-2000米的干燥山坡。产河北、山西、内蒙古、陕西、甘肃和青海。

Herbs, 5-20 cm tall, pilose. Leaves opposite below, gradually increasing in size and becoming narrower upward, oblong-lanceolate. Bracteoles entire or with 1 or 2 small teeth near tip of pedicels; corolla 2.5-3.5 cm long. Seeds narrowly ovoid. Fl. Apr-Aug. Dry mountain slopes at 800-2000 m. Distributed in Hebei, Shanxi, Neimenggu, Shaanxi, Gansu and Qinghai.

蒙古芯芭 *Cymbaria mongolica*

紫葳科 Bignoniaceae

黄花风铃木
Handroanthus chrysanthus (Jacq.) S. O. Grose

落叶乔木。树冠圆伞形。掌状复叶，小叶4-5，倒卵形，纸质，有锯齿，全叶被褐色细茸毛。花冠漏斗形，花缘皱曲，花色鲜黄。蓇葖果，开裂时果荚多重反卷，具绒毛。种子带薄翅。花期3-4月。中国栽培。原产中美洲、南美洲。

黄花风铃木 *Handroanthus chrysanthus*

炮仗花 *Pyrostegia venusta*

照夜白 *Nyctocalos brunfelsiiflora*

Deciduous trees. Crown round umbrella. Leaf palmately compound, leaflets 4-5, elliptic, papery, serrated, pubscent. Corolla funnel-shaped, flowers edge buckle, bright yellow. Follicles, when cracking pods multiple anti-roll, with hairs. Seeds with a thin wing. Fl. Mar-Apr. Cultivated in China. Native to Central and South America.

炮仗花

Pyrostegia venusta (Ker-Gawl.) Miers

常绿攀援木质藤本。叶对生，小叶2-3枚，卵形，顶生小叶常变3叉的丝状卷须；小叶卵形或长圆状披针形。聚伞圆锥花序生于侧枝的顶端。花萼钟状；花冠筒状，弯曲，橙色。蒴果线形，室间开裂，隔膜与果瓣平行。种子横长圆形，有膜质透明的翅。华南和西南有栽培。原产南美洲。

Evergreen Woody climbers. Leaves 2-3-folialate, with or without terminal trifid tendril; leaflets ovate to oblong-lanceolate. Flowers in terminal thyrses; calyx campanulate to denticulate; corolla tubular club-shaped, curved, orange. Cappsule linear; seeds transverse oblong with membranous hyaline wings. Cultivated in S and SW China. Native to South America.

照夜白

Nyctocalos brunfelsiiflora Teijsm. et Binn.

藤本，无卷须。叶具3小叶；小叶椭圆状披针形至椭圆形或倒卵形，无毛。花序总状，具约17花，顶生；花冠白色；雄蕊5枚。蒴果长椭圆体形，压扁。种子具膜质翅。花期8-10月，果期9-11月。生海拔300-600米的低山丛林中。产云南南部。东南亚亦有。

Lianas, without tendrils. Leaves 3-foliolate; leaflets elliptic-lanceolate to elliptic or obovate, glabrous. Inflorescences racemose, ca. 17-flowered, terminal; corolla white; stamens 5. Capsules long ellipsoid, compressed. Seeds membranous winged. Fl. Aug-Oct. Fr. Sep-Nov. Jungles on low hills at 300-600 m. Distributed in S Yunnan. Also in SE Asia.

猫爪藤

Dolichandra unguis-cati (L.) L. G. Lohmann

木质藤本。叶具2小叶，有或无顶生卷须；小叶长圆状披针形至长圆状卵形。聚伞花序具1-6花；花冠筒状，黄色，二唇形。蒴果长圆形至狭椭圆形，木质。种子横长圆形，有翅。华南和西南有栽培。原产南美洲。

Woody climber. Leaves bifolialate, with or without terminal trifid tendril; leaflets oblong-lanceolate to oblong-ovate. Inforescence a 1-6-flowered cyme; corolla tubular, yellow, obliquely bilabiate. Capsules oblong to narrowly ellipsoid, woody. Seeds transverse oblong, in wing. Cultivated in S and SW China. Native to South America.

猫爪藤 *Dolichandra unguis-cati*

Distributed in S Yunnan. Also in S and SE Asia.

蒜香藤 *Mansoa alliacea*

木蝴蝶 *Oroxylum indicum*

蒜香藤

Mansoa alliacea (Lam.) A. H. Gentry

常绿木质藤本。三出复叶对生，中叶椭圆形，长达15厘米，全缘。圆锥花序腋生；花冠筒状；花瓣前端5裂，初开为紫色，后变至白色。花及叶经过揉搓之后有浓浓的蒜香味。荚果。花期春秋两季。中国南方栽培。原产南美洲。

Evergreen woody vine. Leaves opposite, 3 leaflets, middle leaflet ovate, up to 15 cm long, margin entire. Inflorescence panicles, axillary; corolla tubular; petals 5-lobed at front, flowers purple when opening, and white when launch. Flowers and leaves have a thick smell of garlic after rubbing. Pod. Fl. spring and autumn. Cultivated in south of China. Native to South America.

老鸦烟筒花

Millingtonia hortensis L. f.

乔木。叶对生，二至三回羽状复叶；小叶椭圆形、卵形或卵状长圆形，无毛，边缘全缘。花序聚伞状圆锥形；花萼小，杯状；花冠白色，冠筒极细长。蒴果条形。种子碟状长圆形，压扁，由膜质和透明翅所包围。花期9-12月。生海拔500-1200米的低山坡。产云南南部。南亚和东南亚亦有。

Trees. Leaves opposite, 2-3-pinnately compound; leaflets elliptic, ovate or ovate-oblong, glabrous, margin entire. Inflorescences cymose-paniculate; calyx small, cupular; corolla white, tubes very long. Capsules linear. Seeds discoid-oblong, compressed, surrounded by membranous and transparent wings. Fl. Sep-Dec. Low altitude slopes at 500-1200 m.

老鸦烟筒花 *Millingtonia hortensis*

木蝴蝶

Oroxylum indicum (L.) Benth. ex Kurz

小乔木。叶为二至四回羽状复叶。花常夜间开放，具难闻气味；花萼紫色，钟形，果期近木质；花冠紫红色。蒴果木质，约1厘米厚。种子圆形，具纸质翅。花期9-12月。生海拔500-900米的热带和亚热带低开阔林、路边或山坡。产中国西南、华南和华东。南亚和东南亚亦有。

Small trees. Leaves 2(-4)-pinnately compound. Flowers usually open at night, with foul smell; calyx purple, campanulate, becoming subwoody in fruit; corolla purple-red. Capsules woody, ca. 1 cm thick. Seeds rounded, with papery wings. Fl. Sep-Dec. Tropical and subtropical low altitude open forests, roadsides or slopes at 500-900 m. Distributed in SW, S and E China. Also in S and SE Asia.

梓 *Catalpa ovata*

梓

Catalpa ovata G. Don

乔木。叶对生或近对生，有时轮生，疏具柔毛或渐无毛，常3裂。花序圆锥状，顶生；花冠钟形，浅黄色，喉部具2黄色条纹和紫斑。蒴果条形，下垂。花期5-6月，果期8-10月。生海拔(500-)1900-2500米的山坡。产中国西南、华中、华东、华北、西北和东北。引种于各地。

Trees. Leaves opposite or nearly so, sometimes whorled, scabrous, sparsely pubescent or glabrescent, usually 3-lobed. Inflorescences paniculate, terminal; corolla campanulate, pale yellow, yellow striate and purple spots at throat. Capsules linear, nodding. Fl. May-Jun. Fr. Aug-Oct. Slopes at (500-)1900-2500 m. Distributed in SW, C, E, N, NW and NE China. Introduced elsewhere.

黄金树

Catalpa speciosa (Warder ex Barney) Engelm.

乔木；树冠伞状。叶长15-30厘米，顶端长渐尖；叶柄长10-15厘米。圆锥花序顶生；苞片2，线形；花萼2裂，裂片舟状，无毛；花冠白色，喉部有2黄色条纹及紫色细斑点。蒴果圆柱形，黑色，长30-55厘米，宽1-2厘米，2瓣开裂。种子椭圆形，长25-35毫米，宽6-10毫米，两端有毛。花期5-6月，果期8-10月。中国各地均有栽培。原产美国中部至东部。

Trees, canopy umbrella. Leaves 15-30 cm, apex acuminate; petiole 10-15 cm. Inflorescences paniculate, terminal; bract 2, linear; calyx 2-lobes, lobe navicular, glabrous; corolla white, yellow 2-striate and purple spotted at throat. Capsules cylindrical, balck, 30-55 × 1-2 cm, 2 valve. Seeds ellipsoid, 25-35 × 6-10 mm, terminal pubescent. Fl. May-Jun. Fr. Aug-Oct. Cultivated in China. Native to C to E USA.

黄金树 *Catalpa speciosa*

楸 *Catalpa bungei*

楸
Catalpa bungei C. A. Mey.

乔木。叶三角状卵形或卵状长圆形，下面无毛，上面深绿色。伞房状总状花序顶生，具花2-12朵；花冠浅红色，喉部具2条黄色条纹和深紫色斑点。蒴果条形。花期5-6月，果期6-10月。生路边。产华中、华东、华北和西北，栽培于云南、贵州和广西。

Trees. Leaves triangular-ovate or ovate-oblong, glabrous abaxially, dark green adaxially. Inflorescences corymbose-racemose, terminal, 2-12-flowered; corolla pale red, yellow 2-striate and dark purple spotted at throat. Capsules linear. Fl. May-Jun. Fr. Jun-Oct. Roadsides. Distributed in C, E, N and NW China, cultivated in Yunnan, Guizhou and Guangxi.

灰楸 *Catalpa fargesii*

灰楸
Catalpa fargesii Bureau

乔木，高约25米。叶卵形或三角状心形，厚革质。花序聚伞状总状，具7-15花，无毛或具柔毛；花冠钟形，淡红色至淡紫色，上面具紫斑。蒴果圆筒形，下垂。花期3-5月，果期6-11月。生海拔700-1300(-2500)米的路旁或山坡，通常栽培。产中国西南、华南、华中、华东、华北和西北。

Trees, ca. 25 m tall. Leaves ovate or triangular-cordate, thick leathery. Inflorescences corymbose-racemose, 7-15-flowered, glabrous or pubescent; corolla campanulate, pale red to pale purple, purple spotted adaxially. Capsules terete, nodding. Fl. Mar-May. Fr. Jun-Nov. Roadsides or slopes at 700-1300(-2500) m, frequently cultivated. Distributed in SW, S, C, E, N and NW China.

火焰树
Spathodea campanulata Beauv.

乔木，高达20米。偶数羽状复叶；小叶4-9对；长圆形或长圆状卵形。总状花序顶生；花冠猩红至血红色，钟状。蒴果长圆状披针形，木质。种子横长圆形，具白色、有光泽的透明的翅。广泛栽培与热带和亚热带地区。原产热带非洲。

Tree, to 20 m. Leaves imparipinnate; leaflets 4-9 pairs, oblong or oblong-ovate. Flowers in terminal racemes. Corolla scarlet to

火焰树 *Spathodea campanulata*

羽叶楸 *Stereospermum colais*

小萼菜豆树 *Radermachera microcalyx*

blood-red, campanulate. Capsules oblong-lanceolate, woody. Seeds transverse oblong in white shining hyaline wings. Widely cultivated in tropical and subtropical regions. Native to tropical Africa.

羽叶楸

Stereospermum colais (Buch.-Ham. ex Dillwyn) Mabb.

落叶乔木。小叶在中脉两侧各3-6片。花序圆锥状，顶生；花数朵；花萼钟形，紫色；花冠黄白色；口部具绒毛；冠筒圆柱形。蒴果长柱形，有明显4棱。花期5-7月，果期9-11月。生海拔400-1800米的林中或山坡。产云南、贵州、广西和海南。南亚和东南亚亦有。

Deciduous trees. Leaflets 3-6 on each side of midrib. Inflorescences paniculate, terminal; flowers many; calyx campanulate, purple; corolla yellow-white, tomentose at mouth, tubes terete. Capsules long cylindric, distinctly 4-ridged. Fl. May-Jul. Fr. Sep-Nov. Forests or slopes at 400-1800 m. Distributed in Yunnan, Guizhou, Guangxi and Hainan. Also in S and SE Asia.

小萼菜豆树

Radermachera microcalyx C. Y. Wu et W. C. Yin

乔木。1回羽状复叶；小叶5-7，下面具黑色腺点，无毛。聚伞状圆锥花序；花萼钟形，宿存；萼齿5，小；花冠黄色。蒴果绿色，长圆柱形，下垂；外果皮薄革质，2瓣裂。花期1-3月，果期4-12月。生海拔340-1600米的林下。产云南南部和广西。

Trees. Leaves 1-pinnately compound; leaflets 5-7, black glandular abaxially, glabrous. Cymose panicle; calyx campanulate, persistent; teeth 5, minute; corolla yellow. Capsules green, long terete, nodding; pericarps thin leathery, dehiscing by 2 valves. Fl. Jan-Mar. Fr. Apr-Dec. Forests at 340-1600 m. Distributed in S Yunnan and Guangxi.

菜豆树

Radermachera sinica (Hance) Hemsl.

小乔木。二至三回羽状复叶，侧生小叶卵形至卵状披针形，近基部外侧有白色腺点。花序圆锥状，顶生，直立；花冠白色至浅黄色，钟状漏斗形。蒴果圆柱状，下垂，具棱。花期5-9月，果期10-12月。生海拔300-800米的山坡或林中。产中国西南、华南和华东。印度、不丹、缅甸北部和越南亦有。

Small trees. Leaves 2(-3)-pinnately compound, lateral leaflets ovate to ovate-lanceolate, dense white glandular dots near base on outer side. Inflorescences paniculate, terminal, erect; corolla white to pale yellow, campanulate-funnelform. Capsules terete, nodding, angular. Fl. May-Sep. Fr. Oct-Dec. Slopes or forests at 300-800 m. Distributed in SW, S and E China. Also in India, Bhutan, N Myanmar and Vietnam.

菜豆树 *Radermachera sinica*

海南菜豆树 *Radermachera hainanensis*

海南菜豆树
Radermachera hainanensis Merr.

乔木。叶二至三回羽状分裂；小叶长圆状卵形或卵形，纸质，无毛。花序总状，顶生，具少量花；花萼浅红色；花冠金黄色，钟形。种子卵球形，膜质。花期4月。生海拔300-600米的山坡或林中。产云南、广东和海南。老挝、泰国和柬埔寨亦有。

Trees. Leaves 2- or 3-pinnately compound; leaflets oblong-ovate or ovate, papery, glabrous. Inflorescences racemose, terminal, few-flowered; calyx pale red; corolla golden yellow, campanulate. Seeds ovoid, membranous. Fl. Apr. Slopes or forests at 300-600 m. Distributed in Yunnan, Guangdong and Hainan. Also in Laos, Thailand and Cambodia.

凌霄
Campsis grandiflora (Thunb.) Schum.

攀援藤本。小叶7-9枚，边缘具锯齿。花序短圆锥状，顶生；花萼钟形，裂至中部，裂片披针形；花冠外面橙红色，里面红色；花药二叉，黄色。蒴果先端钝。花期5-8月。栽培于华南、华东和华北。印度、巴基斯坦、越南和日本亦有。

Climbing vines. Leaflets 7-9, margin serrate. Inflorescences short paniculate, terminal; calyx campanulate, divided halfway, lobes lanceolate; corolla red adaxially, orange-red abaxially; anthers divergent, yellow. Capsules apex obtuse. Fl. May-Aug. Cultivated in S, E and N China. Also in India, Pakistan, Vietnam and Japan.

角蒿
Incarvillea sinensis Lam.

一年生或多年生草本。叶二至三回羽状分裂。花序疏总状；小苞片绿色，条形；花萼绿色，带紫红色，钟形，萼齿近钻形，基部膨大成腺点；花冠钟状漏斗形，基部收缩成管状。蒴果浅绿色，狭圆筒形。花期5-9月，果期10-11月。生海拔500-2500(-3900)米的山坡或田野。产中国西南、华北、西北、河南、山东和黑龙江。

凌霄 *Campsis grandiflora*

角蒿 *Incarvillea sinensis*

黄花角蒿 *Incarvillea sinensis* var. *przewalskii*

Annual or perennial herbs. Leaves 2- or 3-pinnately divided. Inflorescences sparsely racemose; bractlets green, linear; calyx green with purple-red, campanulate, teeth subulate, base enlarged into glands; corolla campanulate-funnelform, constricted into tubular base. Capsules pale green, narrowly terete. Fl. May-Sep. Fr. Oct-Nov. Slopes or fields at 500-2500(-3900) m. Distributed in SW, N and NW China, Henan, Shandong and Heilongjiang.

黄花角蒿

Incarvillea sinensis var. **przewalskii** Batalin

一年生至多年生草本。茎分枝。叶互生，二至三回羽状细裂，形态多变异。顶生总状花序；花萼钟状，萼齿钻状；花冠淡黄色，钟状漏斗形，基部收缩成细筒，裂片圆形；雄蕊4，2强，花药成对靠合；花柱淡黄色。蒴果圆柱形。种子扁圆形，具透明的膜质翅。花期7-9月，果期10-11月。生海拔2000-2600米的山坡。产甘肃、陕西、青海和四川西北部。

Annual or perennial herbs. Stems branched. Leaves alternate, 2- or 3-pinnately, shape variety. Inflorescences racemose, terminal; calyx campanulate; corolla pale yellow, campanulate-funnelform, constricted into tubular base, lobes rounded; stamens 4, didynamous, anthers adherent; style pale yellow. Capsules cylindrical. Seeds compressed globose, surrounded by transparent and membranous wings. Fl. Jul-Sep. Fr. Oct-Nov. Slopes at 2000-2600 m. Distributed in Gansu, Shaanxi, Qinghai and NW Sichuan.

两头毛

Incarvillea arguta (Royle) Royle

多年生草本。1回羽状复叶；小叶5-11个，卵状披针形，边缘具细齿。花序总状；苞片钻形；花冠浅红色或紫红色，钟状漏斗形。种子多，两端被丝状短尖柔毛。花期3-7月，果期9-12月。生海拔1400-2700(-3400)米的山坡或灌丛中。产中国西南和甘肃。印度和尼泊尔亦有。

Perennial herbs. 1-pinnately compound; leaflets 5-11, ovate-lanceolate, margin serrate. Inflorescences racemose; bracts subulate; corolla pale red or purple-red, campanulate-funnelform. Seeds many, acute and filiform pubescent at both ends. Fl. Mar-Jul. Fr. Sep-Dec. Slopes or shrubbery at 1400-2700(-3400) m. Distributed in SW China and Gansu. Also in India and Nepal.

两头毛 *Incarvillea arguta*

单叶波罗花 *Incarvillea forrestii*

黄波罗花 *Incarvillea lutea*

单叶波罗花

Incarvillea forrestii Fletcher

多年生草本，渐无毛。叶互生，单叶，卵状椭圆形，纸质，边缘具圆齿。花序总状，具6-12花；花冠红色，内面有紫红色条纹和斑点。蒴果披针形，压扁，具4棱。花期5-7月，果期8-11月。生海拔3000-3500米草地或灌丛中。产云南西北部和四川西南部。

Perennial herbs, glabrescent. Leaves alternate, simple, ovate-elliptic, papery, margin crenate. Inflorescences racemose, 6-12-flowered; corolla red, with purple-red streaks and dots inside. Capsules lanceolate, compressed, 4-angled. Fl. May-Jul. Fr. Aug-Nov. Grasslands or thickets at 3000-3500 m. Distributed in NW Yunnan and SW Sichuan.

黄波罗花

Incarvillea lutea Bureau et Franch.

多年生草本。叶一回羽状分裂。花序总状，具5-12花；花萼钟形，绿色，具紫色斑点，脉深紫色；花冠黄色，基部具紫色斑点和条纹。蒴果浅棕色，披针形。花期5-8月，果期9-11月。生海拔2000-3400米的林下、山坡或草地。产云南西北部、四川西部和西藏中部。

Perennial herbs. Leaves 1-pinnately divided. Inflorescences racemose, 5-12-flowered; calyx campanulate, green with purple spots, veins dark purple; corolla yellow, purple spotted and striate at base. Capsules pale brown, lanceolate. Fl. May-Aug. Fr. Sep-Nov. Forests, slopes or grasslands at 2000-3400 m. Distributed in NW Yunnan, W Sichuan and C Xizang.

红波罗花

Incarvillea delavayi Bureau et Franch.

多年生无茎草本。叶基生，一回羽状分裂；侧生小叶4-11对，椭圆状披针形。花序总状，具2-6花，顶生；花萼钟形；花冠红色，钟形。蒴果灰褐色，具4棱。花期7月。生海拔2400-3900米的草地和山坡。产云南西北部和四川西部。

Perennial acaulescent herbs. Leaves basal, 1-pinnately divided; lateral leaflets 4-11 pairs, elliptic-lanceolate. Inflorescences racemose, 2-6-flowered, terminal; calyx campanulate; corolla red, campanulate. Capsules gray-brown, 4-angled. Fl. Jul. Grasslands or slopes at 2400-3900 m. Distributed in NW Yunnan and W Sichuan.

鸡肉参

Incarvillea mairei (H. Lévl.) Griers.

无茎多年生草本。叶基生，一回羽状复叶；侧生小叶2或3对，无毛，边缘具锯齿。花序总状，具2-4花，近顶生；花冠紫红色或粉红色，基部带黄色。蒴果圆锥形。花期5-7月，果期9-11月。生海拔2400-4500米的山坡或路边。产云南北部和西北部、四川西部和西藏东部。

红波罗花 *Incarvillea delavayi*

鸡肉参 *Incarvillea mairei*

多小叶鸡肉参 *Incarvillea mairei* var. *multifoliolata*

Acaulescent perennial herbs. Leaves basal, 1-pinnately compound; lateral leaflets 2 or 3 pairs, glabrous, margin serrate. Inflorescences racemose, 2-4-flowered, subterminal; corolla purple-red or pink, tinged with yellow at base. Capsules conical. Fl. May-Jul. Fr. Sep-Nov. Slopes or roadsides at 2400-4500 m. Distributed in N and NW Yunnan, W Sichuan and E Xizang.

多小叶鸡肉参

Incarvillea mairei (H. Lévl.) Grierson var. **multifoliolata** (C. Y. Wu et W. C. Yin) C. Y. Wu et W. C. Yin

多年生无茎草本。羽状复叶，具4-8对小叶，边缘具锯齿至近全缘，先端锐尖。花序总状，具2-4花，近顶生；花冠紫红色或浅红色。花期6-8月，果期8-10月。生海拔3200-4200米的林中、山坡或草地。产云南西北部和四川西北部。

Perennial herbs, stemless. Leaves pinnately compound, leaflets 4-8 pairs, margin serrulate to subentire, apex acuminate. Inflorescences racemose, 2-4-flowered, subterminal; corolla purple-red or pale red. Fl. Jun-Aug. Fr. Aug-Oct. Forests, slopes or grasslands at 3200-4200 m. Distributed in NW Yunnan and NW Sichuan.

藏波罗花

Incarvillea younghusbandii Sprague

无茎草本。根肉质。叶基生，一回羽状复叶，侧生小叶2-5对，无柄，卵状椭圆形，粗糙，边缘具锯齿。花序短总状，具3-6花或单生；花冠漏斗形；花冠管橘黄色。蒴果近木质，强烈弯曲。花期5-8月，果期8-10月。生海拔4000-5500米的砾石、山坡或灌丛。产西藏和青海。

Herbs stemless. Roots fleshy. Leaves basal, 1-pinnately compound, lateral leaflets 2-5 pairs, sessile, ovate-elliptic, scabrous, margin serrate. Inflorescences short racemose, 3-6-flowered or solitary; corolla funnelform; tubes orange-yellow. Capsules subwoody, strongly curved. Fl. May-Aug. Fr. Aug-Oct. Gravel, slopes or thickets at 4000-5500 m. Distributed in Xizang and Qinghai.

藏波罗花 *Incarvillea younghusbandii*

密生波罗花 *Incarvillea compacta*

密生波罗花 *Incarvillea compacta*

密生波罗花

Incarvillea compacta Maxim.

多年生草本。叶一回羽状复叶，簇生于茎顶。花序密总状，簇生于茎顶；花萼绿色或紫红色，具深紫色斑点，钟形；花冠红色或紫红色。蒴果狭披针形，长11厘米，具明显的4棱。花期5-7月，果期8-12月。生海拔2600-4100米的山坡、草地或灌丛中。产云南西北部、四川西南部、西藏、甘肃南部和青海。

Perennial herbs. Leaves 1-pinnately compound, clustered at stem base. Inflorescences densely racemose, clustered at stem apex; calyx green or purple-red, with dark purple spots, campanulate; corolla red or purple-red. Capsules narrowly lanceolate, 11 cm long, with 4 distinct ridges. Fl. May-Jul. Fr. Aug-Dec. Slopes, grasslands or thickets at 2600-4100 m. Distributed in NW Yunnan, SW Sichuan, Xizang, S Gansu and Qinghai.

中甸角蒿 *Incarvillea zhongdianensis*

中甸角蒿

Incarvillea zhongdianensis Grey-Wils.

草本。叶较小，侧生小叶4-8对，卵状披针形；顶生的1枚小叶较大。花萼绿色带紫红色，钟形；花冠钟状漏斗形，紫红色，喉部黄色，基部收缩成管状。花期6-8月，果期8-10月。生海拔约3200米的草坡或云杉林缘。产云南(香格里拉)。广泛栽培于世界植物园中。

Herbs. Leaves small, lateral leaflets 4-8, ovate-lanceolate; terminal leaflets larger. Inflorescences sparsely racemose or solitary; calyx green with purple-red, campanulate; corolla campanulate-funnelform, purple-red, yellow in throat, constricted into tubular base. Fl. Jun-Aug. Fr. Aug-Oct. Grassy slopes or *Picea* forests edges at ca. 3200 m. Distributed in Yunnan (Shangri-La). Widely cultivated in botanical gardens worldwide.

西南猫尾木

Markhamia stipulata Seem. ex Schum.

乔木。小叶具柔毛至渐无毛，边缘具锯齿至近全缘。花序为顶生总状，具黄色锈色绒毛，具4-10花；花萼和蒴果具卷曲毛；花冠黄色至污黄色，约10厘米。花期9-12月，果期翌年2-3月。生海拔300-1700米的林中。产云南、广西、广东和海南。东南亚亦有。

Trees. Leaflets pubescent to glabrescent, margin serrulate to subentire. Inflorescences terminal racemes, rusty yellow pubescent, 4-10-flowered; calyx and capsules floccose; corolla yellow to dirty yellow, ca. 10 cm long. Fl. Sep-Dec. Fr. next Feb-Mar. Forests at 300-1700 m. Distributed in Yunnan, Guangxi, Guangdong and Hainan. Also in SE Asia.

西南猫尾木 *Markhamia stipulata*

西南猫尾木 *Markhamia stipulata*

毛叶猫尾木 *Markhamia stipulata* var. *kerrii*

火烧花 *Mayodendron igneum*

毛叶猫尾木

Markhamia stipulata Seem. ex K. Schum. var. **kerrii** Sprague

乔木。小叶幼时具锈黄色短柔毛，上面几无毛，下面有短柔毛至微柔毛。花序为顶生总状，密具锈色柔毛，具4-10花；花萼和蒴果疏被至密被绵毛。花期9-12月，果期翌年1-4月。生海拔900-1200米的疏林或潮湿处。产云南、华南和福建。东南亚亦有。

Trees. Leaflets rusty yellow pubescent when young, glabrescent adaxially, pubescent to puberulent abaxially. Inflorescences terminal racemes, rusty yellow pubescent, 4-10-flowered; calyx and capsules sparsely to densely lanate-woolly. Fl. Sep-Dec. Fr. next Jan-Apr. Sparse forests or humid places at 900-1200 m. Distributed in Yunnan, S China and Fujian. Also in SE Asia.

火烧花

Mayodendron igneum (Kurz) Kurz

常绿乔木。叶二回羽状复叶。花序短总状，生于老茎或短侧枝上，具5-13花；花萼杯状，扇形；花冠橙黄色至金黄色，管状。蒴果长条形，下垂。花期2-5月，果期5-9月。生海拔100-1900米的山坡或林中。产云南南部、广西、广东和台湾。南亚和东南亚亦有。

Trees evergreen. Leaves 2-pinnately compound. Inflorescences short racemose, borne on old stems or short lateral branches, 5-13-flowered; calyx tubular, spathelike; corolla orange-yellow to golden yellow, tubular. Capsules long linear, nodding. Fl. Feb-May. Fr. May-Sep. Slopes or forests at 100-1900 m. Distributed in S Yunnan, Guangxi, Guangdong and Taiwan. Also in S and SE Asia.

蓝花楹

Jacaranda mimosifolia D. Don.

落叶乔木。叶对生，二回羽状复叶，羽片通常在16对以上，每1羽片有小叶16-24对；小叶顶端急尖，基部楔形，全缘，无毛。花蓝色，花序长达30厘米，直径约18厘米；花萼筒状，萼齿5。花冠筒细长，蓝色，长约18厘米，裂片圆形；雄蕊4，2强，花丝着生于花冠筒中部。子房圆柱形，无毛。朔果木质，扁卵圆形。花期5-6月。中国栽培。原产热带美洲。

Deciduous trees. Leaves opposite, 2 pinnate leaves, above 16, leaflets 16-24 per pinna pairs; leaflets apex acute, base cuneate, entire, glabrous. Flower blue, inflorescence up to 30 cm diam, ca. 18 cm diam; calyx tubular, lobes 5. Corolla tube slender, blue, ca. 18 cm long; lobes rounded; stamen 4, filaments in the middle of the tube. Ovary cylindrical, glabrous. Capsules woody, flat oval. Fl. May-Jun. Cultivated in China. Native to tropical America.

蓝花楹 *Jacaranda mimosifolia*

硬骨凌霄 *Tecoma capensis*

硬骨凌霄
Tecoma capensis (Thunb.) Lindl.

常绿半藤状或近直立灌木。枝绿褐色，有痂状凸起。叶对生，奇数羽状复叶；总叶柄长，小叶柄短；小叶多为7枚，边缘有锯齿。总状花序顶生；花萼钟状，5齿裂；花冠漏斗状，有深红色的纵纹上唇凹入；雄蕊突出。蒴果线形，略扁。花期春季和秋季。中国栽培。原产非洲南部。

Evergreen scrambling or suberect shrubs. Branch green-brown, scablike protrusions. Leaves opposite, odd-pinnate; petiole long than leaflet petiole; leaflets usually 7, serrated. Infloresceneswracemes, terminal; calyx campanulate, 5-lobed; corolla funnelform, upper lip dark red, stripes vertical concave; stamens protruding. Capsules linear, slightly flat. Fl. spring and autumn. Cultivated in China. Native to S Africa.

胡麻科 Pedaliaceae

芝麻
Sesamum indicum L.

一年生草本。茎四棱形。叶对生或互生，披针形至卵形，具不同程度的三裂。花单生或2-3朵生于叶腋；花白色、粉色或淡粉紫色具深色斑块。蒴果椭圆体形。中国大部分地区有栽培。

Annual herbs. Stems tetragonal. Leaves opposite or alternate, lanceolate to ovate, variously 3-parted. Flowers solitary or 2-3 axillary; flowers white, pink, or mauve-pink with darker markings. Capsules ellipsoid. Cultivated in most parts of China.

芝麻 *Sesamum indicum*

角胡麻科 Martyniaceae

角胡麻
Martynia annua L.

一年生或多年生草本，全株被黏质柔毛。叶对生，阔卵形。总状花序顶生，萼基部具膜质小苞片2枚；花冠二唇形，檐部裂片5，不等大；雄蕊2。蒴果，顶端具2枚短钩状凸起。花期全年。生海拔500-1500米的林下或路边荒地。云南南部有归化。南亚和东南亚亦有归化。原产中美洲。

Annual or perennial herbs, usually viscid pubescent with uniseriate glandular hairs. Leaves opposite, broadly ovate. Inflorescences racemose, terminal; bracts 2, membranous; corolla bilabiate, dark red, lobes 5, semirounded; stamen 2. Capsules ovoid, with hook-like apical beak. Fl. throughout year. Forests or roadsides at 500-1500 m. Naturalized in S Yunnan. Also naturalized in S and SE Asia. Native to Central America.

角胡麻 *Martynia annua*

列当科 Orobanchaceae

草苁蓉

Boschniakia rossica (Cham. et Schltdl.) B. Fedtsch.

寄生肉质草本。根茎横生，圆柱形。茎常2或3个直立。花序穗状，长7-22厘米；花萼杯形，顶部不规则3-5齿裂；花冠深紫红色，宽钟形；冠筒膨大。蒴果近球形，2瓣开裂。花期5-7月，果期7-9月。生海拔1500-1800米的山坡、林中或溪边，寄生于桤木属植物上。产内蒙古、辽宁、吉林和黑龙江。俄罗斯、朝鲜半岛、日本和美国(阿拉斯加)亦有。

Parasitic fleshy herbs. Rootstocks horizontal, cylindric. Stems usually 2 or 3, erect. Inflorescences spicate, 7-22 cm long; calyx cupular, apex irregularly 3-5-toothed; corolla dark purple-red, broadly campanulate; tubes enlarged. Capsules subglobose, dehiscing by 2 valves. Fl. May-Jul. Fr. Jul-Sep. Slopes, forests or riversides at 1500-1800 m, parasitic on species of *Alnus*. Distributed in Neimenggu, Liaoning, Jilin and Heilongjiang. Also in Russia, Korean Peninsula, Japan and USA (Alaska).

丁座草

Boschniakia himalaica Hook. f. et Thoms.

寄生肉质草本，通常仅有1条直立的茎。花序总状；花萼短杯状；花冠黄褐色或淡紫色；花冠管稍膨大；上唇凹缺或全缘；下唇3裂。蒴果近球形至卵状长圆形。花期4-6月，果期6-9月。生海拔2500-4400米的山坡、林下或灌丛，寄生于杜鹃植物根上。产台湾、湖北、西南和西北。印度北部、尼泊尔和不丹亦有。

Parasitic fleshy herbs, stems usually 1, erect. Inflorescences racemose; calyx short cupular; corolla yellow-brown or pale purple; tubes slightly enlarged; upper lips emarginate or entire; lower lips 3-lobed. Capsules subglobose or ovoid-oblong. Fl. Apr-Jun. Fr. Jun-Sep. Slopes, forests or thickets at 2500-4400 m, parasitic on roots of *Rhododendron*. Distributed in Taiwan, Hubei, and SW and NW China. Also in N India, Nepal and Bhutan.

丁座草 *Boschniakia himalaica*

野菰

Aeginetia indica L.

寄生肉质草本。茎黄褐色或紫红色。叶红色，卵状披针形或披针形。花常单生茎端，稍俯垂；花冠具紫红色条纹，不明显二唇形，管状钟形。花期4-8月，果期8-10月。生海拔200-1800米山坡或路边，寄生于芒属和甘蔗属植物上。产中国西南、华南和华东。南亚、东南亚和东亚亦有。

Parasitic fleshy herbs. Stems yellow-brown or purple-red. Leaves red, ovate-lanceolate or lanceolate. Flowers solitary, terminal, slightly nodding; corolla purple-red striate, indistinctly bilabiate, tubular-campanulate. Fl. Apr-Aug. Fr. Aug-Oct. Slopes or roadsides at 200-1800 m, parasitic on species of *Miscanthus* and *Saccharum*. Distributed in SW, S and E China. Also in S, SE and E Asia.

草苁蓉 *Boschniakia rossica*

野菰 *Aeginetia indica*

中国野菰 *Aeginetia sinensis*

中国野菰
Aeginetia sinensis Beck

寄生肉质草本。茎常自基部分枝。叶生茎基，卵状披针形或披针形。单花顶生；花萼佛焰苞状；花冠紫红色，有时上半部分紫红色而下半部分白色，近二唇形。蒴果圆锥形。花期4-6月，果期6-9月。生海拔800-900米的路边，常寄生于禾本科植物的根部。产福建、浙江、安徽和江西。日本亦有。

Parasitic fleshy herbs. Stems usually branched from base. Leaves at stem base, ovate-lanceolate or lanceolate. Flowers solitary, terminal; calyx spathelike; corolla purple-red, sometimes purple-red on upper part and white on lower part, nearly bilabiate. Capsules conical. Fl. Apr-Jun. Fr. Jun-Sep. Roadsides at 800-900 m, frequently parasitic on roots of species of the Poaceae. Distributed in Fujian, Zhejiang, Anhui and Jiangxi. Also in Japan.

肉苁蓉
Cistanche deserticola Ma

寄生草本，高0.4-1.6米。茎不分枝或具2-4分枝。茎下部叶卵形或三角状卵形，上部者披针形或条状披针形。花序穗状，长15-50厘米；花冠淡黄白色或淡紫色，干时褐色，管状钟形，长3-4厘米。蒴果卵球形。花期5-6月，果期6-8月。生海拔200-1200米的沙地。产内蒙古、甘肃、宁夏和新疆。蒙古亦有。

Parasitic herbs, 0.4-1.6 m tall. Stems unbranched or 2-4-branched. Leaves on lower part of stem ovate or triangular-ovate, on upper part lanceolate or linear-lanceolate. Inflorescences spicate, 15-50 cm long; bract subequalling corolla; calyx campanulate; corolla pale yellow-white or pale purple, becoming brown when dry, tubular-campanulate, 3-4 cm long. Capsules ovoid. Fl. May-Jun. Fr. Jun-Aug. Sandy places at 200-1200 m. Distributed in Neimenggu, Gansu, Ningxia and Xinjiang. Also in Mongolia.

肉苁蓉 *Cistanche deserticola*

矮生豆列当
Mannagettaea hummelii Harry Sm.

寄生草本，高3-5厘米。叶数个，宽卵状三角形，无毛，先端钝。花序聚伞状；花冠紫色，稍弯曲；冠筒先端及上面密具柔毛；上唇全缘或2裂；下唇3裂。蒴果长圆形或卵球形。花期6-7月，果期8-9月。生海拔3200-3700米的山坡、灌丛或林中。产四川西部、西藏东部、甘肃西南部和青海东南部。俄罗斯亦有。

Parasitic herbs, 3-5 cm tall. Leaves several, broadly ovate-triangular, glabrous, apex obtuse. Inflorescences corymbose; corolla purple, slightly curved; tubes densely villous at apex and adaxially; upper lips entire or 2-lobed; lower lips 3-lobed. Capsules oblong or ovoid. Fl. Jun-Jul. Fr. Aug-Sep. Slopes, thickets or forests at 3200-3700 m. Distributed in W Sichuan, E Xizang, SW Gansu and SE Qinghai. Also in Russia.

矮生豆列当 *Mannagettaea hummelii*

列当 *Orobanche coerulescens*

列当
Orobanche coerulescens Steph.

寄生草本，密被长柔毛。叶卵状披针形。花序长10-20厘米；花冠深蓝色、蓝紫色、淡紫色或黄色；花冠管收缩，向上开放；上唇2裂。蒴果卵状长圆形或长圆形。花期4-7月，果期7-9月。生海拔900-4000米的山坡或草地，寄生于蒿属植物的根上。产中国西南、华中、华东、华北、西北和东北。南亚、中亚、东亚、俄罗斯和欧洲亦有。

Parasitic herbs, densely villous. Leaves ovate-lanceolate. Inflorescences 10-20 cm long; corolla dark blue, blue-purple, pale purple or yellow; tubes constricted, open upward; upper lips 2-lobed. Capsules ovoid-oblong or oblong. Fl. Apr-Jul. Fr. Jul-Sep. Slopes or grasslands at 900-4000 m, parasitic on roots of *Artemisia*. Distributed in SW, C, E, N, NW and NE China. Also in S, C and E Asia, Russia and Europe.

西藏列当
Orobanche clarkei Hook. f.

二年生或多年生寄生草本，密被腺状柔毛。茎直立。花序穗状；花冠蓝紫色；花冠近直立，不膨大；上唇2裂；柱头2裂。蒴果长圆形。种子卵状长圆形。花期5-6月，果期7-8月。生海拔2900-3400米的灌丛中。产西藏。克什米尔地区、巴基斯坦和塔吉克斯坦亦有。

Biennial or perennial parasite herbs, densely glandular pubescent. Stems erect. Inflorescences spicate; corolla blue-purple; tubes suberect, not enlarged; upper lips 2-lobed; stigmas 2-lobed. Capsules oblong. Seeds ovoid-oblong. Fl. May-Jun. Fr. Jul-Aug. Thickets at 2900-3400 m. Distributed in Xizang. Also in Kashmir, Pakistan and Tajikistan.

西藏列当 *Orobanche clarkei*

黄花列当
Orobanche pycnostachya Hance

多年生或二年生寄生草本，密具腺毛。茎不分枝，直立。叶卵状披针形或披针形。花序穗状，多花；花冠黄色；花冠管稍弯曲，上部膨大；上唇2裂；下唇稍长。蒴果长圆形。花期4-6月，果期6-8月。生海拔300-2500米的山坡或草地，寄生于蒿属植物上。产华中、华东、华北、西北和东北。俄罗斯、蒙古和朝鲜半岛亦有。

Perennial or biennial parasite herbs, densely glandular pubescent. Stems unbranched, erect. Leaves ovate-lanceolate or lanceolate. Inflorescences spicate, many-flowered; corolla yellow; tubes slightly curved, enlarged upward; upper lips 2-lobed; lower lips longer than upper. Capsules oblong. Fl. Apr-Jun. Fr. Jun-Aug. Slopes or grasslands at 300-2500 m, parasitic on species of *Artemisia*. Distributed in C, E, N, NW and NE China. Also in Russia, Mongolia and Korean Peninsula.

黄花列当 *Orobanche pycnostachya*

滇列当
Orobanche yunnanensis (Beck) Hand.-Mazz.

二年生或多年生寄生草本，全株密被具腺状柔毛。叶卵状披针形。花序穗状；花萼不规则2裂至基部；花冠常红色，干后红褐色或棕色；花冠管膨大；上唇先端凹缺；下唇长约为上唇的1/2。蒴果椭圆体形。花期5-6月，果期7-8月。生海拔2200-3400米的山坡上，寄生于牛至属和其他唇形科植物根上。产云南北部、四川西南部和贵州西北部。

Biennial or perennial parasite herbs, densely glandular pubescent. Leaves ovate-lanceolate. Inflorescences spicate; calyx irregularly 2-parted to base; corolla usually red, becoming red-brown or brown when dry; tubes enlarged; upper lips apex emarginate; lower lips ca. 1/2 as long as upper. Capsules ellipsoid. Fl. May-Jun. Fr. Jul-Aug. Slopes at 2200-3400 m, parasitic on roots of *Origanum* and other Lamiaceae. Distributed in N Yunnan, SW Sichuan and NW Guizhou.

滇列当 *Orobanche yunnanensis*

苦苣苔科 Gesneriaceae

长瓣马铃苣苔

Oreocharis auricula C. B. Clarke

多年生无茎草本。叶有时镰形，狭至阔椭圆形至卵形或倒卵形。聚伞花序具4-11花；花冠紫色；花冠管圆柱形，口部狭；花冠裂片4-10 × 1.5-4毫米；冠檐二唇形；退化雄蕊1。花期5-9月，果期8-11月。生海拔200-1800米的山谷、沟边或林下阴湿岩石上。产中国西南、华南、华中和华东。

Perennial acaulous herbs. Leaves sometimes slightly falcate, narrowly to broadly elliptic to ovate or obovate. Cymes 4-11-flowered; corolla purple; tubes cylindric, narrowed at mouth; limbs lobes 4-10 × 1.5-4 mm; limbs 2-lipped; staminode 1. Fl. May-Sep. Fr. Aug-Nov. Valleys, along streams or on shady and damp rocks under forests at 200-1800 m. Distributed in SW, S, C and E China.

橙黄马铃苣苔

Oreocharis aurantiaca Franch.

多年生无茎草本。叶基部有时偏斜，边缘具粗圆齿至粗锯齿或细齿。聚伞花序具2-10花；花序梗具透明至紫红色腺状柔毛；花冠深橙色至橙色；花冠管圆柱形，口部狭；退化雄蕊1。花期7-9月，果期8-10月。生海拔1000-3400米的山坡灌丛干燥石灰岩上。产云南西北部。

Perennial acaulous herbs. Leaves base sometimes oblique, margin coarsely crenate to coarsely dentate or serrate. Cymes 2-10-flowered; peduncles translucent to purple-red glandular pubescent; corolla deep orange to orange; tubes cylindric, narrowed at mouth; staminode 1. Fl. Jul-Sep. Fr. Aug-Oct. Dry limestone rocks in thickets on slopes at 1000-3400 m. Distributed in NW Yunnan.

心叶马铃苣苔 *Oreocharis cordatula*

心叶马铃苣苔

Oreocharis cordatula (Craib) Pellegr.

多年生无茎草本。叶卵状披针形至狭卵形，长3-7.5厘米，边缘具粗圆齿至粗锯齿，上面密被贴伏柔毛，下面密被淡褐色绢状绵毛。花序梗、花梗、花萼和雌蕊被淡褐色腺状柔毛；苞片无；花萼裂片披针形；花冠深橘色至黄色，长1.9-2.4厘米，喉部缢缩；花丝无毛。花期6-8月，果期7-9月。生海拔1900-3200米的山顶、沟谷、溪边石灰岩上。产云南和四川。

Perennial acaulous herbs. Leaves ovate-lanceolate to narrowly ovate, 3-7.5 cm long, margin coarsely crenate to coarsely serrate, adaxially densely appressed pubescent, abaxially densely pale brown woolly. Peduncle, pedicel and calyx and pistil brownish glandular pubescent; bract absent; calyx lobes lanceolate; corolla deep orange to yellow, 1.9-2.4 cm long, narrowed at mouth; filaments glabrous. Fl. Jun-Aug. Fr. Jul-Sep. Limestone by streams in valleys and at montane summits at 1900-3200 m. Distributed in Yunnan and Sichuan.

长瓣马铃苣苔 *Oreocharis auricula*

橙黄马铃苣苔 *Oreocharis aurantiaca*

剑川马铃苣苔 *Oreocharis georgei*

毛花马铃苣苔 *Oreocharis dasyantha*

剑川马铃苣苔
Oreocharis georgei Anthony

多年生草本。叶柄、叶背、花序梗、苞片及花萼均被黄褐色长柔毛。叶狭卵形至椭圆形或狭倒卵形，长2-9厘米，宽0.8-3.5厘米，边缘具锯齿，上面被白色短柔毛和锈色柔毛。聚伞花序具1-6花；花冠黄色至橘黄色，筒部细筒状，喉部缢缩，檐部二唇形；花丝近无毛。花期3-6月，果期6-7月。生海拔2300-3400米的林中岩石上及林缘。产云南和四川。

Perennial herbs. Petioles, leaf abaxial, peduncle, bracts and calyx rust-brown villous. Leaves narrowly ovate to elliptic or narrowly obovate, 2-9 × 0.8-3.5 cm, margin serrate, adaxially whitish pubescent, with a few rust-brown hairs. Cymes 1-6-flowered; corolla yellow to orange-yellow, tube cylindric, narrowed at mouth, limb 2-lipped; filaments glabrescent. Fl. Mar-Jun. Fr. Jun-Jul. Rocks in forests and forest margins at 2300-3400 m. Distributed in Yunnan and Sichuan.

毛花马铃苣苔
Oreocharis dasyantha Chun

多年生草本。叶卵状椭圆形至宽卵形，长6-12厘米，基部偏斜，近圆形至心形，边缘具细锯齿，两面被灰白色短柔毛，侧脉每边5-7；叶柄长达14.5厘米，密被淡褐色绒毛。聚伞花序具1-3(-4)花；花萼裂片狭披针形；花冠黄红色，筒部钟状，檐部稀二唇形；花丝被短柔毛，花药宽长圆形，药室2，纵裂；柱头2，近圆形。花期2月。产海南。

Perennial herbs. Leaves ovate-elliptic to broadly ovate, 6-12 cm long, base often oblique, subrounded to cordate, margin serrulate, gray pubescent on both surfaces; lateral veins 5-7 on each side of midrib; petioles to 14.5 cm long, densely pale brown villous. Cymes 1-3(-4)-flowered; calyx segments narrowly lanceolate; corolla orange-red, tube campanulate, limb barely 2-lipped; filaments pubescent; anthers broadly oblong, 2-loculed, dehiscing longitudinally; stigmas 2, suborbicular. Fl. Feb. Distributed in Hainan.

筒花马铃苣苔
Oreocharis tubiflora K. Y. Pan

多年生草本。叶片下面仅在脉上被锈色绵毛。聚伞花序具3-8花；花冠紫色；花冠管宽管状，自基部至口部渐扩大；花药阔长圆形，2室，纵向开裂；退化雄蕊1。蒴果倒披针状。花期4-9月。生海拔500-700米的岩石上。产福建。

Perennial herbs. Leaves covered with rubiginous tomentum only at veins of abaxial sides. Cymes 3-8-flowered; corolla purple; tubes broadly tubular, gradually ampliate from base to mouth; anthers broadly oblong, 2-loculed, dehiscing longitudinally; staminode 1. Capsules oblanceolate. Fl. Apr-Sep. Rocks at 500-700 m. Distributed in Fujian.

筒花马铃苣苔 *Oreocharis tubiflora*

大花石上莲 *Oreocharis maximowiczii*

大花石上莲

Oreocharis maximowiczii C. B. Clarke

多年生草本。叶片下面密被锈色绵毛。聚伞花序具(1-)3-15花；花冠粉色至紫色，长1.5-2.5厘米；花冠管自基部向口部扩大；雄蕊内藏；花药阔长圆形，退化雄蕊1。花期4-6月，果期6月。生海拔200-800米的山谷岩石、溪边石上、路边或林下。产福建、江西和广东。

Perennial herbs. Leaves covered with rubiginous tomentum at abaxial sides. Cymes (1-) 3-15-flowered; corolla pink to lavender, 1.5-2.5 cm long; tubes ampliate from base to mouth; stamens included; anthers broadly oblong, staminode 1. Fl. Apr-Jun. Fr. Jun. Rocks of mountain valleys, rocks by streams, roadsides or under forests at 200-800 m. Distributed in Fujian, Jiangxi and Guangdong.

大叶石上莲

Oreocharis benthamii C. B. Clarke

多年生草本，密具浅褐色绵毛至毡毛。叶片长圆形至卵形，长5.5-14厘米，宽3-8厘米。聚伞花序具8-15花；花冠紫色至蓝色，自基部向口部渐扩大；退化雄蕊1。花期4-10月，果期8-11月。生海拔200-1400米的沟谷和林中岩石上或悬崖上。产广东、广西、湖南西南部和江西东南部。

Perennial herbs, densely light brown woolly to pannose. Leaves oblong to ovate, 5.5-14 × 3-8 cm. Cymes 8-15-flowered; corolla purple to blue, gradually ampliate from base to mouth; staminode 1. Fl. Apr-Oct. Fr. Aug-Nov. Rocks in valleys and forests or cliffs at 200-1400 m. Distributed in Guangdong, Guangxi, SW Hunan and SE Jiangxi.

黄花马铃苣苔

Oreocharis flavida Merr.

多年生草本。叶卵形至宽卵形，稀椭圆形至倒卵形，长4-10厘米，边缘近全缘或有浅钝齿，上面密被柔毛，下面密被褐色绵毛；叶柄长达10厘米，密被褐色绵毛。聚伞花序具3-7花；花萼裂片披针形；花冠淡黄色至橘黄色，筒部钟状；花药马蹄形，1室，横裂。蒴果长1.5-3(-4)厘米。花期10-12月，果期11-12月。生海拔1000-1900米的山坡林下。产海南。

Perennial herbs. Leaves ovate to broadly ovate, rarely elliptic or obovate, 4-10 cm long, margin nearly entire to shallowly crenate, adaxially densely pubescent, abaxially densely brown woolly; petioles to 10 cm long, densely brown woolly. Cymes 3-7-flowered; calyx segments lanceolate; corolla pale yellow to orange-yellow, tube campanulate; anthers horseshoe-shaped, 1-loculed, dehiscing transversely. Capsules 1.5-3(-4) cm long. Fl. Oct-Dec. Fr. Nov-Dec. Under forests on slopes at 1000-1900 m. Distributed in Hainan.

大叶石上莲 *Oreocharis benthamii*

黄花马铃苣苔 *Oreocharis flavida*

龙胜金盏苣苔 *Isometrum lungshengense*

龙胜金盏苣苔

Isometrum lungshengense (W. T. Wang) W. T. Wang et K. Y. Pan

多年生无茎草本。叶柄长达8厘米，密被白色短柔毛；叶片椭圆状倒卵形至卵形，上面具白色贴伏长柔毛，下面有短柔毛；花冠粉红色至紫红色，外面疏具腺状柔毛。花期9-10月。生海拔700-1500米的路旁或林下岩石上。产广西(龙胜县和临桂县)。

Perennial acaulous herbs. Petioles to 8 cm long, densely white pubescent; leaves elliptic-obovate to ovate, adaxially white appressed villous, abaxially pubescent; corolla pink to purple-red, outside sparsely glandular puberulent. Fl. Sep-Oct. Rocks near roads or in forests at 700-1500 m. Distributed in Guangxi (Longsheng County and Lingui County).

羽裂金盏苣苔

Isometrum primuliflorum (Batal.) B. L. Burtt

多年生无茎草本，具极密褐色绵毛。叶柄长达4厘米，密被褐色绵状毛；叶狭椭圆形至卵形，边缘具小裂片，裂片具圆锯齿。花冠淡紫色，0.9-1.4厘米，外面有具腺微柔毛。花期7月。生海拔2000-2800米的阴湿岩石上。产四川西北部。

Perennial acaulous herbs, very densely brown woolly. Petioles to 4 cm long, very densely brown woolly. Leaves narrowly elliptic to ovate, margin lobulate, lobes crenate-dentate. Corolla pale purple, 0.9-1.4 cm long, outside glandular puberulent. Fl. Jul. Shady and damp rocks at 2000-2800 m. Distributed in NW Sichuan.

羽裂金盏苣苔 *Isometrum primuliflorum*

弥勒苣苔 *Paraisometrum mileense*

弥勒苣苔

Paraisometrum mileense W. T. Wang

多年生草本。叶椭圆形至长圆状椭圆形，长2-4.8厘米，边缘具钝牙齿，上面密被贴伏白色短柔毛，下面密被淡褐色绵毛；叶柄被淡褐色绵毛。花序梗长6.5-12厘米，密被淡褐色短柔毛；花萼裂片披针状条形，外面密被绒毛；花冠长1.6-1.8厘米，筒长1.3-1.4厘米，上唇裂片三角形，下唇三角形，长约2.2毫米；子房长约9毫米。花期2月。产云南(弥勒)。

Perennial herbs. Leaves elliptic to oblong elliptic, 2-4.8 cm long, margin obtusely denticulate, adaxially densely appressed white pubescent, abaxially densely brownish woolly; petioles brownish woolly. Peduncle 6.5-12 cm long, densely brownish pubescent; calyx segments lanceolate-linear, outside densely tomentose; corolla 1.6-1.8 cm long, tube 1.3-1.4 cm long, adaxial lip lobes triangular, abaxial lip deltoid, ca. 2.2 mm long; ovary ca. 9 mm long. Fl. Feb. Distributed in Yunnan (Mile).

凸瓣苣苔

Ancylostemon convexus Craib

多年生草本。叶柄长达12厘米；叶卵形，侧脉4-7，明显。花序梗、叶柄与叶被锈色绒毛与白色柔毛；花萼自中部或以上5裂；花冠橘黄色至橙黄色；冠檐二唇形，上唇圆形。蒴果长3.5-5厘米。花期5-8月，果期7-10月。生海拔2500-3400米的阴湿悬崖上、岩石上或树上。产云南(大理)。

Perennial herbs. Petioles to 12 cm long; leaves ovate, lateral veins 4-7 on each side of midrib, conspicuous. Peduncles, petioles and leaves with rust-brown villous and white puberulent; calyx 5-lobed from below to above the middle; corolla orange to orange-yellow; limbs 2-lipped, adaxial lips rounded. Capsules 3.5-5 cm long. Fl. May-Aug. Fr. Jul-Oct. Shady, damp cliffs, rocks or trees at 2500-3400 m. Distributed in Yunnan (Dali).

粗筒苣苔 *Briggsia kurzii*

粗筒苣苔

Briggsia kurzii (C. B. Clarke) W. E. Evans

多年生草本。叶倒披针形至倒卵形或披针形至卵形，长(0.5-)4-14厘米，上面疏被贴伏短柔毛，基部有时歪斜，狭楔形至圆形，边缘具不规则锯齿。总花梗长1.2-6.2厘米，具短柔毛；花萼裂片披针形至狭三角形；花冠黄色至橘色，稀白色，上唇裂片半圆形。花期6-9月，果期8-10月。生海拔1800-3500米的草坡岩石上。产四川和云南。不丹、缅甸和印度东北部亦有。

Perennial herbs. Leaves oblanceolate to obovate or lanceolate to ovate, (0.5-)4-14 cm long, adaxially sparsely appressed pubescent, base sometimes oblique, narrowly cuneate to rounded,

凸瓣苣苔 *Ancylostemon convexus*

革叶粗筒苣苔 *Briggsia mihieri*

盾叶粗筒苣苔 *Briggsia longipes*

margin irregularly serrate. Peduncle 1.2-6.2 cm long, pubescent; calyx segments lanceolate to narrowly triangular; corolla yellow to orange, rarely white, adaxial lip lobes semiorbicular. Fl. Jun-Sep. Fr. Aug-Oct. Rocks of grassy slopes at 1800-3500 m. Distributed in Sichuan and Yunnan. Also in Bhutan, Myanmar and NE India.

盾叶粗筒苣苔

Briggsia longipes (Hemsl. ex Oliv.) Craib

多年生草本，无茎。植株无茎或茎高6厘米，无毛。叶基生或簇生，狭倒卵形至椭圆形至宽卵形；基部盾状或圆形至近楔形。聚伞花序具1-5(-7)花；花冠浅紫色，内面常有斑点。花期9-12月，果期12月。生海拔1000-1800米的林内潮湿岩石上或岩缝中。产云南东南部和广西西北部。

Perennial herbs, stemless. Plants stemless or stems to 6 cm tall, glabrous. Leaves basal or crowded, narrowly obovate to elliptic to broadly ovate, base peltate or rounded to nearly cuneate. Cymes 1-5(-7)-flowered; corolla pale purple, inside often spotted. Fl. Sep-Dec. Fr. Dec. Forests on damp rocks or crevices of rocks at 1000-1800 m. Distributed in SE Yunnan and NW Guangxi.

革叶粗筒苣苔

Briggsia mihieri (Franch.) Craib

多年生草本，无毛。叶基部或簇生；叶无毛；侧脉3或4，在叶两面不明显。聚伞花序具1-5花；花冠蓝紫色或浅紫色至浅黄色；内常具浅褐色斑点。蒴果长3.4-7厘米，渐无毛。花期9-10月，果期11月。生海拔600-1710米的遮阴或潮湿岩石上。产四川、贵州和广西(隆林县)。

Perennial herbs, glabrous. Leaves basal or crowded; leaves glabrous; lateral veins 3 or 4 on each side of midrib, inconspicuous on both surfaces. Cymes 1-5-flowered; corolla blue-purple or pale purple to pale yellow, inside usually brownish spotted. Capsules 3.4-7 cm long, glabrescent. Fl. Sep-Oct. Fr. Nov. Shady or damp rocks at 600-1710 m. Distributed in Sichuan, Guizhou and Guangxi (Longlin County).

黄花粗筒苣苔

Briggsia aurantiaca Burtt

多年生无茎草本，被锈色绵毛和长柔毛。叶椭圆形至卵形或倒卵形，稀近菱形，长2-12厘米，基部楔形，边缘具不规则圆齿至锯齿；侧脉每边5-7。聚伞花序具1-8花；花萼裂片披针形至卵形；花冠橘色至黄色，下唇内面被紫红色至粉色、橘色或褐色斑点；雌蕊无毛。蒴果倒披针形。花果期6-7月。生海拔2800-3700米的岩石缝中。产西藏东南部。

Perennial stemless herbs, rust woolly and villous. Leaves elliptic to ovate or obovate, rarely subrhombic, 2-12 cm long, base cuneate, margin irregularly crenate to serrate; lateral veins 5-7 on each side of midrib. Cymes 1-8-flowered; calyx segments lanceolate to ovate; corolla orange to yellow, inside abaxial lip purple-red to pink, orange or brown spotted; pistil glabrous. Capsules onlanceolate. Fl. and fr. Jun-Jul. Rock crevices at 2800-3700 m. Distributed in SE Xizang.

黄花粗筒苣苔 *Briggsia aurantiaca*

鄂西粗筒苣苔 *Briggsia speciosa*

鄂西粗筒苣苔

Briggsia speciosa (Hemsl.) Craib

多年生草本，无茎。叶具白色平伏毛，基部常倾斜。聚伞花序具1-5花；苞片2枚，条形至披针形；花冠紫红色，下唇内部具2个黄色或紫色斑点；子房具腺状柔毛。蒴果长6-6.8厘米，渐无毛。花期6-8月，果期7-9月。生海拔300-1600米的遮阴或潮湿山坡。产四川、湖北和湖南。

Perennial herbs, stemless. Leaves white appressed pubescent, base often oblique. Cymes 1-5-flowered; bracts 2, linear to lanceolate; corolla purple-red, inside abaxial lips with 2 yellow or purple spots; ovary glandular pubescent. Capsules 6-6.8 cm long, glabrescent. Fl. Jun-Aug. Fr. Jul-Sep. Shady or damp rocks on slopes at 300-1600 m. Distributed in Sichuan, Hubei and Hunan.

川鄂粗筒苣苔

Briggsia rosthornii (Diels) Burtt

多年生草本。叶卵形至披针形，稀椭圆形，长2-13厘米，基部有时歪斜，心形至楔形，边缘具圆齿至锯齿。总花梗长3.5-20厘米；花萼裂片披针形，稀椭圆形至倒卵形或浅裂；花冠淡粉色至紫色或紫红色，内部深红色或具紫红色斑点，上唇2浅裂，裂片半圆形，先端钝。花期8-9月，果期9-10月。生海拔1000-2000米的林下潮湿岩石上。产四川、湖北和贵州。

Perennial herbs. Leaves ovate to lanceolate, rarely elliptic, 2-13 cm long, base sometimes oblique, cordate to cuneate, margin crenate to serrate. Peduncle 3.5-20 cm long; calyx segments lanceolate, rarely elliptic to obovate or lobed; corolla pale pink to purple or purple-red, inside deep red or purple-red spotted, adaxial lip 2-lobed, lobes semiorbicular, apex obtuse. Fl. Aug-Sep. Fr. Sep-Oct. Shady, damp rocks under forests at 1000-2000 m. Distributed in Sichuan, Hubei and Guizhou.

平伐粗筒苣苔

Briggsia pinfaensis (Lévl.) Craib

多年生无茎草本。叶椭圆状至卵形，长3.2-9厘米，宽2-4厘米，基部楔形，边缘波状，具钝锯齿，两面均被灰色或褐色贴伏短柔毛；叶柄长1.5-5.5厘米。聚伞花序具1-4花；苞片长4-6毫米；花萼裂片披针形；花冠蓝紫色，内面具紫色斑点，长2.8-4厘米，下唇边缘具细尖齿；雌蕊被腺状短柔毛。幼果长4-5厘米，被腺状短柔毛。花期8月。产贵州(贵定县)。

Perennial stemless herbs. Leaves elliptic to ovate, 3.2-9 × 2-4 cm, base cuneate, margin undulate, crenate-serrate, gray or brown appressed puberulent; petioles 1.5-5.5 cm long. Cymes 1-4-flowered; bracts 4-6 mm long; calyx segments lanceolate; corolla blue-purple, inside purple spotted, 2.8-4 cm long, abaxial lip erose at margine; pistil glandular pubescent. Immature capsules 4-5 cm long, glandular pubescent. Fl. Aug. Distributed in Guizhou (Guiding County).

合萼漏斗苣苔

Didissandra petelotii Pellegr.

亚灌木。茎高20-30厘米。叶片长圆状披针形至卵形，稍偏斜。聚伞花序1-2花；花序梗1.5-3厘米，被长柔毛；花萼合生；花冠淡黄色，下喉部具2个红紫色斑。花期1月，果期2月。生海拔约650米的山林下。产广西(那坡县、靖西县)。越南北部亦有。

Shrubs. Stems 20-30 cm. Leaf blade oblong-elliptic lanceolate to ovate, slightly oblique. Cymes

川鄂粗筒苣苔 *Briggsia rosthornii*

平伐粗筒苣苔 *Briggsia pinfaensis*

合萼漏斗苣苔 *Didissandra petelotii*

1-2-flowered; peduncles 1.5-3 cm long, villous; calyxs accrete; corollas pale yellow, under the throat with 2 red-purple spots. Fl. Jan. Fr. Feb. Under the forests at ca. 650 m. Distributed in Guangxi (Napo County, Jingxi County). Also in N Vietnam.

粉绿异裂苣苔

Pseudochirita guangxiensis var. **glauca** Y. G. Wei et Yan Liu

多年生草本。茎和叶密被近贴伏的绒毛。每对叶不等大；叶卵形，长11-27(-30)厘米，基部稍斜，宽楔形至圆形，边缘近全缘或有不明显的钝锯齿。花序梗长3-9(-12)厘米，被短柔毛；花萼钟状，5浅裂；花冠白色，长3.2-4.3厘米，外面疏被腺毛，上唇2裂，下唇3浅裂。蒴果线形，长3-4.5厘米，近无毛。生石灰岩山林中。产广西。越南亦有。

Perennial herbs. Stems and leaves densely subappressed tomentose. Leaves unequal in each pair; leaves ovate, 11-27(-30) cm long, base slightly oblique, broadly cuneate to rounded, margin subentire or indistinctly crenate. Peduncle 3-9(-12) cm long, velutinous; calyx campanulate, 5-lobed; corolla white, 3.2-4.3 cm long, sparsely glandular-hairy outside, adaxial lip 2-lobed, abaxial lip 3-lobed. Capsules linear, 3-4.5 cm long, nearly glabrous. Forests on limestone hills. Distributed in Guangxi. Also in Vietnam.

异片苣苔 *Allostigma guangxiense*

异片苣苔

Allostigma guangxiense W. T. Wang

多年生草本，密具浅灰色至浅褐色的2种长度的毛。同1对叶稍不等大，卵形至椭圆形；叶片两侧不对称，具柔毛，黄色腺点。花冠长约3.8厘米。蒴果长约4厘米。花果期9月。生石灰岩山。产广西西南部。

Perennial herbs, with dense grayish to brownish hairs of 2 lengths. Leaves unequal in each pair, ovate to elliptic; leaves asymmetrical in both sides, puberulent, yellow glandular. Corolla ca. 3.8 cm long. Capsules ca. 4 cm long. Fl. and fr. Sep. Limestone hills. Distributed in SW Guangxi.

粉绿异裂苣苔 *Pseudochirita guangxiensis* var. *glauca*

大苞漏斗苣苔 *Didissandra begoniifolia*

大苞漏斗苣苔
Didissandra begoniifolia Lévl.

草本。聚伞花序5-10花；花序梗3-10(-12)厘米，有长柔毛；苞片2，脱落，宽卵形至圆形，1.5-2.5 × 1.5-3厘米，边缘具不整齐锯齿至近全缘。花期8-9月，果期9-10月。生海拔1200-2100米的山坡灌丛下石缝。产云南东南部、贵州、广西西部和湖北西部。

Herbs. Cymes 5-10-flowered; peduncles 3-10(-12) cm long, villous; bracts 2, deciduous, broadly ovate to orbicular, 1.5-2.5 × 1.5-3 cm, margin irregularly dentate to subentire. Fl. Aug-Sep. Fr. Sep-Oct. Crevices of rocks under thickets on slopes at 1200-2100 m. Distributed in SE Yunnan, Guizhou, W Guangxi and W Hubei.

卷丝苣苔
Corallodiscus kingianus (Craib) Burtt

多年生草本。叶菱状披针形至披针形，长1.6-11厘米，上面无毛，稀于中脉基部具锈色绵毛。聚伞花序具(5-)7-20花；总花梗密具锈色绵毛；花冠淡紫色至蓝色及白色，内面下唇具2深褐色的斑纹。蒴果卵球形至长圆形。花果期6-9月。生海拔2800-4800米的林中或岩石山坡上。产云南西北部、西藏南部、四川西南部和青海南部。印度北部和不丹亦有。

Perennial herbs. Leaves rhombic-lanceolate to lanceolate, 1.6-11 cm long, adaxially glabrous, rarely rust-brown woolly near base of midrib. Cymes(5-)7-20-flowered; peduncles densely rust-brown woolly; corolla purplish to blue and white, inside with 2 deep brown spotted striae on abaxial lips. Capsules ovoid to oblong. Fl. and fr. Jun-Sep. Rocks on slopes or forests at 2800-4800 m. Distributed in NW Yunnan, S Xizang, SW Sichuan and S Qinghai. Also in N India and Bhutan.

卷丝苣苔 *Corallodiscus kingianus*

西藏珊瑚苣苔 *Corallodiscus lanuginosus*

西藏珊瑚苣苔
Corallodiscus lanuginosus (Wall. ex R. Brown) B. L. Burtt

多年生草本。叶平坦、具折皱或泡状隆起。聚伞花序(1-)4-15(-30)花；花冠蓝色或紫色至白色或黄色，下唇内部有或无斑点。蒴果狭长圆形，长0.6-2.5厘米。花期4-10月，果期6-12月。生海拔700-4300米的石坡、峭壁、林缘或林内。产中国西南、华南、华中、华北和西北。印度、尼泊尔、不丹和泰国亦有。

Perennial herbs. Leaves flat, plicate or bullate. Cymes(1-)4-15(-30)-flowered; corolla blue or purple to white or yellow, inside with or without spots on abaxial lips. Capsules narrowly oblong, 0.6-2.5 cm long. Fl. Apr-Oct. Fr. Jun-Dec. Rocky slopes, steep cliffs, forest edges, forests at 700-4300 m. Distributed in SW, S, C, N and NW China. Also in India, Nepal, Bhutan and Thailand.

堇叶苣苔
Platystemma violoides Wall.

矮小草本。叶宽卵形，长 1.5-8 厘米，基部心形或具耳，边缘具粗牙齿，上面被白色贴伏柔毛，下面疏被柔毛。花序梗长 1-4 厘米，疏被腺毛；苞片钻形；花萼裂片卵

堇叶苣苔 *Platystemma violoides*

横蒴苣苔 *Beccarinda tonkinensis*

状长圆形；花冠紫红色，筒部长约3毫米，檐部明显二唇形。蒴果部分包被于花萼内，长5-7毫米。生海拔2300-3200米的沟谷阴湿岩石上或干燥的峭壁。产西藏。印度、尼泊尔和不丹亦有。

Small herbs. Leaves broadly ovate, 1.5-8 cm long, base cordate to auriculate, margin coarsely dentate, adaxially white appressed pubescent, abaxially sparsely white pubescent. Peduncle 1-4 cm long, sparsely glandular pubescent; bracts subulate; calyx segments ovate-oblong; corolla purple-red, tube ca. 3 mm long, limb distinctly 2-lipped. Capsules partially enclosed in persistent calyx, 5-7 mm long. Shady and damp rocks in valleys or dry cliffs at 2300-3200 m. Distributed in Xizang. Also in India, Nepal and Bhutan.

横蒴苣苔

Beccarinda tonkinensis

(Pellegr.) B. L. Burtt

多年生无茎草本。叶基生，卵形至圆形。花序梗长(5-)7-17厘米，具褐色硬毛；花萼裂片卵形至倒卵形，外面疏具柔毛；花冠蓝紫色，外面疏被微柔毛。蒴果长1.5-2.5厘米。花期4-6月，果期5-9月。生海拔700-2400米的林坡岩石上。产云南东南部、四川南部、贵州南部和广西。越南北部亦有。

Perennial herbs stemless. Leaves basal, ovate to orbicular. Peduncles (5-)7-17 cm long, brownish hirsute; calyx segments ovate to obovate, outside sparsely pubescent; corolla blue-purple, outside sparsely puberulent. Capsules 1.5-2.5 cm long. Fl. Apr-Jun. Fr. May-Sep. Rocks on forested slopes at 700-2400 m. Distributed in SE Yunnan, S Sichuan, S Guizhou and Guangxi. Also in N Vietnam.

饰岩横蒴苣苔

Beccarinda argentea

(Anthony) Burtt

多年生草本。无茎或茎高10厘米。叶基生或于茎上开展互生；叶具长2-8厘米的柄；叶椭圆形至卵形。花冠紫色至紫红色；雄蕊4，内藏，花药宽卵形，全部连着。花期3月，果期6月。生海拔1200-1600米的林下岩石上。产云南东南部。

Perennial herbs. Stemless or stems to 10 cm tall. Leaves basal or alternate and spread along stem; petioles 2-8 cm long; leaves elliptic to ovate. Corolla purple to purple-red; stamens 4, included, anthers broad-ovate, all coherent together. Fl. Mar. Fr. Jun. Rocks under forests at 1200-1600 m. Distributed in SE Yunnan.

饰岩横蒴苣苔 *Beccarinda argentea*

单座苣苔 *Metabriggsia ovalifolia*

纤细半蒴苣苔 *Hemiboea gracilis*

单座苣苔

Metabriggsia ovalifolia W. T. Wang

多年生草本。叶少，同1对叶不等大；叶片基部偏斜。花序梗被褐色腺毛；聚伞花序3-12花；花萼裂片披针状条形；花冠白色，带黄绿色。花果期10月。生海拔约1100米的石灰岩山林下。产广西西南部。

Perennial herbs. Leaves few, unequal in a pair; leaves base oblique. Peduncles brown glandular pubescent; cymes 3-12-flowered; calyx segments lanceolate-linear; corolla white, suffused yellow-green. Fl. and fr. Oct. Forests on limestone hills at ca. 1100 m. Distributed in SW Guangxi.

贵州半蒴苣苔

Hemiboea cavaleriei Lévl.

多年生草本，有时基部木质化。叶片稍肉质，干后膜质。聚伞花序具(1-)3-12花；花冠外部浅黄色至白色，内部具紫色斑点。蒴果条状披针形。花期8-10月，果期10-12月。生海拔300-1600米的山谷阴处或石灰质山地林中。产中国西南、华南、华中和华东。越南北部亦有。

Perennial herbs, sometimes woody at base. Leaflets slightly fleshy, membranous after dry. Cymes (1-)3-12-flowered; corolla outside pale yellow to white, inside purple spotted. Capsules linear-lanceolate. Fl. Aug-Oct. Fr. Oct-Dec. Shaded areas in montane valleys or forests on calcareous mountains at 300-1600 m. Distributed in SW, S, C and E China. Also in N Vietnam.

贵州半蒴苣苔 *Hemiboea cavaleriei*

纤细半蒴苣苔

Hemiboea gracilis Franch.

多年生草本。茎疏被紫褐色的斑点；蠕虫状石细胞少量嵌生于维管束附近的基本组织中。侧脉4-6。聚伞花序具1-3花；花萼自基部5深裂；花冠外面粉色至紫色或浅蓝色，里面具黑色或黄色的线条或斑点及一毛环。蒴果长1.7-2.5厘米。花期8-10月，果期10-11月。生海拔300-1300米的山地山谷林中岩石上、溪边岩石上或林缘。产四川、贵州、湖北、湖南和江西。

半蒴苣苔 *Hemiboea subcapitata*

龙州半蒴苣苔 *Hemiboea longzhouensis*

Perennial herbs. Stems sparsely purple-brown spotted; vermiform sclereids surrounding vascular bundles. Lateral veins 4-6 on each side of midrib. Cymes 1-3-flowered; calyx 5-sect from base; corolla outside pink to purple or bluish, inside with darker or yellow lines and spots, inside with a ring of hairs. Capsules 1.7-2.5 cm long. Fl. Aug-Oct. Fr. Oct-Nov. Rocks in montane valley forests, rocky streamsides or forest edges at 300-1300 m. Distributed in Sichuan, Guizhou, Hubei, Hunan and Jiangxi.

半蒴苣苔
Hemiboea subcapitata Clarke

多年生草本。茎疏具紫褐色或紫色斑点。叶片稍肉质，干时草质，蠕虫状石细胞散生于叶肉中。聚伞花序具(1-)3-10至更多花；花冠外部白色，内部具紫色斑点，内部具一圈毛。花期8-10月，果期10-12月。生海拔100-2100米的山谷林下石上或林下阴湿处。产中国西南、华南、华中、华东和西北。

Perennial herbs. Stems sparsely purple-brown or purple spotted. Leaves slightly carnose, dry blades herbaceous, vermiform sclereids dispersed in leaf mesophyll. Cymes (1-)3-10- or more flowered; corolla outside white, inside purple spotted, inside with a ring of hairs. Fl. Aug-Oct. Fr. Oct-Dec. Rocks in montane valley forests or shaded damp forests at 100-2100 m. Distributed in SW, S, C, E and NW China.

龙州半蒴苣苔
Hemiboea longzhouensis W. T. Wang ex Z. Y. Li

多年生草本。茎高20-40厘米或更高，有紫斑，无毛。花冠外面白色，里面有紫斑，3.7-4.4厘米，外面疏被具腺微柔毛，里面有1纵列毛。花期11-12月，果期翌年1-2月。生海拔300-400米的山谷林中石上。产广西西部。

Perennial herbs. Stems 20-40 cm tall or more, purple spotted, glabrous. Corolla outside white, inside purple spotted, 3.7-4.4 cm long, outside sparsely glandular puberulent, inside with a ring of hairs. Fl. Nov-Dec. Fr. next Jan-Feb. Rocks in montane valley forests at 300-400 m. Distributed in W Guangxi.

柔毛半蒴苣苔
Hemiboea mollifolia W. T. Wang

多年生草本。茎具紫褐色斑点，密被开展绒毛；蠕虫状石细胞少量嵌生于束附近的基本组织中。侧脉6-11对。聚伞花序具约3花；花冠外面粉色，外面疏被腺柔毛，里面具一毛环。蒴果长2.2-2.4厘米。花期8-10月，果期9-11月。生海拔600-900米的遮阴山谷中岩石上。产贵州、湖北和湖南。

Perennial herbs. Stems purple-brown spotted, densely spreading villous; vermiform sclereids surrounding vascular bundles. Lateral veins 6-11 on each side of midrib. Cymes ca. 3-flowered; corolla outside pink, outside sparsely glandular puberulent, inside with a ring of hairs. Capsules 2.2-2.4 cm long. Fl. Aug-Oct. Fr. Sep-Nov. Rocks in shaded montane valleys at 600-900 m. Distributed in Guizhou, Hubei and Hunan.

柔毛半蒴苣苔 *Hemiboea mollifolia*

华南半蒴苣苔 *Hemiboea follicularis*

扁圆石蝴蝶 *Petrocosmea oblata*

华南半蒴苣苔

Hemiboea follicularis C. B. Clarke

多年生草本。茎疏具紫色斑点。叶椭圆形至狭卵形或镰形，无毛。聚伞花序具7-20或更多花；总苞球形；花冠隐藏于总苞内，白色，内部具一毛环。蒴果长椭圆状披针形。花期6-8月，果期9-11月。生海拔200-1500米的林下阴湿石上或沟边石缝中。产贵州南部、广西和广东北部。

Perennial herbs. Stems sparsely purple spotted. Leaves elliptic to narrowly ovate or falcate, glabrous. Cymes 7-20- or more flowered; involucres globose; corolla within involucres, white, inside with a ring of hairs. Capsules ellipsoidal-lanceolate. Fl. Jun-Aug. Fr. Sep-Nov. Shady, damp rocks under forests or rock crevices along streamsides at 200-1500 m. Distributed in S Guizhou, Guangxi and N Guangdong.

扁圆石蝴蝶

Petrocosmea oblata Craib

多年生草本。聚伞花序具1花；花萼辐射状，自基部5深裂；花冠蓝色，无毛；冠筒上唇明显2浅裂；花药无喙；退化雄蕊3。蒴果长约4毫米。花果期8-9月。生海拔2200-3000米的山地或石灰岩土壤中。产云南东北部和四川。

Perennial herbs. Cymes 1-flowered; calyx actinomorphic, 5-sect from base; corolla blue, glabrous; tubes adaxial lips distinctly 2-lobed; anthers beakless; staminodes 3. Capsules ca. 4 mm long. Fl. and fr. Aug-Sep. Montane regions or limestone soils at 2200-3000 m. Distributed in NE Yunnan and Sichuan.

中华石蝴蝶 *Petrocosmea sinensis*

中华石蝴蝶

Petrocosmea sinensis Oliv.

多年生草本。叶柄具柔毛至开展短毛；叶阔菱形至阔菱状倒卵形或近圆形。聚伞花序具1-3花；花冠紫色至蓝色，外部具柔毛至疏具柔毛，内部无毛；退化雄蕊3。花期8-11月，果期8-12月。生海拔400-1700米的多山区域阴处石上。产四川、云南北部和湖北西部。

Perennial herbs. Petioles puberulent to spreading pilose; leaves broadly rhombic to broadly rhombic-obovate or nearly orbicular. Cymes 1-3-flowered; corolla purple to blue, outside puberulent to sparsely puberulent, inside glabrous; staminodes 3. Fl. Aug-Nov. Fr. Aug-Dec. Shaded rocks in hilly regions at 400-1700 m. Distributed in Sichuan, N Yunnan and W Hubei.

秦岭石蝴蝶

Petrocosmea qinlingensis W. T. Wang

多年生草本。叶宽卵形至菱状卵形或近圆形，长0.7-3厘米，宽0.7-2.8厘米，边缘浅波状或有不明显圆齿，两面疏被贴伏短柔毛；侧脉不明显。聚伞花序具1花；花萼辐射对称，5裂达基部，外面疏被短柔毛；花冠淡紫色，外面疏被贴伏短柔毛；花药长圆状卵形；子房与花柱被开展的短柔毛。花期8-9月。生海拔700-1100米的山地岩石上。产陕西。

Perennial herbs. Leaves broadly ovate to rhombic-ovate or nearly orbicular, 0.7-3 × 0.7-2.8 cm, margin repand to indistinctly crenate, sparsely appressed puberulent on both surfaces; lateral veins indistinct. Cymes 1-flowered; calyx actinomorphic, 5-sect from base, outside puberulent; corolla purplish, outside sparsely appressed puberulent; anthers oblong ovoid; ovary and style spreading pubescent. Fl.

秦岭石蝴蝶 *Petrocosmea qinlingensis*

生海拔1100-2500米的山地林中或阴处石崖上。产云南东南部和广西西南部。

Perennial herbs. Petioles matted hirsute; leaves ovate to nearly orbicular. Cymes (1 or)2-4-flowered; calyx actinomorphic, 5-sect from base; corolla blue-purple, outside sparsely puberulent, inside glabrous; filaments puberulent; staminodes 3. Fl. May-Nov. Fr. Nov. Monta forests or shady cliffs at 1100-2500 m. Distributed in SE Yunnan and SW Guangxi.

Aug-Sep. Rocks in hilly regions at 700-1100 m. Distributed in Shaanxi.

石蝴蝶

Petrocosmea duclouxii Craib

多年生草本。叶卵形至近圆形。聚伞花序具1(或2)花；花冠蓝紫色，外部疏具柔毛，内部无毛；花冠上唇明显2裂；花丝具褐锈色柔毛；退化雄蕊3。花期5-6月，果期6月。生海拔2000-2600米的多山区域阴处石上。产云南中部。

Perennial herbs. Leaves ovate to nearly orbicular. Cymes 1(or 2)-flowered; corolla blue-purple, outside sparsely puberulent, inside glabrous; tubes adaxial lip distinctly 2-lobed; filaments rust-brown puberulent; staminodes 3. Fl. May-Jun. Fr. Jun. Shaded rocks in hilly regions at 2000-2600 m. Distributed in C Yunnan.

蒙自石蝴蝶

Petrocosmea iodioides Hemsl.

多年生草本。叶柄具垫状柔毛；叶卵形至近圆形。聚伞花序具(1或)2-4花；花萼辐射状，自基部5深裂；花冠蓝紫色，外部疏具柔毛，内部无毛；退化雄蕊3。花期5-11月，果期11月。

石蝴蝶 *Petrocosmea duclouxii*

蒙自石蝴蝶 *Petrocosmea iodioides*

盾叶苣苔 *Metapetrocosmea peltata*

报春苣苔 *Primulina tabacum*

盾叶苣苔

Metapetrocosmea peltata (Merr. et W. Y. Chun) W. T. Wang

多年生小草本。叶片卵形至宽卵形，边缘波状至全缘，先端钝至圆形。苞片狭三角形至条形；花萼裂片披针形至狭卵形，外面具柔毛；花冠约8毫米。蒴果直径约3毫米。花期12月至翌年2月。生海拔300-700米的山区林中溪边岩石上。产海南。

Perennial dwarf herbs. Leaves ovate to broadly ovate, margin repand to entire, apex obtuse to rounded. Bracts narrowly triangular to linear; calyx segments lanceolate to narrowly ovate, outside pubescent; corolla ca. 8 mm long. Capsules ca. 3 mm diam. Fl. Dec to next Feb. Streamside rocks in forests in hilly regions at 300-700 m. Distributed in Hainan.

报春苣苔

Primulina tabacum Hance

多年生草本。叶柄具宽的波状翅，疏被柔毛；叶宽卵形，上面被平伏柔毛，下面疏被毛，边缘深裂，具近掌状脉。聚伞花序具3-7花；花序梗长7-11厘米，具柔毛；苞片狭披针形；花冠紫色，檐部5裂，裂片均为卵形。花期8-10月。生海拔100-300米的多石地或河边悬崖。产广西(贺州)、广东(连县和阳山县)。

Perennial herbs. Petioles with wide, undulate marginal wings, sparsely puberulent; leaves broadly ovate, adaxially appressed puberulent, abaxially sparsely puberulent, margin deeply lobed, subpalmately veined. Cymes 3-7-flowered; peduncles 7-11 cm long, puberulent; bracts narrowly lanceolate; corolla purple, 5-lobed, all lobes ovate. Fl. Aug-Oct. Stony or riverside cliffs at 100-300 m. Distributed in Guangxi (Hezhou), Guangdong (Lian County and Yangshan County).

多葶唇柱苣苔

Chirita polycephala (W. Y. Chun) W. T. Wang

多年生无茎草本。叶基生，薄纸质；侧脉3或4，不明显。聚伞花序具2-7(-15)花；苞片2，离生；花萼自基部5深裂；花冠浅紫色，冠筒内基部具纵向条状分布的柔毛；柱头楔形，2裂。蒴果直立。花期6月，果期10月。生海拔600-800米的山谷林中岩石上。产广东(乳源县和阳山县)。

Perennial stemless herbs. Leaves basal, thin papery; lateral veins 3 or 4 on each side of midrib, inconspicuous. Cymes 2-7(-15)-flowered; bracts 2, free; calyx 5-sect from base; corolla purplish, inside puberulent in longitudinal lines near base of tubes; stigmas cuneate, 2-lobed. Capsules erect. Fl. Jun. Fr. Oct. Rocks in forested valleys at 600-800 m. Distributed in Guangdong (Ruyuan County and Yangshan County).

多葶唇柱苣苔 *Chirita polycephala*

三苞唇柱苣苔 *Chirita tribracteata*

康定唇柱苣苔 *Chirita tibetica*

三苞唇柱苣苔
Chirita tribracteata W. T. Wang

多年生无茎草本。叶基生，对生，基部倾斜；苞片3，轮生。聚伞花序具约7花；花冠蓝色；花冠管狭钟形；花药上部完全合生，无毛；退化雄蕊3。花期6月，果期7月。生石灰岩山岩洞中稍阴处。产广西西北部。

Perennial stemless herbs. Leaves basal, opposite, base oblique; bracts 3, whorled. Cymes ca. 7-flowered; corolla blue; tubes narrowly funnelform; anthers fused by entire adaxial surfaces, glabrous; staminodes 3. Fl. Jun. Fr. Jul. Near caves slightly shaded places in limestone hills. Distributed in NW Guangxi.

蚂蝗七
Chirita fimbrisepala Hand.-Mazz.

多年生无茎草本。叶基生，草质；侧脉3-5，不明显。聚伞花序具(1或)2-5花；苞片2，离生；花萼自基部5深裂；花冠蓝色、紫色至粉绿色，内部具柔毛，上唇内具2紫色线；冠筒狭漏斗状；柱头倒四边形，2裂。花期3-4月，果期4-6月。生海拔400-1000米的林中或山地溪边的岩石或悬崖上。产中国西南、华南、华中和华东。

Perennial stemless herbs. Leaves basal, herbaceous; lateral veins 3-5 on each side of midrib, inconspicuous. Cymes (1 or)2-5-flowered; bracts 2, free; calyx 5-sect from base; corolla blue, purple to pinkish green, inside puberulent, adaxial lips with 2 purple lines; tubes narrowly funnelform; stigmas obtrapeziform, 2-lobed. Fl. Mar-Apr. Fr. Apr-Jun. Rocks or cliffs in forests or montane streamsides at 400-1000 m. Distributed in SW, S, C and E China.

蚂蝗七 *Chirita fimbrisepala*

康定唇柱苣苔
Chirita tibetica (Franch.) Burtt

多年生草本。根茎瘤状。茎生叶4-10，对生。聚伞花序具1-4花；花冠白色，喉部黄色且有紫色条纹；花冠管近管状；花药在上表面完全合生，无毛。花期7-9月。生海拔1400-3200米的岩石、林中或崖壁。产云南东北部和四川西南部。

Perennial herbs. Rhizome tuber-like. Stem leaves 4-10, opposite. Cymes 1-4-flowered; corolla white, throat yellow with purple stripes; tubes nearly tubular; anthers fused by entire adaxial surfaces, glabrous. Fl. Jul-Sep. Rocks, forests or cliffs at 1400-3200 m. Distributed in NE Yunnan and SW Sichuan.

寿城唇柱苣苔 *Chirita shouchengensis*

齿萼唇柱苣苔 *Chirita verecunda*

桂林唇柱苣苔

Chirita gueilinensis W. T. Wang

多年生无茎草本。叶基生，对生；叶基部斜楔形，两侧不对称，纸质，密具柔毛，无腺体，边缘具浅牙齿。聚伞花序具1-5花；花冠紫色，外面具柔毛，内部仅上唇具柔毛。蒴果直立。花期3-4月，果期5月。生海拔高达800米的石灰岩山林下背阴处。产广西东部与东北部和广东西部。

Perennial stemless herbs. Leaves basal, opposite; leaves base oblique-cuneate, both sides unsymmetrical, papery, densely puberulent, eglandular, margin shallowly crenate. Cymes 1-5-flowered; corolla purple, outside puberulent, inside puberulent only on adaxial lips. Capsules erect. Fl. Mar-Apr. Fr. May. Shaded forests areas in limestone hills to 800 m. Distributed in E and NE Guangxi, and W Guangdong.

寿城唇柱苣苔

Chirita shouchengensis Z. Y. Li

多年生无茎草本。叶纸质，倒披针形，长2-3厘米，先端急尖，基部下延，边缘全缘，上面被淡绿色微柔毛，下面被白色贴伏短绒毛。聚伞花序具1花；花序梗、苞片和花梗均密被微柔毛；花萼5裂达基部；花冠淡紫色，长约4.5厘米，外面疏被微柔毛，内面无毛，筒部漏斗形。花期4-6月。生海拔约300米的石灰岩山上，亦有栽培。引自广西。

Perennial stemless herbs. Leaves papery, oblanceolate, 2-3 cm long, apex acute, base decurrent, margin entire, adaxially light green puberulent, abaxially appressed white velutinous. Cymes 1-flowered; peduncle, bracts and pedicel densely puberulent; calyx 5-sect from base; corolla purplish, ca. 4.5 cm long, outside sparsely puberulent, inside glabrous, tube funnelform. Fl. Apr-Jun. Limestone hills at ca. 300 m, also cultivated. Introduced from Guangxi.

齿萼唇柱苣苔

Chirita verecunda (W. Y. Chun) W. T. Wang

多年生草本。叶基生，对生，倾斜，菱状椭圆形，纸质，疏具柔毛，无腺。花萼自基部5等裂，裂片条状披针形，外面有微柔毛，里面有贴伏微柔毛，边缘有细锯齿，先端渐尖。花期12月。生海拔1000-1100米的林下或山谷多石溪边。产广西(蒙山县和荔浦县)。

Perennial herbs. Leaves basal, opposite, oblique, rhombic-elliptic, papery, sparsely puberulent, eglandular. Calyx 5-sect from base, segments equal, linear-lanceolate, outside puberulent, inside appressed puberulent, margin denticulate, apex attenuate. Fl. Dec. Forests or rocky streamsides in valleys at 1000-1100 m. Distributed in Guangxi (Mengshan County and Lipu County).

桂林唇柱苣苔 *Chirita gueilinensis*

烟叶唇柱苣苔

Chirita heterotricha Merr.

多年生草本。根状茎顶部节间长达2.2厘米。叶草质或纸质，狭椭圆形至长圆形、倒卵形或卵形，长3-23厘米，边缘全缘或有不明显小齿，两面被疏柔毛；叶柄长0.5-11厘米。苞片2，椭圆形至卵形或狭三角形，宽0.2-9毫米；花萼5裂至基部；花冠淡紫色或白色，上唇下部具紫斑。花期4-10月，果期5-10月。生海拔400-600米的山谷林中或溪边石上。产海南。

Perennial herbs. Rhizome apical internodes to 2.2 cm long. Leaves herbaceous to papery, narrowly elliptic to oblong, obovate or ovate, 3-23 cm long, margin entire to indistinctly denticulate, sparsely pubescent on both surfaces; petioles

烟叶唇柱苣苔 *Chirita heterotricha*

0.5-11 cm long. Bracts 2, elliptic to ovate or narrowly triangular, 0.2-9 mm broad; calyx 5-sect from or near base; corolla purplish to white with a purple spot below adaxial lip. Fl. Apr-Oct. Fr. May-Oct. Rocky streamsides or forested valleys at 400-600 m. Distributed in Hainan.

粉花唇柱苣苔

Chirita roseoalba W. T. Wang

多年生无茎草本。叶基生，草质；侧脉3或4，不明显。聚伞花序具3-6花；苞片2，离生；花萼自基部5深裂；花冠白色至淡粉色；冠筒漏斗管状；花药上表面完全合生；柱头楔形，2裂。花期7月。生溪边灌丛沟谷。产湖南(大庸县)。

Perennial stemless herbs. Leaves basal, herbaceous; lateral veins 3 or 4 on each side of midrib, inconspicuous. Cymes 3-6-flowered; bracts 2, free; calyx 5-sect from base; corolla white to pinkish; tubes funnelform-tubular; anthers fused by entire adaxial surfaces; stigmas cuneate, 2-lobed. Fl. Jul. Streamsides thickets in valleys. Distributed in Hunan (Dayong County).

大齿唇柱苣苔

Chirita juliae Hance

多年生无茎草本。叶基部，对生；叶柄长3-17厘米；叶卵形至椭圆形。聚伞花序具2-12花；花梗长1-15毫米；花冠蓝色至紫色，上唇长7-8毫米；雌蕊长3-3.2厘米。蒴果直立。花期7-10月，果期10月。生海拔300-600米的丘陵地区的阴影溪边或岩石山坡上。产广东北部、福建西部、湖南东南部和江西东北部。

Perennial stemless herbs. Leaves basal, opposite; petioles 3-17 cm long; leaves ovate to elliptic. Cymes 2-12-flowered; pedicels 1-15 mm long; corolla blue to purple, adaxial lips 7-8 mm long; pistils 3-3.2 cm long. Capsules erect. Fl. Jul-Oct. Fr. Oct. Shaded streamside rocks in hilly regions at 300-600 m. Distributed in N Guangdong, W Fujian, SE Hunan and NE Jiangxi.

粉花唇柱苣苔 *Chirita roseoalba*

大齿唇柱苣苔 *Chirita juliae*

短毛唇柱苣苔 *Chirita brachytricha*

短毛唇柱苣苔
Chirita brachytricha W. T. Wang et D. Y. Chen

多年生无茎草本。叶基生，对生，纸质，疏具柔毛。聚伞花序具3-6花；花萼裂片和花序梗密具紫色柔毛；花冠紫色；花冠管漏斗形管状；退化雄蕊3。花期4-5月。生海拔400-1000米的山区林下潮湿岩石缝或林中沟谷岩石上。产贵州(荔波县)。

Perennial stemless herbs. Leaves basal, opposite, papery, sparsely puberulent. Cymes 3-6-flowered; calyx segments and peduncles densely purple puberulent; corolla purple; tubes funnelform-tubular; staminodes 3. Fl. Apr-May. Damp rocky crevices in forests of hilly regions or rocks in forested valleys at 400-1000 m. Distributed in Guizhou (Libo County).

永福唇柱苣苔
Chirita yungfuensis W. T. Wang

多年生无茎草本。叶基生，对生，革质，具紫色柔毛至短柔毛。聚伞花序具2-4花；苞片2，具紫色柔毛；花冠淡紫色；花冠管管状；花药具髯毛。花期5月。生石灰岩山阴处。产广西东北部。

Perennial stemless herbs. Leaves basal, opposite, leathery, purple puberulent to pilose. Cymes 2-4-flowered; bracts 2, purple pubescent; corolla purplish; tubes tubular; anthers bearded. Fl. May. Shaded places on limestone hills. Distributed in NE Guangxi.

永福唇柱苣苔 *Chirita yungfuensis*

弄岗唇柱苣苔
Chirita longgangensis W. T. Wang

多年生无茎草本。叶基生，对生或轮生，纸质至革质，密具平伏柔毛和伏毛，无腺体。聚伞花序具2-15花；花萼裂片披针状条形；花冠白色至紫红色。蒴果直立。花期9-12月。生海拔200-300米的石灰岩山林边石上。产广西西南部。

Perennial stemless herbs. Leaves basal, opposite or ternate, papery to leathery, densely appressed puberulent and pilose, eglandular. Cymes 2-15-flowered; calyx segments lanceolate-linear; corolla white to purple-red. Capsules erect. Fl. Sep-Dec. Rocks on forest edges on limestone hills at 200-300 m. Distributed in SW Guangxi.

弄岗唇柱苣苔 *Chirita longgangensis*

羽叶唇柱苣苔
Chirita pinnatifida (Hand.-Mazz.) Burtt

多年生无茎草本。叶基生，对生，边缘不规则羽状分裂。聚伞花序具1-4花；花梗密被柔毛和腺毛；花冠紫色至白色，带紫色；花冠管管状漏斗形花期

羽叶唇柱苣苔 *Chirita pinnatifida*

复叶唇柱苣苔 *Chirita pinnata*

5-9月，果期8-11月。生海拔600-2100米的山谷林中石上或溪边。产湖南、华南和华东。

Perennial stemless herbs. Leaves basal, opposite, margin irregularly pinnately lobed. Cymes 1-4-flowered; pedicels dense pubescent and glandular puberulent; corolla purple to white, tinged purple; tubes tubular-funnelform. Fl. May-Sep. Fr. Aug-Nov. Rocks or streamsides in forested valleys at 600-2100 m. Distributed in Hunan, S and E China.

复叶唇柱苣苔

Chirita pinnata W. T. Wang

多年生无茎草本。叶基生，对生；叶深裂成羽状，几成复叶，边缘具深锯齿或细齿至全缘。聚伞花序具1-3花；花冠紫色；花冠管狭至宽钟形，下面常凸起；退化雄蕊3。花果期5-9月。生海拔700-1300米的山区阴处岩石上。产广西(融水县)。

Perennial stemless herbs. Leaves basal, opposite; leaves deeply pinnately lobed, nearly compound, margin deeply crenate or serrate to entire. Cymes 1-3-flowered; corolla purple; tubes narrowly to broadly funnelform, often gibbous abaxially; staminodes 3. Fl. and fr. May-Sep. Shaded rocks in montane regions at 700-1300 m. Distributed in Guangxi (Rongshui County).

神农架唇柱苣苔

Chirita tenuituba (W. T. Wang) W. T. Wang

多年生无茎草本。叶基生，纸质；侧脉约3，不明显。聚伞花序具1-3花；苞片2，离生；花萼自基部5深裂；花冠紫色；冠筒圆柱形；花药上面完全合生；柱头倒三角形，2裂。蒴果直立。花期3-5月。生海拔300-1000米的岩缝或林中悬崖上。产重庆、贵州、湖北和湖南。

Perennial stemless herbs. Leaves basal, papery; lateral veins ca. 3 on each side of midrib, inconspicuous. Cymes 1-3-flowered; bracts 2, free; calyx 5-sect from base; corolla purple; tubes cylindric; anthers fused by entire adaxial surfaces; stigmas obdeltoid, 2-parted. Capsules erect. Fl. Mar-May. Rocky crevices or cliffs in forests at 300-1000 m. Distributed in Chongqing, Guizhou, Hubei and Hunan.

神农架唇柱苣苔 *Chirita tenuituba*

刺齿唇柱苣苔 *Chirita spinulosa*

刺齿唇柱苣苔
Chirita spinulosa D. Fang et W. T. Wang

多年生无茎草本。叶基生，对生，无柄，革质，无毛，无腺体，边缘密具刺齿。聚伞花序具约9花；花冠蓝紫色；花药上面完全合生，基部疏具柔毛；退化雄蕊2。花期11月。生海拔约100米的石灰岩山阴处。产广西(扶绥县)。

Perennial herbs, stemless. Leaves basal, opposite, sessile, leathery, glabrous, eglandular, margin finely spiny denticulate. Cymes ca. 9-flowered; corolla blue-purple; anthers fused by entire adaxial surfaces, basally sparsely puberulent; staminodes 2. Fl. Nov. Shaded areas on limestone hills at ca. 100 m. Distributed in Guangxi (Fusui County).

美丽唇柱苣苔
Chirita speciosa Kurz

多年生草本。茎具红褐色长柔毛。叶4-6，基生或茎生，茎生则簇生于茎顶，互生。聚伞花序具1-6花；花冠蓝紫色，花冠管带黄色；花药密被褐色长柔毛。花期3-9月，果期5-7月。生海拔700-3100米的湿沟谷岩石上。产云南西部和南部。印度东北部、缅甸、泰国和越南北部亦有。

Perennial herbs. Stems with red-brown villose. Leaves 4-6, basal or along stem, crowded at apex, alternate. Cymes 1-6-flowered; corolla blue-purple with yellow in tube; anthers densely brown villose. Fl. Mar-Sep. Fr. May-Jul. Rocks in wet valleys at 700-3100 m. Distributed in W and S Yunnan. Also in NE India, Myanmar, Thailand and N Vietnam.

大叶唇柱苣苔
Chirita macrophylla Wall.

具地上茎多年生草本。叶卵形至椭圆形，9.5-19 × 6-14厘米，草质，上面具平伏柔毛，无腺体，下面疏具柔毛。聚伞花序具(1或)2-6花；花冠黄色至白色，有时浅紫色至浅蓝色；花药上面完全合生；退化雄蕊3。花期5-9月，果期8-12月。生海拔1300-3100米的山地林下岩上。产云南南部和贵州西南部。南亚和东南亚亦有。

Caulescent, perennial herbs. Leaves ovate to elliptic, 9.5-19 × 6-14 cm, herbaceous, adaxially appressed puberulent, eglandular, abaxially sparsely pubescent. Cymes (1 or) 2-6-flowered; corolla yellow to white, sometimes purplish to bluish; anthers fused by entire adaxial surfaces; staminodes 3. Fl. May-Sep. Fr. Aug-Dec. Rocks in montane forests at 1300-3100 m. Distributed in S Yunnan and SW Guizhou. Also in S and SE Asia.

光萼唇柱苣苔
Chirita anachoreta Hance

一年生草本。叶倾斜，薄草质。聚伞花序具1-3(-9)花；花萼外面疏具柔毛至伏毛或无毛，有时具腺点，内部无毛；花冠白色至黄色，具黄色具紫色斑块或蓝紫色；退化雄蕊2或3。蒴果直立。花期7-9月，果期8-11月。生海拔200-2300米的林中石上或山谷溪边。产中国西南、华南、华中和华东。南亚和东南亚亦有。

Annual herbs. Leaves oblique,

美丽唇柱苣苔 *Chirita speciosa*

大叶唇柱苣苔 *Chirita macrophylla*

光萼唇柱苣苔 *Chirita anachoreta*

thin herbaceous. Cymes 1-3(-9)-flowered; calyx outside sparsely puberulent to pilose or glabrous, sometimes glandular, inside glabrous; corolla white to yellow with yellow or purple markings or blue-purple; staminodes 2 or 3. Capsules erect. Fl. Jul-Sep. Fr. Aug-Nov. Rocks in forests or valley streamsides at 200-2300 m. Distributed in SW, S, C and E China. Also in S and SE Asia.

斑叶唇柱苣苔

Chirita pumila D. Don

具茎一年生草本。叶绿色带紫色斑。聚伞花序具(1或)2-7花；花萼外被长柔毛，先端突钻状渐尖，常向外弯曲；花冠白色至紫色，具黄色或紫色斑块。花期7-9月，果期7-10月。生海拔800-2800米的林内、沟边、岩石上、崖壁或草丛。产云南西北部和南部、西藏东南部、贵州西南部和广西西北部。南亚和东南亚亦有。

Caulescent annual herbs. Leaves green, with purple markings. Cymes (1 or)2-7-flowered; calyx villose outside, apex subulate-acuminate, always curved outward; corolla white to purple with yellow or purple markings. Fl. Jul-Sep. Fr. Jul-Oct. Forests, by streams, rocks, cliffs or among grassy clumps at 800-2800 m. Distributed in NW and S Yunnan, SE Xizang, SW Guizhou and NW Guangxi. Also in S and SE Asia.

合苞唇柱苣苔

Chirita infundibuliformis W. T. Wang

多年生草本。茎直立，密被褐色柔毛。茎生叶4-8枚对生；叶倾斜，纸质；侧脉约7。聚伞花序具1或2花；苞片2，贴生成苞片状，宽漏斗形；花萼自中部5裂；花冠紫红色；冠筒近管状；柱头倒四边形，2裂。花期8月。生海拔900-1700米的阔叶林缘或山地溪边。产西藏(墨脱县)。

Perennial herbs. Stems erect, densely brown pubescent. Stem leaves 4-8, opposite; leaves oblique, papery; lateral veins ca. 7 on each side of midrib. Cymes 1- or 2-flowered; bracts 2, connate into an involucre, broadly funnelform; calyx 5-lobed from middle; corolla purple-red; tubes nearly tubular, stigmas obtrapeziform, 2-lobed. Fl. Aug. Broad-leaved forest edges or montane streamsides at 900-1700 m. Distributed in Xizang (Mêdog County).

斑叶唇柱苣苔 *Chirita pumila*

合苞唇柱苣苔 *Chirita infundibuliformis*

百寿唇柱苣苔 *Chirita baishouensis*

百寿唇柱苣苔

Chirita baishouensis Y. G. Wei, H. Q. Wen et S. H. Zhong

多年生草本。叶基生，纸质，椭圆形或卵状椭圆形，被贴伏的短柔毛。聚伞花序8-15个，腋生，有1-4花；花萼中部之上常有2-3枚不明显的小钝齿；花冠淡紫色，内面具紫色斑纹和4条紫色条纹；退化雄蕊3，散生腺状短柔毛。花期3-4月。生海拔约100米处。产广西(永福县)。

Perennial herbs. Leaves all basal, papery, elliptic or ovate-elliptic, with appressed short pubescent. Cymes 8-15, axillary, 1-4-flowered; calyx with 2-3 obtuse-serrate above middle; corolla pale purple, purple markings and 4 purple stripes inside; staminodes 3; sparsely shortly glandular puberulent. Fl. Mar-Apr. ca. 100 m. Distributed in Guangxi (Yongfu County).

休宁小花苣苔

Chiritopsis xiuningensis X. L. Liu et X. H. Guo

多年生草本。叶卵形至宽卵形或椭圆形至近圆形，长2-9厘米，基部宽楔形至近心形，边缘具波状细牙齿至近全缘，疏被微柔毛；叶柄、花序梗和花梗密被开展微柔毛。花萼裂片条形至狭条状披针形；花冠淡黄色，内面无毛，檐部不明显二唇形；子房被微柔毛。蒴果狭卵形。花期7-8月，果期8月。生海拔400-500米的岩石峭壁。产安徽(休宁县)。

Perennial herbs. Leaves ovate to broadly ovate or elliptic to nearly orbicular, 2-9 cm long, base broadly cuneate to nearly cordate, margin repand-denticulate to nearly entire, sparsely puberulent; petioles, peduncle and pedicel densely spreading puberulent. Calyx segments linear to narrowly linear-lanceolate; corolla yellowish, inside glabrous, limb indistinctly 2-lipped; ovary puberulent. Capsules narrowly ovoid. Fl. Jul-Aug. Fr. Aug. Stony cliffs or rocks at 400-500 m. Distributed in Anhui (Xiuning County).

石山苣苔

Petrocodon dealbatus Hance

多年生草本。叶倒披针形、椭圆形，有时镰形，具短糙毛，边缘近全缘至具牙齿、小齿或细圆齿。花萼裂片条形，外部疏具短硬毛；花冠坛状筒形，筒比檐部长，檐部不明显二唇形。花期6-9月，果期10月。生海拔200-1000米的石山林中或山谷阴处石上。产贵州、广东、广西、湖北和湖南。

Perennial herbs. Leaves oblanceolate, elliptic, sometimes falcate, short strigose, margin nearly entire to dentate, denticulate or crenulate. Calyx segments linear, outside sparsely short strigose; corolla urceolate-tubular, tubes longer than limb, limbs indistinctly 2-lipped. Fl. Jun-Sep. Fr. Oct. Forests on limestone hills or shaded rocks in valleys at 200-1000 m. Distributed in Guizhou, Guangdong, Guangxi, Hubei and Hunan.

藏南长蒴苣苔

Didymocarpus primulifolius D. Don

多年生草本。茎高5-20厘米。

休宁小花苣苔 *Chiritopsis xiuningensis*

石山苣苔 *Petrocodon dealbatus*

叶3-4枚生茎顶端；叶片卵形或宽卵形。聚伞花序具5-15花；花冠深紫色，长约1.9厘米，无毛；筒细筒状，长约1.1厘米，上唇2浅裂，下唇3裂近中部。雄蕊无毛，花丝狭线形；退化雄蕊1；雌蕊长约1.1厘米，无毛。花期6-7月，果期7-8月。生海拔2100-2700米的林下石上或陡崖上。产西藏南部。

藏南长蒴苣苔 *Didymocarpus primulifolius*

Perennial herbs. Stem 5-20 cm tall. Leaves 3-4 at the top of the stem; Ovate or broad ovate. Cymes 5-15-flowered; corolla dark purple, ca. 1.9 cm, glabrous; tubes thin tubular, ca. 1.1 cm; upper lip 2-lobed, lower lip 3-lobed near middle. Stamaens glabrous, filaments narrowly linear; staminode 1; pistil ca. 1.1 cm, glabrous. Fl. Jun-Jul. Fr. Jul-Aug. Valleys on the rock or on the steep cliff at 2100-2700 m. Distributed in S Xizang.

云南长蒴苣苔

Didymocarpus yunnanensis (Franch.) W. W. Sm.

多年生草本。叶草纸，狭卵形至卵形，长圆形或倒卵形，长1-14厘米，上面被贴伏短柔毛，下面只沿脉被短柔毛。花序梗与花梗有疏腺毛；花萼稍对称，5浅裂，檐部不明显二唇形；花冠紫色至紫红色，具深色条纹，外面有疏柔毛，筒部狭漏斗形。花期6-9月，果期7-10月。生海拔1500-3400米的山谷石上或石崖上。产云南、四川和西藏。印度亦有。

东南长蒴苣苔 *Didymocarpus hancei*

Perennial herbs. Leaves herbaceous, narrowly ovate to ovate, oblong, or obovate, 1-14 cm long, adaxially appressed puberulent, abaxially puberulent along veins. Peduncle and pedicel sparsely glandular puberulent; calyx slightly zygomorphic, 5-lobed, limb indistinctly 2-lipped; corolla purple to reddish purple with darker stripes, outside sparsely pubescent, tube narrowly funnelform. Fl. Jun-Sep. Fr. Jul-Oct. Rocks, cliffs in valleys at 1500-3400 m. Distributed in Yunnan, Sichuan and Xizang. Also in India.

东南长蒴苣苔

Didymocarpus hancei Hemsl.

无茎草本。叶基生，纸质，具短糙毛，边缘具细圆齿至锯齿。聚伞花序具4至多花；花序梗疏具柔毛至糙毛；花萼辐射状；花冠浅紫色；花冠筒漏斗形；退化雄蕊2。蒴果条形。花期4月，果期5月。生海拔400-1000米的林中、石上或岩壁上。产广东北部、湖南南部、福建和江西。

Plants stemless. Leaves basal, papery, short strigose, margin denticulate to serrate. Cymes 4- to many-flowered, umbellike; peduncles sparsely puberulent to strigose; calyx actinomorphic, umbellike; corolla pale purple; tubes funnelform; staminodes 2. Capsules linear. Fl. Apr. Fr. May. Forests, rocks or cliffs at 400-1000 m. Distributed in N Guangdong, S Hunan, Fujian and Jiangxi.

云南长蒴苣苔 *Didymocarpus yunnanensis*

长檐苣苔

Dolicholoma jasminiflorum D. Fang et W. T. Wang

多年生小草本。叶具白色柔毛，边缘近全缘，具腺点和缘毛；叶柄具开展白色柔毛和腺毛。花小；花冠深5裂，檐部略长于筒部，裂片狭三角形；花冠浅紫色，中心浅黄色。蒴果狭椭圆体形。花期4月，果期5月。生石灰岩山陡崖阴处。产广西西南部。

Perennial small herbs. Leaves white pubescent, margin subentire, glandular and ciliate; petioles spreading white puberulent and glandular puberulent. Flowers small; corolla deeply 5-lobed, limbs slightly longer than tube, lobes narrowly triangular; corolla purplish, yellowish in center. Capsules narrowly ellipsoidal. Fl. Apr. Fr. May. Shady cliffs in limestone hills. Distributed SW Guangxi.

稀裂圆唇苣苔

Gyrocheilos retrotrichus W. T. Wang var. **oligolobus** W. T. Wang

多年生草本。叶上混生有0.2-0.5毫米和1-2毫米的毛。聚伞花序5至多花；花萼4裂或二唇形；裂片大小不等，宽1-3毫米，先端钝至圆形。花期4-7月，果期5-9月。生海拔400-1000米的林下或山谷阴处岩石上。产贵州东南部、广西北部和广东西南部。

Perennial herbs. Leaves adaxially with mixed hairs 0.2-0.5 mm long and 1-2 mm long. Cymes 5- to many-flowered; calyx 4-lobed or 2-lipped; lobes unequal, 1-3 mm wide, apex obtuse to rounded. Fl. Apr-Jul. Fr. May-Sep. Forests or shaded rocks in valleys at 400-1000 m. Distributed in SE

闽赣长蒴苣苔 *Didymocarpus heucherifolius*

闽赣长蒴苣苔

Didymocarpus heucherifolius Hand.-Mazz.

多年生无茎草本。叶基生，多裂，纸质；侧脉3或4。聚伞花序具3至多花；苞片离生；冠檐

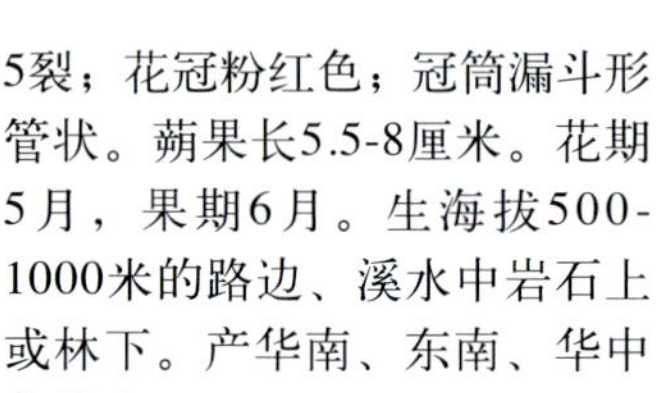

5裂；花冠粉红色；冠筒漏斗形管状。蒴果长5.5-8厘米。花期5月，果期6月。生海拔500-1000米的路边、溪水中岩石上或林下。产华南、东南、华中和华东。

Perennial herbs, stemless. Leaves basal, multilobed, papery; lateral veins 3 or 4 on each side of midrib. Cymes 3- to many-flowered; bracts free; limbs 5-lobed; corolla pink; tubes funnelform-tubular. Capsules 5.5-8 cm long. Fl. May. Fr. Jun. Waysides, streamsides rocks or forests at 500-1000 m. Distributed in S, SE, C and E China.

长檐苣苔 *Dolicholoma jasminiflorum*

稀裂圆唇苣苔 *Gyrocheilos retrotrichus* var. *oligolobus*

朱红苣苔 *Calcareoboea coccinea*

Guizhou, N Guangxi and SW Guangdong.

朱红苣苔

Calcareoboea coccinea C. Y. Wu ex H. W. Li

多年生草本，具平伏柔毛。叶柄长3-14.5厘米；叶卵形，密具柔毛至绢毛，边缘具锯齿。花序似伞形花序，具总苞；能育雄蕊2，退化雄蕊2，极小。花期4-6月，果期7月。生海拔1000-1500米的石灰岩山林中石上。产云南东南部和广西西部。越南北部亦有。

Perennial herbs, appressed pubescent. Petioles 3-14.5 cm long; leaves ovate, densely puberulent to sericeous, margin dentate. Inflorescences umbellike, involucrate; stamens 2, staminodes 2, very small. Fl. Apr -Jun. Fr. Jul. Rocks in forests on limestone hills at 1000-1500 m. Distributed in SE Yunnan and W Guangxi. Also in N Vietnam.

锥序蛛毛苣苔

Paraboea swinhoei (Hance) B. L. Burtt

亚灌木。茎密被浅褐色绵毛，渐无毛。叶对生，纸质，下面被浅灰色绵毛，渐无毛，上面具浅褐色毡毛。聚伞花序近顶生；花冠白色；退化雄蕊2或3。蒴果螺旋状扭曲，长2-2.5厘米，无毛。花期6-9月，果期7-9月。生海拔300-1000米的林下阴湿岩石上。产贵州、广西和台湾。泰国、越南和菲律宾亦有。

Subshrubs. Stems densely brownish woolly, glabrescent. Leaves opposite, papery, adaxially grayish woolly, glabrescent, abaxially brownish pannose. Cymes appearing terminal; corolla white; staminodes 2 or 3. Capsules spirally twisted, 2-2.5 cm long, glabrous. Fl. Jun-Sep. Fr. Jul-Sep. Shady and damp rocks under forests at 300-1000 m. Distributed in Guizhou, Guangxi and Taiwan. Also in Thailand, Vietnam and the Philippines.

蛛毛苣苔

Paraboea sinensis (Oliv.) B. L. Burtt

亚灌木。叶对生，沿茎排列，纸质，上面具浅灰色至褐色毡毛至柔毛，近无毛，下面具浅褐色毡毛。聚伞花序腋生，生近枝顶，具褐色毡毛；花冠蓝紫色至紫红色。蒴果螺旋状扭曲，无毛。花期5-7月，果期5-11月。生海拔600-2500米的石缝或林中崖壁上。产云南西南部和东南部、四川东南部、贵州、广西西南部和湖北西部。缅甸、泰国和越南亦有。

Subshrubs. Leaves opposite, spread along stem, papery, adaxially grayish to brown pannose to puberulent, subglabrescent, abaxially brownish pannose. Cymes axillary, near branch apices, brown pannose; corolla blue-purple to purple-red. Capsules spirally twisted, glabrous. Fl. May-Jul. Fr. May-Nov. Crevices of rocks or cliffs in forests at 600-2500 m. Distributed in SW and SE Yunnan, SE Sichuan, Guizhou, SW Guangxi and W Hubei. Also in Myanmar, Thailand and Vietnam.

锥序蛛毛苣苔 *Paraboea swinhoei*

蛛毛苣苔 *Paraboea sinensis*

锈色蛛毛苣苔
Paraboea rufescens

锈色蛛毛苣苔

Paraboea rufescens (Franch.) Burtt

多年生草本或亚灌木。根状茎木质化，粗壮。叶对生，多簇生于近茎顶。聚伞花序腋生，具锈色毡毛；花冠白色至浅紫色，极少紫红色。蒴果螺旋扭曲。花期6-9月，果期7-10月。生海拔700-1500米的石灰山岩石缝中。产中国西南和华南。泰国和越南亦有。

Perennial herbs or subshrubs. Rhizomes woody, robust. Leaves opposite, mostly crowded near stem apex. Cymes axillary, rust-brown pannose; corolla white to purplish, seldom purple-red. Capsules spirally twisted. Fl. Jun-Sep. Fr. Jul-Oct. Crevices of rocks on limestone hills at 700-1500 m. Distributed in SW and S China. Also in Thailand and Vietnam.

海南蛛毛苣苔

Paraboea hainanensis (Chun) Burtt

多年生草本。叶基生或密集于根状茎顶端；叶倒披针形至倒卵形，长5-18厘米，宽1.2-6厘米，基部渐狭至楔形，边缘具细圆齿至锯齿，上面无毛至近无毛，下面密被黄褐色蛛丝状绵毛。花萼裂片披针形至卵形，外面被腺状柔毛；花冠蓝色至深紫色。蒴果螺旋卷曲，长3-4.5厘米，无毛。花果期7-9月。生海拔约800米的阴湿混交林下岩石上。产海南。

Perennial herbs. Leaves basal or crowded near stem apex; leaf blade oblanceolate to obovate, 5-18 × 1.2-6 cm, base attenuate to cuneate, margin crenulate to serrate, adaxially glabrous to glabrescent, abaxially brown cobwebby-woolly. Calyx segments lanceolate to ovate, outside glandular pubescent; corolla bluish to deep purple. Capsules spirally twisted, 3-4.5 cm long, glabrous. Fl. and fr. Jul-Sep. Shady and damp rocks under mixed forest at ca. 800 m. Distributed in Hainan.

地胆旋蒴苣苔

Boea philippensis C. B. Clarke

纤细草本。叶倒卵形至狭椭圆状匙形或菱形，上面具浅灰色绒毛至柔毛，下面密具绒毛。花冠淡紫色至深红色或白色，7-10毫米，外面被微柔毛，里

海南蛛毛苣苔 *Paraboea hainanensis*

地胆旋蒴苣苔 *Boea philippensis*

旋蒴苣苔 *Boea hygrometrica*

大花旋蒴苣苔 *Boea clarkeana*

面无毛。花期5-6月，果期5-7月。生海拔100-800米的林中阴湿岩石上。产贵州、湖南和华南。越南和菲律宾亦有。

Slender herbs. Leaves obovate to narrowly elliptic-spatulate or rhombic, adaxially grayish villous to pubescent, abaxially densely villous. Corolla purplish to dark red or white, 7-10 mm long, outside puberulent, inside glabrous. Fl. May-Jun. Fr. May-Jul. Shady and damp rocks in forests at 100-800 m. Distributed in Guizhou, Hunan and S China. Also in Vietnam and the Philippines.

旋蒴苣苔

Boea hygrometrica (Bunge) R. Br.

纤细草本。叶全部基生，密集，莲座状，叶片肉质，两面被白色长柔毛，脉纹掌状，中脉不明显。花冠蓝紫色至蓝色至白色或粉红色，外部疏具柔毛。蒴果螺旋状扭曲。花期4-8月，果期4-9月。生海拔100-1500米的河谷岩石上、山坡或林中。产中国西南、华南、华中、华北、华东、西北和东北。

Slender herbs. Leaves all basal, crowded, rosette, leaves fleshy, white villose, venation palmate, midrib indistinct. Corolla blue-purple or blue to white or pink, outside sparsely puberulent. Capsules spirally twisted. Fl. Apr-Aug. Fr. Apr-Sep. Rocks in valleys, slopes or forests at 100-1500 m. Distributed in SW, S, C, N, E, NW and NE China.

大花旋蒴苣苔

Boea clarkeana Hemsl.

多年生草本。叶柄长达10厘米；叶卵形至阔卵形、椭圆形或倒卵形。花萼自中部5裂；花冠白色至蓝色或蓝紫色，长1.4-2.2厘米；退化雄蕊2或3。蒴果长1.8-4.5厘米。花期6-8月，果期6-10月。生海拔500-3100米的山坡岩缝。产中国西南、华中、华北和华东。

Perennial herbs. Petioles to 10 cm long; leaves ovate to broadly ovate, elliptic or obovate. Calyx 5-lobed from middle; corolla white to blue or blue-purple, 1.4-2.2 cm long; staminodes 2 or 3. Capsules 1.8-4.5 cm long. Fl. Jun-Aug. Fr. Jun-Oct. Crevices of rocks on slopes at 500-3100 m. Distributed in SW, C, N and E China.

滇桂喜鹊苣苔

Ornithoboea wildeana Craib

多年生草本。茎高20-40厘米，被长柔毛至短柔毛。叶缘具牙齿状锯齿，有时为重齿。花冠白色，带紫色；下唇具黄色髯毛，裂片长圆形；花序梗和花梗具长柔毛；退化雄蕊2。花果期7-9月。生海拔300-1300米的林缘岩石上。产云南南部和广西西南部。泰国西北部亦有。

Perennial herbs. Stems 20-40 cm tall, villous to pubescent. Leaves margin dentate-serrate, sometimes double. Corolla white, flushed purplish; abaxial lips yellowish bearded, lobes oblong; peduncles and pedicels villous; staminodes 2. Fl. and fr. Jul-Sep. Rocks at forest edges at 300-1300 m. Distributed in S Yunnan and SW Guangxi. Also in NW Thailand.

滇桂喜鹊苣苔 *Ornithoboea wildeana*

长冠苣苔 *Rhabdothamnopsis sinensis*

长冠苣苔

Rhabdothamnopsis sinensis Hemsl.

小灌木。叶柄近无至9毫米；叶有时簇生于近茎顶；叶具柔毛，叶缘具锯齿至圆齿。花冠钟状管形，外面紫色，内面白色带紫色条带；冠筒长1.3-3厘米；花药被髯毛；雌蕊具腺毛。花期5-8月，果期6-10月。生海拔1600-4600米的石灰岩林中。产云南中部和北部、四川西部与西南部和贵州西部。

Small shrubs. Petioles subsessile to 9 mm long; leaves sometimes crowded near stem apex; leaves puberulent, margin serrulate to crenulate. Corolla campanulate-tubular, purple outside, white and with purple stripes inside; tubes 1.3-3 cm long; anthers bearded; pistils glandular puberulent. Fl. May-Aug. Fr. Jun-Oct. Limestone in forests at 1600-4600 m. Distributed in C and N Yunnan, W and SW Sichuan, and W Guizhou.

软叶大苞苣苔

Anna mollifolia (W. T. Wang) W. T. Wang et K. Y. Pan

亚灌木。叶片两侧不对称，具柔毛或稍密，边缘近全缘至具浅锯齿。花序梗密具柔毛；花萼红色；花冠白色，外部渐无毛，裂片近圆形。种子附属物长0.1-0.2毫米。花期8月，果期9月。生海拔1100-1500米的石灰山岩隙中。产云南东南部和广西西南部。

Subshrubs. Leaves asymmetrical in both sides, puberulent to densely so, margin subentire to shallowly serrulate. Peduncles densely puberulent; calyx red; corolla white, outside glabrescent, lobes suborbicular. Seeds appendages 0.1-0.2 mm long. Fl. Aug. Fr. Sep. Rock crevices in limestone hills at 1100-1500 m. Distributed in SE Yunnan and SW Guangxi.

滇黔紫花苣苔

Loxostigma cavaleriei (Lévl. et Van.) Burtt

亚灌木。叶片膜质。聚伞花序不分枝或具3分枝，具1-7花；花冠淡红色或白色，内部常具红色、紫色或棕色斑点。种子两端各具1毛状附属物。花期7-9月，果期10-11月。生海拔600-1600米的林中树干上。产云南东南部、贵州南部和广西。

Subshrubs. Leaves membranous. Cymes unbranched to branched 3, 1-7-flowered; corolla pink to white, inside sometimes red, purple or brown spotted. Seeds with 1 hairlike appendage at each end. Fl. Jul-Sep. Fr. Oct-Nov. Tree trunks in forests at 600-1600 m. Distributed in SE Yunnan, S Guizhou and Guangxi.

齿萼紫花苣苔

Loxostigma fimbrisepalum K. Y. Pan

亚灌木。茎60-100厘米，疏具柔毛，近无毛。叶对生，上面疏具平伏柔毛，下面无毛至疏具平伏柔毛。聚伞花序具2-5分枝，5-18花；花萼自近基部5等裂，裂片卵形至宽三角形；花冠白色至淡紫色，里面具紫斑。花期9-11月。生海拔900-1600米的石缝、石灰岩上或林中附生植物。产云南东南部和广西。

Subshrubs. Stems 60-100 cm tall, sparsely puberulent, glabrescent.

软叶大苞苣苔 *Anna mollifolia*

滇黔紫花苣苔 *Loxostigma cavaleriei*

齿萼紫花苣苔 *Loxostigma fimbrisepalum*

Leaves opposite, adaxially sparsely appressed puberulent, abaxially glabrous to sparsely appressed puberulent. Cymes branched 2-5, 5-18-flowered; calyx 5-sect from near base, segments equal, ovate to broadly triangular; corolla white to lavender, purple spotted inside. Fl. Sep-Nov. Rock crevices, limestone or epiphytic in forests at 900-1600 m. Distributed in SE Yunnan and Guangxi.

红花芒毛苣苔

Aeschynanthus moningeriae (Merr.) W. Y. Chun

小灌木。叶狭椭圆形至卵形或倒卵形。聚伞花序腋生或假顶生，具2-4(-7)花；花冠红色，2.8-3厘米，外面无毛，里面疏被短柔毛，口部强烈偏斜。种子每端各有1毛状附属物。花期9月至翌年2月，果期翌年1-5月。生海拔300-1200米的林中树上或山谷溪边岩石上。产广东和海南。

Small shrubs. Leaves narrowly elliptic to ovate or obovate. Cymes axillary or pseudoterminal, 2-4(-7)-flowered; corolla red, 2.8-3 cm long, outside glabrous, inside sparsely puberulent, mouth strongly oblique. Seeds with 1 hairlike appendage at each end. Fl. Sep to next Feb. Fr. next Jan-May. Trees in forests or streamside rocks in valleys at 300-1200 m. Distributed in Guangdong and Hainan.

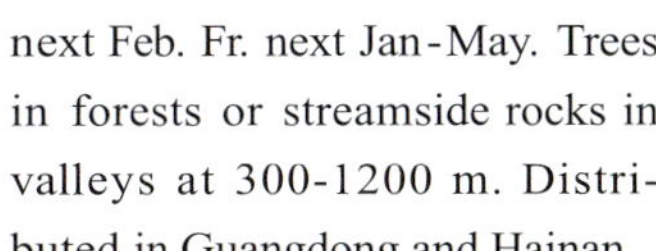

红花芒毛苣苔 *Aeschynanthus moningeriae*

尾叶芒毛苣苔

Aeschynanthus stenosepalus Anthony

小灌木。叶对生，狭椭圆形至披针形或卵形，革质至纸质，无毛，先端长渐尖至尾尖。聚伞花序腋生，具1-3花；苞片常早落，绿色；花萼绿色，有时带红色，自基部5裂；花冠红色。种子两边各具1毛状附属物。花期7-10月，果期10月。在海拔1500-2500米的林缘树生或岩生。产云南西北部和西藏(墨脱县)。缅甸北部亦有。

Small shrubs. Leaves opposite, narrowly elliptic to lanceolate or ovate, leathery to papery, glabrous, apex long acuminate to caudate. Cymes axillary, 1-3-flowered; bracts usually deciduous, green; calyx green, sometimes tinged red, 5-sect from base; corolla red. Seeds with 1 hairlike appendage at each end. Fl. Jul-Oct. Fr. Oct. Trees or rocks in forest edges at 1500-2500 m. Distributed in NW Yunnan and Xizang (Mêdog County). Also in N Myanmar.

尾叶芒毛苣苔 *Aeschynanthus stenosepalus*

狭花芒毛苣苔 *Aeschynanthus wardii*

显苞芒毛苣苔 *Aeschynanthus bracteatus*

狭花芒毛苣苔

Aeschynanthus wardii Merr.

附生灌木。叶片革质至纸质，无毛，下面具少量腺点。聚伞花序腋生，具1-4花；花冠红色至橘红色，外面无毛；口部倾斜。种子两侧各具1毛状附属物。花果期9-11月。生海拔900-2800米的林中或河谷石上。产云南西北部和西部。缅甸北部亦有。

Epiphyte shrubs. Leaflets leathery to papery, glabrous, abaxially few punctate. Cymes axillary, 1-4-flowered; corolla red to orange, outside glabrous, mouths oblique. Seeds with 1 hairlike appendage at each end. Fl. and fr. Sep-Nov. Forests or rocks in river valleys at 900-2800 m. Distributed in NW and W Yunnan. Also in N Myanmar.

显苞芒毛苣苔

Aeschynanthus bracteatus Wall.

附生小灌木。茎高25-150厘米，无毛。叶对生；叶革质至纸质。花萼红色，自基部5裂；花冠红色至粉红色或紫色，口部强烈歪斜；冠檐不明显二唇形。蒴果长7-16(-21)厘米。种子每端各有1毛状附属物。花期6-10月，果期7至翌年1月。生海拔900-3200米的山谷林中树上或溪边崖壁上。产云南、西藏东南部和广西西北部。印度东北部、不丹和缅甸亦有。

Epiphytic small shrubs. Stems 25-150 cm tall, glabrous. Leaves opposite; leaves leathery to papery. Calyx red, 5-sect from base; corolla red to pink or purple, mouth strongly oblique; limbs indistinctly 2-lipped. Capsules 7-16(-21) cm long. Seeds with 1 hairlike appendage at each end. Fl. Jun-Oct. Fr. Jul to next Jan. Trees in forested valleys or streamside cliffs at 900-3200 m. Distributed in Yunnan, SE Xizang and NW Guangxi. Also in NE India, Bhutan and Myanmar.

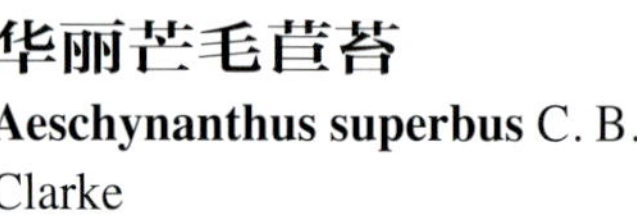

华丽芒毛苣苔

Aeschynanthus superbus C. B. Clarke

附生小灌木。叶对生，全缘，纸质至革质，无毛，下面疏具腺点。聚伞花序腋生，具5-15花；花冠橘红色至深红色，具深色条纹。种子两侧各具1毛状附属物。蒴果条形。花期8-9月，果期10月。生海拔1000-2500米的林中树上。产云南东南部和西

华丽芒毛苣苔 *Aeschynanthus superbus*

条叶芒毛苣苔 *Aeschynanthus linearifolius*

部、西藏东南部。印度东北部、不丹和缅甸北部亦有。

Epiphytic small shrubs. Leaves opposite, margin entire, papery to leathery, glabrous, abaxially sparsely punctate. Cymes axillary, 5-15-flowered; corolla orange-red to dark red with darker stripes. Seeds with 1 hairlike appendage at each end. Capsules linear. Fl. Aug-Sep. Fr. Oct. Trees in forests at 1000-2500 m. Distributed in SE and W Yunnan, and SE Xizang. Also in NE India, Bhutan and N Myanmar.

条叶芒毛苣苔

Aeschynanthus linearifolius C. E. C. Fisch.

小灌木。叶对生，狭至宽倒披针形，革质，无毛。聚伞花序腋生，具1-4花；苞片宿存，红色，披针形；花萼红色，自基部5裂；花冠红色。种子两端各具1毛状附属物。花期7-9月，果期10月。在海拔1900-3100米的山谷林中树生或岩生。产云南和西藏东南部。印度东北部和缅甸北部亦有。

Small shrubs. Leaves opposite, narrowly to broadly oblanceolate, leathery, glabrous. Cymes axillary, 1-4-flowered; bracts persistent, red, lanceolate; calyx red, 5-sect from base; corolla red. Seeds with 1 hairlike appendage at each end. Fl. Jul-Sep. Fr. Oct. Tree or rocks in forested valleys at 1900-3100 m. Distributed in Yunnan and SE Xizang. Also in NE India and N Myanmar.

广西芒毛苣苔 *Aeschynanthus austroyunnanensis* var. *guangxiensis*

广西芒毛苣苔

Aeschynanthus austroyunnanensis W. T. Wang var. **guangxiensis** (W. Y. Chun ex W. T. Wang et K. Y. Pan) W. T. Wang

附生小灌木。茎高达1米，无毛。叶对生，薄革质至纸质，无毛，全缘。花萼外部无毛至近无毛；花冠红色，长2-2.3厘米，外部无毛至近无毛。蒴果长8-20厘米。花期7月，果期12月。生海拔400-1000米的树上、岩石上或石灰岩山崖壁上。产贵州(贞丰县)和广西西部。

Small shrubs. Epiphytic. Stems to 1 m tall, glabrous. Leaves opposite, thin leathery to papery, glabrous, margin entire. Calyx outside glabrous to nearly glabrous; corolla red, 2-2.3 cm long, outside glabrous to nearly glabrous. Capsules 8-20 cm long. Fl. Jul. Fr. Dec. Trees, rocks or cliffs on limestone hills at 400-1000 m. Distributed in Guizhou (Zhenfeng County) and W Guangxi.

矮芒毛苣苔

Aeschynanthus humilis Hemsl.

附生小灌木。叶对生或3枚轮生，全缘。聚伞花序假顶生，具1-3花；花序梗缺无；花萼常带紫红色；花冠红色；冠檐明显二唇形。种子两侧各具1毛状附属物。花期9至翌年1月，果期翌年2月。生海拔1300-2100米的山谷林中树上。产云南南部。

Small shrubs, epiphytic. Leaves opposite or in whorls of 3, leaves margin entire. Cymes pseudoterminal, 1-3-flowered; peduncles absent; calyx often tinged purple-red; corolla red; limbs distinctly 2-lipped. Seeds with 1 hairlike appendage at each end. Fl. Sep to next Jan. Fr. next Feb. Trees in forested valleys at 1300-2100 m. Distributed in S Yunnan.

矮芒毛苣苔 *Aeschynanthus humilis*

黄杨叶芒毛苣苔 *Aeschynanthus buxifolius*

小齿芒毛苣苔 *Aeschynanthus denticuliger*

黄杨叶芒毛苣苔

Aeschynanthus buxifolius Hemsl.

附生小灌木。叶对生或3枚轮生；叶椭圆形至卵形，革质，无毛。聚伞花序腋生，具1花；花冠紫红色，下唇有深红色条纹，内面疏被短腺毛，口部斜，檐部明显二唇形；雄蕊伸出。种子每端各有1附属物。花期6-11月，果期11月。生海拔1300-2200米的林中树上或岩石上。产云南、贵州西南部和广西。越南亦有。

Epiphytic small shrubs. Leaves opposite or in whorls of 3; leaves elliptic to ovate, leathery, glabrous. Cymes axillary, 1-flowered; corolla purple-red, dark red striate on lower lip, sparsely glandular hairy inside, mouth oblique, limbs conspicuously bilabiate; stamens exserted. Seeds each with 1 appendages on both ends. Fl. Jun-Nov. Fr. Nov. Trees or rocks in forests at 1300-2200 m. Distributed in Yunnan, SW Guizhou and Guangxi. Also in Vietnam.

小齿芒毛苣苔

Aeschynanthus denticuliger W. T. Wang

小攀援亚灌木，附生。叶对生或3枚轮生，边缘有小齿。聚伞花序腋生，具1花；花萼绿色，深5裂；花冠黄色或白色，口部不斜；冠檐明显二唇形。种子两端各具1毛状附属物。花期2月，果期3月。生海拔1200-1500米的密林中树上。产云南东南部。老挝和越南北部亦有。

Climbing small subshrubs, epiphytic. Leaves opposite or in whorls of 3, leaves margin denticulate. Cymes axillary, 1-flowered; calyx green, deeply 5-lobed; corolla yellow or white, mouth not oblique; limbs distinctly 2-lipped. Seeds with 1 hairlike appendage at each end. Fl. Feb. Fr. Mar. Trees in dense forests at 1200-1500 m. Distributed in SE Yunnan. Also in Laos and N Vietnam.

大花芒毛苣苔

Aeschynanthus mimetes Burtt

附生小灌木。叶狭椭圆形至宽卵形或倒卵形。花冠橘红色，裂片中央有紫斑，外面上部背面、内面下部有短柔毛。种子一端有1根毛，另一端有2根毛。花期6-9月，果期10-11月。生海拔1000-2500米的林中树上。产云南南部和西部、西藏东南部。印度东北部亦有。

Epiphytic small shrubs. Leaves narrowly elliptic to broadly ovate or obovate; corolla orange-red, with purple spot on the middle of lobes, pubescent on upper part abaxially outside and on lower part inside. Seeds with 1 hair on one end and 2 hairs on another end. Fl. Jun-Sep. Fr. Oct-Nov. Trees in forests at 1000-2500 m. Distributed in S and W Yunnan, and SE Xizang. Also in NE India.

大花芒毛苣苔 *Aeschynanthus mimetes*

束花芒毛苣苔
Aeschynanthus hookeri

束花芒毛苣苔

Aeschynanthus hookeri C. B. Clarke

附生小灌木。叶狭椭圆形至长圆形。聚伞花序假顶生，具4-10花；花冠红色，裂片中央有暗红色条纹。种子一端有1根毛，另一端有2根毛。花期7月。生海拔1200-2100米的山区林中树上。产云南西北部和南部。印度北部、不丹和缅甸北部亦有。

Epiphytic small shrubs. Leaves narrowly elliptic to oblong. Cymes pseudoterminal, 4-10-flowered; corolla red, with dark red lines on the middle of lobes. Seeds with 1 hair on one end and 2 hairs on another end. Fl. Jul. Trees in forests montane regions at 1200-2100 m. Distributed in NW and S Yunnan. Also in N India, Bhutan and N Myanmar.

长圆吊石苣苔

Lysionotus oblongifolius W. T. Wang

亚灌木。茎直立，顶部密被锈褐色平伏柔毛。叶柄长1-4厘米；叶纸质。聚伞花序具4-7花；花萼自基部5深裂；花冠紫红色；冠筒管状；退化雄蕊3。蒴果长3-4厘米。种子附属物近钻形。花期9-10月，果期10月。生海拔约300米的石灰石山林下土地上。产广西西南部。

Subshrubs. Stems erect, apically densely rust-brown appressed puberulent. Petioles 1-4 cm long; leaves papery. Cymes 4-7-flowered; calyx 5-sect from base; corolla purple-red; tubes tubular; staminodes 3. Capsules 3-4cm long. Seed appendages subulate. Fl. Sep-Oct. Fr. Oct. Terrestrial in forests on limestones hills at ca. 300 m. Distributed in SW Guangxi.

短柄吊石苣苔

Lysionotus sessilifolius Hand.-Mazz.

亚灌木，有时攀援。叶3枚轮生，无柄或近无柄，薄革质至纸质，无毛，边缘具细齿。聚伞花序具2-7花；花梗长4-10毫米；花冠紫色，花冠管细，漏斗状。蒴果条形。种子附属物毛状。花期8-10月，果期9-10月。地生或附生或海拔1200-2800米的林中石上或溪边石上。产云南东北部。

Subshrubs, sometimes climbing. Leaves 3 whorled, sessile or subsessile, thin leathery to papery, glabrous, margin serrate. Cymes 2-7-flowered; pedicels 4-10 mm long; corolla purple, tubes slender, funnelform. Capsules linear. Seeds appendages hairlike. Fl. Aug-Oct. Fr. Sep-Oct. Terrestrial or epiphytic in forests and streamsides in valleys at 1200-2800 m. Distributed in NE Yunnan.

长圆吊石苣苔 *Lysionotus oblongifolius*

短柄吊石苣苔 *Lysionotus sessilifolius*

纤细吊石苣苔 *Lysionotus gracilis*

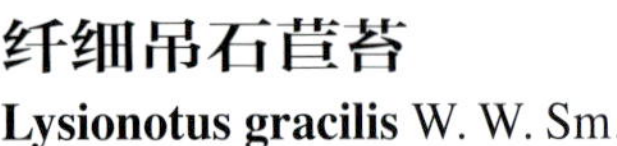

纤细吊石苣苔

Lysionotus gracilis W. W. Sm.

亚灌木。叶3枚轮生，稀对生，革质，上面无毛，下面无毛，中脉有时具柔毛；花序梗纤细。聚伞花序具1-6花；花冠白色，具浅紫色条纹；退化雄蕊3。蒴果条形。花期7-8月，果期8月。生海拔2100-2400米的常绿阔叶林中树上。产云南西部。缅甸北部亦有。

Subshrubs. Leaves 3 whorled, rarely opposite, leathery, adaxially glabrous, abaxially glabrous, midrib sometimes puberulent; peduncles gracile. Cymes 1-6-flowered; corolla white, purplish striate; staminodes 3. Capsules linear. Fl. Jul-Aug. Fr. Aug. Trees evergreen broad-leaved forests at 2100-2400 m. Distributed in W Yunnan. Also in N Myanmar.

墨脱吊石苣苔 *Lysionotus metuoensis*

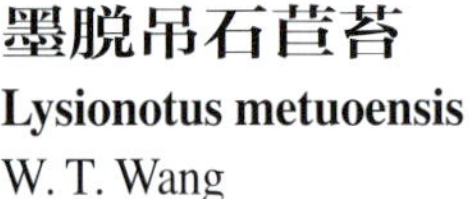

墨脱吊石苣苔

Lysionotus metuoensis W. T. Wang

亚灌木。茎密被开展柔毛。叶近无柄，革质。聚伞花序具1花；花萼5浅裂；花冠白色；冠筒漏斗状管形；退化雄蕊2；花盘环状。蒴果长6.5-13厘米。种子附属物毛发状。花果期8-9月。树生于海拔约1300米的阔叶林中。产西藏（墨脱县）。

Subshrubs. Stems densely spreading pubescent. Leaves nearly sessile, leathery. Cymes 1-flowered; calyx 5-lobed; corolla white; tubes funnelform-tubular; staminodes 2; disc ringlike. Capsules 6.5-13 cm long. Seed appendages hairlike. Fl. and fr. Aug-Sep. Trees in broad-leaved forests at ca. 1300 m. Distributed in Xizang (Mêdog County).

吊石苣苔 *Lysionotus pauciflorus*

吊石苣苔

Lysionotus pauciflorus Maxim.

亚灌木。叶对生或3-4枚轮生，叶片革质，形状多变化。聚伞花序具1-3(-12)花；花冠白色至浅紫色或粉红色，内部具紫色线且有时喉部黄色。种子附属物毛状。花期6-12月，果期8月至翌年1月。生海拔300-2200米的林中、岩石、崖壁、山丘或山区树上。产中国西南、华南、华中和华东。越南北部和日本亦有。

Subshrubs. Leaves opposite or 3-4 whorled, leaves leathery, varies much in shape. Cymes 1-3 (-12)-flowered; corolla white to light purple or pink, inside with purple lines and sometimes yellow throat. Seeds appendages hairlike. Fl. Jun-Dec. Fr. Aug to next Jan. Forests, rocks, cliffs, trees in hilly or montane regions at 300-2200 m. Distributed in SW, S, C and E China. Also in N Vietnam and Japan.

齿叶吊石苣苔

Lysionotus serratus D. Don

附生亚灌木。茎不弯曲，无翅；叶草质，边缘具锯齿至细齿或小圆齿。聚伞花序具3-15

齿叶吊石苣苔 *Lysionotus serratus*

花；花冠淡紫色或白色，有疏柔毛和短腺毛；花盘环状，近全缘。花期7-9月，果期9月至翌年1月。生海拔900-2200米的山地林中树上或石上、溪边或高山草地。产中国西南和广西西北部。南亚和东南亚亦有。

Epiphytic subshrubs. Stems not angled, wingless; leaves herbaceous, margin dentate to serrate or crenulate. Cymes 3-15-flowered; corolla pale purple or white, sparsely pilose and short glandular-hairy; disc annular, subentire. Fl. Jul-Sep. Fr. Sep to next Jan. Trees or stones of montane forests, by streams or alpine grasslands at 900-2200 m. Distributed in SW China and NW Guangxi. Also in S and SE Asia.

合萼吊石苣苔

Lysionotus gamosepalus W. T. Wang

亚灌木。叶纸质，长椭圆形或卵形，长5-13厘米，宽2.8-6厘米，无毛，基部宽楔形至圆形，边缘有锯齿或牙齿；侧脉每侧4-6条，平坦。聚伞花序有2-11花；花萼5裂至中部至中部以上，筒部长6-10毫米；花冠白色或带淡紫色，无毛；药隔凸起长1.2-1.8毫米。花期6-7月，果期8月。生海拔800-1600米的常绿阔叶林中或石崖上或路边石上。产西藏东南部。

Subshrubs. Leaves papery, narrowly elliptic to ovate, 5-13 × 2.8-6 cm, glabrous, base broadly cuneate to rounded, margin serrate to dentate; lateral veins 4-6 on each side of midrib, flat. Cymes 2-11-flowered; calyx 5-lobed from middle to above middle, tube 6-10 mm long; corolla white or tinged purple, glabrous; anthers connective appendage 1.2-1.8 mm long. Fl. Jun-Jul. Fr. Aug. Evergreen, broad-leaved forests, stony cliffs or rocks at waysides at 800-1600 m. Distributed in SE Xizang.

攀援吊石苣苔 *Lysionotus chingii*

攀援吊石苣苔

Lysionotus chingii W. Y. Chun ex W. T. Wang

平卧草本、攀援亚灌木或藤本。茎有时厚软木状。叶纸质，无毛，边缘全缘至具圆锯齿和小腺体。聚伞花序具1花；花冠白色或浅绿色。种子附属物近钻形。花期7-9月，果期9月。生海拔900-1500米的山谷林中树上或岩石上。产云南东南部和广西南部与西部。越南北部亦有。

Prostrate herbs, climbing subshrubs or lianas. Stems sometimes thick corky. Leaves papery, glabrous, margin entire to denticulate with small glands. Cymes 1-flowered; corolla white or tinged greenish. Seeds appendages subulate. Fl. Jul-Sep. Fr. Sep. Trees or rocks in forested valleys at 900-1500 m. Distributed in SE Yunnan, and S and W Guangxi. Also in N Vietnam.

合萼吊石苣苔 *Lysionotus gamosepalus*

椭圆线柱苣苔 *Rhynchotechum ellipticum*

椭圆线柱苣苔

Rhynchotechum ellipticum

(Wall. ex D. Dietr.) A. DC.

亚灌木。叶对生；叶柄具平伏棕锈色卷毛。聚伞花序具15-70花；花冠白色至粉红色，钟形，外面具乳头状凸起；花丝着生于花冠基部。浆果白色，无毛。花期6-10月，果期8月至翌年1月。生海拔100-1800米的林中或沟边阴处。产中国西南、华南和华东。南亚和东南亚亦有。

Subshrubs. Leaves opposite; petioles appressed rust-brown to brown woolly. Cymes 15-70-flowered; corolla white or tinged pink, campanulate, papillate outside; filaments inserted on corolla base. Berries white, glabrous. Fl. Jun-Oct. Fr. Aug to next Jan. Forests or shaded streamsides at 100-1800 m. Distributed in SW, S and E China. Also in S and SE Asia.

毛线柱苣苔

Rhynchotechum vestitum

Wall. ex Clatke

亚灌木。叶柄、花序梗与花梗密被开展的淡黄色长毛。叶对生，长圆形至椭圆形或倒卵形，长13-28.5厘米，基部宽楔形或圆形，边缘有小牙齿。聚伞花序有(3-)5-15花；花萼裂片狭三角形至狭披针形；花冠淡红色。浆果白色，长3-5毫米。花期6-8月，果期10-11月。生海拔800-1300米的山谷林中或溪边阴处。产西藏、云南和广西。不丹和印度北部亦有。

Subshrubs. Petioles, peduncle and pedicel densely spreading yellowish hirsute. Leaves opposite, ovate to elliptic or obovate, 13-28.5 cm long, base broadly cuneate to rounded, margin denticulate. Cymes (3-)5-15-flowered; calyx segments narrowly triangular to narrowly lanceolate; corolla reddish. Berry white, 3-5 mm long. Fl. Jun-Aug. Fr. Oct-Nov. Forests, shaded streamsides in valleys at 800-1300 m. Distributed in Xizang, Yunnan and Guangxi. Also in Bhutan and N India.

尖舌苣苔

Rhynchoglossum obliquum

Blume

一年生草本。无根茎。叶偏斜，一侧椭圆形，另一侧卵形，上面无毛，有时具微柔毛，下面无毛，边缘全缘至波状。聚伞花序具10-30花；花较小，花冠长约1厘米，颜色较淡；下唇3裂至不裂；能育雄蕊2。花期8-9月，果期10-11月。生海拔500-2300米的山地林边、林中或石山洞中。产中国西南和台湾。

Annual herbs. Not rhizomatous. Leaves oblique, one side elliptic, other side ovate, adaxially glabrous, sometimes minutely sparsely puberulent, abaxially glabrous, margin entire to undulate. Cymes 10-30-flowered; flowers small, corolla ca. 1 cm long, with paler color; abaxial lip 3-lobed to undivided; fertile stamens 2. Fl. Aug-Sep. Fr. Oct-Nov. Montane forest edges, forests or stony caves at 500-2300 m. Distributed in SW China and Taiwan.

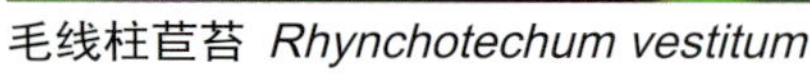

毛线柱苣苔 *Rhynchotechum vestitum*

尖舌苣苔 *Rhynchoglossum obliquum*

盾座苣苔 *Epithema carnosum*

盾座苣苔
Epithema carnosum Benth.

小草本。茎下部叶1片，上部叶2，对生；叶卵状椭圆形，具柔毛，边缘波状细圆齿至具细齿。苞片1，兜状；花冠浅红色至浅紫色或白色。蒴果球形，周裂。花期6-9月，果期9月。生海拔300-1400米的山洞或阴处石上。产云南东南部、贵州西南部、广西、广东北部和台湾。印度东北部、尼泊尔、不丹、缅甸和泰国亦有。

Dwarf herbs. Basal stem leaf 1, upper stem leaves 2, opposite; leaves ovate-elliptic, puberulent, margin undulate-denticulate to serrate. Bract 1, hoodlike; corolla reddish to purplish or white. Capsules subglobose, circumscissile. Fl. Jun-Sep. Fr. Sep. Caves in valleys or shady rocks at 300-1400 m. Distributed in SE Yunnan, SW Guizhou, Guangxi, N Guangdong and Taiwan. Also in NE India, Nepal, Bhutan, Myanmar and Thailand.

台闽苣苔 *Titanotrichum oldhamii*

台闽苣苔
Titanotrichum oldhamii (Hemsl.) Soler.

多年生草本。小叶草质或纸质。能育花的花序总状，不育花的花序穗状；花序多花；花序轴具短毛；苞片披针形至条形；花冠黄色，内部自裂片至基部具紫色斑点。花期7-11月，果期8-12月。生海拔100-1200米的山谷阴处。产台湾和福建。日本亦有。

Perennial herbs. Leaflets herbaceous or papery. Inflorescences of fertile flowers racemose, sterile spicate; inflorescences many-flowered; rachis pilose; bracts lanceolate to linear; corolla yellow, inside purple spotted from lobes to base. Fl. Jul-Nov. Fr. Aug-Dec. Shaded areas in valleys at 100-1200 m. Distributed in Taiwan and Fujian. Also in Japan.

牛耳朵 *Chirita eburnea*

牛耳朵
Chirita eburnea Hance

多年生无茎草本。叶基生，对生，卵形至椭圆形或倒卵形，纸质，具平伏柔毛至短柔毛。聚伞花序(1或)2-13(-17)花；花冠紫色至白色或黄色；花冠管近管状。蒴果直立。花期4-9月，果期5-10月。生海拔0-1900米的石灰岩山林岩石上或林下沟边。产中国西南、华南和华中。

Perennial herbs, stemless. Leaves basal, opposite, ovate to elliptic or obovate, papery, appressed puberulent to pilose. Cymes (1 or) 2-13(-17)-flowered; corolla purple to white or yellow; tubes nearly tubular. Capsules erect. Fl. Apr-Sep. Fr. May-Oct. Rocks in forests or streamsides in forests on limestone hills at 0- to 1900 m. Distributed in SW, S and C China.

圆叶唇柱苣苔
Chirita dielsii (Borza) Burtt

多年生无茎草本。叶均基生，宽卵形至宽倒卵形或圆形至肾形，背面沿脉被长柔毛，侧脉4-5对。聚伞花序具1(或2)花；花萼二唇形；花冠淡紫色或蓝紫色。蒴果直立。花期6-9月，果期8-9月。生海拔1900-3400米的山区阴处岩石上。产云南。

Acaulescent, perennial herbs. Leaves all basal, broadly ovate to broadly obovate or orbicular to reniform, with villose along veins on back surface, lateral nerves 4-5 pairs. Cymes 1(or 2)-flowered; calyx 2-lipped; corolla pale purple or purple-blue. Capsules erect. Fl. Jun-Sep. Fr. Aug-Sep. Shaded rocks in montane regions at 1900-3400 m. Distributed in Yunnan.

钩序唇柱苣苔
Chirita hamosa R. Brown

一年生草本。无根茎。叶狭至宽卵形。聚伞花序蝎尾状，具1-5(-10)花；花冠白色，喉部黄色，常带紫色；花序梗与叶柄合生；花药被髯毛。蒴果直立。花期7-10月，果期7-11月。生海拔300-1500米的阴处林中石上、岩壁或沟边。产云南南部和东南部、广西南部和西部。南亚和东南亚亦有。

Annual herbs. Not rhizomatous. Leaves narrowly to broadly ovate. Cymes scorpioid, 1-5(-10)-flowered; corolla white, yellow at throat, often flushed lavender; peduncles adnate to petioles; anthers bearded. Capsules erect. Fl. Jul-Oct. Fr. Jul-Nov. Shaded rocks in forests, cliffs or streamside at 300-1500 m. Distributed in S and SE Yunnan, and S and W Guangxi. Also in S and SE Asia.

圆叶唇柱苣苔 *Chirita dielsii*

钩序唇柱苣苔 *Chirita hamosa*

狸藻科
Lentibulariaceae

高山捕虫堇 *Pinguicula alpina*

高山捕虫堇
Pinguicula alpina L.

多年生食虫草本，具根状茎。芽卵球形。叶3-13，基生呈莲座状，上面密生分泌黏液的腺毛，边缘内卷。花1-5朵；花梗和花萼无毛；花冠白色，下唇基部具1黄斑，距淡黄色。蒴果卵球状长圆形。花期5-7月，果期7-9月。生海拔1800-4500米的沼泽或山坡湿地。产中国西南和华中。欧洲和亚洲的温带高山地区亦有。

Perennial insectivorous herbs, with rhizomes. Leaves 3-13, basal and in a rosette, adaxially with dense glandular hairs which secreted mucilage, margin involute. Flowers 1-5; pedicel and calyx glabrous; corolla white, with 1 yellow spot at base of lower lip, spur yellowish. Capsules ovoid-oblong. Fl. May-Jul. Fr. Jul-Sep. Bogs or wet places in mountains at 1800-4500 m. Distributed in SW and C China. Also in temperate alpine regions of Europe and Asia.

斜果挖耳草
Utricularia minutissima Vahl

一年生食虫草本。叶狭倒卵形至条形，具1脉。苞片和小苞片基部着生；花冠紫堇色至白色，下唇3浅裂，明显短于距。蒴果斜椭圆体形，沿腹侧纵裂。花期9-11月，果期11-12月。生低海拔的潮湿沙质土或湿草地。产江苏、江西、福建和广西。南亚、东南亚、日本南部和澳大利亚亦有。

Annual insectivorous herbs. Leaves narrowly obovate to linear, 1-nerved. Bracts and bracteoles basifixed; corolla violet to white, lower lip, 3-lobed, much shorter than spur. Capsules obliquely ellipsoid, dehiscing by a longitudinal ventral slit. Fl. Sep-Nov. Fr. Nov-Dec. Wet, sandy soil or grassy places near sea level. Distributed in Jiangsu, Jiangxi, Fujian and Guangxi. Also in S and SE Asia, S Japan and Australia.

毛挖耳草
Utricularia hirta Klein ex Link

多年生食虫草本。叶狭倒卵形，具1脉。花序梗及花序密被开展的毛，毛长0.1-1.5毫米；苞片和小苞片基部着生；花冠紫堇色至白色，下唇3浅裂，明显短于距。蒴果斜卵球形，沿腹侧纵裂。花期9-11月，果期11-12月。生近海潮湿草地和沼泽。产广西南部。南亚和东南亚亦有。

Perennial insectivorous herbs. Leaves narrowly obovate, 1-nerved. Peduncle and inflorescences densely with 0.1-1.5 mm long patent hairs; bracts and bracteoles basifixed; corolla violet to white, lower lip, 3-lobed, much shorter than spur. Capsules obliquely ovoid, dehiscing by a longitudinal ventral slit. Fl. Sep-Nov. Fr. Nov-Dec. Damp or wet, open grassy places and marshes near sea level. Distributed in S Guangxi. Also in S and SE Asia.

斜果挖耳草 *Utricularia minutissima*

毛挖耳草 *Utricularia hirta*

短梗挖耳草
Utricularia caerulea

短梗挖耳草
Utricularia caerulea L.

一年生食虫草本。叶条形至条状倒卵形，具3脉。苞片和小苞片盾状着生；花冠紫色至白色，下唇顶端圆形至微凹，通常短于距。蒴果球形至椭圆体形，腹侧纵裂。花期6月至翌年1月，果期7月至翌年2月。生海拔50-2000米的潮湿岩石、溪边、湿草地和沼泽地。产云南、贵州、华南、福建、湖南和山东。东亚、东南亚、南亚、马达加斯加和澳大利亚亦有。

Annual insectivorous herbs. Leaves linear to linear-obovate, veins 3. Bracts and bracteoles peltate-attached; corolla violet to white, lower lip rounded to emarginated at apex, shorter than spur. Capsules globose to ellipsoid, dehiscing by a longitudinal ventral slit. Fl. Jun to next Jan. Fr. Jul to next Feb. Wet rocks, beside streams, wet grassland or swamps at 50-2000 m. Distributed in Yunnan, Guizhou, S China, Fujian, Hunan and Shandong. Also in E, SE and S Asia, Madagascar and Australia.

钩突挖耳草
Utricularia warburgii K. I. Goebel

一年生食虫草本。叶条状倒卵形，具1脉。苞片和小苞片盾状着生；花冠淡蓝紫色，下唇顶端圆形，短于距，喉凸具2个钩状凸起。蒴果球形至椭圆体形，腹侧纵裂。花期5-9月，果期7-10月。生海拔800-2000米的湿草地和潮湿岩石上。产华东、重庆和湖南。

Annual insectivorous herbs. Leaves linear-obovate, vein 1. Bracts and bracteoles peltate-attached; corolla pale blueish purple, lower lip rounded at apex, shorter than spur, palate with 2 hooked processes. Capsules globose to ellipsoid, dehiscing by a longitudinal ventral slit. Fl. May-Sep. Fr. Jul-Oct. Wet grasslands or wet rocks at 800-2000 m. Distributed in E China, Chongqing and Hunan.

挖耳草 *Utricularia bifida*

挖耳草
Utricularia bifida L.

一年生食虫草本。叶狭条形至条状倒披针形，具1脉。苞片和小苞片基部着生；花冠黄色，下唇顶端圆形，约与距等长。蒴果宽椭圆体形，背腹扁，沿背腹两侧纵裂。花期6-12月，果期7月至翌年1月。生海拔50-1400米的潮湿土壤和石上或稻田中。产山东、河南、西南、华南、华中和华东。亦产东亚、东南亚、南亚和澳大利亚北部。

Annual insectivorous herbs. Leaves narrowly linear to linear-oblanceolate, vein 1. Bracts and bracteoles basifixed; corolla yellow, lower lip rounded at apex, about as long as spur. Capsules broadly ellipsoid, dorsiventrally compressed, denisciing by longitudinal dorsal and ventral slits. Fl. Jun-Dec. Fr. Jul to next Jan. Wet soil, rocks or rice fields at 50-1400 m. Distributed in Shandong, Henan, SW, S, C and E China. Also in E, SE and S Asia, and N Australia.

钩突挖耳草 *Utricularia warburgii*

齿萼挖耳草
Utricularia uliginosa Vahl

一年生食虫草本。叶条形至狭倒卵形，具羽状脉，有时具3脉。苞片和小苞片基部着生；花冠紫堇色至白色，下唇先端圆形或微凹，约与距等长。蒴果卵球形，背腹扁，腹侧纵裂。花期7-10月，果期8-11

齿萼挖耳草 *Utricularia uliginosa*

禾叶挖耳草 *Utricularia graminifolia*

月。生海拔30-400米沼泽地或潮湿的沙地上。产广西、广东、海南和台湾。南亚、东南亚、日本和澳大利亚亦有。

Annual insectiveorous herbs. Leaves linear to narrowly obovate, veins pinnate, sometimes 3-nerved. Bracts and bracteoles basifixed; corolla violet to white, lower lip rounded or emarginated at apex, about as long as spur. Capsules ovoid, dorsiventrally compressed, dehiscing by a longtitudinal ventral slit. Fl. Jul-Oct. Fr. Aug-Nov. Marshes or wet sandy soil at 30-400 m. Distributed in Guangxi, Guangdong, Hainan and Taiwan. Also in S and SE Asia, Japan and Australia.

禾叶挖耳草

Utricularia graminifolia Vahl

多年生或一年生食虫草本。叶条形至狭倒卵形，具羽状脉，有时具3脉。苞片和小苞片基部着生；花冠淡紫色至紫堇色，下唇先端圆形，约与距等长。蒴果椭圆体形，稍背腹扁，沿腹侧纵裂。花期5-12月，果期6月至翌年1月。生海拔100-2100米的沼泽地和潮湿岩面。产云南、广西、广东和福建。南亚、缅甸和泰国亦有。

Perennial or annual insectivorous herbs. Leaves linear to narrowly obovate, veins pinnate, sometimes 3-nerved. Bracts and bracteoles basifixed; corolla mauve or violet, lower lip rounded at apex, about as long as spur. Capsule ellipsoid, slightly dorsiventrally compressed, dehiscing by a longitudinal ventral slit. Fl. May-Dec. Fr. Jun to next Jan. Swamps or wet rooks at 100-2100 m. Distributed in Yunnan, Guangxi, Guangdong and Fujian. Also in S Asia, Myanmar and Thailand.

合苞挖耳草

Utricularia peranomala P. Taylor

一年生食虫草本。叶狭倒卵形至条形，具羽状脉。苞片和小苞片基部着生，基部合生；花冠白色，下唇基部具黄色的凸起，下唇先端圆形至微凹，稍短于距。蒴果球形，腹侧纵裂。花期7-8月，果期8-9月。生海拔约2015米的潮湿岩面苔藓中。特产广西东北部。

Annual insectivorous herbs. Leaves narrowly obovate to linear, veins pinnate. Bracts and bracteoles basifixed, connate at the base; corolla white, with a yellow prominent swelling at base of lower lip, lower lip rounded to retuse at apex, slightly shorter than spur. Capsules globose, dehiscing by a longitudinal ventral slit. Fl. Jul-Aug. Fr. Aug-Sep. Wet rocks among moss at ca. 2015 m. Endemic to NE Guangxi.

合苞挖耳草 *Utricularia peranomala*

肾叶挖耳草 *Utricularia brachiate*

长距挖耳草 *Utricularia forrestii*

肾叶挖耳草

Utricularia brachiate Oliv.

多年生食虫草本，具小块茎。叶肾形，具二歧分枝的脉。苞片和小苞片盾状着生；花冠白色，下唇基部具1(-2)个黄斑，上唇具紫色条纹，下唇5浅裂，长于距。蒴果球形，腹侧纵裂。花期7-9月，果期8-10。生海拔2600-4200米林下岩石苔藓上。产云南西北部、四川西部和西藏东南部。印度北部、尼泊尔、不丹和缅甸北部亦有。

Perennial insectivorous herbs, with a small tuber. Leaves reniform, veins dichotomously branched. Bracts and bracteoles peltate-attached; corolla white, with 1(-2) yellow spots at base of lower lip and violet streaks on upper lip, lower lip 5-lobed, longer than spur. Capsules globose, dehiscing by a longitudinal ventral slit. Fl. Jul-Sep. Fr. Aug-Oct. Among moss on rocks under forests at 2600-4200 m. Distributed in NW Yunnan, W Sichuan and SE Xizang. Also in N India, Nepal, Bhutan and N Myanmar.

长距挖耳草

Utricularia forrestii P. Taylor

多年生食虫草本，具小块茎。叶宽倒卵形至肾形，具二歧分枝的脉。苞片和小苞片盾状着生；花冠淡紫色至紫堇色，下唇3中裂，每1裂片先端微凹至2浅裂，远短于距。蒴果球形，腹侧纵裂。花期7-8月，果期8-9月。生海拔2100-3000米的岩面或树干苔藓中。产云南西北部。缅甸北部亦有。

Perennial insectivorous herbs, with a small tuber. Leaves broadly obovate to reniform, veins dichotomously branched. Bracts and bracteoles peltate-attached; corolla mauve to violet, lower lip 3-lobed to middle, each lobe retuse to 2-lobed at apex, much shorter than spur. Capsules globose, dehiscing by a longitudinal ventral slit. Fl. Jul-Aug. Fr. Aug-Sep. Among moss on rocks or trunks at 2100-3000 m. Distributed in NW Yunnan. Also in N Myanmar.

毛籽挖耳草

Utricularia kumaonensis Oliv.

一年生食虫草本。叶宽倒卵形至肾形，具二歧分枝的脉。苞片和小苞片盾状着生；花冠白色，下唇裂片淡紫红色，而基部具1黄斑，5浅裂，远长于距。蒴果斜卵球形，腹侧纵裂。花期7-8月，果期8-9月。生海拔2600-2700米的岩石、崖壁或倒木的苔藓中。产云南西

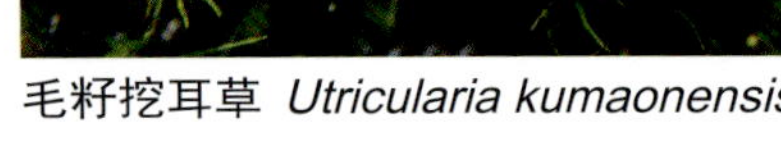

毛籽挖耳草 *Utricularia kumaonensis*

北部。印度北部、尼泊尔、不丹和缅甸北部亦有。

Annual insectivorous herbs. Leaves broadly obovate to reniform, veins dichotomously branched. Bracts and bracteoles peltate-attached; corolla white, with a basal yellow spot and pale mauve lobes on lower lip, lower lip 5-lobed, much longer than spur. Capsules obliquely ovoid, dehiscing by a longitudinal ventral slit. Fl. Jul-Aug. Fr. Aug-Sep. Among moss on rocks, cliffs or fallen trees at 2600-2700 m. Distributed in NW Yunnan. Also in N India, Nepal, Bhutan and N Myanmar.

多序挖耳草 *Utricularia multicaulis*

多序挖耳草
Utricularia multicaulis Oliv.

一年生食虫草本。叶倒卵形，具二歧分枝的脉。苞片和小苞片盾状着生；花冠白色或淡紫红色，下唇3浅裂，基部具1黄斑，明显长于距。蒴果球形至斜卵球形，腹侧纵裂。花期7-9月，果8-10月。生海拔2800-3900米的潮湿石上苔藓中或开阔的沼泽草地。产云南西北部和西藏东南部。印度北部、尼泊尔、不丹和缅甸北部亦有。

Annual insectivorous herbs. Leaves obovate, veins dichotomously branched. Bracts and bracteoles peltate-attached; corolla white or pale mauve, with a yellow spot at base of lower lip, lower lip 3-lobed, much longer than spur. Capsules globose to obliquely ovoid, dehiscing by a longitudinal ventral slit. Fl. Jul-Sep. Fr. Aug-Oct. Wet rocks among moss or open swampy meadows at 2800-3900 m. Distributed in NW Yunnan and SE Xizang. Also in N India, Nepal, Bhutan and N Myanmar.

圆叶挖耳草
Utricularia striatula Sm.

多年生食虫草本。叶圆形，倒卵形或横椭圆形，具二歧分枝的脉。苞片和小苞片盾状着生；花冠白色或紫色，下唇5浅裂而基部具1黄斑，与距约等长。蒴果球形，背腹扁，腹侧纵裂。花期6-10月，果期7-11月。生海拔350-3600米的潮湿岩石或树干上。产中国西南、华南、华中和华东。南亚、东南亚和热带非洲亦有。

Perennial insectivorous herbs. Leaves orbicular, obovate or transversely elliptic, veins dichotomously branched. Bracts and bracteoles peltate-attached; corolla white or violet, lower lip 5-lobed and with a yellow spot at base, about as long as spur. Capsules globose, dorsiventrally compressed, dehiscing by a longitudinal ventral slit. Fl. Jun-Oct. Fr Jul-Nov. Damp rockes and tree trunks at 350-3600 m. Distributed in SW, S, C and E China. Also in S and SE Asia, and tropical Africa.

盾鳞狸藻
Utricularia punctate Wall. ex A. DC.

多年生食虫草本。叶2-3深裂达基部，裂片椭圆形，二至四回二歧状深裂，末回裂片毛发状疏生细刚毛。苞片盾状，无小苞片；花冠淡紫色或白色，下唇全缘或微凹而基部具1黄斑，约与距等长。蒴果椭圆体形，腹背两侧纵裂。花期6-8月，果期7-9月。生低海拔的湖泊、沼泽或稻田。产广西(东兴)和福建(龙海)。东南亚亦有。

Perennial insectivorous herbs. Leaves 2-3 divided to base, segments ellptic, 2-4 times dichotomously divided, ultimate segments capillary, sparsely setulose. Bracts peltate-attached, bracteoles abscent; corolla violet or white, lower lip rounded or retuse at apex, with a yellow blotch at base, about as long as spur. Capsules ellipsoid, dehiscing by longitudinal dorsal and ventral slits. Fl. Jun-Aug. Fr. Jul-Sep. Lakes, swamps or rice fields near sea level. Distributed in Guangxi (Dongxing) and Fujian (Longhai). Also in SE Asia.

圆叶挖耳草
Utricularia striatula

盾鳞狸藻 *Utricularia punctate*

环翅狸藻 *Utricularia gibba*

弯距狸藻 *Utricularia vulgaris* subsp. *macrorhiza*

环翅狸藻

Utricularia gibba L.

一年生或多年生食虫草本。叶不分裂或一至二回二歧状深裂，末回裂片毛发状，无毛或具小刚毛。苞片基部着生，无小苞片；花冠黄色，下唇先端圆形，长于或短于距。蒴果球形，2瓣裂。花期3-11月，果期5-12月。生海拔50-900米的湖泊、沼泽或稻田中。产华东、华中、华南至云南。亚洲、非洲和美洲的热带和亚热带亦有。

Annual or perennial insectivorous herbs. Leaves unlobed or 1-2 times dichotomously divided, ultimate segments capillary, glabrous or sparsely setulose. Bract basifixed, bracteoles absent; corolla yellow, lower lip rounded at apex, longer or shorter than spur. Capsule globose, 2-valvate. Fl. Apr-Nov. Fr. May-Dec. Lakes, swamps or rice fields at 50-900 m. Distributed in E, C and S China to Yunnan. Also in tropical and subtropical Asia, Africa and America.

弯距狸藻

Utricularia vulgaris L. subsp. **macrorhiza** (Leconte) Clausen

多年生食虫草本。叶2裂达基部，裂片羽状深裂，后二至四回二歧状深裂，末回裂片毛发状，具小刚毛。苞片基部着生，无小苞片；花冠黄色，下唇约与距等长，距向上弯曲，背腹面均有内生腺体。蒴果球形，周裂。花期6-8月，果期7-9月。生海拔30-3700米的湖泊、池塘及河流。产中国东北、华北、西北和四川。俄罗斯、蒙古和北美洲亦有。

Perennial insectivorous herbs. Leaves 2-divided to base, segments pinnately divided, secondary segments 2-4 tiems dichotomously divided, ultimate segments capillary, minutely setulose. Bracts basifixed, bracteoles absent; corolla yellow, lower lip about as long as spur, spur markedly curved up ward, with internal glands on both dorsal and ventral surfaces. Capsule globose, circumscissile. Fl. Jun-Aug. Fr. Jul-Sep. Lakes, pools and rivers at 30-3700 m. Distributed in NE, N and NW China, and Sichuan. Also in Russia, Mongolia and North America.

南方狸藻

Utricularia australis R. Br.

多年生食虫草本。叶2深裂达基部，裂片羽状深裂，后一至三回二歧状深裂，末回裂片毛发状，边缘疏生小牙齿和小刚毛。苞片基部着生，无小苞片；花冠黄色，下唇先端圆形或微凹，长于距。蒴果球形，周裂。花期6-11月，果期7-12月。生海拔50-2500米的湖泊、池塘、河流或稻田。产华东、华中、华南和西南。亚洲、非洲、欧洲和大洋洲亦有。

Perennial insectivorous herbs. Leaves 2-divided to base, segments pinnately divided, secondary segments 1-3 times dichotomously divided, ultimate segments capillary, margin sparsely denticulate and setulose. Bracts basifixed, bracteoles absent; corolla yellow, lower lip rounded to retuse at apex, longer than spur. Capsules globose, circumscissile. Fl. Jun-Nov. Fr. Jul-Dec. Lakes, pools, river or rice fields at 50-2500 m. Distributed in E, C, S and SW China. Also in Asia,

南方狸藻 *Utricularia australis*

异枝狸藻
Utricularia intermedia

Africa, Europe and Oceania.

异枝狸藻

Utricularia intermedia Hayne

多年生食虫草本。叶2-3深裂达基部，裂片一至三回二歧状深裂，末回裂片狭条形，边缘具疏小牙齿和小刚毛。苞片基部着生，无小苞片；花冠黄色，下唇先端圆形，稍长于距。蒴果球形，周裂。花期6-9月，果期7-10月。生海拔300-4000米的沼泽、湖泊或池塘。产中国东北、四川西部、西藏南部和东南部。亚洲温带、欧洲和北美洲亦有。

Perennial insectivorous herbs. Leaves 2-3 divided to base, segments 1-3 times dichotomously divided, ultimate segments narrowly linear, margin sparsely denticulate and setulose. Bracts basifixed, bracteoles absent; corolla yellow, lower lip rounded at apex, slightly longer than spur. Capsules globose, circumscissile. Fl. Jun-Sep. Fr. Jul-Oct. Bogs, lakes or ponds at 300-4000 m. Distributed in NE China, W Sichuan, and S and SE Xizang. Also in temperate Asia, Europe and North America.

细叶狸藻

Utricularia minor L.

多年生食虫草本。叶2-3深裂达基部，裂片一至四回二歧状深裂，末回裂片狭条形，边缘无毛或具小刚毛。苞片基部着生，无小苞片；花冠柠檬黄，下唇先端圆形至微凹，长约为距的2倍。蒴果球形，周裂。花期8-9月，果期9-10月。生海拔300-3700米的池塘和沼泽。产中国东北与西南和新疆北部。亚洲温带、欧洲和北美洲亦有。

Perennial insectivorous herbs. Leaves 2-3 divided to base, segments 1-4 times dichotomously divided, ultimate segments narrowly linear, and glabrous, or sparsely setulose. Bracts basifixed, bracteoles absent; corolla lemon-yellow, lower lip rounded to retuse at apex, about 2 × as long as spur. Capsules globose, circumscissile. Fl. Aug-Sep. Fr. Sep-Oct. Ponds or marshes at 300-3700 m. Distributed in NE and SW China, and N Xinjiang. Also in temperate Asia, Europe and North America.

黄花狸藻

Utricularia aurea Lour.

多年生或一年生食虫草本。叶3-4(-5)掌状深裂达基部，裂片羽状深裂，后一至四回二歧式深裂，末回裂片毛发状，具小刚毛。苞片基部着生，无小苞片；花冠淡黄色，下唇先端圆形或微凹，约与距等长，蒴果球形，周裂。花期6-11月，果期7-12月。生海拔50-2700米的湖泊、沼泽、河流或稻田。产山东和长江以南。南亚、东亚至澳大利亚亦有。

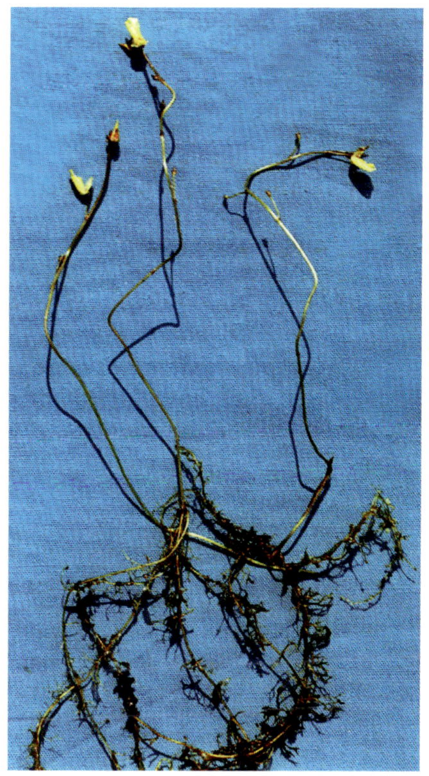

细叶狸藻 *Utricularia minor*

Perennial or annual insectivorous herbs. Leaves 3-4(-5) palmately divided to base, segments pinnately divided, secondary segments 1-4 times dichotomously divided, ultimate segments capillary, setulose. Bracts basifixed, bracteoles absent; corolla pale yellow, lower lip rounded to emarginate at apex, about as long as spur. Capsules globose, circumscissile. Fl. Jun-Nov. Fr. Jul-Dec. Lakes, swamps, rivers or rice fields at 50-2700 m. Distributed in Shandong and the south of Yangtze River. Also in S and E Asia to Australia.

黄花狸藻 *Utricularia aurea*

爵床科
Acanthaceae

红花山牵牛 *Thunbergia coccinea*

直立山牵牛
Thungbergia erecta (Benth.) T. Anders.

直立灌木。枝叶初疏被柔毛，后变无毛。叶柄长2-5毫米；叶卵形至披针形，边缘具波状齿或不规则齿。花单生叶腋；花冠筒白色，冠檐紫堇色。栽培于台湾、广东、广西、海南和云南。原产热带非洲，世界各地栽培。

Erect shrubs. Branches and leaves sparsely pilose, then glabrescent. Petioles 2-5 mm long; leaves ovate to laceolate, margin crenate to irregularly undulate. Flowers axillary; corolla tube white, limbs violet. Cultivated in Taiwan, Guangdong, Guangxi, Hainan and Yunnan. Native to tropical Africa, cultiveated worldwide.

红花山牵牛
Thunbergia coccinea Wall.

藤本。茎具9棱，节间具柔毛。叶卵形至披针形，掌状3-5裂。总状花序腋生或顶生，下垂；花萼退化为1小脊；花冠红色；花丝基部有1毛状体；花药室平行，不等大，基部具矩。花期9月至翌年1月，果期翌年1-5月。生海拔800-1000米的山地林中。产云南和西藏东南部。缅甸、老挝和泰国亦有。

Lianas. Stems 9-angled, pubescent at nodes. Leaves ovate to lanceolate, palmately 3-5-veined; leaves broadly ovate, ovate, or lanceolate, both surfaces pubescent, palmately 3-5-veined, margin undulate or remotely toothed. Racemes axillary or terminal, pendulous; calyx reduced to a minute rim; corolla red; filaments with a tuft of trichomes at base; anther thecae parallel, unequal, spurred at base. Fl. Sep to next Jan. Fr. next Jan-May. Forested montane slopes at 800-1000 m. Distributed in Yunnan and SE Xizang. Also in Myanmar, Laos and Thailand.

山牵牛
Thunbergia grandiflora Roxb.

藤本。茎4棱，具沟槽。叶卵形至三角状卵形，掌状3-7脉。花单生，成对生于叶腋，或每节上2-4花排列成顶生总状花序；花冠淡蓝色，喉部淡黄色；花药室具柔毛，基部具附属物。花期8月至翌年1月，果期11月至翌年3月。生海拔400-1500米的灌丛。产云南、华南和福建。印度、缅甸、泰国和越南亦有。世界热带区域归化。

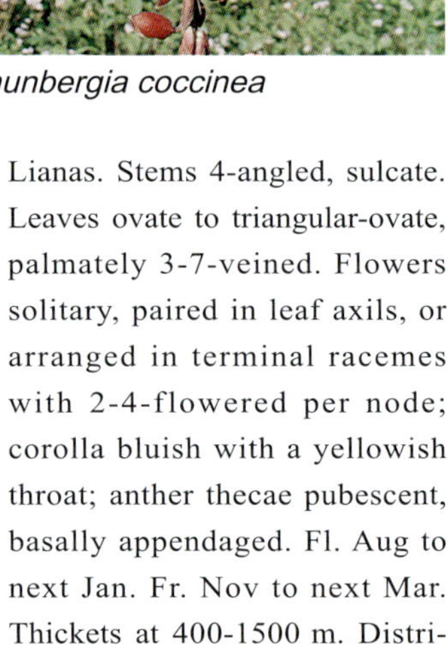

Lianas. Stems 4-angled, sulcate. Leaves ovate to triangular-ovate, palmately 3-7-veined. Flowers solitary, paired in leaf axils, or arranged in terminal racemes with 2-4-flowered per node; corolla bluish with a yellowish throat; anther thecae pubescent, basally appendaged. Fl. Aug to next Jan. Fr. Nov to next Mar. Thickets at 400-1500 m. Distributed in Yunnan, S China and

直立山牵牛 *Thungbergia erecta*

山牵牛 *Thunbergia grandiflora*

羽脉山牵牛 *Thunbergia lutea*

蛇根叶 *Ophiorrhiziphyllon macrobotryum*

Fujian. Also in India, Myanmar, Thailand and Vietnam. Naturalized in tropical regions worldwide.

羽脉山牵牛
Thunbergia lutea T. Anders.

攀援藤本。茎除节间具一圈绒毛外无毛。叶卵形至披针状卵形，下部无毛，上部具柔毛或糙毛。花单生叶腋；花萼小，具10齿；花冠粉红色或白色；花药室具髯毛，基部具矩。蒴果无毛。花期8-12月，果期翌年3-5月。生海拔1000-2500米的林下或灌丛。产云南和西藏东南部。印度、不丹和缅甸亦有。

Climbing vines. Stems glabrous except for a circle of trichomes on nodes. Leaves ovate to lanceolate-ovate, abaxially subglabrous, adaxially pubescent or setose. Flowers axillary, solitary; calyx minute, 10-toothed; corolla pinkish or white; anther thecae barbate, spurred at base. Capsules glabrous. Fl. Aug-Dec. Fr. Mar-May next year. Forests or thickets at 1000-2500 m. Distributed in Yunnan and SE Xizang. Also in India, Bhutan and Myanmar.

碗花草
Thunbergia fragrans Roxb.

草质藤本。茎几具4棱至扁平。叶长圆状卵形至卵形至阔卵形至长圆米状披针形至披针形，两面具柔毛或近无毛，掌状3-5裂。单花腋生；花萼具10-17个不等大小的齿；花冠白色；冠筒阔圆柱形。花期8月至翌年1月，果期11月至翌年3月。生海拔800-2300米的灌丛或路边。产云南、四川、贵州、华南和台湾。南亚和东南亚亦有。

Herbaceous lianas. Stems almost 4-angled to flattened. Leaves oblong-ovate to ovate to broadly ovate to oblong-lanceolate to lanceolate, both surfaces puberulent to subglabrous palmately 3-5-veined. Flowers axillary, solitary; calyx unequally 10-17-dentate; corolla white; tubes basally cylindric. Fl. Aug to next Jan. Fr. Nov to next Mar. Thickets or roadsides at 800-2300 m. Distributed in Yunnan, Sichuan, Guizhou, S China and Taiwan. Also in S and SE Asia.

碗花草 *Thunbergia fragrans*

蛇根叶
Ophiorrhiziphyllon macrobotryum Kurz

草本。叶长圆形至长披针形，基部楔形，有时下延至叶柄，边缘全缘，背面灰白色，沿脉被柔毛。总状花序顶生，被腺毛；花冠黄白色，二唇形，上唇2裂，下唇3裂；能育雄蕊2，伸出花冠外，不育雄蕊2，内藏。蒴果具种子多数；无珠柄钩。花期10月至翌年2月，果期翌年3-5月。生海拔100-1300米的密林、溪边潮湿处。产云南南部。东南亚亦有。

Herbs. Leaves oblong to lanceolate, base cuneate, sometimes decurrent onto the petioles, margin entire, beneath glaucous and pilose along the veins. Raceme terminal, with glandular trichomes; corolla yellowish white, bilabiate, upper lip 2-lobed, lower lip 3-lobed; stamens 2, exserted, staminodes 2, included. Capsules many-seeded, retinacula absent. Fl. Oct to next Feb. Fr. Mar-May next year. Wet places in forest, streamside at 100-1300 m. Distributed in S Yunnan. Also in SE Asia.

瘤子草 *Nelsonia canescens*

瘤子草

Nelsonia canescens (Lam.) Spreng.

一年生草本。叶椭圆形至卵形，两面具绒毛。穗状花序；花冠淡蓝紫色或白色，外部无毛；花冠管圆柱形，近中部收缩，再膨大至喉部。蒴果具8-16粒种子。种子宽椭圆体形，具瘤。花期10月至翌年3月，果期翌年3-5月。生海拔400-2000米的潮湿处和开阔林中。产云南和广西。南亚、东南亚、非洲和马达加斯加。

Annual herbs. Leaves elliptic to ovate, both surfaces villous. Spikes; corolla bluish purple or white, externally glabrous; tubes cylindric, contracted near midpoint then expanded into throat. Capsules 8-16-seeded. Seeds broadly ellipsoid, granulate. Fl. Oct to next Mar. Fr next. Mar-May. Wet places or open forests at 400-2000 m. Distributed in Yunnan and Guangxi. Also in S and SE Asia, Africa and Madagascar.

楠草 *Ruellia repens*

刺苞老鼠簕

Acanthus leucostachyus Wall. ex Nees

直立草本。叶具柄；叶缘稍具刺锯齿，但有时具不明显锯齿，先端锐尖。穗状花序顶生；苞片边缘顶端具刺；小苞片披针形至倒披针形；花冠白色。种子肾形。花期2-5月，果期8-9月。生海拔600-1200米的密林或潮湿处。产云南。南亚和东南亚。

Erect herbs. Leaves petiolate; leaf margins slightly spinose-dentate but sometimes inconspicuously dentate, apex acute. Spikes terminal; bracts margins apically spiny; bracteoles lanceolate to oblanceolate; corolla white. Seeds reniform. Fl. Feb-May. Fr. Aug-Sep. Dense forests or wet places at 600-1200 m. Distributed in Yunnan. Also in S and SE Asia.

刺苞老鼠簕 *Acanthus leucostachyus*

楠草

Ruellia repens L.

多年生披散草本。叶卵形至披针形，边缘全缘；叶柄短。花单生或数朵簇生于叶腋；苞片叶状，长于花萼；花冠白色或粉红色，近5等裂；雄蕊4。蒴果基部具柄，具12-16粒种子。种子近球形，彼此重叠，具增厚的边缘。花期5-8月，果期

地皮消 *Pararuellia delavayana*

6-8月。生海拔100-900米的路边或旷野草地。产广东、广西、海南、云南和台湾。亚洲热带地区亦有。

Perennial creeping herbs. Leaves ovate to lanceolate, margin entire; petioles short. Flowers axillary, solitary or clustered; bracts leaf-like, longer than calyx; corolla white or pinkish-red, subequally 5-lobed; stamens 4. Capsules stipitate, 12-16-seeded. Seeds discoid, with thickened margin. Fl. May-Aug. Fr. Jun-Aug. Roadsides or grasslands at 100-900 m. Distributed in Guangdong, Guangxi, Hainan, Yunnan and Taiwan. Also in tropical regions of Asia.

地皮消

Pararuellia delavayana
(Baill.) E. Hossain

草本。茎极短。穗状花序长约4厘米，具1-3个茎节；苞片椭圆形至卵形；小苞片条形；花冠白色、淡蓝色或粉红色。蒴果具16粒种子。种子密具平伏毛。花期7-9月，果期11月至翌年2月。生海拔700-3000米的林下或山坡。产云南、四川南部和贵州。

Herbs. Stems very short. Spikes ca. 4 cm long, with 1-3 nodes; bracts elliptic to ovate; bracteoles linear; corolla white, light blue or pink. Capsules 16-seeded. Seeds covered with dense appressed trichomes. Fl. Jul-Sep. Fr. Nov to next Feb. Forests or slopes at 700-3000 m. Distributed in Yunnan, S Sichuan and Guizhou.

飞来蓝

Leptosiphonium venustum
(Hance) E. Hossain

草本。叶披针形至倒披针形，基部楔形，下延，边缘浅波状。花单生上部叶腋或数朵集生于枝顶；花冠白色，基部具细长的圆柱形花冠管，近5等裂；雄蕊4。蒴果圆柱形，具种子8-12粒；具珠柄钩。花期8-9月，果期10-11月。生海拔100-800米的林下或溪边。产安徽、福建、广东、广西、湖南、湖北和江西。

Herbs. Leaves laceolate to oblanceolate, base cuneate and decurrent onto the petioles, margin slightly undulate. Flowers borne in leaf axial, solitary or several in dischasial clusters; corolla white, with a long narrow tube at base; subequally 5-lobe; stamens 4. Capsule cylindrical, 8-12-seeded, retinacula present. Fl. Aug-Sep. Fr. Oct-Nov. Forests or streamsides at 100-800 m. Distributed in Anhui, Fujian, Guangdong, Guangxi, Hunan, Hubei and Jiangxi.

飞来蓝 *Leptosiphonium venustum*

喜花草 *Eranthemum pulchellum*

毛冠可爱花
Eranthemum austrosinense var. *pubipetalum*

Guizhou and Yunnan.

假杜鹃

Barleria cristata L.

亚灌木。叶椭圆形至长圆形至卵形。花序腋生，具短且密的聚伞花序，具短梗；花冠蓝紫色；冠筒基部狭圆柱形，逐渐变宽；雄蕊4，二强。蒴果无毛，具4粒种子。花期5月和8-12月，果期5月和10月。生海拔100-2600米及以下的路边、山坡、溪边或旱生植被中。产中国西南、华南和华东。南亚和东南亚亦有。

Subshrubs. Leaves elliptic to oblong to ovate. Inflorescences axillary short and dense cymes, shortly pedunculate; corolla purplish blue; tubes basally narrowly cylindric then gradually widened; stamens 4, didynamous. Capsules glabrous, 4-seeded Fl. May and Aug-Dec. Fr. May and Oct. Roadsides, slopes, along streams or in xeric vegetation below 100-2600 m. Distributed in SW, S and E China. Also in S and SE Asia.

黄花假杜鹃

Barleria prionitis L.

灌木，下部叶腋具刺，具分

喜花草

Eranthemum pulchellum Andrews

灌木，高达2米。叶对生，卵形，有时椭圆形，两面无毛。穗状花序顶生及腋生；苞片大，叶状，白绿色；花萼白色；花冠蓝色或白色，高脚碟形，冠筒长3厘米；雄蕊2。栽培于中国西南和华南。印度和热带喜马拉雅亦有。

Shrubs, to 2 m tall. Leaves opposite, ovate, sometimes elliptic, both sides glabrous. Spikes terminal and axillary; bracts large, leaflike, white-green; calyx white; corolla blue or white, salverform, tubes 3 cm long; stamens 2. Cultivated in SW and S China. Also in India and tropical Himalaya.

毛冠可爱花

Eranthemum austrosinense var. **pubipetalum** (S. Z. Huang ex H. P. Tsui) T. L. Li et Y. F. Deng

多年生草本。叶椭圆形，稀为卵形，边缘全缘；穗状花序顶生或腋生；苞片卵形或椭圆形，每苞片内通常有花1朵；花冠蓝紫色，被短柔毛。子房无毛。蒴果无毛，内有种子4粒。种子密被贴伏长毛。生海拔100-700米的河边灌木丛中或山谷林中阴处。产广西、贵州和云南。

Perennial herbs. Leaves elliptic, rarely ovate, margin entire. Inflorescences axillary or terminal, spikes. Corolla purplish blue, pubescent. Ovary glabrous. Seeds densely appressed tomentose. Forests, thickets in ravines at 100-700 m. Distributed in Guangxi,

假杜鹃 *Barleria cristata*

黄花假杜鹃 *Barleria prionitis*

枝。叶椭圆形至卵形。花簇生于上部叶和/或苞片中；小苞片变为刺；花冠黄色或橙色；冠筒基部较狭，短于喉部。蒴果卵球形。花期10-12月，果期12月至翌年2月。生海拔约600米的路边、灌丛或常绿阔叶林下干燥处。产云南。南亚、东南亚、非洲和马达加斯加亦有。

Shrubs, with spines in lower leaf axils, branched. Leaves elliptic to ovate. Flowers clustered in axils of upper leaves and/or bracts; bractlets spinescent; corolla yellow or orange; tubes with narrow basal portion slightly shorter than throat. Capsules ovoid. Fl. Oct-Dec. Fr. Dec to next Feb. Roadsides, thickets or dry places in evergreen broad-leaved forests at ca. 600 m. Distributed in Yunnan. Also in S and SE Asia, Africa and Madagascar.

水蓑衣

Hygrophila ringens (L.) R. Br. ex Spreng.

多年生草本。叶狭披针形至倒披针形，两面多数钟乳体和无毛或稍具柔毛，中脉两侧具8-11侧脉。花(单生或)2-10朵簇生于叶腋；花冠淡紫色；二强雄蕊。蒴果狭长圆形。花期8-10月，果期12月至翌年2月。生海拔1000米以下的溪边或湿地。产长江以南和河南。南亚、东南亚和日本亦有。

Perennial herbs. Leaves narrowly lanceolate to oblanceolate, both surfaces with numerous cystoliths and glabrous or slightly pubescent, secondary veins 8-11 on each side of midvein. Flowers (solitary or) 2-10 clustered in leaf axils; corolla light purple; stamens didynamous. Capsules narrowly oblong. Fl. Aug-Oct. Fr. Dec to next Feb. Along streams or wet places below 1000 m. Distributed in the south of Yangtze River and Henan. Also in S and SE Asia, and Japan.

肾苞草

Phaulopsis dorsiflora (Retz.) Santapau

草本。叶不等大，卵形至披针形，基部楔形至下延，边缘全缘。穗状花序顶生；苞片圆形或肾形；花冠白色，二唇形，上唇2裂，下唇3裂；雄蕊4。蒴果基部具短柄，具4粒种子，具珠柄钩。花期11月至翌年5月，果期翌年2-6月。生海拔300-800米的路旁、灌丛。产云南。南亚和东南亚亦有。

Herbs. Leaves unequal, ovate to lanceolate, base cuneate to attenuate, margin entire. Spikes terminal; bracts large, orbicular or reniform; corolla white, bilabiate, upper lip 2-lobed, lower lip 3-lobed; stamens 4. Capsule stipitate, 4-seeded; retinacula present. Fl. Nov to next May. Fr. next Feb-Jun. Roadside, thickets at 300-800 m. Distributed in Yunnan. Also in S and SE Asia.

水蓑衣 *Hygrophila ringens*

肾苞草 *Phaulopsis dorsiflora*

长柄恋岩花 *Echinacanthus longipes*

黄花恋岩花 *Echinacanthus lofouensis*

长柄恋岩花

Echinacanthus longipes H.S. Lo et D. Fang

草本。叶基生，长圆形至披针形，基部常偏斜，圆形或楔形，边缘全缘或不明显波状，背面被毛。具伞花序腋生，具2-4花；花冠蓝紫色，近5等裂；雄蕊4，花药基部具芒。蒴果圆柱形，具12-16粒种子；具珠柄钩。花期8-10月，果期9-11月。生海拔500-1200米的石灰岩山地林下。产广西和云南。越南亦有。

Herbs. Leaves basal, oblong to lanceolate, base usually oblique, rounded or cuneate, margin entire or inconspicuously undulate, below pubescent. Cymes axillary, 2-4-flowered; corolla purplish blue, suubequally 5-lobed; stamens 4, anther spurred at base. Capsule cylindrical, 12-16-seeded; retinacula present. Fl. Aug-Oct. Fr. Sep-Nov. Forests on the limestone hills at 500-1200 m. Distributed in Guangxi and Yunnan. Also in Vietnam.

黄花恋岩花

Echinacanthus lofouensis (Lev.) J. R. I. Wood

灌木。叶卵形至狭披针形，基部阔楔形或圆形，边缘全缘，先端渐尖。具伞花序腋生，通常具3花；花冠黄色，被柔毛，5等裂；雄蕊4，花药基部具芒。蒴果棒形，密被柔毛，具8-12粒种子。花期3-5月，果期5-6月。生海拔500-1000米的石灰岩林下或灌丛。特产广西和贵州。

Shrubs. Leaves ovate to narrowly lanceolate, base broadly cuneate or rouanded, margin entire, apex acuminate. Cymes axillary, usually 3-flowered; corolla yellow, pubescent outside, 5-lobed; stamens 4, anther spurred at the base. Capsule clavate, densely pubescent, 8-12-seeded. Fl. Mar-May. Fr. May-Jun. Forests or thickets on limestone hills at 500-1000 m. Endemic to Guangxi and Guizhou.

尖药花

Strobilanthes tomentosa (Nees) J. R. I. Wood

草本或半灌木。幼枝和叶密被白色绒毛。叶片卵形，基部不对称，平截，边缘具不规则圆齿。圆锥花顶生；萼片5等裂至基部；花冠蓝色，5裂；雄蕊4。蒴果长圆形，被柔毛，具6-8粒种子；具珠柄钩。花期8-11月，果期9-12月。生海拔500-2300米的灌丛、路旁或草地。产广西、贵州和云南。南亚和东南亚亦有。

Herbs or subshrubs. Branches and leaves densely covered with white trichomes. Leaves ovate, base oblique, truncate, margin irregularly crenate. Panicles terminal; calyx 5-lobed to the base; corolla blue, 5-lobed; stamens 4. Capsule oblong, pubescent, 6-8-seeded; retinacula present. Fl. Aug-Nov. Fr. Sep-Dec. Thickets, roadside or grassland at 500-2300 m. Distributed in Guangxi, Guizhou and Yunnan. Also in S and SE Asia.

尖药花 *Strobilanthes tomentosa*

日本马蓝 *Strobilanthes japonica*

日本马蓝
Strobilanthes japonica (Thunb.) Miq.

草本。叶披针形，基部楔形，边缘具齿。穗状花序顶生；花冠白色或淡蓝紫色，烟斗状，5裂；雄蕊4。蒴果具4粒种子。花期8-9月，果期10-11月。生海拔500-1100米的溪边、林下阴湿处。产重庆、四川、湖北、湖南和贵州。日本亦有。

Herbs. Leaves lanceolate, base cuneate, margin crenuate. Spikes terminal; corolla white or purplish-blue, funnel-shape, 5-lobed; stamens 4. Capsule 4-seeded. Fl. Aug-Sep. Fr. Oct-Nov. Wet and shade place in streamside or forests at 500-1100 m. Distributed in Chongqing, Sichuan, Hubei, Hunan and Guizhou. Also in Japan.

山一笼鸡
Strobilanthes aprica (Hance) T. Anderson

亚灌木或多年生草本。叶椭圆形至长圆状椭圆形，基部阔楔形，边缘全缘。穗状花序腋生或顶生，密集近头状；花冠淡紫色，基部白色，肿胀，5裂；雄蕊2，外露，花药椭圆形。蒴果具4粒种子。花期8月至翌年1月，果期10月至翌年2月。生海拔2200米以下的灌丛或林下。产广东、广西、江西、贵州、四川和云南。东南亚亦有。

Subshrubs or perennial herbs. Leaves elliptic to oblong-elliptic, base broadly cuneate, margin entire. Spikes axillary or terminal, densely subcapitate; corolla lilac with white base, inflate, 5-lobed; stamens 2, exserted, anther ellipsoid. Capsule 4-seeded. Fl. Aug to next Jan. Fr. Sep to next Feb. Thickets or forests below 2200 m. Distributed in Guangdong, Guangxi, Jiangxi, Guizhou, Sichuan and Yunnan. Also in SE Asia.

曲枝马蓝
Strobilanthes dalzielii (W. W. Sm.) R. Bennoist

草本。茎常之字形曲折。叶卵形至卵状披针形，基部圆形或至楔形，但最顶端一片常心形，边缘具锯齿。穗状花序腋生或顶生，常之字形曲折；花冠蓝紫色，稍弯曲，外面被白色柔毛，5裂；雄蕊4。蒴果具4粒种子。花期10月至翌年1月，果期12月至翌年2月。生海拔400-1200米的溪边林下。产台湾、江西、广东、广西、海南、湖南、贵州和云南。越南、老挝和泰国亦有。

Herbs. Stems usually zigzag. Leaves ovate to ovate-lancaolate, base rounded to cuneate but usually cordate for upmost one, margin serrate. Spikes terminal or axillary, usually zigzag; corolla purplish-blue, slightly curved, white pubescent outside, 5-lobed; stamens 4. Capsule 4-seeded. Fl. Oct to next Jan. Fr. Dec to next Feb. Streamside or forests at 400-1200 m. Distributed in Taiwan, Jiangxi, Guangdong, Guangxi, Hainan, Hunan, Guizhou and Yunnan. Also in Vietnam, Laos and Tailand.

山一笼鸡 *Strobilanthes aprica*

曲枝马蓝 *Strobilanthes dalzielii*

延苞马蓝 *Strobilanthes pteroclada*

延苞马蓝

Strobilanthes pteroclada R. Ben.

亚灌木或多年生草本，茎4棱。叶披针形。花序疏松，聚伞花序，生枝顶或腋生；花互生，藏于两大苞片内；苞片对生；小苞片狭披针形；花冠蓝紫色；子房无毛。蒴果伸长，具4粒种子。种子密被白色贴伏绒毛。花期4-5月，果期6-7月。生海拔300-900米的山谷灌丛或林下。产广西和贵州。越南亦有。

Subshrubs or perennial herbs. Stems 4-angled. Leaf blade lanceolate. Inflorescences axillary or terminal, spikes; bracts lanceolate to narrowly ovate, base gradually narrowed and decurrent onto peduncle; bracteoles narrowly lanceolate; calyx weakly 2-lipped; corolla purplish blue; ovary glabrous. Capsule fusiform, 4-seeded. Seeds white, densely appressed tomentose. Fl. Apr-May. Fr. Jun-Jul. Thickets or forests by streams at 300-900 m. Distributed in Guangxi and Guizhou. Also in Vietnam.

南一笼鸡

Strobilanthes henryi Hemsl.

亚灌木。叶卵形之近圆形，基部阔楔形，稍不对称，常下延至叶柄，边缘具锯齿。穗状花序顶生；花冠淡蓝紫色，烟斗状，5裂；雄蕊2，外露。蒴果具4粒种子。花期9月至翌年2月，果期11月至翌年3月。生海拔1000-2800米的灌丛、疏林下或石上。特产中国，产贵州、湖北、湖南、四川和云南。

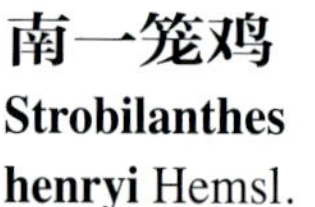

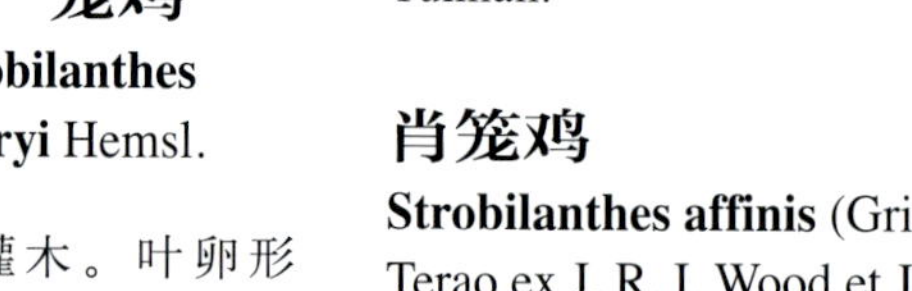

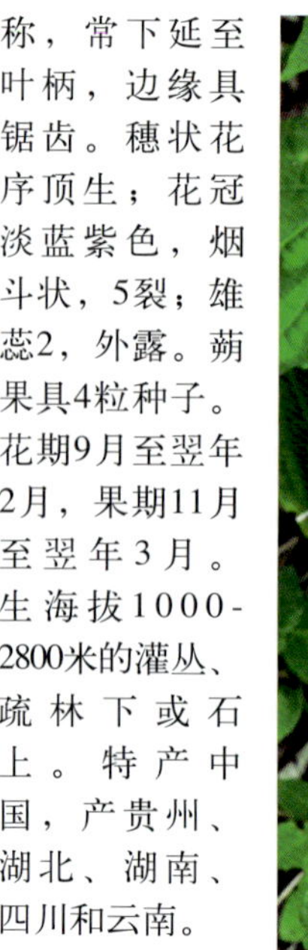

Subshrubs. Leaves ovate to suborbicular, base broadly cuneate, slightly oblique, usually decurrent onto the petioles, margin serrate. Spikes terminal; corolla purplish-blue, funnel-shaped, 5-lobed; stamens 2, exserted. Capsule 4-seeded. Fl. Sep to next Feb. Fr. Nov to next Mar. Thickets, lax forests or rock at 1000-2800 m. Endemic to China, distributed in Guizhou, Hubei, Hunan, Sichuan and Yunnan.

南一笼鸡 *Strobilanthes henryi*

肖笼鸡

Strobilanthes affinis (Griff.) Terao ex J. R. I. Wood et J. R. Bennett

草本。叶卵形至卵状椭圆形，基部圆至楔形，下延至叶柄，边缘具钝齿。穗状花序顶生或腋生；苞片覆瓦状排列；花冠蓝紫色，弯曲，5裂；雄蕊4，长雄蕊外漏，花药长圆形。蒴果具4粒种子。花期9-11月，果期11-12月。生海拔600-1300米的山坡、灌丛。产广西、贵州、湖南和云南。越南、缅甸和印度亦有。

Herbs. Leaves ovate to ovate-elliptic, base rounded to cuneate and decurrent onto the petioles, margin crenuate. Spikes terminal or axillary; bract imbricate; corolla purplish-blue, curved,

肖笼鸡 *Strobilanthes affinis*

板蓝 *Strobilanthes cusia*

红背耳叶马蓝 *Strobilanthes auriculata* var. *dyeriana*

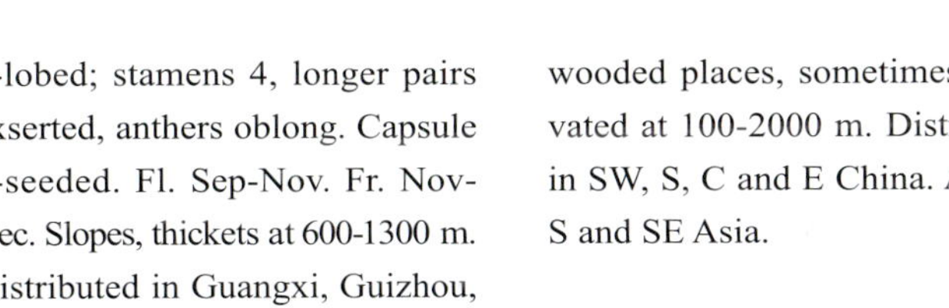

5-lobed; stamens 4, longer pairs exserted, anthers oblong. Capsule 4-seeded. Fl. Sep-Nov. Fr. Nov-Dec. Slopes, thickets at 600-1300 m. Distributed in Guangxi, Guizhou, Hunan and Yunnan. Also in Vietnam, Myanmar and India.

板蓝

Strobilanthes cusia (Nees) Kuntze

直立草本，具分枝，干后黑色。叶椭圆形至卵形。花序顶生或腋生，具苞穗状花序，常聚伞形成1个叶状分枝的圆锥状；花冠蓝色，直伸至稍弯；雄蕊4，内藏。蒴果具4粒种子。种子具平伏毛。花期7月至翌年2月，果期12月至翌年2月。常生海拔100-2000米的林下潮湿处，有时栽培。产中国西南、华南、华中和华东。南亚和东南亚亦有。

Erect herbs, branched, drying blackish. Leaves elliptic to ovate. Inflorescences terminal or axillary, bracteate spikes, often aggregated to form a leafy branched panicle; corolla blue, straight to slightly bent; stamens 4, included. Capsules 4-seeded. Seeds covered with appressed trichomes. Fl. Jul to next Feb. Fr. Dec to next Feb. Usually in moist wooded places, sometimes cultivated at 100-2000 m. Distributed in SW, S, C and E China. Also in S and SE Asia.

红背耳叶马蓝

Strobilanthes auriculata Nees var. **dyeriana** (Mast.) J. R. I. Wood

亚灌木，多分枝。茎4棱，有时之字形。叶下面浅红色，上面绿色，基部渐狭、楔形、圆形或耳状。花序腋生或顶生，穗状；花冠浅紫色至紫色，漏斗形。蒴果狭长圆状倒卵球形。栽培于云南和广东。原产缅甸和越南。

Subshrubs, much branched. Stems 4-angled, sometimes zigzag. Leaves abaxially reddish, adaxially green, base attenuate, cuneate, rounded or auriculate. Inflorescences axillary or terminal, spikes; corolla pale purple to violet, funnel-shaped. Capsules narrowly oblong-obovoid. Cultivated in Yunnan and Guangdong. Native to Myanmar and Vietnam.

翅柄马蓝

Strobilanthes atropurpurea Nees

多年生草本，多分枝。叶卵形至卵状椭圆形，基部具柄；叶柄具翅。花序腋生，穗状，之字形，生一侧，常退化至花单生或于主茎叶腋上成对生；花冠浅紫色、蓝紫色。蒴果纺锤形。花期6-9月。生海拔700-2900米的山坡潮湿处或林下。产中国西南、华南、华中和华东。南亚和东南亚亦有。

Perennial herbs, much branched. Leaves ovate to ovate-elliptic, basally petiolate; petioles winged. Inflorescences axillary, spikes, zigzag, secund, few-flowered, often much reduced to flowers solitary or paired in leaf axil on main stem; corolla pale purple, bluish purple. Capsules fusiform. Fl. Jun-Sep. Moist places on mountain slopes or forests at 700-2900 m. Distributed in SW, S, C and E China. Also in S and SE Asia.

翅柄马蓝 *Strobilanthes atropurpurea*

腺毛马蓝 *Strobilanthes forrestii*

腺毛马蓝

Strobilanthes forrestii Diels

草本。茎4棱，具稀疏柔毛及腺毛，渐无毛。叶近无柄；叶卵形，两面具薄柔毛，上面具钟乳体。花序顶生，穗状，具柔毛及腺毛；花冠蓝色至紫色。蒴果长圆形，具4粒种子。花期7-8月。生海拔约3000米的林下或草坡。产云南。

Herbs. Stems 4-angled, thinly pilose with gland-tipped trichomes, glabrescent. Leaves subsessile; leaves ovate, both surfaces thinly pilose, adaxially with cystoliths. Inflorescences terminal, spikes, pilose with gland-tipped trichomes; bracts leaflike; corolla blue to purple. Capsules oblong, 4-seeded. Fl. Jul-Aug. Forests or grass slopes at ca. 3000 m. Distributed in Yunnan.

阳朔马蓝 *Strobilanthes pseudocollina*

山马蓝

Strobilanthes oresbia W. W. Sm.

草本。叶卵形至圆形，基部楔形至近圆形，常下延至叶柄，边缘具粗齿。穗状花序腋生，偏向一侧；花序轴常“之”字形曲折；花冠蓝紫色，90°弯曲，5裂；雄蕊4。蒴果具4粒种子。花期7-9月，果期10-12月。生海拔1600-3300米的林下。产重庆、四川、西藏和云南。印度和缅甸亦有。

Herbs. Leaves ovate to orbicular, base cuneate to subrounded, usually decurrent onto the petioles, margin dentate. Spikes axillary, secund; rachis usually zigzag; corolla purplish-blue, bent to 90°, 5-lobed; stamens 4. Capsule 4-seeded. Fl. Jul-Sep. Fr. Oct-Dec. Forests at 1600-3300 m. Distributed in Chongqing, Sichuan, Xizang and Yunnan. Also in India and Myanmar.

阳朔马蓝

Strobilanthes pseudocollina K. J. He et D. H. Qin

亚灌木。叶卵形至卵状披针形，基部钝至楔形，边缘近全缘至具疏齿。穗状花序生于上部叶腋；花冠淡紫红色，5裂；雄蕊4。蒴果具4粒种子。花期9-10月，果期11-12月。生海拔100-300米的石灰岩林下石上。产广西。

Subshrubs. Leaves ovate to ovate-lanceolat, base retuse to cuneate, subentire to remotely dentate. Spikes borne from the upper leaf axils; corolla purplish-red, 5-lobed; stamens 4. Capsule 4-seeded. Fl. Sep-Oct. Fr. Nov-Dec. Rock in the forest on limestone hill at 100-300 m. Distributed in Guangxi.

湖南马蓝

Strobilanthes biocullata Y. F. Deng et J. R. I. Wood

亚灌木，丛生。叶密被钟乳体。穗状花序腋生，无叶；苞片下面具2个眼状膨大。花相向对生；花冠蓝紫色；冠筒基部

山马蓝 *Strobilanthes oresbia*

湖南马蓝 *Strobilanthes biocullata*

圆柱形，扭曲约90°，至口部渐增宽，长约2厘米。蒴果棒状，具4粒种子。种子凸透镜状。花期7-10月，果期11月。生海拔200-800米的溪边岩石上。产广西、广东和湖南。

Subshrubs, gregarious. Leaves densely covered with cystoliths. Spikes axillary, leafless; bracts abaxially with 2 swollen bulges resembling eyes. Flowers in opposite pairs; corolla purplish blue; tubes basally cylindric then bent to ca. 90° and gradually widened to ca. 2 cm long at mouth. Capsules clavate, 4-seeded. Seeds lenticular. Fl. Jul- Oct. Fr. Nov. Rocks by streams at 200-800 m. Distributed in Guangxi, Guangdong and Hunan.

云南马蓝

Strobilanthes yunnanensis Diels

半灌木。茎4棱。叶卵形、椭圆形或椭圆状卵形，两面钟乳体丰富。花序腋生或顶生，穗状花序近头状；花常在每个节间有1朵不育；花冠浅蓝色至紫色。蒴果长圆形。花期7-9月，果期10-11月。生海拔800-2800米的灌丛潮湿开阔阴地。产云南、四川、西藏和甘肃。

Subshrubs. Stems 4-angled. Leaves ovate, elliptic or elliptic-ovate, cystoliths abundant on both surfaces. Inflorescences axillary or terminal, subcapitate spikes; flowers usually only one sterile per node; corolla pale blue to pinkish. Capsules oblong. Fl. Jul-Sep. Fr. Oct-Nov. Moist open shady places in thickets at 800-2800 m. Distributed in Yunnan, Sichuan, Xizang and Gansu.

冯氏马蓝

Strobilanthes fengii Y. F. Deng et Y. F. Deng

亚灌木。叶不等大，卵形，基部圆形至阔楔形，边缘波状被睫毛。穗状花序顶生，近头状；花冠红色，稍弯曲，5裂；雄蕊4。蒴果具4粒种子。花期9-12月，果期11月至翌年2月。生海拔1200-1800米的林下。产云南。

Subshrubs. Leaves unequal in each pair, ovate, base rounded to broadly cuneate, margin undulate and ciliate. Spikes terminal, subcapitate; corolla red, slightly curved, 5-lobed; stamens 4. Capsules 4-seeded. Fl. Sep-Dec. Fr. Nov to next Feb. Forest at 1200-1800 m. Distributed in Yunnan.

云南马蓝 *Strobilanthes yunnanensis*

冯氏马蓝 *Strobilanthes fengii*

糯米香 *Strobilanthes tonkinensis*

折苞马蓝 *Strobilanthes brunnescens*

糯米香

Strobilanthes tonkinensis Lindau

草本，高50-100厘米。茎4棱，干后芳香。叶阔长圆状椭圆形，两面疏具柔毛，脉上尤甚，上面具突出的钟乳体。花序顶生，单穗状；花冠粉红色或纯白色。蒴果狭长圆状倒卵球形。花期4-6月和12月，果期6-7月。生海拔200-1500米的湿润林下。产云南和广西。泰国和越南亦有。

Herbs, 50-100 cm tall. Stems 4-angled, fragrant when dry. Leaves broadly oblong-elliptic, both surfaces sparsely pubescent especially on veins, adaxially with prominent cystoliths. Inflorescences terminal, simple spikes; corolla pink to pure white. Capsules narrowly oblong-obovoid. Fl. Apr-Jun, Dec. Fr. Jun-Jul. Moist forests at 200-1500 m. Distributed in Yunnan and Guangxi. Also in Thailand and Vietnam.

折苞马蓝

Strobilanthes brunnescens Benoist

亚灌木。叶卵形至卵状椭圆形，基部楔形，边缘近全缘至具圆齿。穗状花序腋生；苞片覆瓦状排列，先端反折；花冠蓝紫色，稍弯曲，5裂；雄蕊4，花药长圆形。蒴果具种子4粒。花期11-12月，果期翌年2-3月。生海拔300-500米的山谷林下。产广西和云南。越南亦有。

Subshrubs. Leaves ovate to ovate-elliptic, base cuneate, margin subentire to crenuate. Spikes axillary; bracts imbricate, reflexed; corolla purplish-blue, slightly curved, 5-lobed; stamens 4, anther oblong. Capsule 4-seeded. Fl. Nov-Dec. Fr. next Feb-Mar . Forests in valley at 300-500 m. Distributed in Guangxi and Yunnan. Also in Vietnam.

长苞马蓝

Strobilanthes echinata Nees

草本。叶卵形、椭圆形或卵状披针形，基部楔形并下延至叶柄，边缘具锯齿。穗状花序头状；苞片大，卵形，边缘具锯齿；花冠白色，直，5裂；雄蕊4。蒴果具4粒种子。花期4-6月，果期6-9月。生海拔100-2200米的湿润林下。产广东、广西和云南。南亚和东南亚亦有。

Herbs. Leaves ovate, elliptic or ovate-lanceolate, base cuneate and decurrent onto the petioles, margin serrate. Spikes capiate; bracts large, ovate, margin serrate; corolla white, straight, 5-lobed; stamens 4. Capsule 4-seeded. Fl. Apr-Jun. Fr. Jun-Sep. Moist forests at 100-2200 m. Distributed in Guangdong, Guangxi and Yunnan. Also in S and SE Asia.

长苞马蓝 *Strobilanthes echinata*

蒙自马蓝

Strobilanthes lamiifolia (Nees) T. Anderson

草本，多分枝，叶等大。茎细弱，匍匐，有时于节间生根。花序顶生或生于顶部叶叶腋，穗状；花冠紫红色至蓝色，中部膨大；雄蕊4，内藏；子房多毛。蒴果纺锤形，具4粒种子。花期8-11月。生海拔1000-2600米的旱草地或开阔松林中。产云南、四川、西藏和贵州。印度、尼泊尔和不丹亦有。

Herbs, much branched, isophyllous. Stems slender, decumbent and often rooting at nodes. Inflorescences terminal or axillary from apical leaf axils, spikes; corolla purplish red to blue, ventricose; stamens 4, included; ovary hairy. Capsules fusiform, 4-seeded. Fl. Aug-Nov. Dry grasslands or open *Pinus* forests at 1000-2600 m. Distributed in Yunnan, Sichuan,

蒙自马蓝 *Strobilanthes lamiifolia*

Shrubs. Stems 4-angled. Leaves lanceolate to ovate. Inflorescences axillary or terminal, panicles; bracts broadly obovate, caducous; bracteoles oblanceolate to narrowly elliptic. Flowers solitary on each node, distant on rachis. Corolla white, lilac, or deep pink with a white tube. Ovary glabrous; style sparsely pubescent. Capsule 10-15 mm. Seeds ovate, with long trichomes. Fl. Dec to next Jan. Forests on mountain slopes at 800-2000 m. Distributed in S Xizang. Also in Bhutan, India, Myanmar and Nepal.

Xizang and Guizhou. Also in India, Nepal and Bhutan.

球花马蓝

Strobilanthes dimorphotricha Hance

亚灌木。叶不等大，卵形、椭圆形至椭圆状披针形，基部楔形至渐狭。穗状花序头状，组成顶生或腋生圆锥花序；花冠红色，稍弯曲，5裂；雄蕊4，花药球形。蒴果具4粒种子。花期8月至翌年2月，果期10月至翌年4月。生海拔200-2200米的石灰岩林下。产台湾、浙江、福建、江西、广东、广西、海南、湖南、湖北、重庆、四川和云南。越南、老挝、泰国、缅甸和印度亦有。

Subshrubs. Leaves unequal in each pair, ovate, elliptic to elliptic-lanceolate, base cuneate to attenuate. Spikes headlike, forming terminal or axillary panicles; corolla red, slightly curved, 5-lobed; stamens 4, anthers spherical. Capsule 4-seeded. Fl. Aug to next Feb. Fr. Oct to next Apr. Thickets on limestone hills at 200-2200 m. Distributed in Taiwan, Zhejiang, Fujian, Jiangxi, Guangdong, Guangxi, Hainan, Hunan, Hubei, Chongqing, Sichuan and Yunnan. Also in Vietnam, Laos, Thailand, Myanmar and India.

叉花草

Strobilanthes hamiltoniana (Steud.) Bosser et Heine

灌木。茎和枝4棱形。叶片披针形，边缘有细锯齿。穗状花序构成疏松的圆锥花序。苞片倒卵形，早落；小苞片倒披针形至长圆形。花冠堇色。子房无毛，花柱向基部略被刚毛。蒴果长10-15毫米。种子卵形，被长毛。花期12月至翌年1月。生海拔800-2000米的山坡林下。产西藏南部。不丹、印度、缅甸和尼泊尔亦有。

球花马蓝 *Strobilanthes dimorphotricha*

叉花草 *Strobilanthes hamiltoniana*

合页草 *Strobilanthes kingdonii*

合页草

Strobilanthes kingdonii J. R. I. Wood

亚灌木。茎4棱。花序腋生，密头状穗状花序，生于1个短的单或(2或)3叉花梗上；花冠紫色、蓝色或玫瑰红色；冠筒基部圆柱形，口部扩宽至1.2厘米。蒴果长圆形，长约1厘米，具腺端毛，含4粒种子。花期3月、7月和11月。生海拔1500-2800米的溪边常绿阔叶林中。产云南(福贡县)和西藏(墨脱县和察隅县)。

Subshrubs. Stems 4-angled. Inflorescences axillary, dense headlike spikes, borne on a short usually simple or (2 or)3-furcate peduncle; corolla purple, blue or rose; tubes basally cylindric and widened to ca. 1.2 cm long at mouth. Capsules oblong, ca. 1 cm long, gland-tipped pilose, 4-seeded. Fl. Mar, Jul, Nov. Evergreen broad-leaved forests by streams at 1500-2800 m. Distributed in Yunnan (Fugong County) and Xizang (Mêdog County and Zayü County).

华南马蓝

Strobilanthes austrosinensis Y. F. Deng et J. R. I. Wood

草本。叶椭圆形至近圆形，两面被毛，边缘具锯齿。穗状花序顶生，被毛；花冠蓝紫色，烟斗状，稍弯曲，5裂；雄蕊4。蒴果具4粒种子。花期7-9月，果期10-12月。生海拔100-1500米的灌丛、林缘、溪边或路旁。中国特有，产广东、广西、湖南和江西。

华南马蓝 *Strobilanthes austrosinensis*

Herbs. Leaves elliptic to suborbicular, both surfaces pubescent, margin serrate. Spicks terminal, pubescent; corolla purplish-blue, funnel-shape, slingtly curved, 5-lobed; stamens 4. Capsule 4-seeded. Fl. Jul-Sep. Fr. Oct-Dec. Thickets, forests margin, streamside or roadside at 100-1500 m. Endemic to China, distributed in Guangdong, Guangxi, Hunan and Jiangxi.

红毛马蓝

Strobilanthes hossei C. B. Clarke

草本，几乎全株被红褐色粗毛。叶卵形至卵状披针形，基部阔楔形至近圆形，边缘具圆齿。穗状花序腋生；花冠淡蓝色，直，5裂；雄蕊4。蒴果具4粒种子。花期8-12月，果期11月至翌年2月。生海拔1000-1800米的灌丛。产广西和云南。东南亚亦有。

Herbs, almost whole plants covered with reddish brown setose. Leaves ovate to ovate-lanceolate, base broadly cuneate to subrounded, margin crenuate. Spikes axillary; corolla pale blue, straight, 5-lobed; stamens 4. Capsule 4-seeded. Fl. Aug-Dec. Fr. Nov to next Feb. Thickets at 1000-1800 m. Distributed in Guangxi and Yunnan. Also in SE Asia.

红毛马蓝 *Strobilanthes hossei*

急流马蓝

Strobilanthes torrentium Benoist

亚灌木。叶卵形至椭圆形，基部阔楔形并下延至叶柄，边缘

急流马蓝 *Strobilanthes torrentium*

鳞花草 *Lepidagathis incurva*

具锯齿。穗状花序腋生或顶生，密集，头状，密被锈色长丝状毛；花冠蓝紫色，5裂；雄蕊4。蒴果具4粒种子。花期8-10月，果期10-12月。生海拔1900-2300米的林下或溪边。产云南。印度和缅甸亦有。

Subshrubs. Leaves ovate to elliptic, base broadly cuneate and decurrent onto the petioles, margin serrate. Spike axillary or terminal, compact, headlike, densely covered with long silky rufous trichomes; corolla purplish-blue, 5-lobed; stamens 4. Capsule 4-seeded. Fl. Aug-Oct. Fr. Oct-Dec. Forests or streamside at 1900-2300 m. Distributed in Yunnan. Also in India and Myanmar.

台湾鳞花草

Lepidagathis formosensis C. B. Clarke ex Hayata

半灌木状草本。叶长圆状卵形或卵形，先端渐尖，基部渐狭，下延至柄，边缘全缘至波状，上面无毛，背面沿脉疏被柔毛；柄长0.5-2厘米。穗状花序偏向一侧；苞片、小苞片和花萼被短柔毛；花冠白色，外面被短柔毛，冠檐4裂，长圆形，下方裂片最宽，先端2齿或全缘。花期9月至翌年5月，果期翌年1-4月。生海拔100-2300米的次生常绿阔叶林、溪边、水沟和山坡。产台湾和广东。

Suffruticose herbs. Leaves oblong-ovate ovate, apex acuminate, base attenuate and decurrent onto petioles, margin entire to sinuate, adaxially glabrous, abaxially sparsely pubescent along veins; petioles 0.5-2 cm. Spikes secund; bracts, bracteoles and calyx pubescent; corolla white, outside pubescent, limb 4-lobed, oblong, lower lobe broadest, 2-toothed or entire. Fl. Sep to next May. Fr. next Jan-Apr. Secondary evergreen broad-leaved forests, along streams, ditches or slopes at 100-2300 m. Distributed in Taiwan and Guangdong.

台湾鳞花草 *Lepidagathis formosensis*

鳞花草

Lepidagathis incurva Buch.-Ham. ex D. Don

草本，叶常不等大。茎4棱，具沟槽，具柔毛或迅速脱落。叶卵形至椭圆形至狭椭圆形。穗状花序伸长，侧生；花冠白色，具紫色条斑。种子轮廓近圆形。花果期10月至翌年3月。生海拔100-2200米的草地、灌丛、路边、绿篱或溪边。产云南和华南。印度、孟加拉国、缅甸、泰国和越南亦有。

Herbs, often ± anisophyllous. Stems 4-angled, sulcate, pubescent or soon glabrescent. Leaves ovate to elliptic to narrowly elliptic. Spikes elongate, secund; corolla white streaked with purple. Seeds subcircular in outline. Fl. and fr. Oct to next Mar. Grasslands, thickets, roadsides, hedgerows or streamsides at 100-2200 m. Distributed in Yunnan and S China. Also in India, Bangladesh, Myanmar, Thailand and Vietnam.

色萼花 *Chroesthes laceolata*

穿心莲 *Andrographis paniculata*

色萼花

Chroesthes laceolata (T. Anderson) B. Hansen

灌木。叶片椭圆形至披针形，基部楔形，边缘全缘或近全缘，无毛。聚伞圆锥花序顶生；花萼不等大；花冠白色，具粉色或紫色斑点，二唇形，上唇2裂，下唇3裂；雄蕊4，花药基部具芒。蒴果倒卵形，具4粒种子。花期2-3月，果期5-7月。生海拔400-1400米的林下。产广西和云南。东南亚亦有。

Shrubs. Leaves elliptic to lanceolate, base cuneate, margin entire or subentire, glabrous. Thyrses terminal; calyx lobes unequal; corolla white with pink or purple dots, bilabiate, upper lip 2-lobed, upper lip 3-lobed; stamens 4, anthers spurred at base. Capsule obovoid, 4-seeded. Fl. Feb-Mar. Fr. May-Jul. Forests at 400-1400 m. Distributed in Guangxi and Yunnan. Also in SE Asia.

疏花穿心莲

Andrographis laxiflora (Bl.) Lindau

多年生草本。茎四棱形。叶卵形，边缘全缘。总状花序顶生和腋生，花序轴纤细，常波状弯曲，通常不分枝，近无毛；苞片1，披针状线形；小苞片2枚钻形；花萼5裂，裂片线形；花冠白色。蒴果线状长圆形。种子有皱纹。花期8-11月，过期11月至翌年3月。产云南、贵州、广西和海南。南亚和东南亚亦有。

Perennial herbs. Stems 4-angled. Leaf blade ovate, margin entire. Inflorescences axillary or terminal, racemes or panicles of racemes. Bracts 1, lanceolate; bracteoles 2, subulate. Calyx 5-lobed, lobes subulate. Corolla white. Capsule linear-cylindric. Seeds rugose. Fl. Aug-Nov. Fr. Nov to next Mar. Thickets, bamboo forests at 200-1500 m. Distributed in Guangxi, Guizhou, Hainan, Yunnan. Also in S and SE Asia.

穿心莲

Andrographis paniculata (N. L. Burm.) Wall. ex Nees

一年生草本。茎4棱，无毛。叶两面无毛，全缘，基部渐狭且下延至叶柄。总状花序集成大型圆锥花序；花冠白色，二唇形；下唇具紫色斑点。蒴果扁椭圆体形。花果期终年。栽培或归化于中国西南、华南、华中和华东。原产印度和斯里兰卡，栽培或归化于东南亚和加勒比。

Annual herbs. Stems 4-angled, glabrous. Leaves both surfaces glabrous, margin entire, base attenuate and decurrent onto petioles. Racemes formed into large panicles; corolla white, 2-lipped; lower lips with purple dots. Capsules ellipsoid-compressed. Fl. and fr. throughout year. Cultivated or naturalized in SW, S, C and E China. Native to India and Sri Lanka, cultivated or naturalized in SE Asia and Caribbean.

疏花穿心莲 *Andrographis laxiflora*

广西火焰花 *Cystacanthus colaniae*

广西火焰花
Cystacanthus colaniae
(Benoist) Y. F. Deng

灌木。叶卵形至卵状披针形。聚伞圆锥花序长10-20厘米；聚伞花序少花；花冠长约1.2厘米，一侧膨大，弯至90°，中部骤然膨大；雄蕊2；柱头稍2裂。蒴果长1.8-2厘米，无毛。生海拔200-500米的石灰质山丘。产云南、广西和海南。越南亦有。

Shrubs. Leaves ovate to ovate-lanceolate. Thyrses 10- 20 cm long; cymes few-flowered; corolla ca. 1.2 cm long, ventricose, bent to 90° and abruptly inflated at middle; stamens 2; stigmas slightly 2-cleft. Capsules 1.8-2 cm long, glabrous. Limestone hills at 200-500 m. Distributed in Yunnan, Guangxi and Hainan. Also in Vietnam.

火焰花
Phlogacanthus curviflorus
(Wall.) Nees

灌木。叶椭圆形至长圆形，基部阔楔形且下延至叶柄，边缘全缘至近圆齿。聚伞圆锥花序顶生；聚伞花序具(1-)3-5花；花冠紫红色；下唇3裂，上唇2深裂。蒴果具8-10粒种子。花期10月至翌年2月，果期翌年2-5月。生海拔400-1600米的灌丛、林缘或沟壑。产云南和西藏。印度、不丹、缅甸、老挝、泰国和越南亦有。

Shrubs. Leaves elliptic to oblong, base broadly cuneate and decurrent onto petioles, margin entire to subcrenate. Thyrses terminal; cymes (1-)3-5-flowered; corolla purplish red; lower lips 3-lobed, upper lips 2-cleft. Capsules 8-10-seeded. Fl. Oct to next Feb. Fr. next Feb-May. Thickets, forest edges or ravines at 400-1600 m. Distributed in Yunnan and Xizang. Also in India, Bhutan, Myanmar, Laos, Thailand and Vietnam.

宽叶十万错
Asystasia gangetica (L.)
T. Anders.

斜升草本。茎4棱形，具柔毛。叶卵形至椭圆形，3-12 × 1-5厘米。总状花序顶生或腋生；花冠白色或黄色，3-3.5厘米长，口部约1厘米宽；下唇中裂片具紫色或栗色斑块；雄蕊内藏。花期9-12月，果期12月至翌年3月。生林缘。归化于云南、广西和广东。热带亚洲和太平洋岛屿亦有。

Herbs ascending. Stems 4-angled, pilose. Leaves ovate to elliptic, 3-12 × 1-5 cm. Racemes axillary or terminal; corolla white or yellow, 3-3.5 cm long, ca. 1 cm wide at mouth; middle lobe of lower lips with violet or maroon markings; stamens included. Fl. Sep-Dec. Fr. Dec to next Mar. Forest edges. Naturalized in Yunnan, Guangxi and Guangdong. Also in tropical Asia and Pacific Islands.

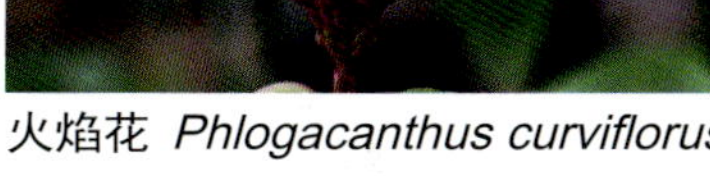

火焰花 *Phlogacanthus curviflorus*

宽叶十万错 *Asystasia gangetica*

白接骨 *Asystasia neesiana*

白接骨

Asystasia neesiana (Wall.) Nees

草本。茎4棱，具槽。花序顶生，穗状或总状，不分枝或具1至数个基生分枝形成圆锥状；花冠粉红色或蓝紫色，外面具腺头状柔毛；冠筒基部圆柱形且狭，伸长，至少为喉部及冠檐合并的2倍长。花期7-9月，果期10月至翌年1月。生海拔100-1800米的常绿阔叶林、潮湿岸堤、溪边、开垦地、沟渠或小道。产中国西南、华南、华中和华东。南亚和东南亚亦有。

Herbs. Stems 4-angled, sulcate. Inflorescences terminal, spikes or racemes, unbranched or with one or more basal branches forming a panicle; corolla pink to purplish blue, outside gland-tipped pubescent; tubes with cylindric basal portion narrow, elongate, at least 2 × as long as throat and limbs combined. Fl. Jul-Sep. Fr. Oct to next Jan. Evergreen broad-leaved forests, moist banks, streamsides, disturbed sites, ditches or trails at 100-1800 m. Distributed in SW, S, C and E China. Also in S and SE Asia.

山壳骨 *Pseuderanthemum latifolium*

海康钩粉草
Pseuderanthemum haikangense

海康钩粉草

Pseuderanthemum haikangense C. Y. Wu et H. S. Lo

灌木。茎圆柱状，无毛。叶椭圆状圆形、长圆形或卵形，宽2-3.5厘米。总状花序顶生，长达30厘米；小苞片长1-2毫米；花冠白色或粉红色，下唇具红色斑点，长约4厘米。蒴果棍棒状。花期6-8月，果期9月。生海拔200-900米的森林。产云南、广东和海南。

Shrubs. Stems terete, glabrous. Leaves elliptic-orbicular, oblong, ovate, 2-3.5 cm width. Racemes terminal, to 30 cm long; bracteoles 1-2 mm long; corolla white or pink and with red spots on lobes of lower lips, ca. 4 cm long. Capsules clavate. Fl. Jun-Aug. Fr. Sep. Forests at 200-900 m. Distributed in Yunnan, Guangdong and Hainan.

山壳骨

Pseuderanthemum latifolium (Vahl) B. Hansen

多年生草本，高达1米。茎4棱，具柔毛。叶椭圆形，下面仅中脉有绒毛，上面无毛。总状花序达30厘米，常有3花簇生于节间；花冠淡紫色，下唇具紫斑。蒴果具柔毛。花期4-6月，果期8-9月。生海拔100-1600米的林下。产云南和华南。南亚和东南亚亦有。

Perennial herbs, to 1 m tall. Stems 4-angled, pubescent. Leaves elliptic, abaxially glabrous except villous on midvein, adaxially glabrous. Racemes to

多花山壳骨 *Pseuderanthemum polyanthum*

太平爵床 *Mackya tapingensis*

30 cm long, usually with 3-flowered clusters at nodes; corolla light purple with purple dots on lower lip. Capsules pubescent. Fl. Apr-Jun. Fr. Aug-Sep. Forests at 100-1600 m. Distributed in Yunnan and S China. Also in S and SE Asia.

多花山壳骨

Pseuderanthemum polyanthum (C. B. Clarke) Merr.

草本。茎浅灰色，近圆柱状，渐无毛。叶宽卵形至长圆形，两面无毛。聚伞圆锥花序长5-12厘米；苞片三角形；花冠蓝紫色；下唇3裂；上唇2裂。蒴果棍棒状。花期4-6月，果期8月。生海拔300-1600米的林下或灌丛。产云南和广西。印度、缅甸、泰国、越南和马来西亚亦有。

Herbs. Stems grayish, subterete, glabrescent. Leaves broadly ovate to oblong, both surfaces glabrous. Thyrses 5-12 cm long; bracts triangular; corolla bluish purple; lower lips 3-lobed; upper lips 2-lobed. Capsules clavate. Fl. Apr-Jun. Fr. Aug. Forests or thickets at 300-1600 m. Distributed in Yunnan and Guangxi. Also in India, Myanmar, Thailand, Vietnam and Malaysia.

太平爵床

Mackya tapingensis (W. W. Sm.) Y. F. Deng et C. Y. Wu

灌木。叶披针形至长圆状披针形，基部楔形，边缘全缘至稍波状。总状花序顶生或腋生，偏向一侧；花冠钟状，玫瑰红色，5裂；雄蕊2。蒴果基部具短柄，具4粒种子。花期10月至翌年1月，果期12月至翌年3月。生海拔600-1800米的林下。产云南。缅甸亦有。

Shrubs. Leaves lanceolate to oblong-lanceolate, base cuneate, margin entire to slightly undulate. Racemes terminal or axillary, second; corolla campanulate, rose, 5-lobed; stamens 2. Capsule stipitate, 4-seeded. Fl. Oct to next Jan. Fr. Dec to next Mar. Forests at 600-1800 m. Distributed in Yunnan. Also in Myanmar.

钟花草

Codonacanthus pauciflorus (Nees) Nees

草本。叶椭圆形至披针形，基部圆形、楔形或下延至叶柄，边缘全缘。总状花序顶生，常数个组成圆锥花序；花冠白色，在下唇基部具紫色斑点，钟形，5裂；雄蕊2。蒴果基部具短柄，具4粒种子。花果期8至翌年3月。生海拔100-1500米的林下、溪边或灌丛。产台湾、福建、江西、广东、广西、海南、贵州和云南。南亚、东南亚和日本亦有。

Herbs. Leaves elliptic to lanceolate, base rounded, cuneate or decurrente onto the petioles, margin entire. Racemes terminal, usually severl forming a panicle; corolla white with purple dots at base of lower lip, campanulate, 5-lobed; stamens 2. Capsule stalked at base, 4-seeded. Fl. and fr. Aug to next Mar. Forests, streamside or thickets at 100-1500 m. Distributed in Taiwan, Fujian, Jiangxi, Guangdong, Guangxi, Hainan, Guizhou and Yunnan. Also in S and SE Asia, and Japan.

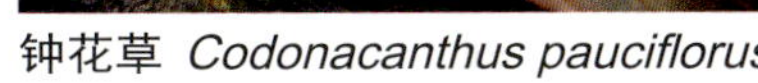

钟花草 *Codonacanthus pauciflorus*

纤穗爵床 *Leptostachya wallichii*

纤穗爵床

Leptostachya wallichii Nees

草本。叶披针形或椭圆形，基部圆形或阔楔形，边缘全缘。穗状花序顶生；花冠白色，二唇形，上唇2裂，下唇3裂；雄蕊2。蒴果具4粒种子。花期8-12月，果期10月至翌年2月。生海拔900-1600米的林下。产广东、广西和海南。南亚和东南亚亦有。

Herbs. Leaves lanceolate or elliptic, base rounded or broadly cuneate, margin entire. Spikes terminal; corolla white, bilabiate, upper lip 2-lobed, lower lip 3-lobed; stamens 2. Capsule 4-seeded. Fl. Aug-Dec. Fr. Oct to next Feb. Moist forests at 900-1600 m. Distributed in Guangdong, Guangxi and Hainan. Also in S and SE Asia.

叉序草 *Isoglossa collina*

叉序草

Isoglossa collina (T. Anderson) B. Hansen

草本，高40-100厘米。叶卵形至卵状椭圆形。花序于上部叶叶腋顶生或腋生，圆锥状；分枝稍扁平；花冠白色，具淡紫色斑点冠筒基部圆柱形，约1/3渐宽为漏斗状喉部。花期8-11月。生海拔300-2200米的常绿阔叶林或溪边湿地。产云南、西藏、广西、广东、湖南和江西。印度、不丹和泰国亦有。

Herbs, 40-100cm tall. Leaves ovate to ovate-elliptic. Inflorescences terminal or axillary from upper leaf axils, panicles; branches slightly flattened; corolla white with purplish dots; tubes basally cylindric for ca. 1/3 of its length then gradually widening into a funnel-shaped throat. Fl. Aug-Nov. Evergreen broad-leaved forests or wetlands by streams at 300-2200 m. Distributed in Yunnan, Xizang, Guangxi, Guangdong, Hunan and Jiangxi. Also in India, Bhutan and Thailand.

双萼观音草 *Peristrophe paniculata*

双萼观音草

Peristrophe paniculata (Forssk.) Brummitt

草本。叶卵形至披针形，基部渐狭至阔楔形，边缘具细齿。具伞花序多数组成顶生圆锥花序；苞片小，小苞片不等大；花冠粉红色，二唇形，上唇2裂，下唇3裂；雄蕊2，花药不等高，药室球形。蒴果基部具柄，具4粒种子。花期8-12月，果期10月至翌年2月。生海拔600-2200米的荒地或路旁。产云南、四川和广西。亚洲、非洲和澳大利亚的热带地区亦有。

Herbs. Leaves ovate to lanceolate, base attenuate to broadly cuneate, margin serrulate. Cymes several forming a terminal panicle; bracts small, bracteoles unequal; corolla pink, bilabiate, upper lip 2-lobed, lower lip 3-lobed; stamens 2, anther-thecae super-

九头狮子草 *Peristrophe japonica*

糙叶山蓝 *Peristrophe strigosa*

posed, spherical. Capsule stipitate, 4-seeded. Fl. Aug-Dec. Fr. Oct to next Feb. Roadside or waste place at 600-2200 m. Distributed in Yunnan, Sichuan and Guangxi. Also in tropical regions of Asia, Africa and Australia.

九头狮子草

Peristrophe japonica (Thunb.) Bremek.

多年生草本。茎4棱。叶两面疏被柔毛。花序由1-4(或更多)具梗，由总苞的顶生或腋生的聚伞花序组成；总苞内具1-3花；花冠白色至浅粉色至亮紫色，冠檐具粉色小点或黑色线条。蒴果具柔毛。种子具瘤。花期7月至翌年2月，果期7-10月。生海拔1500米以下的路边、草地、溪边、山坡或常绿阔叶林中。产中国西南、华南、东南、华中、华北和华东。日本亦有。

Perennial herbs. Stems 4-angled. Leaves both surfaces sparsely pubescent. Cymes terminal or axillary, consisting of 1-4(or more) pedunculate involucres; involucres bearing 1-3 flowers; corolla white to pale pink to light purple and with pink specks or dark lines on limb. Capsules pubescent. Seeds tuberculate. Fl. Jul to next Feb. Fr. Jul-Oct. Roadsides, grasslands, streamsides or montane slopes below 1500 m. Distributed in SW, S, SE, C, N and E China. Also in Japan.

糙叶山蓝

Peristrophe strigosa C. Y. Wu et H. S. Lo

草本。叶卵形至卵状披针形，基部楔形，边缘全缘。聚伞花序顶生或腋生，常具3分枝；苞片小，短于花萼；花冠粉红色，二唇形，上唇2裂，下唇3裂；雄蕊2，花药线形，平行。蒴果具4粒种子。花期10月至翌年2月，果期翌年1-3月。生林下。产海南。

Herb. Leaves ovate to ovate-lanceolat, base cuneate, margin entire. Cymes terminal or axillary, usually 3-branched; bracts small, shorter than calyx; corolla pink, bilabiate, upper lip 2-lobed, lower lip 3-lobed; stamens 2, anther-thecae linear, parallel. Capsule 4-seeded. Fl. Oct to next Feb. Fr. next Jan-Mar. Forests. Distributed in Hainan.

三花枪刀药

Hypoestes triflora Roem. et Schult.

匍匐草本，多分枝。茎4棱，具沟槽和硬毛。聚伞花序在叶腋和/或叶状苞片腋处，或顶生成簇，具(1-)3(-5)花；花萼裂片5，条状披针形；花冠白色至粉红色，边缘栗色。花期7-9月，果期9月至翌年1月。生海拔300-2400米的小路边或林下。产云南。印度、尼泊尔、不丹、缅甸和非洲亦有。

Herbs decumbent, much branched. Stems 4-angled, sulcate, strigose. Cymes pedunculate in leaf axils and/or in axils of subleaflike bracts or in a terminal cluster, (1-)3(-5)-flowered; calyx lobes 5, linear-lanceolate; corolla white to pink with maroon markings. Fl. Jul-Sep. Fr. Sep to next Jan. Trailsides or forests at 300-2400 m. Distributed in Yunnan. Also in India, Nepal, Bhutan, Myanmar and Africa.

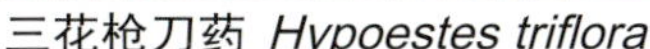
三花枪刀药 *Hypoestes triflora*

鳄嘴花 *Clinacanthus nutans*

鳄嘴花
Clinacanthus nutans (Burm. f.) Lindau

灌木。叶披针形，基部楔形至圆形，边缘近全缘。总状花序1至数个生于枝顶，常组成圆锥花序；花冠暗红色，二唇形，上唇凹，下唇3裂；雄蕊2，花药1室。蒴果被毛，具4粒种子。花期10月至翌年1月，果期翌年3月。生海拔700米以下的林中或灌丛。产广东、广西、海南和云南。印度尼西亚、马来西亚、泰国、老挝和越南亦有。

Shrubs. Leaves lanceolate, base cuneate to rounded, margin sunentire. Racemes 1-several borne on the shoot apex, usually forming a panicle; corolla dull red, bilabiate, upper lip emarginaye, lower lip 3-lobed; stamens 2, anther monothecous. Capsule pilose, 4-seeded. Fl. Oct to next Jan. Fr. Mar next. Thickets below 700 m. Distributed in Guangdong, Guangxi, Hainan and Yunnan. Also in Indonesia, Malaysia, Thailand, Laos and Vietnam.

针子草
Justicia vagabunda Benoist

灌木，高达1.3米。叶长圆状披针形，无毛或有时沿脉具柔毛。花萼裂片5，外面被短柔毛；花冠白色带蓝色条纹，筒短，与檐部等长，檐部二唇形，上唇三角形，先端微凹，下唇3裂。生海拔500-800米的林下、灌丛或溪边。产云南。越南和柬埔寨亦有。

Shrubs, up to 1.3 m tall. Leaves oblong-lanceolate, glabrous or sometimes pilose along midvein. Calyx lobes 5, outside pubescent; corolla white with blue streaks, tubes short, equal to limb, limbs 2-labiate, upper lips triangular, apex emarginate, lower lips 3-lobed. Forests, thickets or streamsides at 500-800 m. Distributed in Yunnan. Also in Vietnam and Cambodia.

密花孩儿草
Rungia densiflora H. S. Lo

草本。叶卵形至卵状披针形，基部楔形，稍下延至叶柄，边缘全缘。穗状花序顶生或腋生；苞片4列，均能育；花冠蓝色，二唇形，上唇2裂，下唇3裂；雄蕊2，下方药室基部具芒。蒴果具4粒种子，胎座与珠柄钩与蒴果内壁成熟时分离。花期9-10月，果期11-12月。生海拔400-800米的林下或溪边。中国特有，产安徽、浙江、江西和广东。

Herbs. Leaves ovate to ovate-lanceolate, base cuneate and slightly decurrent onto the petioles, margin entire. Spikes terminal or axillary; bracts 4-ranked, all fertile; corolla blue, bilabiate, upper lip 2-lobed, lower lip 3-lobed; stamens 2, lower anther-theca spurred at base. Capsule 4-seeded, septa with attached retinacula separating from inner wall of mature capsule. Fl. Sep-Oct. Fr. Nov-Dec. Forests or streamside at 400-800 m. Endemic to China, distributed in Anhui, Zhejiang, Jiangxi and Guangdong.

孩儿草
Rungia pectinata (L.) Nees

一年或多年生草本。茎基部匍匐，节间生根，然后直立。穗状花序腋生或顶生于一侧，单个或有时2至3个组成复合花序；苞片二型；小苞片顶端2裂，稍尖；花萼无色；花冠蓝色或白色；下唇3裂。蒴果椭圆体形，具2-4粒种子。种子稍具皱。花期11月至翌年1月，果期翌年1-4月。生弃荒地。产云南、广西、广东和海南。南亚和东南亚亦有。

Annual or perennial herbs. Stems basally prostrate and rooting at nodes then erect. Spikes axillary or terminal, 1-sided, solitary or sometimes 2 or 3 compound; bracts dimorphic; bracteoles apex 2-cleft and subacute; calyx color-

针子草 *Justicia vagabunda*

密花孩儿草 *Rungia densiflora*

孩儿草 *Rungia pectinata*

less; corolla blue or white; lower lips 3-lobed. Capsules ellipsoid, 2-4-seeded. Seeds minutely verrucose. Fl. Nov to next Jan. Fr. next Jan-Apr. Wastelands. Distributed in Yunnan, Guangxi, Guangdong and Hainan. Also in S and SE Asia.

云南孩儿草

Rungia yunnanensis H. S. Lo

直立草本，高达1米。叶椭圆形至宽椭圆形，钟乳体白色，两面无毛。穗状花序顶生，长达9厘米，偏向一侧。蒴果长约1.2厘米，被柔毛。果期4-6月。生海拔900-1900米的溪边。产云南。

Herbs, up to 1 m tall, erect. Leaves elliptic to broadly elliptic, cystoliths white, glabrous on both surfaces. Spikes terminal, to 9 cm long, 1-sided. Capsules ca. 1.2 cm long, pubescent. Fr. Apr-Jun. Streamsides at 900-1900 m. Distributed in Yunnan.

云南孩儿草 *Rungia yunnanensis*

滇灵枝草

Rhinacanthus beesianus Diels

灌木，茎高0.9-1.5米，近无毛。叶狭倒卵形或倒披针形。圆锥花序顶生。苞叶叶状，披针形。花萼5裂，裂片线形，被柔毛。花冠白色，芳香。上唇极短2裂，下唇裂片椭圆形。花期8-9月。生海拔2100-2400米的干山坡。产云南。

Shrubs, to 0.9-1.5 m tall. Stems 4-angled, subglabrous. Leaf blade narrowly obovate to oblanceolate, margin slightly undulate to subentire, apex acute. Panicles terminal. Bracts leaflike, lanceolate. Calyx 5-lobed; lobes linear, pubescent. Corolla white, fragrant; lower lip lobes elliptic; upper lip apex 2-cleft. Fl. Aug-Sep. Open dry situations on mountain slopes at 2100-2400 m. Distributed in Yunnan.

滇灵枝草 *Rhinacanthus beesianus*

灵枝草（白鹤灵芝）*Rhinacanthus nasutus*

灵枝草（白鹤灵芝）

Rhinacanthus nasutus (L.) Kurz

亚灌木或多年生草本，高达1.5米。叶椭圆形或卵状椭圆形，下面密具柔毛，上面疏具柔毛至近无毛。花冠浅绿色，外面被短柔毛和无腺体的毛状体，先端具腺体。蒴果具柔毛及腺毛。花期10-12月。生海拔700米以下的灌丛、林下或村边栽培。产云南、广东和海南。南亚、东南亚和马达加斯加亦有。

Subshrubs or perennial herbs, to 1.5 m tall. Leaves elliptic or ovate-elliptic, abaxially densely pubescent, adaxially sparsely pubescent to subglabrous. Corolla greenish white, outside pubescent with gland-tipped and non-glandular trichomes. Capsules pubescent with gland-tipped trichomes. Fl. Oct-Dec. Thickets, forests or cultivated around villages below 700 m. Distributed in Yunnan, Guangdong and Hainan. Also in S and SE Asia, and Madagascar.

广西秋英爵床

Cosmianthemum guangxiensis H. S. Lo et D. Fang

草本。叶卵状椭圆形至披针形，基部圆至阔楔形，边缘全缘。聚伞圆锥花序顶生；花冠白色或黄绿色，长1-1.5厘米，二唇形，上唇2裂，下唇3裂；雄蕊2。蒴果具4粒种子。花期9-10月，果期11-12月。生海拔约400米的林下。产广西。

Herbs. Leaves ovate-elliptic to lanceolate, base rounded to broadly cuneate, margin entire. Thyrses terminal; corolla white or yellowish green, 1-1.5 cm long, bilabiate, upper lip 2-lobed, lower lip 3-lobed; stamens 2. Capsule 4-seeded. Fl. Sep-Oct. Fr. Nov-Dec. Forests at ca. 400 m. Distributed in Guangxi.

广西秋英爵床 *Cosmianthemum guangxiensis*

虾衣花

Justicia brandegeana Wassh. et L. B. Sm.

草本。叶卵形，基部渐狭而下延至叶柄，边缘全缘。穗状花序紧密；苞片大，各种颜色；花冠白色，具红色斑点；雄蕊2，花药下方1室基部具芒。蒴果具4粒种子。中国南方各地栽培。原产墨西哥。

Herbs. Leaves ovate, base attenuate and decurrent onto the petioles, margin entire. Spikes congest-

虾衣花 *Justicia brandegeana*

鸭嘴花 *Justicia adhatoda*

绵毛杜根藤 *Justicia alboveltata*

ed; bracts large, variable in colour; corolla white with red dots; stamens 2, lower thecae spurred at base. Capsule 4-seeded. Widely cultivated in south of China. Native to Mexico.

鸭嘴花

Justicia adhatoda L.

灌木。叶卵形至椭圆形状卵形。穗状花序；花萼裂片5；花冠白色或粉红色，外面带淡紫色或粉红色条纹，宽管状；冠筒基部圆柱形，稍膨大向上弯曲。蒴果倒卵球形。花期1-3月，果期6-7月。生海拔850-1800米的路边或灌丛。归化或引种于云南、广西、广东和海南。可能原产南亚和东南亚，但广泛栽植和归化于热带地区。

Shrubs. Leaves ovate to elliptic-ovate. Spikes; calyx lobes 5; corolla white or pink with purplish or pinkish stripes outside, broadly tubular; tubes basally cylindric and then slightly inflated and bent upward. Capsules obovoid. Fl. Jan-Mar. Fr. Jun-Jul. Roadsides or thickets at 850-1800 m. Naturalized or cultivated in Yunnan, Guangxi, Guangdong and Hainan. Probably native to S and SE Asia, but widely cultivated and naturalized in tropics.

绵毛杜根藤

Justicia alboveltata W. W. Sm.

草本，多分枝。叶小，披针形，基部楔形，边缘全缘，两面被白色绵毛。穗状花序短缩成簇；花冠黄白色或黄色，二唇形，上唇2裂，下唇3裂；雄蕊2，下方1室基部具芒。蒴果具4粒种子。花期8-10月，果期11月。生海拔约2700米的灌丛或溪边。特产云南。

Herbs, much branched. Leaves small, lanceolate, base cuneate, margin entire, villous. Spikes abbreviated; corolla yellowish white or yellow, bilabiate, upper lip 2-lobed, lower lip 3-lobed; stamens 2, lower thecae spurred at base. Capsule 4-seeded. Fl. Aug-Oct. Fr. Dec. Thickets or streamside at ca. 2700 m. Endemic to Yunnan.

杜根藤

Justicia quadrifaria (Nees) T. Anderson

草本。叶长圆形至披针形，基部楔形，下延至叶柄，边缘全缘或稍具齿。穗状花序短缩；苞片叶状；花冠黄白色，具淡紫色斑点，二唇形，上唇2裂，下唇3裂；雄蕊2，花药下方1室基部具芒。蒴果基部具短柄，具4粒种子。花期6-9月，果期9-11月。生海拔800-1600米的林下石上。产中国西南、华南、东南、华东和华中。南亚和东南亚亦有。

Herbs. Leaves oblong to lanceolate, base cuneate and decurrent onto the petioles, margin entire or slightly serrate. Spikes abbreviated; bracts leaflike; corolla yellowish white, with purplish dots, bilabiate, upper lip 2-lobed, lower lip 3-lobed; stamens 2, lower thecae spurred at base. Capsule stipitate, 4-seeded. Fl. Jun-Sep. Fr Sep-Nov. Rock in forests at 800-1600 m. Distributed in SW, S, SE, E and C China. Also in S and SE Asia.

杜根藤 *Justicia quadrifaria*

紫苞爵床 *Justicia latiflora*

紫苞爵床

Justicia latiflora Hemsl.

直立草本或灌木。茎4棱。叶两面沿脉具硬毛。穗状花序顶生，密具花；苞片基部浅绿色，中部以上紫红色；花冠白色，裂片上具紫红色点或条纹；花冠下唇宽3齿；子房无毛。蒴果倒卵球形棒状，无毛。种子具皱。花期2-4月，果期5-6月。生海拔600-1800米的林中或溪边。产贵州、重庆、湖北和湖南。

Herbs or shrubs, erect. Stems 4-angled. Leaves both surfaces hispid along veins. Spikes terminal, densely-flowered; bracts basally pale green and purplish red above middle; corolla pale white with purplish red dots or stripes on lobes; lower lips of tubes broadly 3-lobed; ovary glabrous. Capsules obovoid-clavate, glabrous. Seeds rugose. Fl. Feb-Apr. Fr. May-Jun. Forests or streamsides at 600-1800 m. Distributed in Guizhou, Chongqing, Hubei and Hunan.

琴叶爵床

Justicia panduriformis R. Ben.

灌木。茎4棱。叶无柄，倒卵形，无毛，基部狭，耳形抱茎，边缘全缘。穗状花序顶生；花冠白色，具淡紫色斑点，二唇形，上唇2裂，下唇3裂；雄蕊2，花药下方1室基部具芒。蒴果基部具柄，具4粒种子。花期4-6月，果期8-9月。生石山林下。产云南和广西。越南亦有。

Shrubs. Stems 4-angled. Leaves sessile, obovate, glabrous, base attenuate and auriculate, margin entire. Spikes terminal; corolla white, with purplish dots, bilabiate, upper lip 2-lobed, lower lip 3-lobed; stamens 2, lower thecae spurred at base. Capsule stipitate, 4-seeded. Fl. Apr-Jun. Fr. Aug-Sep. Rock in forests. Distributed in Yunnan and Guangxi. Also in Vietnam.

桂南爵床

Justicia austroguangxiensis D. Fang et H. S. Lo

草本。叶基生，莲座状，倒卵形至倒卵状椭圆形，基部狭下延至叶柄，边缘全缘。穗状花序顶生或生于叶腋；花冠黄绿色，具淡紫色斑块，二唇形，上唇2裂，下唇3裂；雄蕊2，花药下方1室基部具芒。蒴果基部具柄，具4粒种子。花期3-7月，果期6-8月。生海拔300-500米的密林下石上。特产广西。

Herbs. Leaves in a basal rosette, obovate to obovate-elliptic, base attenuate and decurrent onto the petioles, margin entire. Spikes terminal or in leaf axil; corolla light yellowish green with purplish stripes, bilabiate, upper lip 2-lobed, lower lip 3-lobed; stamens 2, lower thecae spurred at base. Capsule stipitate, 4-seeded. Fl. Mar-Jul. Fr. Jun-Aug. Dense forests on rocks at 300-500 m. Endemic to Guangxi.

南岭爵床

Justicia leptostachya Hemsl.

草本。茎4棱，具二叉状毛。叶下沿脉具硬毛，上面疏被硬毛。穗状花序顶生，常分枝；花序轴具硬毛，每节有数花；苞片条形；花冠下唇3裂；子房无毛。

琴叶爵床 *Justicia panduriformis*

南岭爵床 *Justicia leptostachya*

桂南爵床
Justicia austroguangxiensis

蒴果棒状，具柔毛，具4粒种子。种子深褐色，稍多刺。花期3-6月，果期7-8月。生岩石上。产广西、广东和湖南。

Herbs. Stems 4-angled, bifariously pubescent. Leaves abaxially strigose along veins, adaxially sparsely strigose. Spikes terminal, often branched; rachis hispid, with several flowers per node; bracts linear; corolla lower lips 3-lobed; ovary glabrous. Capsules clavate, puberulent, 4-seeded. Seeds dark brown, slightly muricate. Fl. Mar-Jun. Fr. Jul-Aug. Rocks. Distributed in Guangxi, Guangdong and Hunan.

野靛棵

Justicia patentiflora Hemsl.

多年生草本。叶卵形至长圆状披针形，膜质，无毛。穗状花序；花萼裂片5，被微柔毛；花冠紫红色具淡紫色斑点；冠筒基部以上急反折，外具柔毛。蒴果狭倒卵球形。花期11月至翌年1月，果期翌年4-5月。生海拔500-1800(-2400)米的林中、溪边或石灰岩丘陵。产云南。越南亦有。

Perennial herbs. Leaves ovate to oblong-lanceolate, membranous, glabrous. Spikes; calyx lobes 5, puberulent; corolla purplish red with purplish spots; tubes abruptly recurved above base, outside puberulent. Capsules narrowly obovoid. Fl. Nov to next Jan. Fr. next Apr-May. Forests, by streams or limestone hills at 500-1800(-2400) m. Distributed in Yunnan. Also in Vietnam.

野靛棵 *Justicia patentiflora*

黑叶小驳骨 *Justicia ventricosa*

黑叶小驳骨
Justicia ventricosa Wall. ex Sims.

亚灌木或多年生草本。叶椭圆形至倒卵形，无毛，基部渐狭，边缘全缘或稍波状。穗状花序顶生；苞片叶状，绿色或栗色；花冠黄白色，具淡紫色条纹，二唇形，上唇2裂，下唇3裂；雄蕊2，花药下方1室具芒。蒴果具4粒种子。花期11月至翌年1月，果期翌年4-5月。栽培或归化于广东、广西、海南和云南等地。原产东南亚。

Subshrubs or perennial herbs. Leaves elliptic to obovate, glabrous, base attenuate, margin entire or slightly undulate. Spikes terminal; bract leafy, green or mroom; corolla yellowish white, with purplish stripes, bilabiate, upper lip 2-lobed, lower lip 3-lobed; stamens 2, lower theca spurred at base. Capsule 4-seeded. Fl. Nov to next Jan. Fr. next Apr-May. Cultivated or naturalized in Guangdong, Guangxi, Hainan and Yunnan. Native in SE Asia.

小驳骨 *Justicia gendarussa*

小驳骨
Justicia gendarussa Burm. f.

亚灌木，多分枝。茎于节间膨大。叶狭披针形。穗状花序顶生或腋生，间断，常形成叶状圆锥状；花冠乳白色，二唇形；下唇基部具紫色斑点；上唇具紫斑；雄蕊2。蒴果棍棒状。花期1-4月。生路边或灌丛。归化或栽植于云南、广西、广东、海南和台湾。原产或归化于南亚、东南亚和巴布亚新几内亚。

Subshrubs, much branched. Stems swollen at nodes. Leaves narrowly lanceolate. Spikes terminal or axillary, interrupted, usually in a leafy panicle; corolla creamy white, 2-lipped; lower lips violet dotted basally; upper lips violet blotched; stamens 2. Capsules clavate. Fl. Jan-Apr. Roadsides or thickets. Naturalized or cultivated in Yunnan, Guangxi, Guangdong, Hainan and Taiwan. Native to or naturalized in S and SE Asia, and Papua New Guinea.

爵床
Justicia procumbens L.

匍匐草本。茎4棱，具沟槽，具柔毛。叶近无毛至疏具硬毛，钟乳体多数。穗状花序圆柱状，密被柔毛；花萼裂片4，条形，边缘黄白色，具缘毛；花冠粉红色或白色，下唇具红色斑点。花果期终年。生海拔1500米以下的荒地、路边、草坪、旷野或海滨林地。产中国西南、华南、华中、华东、华北和西北。南亚、东南亚和东亚亦有。

Herbs procumbent. Stems 4-angled, sulcate, pubescent. Leaves subglabrous to sparsely hispid, cystoliths numerous. Spikes cylindric, densely pilose; calyx lobes 4, linear, margin yellowish white and ciliate; corolla pink or white and red-spotted on lower lip. Fl. and fr. all year. Wastelands, roadsides, lawns, open fields or littoral forests below 1500 m. Distributed in SW, S, C, E, N and NW China. Also S, SE and E Asia.

爵床 *Justicia procumbens*

苦槛蓝科 Myoporaceae

苦槛蓝

Pentacoelium bontioides (Sieb. et Zucc.) A. Gray

灌木。叶互生，无毛；叶软革质。花腋生，具1或2(-4)成簇；花冠白色至浅粉紫色，具紫色斑点或条纹，漏斗形至钟形。果实紫红色但是干后浅棕色，卵状球形，肉质，具5-8粒种子。花期4-6月，果期5-7月。生近海平面的沙地或海岸边多石灌丛。产广西、广东、海南、台湾、福建和浙江。越南和日本亦有。

Shrubs. Leaves alternate, glabrous; leaves softly leathery. Flowers axillary, 1 or 2(-4) in fascicles. Corolla white to pale pinkish purple with purple spots or streaks, funnel-shaped to campanulate. Fruits purplish red but drying pale brown, ovoid-spherical, fleshy, with 5-8 seeds. Fl. Apr-Jun. Fr. May-Jul. Sandy sitcs or stony thickets along coasts at near sea level. Distributed in Guangxi, Guangdong, Hainan, Taiwan, Fujian and Zhejiang. Also in Vietnam and Japan.

苦槛蓝 *Pentacoelium bontioides*

透骨草科 Phrymaceae

透骨草

Phryma leptostachya L. subsp. **asiatica** (Hara) Kitamura

多年生草本。茎4棱。叶两面具柔毛。花序轴长10-30厘米；苞片钻形至条形；花梗花期直立，后反折；花冠亮紫色、浅红色至白色，管状漏斗形；冠檐二唇形。瘦果长椭圆体形，包于宿存棒状花萼中。花期6-10月，果期8-12月。生海拔300-2800米的湿润沟谷、林中、路边、山坡或山腰。产中国西南、华南、东南、华中、华北、华西、华东和东北。南亚、东南亚和东北亚亦有。

Perennial herbs. Stems 4-angular. Both surfaces of leaves puberulent. Inflorescences axises 10-30 cm long; bracts subulate to linear; pedicels erect at anthesis but later reflexed; corolla light purplish, pale red to white, tubular-funnelform; limbs 2-lipped. Achenes long ellipsoid, enveloped in persistent clavate calyx. Fl. Jun-Oct. Fr. Aug-Dec. Moist ravines, forests, roadsides, mountain slopes or hillsides at 300-2800 m. Distributed in SW, S, SE, C, N, W, E and NE China. Also in S, SE and NE Asia.

透骨草 *Phryma leptostachya* subsp. *asiatica*

车前科
Plantaginaceae

北车前
Plantago media L.

多年生草本，具直根。叶椭圆形至倒卵形，具7或9脉，被白色柔毛。花冠银白色；雄蕊着生于冠筒近基部。盖果卵状椭圆体形，于中部稍下方周裂，具(2-)4粒种子。种子长1.5-2毫米，腹面隆起。花期6-8月，果期7-9月。生海拔1360-2000米的草甸、河滩、沟谷或山坡台地。产内蒙古和新疆。欧洲和中亚和北业亦有。

北车前 *Plantago media*

Perennial herbs, with a taproot. Leaves elliptic to obovate, veins 7 or 9, white pilose. Corolla silver-white; stamens adnate to near base of corolla tube. Pyxis ovoid-ellipsoid, circumscissile slightly below middle, with (2-)4 seeds. Seeds 1.5-2 mm long, ventral face prominent, shiny. Fl. Jun-Aug. Fl. Jul-Sep. Meadows, wet banks, ravines, mountain slopes or terraces at 1360-2000 m. Distributed in Neimenggu and Xinjiang. Also in Europe, C and N Asia.

大车前
Plantago major L.

多年生草本，具须根。叶宽卵形至宽椭圆形，具(3-)5或7脉。花无梗；花冠白色；雄蕊着生于冠筒近基部。盖果于中部或稍下周裂，具(8-)12-24(-34)粒种子。种子长0.8-1.2毫米，具角，黄褐色。花期6-8月，果期7-9月。生海拔2800米以下的草甸、湿地或荒地。广布中国。欧亚温带和寒温带亦有。

Perennial herbs, with fibrous roots. Leaves broadly ovate to broadly elliptic, veins (3-)5 or 7. Flowers sessile; corolla white; stamens adnate to near base of corolla tube, anthers purplish or white. Pyxis circumscissile at or just below middle, with (8-)12-24(-34) seeds. Seeds 0.8-1.2 mm long, angled, yellowish brown. Fl. Jun-Aug. Fr. Jul-Sep. Meadows, wet places or waste places below 2800 m. Widely distributed in China. Also in temperate and cold temperate areas of Asia and Europe.

车前
Plantago asiatica L.

多年生草本，具须根。叶宽卵形至宽椭圆形，具5-7脉。花具梗；龙骨突不延至萼片先端；花冠白色；雄蕊着生于冠筒近基部；盖果近基部周裂，具5-6(-12)粒种子。种子长(1.2-)1.5-2毫米，具角，褐黑色。花期4-8月，果期6-9月。生海拔3800米以下的山坡、河岸、路边或草地。广布中国。南亚和东北亚亦有。

Perennial herbs, with fibrous roots. Leaves broadly ovate to broadly elliptic, veins 5-7.

大车前 *Plantago major*

疏花车前 *Plantago asiatica* subsp. *erosa*

裂，具3或5脉。花冠淡黄色；雄蕊着生于冠筒近基部。盖果纺锤状椭圆体形，于中部稍下方周裂，具2-3粒种子。种子长2.1-2.8毫米，腹面具1浅沟，黑褐色。花果期4-5月。生海拔140-160米的江心冲击岛。特产重庆。

Perennial herbs, with fibrous roots. Leaves lanceolate to linear-lanceolate, margin usually dentate to incised, veins 3 or 5. Corolla yellowish; stamens adnate to near base of corolla tube. Pyxis fusiform-ellipsoid, circumscissile just below middle, with 2-3 seeds. Seeds 2.1-2.8 mm long, with a shallow groove on ventral face, blackish brown. Fr. and fr. Apr-May. Alluvial islets at 140-160 m. Endemic in Chongqing.

车前 *Plantago asiatica*

Flower shortly pedicellate; keel not extending to apex of sepal; corolla white; stamens adnate to near base of corolla tube; pyxis circumscissile near base, with 5-6(-12) seeds. Seeds (1.2-)1.5-2 mm long, angled, blackish brown. Fl. Apr-Aug. Fr. Jun-Sep. Mountain slopes, river banks, roadsides or lawns below 3800 m. Widely distributed in China. Also in S and NE Asia.

疏花车前

Plantago asiatica L. subsp. **erosa** (Wall.) Z. Y. Li

多年生草本，具须根。叶宽卵形至宽椭圆形，具3或5脉。龙骨突延至萼片先端；花冠白色；雄蕊着生于冠筒近基部。盖果圆锥状卵球形，具6-15粒种子。种子长1.2-1.7(-2)毫米，褐黑色。花期5-7月，果期8-9月。生海拔400-3800米的山坡、河岸、田中或路边。产中国西南、华南、华中、华东和青海。南亚亦有。

Perennial herbs, with fibrous roots. Leaves broadly ovate to broadly elliptic, veins 3 or 5. Keel extending to apex of sepal; corolla white; stamens adnate to near base of corolla tube. Pyxis conic-ovoid, with 6-15 seeds. Seeds 1.2-1.7 (-2) mm long, blackish brown. Fl. May-Jul. Fr. Aug-Sep. Mountain slopes, river banks, fields or roadsides at 400-3800 m. Distributed in SW, S, C and E China, and Qinghai. Also in S Asia.

丰都车前

Plantago fengdouensis (Z. E. Chao et Yong Wang) Yong Wang et Z. Y. Li

多年生草本，具须根。叶披针形至条状披针形，边缘常具牙齿或锐

丰都车前 *Plantago fengdouensis*

北美车前 *Plantago virginica*

北美车前
Plantago virginica L.

一年生或二年生草本，具直根。叶倒披针形至倒卵状披针形，具3或5脉。花冠淡黄色；雄蕊着生于冠筒顶端。盖果卵球形，于下部周裂，具2粒种子。种子长(1-)1.4-1.8毫米，腹面具1浅沟，淡褐色。花期4-5月，果期5-6月。生海拔800米以下的草地、路边或湖畔。归化于中国西南、华南、华东和华中。原产北美洲。

Annual or biennial herbs, with a taproot. Leaves oblanceolate to obovate-lanceolate, veins 3 or 5. corolla yellowish; stamens adnate to near apex of corolla tube. Pyxis ovoid, circumscissile near base, with 2 seeds. Seeds (1-)1.4-1.8 mm long, with a shallow groove on ventral face, light brown. Fl. Apr-May. Fr. May-Jun. Grasslands, roadsides or lake banks below 800 m. Naturalized in SW, S, E and C China. Native to North America.

平车前 *Plantago depressa*

平车前
Plantago depressa Willd.

一年生、冬性一年生或多年生草本，具直根。叶椭圆形至卵状披针形；具5或7脉。花冠白色；雄蕊着生于冠筒近顶端。盖果圆锥状卵球形，近基部周裂，具4粒种子。种子长1.2-1.8毫米，腹面隆起至稍扁平，黄褐色至黑色。花期5-7月，果期7-9月。生海拔4500米以下的草甸、湿地或路边。除华南外中国广布。南亚、中亚、西亚和东北亚亦有。

Annual herbs, winter annual or perennial herbs with a taproot. Leaves elliptic to ovate-lanceolate, white pubescent; veins 5 or 7. Corolla white; stamens adnate to near apex of corolla tube. Pyxis conic-ovoid, circumscissile near

长叶车前 *Plantago lanceolata*

base, with 4 seeds. Seeds 1.2-1.8 mm long, ventral face prominent to slightly flat, yellowish brown to black. Fl. May-Jul. Fr. Jul-Sep. Meadows, wet places or roadsides below 4500 m. Widespread in China, except S China. Also in S, C, W and NE Asia.

长叶车前
Plantago lanceolata L.

多年生草本，具直根。叶条状披针形至椭圆状披针形，具(3-)5(-7)脉。花冠白色；雄蕊贴生于冠筒中部。盖果狭卵球形，于基部周裂，具(1或)2粒种子。种子长2-2.6毫米，腹面具1宽沟，褐色至深褐色。花期5-6月，果期7-8月。生海拔900米以下的海边、草地或沼泽地。广布中国。北温带亦有。

Perennial herbs, with a taproot Leaves linear-lanceolate to elliptic-lanceolate, veins (3-)5(-7). Corolla white; stamens adnate to near middle of corolla tubeh. Pyxis narrowly ovoid, circumscissile near base, with (1 or)2 seeds. Seeds 2-2.6 mm long, with a broad groove on ventral face, brown to dark brown. Fl. May-Jun. Fr. Jul-Aug. Seasides, meadows or boggy places below 900 m. Widespread in China. Distributed in the north temperate zone.

芒苞车前
Plantago aristata Michaux.

一年生或越年生草本，具直根。叶椭圆状条形至条形，具3脉，被长柔毛。苞片远长于花，具狭条形至钻形的芒尖；花冠淡黄色；雄蕊着生于冠筒近顶端。盖果椭圆体形，于中部稍下方周裂，具2粒种子。种子长(1.9-)2.3-2.7毫米，腹面具1宽沟。花期5-6月，果期6-7月。生海拔3-20米的海边、荒地及路旁。归化于山东和江苏。原产北美洲。

Annual or winter annual herbs, with a taproot. Leaves elliptic-linear to linear, veins 3, villose. Bracts much longer than flowers, with linear to subulate arista; corolla yellowish; stamens adnate to near apex of corolla tube. Pyxis ellipsoid, circumscissile slightly below middle, with 2 seeds. Seeds (1.9-)2.3-2.7 mm long, with a broad groove on ventral face. Fl. May-Jun. Fr. Jun-Jul. Seaside, wastelands and roadsides at 3-20 m. Naturalized in Shandong and Jiangsu. Native to North America.

芒苞车前 *Plantago aristata*

小车前 *Plantago minuta*

小车前
Plantago minuta Pall.

一年生或多年生草本，具直根。叶狭条形至条状披针形，具3脉。花冠白色；雄蕊着生于冠筒近顶端。盖果卵球形，于近基部周裂，具2粒种子。种子长(2.5-)3-4毫米，腹面具1宽沟，黄褐色。花期6-8月，果期7-9月。生海拔400-4300米的盐碱地、戈壁滩、沟谷或河滩。产内蒙古、山西、西北和西藏。俄罗斯南部、蒙古和哈萨克斯坦。

Annual or perennial herbs, with a taproot. Leaves basal, narrowly linear to linear-lanceolate, veins 3. Corolla white; stamens adnate to near apex of corolla tube. Pyxis ovoid, circumscissile near base, with 2 seeds. Seeds (2.5-)3-4 mm long, with a broad groove on ventral face, yellowish brown. Fl. Jun-Aug. Fr. Jul-Sep. Saline places, gravel beds, or ravines at 400-4300 m. Distributed in Neimenggu, Shanxi, NW China and Xizang. Also in S Russia, Mongolia and Kazakhstan.

盐生车前
Plantago maritima L. subsp. **ciliata** Printz

多年生草本，具直根。叶条形，革质，具(1-)3或5脉。花冠白色至淡黄色，筒部具柔毛，裂片具缘毛；雄蕊着生于冠筒中部。盖果圆锥状卵球形，约于中部周裂，具1或2粒种子。种子长1.6-2.3毫米，腹面平坦，黄褐色至深褐色。花期6-7月，果期7-8月。生海拔100-3800米的盐质沼泽或盐质草甸。产华北和西北。中亚亦有。

Perennial herbs, with a taproot. Leaves linear, leathery, veins (1-)3 or 5. Corolla white to yellowish, tubes pubescent, lobes ciliolate; stamens adnate to middle of corolla tube. Pyxis conic-ovoid, circumscissile at about middle, with 1 or 2 seeds. Seeds 1.6-2.3 mm long, ventral face flat, yellowish brown to dark brown. Fl. Jun-Jul. Fr. Jul-Aug. Salt marshes or saline meadows at 100-3800 m. Distributed in N and NW China. Also in C Asia.

盐生车前 *Plantago maritima* subsp. *ciliata*

茜草科 Rubiaceae

岩上珠
Clarkella nana (Edgew.) Hook. f.

矮小草本，高10厘米。块根长球形。叶片薄纸质，卵形至阔卵形，两面近无毛。聚伞花序顶生；花冠管状漏斗形，白色，被短柔毛。果倒圆锥形，被微柔毛。花期5-7月，果期7-9月。生海拔1000-1400米的潮湿岩石上。产云南南部及东南部和西南部、贵州中部、广西北部和广东北部。印度北部、缅甸北部和泰国亦有。

Low herbs, to 10 cm tall. Tubers ellipsoid-oblong. Leaves drying thinly papery, ovate to broadly ovate, both surfaces subglabrous. Inflorescences terminal, cymose; corolla funnelform, white, outside

岩上珠 *Clarkella nana*

岩黄树 *Xanthophytum kwangtungense*

Fruits indehiscent, globose to ovoid. Seeds foveolate. Fl. and fr. Mar-Dec. Forest edges, open fields, streamsides, thickets and grasslands at 100-1500 m. Distributed in SW and S China. Also in Japan, S and SE Asia, and Australia.

puberulent. Fruit obconical, villosulous. Fl. May-Jul. Fr. Jul-Sep. Wet rocks at 1000-1400 m. Distributed in S, SE and SW Yunnan, C Guizhou, N Guangxi and N Guangdong. Also in N India, N Myanmar and Thailand.

岩黄树
Xanthophytum kwangtungense (Chun et F. C. How) H. S. Lo

直立灌木，高0.5-1米；小枝密被柔毛。叶纸质，椭圆形或椭圆状长圆形，上面脉上常被疏柔毛，下面密被长柔毛。聚伞花序紧缩成头状；花萼裂片近狭匙形；花冠浅黄色，钟状漏斗形。蒴果近球形，被短硬毛。花期5月，果期7-10月。生林下潮湿地方。产广西东南部和云南南部。越南亦有。

Shrubs, 0.5-1 m tall; branches densely pilose. Leaves elliptic or elliptic-oblong, papery, adaxially usually sparsely pilose along veins, abaxially densely villous. Inflorescences subcapitate to congested-cymose; calyx lobes subspatulate; corolla pale yellow, campanulate-funnelform. Capsules subglobose, strigillose. Fl. May. Fr. Jul-Oct. Wet places in forests. Distributed in SE Guangxi and S Yunnan. Also in Vietnam.

耳草
Hedyotis auricularia L.

多年生草本。叶披针形至椭圆形；托叶具3-9条形或刚毛状裂片。花序腋生，团伞状至密集聚伞状；花冠白色，管状或管状漏斗形。果实开裂，球形至卵球形。种子具凹坑。花果期3-12月。生海拔100-1500米的林缘、开阔地、溪边、灌丛和草地。产中国西南和华南。日本、南亚、东南亚和澳大利亚亦有。

Perennial herbs. Leaves lanceolate to elliptic; stipules with 3-9 linear or setiform lobes. Inflorescences axillary, glomerulate to congested-cymose; corolla white, tubular or tubular-funnelform.

金毛耳草
Hedyotis chrysotricha (Palib.) Merr.

铺散草本，长约30厘米，被金黄色硬毛。叶薄纸质，阔披针形、椭圆形或卵形。聚伞花序腋生，具1-3花，近无梗；花4数；花冠白色或紫色，漏斗状。蒴果近球形。花果期几乎全年。生海拔100-900米山谷阔叶林中或山坡灌木丛中。产中国西南、华南、华中和华东。菲律宾和日本亦有。

Diffuse herbs, ca. 30 cm long, with golden yellow hairs. Leaves thinly papery, broadly lanceolate, elliptic or ovate. Cymes axillary, 1-3-flowered, subsessile; flowers 4-merous; corolla white or purple, funnelform. Capsules subglobose. Fl. and fr. nearly year-round. Broad-leaved forests in valleys or thickets on montane slopes at 100-900 m. Commonly distributed in SW, S, C and E China. Also in the Philippines and Japan.

耳草 *Hedyotis auricularia*

金毛耳草 *Hedyotis chrysotricha*

鼎湖耳草 *Hedyotis effusa*

剑叶耳草 *Hedyotis caudatifolia*

鼎湖耳草

Hedyotis effusa Hance

直立草本，高可达1米，几全株无毛。茎圆柱形。叶纸质，卵形或卵状披针形。聚伞花序排列成圆锥花序状，顶生；花4数；花冠漏斗形，白色，外面无毛，里面喉部被毛。蒴果近球形。花期7-9月，果期8月至翌年3月。生海拔200-500米的林下或山谷溪旁。产广东和广西。

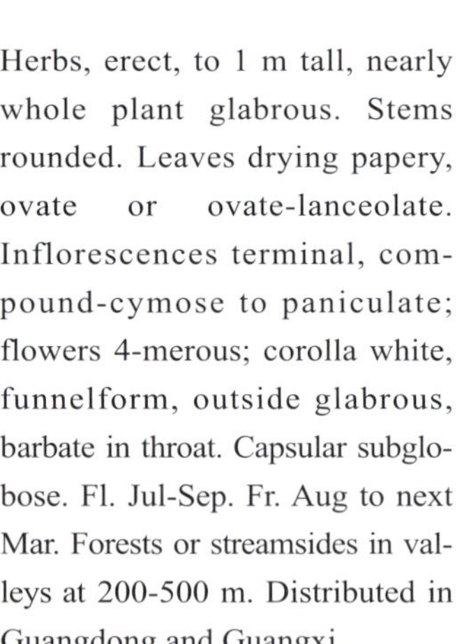

Herbs, erect, to 1 m tall, nearly whole plant glabrous. Stems rounded. Leaves drying papery, ovate or ovate-lanceolate. Inflorescences terminal, compound-cymose to paniculate; flowers 4-merous; corolla white, funnelform, outside glabrous, barbate in throat. Capsular subglobose. Fl. Jul-Sep. Fr. Aug to next Mar. Forests or streamsides in valleys at 200-500 m. Distributed in Guangdong and Guangxi.

剑叶耳草

Hedyotis caudatifolia Merr. et Metcalf

直立亚灌木，高达90厘米，全株无毛。叶革质，通常披针形。聚伞花序排成圆锥花序式；花4数；花冠管状，白色或粉红色。蒴果长圆形或椭圆形。花期5-6月，果期6-8月。生海拔300-700米的丛林下比较干旱的沙质土壤上。产福建、广东、广西、湖南、江西和浙江。

Subshrubs, erect, to 90 cm tall, whole plant glabrous; stems glabrous. Leaves drying leathery, usually lanceolate. Inflorescences compound-cymose to paniculate; flowers 4-merous; corolla white or pink, tubular. Capsules ellipsoid-oblong or ellipsoid. Fl. May-Jun. Fr. Jun-Aug. Dry soil in jungles or thickets at 300-700 m. Distributed in Fujian, Guangdong, Guangxi, Hunan, Jiangxi and Zhejiang.

粗毛耳草

Hedyotis mellii Tutch.

多年生直立草本。茎近方柱形，被粗毛至无毛。叶纸质，卵状披针形，两面均被疏短毛。花序生顶端及常生于最上部叶腋，聚伞状或圆锥状；花冠白色，钟形，喉部和裂片密被绒毛。花期6-11月，果期8-11月。生海拔400-1100米的山地或山坡上的热带丛林或灌丛下。产广西、广东、福建、湖南和江西。

Perennial herbs, erect. Stems subquadrangular, hirtellous to glabrous. Leaves papery, ovate-lanceolate, both surfaces sparsely puberulent. Inflorescences terminal and usually also in axils of uppermost leaves, cymose or paniculate; corolla white, funnelform, inside densely tomentose in throat and onto lobes. Fl. Jun-Nov. Fr. Aug-Nov. Jungles or thickets on mountains or mountain slopes at 400-1100 m. Distributed in Guangxi, Guangdong, Fujian, Hunan and Jiangxi.

粗毛耳草 *Hedyotis mellii*

corolla white or yellow, tabular. Capsules compressed globose. Fl. Jul-Sep. Fr. Aug-Sep. Montane forests or damp sites in valleys at 1000-1800 m. Distributed in Yunnan. Also in S and SE Asia.

长节耳草 *Hedyotis uncinella*

牛白藤 *Hedyotis hedyotidea*

长节耳草

Hedyotis uncinella Hook. et Arn.

直立多年生草本，无毛。茎单生，4棱，节间长。叶纸质，卵状长圆形或长圆状披针形。花序顶生或腋生，密集成头状；花4数；花冠白色或紫色。蒴果卵球形，熟时开裂为2果片。花果期4-9月。生海拔200-1200米的干旱旷地上。产贵州、广东、台湾、海南、香港和湖南。印度和缅甸亦有。

Erect perennial herbs, glabrous. Stems single, 4-angled, internodes long. Leaves papery, ovate-oblong or oblong-lanceolate. Inflorescences terminal or axillary, capitate; flowers 4-merous; corolla white or purple. Capsules ovoid, 2-valved at maturity. Fl. and fr. Apr-Sep. Drought open field at 200-1200 m. Distributed in Guizhou, Guangdong, Taiwan, Hainan, Hong Kong and Hunan. Also in India and Myanmar.

攀茎耳草

Hedyotis scandens Roxb.

多年生灌木或草本，藤状，攀援或铺散，无毛。叶近革质，长圆状披针形或窄椭圆形，侧脉4-5对。聚伞花序排成开展的圆锥花序；花4数；花冠白色或黄色，筒状。蒴果扁球形。花期7-9月，果期8-9月。生海拔1000-1800米的山地林中或山谷潮湿处。产云南。南亚和东南亚亦有。

Shrubs or herbs, perennial, lianescent, climbing or scandent, glabrous. Leaves subleathery, oblong-lanceolate or narrowly elliptic, lateral veins 4-5 pairs. Cymes arranged into spreading panicles; flowers 4-merous;

攀茎耳草 *Hedyotis scandens*

牛白藤

Hedyotis hedyotidea (DC.) Merr.

灌木或亚灌木。茎密被乳突至粉末状柔毛。近头状花序生顶端及最上部叶腋，由1-10个聚伞花序组成；花冠白色至灰黄色，管状至漏斗状。果实先腹部开裂，然后室间开裂。种子数粒，具棱。花果期4-12月。生海拔200-1000米的沟谷灌丛或山坡。产中国西南、华南和东南。东南亚亦有。

Shrubs or subshrubs. Stems densely papillose to farinose puberulent. Inflorescences terminal and in axils of uppermost leaves, subcapitate, with 1-10 cymose groups of flowers; corolla white to pale yellow, tubular to funnelform. Fruits loculicidally then septicidally dehiscent. Seeds several, angled. Fl. and fr. Apr-Dec. Thickets in ravines or hill slopes at 200-1000 m. Distributed in SW, S and SE China. Also in SE Asia.

伞房花耳草 *Hedyotis corymbosa*

伞房花耳草

Hedyotis corymbosa (L.) Lam.

柔弱草本，高可达40厘米；茎4棱。叶膜质，线形至狭披针形。聚伞花序腋生；花4数；花冠白色或粉红色，管形。蒴果近球形，膜质。花果期几乎全年。生海拔300-900米的沟边或湿润的草地上。产云南和四川向东至华东。亚洲热带地区、非洲和美洲等地亦有。

Slender herbs, to 40 cm tall; stems 4-angled. Leaves linear to narrowly lanceolate, membranous. Inflorescences axillary, cymose; flowers 4-merous; corolla white or pink, tubular. Capsules subglobose, membranous. Fl. and fr. almost year-round. Streamsides or humid grasslands at 300-900 m. Spreading from Yunnan and Sichuan eastward to E China. Also in tropical Asia, Africa and America.

白花蛇舌草

Hedyotis diffusa Willd.

铺散草本，无毛，长20-50厘米。叶无柄，膜质，条形，侧脉不明显。花腋生，单生或双生，近无梗，4数；花冠白色，管状。蒴果膜质，扁球形。花果期5-10月。生海拔900米以下的梯田、田埂或湿润旷野中。产中国西南、华南和华东。日本、南亚和东南亚亦有。

Diffuse herbs, glabrous, 20-50 cm long. Leaves sessile, membranous, linear, lateral veins indistinct. Flowers axillary, single or binate, subsessile, 4-merous; corolla white, tubulate. Capsules membranous, compressed globose. Fl. and fr. May-Oct. Paddy fields, farmland ridges or humid open fields below 900 m. Distributed in SW, S and E China. Also in Japan, S and SE Asia.

白花蛇舌草 *Hedyotis diffusa*

臭味新耳草

Neanotis ingrata (Wall. ex Hook. f.) W. H. Lewis

草本，有臭味，被长毛。茎显著具棱或槽。叶卵状披针形。多歧聚伞花序顶生或近顶生，具梗；花近无梗；花冠白色，管状漏斗形。蒴果近扁球状。花期6-9月。生海拔500-1500米的河岸草坡上或山坡林下。产中国西南、华南、华中和华东。不丹、印度和尼泊尔亦有。

Herbs, smelly, pilose. Stems distinctly ridged or sulcate. Leaves ovate-lanceolate. Pleiochasiums terminal or subterminal, pedunculate; flowers subsessile; corolla white, tubular-funnelform. Capsules nearly oblate. Fl. Jun-Sep. Grassy slopes on river banks or forests on montane slopes at 500-1500 m. Distributed in SW, S, C and E China. Also in Bhutan,

臭味新耳草 *Neanotis ingrata*

薄叶新耳草 *Neanotis hirsuta*

India and Nepal.

薄叶新耳草

Neanotis hirsuta (L. f.) W. H. Lewis

匍匐草本。茎具纵棱。叶卵形或椭圆形，两面近无毛。花序假腋生或顶生，常聚集成头状；花白色，漏斗形，外面无毛。蒴果扁球形，有毛或无毛。种子平凸。花果期6-10月。生海拔500-1500米的林下或溪旁湿地上。产华东、东南、华南和西南。南亚、东南亚和日本亦有。

Creeping herbs. Stems angles. Leaves drying papery, ovate or elliptic, both surfaces subglabrous. Inflorescences terminal or pseudoaxillary, capitate; corolla white. Capsules compressed globose, glabrous. Seeds plano-convex. Fl. and fr. Jun-Oct. Wet sites at streamsides or in forests at 500-1500 m. Distributed in E, SE, S and SW China. Also in S and SE Asia, and Japan.

西南新耳草

Neanotis wightiana (Wall. ex Wight et Arn.) W. H. Lewis

多年生匍匐草本，节上生根。茎显著具棱和槽。叶卵形至披针形，脉不明显。聚伞花序顶生，无梗，团聚状，常具2叶状苞片；花近无梗；花冠白色，稀淡红色。蒴果扁球形，室间具深沟。花期5-7月，果期6-10月。生海拔900-1900米的草坡、路边或溪边。产云南、四川、广西和贵州。不丹、印度和越南亦有。

Perennial procumbent herbs, rooting at nodes. Stems distinctly ridged and sulcate. Leaves ovate to lanceolate, veins indistinct. Cymes terminal, sessile, glomerate, usually with 2 foliaceous bracts; flowers subsessile; corolla white, rarely pale red. Capsules oblate, deeply grooved between cells. Fl. May-Jul. Fr. Jun-Oct. Grassy slopes, roadsides or streamsides at 900-1900 m. Distributed in Yunnan, Sichuan, Guangxi and Guizhou. Also in Bhutan, India and Vietnam.

伞花螺序草

Spiradiclis umbelliformis H. S. Lo

匍匐草本。茎密被多细胞长柔毛。叶纸质，卵状圆形，两面无毛或近无毛；侧脉4-6对。聚伞花序伞形状；花二型；花冠白色或微红紫，漏斗状，外面无毛。蒴果近球状，无毛。花期4-5月，果期6-9月。生海拔900-1100米林下的岩石隙缝中。产广西(那坡县)。

Creeping herbs. Stems densely multicellular villosulous. Leaves ovate-orbicular, papery, both surfaces glabrescent or subglabrous; secondary veins 4-6 pairs. Inflorescences cymose, umbelliform; flowers distylous; corolla white to pale purplish red, funnelform, glabrous outside. Capsules subglobose, glabrescent. Fl. Apr-May. Fr. Jun-Sep. Rock crevices in forests at 900-1100 m. Distributed in Guangxi (Napo County).

西南新耳草 *Neanotis wightiana*

伞花螺序草 *Spiradiclis umbelliformis*

糙边螺序草 *Spiradiclis scabrida*

心叶螺序草 *Spiradiclis cordata*

心叶螺序草

Spiradiclis cordata H. S. Lo et W. L. Sha

草本。无茎或具短茎。叶簇生短茎上，叶片纸质，椭圆状卵形或椭圆状长圆形，基部心形，上面被疏短糙毛，下面被长柔毛；侧脉15-19对。花序聚伞状至圆锥状；花冠白色，管状漏斗形。蒴果近球形，被柔毛。花期8月，果期8-11月。生海拔200-550米的岩石上。产广西西南部。

Herbs. Acaulescent or stems short. Leaves clustered at base of plant; blade elliptic-ovate to elliptic-oblong, papery, base cordate; adaxially sparsely hispidulous, abaxially villous, secondary veins 15-19 pairs. Inflorescences cymose to paniculate; corolla white, tubular-funnelform. Capsules subglobose, puberulent. Fl. Aug. Fr. Aug-Nov. Rocks at 200-550 m. Distributed in SW Guangxi.

糙边螺序草

Spiradiclis scabrida D. Fang et D. H. Qin

草本，高40-50厘米。嫩枝被短柔毛。叶片纸质，卵形或狭卵形，常疏被短糙伏毛；侧脉6-9对。伞房状聚伞花序；花冠淡粉红色，高脚碟状，无毛，长25-26毫米。蒴果近球形，无毛。花期11-12月，果期翌年3-4月。生海拔800-1200米的林中岩石上。产广西(那坡县)。

Herbs, 40-50 cm tall. Branches puberulent when young. Leaves ovate or narrowly ovate, papery, usually sparsely scabridulous; secondary veins 6-9 pairs. Inflorescences corymbose-cymose; corolla pale pink, salverform, both sides glabrous; tube 25-26 mm. Capsules subglobose, glabrescent. Fl. Nov-Dec. Fr. next Mar-Apr. Rocks in forests at 800-1200 m. Distributed in Guangxi (Napo County).

紫花螺序草

Spiradiclis purpureocaerulea H. S. Lo

草本。茎密被柔毛。叶纸质，卵形，上面密被糙毛状短硬毛，下面密被柔毛。聚伞花序紧缩成近头状，总梗极短；花冠蓝紫色，高脚碟形，冠管纤细，长2.5-2.7厘米。蒴果近球形，被柔毛。花期7-8月，果期9-10月。生海拔150-500米密林下的石灰岩石上。产广西南部。

Herbs. Stems densely pubescent. Leaves drying papery, ovate, adaxially densely strigose-hispidulous, abaxially densely pubescent. Inflorescences subcapitate to congested-cymose; puduncles very short; corolla bluish purple, slenderly salverform, 2.5-2.7 cm. Capsules subglobose, pubescent. Fl. Jul-Aug. Fr. Sep-Oct. Dense forests on limestone hills at 150-500 m. Distributed in S Guangxi.

紫花螺序草 *Spiradiclis purpureocaerulea*

红花螺序草 *Spiradiclis coccinea*

红花螺序草
Spiradiclis coccinea H. S. Lo

直立草本，高约40厘米。茎常无毛。叶纸质，狭椭圆状长圆形至长圆形，两面无毛；侧脉6-9对。聚伞花序几无总梗；花冠深红色，管状高脚碟形，外面无毛或被短柔毛，里面在雄蕊之上被1环白色长毛或散生柔毛。蒴果近球形，近无毛。花期5-8月，果期7-10月。生海拔350-600米的密林下的岩石上。产广西南部。

Erect herbs, ca. 40 cm tall. Stems often glabrous. Leaves drying papery, narrowly elliptic-oblong to oblong, glabrous on both surfaces; secondary veins 6-9 pairs. Inflorescences cymose; peduncle very short; corolla dark red, tubular-salverform, outside glabrous or puberulent, inside with pilose ring above stamens or spasely pubescent. Capsules subglobose, subglabrous. Fl. May-Aug. Fr. Jul-Oct. Rocks in dense forests at 350-600 m. Distributed in S Guangxi.

长苞螺序草
Spiradiclis longibracteata S. Y. Liu et S. J. Wei

草本。茎密被长柔毛。叶茎生或基生成莲座状，叶片倒卵状披针形或倒披针形；侧脉18-22对。聚伞花序稠密多花；花冠紫红色，高脚蝶状，外面被短柔毛，近中部或基部被1环白色长柔毛。蒴果圆球形，被柔毛。花期4月，果期8-10月。生海拔150-550米的林下湿处。产广西(东兰县和隆安县)。

Herbs. Stems densely villous. Leaves cauline or clustered at base of plant; leaves obovate-lanceolate or oblanceolate; secondary veins 18-22 pairs. Inflorescences cymose, densely many flowered; corolla salverform, purplish red, sparsely puberulent outside, inside with villous ring near middle or at base. Capsules globose, pilose. Fl. Apr. Fr. Aug-Oct. Wet places in forests at 150-550 m. Distributed in Guangxi (Donglan County and Long'an County).

大叶螺序草
Spiradiclis bifida Kurz

直立草本，高达0.5米。茎被柔毛。叶片薄纸质，椭圆形或长圆形；侧脉15-19对。花序圆锥状，长达20厘米；花白色，短管形，外面无毛。果近球形。生海拔150-700米密林下的潮湿地方。产云南西南部和南部。印度东北部亦有。

长苞螺序草 *Spiradiclis longibracteata*

Erect herbs, to 0.5 m tall. Stems pubescent. Leaves drying thinly papery, elliptic or oblong; secondary veins 15-19 pairs. Inflorescences paniculate, to 20 cm; flowers white, shortly tubular, outside glabrous. Capsules subglobose. Wet places in forests at 150-700 m. Distributed in SW and S Yunnan. Also in NE India.

大叶螺序草 *Spiradiclis bifida*

两广螺序草 *Spiradiclis fusca*

两广螺序草

Spiradiclis fusca H. S. Lo

草本，高30-80厘米。茎无毛。叶薄纸质，长圆状披针形至近椭圆形；侧脉10-13对。花序圆锥状；花冠白色或淡紫色，漏斗状，外面无毛。果近球形，近无毛。花期2-3月，果期4-6月。生海拔150-450米的石山上。产广西(桂林)和广东(连县)。

Herbs, 30-80 cm tall. Stems glabrous. Leaves drying thinly papery, oblong-lanceolate to subelliptic, secondary veins 10-13 pairs. Inflorescences paniculiform; corolla white or pale purple, funnelform, glabrous outside. Capsules subglobose, subglabrous. Fl. Feb-Mar. Fr. Apr-Jun. Limestone hills at 150-450 m. Distributed in Guangxi (Guilin) and Guangdong (Lian County).

疏花螺序草 *Spiradiclis laxiflora*

滇南螺序草

Spiradiclis malipoensis H. S. Lo

草本，高20-25厘米。叶纸质，披针状长圆形或卵状椭圆形，两面无毛。聚伞花序伞形；花白色至粉红色，管状漏斗形，外面无毛，里面喉部被较密的柔毛。蒴果近椭圆形，无毛。花期4-6月，果期7-12月。生海拔1100-1350米处的密林中。产云南(麻栗坡县)。

Herbs, 20-25 cm tall. Leaves papery, lanceolate-oblong or ovate-elliptic, glabrous. Inflorescences cymose, umbelliform; corolla white to pink, tubular-funnelform, glabrous outside, inside densely pubescent at throat. Capsules ellipsoid, glabrous. Fl. Apr-Jun. Fr. Jul-Dec. Dense forests at 1100-1350 m. Distributed in Yunnan (Malipo County).

滇南螺序草 *Spiradiclis malipoensis*

疏花螺序草

Spiradiclis laxiflora W. L. Sha et X. X. Chen

多年生草本，高10-25厘米。茎无毛。叶纸质，椭圆形或倒卵状椭圆形，两面无毛；侧脉9-11对。聚伞花序圆锥状；花淡黄色或白色，管状漏斗形。蒴果椭圆状，无毛。花期4-5月，果期6月。生海拔150-500米疏林下的石灰岩石上。产广西南部。

Perennial herbs, 10-25 cm tall. Stems glabrous. Leaves drying papery, elliptic or obovate-elliptic, both surfaces glabrous, secondary veins 9-11 pairs. Inflorescences cymose, paniculiform; corolla slight yellow or white,

along streams at 50-1200 m. Distributed in Yunnan and Guangxi. Also in Myanmar and N Vietnam.

峨眉螺序草 *Spiradiclis emeiensis*

tubular-funnelform. Capsules ellipsoid, glabrous. Fl. Apr-May. Fr. Jun. Sparse forests on limestone hills at 150-500 m. Distributed in S Guangxi.

峨眉螺序草

Spiradiclis emeiensis H. S. Lo

草本，高达25厘米。茎密被柔毛。叶薄纸质，卵形至椭圆形，两面疏生短柔毛，下面脉上毛较密。聚伞花序稠密多花；花冠白色，管状漏斗形，外面被长柔毛。蒴果椭圆形，密被长柔毛。花期6月，果期8月。生海拔800-1200米的密林下。产四川和云南。

Herbs, up to 25 cm tall. Stems densely pubescent. Leaves drying thinly papery, ovate to elliptic, both surfaces sparsely pubescent or densely pubescent along abaxial veins. Inflorescences cymose, densely many flowered; corolla white, tubular funnelform, outside villous. Capsules narrowly ellipsoid, densely villous. Fl. Jun. Fr. Aug. Dense forests at 800-1200 m. Ditributed in Sichuan and Yunnan.

螺序草

Spiradiclis caespitosa Blume

多年生草本。茎被短柔毛。叶纸质，椭圆形或椭圆状卵形，两面近无毛至无毛。花序聚伞状；花冠白色，短管状。蒴果线状长圆形。常生海拔50-1200米林下的沟谷边，有时亦见于林缘的稻田中。产云南和广西。缅甸和越南北部亦有。

Perennial herbs. Stems puberulent. Leaves drying papery, elliptic to elliptic-ovate, puberulent to glabrescent on both surfaces. Inflorescences cymose; corolla white, shortly tubular, outside glabrous. Capsules linear-oblong. Moist shady sites, often

腺叶螺序草

Spiradiclis glandulosa L. Wu et Q. R. Liu

匍匐草本，高10厘米。茎密被短柔毛。叶草质，圆形或卵状圆形，下面密被腺状的小点；侧脉3-5对。花序伞形状；花萼裂片长圆状披针形；花冠粉红色，漏斗状，外面无毛。蒴果近球状，被微柔毛。花期1-3月，果期2-4月。生海拔150-450米的石灰岩山上。产广东(乳源县)、广西东部和湖南南部。

Creeping herbs, 10 cm tall. Stems densely villosulous. Leaves coriaceous, orbicular or ovate-orbicular, abaxially densely glandule-like spots, secondary veins 3-5 pairs. Inflorescences umbelliform; calyx lobes oblong-lanceolate. Corolla pink, funnelform, glabrous outside. Capsules subglobose, villosulous. Fl. Jan-Mar. Fr. Feb-Apr. Limestone hills at 150-450 m. Distributed in Guangdong (Ruyuan County), E Guangxi and S Hunan.

螺序草 *Spiradiclis caespitosa*

腺叶螺序草 *Spiradiclis glandulosa*

罗氏螺序草 *Spiradiclis loana*

罗氏螺序草

Spiradiclis loana R. J. Wang

草本，高达15厘米。茎短，被柔毛。叶莲座状，叶片倒披针形、椭圆形或倒卵形，纸质，上面被刺状毛；侧脉7-11对。聚伞花序圆锥状；花冠漏斗形，白色，外面被短柔毛。蒴果近球形。花期7-8月，果期8-11月。生海拔150-350米密林下的岩石上。产广西(龙州县)。

Herbs, up to 15 cm tall. Stems short, pubescent. Leaves rosette-like, blade oblanceolate, elliptic or obovate, papery, spiny pilose adaxially; secondary veins 7-11 pairs. Inflorescences cymose, paniculiform; corolla funnelform, white, puberulous outside. Capsules subglobose. Fl. Jul-Aug. Fr. Aug-Nov. Dense forests on limestone hills at 150-350 m. Distributed in Guangxi (Longzhou County).

少花螺序草

Spiradiclis pauciflora L. Wu et Q. R. Liu

少花螺序草 *Spiradiclis pauciflora*

草本，高10厘米。枝密被柔毛。叶纸质，常卵形或椭圆形，两面被柔毛；侧脉3-4对。聚伞花序有花3-5朵；花冠淡紫红色或近白色，漏斗形，外面近无毛，里面被柔毛或中部有1圈白色长毛。蒴果近球形，被柔毛。花期4-5月，果期5-6月。生海拔650-1350米密林下的岩石上。产广西西部。

Herbs, up to 10 cm tall. Stems densely pubescent. Leaves drying papery, usually ovate to elliptic-ovate, both surfaces pubescent; secondary veins 3-4 pairs. Inflorescences cymose, 3-5-flowered; corolla white to pale purplish red, funnelform, subglabrous outside, inside pubescent or with a pubescent ring of long hairs near middle. Capsules subglobose, puberulent. Fl. Apr-May. Fr. May-Jun. Dense forests on limestone hills at 650-1350 m. Distributed in W Guangxi.

独龙蛇根草 *Ophiorrhiza dulongensis*

独龙蛇根草

Ophiorrhiza dulongensis H. S. Lo

匍匐草本。茎被短柔毛。叶薄纸质，阔卵形或卵形，侧脉4-6对。花序有花3-4朵，簇生；花二型；苞片线形；花冠白色，漏斗状，外面近无毛，里面被密长柔毛。果未见。花期7月。生海拔2300-2400米的常绿阔叶林中。产云南西北部。

Procumbent herbs. Stems puberulent. Leaves drying thinly papery, broadly ovate or ovate, secondary veins 4-6 pairs. Inflorescences fa-

垂花蛇根草 *Ophiorrhiza nutans*

sciculate, 3- or 4-flowered; flowers distylous; bracts linear; corolla white, funnelform, glabrous outside, densely villous inside. Capsules unknown. Fl. Jul. Evergreen broad-leaved forests at 2300-2400 m. Distributed in NW Yunnan.

垂花蛇根草

Ophiorrhiza nutans C. B. Clarke ex Hook. f

草本，高可达70厘米。茎被多细胞长柔毛。叶片纸质，卵形、椭圆形或长圆形，侧脉9-13对。花序常密集成聚伞状；花冠白色，近管状，外面无毛，里面中部以上疏被柔毛。蒴果僧帽状，被短硬毛。生海拔700-2400米的密林下。产云南和西藏(墨脱县)。印度东北部、尼泊尔和缅甸亦有。

Herbs, up to 70 cm tall. Stems densely villous with multicellular trichomes. Leaves drying papery, ovate, elliptic or oblong; secondary veins 9-13 pairs. Inflorescences congested-cymose; corolla white, tubular-funnelform, glabrous outside, inside sparsely pubescent. Capsules mitriform, hispidulous. Moist forests at 700-2400 m. Distributed in Yunnan and Xizang (Mêdog County). Also in NE India, Nepal and Myanmar.

大齿蛇根草

Ophiorrhiza macrodonta H. S. Lo

高大草本或半灌木。叶纸质，长圆形至心形，侧脉14-16对，托叶卵形至卵状披针形。聚伞花序顶生，螺状，初时俯垂，后变直，具多花；花二型，花柱异长；花冠淡红色，管状漏斗形。蒴果倒心形。花期9月。生海拔约1500米的林下湿地。产云南。

Large herbs or subshrubs. Leaves papery, oblong to ovate, lateral veins 14-16 pairs, stipules ovate to lanceolate-ovate. Cymes terminal, heliciform, pendulous first, erect later, many-flowered; flowers dimorphic, heterostyled; corolla reddish, tubular-funnelform. Capsules obcordate. Fl. Sep. Wet places under forests at ca. 1500 m. Distributed in Yunnan.

大齿蛇根草 *Ophiorrhiza macrodonta*

那坡蛇根草

Ophiorrhiza napoensis H. S. Lo

草本，近直立，高30厘米或过之。茎无毛。叶纸质，狭披针形至近卵形，无毛。花序密集成聚伞状；花冠白色，高脚碟形，外面无毛，里面近无毛。果倒心形，无毛。花期10-11月，果期11月至翌年1月。生海拔1100-1200米的山坡林下。产广西(那坡县)。

Herbs, suberect, to 30 cm or taller. Stems glabrous. Leaves drying papery, narrowly lanceolate to subovate, glabrous on both surfaces. Inflorescences congested-cymose; corolla white, salverform, glabrous outside, subglabrous inside. Carpsules obcordate, glabrous. Fl. Oct-Nov. Fr. Nov to next Jan. Forests on hill slopes at 1100-1200 m. Distributed in Guangxi (Napo County).

那坡蛇根草 *Ophiorrhiza napoensis*

延翅蛇根草 *Ophiorrhiza alatiflora*

日本蛇根草 *Ophiorrhiza japonica*

延翅蛇根草

Ophiorrhiza alatiflora H. S. Lo

草本或亚灌木，近直立。茎无毛。叶片纸质，卵形或长圆状卵形，两面无毛。花序近簇生至近圆锥状；花二型；花冠白色带紫色，近管状，外面无毛，背部有阔翅，下延至冠管基部。蒴果僧帽状，被微柔毛。产海拔1000-1500米的密林下。产云南东南部。

Herbs or subshrubs, suberect. Stems glabrous. Leaves papery, ovate or oblong-ovate, glabrous on both surfaces. Inflorescences subfasciculate to paniculate; flowers distylous; corolla white striped with purple, outside glabrous and winged along entire length. Capsules mitriform, pilosulous. Forests at 1000-1500 m. Distributed in SE Yunnan.

日本蛇根草

Ophiorrhiza japonica Blume

多年生草本。茎上部直立，下部匍地生根。叶纸质，卵形至披针形，侧脉6-8对，叶柄压扁，托叶早落。聚伞花序顶生，多花；花二型，花柱异长；花冠近漏斗状，白色或粉红色。蒴果近僧帽状。花期冬季，果期春夏季。生海拔100-2400米肥沃的山谷林下。产中国西南、华南、华中和华东。越南和日本亦有。

Perennial herbs. Upper part of stems erect, lower part of stems creeping and rooting. Leaves papery, ovate to lanceolate, lateral veins 6-8 pairs, petioles flattened, stipules caducous. Cymes terminal, heliciform, many-flowered; flowers dimorphic, heterostyled; corolla subfunnelform, white or pink. Capsules submitriform. Fl. winter. Fr. spring to summer. Ravine forests with fertile soil at 100-2400 m. Distributed in SW, S, C and E China. Also in Vietnam and Japan.

广州蛇根草

Ophiorrhiza cantonensis Hance

草本或亚灌木，高可达1.2米。叶片纸质，通常长圆状椭圆形，两面无毛至被较密的短柔毛。花序圆锥状或伞房状；花二型；小苞片钻形或线形；花冠白色或微红，管状漏斗形，外面近无毛。蒴果僧帽状，密被短柔毛至近无毛。花期冬春季，果期春夏季。常生海拔300-900米的密林下沟谷边。产广东、海南、广西、云南和贵州。

Herbs or subshrubs, to 1.2 m tall. Leaves drying papery, usually oblong-elliptic, both surfaces glabrous to densely puberulent. Inflorescences paniculiform to corymbose; flowers distylous; bracts subulate or linear; corolla white to pink, tubular-funnelform, subglabrous outside.

广州蛇根草 *Ophiorrhiza cantonensis*

峨眉蛇根草 *Ophiorrhiza chinensis* f. *emeiensis*

Capsules mitriform, densely puberulent to subglabrous. Fl. winter and spring. Fr. spring and summer. Ravines and watersides in forests at 300-900 m. Distributed in Guangdong, Hainan, Guangxi, Yunnan and Guizhou.

峨眉蛇根草

Ophiorrhiza chinensis H. S. Lo f. **emeiensis** H. S. Lo

草本或亚灌木，高30-65厘米。茎无毛。叶厚纸质，长圆形或狭长圆形，两面无毛。花序开展，多花；花二型；苞片线形或钻状线形，果期脱落；花冠粉红色或紫红色，管状漏斗形，外面无毛。蒴果近倒心形，近无毛。花期3-5月，果期5-9月。生海拔600-1000米的阔叶密林下。产四川(峨眉山)。

Herbs or subshrubs, 30-65 cm tall. Stems glabrous. Leaves thickly papery, oblong or narrowly oblong, glabrous on both surfaces. Inflorescences cymose to paniculiform, many flowered; flowers distylous; bracts linear or subulate linear, deciduous in fruit; corolla pink or purplish red, tubular-funnelform, outside glabrous. Capsules obcordate-mitriform, subglabrous. Fl. Mar-May. Fr. May-Sep. Broad-leaved forests at 600-1000 m. Distributed in Sichuan (Emei Mountain.).

短小蛇根草

Ophiorrhiza pumila Champ. ex Benth.

草本，可达30厘米。茎密被短柔毛。叶纸质，卵状，椭圆形或椭圆状长圆形，上面被短糙毛或无毛，下面被短柔毛。聚伞花序，近簇生；花冠白色，近管状，外面被短柔毛。蒴果僧帽状，被短硬毛。花期4-9月，果期6-10月。生海拔200-700米的林下沟溪边或湿地上阴处。产广西、广东、香港、海南、江西、福建和台湾。日本和越南北部亦有。

Herbs to 30 cm tall. Stems densely puberulent. Leaves drying papery, ovate, elliptic or elliptic-oblong, adaxially sparsely scaberulous or glabrous, abaxially densely puberulent. Inflorescences cymose, subfasciculate; bracts subulate; flowers homostylous; corolla white, tubular, outside puberulent. Capsules mitriform or somewhat obcordate, hispidulous. Fl. Apr-Sep. Fr. Jun-Oct. Shady places on wet lands, streamsides or riversides in forests at 200-700 m. Distributed in Guangxi, Guangdong, Hong Kong, Hainan, Jiangxi, Fujian and Taiwan. Also in Japan and N Vietnam.

短小蛇根草 *Ophiorrhiza pumila*

香茜 *Carlemannia tetragona*

香茜

Carlemannia tetragona Hook. f.

草本或半灌木，干时芳香。小枝4棱。叶对生，椭圆形或卵形，显著具锯齿，无托叶。聚伞花序伞房状；花4数；花冠白色，喉部黄色，狭漏斗形。蒴果阔金字塔形，基部截平，星状开裂成4果瓣。花期6-9月，果期10-12月。生海拔600-1500米的密林中。产云南和西藏。印度东北部、缅甸、泰国北部和印度尼西亚亦有。

Herbs or semishrubs, fragrant when dry. Branchlets tetragonous. Leaves opposite, elliptic or ovate, distinctly

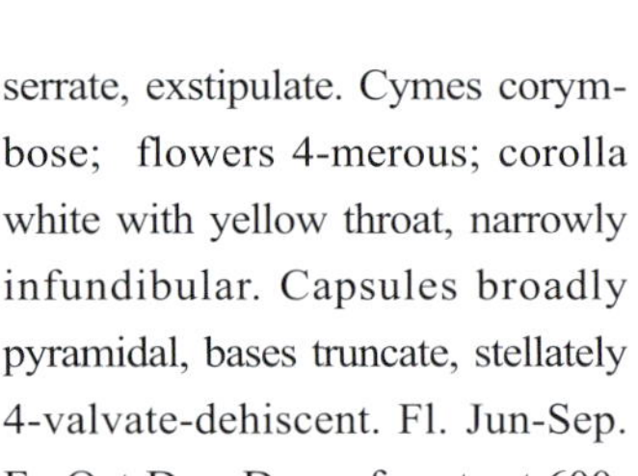

serrate, exstipulate. Cymes corymbose; flowers 4-merous; corolla white with yellow throat, narrowly infundibular. Capsules broadly pyramidal, bases truncate, stellately 4-valvate-dehiscent. Fl. Jun-Sep. Fr. Oct-Dec. Dense forests at 600-1500 m. Distributed in Yunnan and Xizang. Also in NE India,

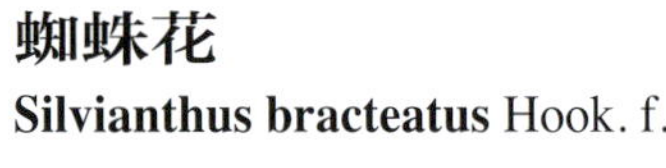

Myanmar, N Thailand and Indonesia.

蜘蛛花

Silvianthus bracteatus Hook. f.

灌木，高0.5-1米。叶对生，具柄，椭圆形，膜质，边缘具波状牙齿，无托叶。聚伞花序具短梗或无梗，近球形，多花；花5数；花冠白色，漏斗状钟形，裂片镊合状排列。蒴果近球形，肉质，5瓣裂。花期春季，果期秋季。生海拔700-900米的林中。产云南。印度东北部和缅甸亦有。

Shrubs, 0.5-1 m tall. Leaves opposite, petiolate, elliptic, membranous, margin undulate-dentate, exstipulate. Cymes shortly pedunculate or sessile, subglobose, many flowered; flowers 5-merous; corolla white, infundibular-campanulate, lobes valvate. Capsules subglobose, fleshy, 5-valved. Fl. spring. Fr. autumn. Forests at 700-900 m. Distributed in Yunnan. Also in NE India and Myanmar.

岩雪花

Argostemma saxatile Chun et F. C. How ex W. C. Ko

草本，高5-15厘米。茎被长柔毛。叶对生，一大一小，长圆状椭圆形或近卵形；侧脉5-7对。花序顶生，伞形状或短总状，有花2-4朵；花冠白色，辐射状，无毛或近无毛；花药合生。果未见。花期3-4月。生海拔300-600米的密林下潮湿地。产广西西南部。

Herbs, 5-15 cm tall. Stems villous. Leaves opposite, anisophyl-

蜘蛛花 *Silvianthus bracteatus*

岩雪花 *Argostemma saxatile*

lous, oblong-elliptic or subovate; secondary veins 5-7 pairs. Inflorescences terminal, umbelliform or short racemiform, 2-4-flowered; corolla white, rotate, glabrous or subglabrous; anthers coherent. Capsules unknown. Fl. Mar-Apr. Wet sites in dense forests at 300-600 m. Distributed in SW Guangxi.

毛花报春茜

Leptomischus erianthus H. S. Lo

草本，高达1米。茎方柱形，密被短硬毛。叶纸质，狭椭圆形至近椭圆形，上面近无毛或散生糙伏毛，下面脉上被多细胞长柔毛。花序顶生，近头状；花萼裂片狭三角形；花冠白色，管状漏斗形，外面密被绒毛。蒴果倒圆锥状，被多细胞长柔毛。花期4-6月，果期6-9月。生海拔1500-1700米的密林下的潮湿山谷。产云南东南部。

Herbs, up to 1 m tall. Stems quadrangular, densely hirtellous. Leaves drying papery, narrowly elliptic to subelliptic, adaxially sparsely strigose to glabrescent, abaxially glabrescent except multicellular villous along principal veins. Inflorescences terminal, subcapitate; calyx lobes narrowly triangular; corolla white, tubular-funnelform, densely tomentose outside, inside white villous. Capsules obconic, multicellular villous. Fl. Apr-Jun. Fr. Jun-Sep. Dense forests in moist valleys at 1500-1700 m. Distributed in SE Yunnan.

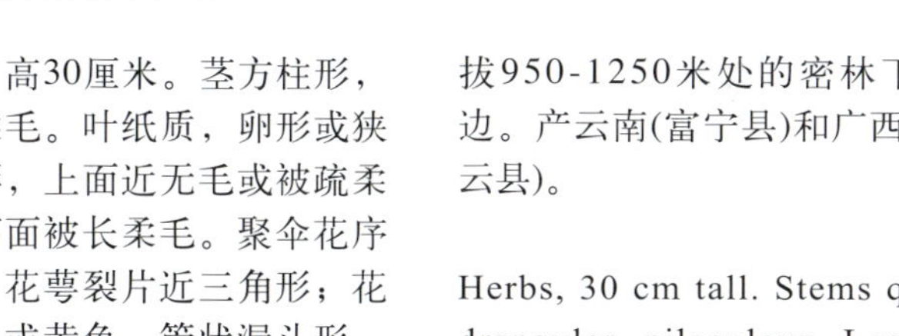

毛花报春茜 *Leptomischus erianthus*

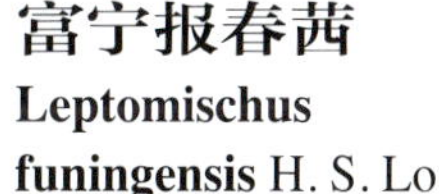

富宁报春茜

Leptomischus funingensis H. S. Lo

草本，高30厘米。茎方柱形，被短柔毛。叶纸质，卵形或狭椭圆形，上面近无毛或被疏柔毛，下面被长柔毛。聚伞花序顶生；花萼裂片近三角形；花冠白色或黄色，管状漏斗形，里外均被长柔毛。蒴果被疏柔毛，倒圆锥形或半球形。花期5-7月，果期7-10月。生海拔950-1250米处的密林下溪边。产云南(富宁县)和广西(凌云县)。

Herbs, 30 cm tall. Stems quadrangular, pilosulous. Leaves drying papery, adaxially glabrescent to pilose, abaxially villous; stipules suborbicular, usually reflexed. Inflorescences terminal, cymose; calyx lobes subtriangular; flowers distylous; calyx obconic densely multicellular villous, lobes subtriangular; corolla white or yellow, tubular-funnelform, both sides villous. Capsules obconic or subglobose, pilose. Fl. May-Jul. Fr. Jul-Oct. Streamsides in dense forests at 950-1250 m. Distributed in Yunnan (Funing County) and Guangxi (Lingyun County).

富宁报春茜 *Leptomischus funingensis*

水锦树 *Wendlandia uvariifolia*

龙州水锦树
Wendlandia oligantha W. C. Chen

灌木或乔木，高3-10米。小枝灰褐色。叶革质，椭圆形、卵形或卵状长圆形，两面无毛或有时在下面中脉上有疏短柔毛。花序圆锥状，顶生；花冠高脚碟状，白色，长约4.5毫米，外面无毛，喉部有白色柔毛。蒴果近球形，无毛。花期7-8月，果期8-12月。生海拔300-1000米的山谷林中或灌丛。产广西。

Shrubs or trees, 3-10 m tall. Branches grayish brown. Leaves leathery, elliptic, ovate or ovate-oblong, adaxially glabrous, abaxially glabrous or sparsely puberulent along principal veins. Inflorescences paniculate, terminal; corolla white, salverform, ca. 4.5 mm long, outside glabrous, inside villous at throat. Capsules subglobose, glabrous. Fl. Jul-Aug. Fr. Aug-Dec. Forests or thickets in valleys at 300-1000 m. Distributed in Guangxi.

水锦树
Wendlandia uvariifolia Hance

灌木或乔木。小枝密被锈色毛。叶宽椭圆形，长圆形，卵形或披针形，纸质。圆锥状聚伞花序顶生，多花；花小，无梗；花冠漏斗状，白色。蒴果球形，被柔毛。花期1-5月，果期4-10月。生海拔100-1200米的灌丛、林中、林缘、山坡或溪边。产云南、贵州、广西、广东、海南和台湾。越南亦有。

Shrubs or trees. Twigs densely rusty hirsute. Leaves broadly elliptic, oblong, ovate or lanceolate, papery. Paniculate-cymes terminal, many-flowered; flowers small, sessile; corolla funnelform, white. Capsules globose, pubescent. Fl. Jan-May. Fr. Apr-Oct. Thickets, forests, forest edges, montane slopes or stream sides at 100-1200 m. Distributed in Yunnan, Guizhou, Guangxi, Guangdong, Hainan and Taiwan. Also in Vietnam.

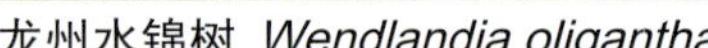

龙州水锦树 *Wendlandia oligantha*

土连翘 *Hymenodictyon flaccidum*

土连翘
Hymenodictyon flaccidum Wall.

落叶乔木，高6-20米。叶纸质或薄革质，倒卵形、卵形或长圆形，两面通常无毛。总状花序腋生；花冠漏斗形，黄色至红色，外面被短柔毛。蒴果倒圆锥体形，有灰白色斑点。花期5-7月，果期8-12月。生海拔300-3000米的山谷或溪边的林中或灌丛中。产广西、四川和云南。印度北部、尼泊尔、不丹和越南北部亦有。

Trees, deciduous, 6-20 m tall. Leaves drying papery or thinly

石丁香 *Neohymenopogon parasiticus*

leathery, obovate, ovate or oblong, both surfaces often glabrous to glabrescent. Inflorescences raceme, axillary; corolla funnelform, yellow to red, densely puberulent outside. Capsules obovoid, with several prominent whitened lenticels. Fl. May-Jul. Fr. Aug-Dec. Forests or thickets at streamsides or in valleys at 300-3000 m. Distributed in Guangxi, Sichuan and Yunnan. Also in N India, Nepal, Bhutan and N Vietnam.

石丁香

Neohymenopogon parasiticus (Wall.) Bennet

小灌木，附生。叶大，椭圆状披针形至倒卵形，侧脉多而密，15-28对。伞房状聚伞花序顶生，疏散，大型，三歧分枝，苞片大型，叶状，白色，具长柄；花冠白色，高脚碟状。蒴果长圆形。花期6-8月，果期9-12月。生海拔1200-2700米的山谷灌丛或林中，常附生在岩石上或树上。产云南和西藏。南亚和东南亚亦有。

Small shrubs, epiphytic. Leaves large, elliptic-lanceolate to obovate, lateral veins 15-28 pairs. Corymbose-cymes terminal, lax, large, trichotomous, bracts large, foliaceous, white, oblong, long petiolate; corolla white, salverform. Capsules oblong. Fl. Jun-Aug. Fr. Sep-Dec. Thickets or forests in valleys at 1200-2700 m, often on rocks or trees. Distributed in Yunnan and Xizang. Also in S and SE Asia.

滇丁香 *Luculia pinceana*

流苏子

Coptosapelta diffusa (Champ. ex Benth.) van Steenis

流苏子 *Coptosapelta diffusa*

藤本或攀援灌木，高2-5米。枝圆柱状。叶纸质至革质，卵形至披针形。花单生叶腋，具花梗；花冠白色或黄色，高脚碟状。蒴果压扁球形。花期5-7月，果期6-12月。生海拔100-1500米的山地林中或灌丛中。产中国西南、华南、华中和华东。日本亦有。

Lianas or scandent shrubs, 2-5 m tall. Branches terete. Leaves papery to leathery, ovate to lanceolate. Flowers solitary, axillary, pedicellate; corolla white or yellow, salverform. Capsules compressed globose. Fl. May-Jul. Fr. Jun-Dec. Montane forests or thickets at 100-1500 m. Distributed in SW, S, C and E China. Also in Japan.

滇丁香

Luculia pinceana Hook.

灌木或乔木，高2-10米。叶纸质，长圆形至卵形。伞房状聚伞花序顶生，多花，具苞片；花芳香，5数；花冠粉红色至白色，高脚碟状，在裂片之间内面基部具2个片状附属物。蒴果近圆柱状，具肋。花果期3-12月。生海拔600-3000米的山坡、林下、灌丛、溪边或山谷。产云南、西藏、贵州和广西。印度、尼泊尔、缅甸和越南亦有。

Shrubs or trees, 2-10 m tall. Leaves papery, oblong to ovate. Corymbose-cymes terminal, many-flowered, bracteate; flowers fragrant, 5-merous; corolla pink to white, salverform, with 2 lamellate appendages at base between lobes inside. Capsules subcylindric, ribbed. Fl. and fr. Mar-Dec. Montane slopes, forests, thickets, streamsides or valleys at 600-3000 m. Distributed in Yunnan, Xizang, Guizhou and Guangxi. Also in India, Nepal, Myanmar and Vietnam.

鸡冠滇丁香 *Luculia yunnanensis*

鸡冠滇丁香

Luculia yunnanensis S. Y. Hu

灌木或乔木，高3.5-10米。叶倒披针形至长圆形，托叶披针形，早落。伞房状聚伞花序顶生，多花，具苞片；花芳香，5数；花冠红色，高脚碟状，在裂片之间内面基部具2个片状附属物。果实倒卵球形；花序梗、花梗和萼管被绒毛，果被柔毛。花果期3-11月。生海拔1200-3200米的山地林下或灌丛中。产云南。

Shrubs or trees, 3.5-10 m tall. Leaves oblanceolate to oblong, stipules lanceolate, caducous. Corymbose-cymes terminal, many-flowered, bracteate; flowers fragrant, 5-merous; corolla red, salverform, with 2 lamellate appendages at base between lobes inside. Capsules obovoid; peduncles, pedicels and calyx tubes tomentose, capsules villose. Fl. and fr. Mar-Nov. Montane forests or thickets at 1200-3200 m. Distributed in Yunnan.

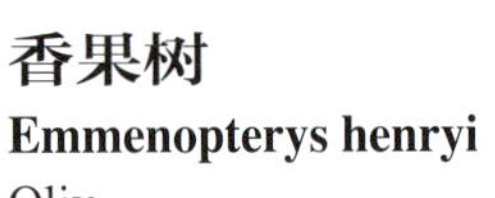

香果树

Emmenopterys henryi Oliv.

落叶大乔木，高达30米。树皮鳞片状。叶宽椭圆形至卵状长圆形。聚伞花序顶生，圆锥状；花芳香，具梗；变态叶状萼裂片白色，匙形或广椭圆形；花冠白色或黄色，漏斗状。蒴果长圆状卵球形或近纺锤形，具纵肋。花期6-8月，果期8-11月。生海拔400-1600米的山谷林中。产秦岭以南。

Deciduous large trees, to 30 m tall. Bark scaly. Leaves broadly elliptic to ovate-oblong. Cymes terminal, paniculate; flowers fragrant, pedicellate; petaloid calycophylls white, spathulate or broadly elliptic; corolla white or yellow, funnelform. Capsules oblong-ovoid or subfusiform, longitudinally ribbed. Fl. Jun-Aug. Fr. Aug-Nov. Forests in valleys at 400-1600 m. Distributed in the south of Qinling Mountain.

香果树 *Emmenopterys henryi*

大叶钩藤 *Uncaria macrophylla*

大叶钩藤

Uncaria macrophylla Wall.

大藤本，具钩刺。嫩枝被硬毛。叶大型，卵形或阔椭圆形，近革质。头状花序不计花冠直径15-20毫米，总梗长达7厘米；花冠高脚蝶状。小蒴果被苍白色柔毛。花期7月、9月和

12月，果期翌年3-4月和9-11月。生海拔300-900米的山谷林中、林缘或灌丛，常攀援在次生林树冠上。产云南、广西、广东和海南。南亚和东南亚亦有。

Large lianas, with hooked thorns. Twigs hispid. Leaves large, ovate or broadly elliptic, subleathery. Flowering heads across calyces 15-20 mm diam, peduncles to 7 cm long; corolla salverform. Small capsules pale pubescent. Fl. Jul, Sep and Dec. Fr. next Mar-Apr and Sep-Nov. Valley forests, forest edges or thickets at 300-900 m, often on canopy of secondary forests. Distributed in Yunnan, Guangxi, Guangdong and Hainan. Also in S and SE Asia.

华钩藤 *Uncaria sinensis*

华钩藤

Uncaria sinensis (Oliv.) Havil.

藤本。幼茎细弱，四方形，无毛。叶干薄纸质，椭圆形至卵形。花序腋生，单生头状或有时顶生3-5组花序，无毛；花冠高脚碟状。头状果序椭圆形，具糙伏毛至硬毛。花果期6-10月。生海拔900-1100米疏林中或湿润次生林中。产中国西南、华中和华西。

Lianas. Young stems slender, quadrangular, glabrous. Leaves drying thinly papery, elliptic to ovate. Inflorescences axillary, heads solitary or sometimes in terminal groups of 3-5, glabrous; corolla salverform. Fruiting heads ellipsoid, strigillose to strigose. Fl. and fr. Jun-Oct. Sparse forests or wet secondary forests at 900-1100 m. Distributed in SW, C and W China.

白钩藤

Uncaria sessilifructus Roxb.

大藤本。嫩枝被柔毛。叶卵形、椭圆形或长圆形，近革质。头状花序不计花冠直径5-10毫米，总梗长达15厘米；小花无梗；花冠黄白色，高脚碟状，外面被绸毛。小蒴果稍被柔毛。花果期3-12月。生海拔300-1500米的山谷溪边密林或灌丛中。产云南和广西。南亚和东南亚亦有。

Large lianas. Twigs pubescent. Leaves ovate, elliptic or oblong, subleathery. Flowering heads across calyces 5-10 mm diam, peduncles to 15 cm long; flowers sessile; corolla yellowish-white, salverform, sericeous outside. Small capsules slightly pubescent. Fl. and fr. Mar-Dec. Dense forests or thickets in valleys at 300-1500 m. Distributed in Yunnan and Guangxi. Also in S and SE Asia.

平滑钩藤

Uncaria laevigata Wall. ex G. Don

藤本，具钩刺。嫩枝纤细，四棱形或稍具棱，被短柔毛。叶椭圆形或长圆形，近革质。头状花序单生叶腋或3个组成腋生聚伞花序，总梗长达8厘米；花无梗。小蒴果纺锤状，被柔毛。花果期5-11月。生海拔600-1300米的山谷溪边林中。产云南、台湾和广西。南亚和东南亚亦有。

Lianas, with hooked thorns. Twigs slender, tetragonous or slightly angulate, puberulous. Leaves elliptic or oblong, subleathery. Flowering heads solitary in axils or 3 in axillary cyme, pubuncles to 8 cm long; flowers sessile. Small capsules fusiform, pubescent. Fl. and fr. May-Nov. Forests by streams in valleys at 600-1300 m. Distributed in Yunnan, Taiwan and Guangxi. Also in S and SE Asia.

白钩藤 *Uncaria sessilifructus*

平滑钩藤 *Uncaria laevigata*

毛钩藤 *Uncaria hirsuta*

毛钩藤
Uncaria hirsuta Havil.

藤本。幼茎细弱、具毛。叶卵形、披针状长圆形或椭圆形。花序腋生或常成对和顶生于茎顶，头状花序直径18-25毫米；花近无柄；花冠浅黄色或浅红色、高脚碟状。果实近无柄、倒卵球形。花果期1-12月。生海拔低于100-500米的沟谷溪边灌丛或林中。产贵州、广西、广东、台湾和福建。

Lianas. Young stems slender, hirsute. Leaves ovate, lanceolate-oblong or elliptic. Inflorescences axillary or frequently paired and terminal on stems heads 18-25 mm diam; flowers subsessile; corolla pale yellow or pale red, salverform. Fruits subsessile, obovoid. Fl. and fr. Jan-Dec. Thickets or forests at streamsides in valleys below 100-500 m. Distributed in Guizhou, Guangxi, Guangdong, Taiwan and Fujian.

钩藤 *Uncaria rhynchophylla*

钩藤
Uncaria rhynchophylla (Miq.) Miq. ex Havil.

藤本。幼茎无毛，有时灰色。叶常红褐色或深红色，椭圆形、披针形或椭圆状长圆形，两面无毛。花序腋生或顶生、单生或常7-11个顶生成簇；花冠高脚碟状。果实有柄或近无柄、倒卵球形至纺锤形。花果期5-12月。生海拔1000米以下的沟谷溪边的疏林或灌丛中。产中国西南、华南、东南、华中和华东。日本亦有。

Lianas. Young stems glabrous, sometimes glaucous. Leaves drying papery and often red-brown or dark red, elliptic, lanceolate or elliptic-oblong, glabrous on both surfaces. Inflorescences axillary and terminal, solitary or usually in terminal groups of 7-11; corolla salverform. Fruits sessile or subsessile, obovoid to fusiform. Fl. and fr. May-Dec. Sparse forests or thickets at streamsides in valleys below 1000 m. Distributed

攀茎钩藤 *Uncaria scandens*

in SW, S, SE, C and E China. Also in Japan.

攀茎钩藤

Uncaria scandens (Sm.) Hutchins.

大藤本。嫩枝密被锈色柔毛。叶卵形、椭圆形或长圆形，纸质。头状花序不计花冠直径25毫米；花无梗，花萼密被灰白色硬毛或柔毛，裂片条形；花冠淡黄色，裂片长倒卵形。小蒴果倒披针形，被疏柔毛。花期2-4月，果期7-11月。生海拔100-1500米的山谷溪边阔叶疏林中。产中国西南和华南。

Large lianas. Twigs densely rusty pubescent. Leaves ovate, elliptic or oblong, papery. Flowering heads across calyces 25 mm diam; flowers sessile, calyx densely greyish white hirsute or pubescent, lobes linear; corolla pale yellow, lobes long obovate. Small capsules oblanceolate, sparsely villose. Fl. Feb-Apr. Fr. Jul-Nov. Broad-leaved sparse forests by streams in valley at 100-1500 m. Distributed in SW and S China.

团花

Neolamarckia cadamba (Roxb.) Bosser

落叶大乔木，高达30米。叶椭圆形或长圆状椭圆形，薄革质。头状花序单个顶生，不计花冠直径4-5厘米，总梗粗壮，长2-4厘米；花冠黄白色，漏斗状。花果期6-11月。生海拔600-1000米的阔叶林或溪边。产云南、广西和广东。南亚和东南亚亦有。

Deciduous large trees, to 30 m tall. Leaves elliptic or oblong-elliptic, thinly leathery. Flowering heads solitary, terminal, 4-5 cm diam, peduncles stout, 2-4 cm long; corolla yellowish-white, funnelform. Fl. and fr. Jun-Nov. Broad-leaved forests or streamsides in valleys at 600-1000 m. Distributed in Yunnan, Guangxi and Guangdong. Also in S and SE Asia.

团花 *Neolamarckia cadamba*

黄棉木 *Metadina trichotoma*

黄棉木

Metadina trichotoma (Zoll. et Mor.) Bakh. f.

乔木，高5-10米。叶厚纸质或近革质，长披针形、椭圆状倒披针形或椭圆状长圆形，两面无毛或下面脉上有绒毛。数个头状花序作聚伞花序式排列，顶生或腋生；花冠窄漏斗状，无毛。蒴果倒圆锥体形。花果期4-12月。生海拔300-1400米的林谷溪畔。产广东、广西、云南和湖南等地。南亚和东南亚亦有。

Trees, 5-10 m tall. Leaves drying stiffly papery to subleathery, lanceolate, elliptic-lanceolate or ovate-oblong, both surfaces glabrous or abaxially tomentulose along veins. Inflorescences terminal and leaves, capitate with several globose heads in cymes; corolla narrowly funnelform, glabrous. Capsules obconic. Fl. and fr. Apr-Dec. Forests at streamsides in valleys at 300-1400 m. Distributed in Guangdong, Guangxi, Yunnan and Hunan. Also in S and SE Asia.

鸡仔木 *Sinoadina racemosa*

鸡仔木

Sinoadina racemosa (Siebold et Zucc.) Ridsdale

乔木，高4-12米。小枝无毛。叶薄革质，卵形、卵状长圆形或椭圆形，两面通常无毛或下面有短柔毛。头状花序常约10个排成圆锥状花序；花冠淡黄色。蒴果倒卵状楔形，有稀疏的毛。花果期5-12月。生海拔300-1500米的山林中或水边。产长江流域及以南各省区。日本、泰国和缅甸亦有。

Trees, 4-12 m tall. Branches glabrous. Leaves thinly leathery, ovate, ovate-oblong or elliptic, both surfaces often glabrous or abaxially puberulent. Inflorescences paniculiform with ca. 10 subglobose heads; corolla pale yellow. Capsules obovoid-cuneate, sparsely hirtellous. Fl. and fr. May-Dec. Forests or watersides at 300-1500 m. Distributed in provinces south of Yangtze River. Also in Japan, Thailand and Myanmar.

海南槽裂木 *Pertusadina metcalfii*

海南槽裂木

Pertusadina metcalfii (Merrill ex H. L. Li) Y. F. Deng et C. M. Hu

乔木或灌木，高达30米。小枝近无毛。叶厚纸质，椭圆形至椭圆状长圆形；侧脉7-10对。花序头状，单生；花冠黄色，高脚碟状，内外均无毛。蒴果倒圆锥体形，疏被短柔毛。花期5-6月，果期9-12月。生海拔100-900米的密林中。产广东、海南、广西、福建、浙江和湖南。泰国亦有。

Shrubs or large trees, to 30 m tall. Branches subglabrous. Leaves drying thickly papery, elliptic to elliptic-oblong or oblanceolate; second veins 7-10 pairs. Inflorescences capitate, solitary; corolla yellow, salverform, glabrous on both sides. Capsules obconic, sparsely puberulent. Fl. May-Jun. Fr. Sep-Dec. Dense forests at 100-900 m. Distributed in Guangdong, Hainan, Guangxi, Fujian, Zhejiang and Hunan. Also in Thailand.

水团花

Adina pilulifera (Lam.) Franch. ex Drake

常绿灌木或小乔木，高1-5米。叶椭圆形、椭圆状披针形至长圆形。头状花序腋生，总梗长2-5厘米；花冠白色，狭漏斗形，花柱伸出。蒴果倒楔形。花期6-9月，果期7-12月。生海拔200-400米的灌丛、疏林、溪边或沟谷中。产中国西南、华中、华南和华东。越南和日本亦有。

Evergreen shrubs or small trees, 1-5 m tall. Leaves elliptic, elliptic-lanceolate to oblong. Flowering heads axillary, peduncles 2-5 cm long; corolla white, narrowly funnelform, style exserted. Capsules obcuneate. Fl. Jun-Sep. Fr. Jul-Dec. Thickets, sparse forests, streamsides or valleys at 200-400 m. Distributed in SW, C, S and E China. Also in Vietnam and Japan.

水团花 *Adina pilulifera*

细叶水团花 *Adina rubella*

细叶水团花
Adina rubella Hance

落叶小灌木，高1-3米。叶卵状披针形或卵状长圆形，革质。头状花序单独顶生，总花梗长2-5厘米；花冠紫红色，花柱伸出。蒴果长圆状倒楔形。花果期5-12月。生海拔100-600米的溪边、河边或沙滩。产华南、华中和华东。朝鲜半岛亦有。

Deciduous dwarf shrubs, 1-3 m tall. Leaves ovate-lanceolate or ovate-oblong, leathery. Flowering heads solitary, terminal, peduncles 2-5 cm long; corolla purplish red, style exserted. Capsules oblong-obcuneate. Fl. and fr. May-Dec. Streamsides, riversides or sandy coasts at 100-600 m. Distributed in S, C and E China. Also in Korean Peninsula.

心叶木
Haldina cordifolia (Roxb.) Ridsd.

高大落叶乔木，7-40米高，具板根。树皮红褐色。叶具柄，宽卵形，基部心形，上面疏被长硬毛，下面密被短柔毛，托叶有显著龙骨。头状花序淡黄色，腋生，每节2-4(-10)个，总花梗长达10厘米。蒴果基部密生硬毛或近无毛。花期春夏季。生海拔300-1000米的热带雨林中。产云南。南亚和东南亚亦有。

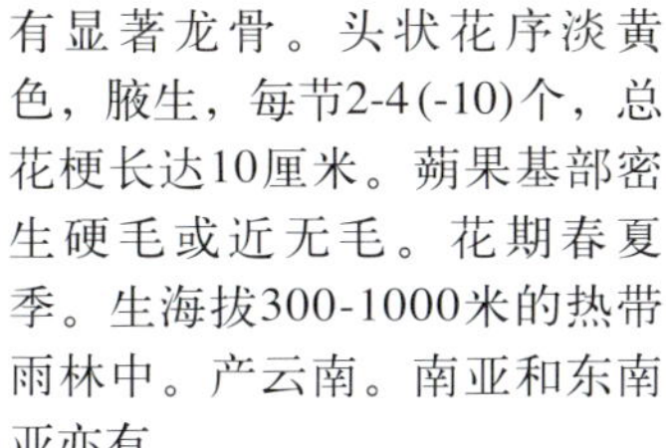

Large deciduous trees, 7-40 m tall, with plank buttresses. Bark reddish brown. Leaves petiolate, broadly ovate, bases cordate, sparsely long hirsute adaxially, densely puberulent abaxially, stipules strongly keeled. Flowering heads yellowish, axillary, 2-4(-10) at each node, peduncles up to 10 cm long. Capsules densely strigillose or glabrescent near base. Fl. spring to summer. Tropical rain forests at 300-1000 m. Distributed in Yunnan. Also in S and SE Asia.

风箱树
Cephalanthus tetrandrus (Roxb.) Ridsd. et Bakh. f.

落叶灌木或小乔木，高1-5米。叶对生或轮生，近革质，卵状披针形至椭圆状长圆形，两面常无毛。花序头状；花冠白色，高脚蝶状，外面无毛。坚果倒圆锥体形。花期6-9月，果期7-10月。生海拔700-800米处略荫蔽的水沟旁或溪畔。产长江流域及以南各省区。印度、孟加拉国、缅甸、泰国、老挝和越南北部亦有。

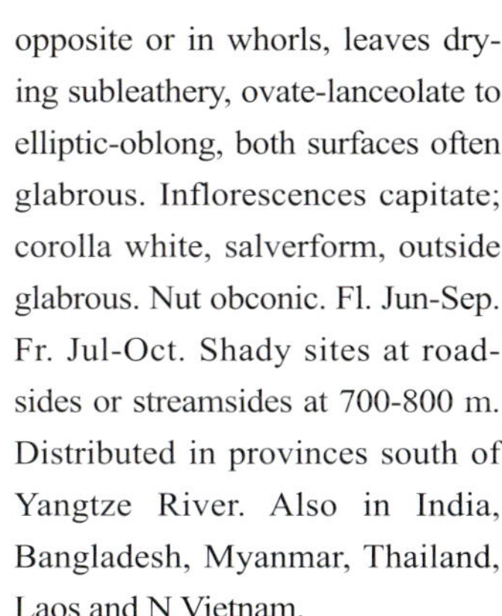

Deciduous shrubs or small trees, 1-5 m tall. Leaves opposite or in whorls, leaves drying subleathery, ovate-lanceolate to elliptic-oblong, both surfaces often glabrous. Inflorescences capitate; corolla white, salverform, outside glabrous. Nut obconic. Fl. Jun-Sep. Fr. Jul-Oct. Shady sites at roadsides or streamsides at 700-800 m. Distributed in provinces south of Yangtze River. Also in India, Bangladesh, Myanmar, Thailand, Laos and N Vietnam.

心叶木 *Haldina cordifolia*

风箱树 *Cephalanthus tetrandrus*

中华尖药花 *Acranthera sinensis*

中华尖药花

Acranthera sinensis C. Y. Wu

草本或亚灌木，高达1米。茎钝方柱形。叶椭圆形或倒卵形，薄纸质至膜质，下面及中脉密被毛。花单生腋生短枝顶端；花萼裂片线状披针形；花冠漏斗状，淡红色或紫红色。浆果扁圆柱状。花期4-7月，果期7-10月。生海拔1000-1800米的山地林下。产云南东南部。

Herbs or subshrubs, up to 1 m tall. Branches generally quadrangular. Leaves elliptic or obovate, thinly papery to membranous, abaxially strigillose to strigose or hispid with pubescence denser along principal veins. Inflorescences generally pseudoaxillary, 1-flowered; calyx lobes linear-lanceolate; corolla funnelform, purple outside and pink inside. Berry compressed cylindrical. Fl. Apr-Jul. Fr. Jul-Oct. Forests on mountain slopes at 1000-1800 m. Distributed in SE Yunnan.

大叶玉叶金花

Mussaenda macrophylla Wall.

直立或攀援灌木。枝4棱。叶大，长圆形至卵形。聚伞花序有短总梗，苞片大；花叶宽卵形或菱形，白色，7脉；花大，近无梗，密被棕色柔毛；萼钟形，裂片近叶状，披针形；花冠橙黄色，裂片卵形。浆果暗紫色，椭圆体形。花期6-7月，果期8-11月。生海拔1300米以下的山谷林中或灌丛中。产广西、广东和台湾。东南亚亦有。

Erect or climbing shrubs. Branches tetragonous. Leaves large, oblong to ovate. Cymes shortly pedunculate, bracts large; petaloid calycophylls broadly ovate or rhombic, white, 7-veined; flowers large, subsessile, densely brown-villose; calyx campanulate, lobes subfoliaceous, lanceolate; corolla orange-yellow, lobes ovate. Berries dark purple, ellipsoid. Fl. Jun-Jul. Fr. Aug-Nov. Valley forests or thickets below 1300 m. Distributed in Guangxi, Guangdong and Taiwan. Also in SE Asia.

黐花 (大叶白纸扇)

Mussaenda shikokiana Makino

直立或攀援灌木，高1-3米。枝密被短柔毛。叶薄纸质，宽卵形至椭圆形。聚伞花序顶生，花疏散；花叶倒卵形；萼裂片近叶状，白色，披针形，长达1厘米；花冠黄色，裂片卵形。浆果近球形。花期5-7月，果期7-10月。生海拔100-1000米的山地疏林下或路边。产中国西南、华南、华中和华东。日本亦有。

Erect or climbing shrubs, 1-3 m tall. Branches densely pubescent. Leaves thin papery, broadly ovate or elliptic. Cymes terminal, flowers sparse; petaloid calycophylls obovate; calyx lobes subfoliaceous, white, lanceolate, to 1 cm long; corolla yellow, lobes ovate. Berries subglobose. Fl. May-Jul. Fr. Jul-Oct. Montane sparse forests or roadsides at 100-1000 m. Distributed in SW, S, C and E China. Also in Japan.

大叶玉叶金花 *Mussaenda macrophylla*

黐花 (大叶白纸扇) *Mussaenda shikokiana*

红毛玉叶金花 *Mussaenda hossei*

红毛玉叶金花
Mussaenda hossei Craib

亚灌木。枝条密具发白、淡红色或褐色黏毛、绒毛或长柔毛。叶厚纸质，倒披针形或长圆状倒披针形。聚伞花序顶生，总梗分枝，有绒毛；花近无梗，5数；花叶白色；花萼裂片披针形；花冠橙黄色。浆果长圆状椭圆体形。花期11月至翌年3月。生海拔600-1600米的林中。产云南南部。东南亚亦有。

Subshrubs. Branches densely whitened, reddish or brownish villosulous, tomentulose or villous. Leaves thick leathery, oblanceolate or oblong-oblanceolate. Cymes terminal, peduncles branched, tomentose; flowers subsessile, 5-merous; petaloid calycophylls white; calyx lobes lanceolate; corolla orange-yellow. Berries oblong-ellipsoid. Fl. Nov to next Mar. Forests at 600-1600 m. Distributed in S Yunnan. Also in SE Asia.

楠藤
Mussaenda erosa Champ. ex Benth.

攀援灌木，高3米；小枝无毛。叶纸质，卵形至长圆状椭圆形，两面无毛或沿脉有稀疏的贴伏毛。伞房状多歧聚伞花序；花冠橙黄色，外面密被柔毛。浆果近球形或阔椭圆形，无毛。花期4-7月，果期9-12月。生海拔300-800米疏林下的路边。产广东、香港、广西、云南、四川、贵州、福建、海南和台湾。日本和越南亦有。

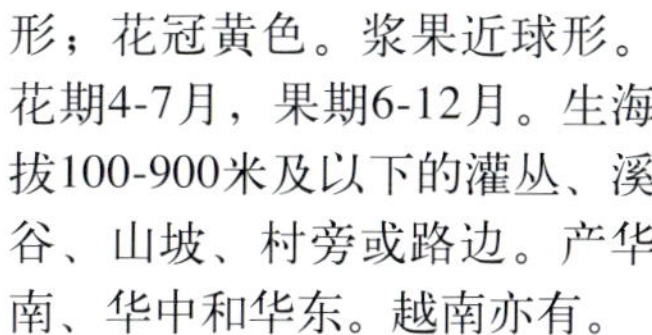

Climbing shrubs, 3 m tall; branches glabrous. Leaves thickly papery, ovate to oblong-elliptic, both surfaces glabrous or sparsely strigose on principal veins. Inflorescences compound-cymose to compound-corymbiform; corolla yellow to orange-yellow, outside densely pubescent. Berry ellipsoid to subglobose, glabrous. Fl. Apr-Jul. Fr. Sep-Dec. Sparse evergreen forests, streamsides, along roads at 300-800 m. Distributed in Guangdong, Hong Kong, Guangxi, Yunnan, Sichuan, Guizhou, Fujian, Hainan and Taiwan. Also in Japan and Vietnam.

玉叶金花
Mussaenda pubescens Ait. f.

楠藤 *Mussaenda erosa*

攀援灌木，被柔毛。叶对生或轮生，膜质或薄纸质，卵状长圆形或卵状披针形。聚伞花序顶生，密花；花叶宽椭圆形，5-7脉；花萼裂片条形；花冠黄色。浆果近球形。花期4-7月，果期6-12月。生海拔100-900米及以下的灌丛、溪谷、山坡、村旁或路边。产华南、华中和华东。越南亦有。

Climbing shrubs, pubescent. Leaves opposite or verticillate, membranous or thinly papery, ovate-oblong or ovate-lanceolate. Cymes terminal, many-flowered; petaloid calycophylls broadly elliptic, 5-7-veined; calyx lobes linear; corolla yellow. Berries subglobose. Fl. Apr-Jul. Fr. Jun-Dec. Thickets, valleys, montane slopes, village margins or roadsides below 100-900 m. Distributed in S, C and E China. Also in Vietnam.

玉叶金花 *Mussaenda pubescens*

广东玉叶金花 *Mussaenda kwangtungensis*

粗毛玉叶金花
Mussaenda hirsutula

粗毛玉叶金花
Mussaenda hirsutula Miq.

攀援灌木。小枝密被锈色或灰色柔毛。叶纸质，长圆形至倒卵形。聚伞花序顶生或腋生；花近无梗；花叶宽椭圆形，被柔毛，7脉；花萼裂片条形；花冠黄色，裂片椭圆形。浆果椭圆体形至近球形。花期4-6月，果期7月至翌年1月。生海拔300-800米的灌丛、溪边或山谷中，常在树冠上。产云南、贵州、广东、海南和湖南。

Climbing shrubs. Branchlets densely rusty or gray villose. Leaves papery, oblong to subovate. Cymes terminal or axillary; flowers subsessile; petaloid calycophylls broadly elliptic, pubescent, 7-veined; calyx lobes linear; corolla yellow, lobes elliptic. Berries ellipsoid to subglobose. Fl. Apr-Jun. Fr. Jul to next Jan. Thickets, streamsides or valleys at 300-800 m, often on tree crowns. Distributed in Yunnan, Guizhou, Guangdong, Hainan and Hunan.

贡山玉叶金花
Mussaenda treutleri Stapf

直立灌木，被柔毛或长柔毛。叶纸质，卵形至卵状椭圆形，基部截形。花序顶生或腋生，多花，密集；花叶白色，卵形，被长柔毛，7脉；花萼裂片条形；花冠橙红色，裂片卵形。浆果球形。花期7-9月。生海拔600-2000米密林中的湿土上。产云南。印度东北部、尼泊尔和不丹亦有。

Erect shrubs, pubescent or villose. Leaves papery, ovate or ovate-elliptic, base truncate. Cymes terminal or axillary, many-flowered, compact; petaloid calycophylls white, ovate, villose, 7-veined; calyx lobes linear; corolla orange-red, lobes ovate. Berries globose. Fl. Jul-Sep. Wet soils in dense forests at 600-2000 m. Distributed in Yunnan. Also in NE India, Nepal and Bhutan.

广东玉叶金花
Mussaenda kwangtungensis H. L. Li

攀援灌木，高1-2.5米。分枝棕色，被柔毛。叶薄纸质，披针状椭圆形。聚伞花序顶生，紧密；花近无梗；花叶长圆状卵形，具5脉；花萼裂片条形；花冠黄色，裂片卵形。花期5-9月。生山地丛林中，常在树冠上。产广东南部。

Climbing shrubs, 1-2.5 m tall. Branches brown, pubescent. Leaves thin papery, lanceolate-elliptic. Cymes terminal, compact; flowers subsessile; petaloid calycophylls oblong-ovate, 5-veined; calyx lobes linear; corolla yellow, lobes ovate. Fl. May-Sep. Montane forests, often on tree crowns. Distributed in S Guangdong.

贡山玉叶金花 *Mussaenda treutleri*

展枝玉叶金花 *Mussaenda divaricata*

展枝玉叶金花

Mussaenda divaricata Hutchins.

直立攀援灌木。小枝被稀疏短柔毛。叶薄纸质至近膜质，被短柔毛，椭圆形至卵状椭圆状。聚伞花序顶生，具多花，疏松；花叶大型，宽椭圆形或卵形，具7脉；花冠小，黄色，被黄色棒状毛。浆果椭圆状。花期5-9月，果期10月。生海拔1400米的峡谷林下、林缘或灌丛中。产中国西南、华中和华南。越南亦有。

Erect climbing shrubs. Branchlets sparsely pubescent. Leaves thin papery or submembranous, pubescent, elliptic or ovate-elliptic. Cymes terminal, many-flowered, lax; petaloid calycophylls large, broadly elliptic or ovate, 7-veined; corolla small, yellow, clavate-hairy. Berries ellipsoid. Fl. May-Sep. Fr. Oct. Valley forests, forest edges or thickets at 1400 m. Distributed in SW, C and S China. Also in Vietnam.

长瓣玉叶金花

Mussaenda longgipetala H. L. Li

攀援灌木，高达5米。枝圆密被长柔毛。叶片近膜质，长圆状卵形或椭圆状卵形，两面疏被灰白色短柔毛。聚伞花序顶生；花冠金黄色，高脚蝶状，外面密被灰白色长柔毛，内面密被金黄色棒状毛。果未见。花期5-8月。生海拔200米以下的低矮丘陵疏林下。产广西南部。越南亦有。

Climbing shrubs, to 5 m tall. Branches densely villous. Leaves submembranous, oblong-ovate or elliptic-ovate, both surfaces sparsely white puberulent. Inflorescences terminal, cymose; corolla orange-yellow, outside densely gray-white villous, inside densely barlike hairs. Fruit unknown. Fl. Apr-Aug. Sparse evergreen forests or thickets on hills below 200 m. Ditributed in S Guangxi. Also in Vietnam.

长瓣玉叶金花 *Mussaenda longgipetala*

裂果金花

Schizo mussaenda henryi (Hutch.) X. F. Deng et D. X. Zhang

灌木，高达3米。叶膜质，被短柔毛，长圆形，托叶条形。聚伞花序顶生，苞片条状披针形；花叶卵状披针形，3-5脉；花萼裂片长1毫米；花冠黄色。蒴果倒卵球形或陀螺状，顶部室背开裂。花期春夏季。生海拔100-1000米的山地阔叶林下或路边灌丛。产云南南部和广西。东南亚亦有。

Shrubs, to 3 m tall. Leaves membranous, pubescent, oblong, stipules linear. Cymes terminal, bracts linear-lanceolate; petaloid calycophylls ovate-lanceolate, veins 3-5; calyx lobes 1 mm long; corolla yellow. Capsules obovoid or turbinate, loculicidally dehiscent at apex. Fl. spring to summer. Montane broad-leaved forests or thickets on roadsides at 100-1000 m. Distributed in S Yunnan and Guangxi. Also in SE Asia.

裂果金花 *SchizoMussaenda henryi*

密脉木 *Myrioneuron faberi*

密脉木

Myrioneuron faberi Hemsl.

高大草本或有时灌木状。叶有时聚生于上部枝上，干纸质，倒卵形、椭圆形或长圆状倒卵形；次级叶脉9-15对。花序全顶生，球形；花冠黄色，管状。浆果近球形，具梗，常伸长至8毫米。花期8月，果期10-12月。生海拔500-1500米的林中或溪边。产广西、西南和华中。

Large to suffruticose herbs. Leaves sometimes crowded at upper parts of branches, drying papery, obovate, elliptic or oblong-obovate; secondary veins 9-15 pairs. Inflorescences terminal, globose; corolla yellow, tubular. Berries subglobose, with pedicels often elongating, to 8 mm long. Fl. Aug. Fr. Oct-Dec. Forests or often by streams at 500-1500 m. Distributed in Guangxi, SW and C China.

越南密脉木

Myrioneuron tonkinense Pitard

多年生草本或半灌木，高1-2米。树皮灰白色。叶具柄，纸质，倒卵形至椭圆形，侧脉15-18对，平行。花序对生于叶腋，短穗状或总状；花萼裂片显著钻形；花冠黄色，管状。浆果近球形，白色。花期6-8月，果期10-12月。生海拔100-1700米的山谷溪边密林中。产云南、广西、广东和海南。越南亦有。

Perennial herbs or semishrubs, 1-2 m tall. Bark grayish white. Leaves petiolate, papery, obovate to elliptic, lateral veins 15-18 pairs, parallell. Inflorescences axillary, opposite, shortly spicate or racemose; calyx lobes linear-subulate; corolla yellow, tubular. Berries subglobose, white. Fl. Jun-Aug. Fr. Oct-Dec. Dense forests by streams in valleys at 100-1700 m. Distributed in Yunnan, Guangxi, Guangdong and Hainan. Also in Vietnam.

毛腺萼木

Mycetia hirta Hutchins.

灌木，高1-2米，被长硬毛。树皮黄白色，光亮。对生叶稍不等大，纸质，长圆状椭圆形至宽披针形，侧脉18-23对。聚伞花序顶生，多花；花萼钟形，边缘具带柄腺毛或撕裂状；花冠黄色，狭管状。果近球形，白色，被毛。花期6-7月，果期

毛腺萼木 *Mycetia hirta*

越南密脉木 *Myrioneuron tonkinense*

腺萼木 *Mycetia glandulosa*

9-10月。生海拔500-1600米的山谷溪边林中。产云南和海南。

Shrubs, 1-2 m tall, hirtous. Bark yellowish white, shiny. Opposite leaves slightly unequal in size, papery, oblong elliptic or broadly lanceolate, lateral veins 18-23 pairs. Cymes terminal, many-flowered; calyx campanulate, margin stipitate-glandulose or lacerate; corolla yellow, narrowly tubular. Capsules subglobose, white, hairy. Fl. Jun-Jul. Fr. Sep-Oct. Forests by streams in valleys at 500-1600 m. Distributed in Yunnan and Hainan.

腺萼木

Mycetia glandulosa Craib

灌木，被柔毛。树皮光亮。对生叶不等大，纸质，倒披针形至长圆形，镰状弯曲，偏斜，侧脉显著。聚伞花序顶生，多花，具梗；花萼裂片边缘有流苏状腺体；花冠黄色，狭管状。果近球形，浆果状或蒴果状。花期5月，果期秋季。生海拔900-1500米的山谷溪边林中。产云南。泰国亦有。

Shrubs, pubescent. Bark shiny. Opposite leaves unequal in size, papery, oblanceolate to oblong, falcate bent, oblique, lateral veins distinct. Cymes terminal, many-flowered, pedunculate; calyx lobes with fimbriate glands at margin; corolla yellow, narrowly tubular. Fruits subglobose, baccate or capsular. Fl. May. Fr. autumn. Forests by streams in valleys at 900-1500 m. Distributed in Yunnan. Also in Thailand.

华腺萼木

Mycetia sinensis (Hemsl.) Craib

灌木或亚灌木，高可达0.7米。嫩枝被皱卷柔毛。叶近膜质，长圆状披针形至椭圆状长圆形。聚伞花序顶生，开展；花冠白色，狭管状，内外均无毛。果近球形，无毛。花期7-8月，果期9-11月。生海拔200-1000米密林下的沟溪边或林中路旁。产湖南、江西、福建、云南、广西、广东和海南。

Shrubs or subshrubs, up to 0.7 m tall. Branches densely hirtellous when young. Leaves drying submembranous, oblong-lanceolate to elliptic-oblong. Inflorescences terminal, laxly cymose; corolla white, narrowly tubular, glabrous on both side. Berries subglobose, glabrous. Fl. Jul-Aug. Fr. Sep-Nov. Streamsides or roadsides in dense forests at 200-1000 m. Distributed in Hunan, Jiangxi, Fujian, Yunnan, Guangxi, Guangdong and Hainan.

华腺萼木 *Mycetia sinensis*

纤梗腺萼木

Mycetia gracilis Craib

灌木。树皮光亮。对生叶极不等大，薄革质，宽倒披针形至窄披针形，侧脉10-14对。聚伞花序顶生，疏松，5-7花，分枝和小花梗纤细，有时线状；花萼裂片条形；花冠黄色，狭管状。果球形，浆果状或蒴果状。花期8-9月，果期11-12月。生海拔600-1300米的山谷溪边林中。产云南。泰国和越南亦有。

Shrubs. Bark shiny. Opposite leaves very unequal in size, thin leathery, broadly oblanceolate to narrowly lanceolate, lateral veins 10-14 pairs. Cymes terminal, lax, 5-7-flowered, branches and pedicels slender, sometimes filiform; calyx lobes linear; corolla yellow, narrowly tubular. Fruits globose, baccate or capsular. Fl. Aug-Sep. Fr. Nov-Dec. Forests by streams in valleys at 600-1300 m. Distributed in Yunnan. Also in Thailand and Vietnam.

纤梗腺萼木 *Mycetia gracilis*

长叶腺萼木 *Mycetia longifolia*

长叶腺萼木
Mycetia longifolia (Wall.) Kuntze

灌木；嫩枝被柔毛。叶纸质，椭圆状披针形或椭圆形。聚伞花序顶生或腋生，开展；花冠黄色，狭管状，外面无毛或被微柔毛。果近球形，无毛。花期夏秋季。生林下。产云南南部和西部、西藏(墨脱县)。南亚和东南亚亦有。

Shrubs; branches pubescent. Leaves drying papery, elliptic-lanceolate or elliptic. Inflorescences terminal or pseudoaxillary, laxly cymose; corolla yellow, narrowly tubular, outside glabrous to villosulous. Berries subglobose, glabrescent. Fl. summer to autumn. Forests. Distributed in S and W Yunnan and Xizang (Mêdog County). Also in S and SE Asia.

狭叶栀子
Gardenia stenophylla Merr.

灌木。叶对生或三叶轮生，狭披针形或条状披针形，基部常下延。花单生，假腋生或顶生；花冠外部无毛；花冠管圆柱柱，裂片5-8。浆果黄色或橘红色，具5-8明显至不明显的纵棱。花期4-8月，果期5月至翌年1月。生海拔100-800米的沟谷溪边林中或灌丛中或河边。产华南和华东。越南亦有。

Shrubs. Leaves opposite or ternate, narrowly lanceolate or linear-lanceolate, base often decurrent. Flowers solitary, pseudoaxillary or terminal; corolla outside glabrous; tubes cylindrical, lobes 5-8. Berries yellow or orange-red, with 5-8 weak to developed longitudinal ridges. Fl. Apr-Aug. Fr. May to next Jan. Forests or thickets at streamsides in valleys or riversides at 100-800 m. Distributed in S and E China. Also in Vietnam.

狭叶栀子 *Gardenia stenophylla*

海南栀子
Gardenia hainanensis Merr.

乔木。叶倒卵状长圆形、椭圆状长圆形或倒披针形，长5-19.5厘米，先端急尖或短渐尖，尖端常稍钝，两面无毛。花芳香，单生，直径4-5厘米；花萼裂片长4-5毫米；花冠白色，高脚碟状。果黄色，具5条不明显或发达纵棱，顶部具宿存的萼檐。花期4月，果期5-10月。生海拔100-1200米的山谷溪边林中或山坡。产广西和海南。

Trees. Leaves obovate-oblong, elliptic-oblong or oblanceolate, 5-19.5 cm long, apex acute or shortly acuminate with tip often slightly obtuse, both surfaces glabrous. Flowers fragrant, solitary, 4-5 cm diam; calyx lobes 4-5 mm long; corolla white, salverform. Fruits yellow, with 5 weak to developed longitudinal ridges, with persistent calyx limb. Fl. Apr. Fr. May-Oct. Forests at streamsides in valleys, or mountain slopes at 100-1200 m. Distributed in Guangxi and Hainan.

海南栀子 *Gardenia hainanensis*

栀子 *Gardenia jasminoides*

栀子
Gardenia jasminoides J. Ellis

灌木，高0.3-3米，无毛。叶革质，长圆状披针形至倒卵形。花芳香，单朵顶生，萼管具棱，裂片披针形；花冠白色，高脚碟状。蒴果卵球形，具翅状纵棱。花期3-7月，果期5月至翌年2月。生海1500米以下的林下、灌丛和荒野。产中国西南、华南、华中和华东。东北亚、南亚和东南亚亦有。

Shrubs, 0.3-3 m tall, glabrous. Leaves leathery, oblong-lanceolate to obovate. Flowers fragrant, solitary, terminal, calyx tube ridged, lobes lanceolate; corolla white, salverform. Capsules ovoid, with longitudinal winged ridges. Fl. Mar-Jul. Fr. May to next Feb. Forests, thickets or fields below 1500 m. Distributed in SW, S, C and E China. Also in NE, S and SE Asia.

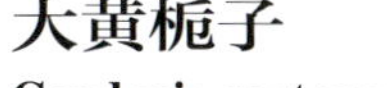

大黄栀子
Gardenia sootepensis Hutchins.

乔木，高7-10米。节显著，节间密集，短。叶倒卵形至长圆形，下面密被绒毛，托叶近膜质。花大，直径约7厘米，芳香，黄色或白色；花冠高脚碟形，具条纹。蒴果绿色，顶端无宿存萼裂片。花期4-8月，果期6月至翌年4月。生海拔700-1600米的山坡、村边或溪边。产云南南部。老挝和泰国亦有。

Trees, 7-10 m tall. Nodes distinct, internodes dense, short. Leaves obovate to oblong, densely tomentose abaxially, stipule submembranous. Flowers large, ca. 7 cm diam, fragrant, yellow or white; corolla salverform, striate. Capsules green, without persistent calyx lobes on top. Fl. Apr-Aug. Fr. Jun to next Apr. Slopes, village margins or streamsides at 700-1600 m. Distributed in S Yunnan. Also in Laos and Thailand.

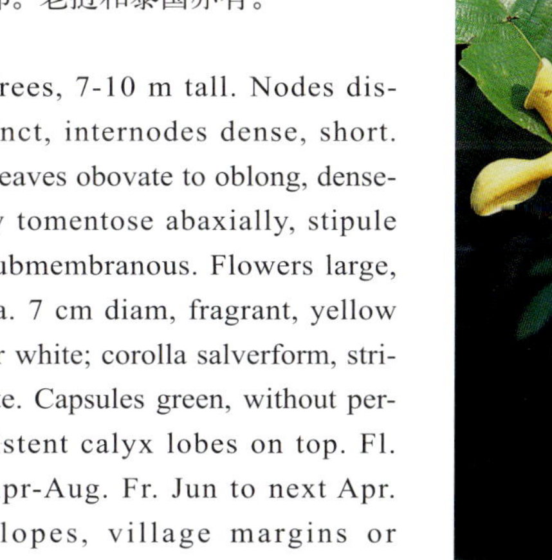

大黄栀子 *Gardenia sootepensis*

山石榴 *Catunaregam spinosa*

山石榴

Catunaregam spinosa (Thunb.) Tirveng.

有刺灌木或小乔木，有时攀援状。叶对生或簇生，倒卵形或长圆状倒卵形，纸质或近革质，侧脉4-7对。花单生或2-3朵簇生；花冠白色或淡黄色，钟状，裂片卵形。浆果大，球形。花期3-6月，果期5月至翌年1月。生海拔1600米以下的灌丛、林中或旷野。产中国西南、华南和华东。南亚、东南亚和东非亦有。

Spinescent shrubs or small trees, sometimes scandent. Leaves opposite or fascicled, obovate or oblong-obovate, papery or subleathery, lateral veins 4-7 pairs. Flowers solitary or 2-3 clustered; corolla white or pale yellow, campanulate, lobes ovate. Berries large, globose. Fl. Mar-Jun. Fr. May to next Jan. Thickets, forests or open fields below 1600 m. Distributed in SW, S and E China. Also in S and SE Asia, and E Africa.

簕茜(鸡爪簕)

Benkara sinensis (Lour.) Ridsdale

灌木或小乔木，多分枝。枝压扁至圆柱形，具刺。叶干纸质至厚纸质，卵状椭圆形、椭圆状长圆形或卵形。花序密伞房状，常伞形，数花至多花；花冠白色或黄色。浆果球形。花期3-12月，果期5月至翌年2月。生海拔1200米以下的林中、林缘或山坡灌丛、山地或田间。产云南、广西、广东、海南、台湾和福建。泰国、越南和日本亦有。

Shrubs or small trees, many branched. Branches compressed to terete, with thorns. Leaves drying papery to thickly papery, ovate-elliptic, elliptic-oblong or ovate. Inflorescences congested-cymose, often umbelliform, several to many flowered; corolla white or yellow. Berries globose. Fl. Mar-Dec. Fr. May to next Feb. Forests, forest edges or thickets on hills, mountains or in fields below 1200 m. Distributed in Yunnan, Guangxi, Guangdong, Hainan, Taiwan and Fujian. Also in Thailand, Vietnam and Japan.

多毛茜树

Aidia pycnantha (Drake) Tirvengadum

灌木或乔木，高2-12米。嫩枝密被柔毛。叶薄革质，长圆形、长圆状披针形或长圆状倒披针形，下面密被柔毛。聚伞花序腋生；花冠白色或淡黄色，高脚碟形，喉部密被长柔毛。浆果球形，有疏毛。花期3-9月，果期4-12月。生海拔20-1000米的林中或灌丛中。产中国东南、西南和华南。越南亦有。

Shrubs or trees, 2-12 m tall. Branches densely pubescence. Leaves drying thinly leathery or papery, oblong, oblong-lanceolate or oblong-oblanceolate, abaxially pubescence. Inflorescences cymose, axillary; corolla white or pale yellow, salverform, inside densely villous in throat. Berry globose, sparsely strigillose. Fl. Mar-Sep. Fr. Apr-Dec. Thickets or forests at 20-1000 m. Distributed in SE, SW and S China. Also in Vietnam.

簕茜(鸡爪簕) *Benkara sinensis*

多毛茜树 *Aidia pycnantha*

茜树 *Aidia cochinchinensis*

香楠 *Aidia canthioides*

香楠

Aidia canthioides (Champ. ex Benth.) Masamu.

灌木或乔木，高1-12米。小枝无毛。叶薄革质，长圆状椭圆形至披针形，两面无毛。花序腋生，紧缩，近无总梗；花冠高脚碟形，白色或黄白色，喉部被长柔毛。浆果球形。花期4-6月，果期5月至翌年2月。生海拔50-1500米的山坡、山谷溪边、丘陵的灌丛中或林中。产中国东南、西南和华南。日本和越南亦有。

Shrubs or trees, 1-12 m tall. Branches glabrous. Leaves drying thinly leathery, oblong-elliptic to lanceolate, both surfaces glabrous. Inflorescences congested, axillary, subsessile; corolla white or yellowish white, salverform. Berry globose. Fl. Apr-Jun. Fr. May to next Feb. Thickets or forests on hills, on mountain slopes or at streamsides in valleys at 50-1500 m. Distributed in SE, SW and S China. Also in Japan and Vietnam.

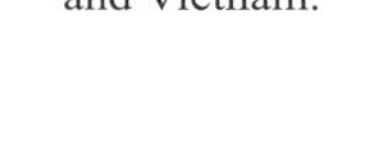

茜树

Aidia cochinchinensis Lour.

灌木或乔木，高2-15米，无毛。叶具柄，椭圆状长圆形至窄椭圆形，纸质或革质，侧脉5-10对。聚伞花序与叶对生，具多花，具苞片；花4-5数；花冠黄色或白色，裂片长圆形。浆果球形。花期4月。生海拔500-1300米的丘陵、山坡、山谷或溪边。产云南和海南。越南亦有。

Shrubs or trees, 2-15 m tall, glabrous. Leaves petiolate, elliptic-oblong to narrowly elliptic, papery or leathery, lateral veins 5-10 pairs. Cymes opposite to leaves, many-flowered, bracteate; flowers 4-5-merous; corolla yellow or white, lobes oblong. Berries globose. Fl. Apr. Hills, slopes, valleys or streamsides at 500-1300 m. Distributed in Yunnan and Hainan. Also in Vietnam.

岭罗麦

Tarennoidea wallichii (Hook. f.) Tirveng. et C. Sastre

乔木，高3-20米，节显著。叶长圆形至披针形，革质。聚伞花序排成圆锥花序状，顶生或腋生，疏散；花5数；花冠黄色或白色，喉部被长柔毛。浆果球形。花期3-6月，果期7月至翌年2月。生海拔400-2200米的丘陵、山坡、山谷溪边的林中或灌丛中。产云南、贵州、广西、广东和海南。南亚和东南亚亦有。

Trees, 3-20 m tall, nodes distinct. Leaves oblong to lanceolate, leathery. Cymes in panicle, terminal or axillary, lax; flowers 5-merous; corolla yellow or white, villose at throat. Berries globose. Fl. Mar-Jun. Fr. Jul to next Feb. Forests or thickets, on hills, mountain slopes by streams or in valleys at 400-2200 m. Distributed in Yunnan, Guizhou, Guangxi, Guangdong and Hainan. Also in S and SE Asia.

岭罗麦 *Tarennoideawallichii*

滇短萼齿木 *Brachytome hirtellata*

须弥茜树 *Himalrandia lichiangensis*

须弥茜树
Himalrandia lichiangensis (W. W. Sm.) Tirveng.

灌木，高0.6-3米。叶纸质或薄革质，倒卵形或倒卵状匙形，两面被糙伏毛。花单朵顶生抑缩的侧生短枝上；花冠高脚蝶状，黄色或白色，外面无毛，内面被白色硬毛。浆果球形。种子1或2粒。花期5月，果期7-11月。生海拔1400-2400米的山坡、山谷沟边的林中或灌丛中。产四川和云南北部。

Shrubs, 0.6-3 m tall. Leaves drying papery to thinly leathery, obovate or obovate-spatulate, both surfaces strigillose. Inflorescences terminal usually on short shoots, 1-flowered; corolla salverform, yellow or white, outside glabrous, inside hirsute. Berry globose. Seeds 1 or 2. Fl. May. Fr. Jul-Nov. Forests or thickets at streamsides in valleys or on mountains at 1400-2400 m. Distributed in Sichuan and N Yunnan.

滇短萼齿木
Brachytome hirtellata Hu

灌木，高约3米，密被短硬毛。叶对生或3叶假轮生，膜质，倒披针形或长圆状披针形，下面密被短硬毛。聚伞花序与叶对生，具5-10花，具苞片；花冠漏斗状，白色或黄白色。浆果球形，被短硬毛。花期3-6月，果期5月至翌年3月。生海拔400-2200米的山谷、溪边、灌丛或林中。产云南和西藏。越南亦有。

Shrubs, ca. 3 m tall, densely hirtellous. Leaves opposite or in pseudo-whorls of 3, membranous, oblanceolate or oblong-lanceolate, abaxially densely hirtellous. Cymes opposite to leaves, 5-10 flowered, bracteate; corolla funnelform, white or yellowish white. Berries globose, hirtellous. Fl. Mar-Jun. Fr. May to next Mar. Valleys, streamsides, thickets or forests at 400-2200 m. Distributed in Yunnan and Xizang. Also in Vietnam.

狗骨柴
Diplospora dubia (Lindley) Masam.

灌木或乔木，高1-12米。小枝无毛。叶革质，卵状长圆形至披针形，两面无毛。花序近头状至稠密簇生；花4数；花冠白色或黄色，高脚碟状。浆果近球形，无毛或近无毛。花期4-8

狗骨柴 *Diplospora dubia*

毛狗骨柴 *Diplospora fruticosa*

月，果期5月至翌年2月。生海拔40-1500米的林中或灌丛中。产云南和四川向东至华东。日本和越南亦有。

Shrubs or trees, 1-12 m. Branches glabrous. Leaves drying leathery, ovate-oblong to lanceolate, both surfaces glabrous. Inflorescences subcapitate to congested-fasciculate; flowers 4-merous; corolla white or yellow, salverform, glabrous outside. Berry subglobose, glabrous or subglabrous. Fl. Apr-Aug. Fr. May to next Feb. Forests or thickets at 40-1500 m. Distributed in Yunnan and Sichuan eastward to E China. Also in Japan and Vietnam.

毛狗骨柴

Diplospora fruticosa
Hemsl.

灌木或乔木，被短柔毛。叶干纸质至薄革质，长圆形、长圆状披针形或窄椭圆形；次级叶脉7-13对。花序聚伞状，近无柄，密被柔毛至糙伏毛；苞片多合生成对；花冠白色，外部无毛。果实红色，近球形。花期3-5月，果期6月至翌年2月。生海拔200-2000米的灌丛或沟谷林中。产中国西南、华中和华东。越南亦有。

Shrubs or trees, puberulent. Leaves drying papery or thinly leathery, oblong, oblong-lanceolate or narrowly elliptic; secondary veins 7-13 pairs. Inflorescences shortly cymose, subsessile, densely puberulent to strigillose; bracts mostly united in pairs; corolla white, glabrous outside. Fruits red, subglobose. Fl. Mar-May. Fr. Jun to next Feb. Thickets or forests in ravines at 200-2000 m. Distributed in SW, C and E China. Also in Vietnam.

白皮乌口树

Tarenna depauperata
Hutchins.

灌木或小乔木，高1-6米。小枝无毛。叶纸质或革质，椭圆状倒卵形至近卵形，两面无毛。伞房状的聚伞花序顶生；花冠白色，外面无毛，内面有长柔毛。浆果球形，无毛。种子1或2粒。花期4-11月，果期4月至翌年1月。生海拔200-1700米的林中和灌丛中。产江苏、广东、广西、贵州和云南。越南亦有。

Shrubs or small trees, 1-6 m tall. Branches glabrous. Leaves papery or leathery, elliptic-obovate to subovate, both surfaces glabrous. Inflorescences corymbiform to pyramidal; corolla white, outside glabrous, villous inside. Berries globose, glabrescent. Seeds 1 or 2. Fl. Apr-Nov. Fr. Apr to next Jan. Forests or thickets at 200-1700 m. Distributed in Jiangsu, Guangdong, Guangxi, Guizhou and Yunnan. Also in Vietnam.

白皮乌口树 *Tarenna depauperata*

白花苦灯笼 *Tarenna mollissima*

白花苦灯笼

Tarenna mollissima (Hook. et Arn.) Robins.

灌木或小乔木，高1-6米，密被软毛或微绒毛。叶披针形至卵状椭圆形，纸质。伞房状聚伞花序顶生，多花；花4-5数；花冠白色，喉部密被长柔毛。浆果近球状，被软毛。花期5-7月，果期5月至翌年1月。生海拔200-1100米的林中、灌丛或山坡。产中国西南、华南、华中和华东。越南亦有。

Shrubs or small trees, 1-6 m tall, densely soft hairy or tomentulose. Leaves lanceolate to ovate-elliptic, papery. Corymbose-cymes terminal, many-flowered; flowers 4-5-merous; corolla white, densely villose at throat. Berries subglobose, soft hairy. Fl. May to next Jul. Fr. May to next Jan. Forests, thickets or montane slopes at 200-1100 m. Distributed in SW, S, C and E China. Also in Vietnam.

尖萼乌口树

Tarenna acutisepala W. C. Chen

灌木，高1-2.5米。叶长圆形或披针形，纸质或近革质。伞房状聚伞花序顶生，密花，具短总梗；花萼裂片三角状披针形；花冠高脚碟状，淡黄色，裂片椭圆形。浆果近球形，顶端具宿存花萼裂片。花期4-9月，果期5-11月。生海拔500-1600米的林中、沟谷灌丛或山坡。产四川、广西、广东、福建、湖北、湖南、江苏和江西。

Shrubs, 1-2.5 m tall. Leaves oblong or lanceolate, papery or subleathery. Corymbose-cymes terminal, densely flowered, shortly pedunculate; calyx lobes triangular-lanceolate; corolla salverform, pale yellow, lobes elliptic. Berries subglobose, crowned by persistent calyx lobes. Fl. Apr-Sep. Fr. May-Nov. Forests, valley thickets or montane slopes at 500-1600 m. Distributed in Sichuan, Guangxi, Guangdong, Fujian, Hubei, Hunan, Jiangsu and Jiangxi.

假桂乌口树

Tarenna attenuata (Hook. f.) Hutchins.

灌木或乔木，高1-8米。小枝无毛。叶常薄革质，长圆状披针形至倒卵形，通常两面无毛。伞房状的聚伞花序顶生；花冠白色或淡黄色，外面无毛，喉部有柔毛。浆果近球形，无毛。种子2粒。花期4-11月，果

尖萼乌口树 *Tarenna acutisepala*

假桂乌口树 *Tarenna attenuata*

藏药木 *Hyptianthera stricta*

期5月至翌年1月。生海拔15-1200米的林中或灌丛中。产广东、广西、海南和云南。印度、泰国、越南和柬埔寨亦有。

Shrubs or trees, 1-8 m tall. Branches glabrous. Leaves often thinly leathery, oblong-lanceolate to obovate, both surfaces glabrous. Inflorescences corymbiform to pyramidal, terminal; corolla white or pale yellow, outside glabrous, villosulous at throat. Berries subglobose, glabrous. Seeds 2. Fl. Apr-Nov. Fr. May to next Jan. Forests or thickets at 15-1200 m. Distributed in Guangdong, Guangxi, Hainan and Yunnan. Also in India, Thailand, Vietnam and Cambodia.

藏药木

Hyptianthera stricta (Roxb.) Wight et Arn.

灌木或小乔木，高2-8米。小枝无毛。叶纸质或薄革质，长圆状披针形或狭长圆形，两面常无毛。花无梗，数至多朵簇生叶腋；花冠白色，短漏斗形，外面无毛。浆果近球形，有微柔毛。花期4-8月，果期8月至翌年2月。生海拔100-1500米的山地或溪边的林中或灌丛中。产云南和西藏。南亚和东南亚北部亦有。

Shrubs or small trees, 2-8 m tall. Branches glabrous. Leaves drying papery or thinly leathery, oblong-lanceolate or narrowly oblong, usually glabrous on both surfaces. Inflorescences fasciculate, sessile; flowers sessile; corolla short funnelform, abaxially glabrous. Berry subglobose, puberulent. Fl. Apr-Aug. Fr. Aug to next Feb. Forests or thickets at streamsides or on mountains at 100-1500 m. Distributed in Yunnan and Xizang. Also in S Asia and north of SE Asia.

猪肚木 *Canthium horridum*

猪肚木

Canthium horridum Blume

具刺灌木，高2-3米，被毛。叶卵形或椭圆形，纸质，侧脉2-3对。花小，无梗或具短梗，单生或数朵簇生于叶腋；花冠白色，近坛状，裂片长圆形。核果黄色，大，卵球形。花期4-6月，果期7-11月。生海拔500米以下的疏林或灌丛中。产中国西南和华南。南亚和东南亚亦有。

Spinescent shrubs 2-3 m tall, hairy. Leaves ovate or elliptic, papery, lateral 2-3 pairs. Flowers small, sessile or shortly pedicellate, solitary or several clustered in axils of leaves; corolla white, suburceolate, lobes oblong. Drupes yellow, large, ovoid. Fl. Apr-Jun. Fr. Jul-Nov. Sparse forests or thickets below 500 m. Distributed in SW and S China. Also in S and SE Asia.

鱼骨木

Canthium dicoccum (Gaertn.) Merr.

灌木至中等乔木，高8-15米。小枝无毛。叶革质，卵形、椭圆形至卵状披针形。聚伞花序具短总花梗，比叶短，偶被微柔毛；花冠绿白色或淡黄色，圆筒形，喉部具绒毛，顶部5裂，偶有4裂。核果倒卵形，略扁。花期1-5月，果期6-9月。常生海拔200-1500米至中海拔疏林或灌丛中。产广东、香港、海南、广西、云南和西藏(墨脱县)。南亚、东南亚和澳大利亚亦有。

Shrubs or trees, 8-15 m tall. Branches glabrous. Leaf blade drying leathery, ovate to ovate-lanceolate; stipules triangular. Inflorescences cymose, not longer than leaf; corolla white, urceolate-tubular, inside villous in throat. Drupes obovoid, laterally compressed, glabrescent. Fl. Jan-May. Fr. Jun-Sep. Broad-leaved forests at low to middle altitude at 200-1500 m. Distributed in Guangdong, Hong Kong, Hainan, Guangxi, Yunnan and Xizang (Mêdog County). Also in S and SE Asia, and Australia.

鱼骨木 *Canthium dicoccum*

海岸桐 *Guettarda speciosa*

海岸桐
Guettarda speciosa L.

常绿小乔木。叶宽倒卵形或宽椭圆形，长11-20厘米，宽8-18厘米，先端急尖，钝或圆形；侧脉7-11对。花序近头状或聚伞状，密被茸毛，花序轴蝎尾状；花无梗，芳香；萼檐管形，截平；花冠白色，外面密被绒毛，先端7-8裂，裂片倒卵形，先端钝至圆。核果成熟后绿色。花期4-7月。生海岸沙地的灌丛边缘。产台湾、海南和广东。广布于热带海岸。

Evergreen small trees. Leaves broadly obovate or broadly elliptic, 11-20 × 8-18 cm, apex acute, obtuse or rounded; lateral veins 7-11 pairs. Inflorescences subcapitate to cymose, densely velutinous, axes scorpioid; flowers sessile, fragrant; calyx limb tubular, truncate; corolla white, outside densely velutinous, lobes 7-8 at apex, lobes obovate, apex obtuse to rounded. Drupes apparently green at maturity. Fl. Apr-Jul. Thickets on sandy coasts. Distributed in Taiwan, Hainan and Guangdong. Also widely in tropical coast.

假鱼骨木
Psydrax dicocca Gaertn.

灌木或乔木，高8-15米。小枝无毛。叶革质，卵形，椭圆形至卵状披针形。聚伞花序具短总花梗，比叶短；花冠绿白色或淡黄色，近管形，喉部具绒毛。核果倒卵形，略扁。花期1-5月，果期6-9月。生海拔200-1500米的疏林或灌丛中。产广东、香港、海南、广西、云南和西藏。南亚、东南亚和澳大利亚亦有。

Shrubs or trees, 8-15 m tall. Branches glabrous. Leaves drying leathery, ovate to ovate-lanceolate; stipules triangular. Inflorescences cymose, not longer than leaf; corolla white, subtubular, inside villous in throat. Drupes obovoid, laterally compressed, glabrescent. Fl. Jan-May. Fr. Jun-Sep. Broad-leaved forests at low to middle elevations at 200-1500 m. Distributed in Guangdong, Hong Kong, Hainan, Guangxi, Yunnan and Xizang. Also in S and SE Asia, and Australia.

假鱼骨木 *Psydrax dicocca*

小粒咖啡 *Coffea arabica*

小粒咖啡
Coffea arabica L.

小乔木或大灌木，高5-8米。叶革质，卵状披针形或披针形，边缘全缘或波状，中脉凸起。聚伞花序数个簇生于叶腋内，每个具2-5花，近无梗或具短梗；花冠白色。浆果椭圆体形，红色。花期3-7月，果期10月至翌年1月。中国西南、华南和华东有栽培。原产东非，世界热带和亚热带地区广泛栽培。

糙叶大沙叶 *Pavetta scabrifolia*

Small trees or large shrubs, 5-8 m tall. Leaves leathery, ovate-lanceolate or lanceolate, margin entire or undulate, midribs convex. Cymes several fascicled in leaf axils, each 2-5-flowered, sessile or shortly pedunculate; corolla white. Berries ellipsoid, red. Fl. Mar-Jul. Fr. Oct to next Jan. Cultivated in SW, S and E China. Native to E Africa, widely cultivated in tropical and subtropical areas of the world.

糙叶大沙叶

Pavetta scabrifolia Bremek.

灌木，被柔毛。叶干膜质，披针形，粗糙，侧脉5-6对，托叶宽三角形。伞房花序疏散，被柔毛；花白色，4数；花冠管长1.7厘米，内面无毛，花柱长3.8厘米。浆果球形，直径约5毫米。花期5-6月。生海拔900-1300米的疏林或沟边。产云南。

Shrubs, pubescent. Leaves drying membranous, lanceolate, scabrous, lateral veins 5-6 pairs, stipules broadly triangular. Corymbs lax, pubescent; flowers white, 4-merous; corolla tube 1.7 cm long, glabrous inside, styles 3.8 cm long. Berries globose, ca. 5 mm diam. Fl. May-Jun. Sparse forests or ditch sides at 900-1300 m. Distributed in Yunnan.

香港大沙叶

Pavetta hongkongensis Bremek.

灌木或小乔木，高1-4米。叶膜质，长圆形至椭圆状倒卵形。伞房花序顶生，大型，疏松，多花；花具长梗，4数；花冠白色，

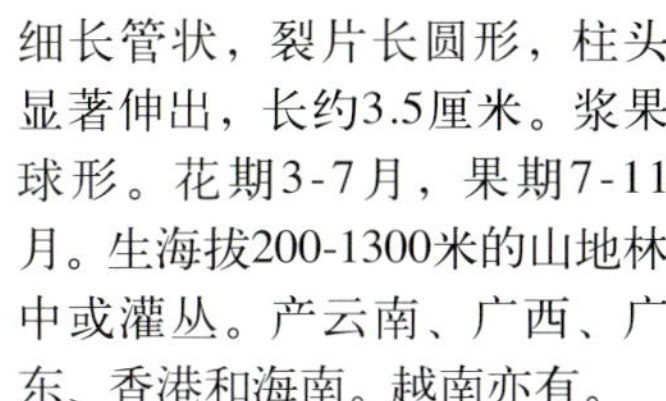

细长管状，裂片长圆形，柱头显著伸出，长约3.5厘米。浆果球形。花期3-7月，果期7-11月。生海拔200-1300米的山地林中或灌丛。产云南、广西、广东、香港和海南。越南亦有。

Shrubs or small trees, 1-4 m tall. Leaves membranous, oblong to elliptic-obovate. Corymbs terminal, large, lax, many-flowered; flowers long pedicellate, 4-merous; corolla white, very narrowly tubular, lobes oblong, style distinctly exserted, ca. 3.5 cm long. Berries globose. Fl. Mar-Jul. Fr. Jul-Nov. Montane forests or thickets at 200-1300 m. Distributed in Yunnan, Guangxi, Guangdong, Hong Kong and Hainan. Also in Vietnam.

香港大沙叶 *Pavetta hongkongensis*

龙山龙船花 *Ixora longshanensis*

龙山龙船花

Ixora longshanensis Tao Chen

小乔木。叶对生，无柄，椭圆形至倒披针形，基部心形且半抱茎，顶端锐尖至渐尖。花序顶生，伞房状，具柔毛，红色至紫红色；萼檐深裂；花冠蕾期紫红色。生溪边沟谷密林。产云南南部。

Small trees. Leaves opposite, sessile, elliptic to oblanceolate, base cordulate and amplexicaul, apex acute to acuminate. Inflorescences terminal, corymbiform, puberulent, red to purple-red; calyx limbs deeply lobed; corolla in bud red-purple. Dense forests in valleys or at streamsides. Distributed in S Yunnan.

云南龙船花

Ixora yunnanensis Hutchins.

灌木，高达1米，无毛。叶对生或近轮生，纸质，披针形或长圆形，基部渐狭下延。花序顶生，多花，近无梗；花芳香，近无梗，4数，萼筒陀螺状，裂片披针形；花冠白色，裂片长圆形。核果长圆形。花期5月。生海拔160-240米的湿润溪边密林中。产云南。

Shrubs, to 1 m tall, glabrous. Leaves opposite or sub-vertixillate, papery, lanceolate or oblong, attenuate and decurrent at base. Inflorescences terminal, many-flowered, subsessile; flowers fragrant, subsessile, 4-merous, calyx tube turbinate, lobes lanceolate; corolla white, lobes oblong. Drupes oblong. Fl. May. Dense forests by humid streams at 160-240 m. Distributed in Yunnan.

云南龙船花 *Ixora yunnanensis*

白花龙船花

Ixora henryi Lévl.

灌木，高1-3米，无毛。叶具柄，纸质，长圆形、披针形或椭圆形。花序顶生，多花，排成三歧伞房式的聚伞花序；花具梗，4数；花冠白色，管纤细，裂片长圆形。核果球形。花期9-12月，果期翌年5-7月。生海拔200-2000米的阔叶林或林缘溪边。产云南、贵州、广西、广东和海南。越南亦有。

Shrubs, 1-3 m tall, glabrous. Leaves petioloate, papery, oblong, lanceolate or elliptic. Inflorescences terminal, many-flowered, arranged in trichotomous, corymbous cymes; flowers pedicellate, 4-merous; corolla white, tube slender, lobes oblong. Drupes globose. Fl. Sep-Dec. Fr. May-Jul next year. Broad-leaved forests or streamsides at forest edges at 200-2000 m. Distributed in Yunnan, Guizhou, Guangxi, Guangdong and Hainan. Also in Vietnam.

龙船花

Ixora chinensis Lam.

灌木，高0.8-2米，无毛。叶近无柄，披针形，托叶合生成鞘。伞房花序顶生，多花；花4数；花冠红色或红黄色。核果近球形，中部具槽。花期5-7月和12月，果期9-10月。生海拔200-800米的灌丛或疏林中，亦栽培于庭院和公路边。产广西、广东、香港和福建，中国西南和华南栽培。东南亚亦有。

Shrubs, 0.8-2 m tall, glabrous.

白花龙船花 *Ixora henryi*

龙船花 *Ixora chinensis*

Leaves subsessile, lanceolate, stipules connate into a sheath. Corymbs terminal, many-flowered; flowers 4-merous; corolla red or reddish yellow. Drupes subglobose, grooved at middle. Fl. May-Jul and Dec. Fr. Sep-Oct. Thickets or sparse forests at 200-800 m, also cultivated in gardens or roadsides. Distributed in Guangxi, Guangdong, Hong Kong and Fujian, cultivated in SW and S China. Also in SE Asia.

橙江龙船花

Ixora coccinea L.

灌木，高1.2-2米。叶近无梗，革质，长圆形，托叶合生成鞘。伞房花序顶生，多花；花4数；花冠红色，裂片长圆形，开展。核果双生，熟时红黑色。中国西南和华南栽培。原

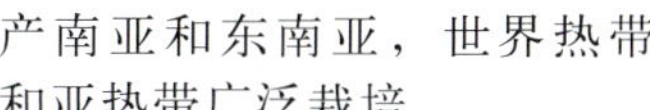

产南亚和东南亚，世界热带和亚热带广泛栽培。

Shrubs, 1.2-2 m tall. Leaves subsessile, leathery, oblong, stipules connate into a sheath. Corymbs terminal, many-flowered; flowers 4-merous; corolla red, lobes oblong, spreading. Drupes binate, reddish black at maturity. Cultivated in SW and S China. Native to S and SE Asia, widely cultivated in tropical and subtropical areas of the world.

长柱山丹

Duperrea pavettifolia (Kurz) Pitard

灌木或小乔木，被短柔毛和短硬毛。叶纸质，长圆状椭圆形、长圆状披针形至倒披针形。花序顶生或腋生；萼筒倒圆锥形，裂片条形；花冠高脚蝶状，白色，冠筒纤细，花柱细长，柱头近椭圆形。浆果扁球形。花期4-6月，果期9-12月。生海拔100-1100米的山地林中或山谷溪边。产云南、广西和海南。东南亚亦有。

橙江龙船花 *Ixora coccinea*

Shrubs or small trees, puberulent and strigillose. Leaves papery, oblong-elliptic, oblong-lanceolate to oblanceolate. Inflorescences terminal or axillary; calyx tube obconical, lobes linear; corolla salverform, white, tube slender, style slender, stigma subellipsoid. Berries compressed globose. Fl. Apr-Jun. Fr. Sep-Dec. Montane forests or streamsides in valleys at 100-1100 m. Distributed in Yunnan, Guangxi and Hainan. Also in SE Asia.

长柱山丹 *Duperrea pavettifolia*

美果九节 *Psychotria calocarpa*

美果九节
Psychotria calocarpa Kurz

亚灌木，高0.2-1米。根茎匍匐。叶纸质，长圆形至披针形；侧脉9-15对，在近边缘处结成一明显的边脉。聚伞花序小，顶生或腋生；花小，4数；花冠外面紫色，内面白色，被长毛。核果椭圆体形，熟时红色或橙色。花期5-7月，果期8月至翌年2月。生海拔800-1700米的山地林中。产云南和西藏。南亚和东南亚亦有。

Subshrubs, 0.2-1 m tall. Rhizomes procumbent. Leaves papery, oblong to lanceolate; secondary veins 9-15 pairs, forming a obvious submarginal vein. Cymes small, terminal or axillary; flowers small, 4-merous; corolla purple outside, white and villose inside. Drupes ellipsoid, red or orange at maturity. Fl. May-Jul. Fr. Aug to next Feb. Montane forests at 800-1700 m. Distributed in Yunnan and Xizang. Also in S and SE Asia.

假九节
Psychotria tutcheri Dunn

灌木。叶干纸质至薄革质；次级叶脉4-8对；托叶早落，1/4-1/2裂。花序顶生及有时假腋生，聚伞状，分枝部分伞房状；花冠白色或浅绿色，管状。核果近球形；果核近光滑或具浅的3或4条棱。花期4-7月，果期6-12月。生海拔200-1000米的沟谷林中或灌木丛中。产云南、广西、广东、海南和福建。越南亦有。

Shrubs. Leaves drying papery to thinly leathery; secondary veins 4-8 pairs; stipules caducous, 2-lobed for 1/4-1/2. Inflorescences terminal and sometimes pseudoaxillary, cymose, branched portion corymbiform; corolla white or greenish white, tubular. Drupes subglobose; pyrenes shallowly 3- or 4-ribbed to subsmooth. Fl. Apr-Jul. Fr. Jun-Dec. Thickets or forests in ravines or hill slopes at 200-1000 m. Distributed in Yunnan, Guangxi, Guangdong, Hainan and Fujian. Also in Vietnam.

云南九节
Psychotria yunnanensis Hutchins.

灌木，高1-4米。小枝无毛，托叶脱落后在节上有红褐色密毛。叶长圆形、椭圆形或倒披针形，侧脉8-16对。聚伞圆锥花序顶生或腋生，三歧分枝；花冠白色。核果长圆形。花期

假九节 *Psychotria tutcheri*

云南九节 *Psychotria yunnanensis*

驳骨九节 *Psychotria prainii*

4-12月。生海拔800-2300米的山谷林或林缘。产西藏、云南和广西。

Shrubs, 1-4 m tall. Branchlets glabrous, densely reddish-brown-hairy on the nodes after stipules falling away. Leaves oblong, elliptic or oblanceolate, lateral veins 8-16 pairs. Thyrses terminal or axillary, trichotomous; corolla white. Drupes oblong. Fl. Apr-Dec. Valley forests or forest edges at 800-2300 m. Distributed in Xizang, Yunnan and Guangxi.

驳骨九节
Psychotria prainii Lévl.

灌木，高0.5-2米，被暗红色皱软毛。叶密集于枝顶，椭圆形、长圆形至卵形。聚伞花序顶生，密集成头状，具短总梗；花冠白色。核果红色，椭圆体形或倒卵形，熟时红色。花期5-8月，果期7-11月。生海拔1000-1700米的山谷林中、灌丛石上或山坡上。产云南、贵州、广西和广东。东南亚亦有。

Shrubs, 0.5-2 m tall, covered with dark red crinkled soft hairs. Leaves densely crowded on apex of brances, oblong to ovate. Cymes terminal, conferted into a head, shortly pedunculate; corolla white. Drupes red, ellipsoid or obovoid, red at maturity. Fl. May-Aug. Fr. Jul-Nov. Valley forests, rocks in thickets or montane slopes at 1000-1700 m. Distributed in Yunnan, Guizhou, Guangxi and Guangdong. Also in SE Asia.

九节
Psychotria asiatica L.

灌木或小乔木。叶椭圆状长圆形或披针状长圆形。花序顶生或有时假腋生，聚伞状至圆锥状，多花；花无柄至具梗组成3-5小聚伞花序；花冠白色，漏斗形。核果红色，近球形至阔椭圆体形。花果期全年。生海拔1500米以下的灌木或沟谷林中、山坡或村庄边缘。产中国西南、华南和东南。南亚、东南亚和日本亦有。

Shrubs or small trees. Leaves elliptic-oblong or lanceolate-oblong. Inflorescences terminal or sometimes pseudoaxillary, cymose to paniculiform, many-flowered; flowers subsessile to pedicellate in dichotomous cymules of 3-5; corolla white, funnelform. Drupes red, subglobose to broadly ellipsoid. Fl. and fr. year-round. Thickets or forests in ravines, on hill slopes, or at village edges below 1500 m. Distributed in SW, S and SE China. Also in S and SE Asia, and Japan.

九节 *Psychotria asiatica*

毛九节 *Psychotria pilifera*

毛九节
Psychotria pilifera Hutchins.

灌木，高0.5-2米。小枝密被棕褐色卷曲长柔毛。叶纸质或膜质，椭圆形，倒卵形或倒披针形，侧脉8-15对。聚伞花序顶生或腋生；花芽漏斗形。果熟时红色，长圆状椭圆体形。花期7月，果期8-12月。生海拔1300-1700米的山谷林中。产云南。

Shrubs, 0.5-2 m tall. Branchlets densely brown crispate villose. Leaves papery or membranous, elliptic, obovate or oblanceolate, lateral veins 8-15 pairs. Cymes terminal or axillary; corolla in bud funnelform. Drupes red at maturity, oblong-ellipsoid. Fl. Jul. Fr. Aug-Dec. Valley forests at 1300-1700 m. Distributed in Yunnan.

蔓九节
Psychotria serpens L.

攀援或匍匐藤本，具气生根。叶形卵形、椭圆形至披针形，侧脉4-10对。聚伞花序顶生，圆锥状或伞房状；花冠白色。核果浆果状，球形或椭圆体形，成熟时白色。花期4-6月，果期全年。生海拔100-1400米的平地、丘陵、山地、灌丛或沟谷林中，常攀援于树干或石头上。产华南和华东。东南亚和东北亚亦有。

Climbing or creeping lianas, with aerial roots. Leaves ovate, elliptic to lanceolate, lateral veins 4-10 pairs. Cymes terminal, paniculate or corymbose; corolla white. Drupes berrylike, globose or ellipsoid, white at maturity. Fl. Apr-Jun. Fr. all around the year. Plains, hills, mountains, thickets or valley forests at 100-1400 m, usually creeping on trunks or rocks. Distributed in S and E China. Also in SE and NE Asia.

蔓九节 *Psychotria serpens*

弯管花
Chassalia curviflora Thwaites

直立小灌木，高1-2米，无毛。叶长圆状披针形或倒披针形，膜质。聚伞花序多花，顶生；花近无梗，具3型；花冠管弯曲，细长，

弯管花 *Chassalia curviflora*

爱地草 *Geophila repens*

斜基粗叶木 *Lasianthus attenuatus*

两面被毛，裂片4-5，顶端肿胀，具浅沟。果实紫色，压扁球状。花期4-6月，果期4月至翌年1月。生海拔100-2000米的山谷溪边林中。产中国西南和华南。南亚和东南亚亦有。

Erect undershrubs, 1-2 m tall, glabrous. Leaves oblong elliptic or oblanceolate, membanaceous. Cymes many flowered, terminal; flowers subsessile, trimorphic; corolla tube curved, slender, hairy on both sides, lobes 4-5, apex swollen, shallowly sulcate. Fruits purple, compressed, globose. Fl. Apr-Jun. Fr. Apr to next Jan. Forests by streams in valleys at 100-2000 m. Distributed in SW and S China. Also in S and SE Asia.

爱地草

Geophila repens (L.) I. M. Johnst.

多年生匍匐草本，长达40厘米。茎纤细，具不定根。叶心状圆形至近圆形，具柄，膜质，托叶卵形。花单生，具梗；花冠白色，裂片5，开展。核果球形，红色，光滑，顶端冠以宿存萼裂片。花期7-9月，果期9-12月。生海拔100-600米的杂木林、沟边或湿地上。产云南、贵州、广西、广东、海南、台湾和福建。世界热带地区广布。

Perennial creeping herbs, up to 40 cm long. Stems slender, adventitious roots present. Leaves cordate-orbicular to suborbicular, petiolate, membranous, stipules ovate. Flowers solitary, pedunculate; corolla white, lobes 5, spreading. Drupes globose, red, smooth, apex coronated with persistent calyx-lobes. Fl. Jul-Sep. Fr. Sep-Dec. Mixed forests, streamsides or other wet places at 100-600 m. Distributed in Yunnan, Guizhou, Guangxi, Guangdong, Hainan, Taiwan and Fujian. Also widely in tropical areas of the world.

染木树

Saprosma ternata (Wall.) Hook. f.

直立灌木或小乔木，高1-6米。小枝黄绿色，具棱角。叶通常3片轮生，长圆状披针形或长圆状椭圆形。花数朵生于腋生长总梗的顶端，具小花梗；花冠白色，漏斗状。核果椭圆体形或近球形。花期夏季，果期9-12月。生海拔540-1640米的山谷林、疏林或灌丛。产云南、西藏和海南。印度、缅甸、越南和马来西亚亦有。

Erect shrubs or small trees, 1-6 m tall. Twigs yellowish-green, angulate. Leaves usually in whorls of 3, oblong-lanceolate or oblong-elliptic. Flowers several terminal on long peduncle in axils, pedicellate; corolla white, funnelform. Drupes ellipsoid or subglobose. Fl. summer. Fr. Sep-Dec. Valley forests, sparse forests or thickets at 540-1640 m. Distributed in Yunnan, Xizang and Hainan. Also in India, Myanmar, Vietnam and Malaysia.

斜基粗叶木

Lasianthus attenuatus Jack.

灌木，高1-2米，被绒毛。叶革质，长圆形、卵形至披针形，基部偏斜，稍心形。聚伞花序腋生，无柄，3-4花；花无梗，5数；花冠白色，内部被长柔毛。核果蓝色，近球形。花期4月，果期8-9月。生海拔200-1800米的林下或阴湿地。产云南、广西、广东、海南、台湾和福建。南亚、东南亚、日本和巴布亚新几内亚亦有。

Shrubs, 1-2 m tall, tomentose. Leaves leathery, oblong, ovate to lanceolate, base oblique, slightly cordate. Cymes axillary, sessile, 3-4-flowered; flowers sessile, 5-merous; corolla white, villous inside. Drupes blue, subglobose. Fl. Apr. Fr. Aug-Sep. Forests or shaded and wet places at 200-1800 m. Distributed in Yunnan, Guangxi, Guangdong, Hainan, Taiwan and Fujian. Also in S and SE Asia, Japan and Papua New Guinea.

染木树 *Saprosma ternata*

黄果粗叶木 *Lasianthus calycinus*

黄果粗叶木

Lasianthus calycinus Dunn

灌木。叶厚纸质，长圆状椭圆形，长13-15厘米，先端急尖至短渐尖。花序团伞状至近头状，无梗，具4-6花；无苞片；萼管长2-2.5毫米，被糙伏毛，檐部5深裂，裂片条状披针形至狭三角形，长2.5-5毫米；花冠长7-9毫米，裂片5，卵形，比冠管略短。果近球形，橙黄色或橙红色。花期4-5月，果期9-10月。生海拔600-700米的林中或阴湿地。产海南。

Shrubs. Leaves thickly papery, oblong-elliptic, 13-15 cm long, apex acute to shortly acuminate. Inflorescences glomerulate to subcapitate, sessile, 4-6-flowered; bracts absent; calyx tube 2-2.5 mm long, strigillose, limb 5-parted, lobes linear-lanceolate to narrowly triangular, 2.5-5 mm long; corolla 7-9 mm long, lobes 5, ovate, slightly shorter than tube. Fruit subglobose, orange or orange-red. Fl. Apr-May. Fr. Sep-Oct. Forests or shaded and wet places at 600-700 m. Distributed in Hainan.

台湾粗叶木

Lasianthus formosensis Matsum.

灌木，高1-2米，被长柔毛。叶纸质，长圆形或卵状椭圆形，基部尖或钝，侧脉6-7对，小脉近平行。聚伞花序无梗，具6-9花，无苞片；花无梗，5数；花冠管状漏斗形。核果蓝色，卵球形。花期10-12月，果期翌年4月。生海拔500-1000米的山地林下或阴湿地。产云南、广西、广东、海南和台湾。泰国、越南和日本亦有。

Shrubs, 1-2 m tall, villous. Leaves papery, oblong or ovate-elliptic, base acute or obtuse, lateral veins 6-7 pairs, veinlets subparallel. Cymes sessile, 6-9-flowered, bracts absent; flowers sessile, 5-merous; corolla tubular-funnelform. Drupes blue, ovoid. Fl. Oct-Dec. Fr. next Apr. Montane forests or shaded and wet places at 500-1000 m. Distributed in Yunnan, Guangxi, Guangdong, Hainan and Taiwan. Also in Thailand, Vietnam and Japan.

台湾粗叶木 *Lasianthus formosensis*

西南粗叶木

Lasianthus henryi Hutchins.

灌木，高1-2(-4)米。小枝密被贴伏硬毛。叶纸质，长圆形披针形，上面无毛，下面脉上被贴伏柔毛。聚伞状花序簇生叶腋，几无梗；花无梗；花冠白色，外面被硬毛，狭管状，里面中部以上被白色长柔毛。核果近球形，成熟时蓝色。花期5-6月，果期7-10月。常生海拔200-1900米的林缘或疏林中。产中国东南、华南和西南。

Shrubs, 1-2(-4) m tall. Branches and branchlets densely appressed strigillose. Leaves subleathery or papery, oblong-lanceolate, glabrous adaxially, appressed pubescent abaxially on nerves. Inflorescences glomerulate to

西南粗叶木 *Lasianthus henryi*

云南鸡矢藤 *Paederia yunnanensis*

耳叶鸡矢藤 *Paederia cavaleriei*

subcapitate, sessile or subsessile; flowers sessile; corolla white, slender tubular, outside strigillose, inside villous. Drupe globose, blue when mature. Fl. May-Jun. Fr. Jul-Oct. Forests, shaded and wet places at 200-1900 m. Distributed in SE, S and SW China.

云南鸡矢藤

Paederia yunnanensis (Lévl.) Rehd.

木质藤本，长3-7米，被绒毛或短粗毛。叶近膜质，卵状心形，侧脉6-8对。圆锥花序腋生或顶生，狭窄，伸长；花5数；花冠狭筒状，紫色，裂片宽三角形。核果卵球形，压扁。花期6-10月，果期7-12月。生海拔300-3000米的沟谷林缘。产云南、四川、贵州和广西。越南亦有。

Woody lianas, 3-7 m long, tomentose or short, scabrous hairy. Leaves submembranous, ovate-cordate, lateral veins 6-8 pairs. Panicles axillary or terminal, narrow, elongate; flower 5-merous; corolla narrowly tubular, purple, lobes broadly triangular. Drupes ovoid, compressed. Fl. Jun-Oct. Fr. Jul-Dec. Forest edges in valleys at 300-3000 m. Distributed in Yunnan, Sichuan, Guizhou and Guangxi. Also in Vietnam.

耳叶鸡矢藤

Paederia cavaleriei H. Lévl.

缠绕藤本。叶对生，近膜质至纸质。花序腋生及/或顶生，圆锥状至总状，圆柱形至狭金字塔形，分枝为2-4部分；花冠管状漏斗形。果实球形；果核平凸至凹凸状。花期4-8月，果期8-11月。生海拔100-3000米的山地灌丛。产中国西南、华南和华中。

Climbing vines. Leaves opposite, submembranous to papery. Inflorescences axillary and/or terminal, paniculate to racemiform, cylindrical to narrowly pyramidal, branched to 2-4 orders; corolla tubular-funnelform. Fruits globose; pyrenes plano-convex to concavo-convex. Fl. Apr-Aug. Fr. Aug-Nov. Thickets on mountains at 100-3000 m. Distributed in SW, S and C China.

鸡矢藤

Paederia foetida L.

藤本，长3-5米。叶纸质，形状多变，卵形、卵状长圆形至披针形，侧脉4-6对，托叶三角形。圆锥状聚伞花序腋生或顶生；花5数；花冠筒状，浅紫色，内面被绒毛。核果球形。花期5-10月，果期7-12月。生海拔200-2000米的林下、林缘、灌丛或荒地上。产华北及以南的大部分地区。南亚、东南亚和东北亚亦有。

Lianas, 3-5 m long. Leaves papery, varied in shape, ovate, ovate-oblong to lanceolate, lateral veins 4-6 pairs, stipules triangular. Paniculate cymes axillary or terminal; flowers 5-merous; corolla tubular, pale purple, tomentose inside. Drupes globose. Fl. May-Oct. Fr. Jul-Dec. Forests, forest edges, thickets or waste fields at 200-2000 m. Distributed in N China and south of N China. Also in S, SE and NE Asia.

鸡矢藤 *Paederia foetida*

聚花野丁香 *Leptodermis glomerata*

聚花野丁香

Leptodermis glomerata Hutchins.

矮灌木，高达0.6米。茎自基部分枝，纤细。叶对生，纸质，披针形至卵形，基部极尖，侧脉4-6对，托叶尾尖。花团聚枝顶，5数，二型；花萼裂片钻形；花冠管上部膨大。蒴果狭椭圆状，成熟时5瓣裂。生海拔1800-2500米的疏林或山坡。产云南。

Dwarf shrubs, up to 0.6 m tall. Stems branched at base, slender. Leaves opposite, papery, lanceolate to ovate, base very pointed, lateral veins 4-6 pairs, stipules cuspidate. Flowers terminal, glomerate, 5-merous, dimorphic; calyx lobes subulate; corolla tubes dilated upwards. Capsules narrowly ellipsoid, 5-valved at maturity. Sparse forests or montane slopes at 1800-2500 m. Distributed in Yunnan.

野丁香

Leptodermis potaninii Batal.

灌木，高0.5-2米，被长柔毛。叶薄，被白色短柔毛，卵形至披针形，下面苍白，侧脉3-4对。聚伞花序顶生，无梗，具3花；花5-6数；花萼裂片窄三角形；花冠漏斗状。蒴果成熟时5瓣裂。花期5月，果期秋冬季。生海拔800-2700米的山坡灌丛中。产云南、四川、贵州、湖北和陕西。

Shrubs, 0.5-2 m tall, pilose. Leaves thin, white pubescent, ovate to lanceolate, abaxially pale, lateral vein 3-4 pairs. Cymes terminal, sessile, 3-flowered; flowers 5-6-merous; calyx lobes narrowly triangular; corolla funnelform. Capsules 5-valved at maturity. Fl. May. Fr. autumn and winter. Thickets on montane slopes at 800-2700 m. Distributed in Yunnan, Sichuan, Guizhou, Hubei and Shaanxi.

吉隆野丁香

Leptodermis kumaonensis R. Parker

灌木，高1-2米。小枝圆柱状，被腺质柔毛。叶纸质，长圆状披针形或卵状披针形，上面被糙伏毛状的柔毛；侧脉6-9对。花常3-5朵簇生侧生短枝之顶；

野丁香 *Leptodermis potaninii*

吉隆野丁香 *Leptodermis kumaonensis*

花冠漏斗状，外面被柔毛或近无毛，里面喉部以下被白色长柔毛。生海拔2800-3000米处的灌丛中或林缘。产西藏(吉隆县)。印度北部、尼泊尔和不丹亦有。

Shrubs, 1-2 m tall. Branchlets terete, glandular pilose. Leaves papery, oblong-lanceolate or ovate-lanceolate, strigose pilose adaxially, pilose on midrib and lateral veins abaxially; lateral veins 6-9 pairs. Flowers 3-5 fascicled on ends of lateral short branches; corolla funnelform, pilose or subglabrous outside, white villous below throat inside. Thickets or forest margins at 2800-3000 m. Ditributed in Xizang (Gyirong County). Also in N India, Nepal and Bhutan.

薄皮木

Leptodermis oblonga Bunge

灌木，高0.2-1米。茎多分枝，纤细，皮薄。叶纸质，披针形或长圆形，侧脉约3对；小苞片透明，卵形。花无梗，3朵簇生，5数，顶生，稀腋生；花萼裂片卵形；花冠紫红色，漏斗状。蒴果椭圆状，成熟时5瓣裂。花期6-8月，果期10月。生海拔100-1200米的向阳山坡、路边或灌丛中。产华北和西北。

Shrubs, 0.2-1 m tall. Stems much branched, slender, epidermis thin. Leaves papery, lanceolate or oblong, lateral veins ca. 3 pairs; bracteoles transparent, ovate. Flowers sessile, 3-fasciculate, 5-merous, terminal, rarely axillary; calyx lobes ovate; corolla purplish-red, funnelform. Capsules ellipsoid, 5-valved at maturity. Fl. Jun-Aug. Fr. Oct. Sunny slopes, roadsides or thickets at 100-1200 m. Distributed in N and NW China.

纤枝野丁香

Leptodermis schneideri H. Winkl.

灌木，高0.2-1.3米。嫩枝被短绒毛或近无毛。叶在长枝上疏生，在极短枝条上极密挤，长圆形或卵形，两面无毛。花于侧枝上顶生或在上部小枝上腋生，均生花1-3朵；花冠狭漏斗形，多少弯，外面被圆锥状绒毛，里面喉部密被长毛。蒴果近椭圆体形。种子的假种皮紧贴种皮。花期6-8月，果期7-9月。产四川、西藏和云南。

Shrubs, 0.2-1.3 m. Young branches tomentulose or subglabrous. Leaves sparsely arranged on long branches, very crowded on very short branches; blade oblong or ovate, glabrous on both surfaces. Flowers terminal on lateral foliate branches or upper nodes axillary, 1-3 flowered; corolla narrowly funnelform, covered with conical woolly hairs outside, densely covered with long hairs on throat inside. Capsules sub-ellipsoid. Seed aril adherent to testa. Fl. Jun-Aug. Fr. Jul-Sep. Distributed in Sichuan, Xizang and Yunnan.

薄皮木 *Leptodermis oblonga*

纤枝野丁香 *Leptodermis schneideri*

高山野丁香 *Leptodermis forrestii*

云南钩毛草 *Kelloggia chinensis*

高山野丁香
Leptodermis forrestii Diels

灌木，高0.6-1.2米。嫩枝密被短柔毛。叶膜状纸质，通常卵形或披针形，上面散生糙伏毛，下面无毛或脉上被皱卷长柔毛。花通常单朵顶生；花冠浅蓝色或微染红色，漏斗状，外面无毛，里面被白色长柔毛。蒴果近椭圆体形。花期6-7月，果期7-8月。生海拔3200-3400米的林中。产四川、云南西北部和西藏东南部。

Shrubs, 0.6-1.2 m tall. Young branches densely covered with short soft hairs. Leaves membranous-papery, often ovate or lanceolate, dispersedly strigose adaxially, glabrous or with crinkled long soft hairs on veins abaxially. Flowers usually solitary, terminal; corolla light blue or slightly reddish, funnelform, glabrous outside, white villous inside. Fl. Jun-Jul. Fr. Jul-Aug. Capsules sub-ellipsoid. Forests at 3200-3400 m. Distributed in Sichuan, NW Yunnan and SE Xizang.

川滇野丁香
Leptodermis pilosa Diels

灌木，被长柔毛和短绒毛。叶纸质，卵形至披针形，边缘具睫毛，侧脉3-5对，托叶三角形。聚伞花序顶生或腋生，具3-7花；花近无梗，5数；花萼裂片小；花冠漏斗状。蒴果成熟时5瓣裂。花期6月，果期9-10月。生海拔600-3800米的向阳山坡、灌丛或路边。产云南、四川、西藏、湖北、山西和甘肃。

Shrubs, pilose and tomentulose. Leaves papery, ovate to lanceolate, margin ciliate, lateral veins 3-5 pairs, stipules triangular. Cymes terminal or axillary, 3-7-flowered; flowers subsessile, 5-merous; calyx lobes small; corolla funnelform. Capsules 5-valved at maturity. Fl. Jun. Fr. Sep-Oct. Sunny slopes, thickets or roadsides at 600-3800 m. Distributed in Yunnan, Sichuan, Xizang, Hubei, Shanxi and Gansu.

云南钩毛草
Kelloggia chinensis Franch.

多年生草本，高达30厘米。茎被短柔毛。叶薄纸质，狭披针形，上面散生小睫毛，下面中脉被短柔毛。聚伞花序有花3-20朵；花冠白色至红色，短漏斗形，外面被微毛。蒴果近球形，密被白色钩毛。花期7月，果期7-9月。生海拔3000-3700米山地的湿润草坡上或林缘。产四川、西藏和云南。不丹亦有。

Perennial herbs, to 30 cm tall. Stems puberulent. Leaves drying thinly papery, narrowly lanceolate, adaxially sparsely strigillose, abaxially glabrous except puberulent along midrib. Inflorescences 3-20-flowered; calyx densely covered with hooked trichomes; corolla white to pink or red, outside puberulent or hispidulous. Mericarps ovoid, densely covered with hooked trichomes. Fl. Jul. Fr. Jul-Sep. Dry to wet mountain grasslands, along trails, forests and thickets openings at 3000-3700 m. Distributed in Sichuan, Xizang and Yunnan. Also in Bhutan.

六月雪
Serissa japonica (Thunb.) Thunb.

常绿小灌木，高60-90厘米，有

川滇野丁香 *Leptodermis pilosa*

六月雪 *Serissa japonica*

臭味。叶对生并簇生于短枝上，革质，卵形至倒披针形。花单生或数朵簇生，顶生或腋生，4-6数；花冠漏斗状，淡红色或白色，裂片开展。核果球形。花期4-10月，果期6-11月。生海拔100-1600米河边或丘陵阔叶林中。产中国西南、华南和华东。越南和日本亦有。

Evergreen small shrubs, 60-90 cm tall, smelly. Leaves opposite and crowded on short branchlets, leathery, ovate to oblanceolate. Flowers solitary or several clustered, terminal or axillary, 4-6-merous; corolla funnelform, pale red or white, lobes spreading. Drupes globose. Fl. Apr-Oct. Fr. Jun-Nov. Riversides or broad-leaved forests on hills at 100-1600 m. Distributed in SW, S and E China. Also in Vietnam and Japan.

白马骨 *Serissa serissoides*

白马骨

Serissa serissoides (DC.) Druce

小灌木，高达1米。嫩枝被微柔毛。叶通常丛生，薄纸质，倒卵形或倒披针形，除下面被疏毛外，其余无毛；侧脉每边2-3条。花无梗，常单生小枝顶部；花冠白色，外面无毛。果未见。花期4-6月。生荒地或草坪。产长江以南各地。日本亦有。

Small shrubs, to 1 m tall. Branches pilosulous when young. Leaves usually in clusters; blade drying thinly papery, obovate or oblanceolate, glabrous except sparsely pubescent abaxially; secondary veins 2 or 3 pairs. Flowers usually solitary, terminal; corolla white, outside glabrous. Fruit not seen. Fl. Apr-Jun. Wastelands or lawns. Distributed in the south of Yangtze River. Also in Japan.

薄柱草

Nertera sinensis Hemsl.

簇生草本，高5-10厘米，无毛。茎纤细，柔弱，稍匍匐，节上生根。叶小，纸质，长圆状披针形，托叶三角形。花小，直径1-3毫米，单朵顶生，无梗，4数；花冠淡绿色。核果深蓝色至黑色，球形。花期7-8月，果期7-11月。生海拔500-1300米的山地林下、路边、沟边或溪边岩石上。产中国西南、华南、华中和华东。

Tufted herbs, 5-10 cm tall, glabrous. Stems slender, Tender, subcreeping, rooting on nodes. Leaves small, papery, oblong-lanceolate, stipules triangular. Flowers small, 1-3 mm diam, solitary, terminal, sessile, 4-merous; corolla pale green. Drupes dark blue to black, globose. Fl. Jul-Aug. Fr. Jul-Nov. Montane forests, roadsides, ditch sides or rocks by streams at 500-1300 m. Distributed in SW, S, C and E China.

薄柱草 *Nertera sinensis*

虎刺 *Damnacanthus indicus*

虎刺

Damnacanthus indicus Gaertn. f.

具刺灌木，高0.3-1米。根肉质，链珠状。小枝被糙硬毛；刺生于关节，针状。叶纸质或薄革质，卵形、心形或圆形，侧脉3-4对。花1-2生于叶腋；花冠白色，漏斗状。核果红色，近球形。花期3-6月，果期3月至翌年1月。生海拔100-1500米的山地林中或石山灌丛中。产中国西南、华南、华中和华东。印度北部、日本和朝鲜半岛亦有。

Spinescent shrubs, 0.3-1 m tall. Roots succulent, moniliform. Branchlets strigose; spines at nodes, acicular. Leaves papery or thin leathery, ovate, cordate or rounded, lateral veins 3-4 pairs. Flowers 1-2 in leaf axils; corolla white, funnelform. Drupes red, subglobose. Fl. Mar-Jun. Fr. Mar to next Jan. Montane forests or rocky thickets at 100-1500 m. Distributed in SW, S, C and E China. Also in N India, Japan and Korean Peninsula.

短刺虎刺

Damnacanthus giganteus Nakai

具刺灌木，高0.5-2米。根肉质，念珠状。叶革质，披针形，侧脉5-7对，托叶生叶柄间，早落。花成对生于腋生短总梗的顶端；花冠白色，漏斗状。核果红色，近球形。花期3-5月，果期10月至翌年1月。生海拔500-1100米的山地林下或灌丛中。产中国西南、华南、华中和华东。日本亦有。

Spiny shrubs, 0.5-2 m tall. Roots succulent, moniliform. Leaves leathery, lanceolate, lateral veins 5-7 pairs, stipules interpetiolar, caducous. Flowers in pairs terminal on an axillary short peduncle; corolla white, funnelform. Drupes red, subglobose. Fl. Mar-May. Fr. Oct to next Jan. Montane forests or thickets at 500-1100 m. Distributed in SW, S, C and E China. Also in Japan.

短刺虎刺 *Damnacanthus giganteus*

柳叶虎刺

Damnacanthus labordei (Lévl.) H. S. Lo

小灌木。根肉质，念珠状。枝4棱至扁平，无刺。叶干纸质，披针形至披针状条形；次级叶脉9-16对；托叶早落。花序无毛；花冠白色至黄色，外部无毛。核果直径约8毫米。花期2-3月和10-12月，果期9-12月。生海拔800-1800米的林中或灌丛。产中国西南、华南和华中。越南亦有。

Small shrubs. Roots fleshy, moniliform. Branches 4-angled to flattened, without spines. Leaves drying papery, lanceolate to lanceolate-linear; secondary veins 9-16 pairs; stipules caducous. Inflorescences glabrous; corolla white to yellow, outside glabrous. Drupes ca. 8 mm diam. Fl. Feb-Mar and Oct-Dec. Fr. Sep-Dec. Forests or thickets at 800-1800 m. Distributed in SW, S and C China. Also in Vietnam.

柳叶虎刺 *Damnacanthus labordei*

四蕊三角瓣花 *Prismatomeris tetrandra*

四蕊三角瓣花

Prismatomeris tetrandra (Roxb.) K. Schum.

灌木至小乔木，高达8米。小枝4棱柱形，无毛。叶近革质，长圆形至披针形，无毛。伞形花序常顶生；花冠高脚碟形，白色，外面无毛。核果近球形，无毛。种子1(-2)粒。花期5-9月，果期9-12月。生海拔300-2400米的疏、密林下或灌丛中。产福建、广东、海南、广西和云南南部。印度、柬埔寨、泰国和越南亦有。

Shrubs to small trees, to 8 m tall. Branches usually quadrangular, glabrous. Leaves drying subleathery, oblong to lanceolate, glabrous. Inflorescences terminal, umbelliform; corolla salverform, white, glabrous outside. Drupes subglobose, glabrous. Seeds 1(-2). Fl. May-Sep. Fr. Sep-Dec. Forests, thickets at 300-2400 m. Distributed in Fujian, Guangdong, Hainan, Guangxi and S Yunnan. Also in India, Cambodia, Thailand and Vietnam.

黄木巴戟

Morinda angustifolia Roxb.

灌木或小乔木，高达6米。枝4棱。叶长圆形、椭圆形或倒披针形，侧脉9-13对。头状花序与顶生叶对生，具总梗；花无梗；花冠白色，高脚碟形，向内弯，裂片卵状披针形。聚花核果浆果状。花期4-5月，果期夏秋季。生海拔500-1400米的山谷林或丘陵林中。产云南。南亚和东南亚亦有。

Shrubs or small trees, to 6 m tall. Branches tetragonous. Leaves oblong, elliptic or oblanceolate, lateral veins 9-13 pairs. Capitulums opposite to one terminal leaf, pedunculate; flowers sessile; corolla white, salverform, incurved, lobes ovate to lanceolate. Drupecetum baccate. Fl. Apr-May. Fr. summer to autumn. Forests in valleys or hills at 500-1400 m. Distributed in Yunnan. Also in S and SE Asia.

黄木巴戟 *Morinda angustifolia*

大果巴戟 *Morinda cochinchinensis*

大果巴戟
Morinda cochinchinensis DC.

木质藤本；幼枝密被长柔毛。叶纸质，椭圆形、长圆形或倒卵状长圆形，上面疏被糙硬伏毛，下面密被柔毛。顶生头状花序3-18排列成伞形；花冠白色，漏斗形至辐射状，外面被短柔毛或无毛。聚花核果近球形，外面被柔毛。花期5-7月，果期7-11月。生海拔100-1200米的林下或灌丛中。产福建、广东、香港、海南和广西。越南亦有。

Lianas; branches when young densely villous. Leaves drying papery, elliptic, oblong or obovate-oblong, adaxially sparsely strigose, abaxially densely pubescence. Inflorescences terminal, fascicled or umbellate with 3-18 heads; corolla white, funnelform to rotate, outside pilosulous or glabrescent. Drupecetum subglobose, puberulent. Fl. May-Jul. Fr. Jul-Nov. Forests or thickets at 100-1200 m. Distributed in Fujian, Guangdong, Hong Kong Hainan and Guangxi. Also in Vietnam.

羊角藤 *Morinda umbellata* subsp. *obovata*

羊角藤
Morinda umbellata L. subsp. **obovata** Y. Z. Ruan

藤本。叶纸质至革质，倒卵状披针形至倒卵状长圆形，下面光亮，侧脉4-5对，托叶钻形，叶柄被疏毛。头状花序具6-12花，顶生，具梗；花无梗，4-5数；花冠白色，钟状，顶端钩状。聚花核果近球形，成熟时红色。花期6-7月，果期10-11月。生海拔300-1200米的山地林中、灌丛、溪边或路边。产华南、华中和华东。

Lianas. Leaves papery or leathery, obovate-lanceolate or obovate-oblong, adaxially shiny, lateral veins 4-5 pairs, stipules tabulate, petioles sparsely hirsute. Capitulums 6-12-flowered, terminal, pedunculate; flowers sessile, 4-5-merous; corolla white, campanulate, apex hooked. Drupecetum subglobose, red at maturity. Fl. Jun-Jul. Fr. Oct-Nov. Montane forests, thickets, streamsides or roadsides at 300-1200 m. Distributed in S, C and E China.

鸡眼藤 *Morinda parvifolia*

鸡眼藤
Morinda parvifolia Bartl. ex DC.

藤本。叶形多变，倒卵形至披针形，被毛或变无毛，侧脉3-6对，托叶钻形。头状花序近球形或圆锥形，具3-17花，顶生，具总梗；花4-5数，无梗；花冠白色。聚花核果近球形，成熟时橙红色。花期4-6月，果期6-8月。生海拔400米以

墨苜蓿 *Richardia scabra*

下的路边、沟边、荒地、灌丛或疏林下。产华南和华东。越南和菲律宾亦有。

Lianas. Leaves various in shape, obovate to lanceolate, hirsute or glabrescent, lateral veins 3-6 pairs, stipules tabulate. Capitulums subglobose or conical, 3-17-flowered, terminal, pedunculate; flowers 4-5-merous, sessile; corolla white. Drupecetum subglobose, orange at maturity. Fl. Apr-Jun. Fr. Jun-Aug. Roadsides, ditchsides, wastelands, thickets or sparse forests below 400 m. Distributed in S and E China. Also in Vietnam and the Philippines.

墨苜蓿
Richardia scabra L.

一年生草本，俯卧或近直立。叶卵形、椭圆形或披针形，纸质，边缘具睫毛，托叶顶端截形，具数个流苏状刚毛。头状花序顶生，近无梗；花5-6数，萼片披针形；花冠白色，漏斗状或高脚碟状。分果爿3-6，长圆形至倒卵形。花果期2-12月。生湿地中。归化于广东、海南、香港和台湾。原产热带美洲。

Annual herbs, decumbent or suberect. Leaves ovate, elliptic or lanceolate, papery, margin ciliate, stipules apex truncate, fringed with several bristles. Flowering heads terminal, subsessile; flowers 5-6-merous, calyx lobes lanceolate; corolla white, funnelform or salverform. Mericarps 3-6, oblong to obovoid. Fl. and fr. Feb-Dec. Waterlands. Naturalized in Guangdong, Hainan, Hong Kong and Taiwan. Native to tropical America.

糙叶丰花草
Spermacoce hispida L.

平卧草本，高可达50厘米。枝被短硬毛。叶纸质至革质，椭圆形、倒卵形或匙形，两面被短糙毛。花序腋生，无总梗；花冠淡红色或白色，漏斗形，里外均无毛。蒴果椭圆形。种子表面有小颗粒。花果期5-11月。生海拔100米以下的空旷沙地上。产福建、台湾、广东、海南和广西。南亚、东南亚和澳大利亚亦有。

Herbs, prostrate, to 50 cm tall. Stems quadrate, hispidulous. Leaves papery to leathery, elliptic, obovate or spatulate, both surfaces scaberulous. Inflorescences axillary, sessile; corolla pale red or white, funnelform, both sides glabrous. Capsules ellipsoid. Sends surface with tiny pits. Fl. and fr. Mar-Dec. Open sandy lands below 100 m. Distributed in Fujian, Taiwan, Guangdong, Hainan and Guangxi. Also in S and SE Asia, and Australia.

糙叶丰花草 *Spermacoce hispida*

阔叶丰花草
Spermacoce alata Aubl.

草本，高可达1米。茎四棱柱形，被短硬毛。叶纸质，椭圆形或卵状长圆形，两面被短硬毛。花序常腋生，无总梗；花冠漏斗形，淡蓝白色，里面疏被柔毛。蒴果椭圆形至近球形。种子表面有小颗粒。花果期5-11月。逸为野生，多生海拔100-800米的废墟和荒地上。产浙江、福建、广东、广西、海南和台湾。美洲、南亚和东南亚、非洲和澳大利亚亦有。

Herbs, 1 m tall. Stems hispidulous. Leaves papery, elliptic or ovate-oblong, both surfaces hispidulous. Inflorescences usually axillary, sessile; corolla funnelform, slightly blue-white, inside sparsely pubescent. Capsules ellipsoid to subglobose. Sends surface with tiny pits. Fl. and fr. May-Nov. Naturalized in disturbed ground and wastelands at 100-800 m. Distributed in Zhejiang, Fujian, Guangdong, Guangxi, Hainan and Taiwan. Also in America, S and SE Asia, Africa and Australia.

阔叶丰花草 *Spermacoce alata*

光叶丰花草 *Spermacoce remota*

丰花草 *Spermacoce pusilla*

丰花草

Spermacoce pusilla Wall.

直立草本，高可达60厘米。茎四棱柱形。叶革质，线状长圆形，两面粗糙。花序顶生或腋生，无总梗；花冠漏斗形，白色，两面无毛或里面疏被短柔毛。蒴果长圆形或近倒卵形。种子表面无毛。花果期8-12月。生海拔100-1500米的草地和草坡。产华东、东南、华南和西南。热带非洲和亚洲亦有。

Herbs, erect, to 60 cm tall. Stems 4-angled. Leaves drying leathery, linear-oblong, scaberulous on both surfaces. Inflorescences terminal and axillary, sessile; corolla funnelform, white, both sides glabrous or sometimes sparsely puberulent inside. Capsules oblong or subobovate to ellipsoid. Seeds glabrous. Fl. and fr. Aug-Dec. Grasslands and grassy slopes at 100-1500 m. Distributed in E, SE, S and SW China. Also in tropical Africa and Asia.

光叶丰花草

Spermacoce remota Lam.

草本或亚灌木，高可达65厘米。茎圆柱形至近四棱柱形。叶纸质，狭椭圆形至披针形，微柔毛至无毛。花序顶生或腋生；花冠漏斗形，白色，外面无毛，里面喉部被短柔毛。蒴果椭圆体形。种子表面具横向的不规则深沟。花果期6月至翌年1月。生海拔100-300米的潮湿地。产广东和台湾。原产新热带地区。

单花拉拉藤 *Galium exile*

Herbs or subshrubs, to 65 cm tall. Stems subterete to subquadrate. Leaves papery, narrowly elliptic to lanceolate, puberulent to glabrescent. Inflorescences terminal and in uppermost leaf axils; corolla white, funnelform, outside glabrous, inside pubescent in throat. Capsules ellipsoid. Seeds transversely ruminate-rugose with irregular deep grooves. Fl. and fr. Jun to next Jan. Wet sites at 100-300 m. Distributed in Guangdong and Taiwan. Native to the Neotropics.

单花拉拉藤

Galium exile Hook. f.

一年生草本，高4-20厘米。茎纤弱，具4棱，疏被倒钩刺至无毛。叶纸质，对生或4片轮生，倒卵形、宽披针形至狭椭圆形，近无毛。花常单生；花冠白色。果爿近球形，密被长钩毛。花期6-7月，果期8-9月。

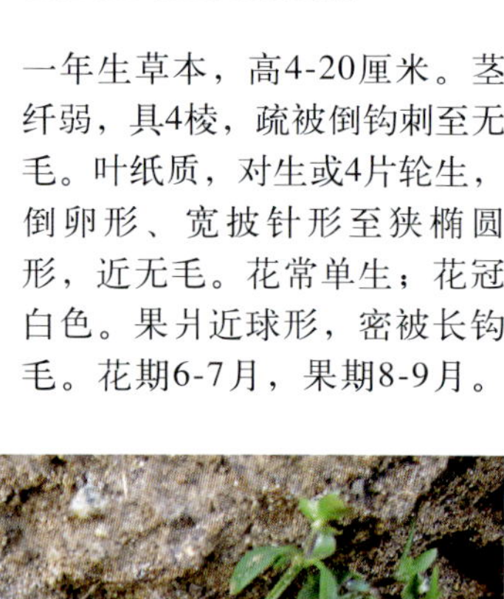

六叶葎 *Galium hoffmeisteri*

生海拔1200-4800米的山坡石隙缝中、沙砾干草坝或河滩草地。产中国西南和西北。印度和尼泊尔亦有。

Annual herbs, 4-20 cm tall. Stems slender, 4-angled, sparsely retrorsely aculeolate to glabrous. Leaves blades drying papery, opposite or in whorls of 4; obovate, oblanceolate to linear-elliptic, subglabrous. Flowers mostly solitary; corolla white, rotate. Mericarps subglobose, with dense, uncinate trichomes. Fl. Jun-Jul. Fr. Aug-Sep. Rock crevices on mountain slopes, sand and gravel drifts on grassy plains at 1200-4800 m. Distributed in SW and NW China. Also in India and Nepal.

六叶葎

Galium hoffmeisteri (Klotzsch) Ehrend. et Schönb.-Tem. ex R. R. Mill

草本，高10-60厘米。茎具4棱。叶片纸质，常6片轮生，狭椭圆状长圆形至阔披针形。聚伞花序；花冠白色或黄绿色，辐状，无毛。果爿近球形，密被钩毛。花期4-8月，果期5-9月。生海拔400-4000米的草丛或灌丛中及林下。产中国除华南以外的大多数地区。印度、巴基斯坦、阿富汗、尼泊尔、不丹、缅甸、日本、朝鲜半岛和俄罗斯亦有。

Herbs, 10-60 cm tall. Stems 4-angled. Leaves papery, usually in whorls of 6, narrowly elliptic-oblong to broadly oblanceolate. Inflorescences cymose; corolla white or yellow-green, rotate, glabrescent. Mericarps subglobose, with dense uncinate trichomes. Fl. Apr-Aug. Fr. May-Sep. Forests on mountain slopes, thickets, meadows at 400-4000 m. Distributed in most parts of China, except S China. Also in India, Pakistan, Afghanistan, Nepal, Bhutan, Myanmar, Japan, Korean Peninsula and Russia.

林猪殃殃 (车叶葎)

Galiumparadoxum Maxim.

多年生草本。茎直立，具4棱及狭翅。叶对生，近圆形、宽卵形至卵状披针形，膜质，边缘具向上的硬纤毛。聚伞花序生顶端及上部叶腋，具3-11花；花冠白色，旋转。果爿卵形，密被棕黄色钩毛。花期5-8月，果期6-9月。生海拔1200-4000米的森林、草地、水边或湿润亚高山岩石上。产中国除新疆以外的大多数地方。南亚和东北亚亦有。

Perennial herbs. Stems erect, 4-angled and narrowly winged. Leaves opposite, suborbicular, broadly ovate to ovate-lanceolate, membranous, margins antrorsely hispidulous-ciliolate. Inflorescences terminal and in axils of upper leaves with 3-11-flowered cymes; corolla white, rotate. Mericarps ovoid, densely covered with uncinate yellowish brown trichomes. Fl. May-Aug. Fr. Jun-Sep. Forests, meadows, near water or shady (sub)alpine rocks at 1200-4000 m. Distributed in most parts of China, except Xinjiang. Also in S and NE Asia.

麦仁珠

Galium tricornutum Dandy

一年生草本，高5-80厘米，少分枝。叶近无梗，纸质，6-8片轮生，倒披针形，柔弱，具1脉。聚伞花序腋生，3-5花，向下弯曲；花小，4数，白色。小坚果近球形，具瘤状凸起。花期4-6月，果期5月至翌年3月。生海拔400-4000米的草甸、旷野、河滩或沟边。产中国西南、华南、华东、华北和西北。南亚、西亚、北非、欧洲和北美洲亦有。

Annual herbs, 5-80 cm tall, few-branched. Leaves subsessile, papery, 6-8-verticilate, oblanceolate, flaccid, 1-veined. Cymes axillary, 3-5-flowered, deflexed; flowers small, 4-merous, white. Nutlets subglobose, tuberculate. Fl. Apr-Jun. Fr. May to next Mar. Meadows, open fields, river beaches or ditch sides at 400-4000 m. Widely distributed in SW, S, E, N and NW China. Also in S and W Asia, N Africa, Europe and North America.

林猪殃殃（车叶葎）*Galiumparadoxum*

麦仁珠 *Galium tricornutum*

猪殃殃（原拉拉藤） *Galium aparine*

猪殃殃（原拉拉藤）

Galium aparine L.

矮小一年生草本，柔软。茎4棱，具倒钩毛。叶近无梗，6-8片轮生，条状倒披针形或长圆状倒披针形，仅1脉。聚伞花序具1花；花小，4数，黄绿色或白色。小坚果小，密被钩状毛。花期3-7月，果期4-11月。生海拔2500米以下的山坡、旷野、沟边、湖边、林缘或草甸。几乎遍布中国各地。南亚、西亚、东北亚和地中海地区亦有。

Dwarf annual herbs, effeminacy. Stems 4-angled, retroverse aculeate. Leaves subsessile, 6-8-verticillate, linear-oblanceolate or oblong-oblanceolate, 1-veined. Cymes 1-flowered; flowers small, 4-merous, yellowish green or white. Nutlets small, densely hooked hairy. Fl. Mar-Jul. Fr. Apr-Nov. Slopes, open fields, ditch or lake sides, forest edges or meadows below 2500 m. Distributed throughout almost all the China. Also in S, W and NE Asia, and Mediterranean regions.

小红参

Galium elegans Wall. ex Roxb.

多年生草本，直立或攀援。茎被毛。叶4叶轮生，近无柄，卵形至披针形，3-5脉。聚伞花序多花，三歧分枝，宽松柔弱；花单性，白色或灰黄色。小坚果小，密被钩状毛。花期4-8(-10)月，果期5-12月。生海拔200-3500米的林中、灌丛、草甸、溪边、旷野或岩石上。产中国西南、东南、华中和华西。南亚和东南亚亦有。

Perennial herbs, erect or climbling. Stems hirsute. Leaves in whorls of 4, subsessile, ovate to lanceolate, 3-5-veined. Cymes many-flowered, trichotomous, lax and effeminacy; flowers unisexual, white or pale yellow. Nutlets small, densely hooked villose. Fl. Apr-Aug(-Oct). Fr. May-Dec. Forests, thickets, meadows, streamsides, open fields or rocks at 200-3500 m. Distributed in SW, SE, C and W China. Also in S and SE Asia.

小红参 *Galium elegans*

线叶拉拉藤

Galium linearifolium Turcz.

多年生草本。叶4枚轮生，条状匙形，常稍镰形，宽1-4毫米。花序顶生，圆锥状，具少花至多花，聚伞花序长1.5-5厘米；花冠白色，旋转。分果爿椭圆体形至近球形，光滑无毛。花期6-8月，果期7-9月。生海拔400-1800米的草坡、林中、灌丛或山地草甸。产湖北、河北和辽宁。朝鲜半岛亦有。

线叶拉拉藤 *Galium linearifolium*

四叶葎 *Galium bungei*

Perennial herbs. Leaves in whorls of 4, linear-spatulate, often slightly falcate, 1-4 mm wide. Inflorescences terminal, paniculiform, with few- to many-flowered, 1.5-5 cm long cymes; corolla white, rotate. Mericarps ellipsoid to subglobose, glabrous and smooth. Fl. Jun-Aug. Fr. Jul-Sep. Grassy slopes, forests, thickets or mountain meadows at 400-1800 m. Distributed in Hubei, Hebei and Liaoning. Also in Korean Peninsula.

四叶葎
Galium bungei Steud.

多年生直立草本。茎常丛生，4棱。叶4枚轮生，长宽比常3-5，纸质，卵状披针形或狭倒披针形。花序聚伞状至圆锥状；花冠黄绿色或白色，4瓣。果爿螺旋状，具皮刺或具平伏钩状毛。花期4-9月，果期5月至翌年1月。生海拔3600米以下的林中、灌丛、山坡草甸或溪边。产中国大部分地区。朝鲜半岛和日本亦有。

Perennial herbs, erect. Stems often caespitose, 4-angled. Leaves in whorls of 4, length/breadth index usually 3-5, papery, ovate-lanceolate or narrowly oblanceolate. Inflorescences cymose to paniculate; corolla yellowish green or white, lobes 4. Mericarps tuberculate, aculeolate or with appressed and uncinate trichomes. Fl. Apr-Sep. Fr. May to next Jan. Forests, thickets, meadows on mountains or streamsides below 3600 m. Distributed in most parts of China. Also in Korean Peninsula and Japan.

喀喇套拉拉藤
Galium karataviense (Pavlov) Pobed.

多年生草本。茎4棱上密具倒皮刺。叶6-10轮生，狭披针形、倒披针形或狭椭圆形。花序顶生及腋生，具数朵至多花聚伞花序；花淡蓝色至紫罗兰色，冠筒长为裂片的1-1.5倍。分果爿近球形至椭圆体形。花果期6-9月。生海拔700-3300米的湿润林中、河边、河岸或湿草地。产中国西南、华北、华西、西北和东北。中亚亦有。

Perennial herbs. Stems densely retrorsely aculeolate on 4 angles. Leaves in whorls of 6-10, narrowly (ob)lanceolate or narrowly elliptic. Inflorescences terminal and axillary, with several- to many-flowered cymes; corolla bluish to violet, tubes 1-1.5 × as long as lobes. Mericarps subglobose to ellipsoid. Fl. and fr. Jun-Sep. Humid forests, riversides, beaches or wet grasslands at 700-3300 m. Distributed in SW, N, W, NW and NE China. Also in C Asia.

喀喇套拉拉藤 *Galium karataviense*

北方拉拉藤 *Galium boreale*

北方拉拉藤
Galium boreale L.

多年生草本，直立，高20-65厘米。4叶轮生，窄披针形，背面光滑，基生脉3条。聚伞花序顶生或生于上部叶腋，密花；花冠白色或灰黄色。小坚果小，密被毛。花期5-8(-9)月，果期(5-)6-10月。生海拔200-4600米的林中、灌丛、沟边、草地或草甸。产中国西南、华北、东北和西北。南亚、西南亚和东北亚、欧洲和北美洲亦有。

Perennial herbs, erect, 20-65 cm tall. Leaves in whorls of 4, narrowly lanceolate, abaxially glabrous, basal veins 3. Cymes terminal or in axils of upper leaves, densely flowered; corolla white or pale yellow. Nutlets small, densely hirsute. Fl. May-Aug(-Sep). Fr. (May-)Jun-Oct. Forests, thickets, ditch sides, grasslands or meadows at 200-4600 m. Widely distributed in SW, N, NE and NW China. Also in S, SW and NE Asia, Europe and North America.

蓬子菜
Galium verum L.

多年生草本，直立。叶6-10片轮生，无梗，纸质，条形。聚伞花序大，多花，排成具叶的圆锥花序；花小，簇生；萼管无毛；花冠黄色。分果爿近球形，无毛。花期4-8月，果期5-10月。生海拔4100米以下的山地、山谷、河滩、旷野、沟边、草地、灌丛或林下。产中国西南、华东、华北、东北和西北。南亚、西南亚、中亚、北亚、东北亚和欧洲亦有。

Perennial herbs, erect, 25-60 cm tall. Leaves in whorls of 6-10, sessile, papery, linear. Cymes large, many-flowered, arranged in foliate panicles; flowers small, clustered; calyx tube glabrous; corolla yellow. Mericarp subglobose, glabrous. Fl. Apr-Aug. Fr. May-Oct. Mountains, valleys, river beaches, open fields, ditch sides, grasslands, meadows, thickets or forests below 4100 m. Distributed in SW, E, N, NE and NW China. Also in S, SW, C, N and NE Asia, and Europe.

滇小叶葎
Galium asperifolium Wall. var. **verrucifructum** Cuf.

多年生草本，直立，纤细柔弱。茎4棱，不具刺。6叶轮生，小，披针形，下面光滑，具1脉。聚伞花序顶生或腋生，多数密集，分歧；花小，4数，绿白色或黄色。小坚果小，近球形，灰白色，具颗粒瘤状凸起。花果期8-10月。生海拔2300-3500米的山坡、草地或灌丛。产四川、西藏和云南西北部。

蓬子菜 *Galium verum*

滇小叶葎 *Galium asperifolium* var. *verrucifructum*

Perennial herbs, erect, slender and fragile. Stems 4-angled, not aculeate. Leaves 6-verticilate, smaller, oblanceolate, abaxially glabrous, 1-veined. Cymes terminal or axillary, many-flowered and dense, divergent; flowers small, 4-merous, greenish-white or yellow. Nutlets small, subglobose, grayish white, appressed granular-tuberculate. Fl. and fr. Aug-Oct. Montane slopes, grasslands or thickets at 2300-3500 m. Distributed in Sichuan, Xizang and NW Yunnan.

车轴草

Galium odoratum (L.) Scopo.

多年生草本。茎直立，具4棱。叶6-10枚轮生，长宽比指数约为4，倒披针形、长圆状披针形或狭椭圆形，边缘具向上的刚毛。聚伞花序顶生；花冠白色或蓝白色，稍漏斗形，4瓣。果爿近球形，密被钩状毛。花果期6-9月。生海拔1500-2800米的山地林中。产中国西南、华北、华西、华东、西北和东北。东北亚、西南亚、非洲西北部和欧洲亦有。北美洲引入栽培。

Perennial herbs. Stems erect, 4-angled. Leaves in whorls of 6-10, length/breadth index ca. 4, oblanceolate, oblong-lanceolate or narrowly elliptic, antrorsely aculeolate on margins. Cymes terminal; corolla white or bluish white, broadly funnelform, lobes 4. Mericarps subglobose, with dense uncinate trichomes. Afforestation plant. Also for water and soil conservation. Fl. and fr. Jun-Sep. Montane forests at 1500-2800 m. Distributed in SW, N, W, E, NW and NE China. Also in NE and SW Asia, NW Africa and Europe. Introduced in North America.

卵叶轮草

Galium platygalium (Maxim.) Pobed.

多年生草本。茎直立，具4角棱。叶4(-6)轮生，干纸质至革质，椭圆形或椭圆状长圆形至卵形，边缘具钩状缘毛。圆锥状聚伞花序伞房状；花冠白色，漏斗形，裂片4。果爿无毛且光滑。花果期7-9月。生海拔约1700米的山坡开阔林中。产山西、黑龙江和吉林。俄罗斯和朝鲜半岛亦有。

Perennial herbs. Stems erect, with 4 thickened angles. Leaves in whorls of 4(-6), drying papery to leathery, elliptic or elliptic-oblong to ovate, margins antrorsely ciliolate. Inflorescences umbel-like thyrsoid; corolla white, funnelform, lobes 4. Mericarps glabrous and smooth. Fl. and fr. Jul-Sep. Open forests on mountain slopes at ca. 1700 m. Distributed in Shanxi, Heilongjiang and Jilin. Also in Russia and Korean Peninsula.

车轴草 *Galium odoratum*

卵叶轮草 *Galium platygalium*

山东茜草 *Rubia truppeliana*

大叶茜草 *Rubia schumanniana*

山东茜草
Rubia truppeliana Loes.

草本，长达2米，匍匐或缠绕。茎四方形，具纵槽和倒向刺。叶4-8片轮生，膜质或近纸质，披针形，基生脉3。圆锥花序顶生，单一，明显具总梗；花具短梗；花冠辐状，裂片卵状三角形。花期7-8月。生海拔100-300米的林下或灌丛。产山东。

Herbs, to 2 m long, creeping or twining. Stems quadrangular, longitudinally sulcate, retrorsely aculeate. Leaves 4-8 in a whorl, membranous or subpapery, lanceolate, basiveins 3. Panicles terminal, solitary, distinctly pedunculate; flowers shortly pedicellate; corolla rotate, lobes ovate-triangular. Fl. Jul-Aug. Forests or thickets at 100-300 m. Distributed in Shandong.

大叶茜草
Rubia schumanniana E. Pritz.

近直立草本，高约1米。茎和分枝4棱，具纵槽。叶4片轮生，厚纸质至革质，粗糙，披针形至卵形，基出脉3。聚伞花序圆锥状，顶生或腋生；花小，5数；花冠白色或绿黄色。核果浆果状，小，球形，熟时黑色。花期5-7月，果期8-10月。生海拔800-3000米的林中。产云南和四川。

Suberect herbs, ca. 1 m tall. Stems and branches quadrangular, longitudinally sulcate. Leaves 4 in a whorl, thickly papery to leathery, coarse, lanceolate to ovate, base veins 3. Cymes paniculate, terminal or axillary; flowers small, 5-merous; corolla white or greenish-yellow. Drupes bacciform, small, globose, black at maturity. Fl. May-Jul. Fr. Aug-Oct. Forests at 800-3000 m. Distributed in Yunnan and Sichuan.

卵叶茜草
Rubia ovatifolia Z. Ying Zhang ex Q. Lin

多年生草质缠绕藤本。叶4片轮生，干薄纸质，卵状心形至近圆心形，长宽比指数1.5-2；主脉5条，掌状。花序聚伞圆锥状；花冠白色或灰黄色，近钟形。浆果成熟时黑色。花期7月，果期10-11月。生海拔1700-2200米的山地疏林中或灌丛中。产中国西南、华东、华中、华北和华西。

Perennial climbing vines, herbaceous. Leaves in whorls of 4, drying thinly papery, ovate-cordiform to suborbicular-cordiform, length/ breadth index 1.5-2; principal veins 5, palmate. Inflorescences thyrsoid; corolla whitish or pale yellow, subcampanulate. Berries black at maturity. Fl. Jul. Fr. Oct-Nov. Sparse forests or thickets on mountains at 1700-2200 m. Distributed in SW, E, C, N and W China.

卵叶茜草 *Rubia ovatifolia*

柄花茜草
Rubia podantha Diels

草质攀援藤本。叶4片轮生，纸质，披针形至卵形，边缘常具睫毛，基出脉3-5。圆锥状聚伞花序腋生或顶生；花5数，萼近球形；花冠紫红色或黄白色，杯状，裂片卵形至披针形。核果浆果状，球形，单生或双生，熟时黑色。花期4-6月，果期6-9月。生海拔700-3000米的林缘、疏林或草地。产四川、云南和广西。

柄花茜草 *Rubia podantha*

Herbaceous climbing lianas. Leaves 4 in a whorl, papery, lanceolate to ovate, margin usually ciliate, basiveins 3-5. Paniculate-cymes axillary or terminal; flowers 5-merous, calyx subglobose; corolla purplish-red or yellowish-white, cotyloid, lobes ovate to lanceolate. Drupes bacciform, globose, solitary or binate, black at maturity. Fl. Apr-Jun. Fr. Jun-Sep. Forest edges, sparse forests or grasslands at 700-3000 m. Distributed in Sichuan, Yunnan and Guangxi.

钩毛茜草

Rubia oncotricha Hand.-Mazz.

藤状草本，常平卧或披散，密被灰色钩状硬毛。叶4片轮生，具短柄，近纸质，披针形或有时卵形。聚伞花序常排成窄圆锥状，顶生或腋生；花5数；花冠白色或黄色。核果浆果状，具淡棕色斑点。花期7-9月，果期9-11月。生海拔500-3200米的山谷林下、林缘、石灰岩山坡或草地。产云南、贵州、四川和广西。

Vinelike herbs, usually prostrate or diffused, densely grey hooked hirsute. Leaves 4 in a whorl, shortly petiolate, subpapery, lanceolate or sometimes ovate. Cymes usually in narrow panicles, terminal or axillary; flowers 5-merous; corolla white or yellow. Drupes bacciform, pale brown maculate. Fl. Jul-Sep. Fr. Sep-Nov. Valley forests, forest edges, limestone montane slopes or grasslands at 500-3200 m. Distributed in Yunnan, Guizhou, Sichuan and Guangxi.

金剑草

Rubia alata Wall.

草质攀援藤本。茎4棱，有倒生皮刺，近无毛。叶4片轮生，薄革质，线状披针形至狭披针形，两面疏被短糙毛。花序腋生或顶生，圆锥花序式；花冠钟状，白色或淡黄色，外面无毛。花期5-8月，果期8-11月。生海拔600-2000米的山坡林缘或灌丛中。产长江以南，东至台湾，西至四川西部，北至河南中部和陕西南部。

Climbing herbs. Stems quadrangular, subglabrous, retrorsely aculeolate. Leaves in whorls of 4; blade drying thinly leathery, linear-lanceolate to narrowly lanceolate, both surfaces sparsely scaberulous. Inflorescences terminal and axillary, paniculate; corolla white or pale yellow, campanulate, glabrous. Fl. May-Aug. Fr. Aug-Nov. Forest margins on mountain slopes or thickets at 600-2000 m. Distributed in the south of Yangtze River, east to Taiwan, west to W Sichuan, north to C Henan and S Shaanxi.

钩毛茜草 *Rubia oncotricha*

金剑草 *Rubia alata*

梵茜草 *Rubia manjith*

多花茜草 *Rubia wallichiana*

梵茜草
Rubia manjith Roxb. ex Flem.

草质攀援藤本。茎、枝有4直棱，棱上有倒生小皮刺。叶4片轮生，纸质，长圆状披针形至卵状披针形，两面粗糙。花序圆锥状；花冠红色、紫红色或橙色，辐状。花期7-8月，果期10月。生海拔700-3600米的林下或灌丛中。产云南、四川、青海和西藏。印度、不丹和尼泊尔亦有。

Climbing herbs. Stems and branches 4-ridged, retrorsely aculeolate. Leaves in whorls of 4; blade drying papery, oblong-lanceolate to ovate-lanceolate, both surfaces scaberulous. Inflorescences paniculate; corolla red, purplish red, or orange, rotate, glabrous. Fl. Jul-Aug. Fr. Oct. Forests and thickets at 700-3600 m. Distributed in Yunnan, Sichuan, Qinghai and Xizang. Also in India, Bhutan and Nepal.

茜草
Rubia cordifolia L.

多年生草质攀援藤本。茎四方形，棱具倒向钩刺。叶4枚轮生，纸质，披针形至长圆状披针形。聚伞花序腋生或顶生，多分枝；花5数；花冠淡黄色，辐状。核果浆果状，球形。花期8-9月，果期10-11月。生海拔300-2800米的疏林、林缘或草地。产中国西南、华北、东北和西北。南亚、西南亚、东南亚、东北亚和非洲亦有。

Perennial herbaceous climbing lianas. Stems quadrangular, ribs retrorsely hooked aculeate. Leaves 4 in a whorl, papery, lanceolate or oblong-lanceolate. Cymes axillary or terminal, many-branched; flowers 5-merous; corolla pale yellow, rotate. Drupes bacciform, globose. Fl. Aug-Sep. Fr. Oct-Nov. Sparse forests, forest edges or grasslands at 300-2800 m. Distributed in SW, N, NE and NW China. Also in S, SW, SE and NE Asia, and Africa.

多花茜草
Rubia wallichiana Decne.

草质攀援藤本。茎和小枝有4钝棱角，常无毛。叶4(或6)片轮生，薄纸质，披针形或卵状披针形，两面粗糙。花序腋生和

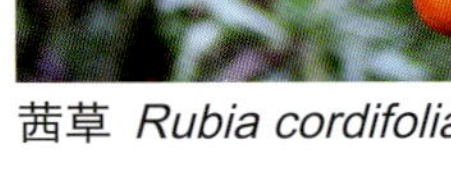

茜草 *Rubia cordifolia*

东南茜草 *Rubia argyi*

顶生，圆锥花序式；花冠紫红色、绿黄色或白色，辐状。花期8-10月，果期8-12月。生海拔300-2600米的林中、林缘和灌丛中。产江西、湖南、广东、海南、广西、四川和云南。印度东北部、不丹和尼泊尔亦有。

Climbing herbs. Stems and branches 4-angled, often glabrous. Leaves in whorls of 4(-6); blade drying thinly papery, lanceolate or ovate-lanceolate, both surfaces sparsely scaberulous. Inflorescences terminal and axillary, paniculate; corolla purplish red, greenish yellow or white, rotate. Fl. Aug-Oct. Fr. Aug-Dec. Forests, forest margins or thickets at 300-2600 m. Distributed in Jiangxi, Hunan, Guangdong, Hainan, Guangxi, Sichuan and Yunnan. Also in NE India, Bhutan and Nepal.

东南茜草

Rubia argyi (Lévl. et Vaniot) H. Hara ex Lauener

多年生草质藤本。茎、枝上四棱形，或具4狭翅，棱上具倒向钩状刺，无毛。叶4-6轮生，具柄，心形。聚伞花序分枝成圆锥花序式；花4-5数，白色。核果浆果状，近球形。花期7-10月，果期8-11月。生海拔300-3400米林缘、灌丛或村边栅栏。产中国西南、华南、华中和华东。日本和朝鲜半岛亦有。

Perennial herbaceous lianas. Stems and branches quadrangular, or narrowly 4-winged, ribs retrorsely hooked aculeate, glabrous. Leaves 4-6 in a whorl, petiolate, cordate. Cymes branched in panicles; flowers 4-5-merous, white. Drupes bacciform, subglobose. Fl. Jul-Oct. Fr. Aug-Nov. Forest edges, thickets or fences at village sides. Distributed in SW, S, C and E China. Also in Japan and Korean Peninsula.

日本粗叶木

Lasianthus japonicus Miq.

灌木，高1-2米，近无毛至被微柔毛。叶革质至纸质，披针形至长圆形，被粗毛。聚伞花序具梗或至无梗，具2-3花；花无梗，4-5数，花萼裂片三角形或披针形；花冠内部被长柔毛。核果蓝色，球形。花期4-5月，果期6-10月。生海拔200-2300米的林下或阴湿地。产中国西南、华南、华中和华东。印度、日本南部和东南亚亦有。

Shrubs, 1-2 m tall, subglabrous to pubescent. Leaves leathery to papery, lanceolate to oblong, hirsute. Cymes pedunculate to subsessile, 2-3-flowered; flowers sessile, 4-5-merous, calyx lobes triangular or lanceolate; corolla villous inside. Drupes blue, globose. Fl. Apr-May. Fr. Jun-Oct. Forests or shaded and wet places at 200-2300 m. Distributed in SW, S, C and E China. Also in India, S Japan and SE Asia.

日本粗叶木 *Lasianthus japonicus*

梗花粗叶木 *Lasianthus biermannii*

梗花粗叶木

Lasianthus biermannii King ex Hook. f.

灌木，高2-5米，被柔毛。叶纸质，椭圆形、卵形或披针形，托叶三角形或披针形。聚伞花序具总梗；花近无梗，4-6数，花萼裂片狭披针形；花冠内部被长柔毛。核果蓝色，球形。花期4月和10-11月，果期7月。生海拔1000-2500米的山地林下或阴湿地。产西藏、云南、贵州和海南。南亚亦有。

Shrubs, 2-5 m tall, pubescent. Leaves papery, elliptic, ovate or lanceolate, stipules triangular or lanceolate. Cymes pedunculate; flowers subsessile, 4-6-merous, calyx lobes narrowly lanceolate; corolla villous inside. Drupes blue, globose. Fl. Apr and Oct-Nov. Fr. Jul. Montane forests or shaded and wet places at 1000-2500 m. Distributed in Xizang, Yunnan, Guizhou and Hainan. Also in S Asia.

长管越南茜
Rubovietnamia aristata Tirveng.

灌木，高达3米。嫩枝近无毛。叶厚纸质，椭圆形或长圆状披针形，上面无毛或疏被短糙毛，下面疏被短糙毛。聚伞花序常假顶生；花冠白色，高脚碟形，外面被短硬毛。浆果近球形，被短硬毛至无毛。花期5-7月，果期7月。生海拔200-1400米处石灰岩山坡的灌丛中或林中。产广西和云南。越南亦有。

Shrubs, up to 3 m tall. Branches subglabrous. Leaves drying thickly papery, elliptic or oblanceolate, adaxially glabrescent to sparsely strigillose, abaxially sparsely strigillose. Inflorescences often pseudoaxillary, cymose; corolla white, salverform, outside densely sericeous-strigose. Berry subglobose, strigillose to glabrescent. Fl. May-Jul. Fr. Jul. Thickets or forests on limestone hills at 200-1400 m. Distributed in Guangxi and Yunnan. Also in Vietnam.

长管越南茜 *Rubovietnamia aristata*

忍冬科 Caprifoliaceae

血满草
Sambucus adnata Wall.

多年生草本。根和根茎红色，具红色汁液。奇数羽状复叶对生，小叶3-5对，长椭圆形、长卵形或披针形，边缘具锯齿，小叶托叶退化成瓶状凸起的腺体。伞状聚伞花序顶生，大型；花小，具恶臭；花冠白色。浆果红色，球形。花期5-7(-9)月，果期9-10月。生海拔1600-3600米的林下、灌丛、山谷、湿地或高山草地。产中国西南和西北。印度和不丹亦有。

Perennial herbs. Roots and rhizomes red, with red juice. Imparipinnate compound leaves opposite, leaflets 3-5 pairs, narrowly elliptic, narrowly ovate or lanceolate, margin serrate, stipules of leaflets reduced to bottlelike glands. Umbellike cymes terminal, large; flowers small, effluvial; corolla white. Berries red, globose. Fl. May-Jul(-Sep). Fr. Sep-Oct. Forests, thickets, valleys, wetlands or alpine grasslands at 1600-3600 m. Distributed in SW and NW China. Also in India and Bhutan.

血满草 *Sambucus adnata*

接骨草
Sambucus javanica Blume

灌木状草本。奇数羽状复叶对生，托叶叶状，有时退化成蓝色腺体，小叶2-3对，边缘具细锯齿。复伞状聚伞花序顶生，大而疏散；不育花杯形，不脱落，可

接骨草 *Sambucus javanica*

育花小；花冠白色，花药黄色或紫色。浆果红色或黑色，近球形。花期4-5(-8)月，果期8-9月。生海拔300-2600米的山坡、林下、沟边或草丛。广布秦岭以南。南亚、东南亚和日本亦有。

Suffrutescent herbs. Imparipinnate compound leaves opposite, stipules foliaceous, sometimes reduced to blue glandulars, leaflets 2-3 pairs, margin serrulate. Compound umbellate cymes terminal, large and lax; some flowers modified into persistent urceolate nectaries, fertile flowers small; corolla white, anthers yellow or purple. Berries red or black, nearly globose. Fl. Apr-May(-Aug). Fr. Aug-Sep. Montane slopes, forests, streamsides or grasslands at 300-2600 m. Widely distributed in the south of Qinling Mountain. Also in S and SE Asia, and Japan.

接骨木

Sambucus williamsii Hance

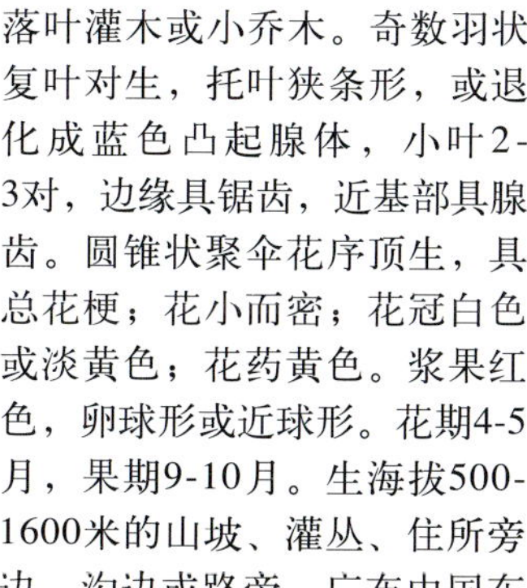

落叶灌木或小乔木。奇数羽状复叶对生，托叶狭条形，或退化成蓝色凸起腺体，小叶2-3对，边缘具锯齿，近基部具腺齿。圆锥状聚伞花序顶生，具总花梗；花小而密；花冠白色或淡黄色；花药黄色。浆果红色，卵球形或近球形。花期4-5月，果期9-10月。生海拔500-1600米的山坡、灌丛、住所旁边、沟边或路旁。广布中国东北到西南的大部分省区。

Deciduous shrubs or small trees. Imparipinnate compound leaves opposite, stipules narrowly linear, or reduced to blue glandulars, leaflets 2-3 pairs, margin serrate, usually glandular serrate nearly base. Cymose panicles terminal, pedunculate; flowers small and dense; corolla white or yellowish; anthers yellow. Berries red, ovoid or subglobose. Fl. Apr-May. Fr. Sep-Oct. Slopes, thickets, beside houses, ditchsides or roadsides at 500-1600 m. Widely distributed throughout the provinces from NE to SW China.

接骨木 *Sambucus williamsii*

聚花荚蒾 *Viburnum glomeratum*

聚花荚蒾

Viburnum glomeratum Maxim.

落叶灌木或小乔木，被黄色或黄白色簇状毛。叶纸质，卵状椭圆形、卵形或宽卵形，边缘具牙齿。复伞状聚伞花序顶生，直径3-6厘米；萼筒筒状倒圆锥形，萼齿三角形；花冠白色，辐状。核果红色，后变黑色，长圆状球形。花期4-6月，果期7-9月。生海拔300-3200米的山谷林中、灌丛中或草坡阴湿处。产中国西南、华中、华东和西北。缅甸北部亦有。

Deciduous shrubs or small trees, yellow or yellowish-white tufted hairy. Leaves chartaceous, ovate-elliptic, ovate or broadly ovate, margin dentate. Compound umbellike cymes terminal, 3-6 cm diam; calyx tube tubular-obconical, lobes triangular; corolla white, rotate. Drupes red and turned black, oblong-globose. Fl. Apr-Jun. Fr. Jul-Sep. Valley forests, thickets or shady palces on grassy slopes at 300-3200 m. Distributed in SW, C, E and NW China. Also in N Myanmar.

绣球荚蒾

Viburnum macrocephalum Fort.

落叶或半常绿灌木，被灰白色或黄白色簇状短毛。叶纸质，卵形、椭圆形或卵状长圆形，边缘具小齿。近伞状聚伞花序顶生，直径8-15厘米，全部由大型不孕花组成(野生者中央具可孕花)；花冠白色，辐状。核果红色，后变黑色。花期4-5月，果期9-10月。生海拔400-1000米的林中或山坡灌丛。产中国东南、华中、华北和华东，亦广泛栽培。

Deciduous or semievergreen shrubs, shortly gray-white or yellowish-white tufted hairy. Leaves chartaceous, ovate, elliptic or ovate-oblong, margin serrulate. Compound umbellike cymes terminal, 8-15 cm diam, totally composed of large sterile flowers (with fertile flowers at center in wild form); corolla white, rotate. Drupes red and turned black later. Fl. Apr-May. Fr. Sep-Oct. Forests or thickets on mountain slopes at 400-1000 m. Distributed in SE, C, N and E China, also commonly cultivated.

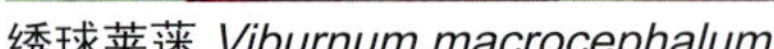

绣球荚蒾 *Viburnum macrocephalum*

烟管荚蒾

Viburnum utile

Hemsl.

常绿灌木。叶上面无毛或疏被星状毛，下面脉上常被锈色星状毛。复伞形式聚伞花序第1辐射枝具5分枝；花无香气，常生于第2和第3辐射枝上。果实最初变红，后为黑色；果核具2条较浅背沟和3条腹沟。花期3-4月，果期8月。生海拔500-1800米的林缘或灌丛。产中国西南、华中、华北和华西。

Evergreen shrubs. Leaves glabrous or sparsely stellate-hairy adaxially, abaxially veins sometimes with rusty stellate hairs. Inflorescences a compound umbellike cyme, 1st node of inflorescences with 5 rays; flowers usually on rays of 2nd and 3rd orders, not fragrant. Fruits initially tur-

烟管荚蒾 *Viburnum utile*

ning red, maturing black; pyrenes with 2 very shallow dorsal grooves and 3 ventral grooves. Fl. Mar-Apr. Fr. Aug. Forest edges or thickets at 500-1800 m. Distributed in SW, C, N and W China.

修枝荚蒾

Viburnum burejaeticum Regel et Herder

落叶灌木，被簇状短毛。小枝黄白色，无毛。叶纸质，宽卵形或椭圆形，基部钝或圆形，常不等大。复伞状聚伞花序顶生，直径4-5厘米；花冠白色，辐状，裂片宽卵形。核果红色，后变黑色，椭圆体形至长圆形。花期5-6月，果期8-9月。生海拔600-1400米的针阔叶混交林中。产黑龙江、吉林和辽宁。俄罗斯、蒙古和朝鲜半岛亦有。

Deciduous shrubs, shortly tufted hairy. Branchlets yellowish-white, glabrous. Leaves chartaceous, broadly ovate or elliptic, base obtuse or rotundate, usually oblique. Compound umbellike cymes terminal, 4-5 cm diam; corolla white, rotate, lobes broadly ovate. Drupes red and turning black later, ellipsoid to oblong. Fl. May-Jun. Fr. Aug-Sep. Needle-leaved and broad-leaved mixed forests at 600-1400 m. Distributed in Heilongjiang, Jilin and Liaoning. Also in Russia, Mongolia and Korean Peninsula.

修枝荚蒾 *Viburnum burejaeticum*

陕西荚蒾

Viburnum schensianum Maxim.

落叶灌木或小乔木，被绒毛。小枝稍四角状。叶纸质，卵状椭圆形、宽卵形或近圆形，边缘具密小尖齿。复伞状聚伞花序顶生，直径4-8厘米；花冠白色，辐状，裂片圆卵形；花药黄色。核果红色，后变黑色，椭圆体形。花期5-7月，果期8-9月。生海拔500-3200米的林下或灌丛中。产四川、湖北、河南、河北、山西、陕西、甘肃、江苏和山东。

Deciduous shrubs or small trees, tomentose. Branchlets somewhat tetragonous. Leaves chartaceous, ovate-elliptic, broadly ovate or suborbicular, margin densely sharp-serrulate. Compound umbellike cymes terminal, 4-8 cm diam; corolla white, rotate, lobes orbicular-ovate; anthers yellow. Drupes red and turned black, ellipsoid. Fl. May-Jul. Fr. Aug-Sep. Forests or thickets at 500-3200 m. Distributed in Sichuan, Hubei, Henan, Hebei, Shanxi, Shaanxi, Gansu, Jiangsu and Shandong.

陕西荚蒾 *Viburnum schensianum*

蒙古荚蒾 *Viburnum mongolicum*

蒙古荚蒾

Viburnum mongolicum (Pall.) Rehd.

落叶灌木，被簇状短毛。二年生小枝黄白色。叶宽卵形至椭圆形，边缘具波状浅齿，上面被簇状或叉状毛，侧脉4-5对。复伞状聚伞花序顶生，具少数花；花冠淡黄白色，筒状钟形；花药黄色。核果红色，后变黑色，椭圆体形。花期5-7月，果期7-9月。生海拔800-2700米的山坡疏林下。产华北和西北。俄罗斯和蒙古亦有。

Deciduous shrubs, shortly tufted hairy. Branchlets of previous year yellowish-white. Leaves broadly ovate to elliptic, margin shallowly undulate-serrate, adaxially stellate-pubescent or forked-hairy, lateral veins 4-5 pairs. Compound umbellike cymes terminal, few flowered; corolla lightly yellowish-white, tubular-campanulate; anthers yellow. Drupes red and turned black, ellipsoid. Fl. May-Jul. Fr. Jul-Sep. Sparse forests on slopes at 800-2700 m. Distributed in N and NW China. Also in Russia and Mongolia.

壶花荚蒾

Viburnum urceolatum Siebold et Zucc.

落叶灌木。叶纸质，卵状披针形或卵状长圆形，顶端渐尖至长渐尖，除基部为全缘外常有细齿或不整齐锯齿，侧脉通常4-6对，近缘前互相网结；叶柄长1-4厘米。花冠外面紫红色，筒状钟形，裂片长为筒的1/5-1/4。果实先红色，后变黑色；核扁，有2条浅背沟和3条腹沟。花期6-7月，果期9-10月。生海拔600-2600米的林中。产中国中南和东南。日本亦有。

Deciduous shrubs. Leaves papery, ovate-lanceolate or ovate-oblong, apex acuminate to long-acuminate, margin usually often serrulate or irregularly serrate except at base, lateral veins 4-6-paired, anastomosing near margin; petioles 1-4 cm long. Corolla purple-red outside, tubular-campanulate, lobes 1/5-1/4 as long as tube. Fruits initially turning red, maturing black; pyrenes compressed, with 2 shallow dorsal grooves and 3 ventral grooves. Fl. Jun-Jul. Fr. Sep-Oct. Forests at 600-2600 m. Distributed in SC and SE China. Also in Japan.

密花荚蒾

Viburnum congestum Rehd.

常绿灌木，被绒毛。小枝灰褐色。叶椭圆状卵形或椭圆形，全缘。近伞状聚伞花序小而密；花芳香，无柄；花冠白色，钟状漏斗形，裂片圆卵形。核果红色，后变黑色，圆球形。花期1-9月，果期8-10月。生海拔1000-2800米的林中、林缘或灌丛。产云南、四川、贵州和甘肃。

Evergreen shrubs, tomentose. Branchlets grayish-brown. Leaves elliptic-ovate or elliptic, margin entire. Compound umbellike cymes small and dense; flowers fragrant, sessile; corolla white, campanulate-infundibular, lobes orbicular-ovate. Drupes red, turned black later, globose. Fl. Jan-Sep. Fr. Aug-Oct. Forests, forest edges or thickets at 1000-2800 m. Distributed in Yunnan, Sichuan, Guizhou and Gansu.

壶花荚蒾 *Viburnum urceolatum*

密花荚蒾 *Viburnum congestum*

金佛山荚蒾 *Viburnum chinshanense*

金佛山荚蒾

Viburnum chinshanense Graebn.

半常绿灌木。叶下面被星状毛，上部无毛。复伞形式聚伞花序第一辐射枝5-7条；花无香味，通常生于第二辐射枝上。果实最初变为红色，成熟时黑色，疏被星状柔毛；核具2条背沟和3条腹沟。花期4-5月，果期7月。生海拔100-1900米的疏林中或灌丛。产中国西南和华西。

Semievergreen shrubs. Leaves abaxially stellate-pubescent, adaxially glabrous. Inflorescences a compound umbellike cyme, 1st node of inflorescences with 5-7 rays; flowers usually on rays of 2nd order, not fragrant. Fruits initially turning red, maturing black, sparsely stellate-pubescent; pyrenes with 2 dorsal grooves and 3 ventral grooves. Fl. Apr-May. Fr. Jul. Sparse forests or thickets at 100-1900 m. Distributed in SW and W China.

皱叶荚蒾

Viburnum rhytidophyllum Hemsl.

常绿灌木或小乔木，密被绒毛。小枝粗壮，稍有棱角。叶极皱，卵状长圆形至卵状披针形，全缘或具不明显小齿。复伞状聚伞花序顶生，直径7-12厘米，花稠密；花冠白色，辐状。核果红色，后变黑色，阔椭圆体形。花期4-5月，果期9-10月。生海拔700-2400米的山坡林下或灌丛中。产四川、贵州、湖北和陕西。欧洲有栽培。

Evergreen shrubs or small trees, densely tomentose. Branchlets thick, somewhat tetragonous. Leaves very rugose, ovate-oblong to ovate-lanceolate, margin or inconspicuously serrulate. Compound umbellike cymes terminal, 7-12 cm diam, flowers dense and closed; corolla white, rotate. Drupes red and turned black, broadly ellipsoid. Fl. Apr-May. Fr. Sep-Oct. Forests or thickets on slopes at 700-2400 m. Distributed in Sichuan, Guizhou, Hubei and Shaanxi. Also cultivated in Europe.

皱叶荚蒾 *Viburnum rhytidophyllum*

显脉荚蒾 *Viburnum nervosum*

显脉荚蒾

Viburnum nervosum D. Don

落叶灌木或小乔木。叶常簇生于枝顶，下面常稍星状毛，上面无毛或近无毛，中脉在上面凸起。复伞形式聚伞花序第一辐射枝5-7条；花芳香，生于第二和第三个辐射枝上。果实最初黄色，后红色，成熟时紫黑色。花期4-6月，果期9-10月。生海拔(1800-)2100-4500米的灌丛或林中(常为冷杉林)。产云南、四川和西藏。印度、尼泊尔、不丹、缅甸北部和越南北部亦有。

Deciduous shrubs or small trees. Leaves often clustered at apices of branchlets, abaxially often ± stellate-pubescent, adaxially glabrous or subglabrous, midvein raised abaxially. Inflorescences a compound umbellike cyme, 1st node of inflorescences with 5-7 rays; flowers on rays of 2nd and 3rd orders, fragrant. Fruits initially turning yellow, then red, maturing purplish black. Fl. Apr-Jun. Fr. Sep-Oct. Forests (usually *Abies* forests) or thickets at (1800-) 2100-4500 m. Distributed in Yunnan, Sichuan and Xizang. Also in India, Nepal, Bhutan, N Myanmar and N Vietnam.

合轴荚蒾

Viburnum sympodiale Graebn.

落叶灌木或小乔木。叶常簇生于枝顶，下面具棕黄色鳞片或糠状星状毛，上面无毛或于叶脉上具星状毛。复伞形状聚伞花序第一辐射枝5条，边缘具大的不育辐射状花；花芳香。果实最初变为黄色，后转红色，成熟后紫黑色。花期4-5月，果期8-9月。生海拔800-2600米的林中或灌丛。产中国西南、华南、东南、华中、华北、华西和华东。

Deciduous shrubs or small trees. Leaves often clustered at apices of branchlets, abaxially yellow-brownish lepidote or furfuraceous stellate-pubescent, adaxially glabrous or stellate-pubescent on veins. Inflorescences a compound umbellike cyme, 1st node of inflorescences usually with 5 rays, with large white sterile radiant flowers on margins; flowers fragrant. Fruits initially turning yellow, then red, maturing pur-

合轴荚蒾 *Viburnum sympodiale*

三脉叶荚蒾 *Viburnum triplinerve*

ple-nigrescent. Fl. Apr-May. Fr. Aug-Sep. Forests or thickets at 800-2600 m. Distributed in SW, S, SE, C, N, W and E China.

三脉叶荚蒾

Viburnum triplinerve
Hand.-Mazz.

常绿灌木，全体无毛。幼枝纤细。叶革质，椭圆形或近圆形，全缘，具离基三出脉。聚伞花序顶生；总花梗纤细；萼齿宽三角形或宽卵形；花冠辐状，裂片近圆形。核果近圆形，熟时紫褐色。花期4-5月，果期6-10月。生海拔约550米的山地。产广西。

Evergreen shrubs, wholly glabrous. Branchlets slender. Leaves coriaceous, elliptic or sub-orbicular, margin entire, triplicostate. Cymes terminal; peduncles slender; calyx lobes broadly triangular or ovate; corolla rotate, lobes sub-orbicular. Drupes subglobose, maturing purplish-brown. Fl. Apr-May. Fr. Jun-Oct. Mountains at ca. 550 m. Distributed in Guangxi.

球核荚蒾

Viburnum propinquum
Hemsl.

常绿灌木。叶常对生，革质，幼时紫色，卵形至卵状披针形、或椭圆形至椭圆状长圆形。花序为一复合伞房状聚伞花序，顶生；伞幅轮生；花冠浅绿色、旋转。果实不变红，成熟后蓝黑色，有光泽，近球形或卵球形；分核球形。生海拔500-1300米的林中或灌丛中。产中国西南、华南、东南、华中、华西和华东。

Evergreen shrubs. Leaves always opposite, coriaceous, purplish when young, ovate to ovate-lanceolate, or elliptic to elliptic-oblong. Inflorescences a compound umbellike cyme, terminal; rays whorled; corolla greenish white, rotate. Fruits not turning red, maturing blue-blackish, lustrous, subglobose or ovoid; pyrenes globose. Forests or thickets at 500-1300 m. Distributed in SW, S, SE, C, W and E China.

狭叶球核荚蒾

Viburnum propinquum
Hemsl. var. **mairei** W. W. Sm.

常绿灌木。叶条状披针形或倒披针形，革质，下面常于叶腋具星状毛，上面无毛，基部楔形。复伞形式聚伞花序第一辐射枝7条；花无香气，生于第三个辐射枝上。果实成熟蓝黑色，光亮；果核具1极小及浅的腹沟或无。花期3-5月，果期5-10月。生海拔400-500米的灌丛。产云南、四川、贵州和湖北。

Evergreen shrubs. Leaves linear-lanceolate or oblanceolate, coriaceous, abaxially sometimes stellate-pubescent in axils of veins, adaxially glabrous, base cuneate. Inflorescences a compound umbellike cyme, 1st node of inflorescences usually with 7 rays; flowers on rays of 3rd order, not fragrant. Fruits maturing blue-blackish, lustrous; pyrenes with 1 very small and shallow ventral groove or without groove. Fl. Mar-May. Fr. May-Oct. Thickets at 400-500 m. Distributed in Yunnan, Sichuan, Guizhou and Hubei.

球核荚蒾 *Viburnum propinquum*

狭叶球核荚蒾 *Viburnum propinquum* var. *mairei*

樟叶荚蒾 *Viburnum cinnamomifolium*

川西荚蒾 *Viburnum davidii*

樟叶荚蒾

Viburnum cinnamomifolium Rehd.

常绿灌木或小乔木。小枝紫褐色。幼叶紫色，后变绿色，革质，椭圆状长圆形，全缘，具离基三出脉。近伞状聚伞花序顶生，大而疏散；花冠淡黄绿色，辐状，裂片宽卵形，反曲。核果蓝黑色，近球形。花期5月，果期6-7月。生海拔1000-1800米的山坡灌丛中。产云南和四川。

Evergreen shrubs or small trees. Branchlets purplish-brown. Leaves purple when young, turned green later, coriaceous, elliptic-oblong, margin entire, triplicostate. Compound umbel-like cymes terminal, large and lax; corolla yellowish-green, rotate, lobes broadly ovate, revolute. Drupes bluish-black, subglobose. Fl. May. Fr. Jun-Jul. Slope thickets at 1000-1800 m. Distributed in Yunnan and Sichuan.

川西荚蒾

Viburnum davidii Franch.

常绿灌木或小乔木。小枝紫褐色。叶厚革质，椭圆状倒卵形至椭圆形，显著皱纹状，全缘，具离基三出脉。近伞状聚伞花序顶生，小而稠密；萼齿披针形；花冠暗红色，辐状，裂片圆形。核果蓝黑色，卵圆形或椭圆状矩圆形。花期6月，果期9-10月。生海拔1800-2400米的山地。产四川。

Evergreen shrubs or small trees. Branchlets purplish-brown. Leaves thickly coriaceous, elliptic-obovate to elliptic, distinctly rugose, margin entire, triplicostate. Compound umbellike cymes terminal, small and dense; calyx lobes lanceolate; corolla darkly red, rotate, lobes rotundate. Drupes bluish-black, ovoid or ellipsoid-oblong. Fl. Jun. Fr. Sep-Oct. Mountains at 1800-2400 m. Distributed in Sichuan.

香荚蒾

Viburnum farreri Stearn

落叶灌木。叶长大后仅于脉腋具星状毛。花序圆锥状；伞幅交互对生；第一辐射枝2条；花

香荚蒾 *Viburnum farreri*

芳香，生于第一至第三个辐射枝上。果实最初变为黄色，成熟后转为紫红色，长圆形；果核具1深腹沟。花期4-5月，果期6-7月。生海拔1600-2800米的林中。产华北、华西、华东和西北。

Deciduous shrubs. Leaves glabrous on both surfaces but stellate-pubescent in axils of veins when matured. Inflorescences paniculate; rays opposite, decussate; 1st node of inflorescences with 2 rays; flowers on rays from 1st to 3rd orders, fragrant. Fruits initially turning yellow, maturing purple-reddish, oblong; pyrenes with 1 deep ventral groove. Fl. Apr-May. Fr. Jun-Jul. Forests at 1600-2800 m. Distributed in N, W, E and NW China.

红荚蒾 *Viburnum erubescens*

红荚蒾

Viburnum erubescens Wall.

落叶灌木或小乔木。小枝紫褐色。叶对生，厚纸质，椭圆形或长圆形。圆锥花序常下垂；苞片绿色，叶状，条形；萼筒筒状，萼齿卵状三角形；花冠白色或粉红色，高脚碟状。核果紫红色，后变黑色，卵球形。花期4-6月，果期8月。生海拔(1500-)2400-3500米的林中或灌丛。产中国西南、华中和华西。印度北部、尼泊尔、不丹和缅甸北部亦有。

Deciduous shrubs or small trees. Branchlets purplish-brown. Leaves opposite, thick chartaceous, elliptic or oblong. Panicles usually nodding; bracts green, leaflike, linear; calyx tube tubular, lobes ovate-tri angular; corolla white or pink, hypocrateriform. Drupes purplish-red, later nigrescent, ellipsoid. Fl. Apr-Jun. Fr. Aug. Forests or thickets at (1500-)2400-3500 m. Distributed in SW, C and W China. Also in N India, Nepal, Bhutan and N Myanmar.

少花荚蒾

Viburnum oliganthum Batal.

常绿灌木或小乔木。叶常亚革质至革质，常倒披针形至条状倒披针形或倒卵状长圆形至长圆形，边缘离基部1/3-1/2及以上具疏浅锯齿，齿顶细尖而弯向内或向前，上面中脉尤为明显。圆锥花序顶生；花冠漏斗状；花药紫红色。核扁，有1条宽广的深腹沟。花期4-6月，果期6-8月。生海拔1000-2200米的林中或灌丛。产湖北、四川、贵州、云南和西藏。

Evergreen shrubs or small trees. Leaves usually subleathery or leathery, usually oblanceolate to linear-oblanceolate or obovate-oblong to oblong, margin remotely and shallowly serrate above 1/3-1/2 from base, apex apiculate and bend inward or forward, midvein particularly raised adaxially. Panicles terminal; corolla funnelform; anthers purple-red. Pyrenes compressed, with 1 broad and deep ventral groove. Fl. Apr-Jun. Fr. Jun-Aug. Forests and scrubs at 1000-2200 m. Distributed in Hubei, Sichuan, Guizhou, Yunnan and Xizang.

少花荚蒾 *Viburnum oliganthum*

巴东荚蒾 *Viburnum henryi*

巴东荚蒾
Viburnum henryi Hemsl.

灌木或小乔木。叶革质，下面脉腋具星状毛，上面无毛，光亮。圆锥花序顶生；伞幅交互对生；第一辐射枝2条；花芳香，生第二和第三辐射枝上。果实最初变为红色，成熟时紫黑色，椭圆形；果核具1条深腹沟。花期6月，果期8-9月。生海拔900-2600米的密林中或湿草坡。产中国西南、东南、华中、华西和华东。

Shrubs or small trees. Leaves subcoriaceous, abaxially stellate-pubescent in axils of veins, adaxially glabrous and lustrous. Inflorescences paniculate, terminal; rays opposite, decussate; 1st node of inflorescences with 2 rays; flowers on rays of 2nd and 3rd orders, fragrant. Fruits initially turning red, maturing purple-blackish, ellipsoid; pyrenes with a deep ventral groove. Fl. Jun. Fr. Aug-Sep. Dense forests or moist grassy slopes at 900-2600 m. Distributed in SW, SE, C, W and E China.

珊瑚树 *Viburnum odoratissimum*

珊瑚树
Viburnum odoratissimum Ker Gawl.

常绿灌木或小乔木，高可达15米。枝灰色。叶革质，椭圆形、长圆形或倒卵形，上面有光泽，侧脉5-6对。圆锥花序顶生，宽尖塔形，总花梗长可达10厘米；花芳香；花冠白色，辐状，裂片反折。核果红色，后变黑色，卵球形。花期3-5月，果期6-9月。生海拔2500米以下的林下或灌丛。产华南、华中和华东，亦广泛栽培。印度、缅甸、泰国、越南、菲律宾、朝鲜半岛和日本亦有。

Evergreen shrubs or small trees, up to 10 15 m tall. Branches gray. Leaves coriaceous, elliptic, oblong or obovate, upper surfaces lustrous, lateral veins 5-6 pairs. Panicles terminal, broadly pyramidal, peduncles up to 10 cm long; flowers fragrant; corolla white, rotate, lobes revolute. Drupes red and turned black, ovoid. Fl. Mar-May. Fr. Jun-Sep. Forests or thickets below 2500 m.

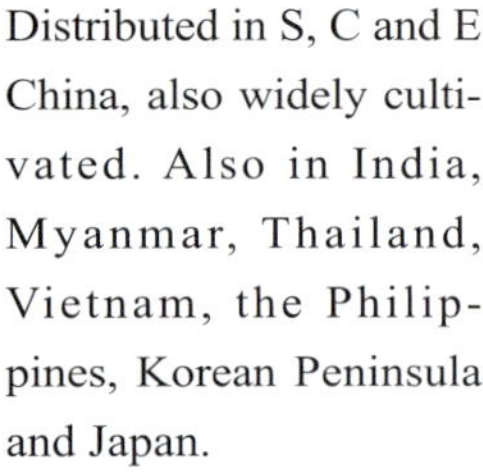
Distributed in S, C and E China, also widely cultivated. Also in India, Myanmar, Thailand, Vietnam, the Philippines, Korean Peninsula and Japan.

短序荚蒾
Viburnum brachybotryum Hemsl.

常绿灌木或小乔木，被黄褐色簇状毛。小枝黄白色。叶革质，倒卵形或长圆形，边缘疏生尖锯齿或近全缘。圆锥花序尖形，顶生；苞片和小苞片宿存；花冠白色，辐状，裂片卵形；花药黄白色。核果鲜红色，卵球形，先端尖。花期1-3月，果期7-8月。生海拔(400-)600-1900米的山谷密林或山坡灌丛中。产云南、四川、贵州、广西、湖北、湖南和江西。

Evergreen shrubs or small trees, yellowish tufted hairy. Branchlets yellowish-white. Leaves coriaceous, obovate or oblong, margin remotely sharp-serrate or nearly entire. Panicles pyramidal, terminal; bracts and bracteoles persistent; corolla white, rotate, lobes ovate; anthers yellowish-white. Drupes red, ovoid, apex acuminate. Fl. Jan-Mar. Fr. Jul-Aug. Dense valley forests or slope thickets at (400-)600-1900 m. Distributed in Yunnan, Sichuan, Guizhou, Guangxi, Hubei, Hunan and Jiangxi.

伞房荚蒾
Viburnum corymbiflorum P. S. Hsu et S. C. Hsu

灌木或小乔木。叶上面光亮，下部于中脉无毛或极稀被星状毛。花序轴缩短，花序呈圆锥状或伞房状；伞幅交互对生；第一辐射枝2条；花生于第三辐射枝上。果实成熟时红

短序荚蒾 *Viburnum brachybotryum*

色，椭圆体形；果核具1深腹沟。花期4-5月，果期6-7月。生海拔1000-2400米的林中或灌丛。产中国西南、华南、东南、华中和华东。

Shrubs or small trees. Leaves lustrous adaxially, abaxially glabrous or very sparsely stellate-pubescent on veins. Inflorescences paniculate, corymblike due to shortening of inflorescences axis; rays opposite, decussate; 1st node of inflorescences with 2 rays; flowers on rays of 3rd order. Fruits maturing red, ellipsoid; pyrenes with a deep ventral groove. Fl. Apr-May. Fr. Jun-Jul. Forests or thickets at 1000-2400 m. Distributed in SW, S, SE, C and E China.

粉团

Viburnum plicatum Thunb.

落叶灌木。叶对生，纸质，宽卵形或矩圆状卵形，侧脉10-12对。复伞形式聚伞花序第一辐射枝6-8条，完全由大型不育花组成；花无香气，生于第四辐射枝上；不孕花白色，辐状，直径达4厘米。花期4-5月。产湖北和贵州，各地常有栽培。日本亦有。

Deciduous shrubs. Leaves opposite, chartaceous, broadly ovate or oblong-ovate, lateral veins 10-12-jugate. Inflorescences a compound umbellike cyme, 1st node of inflorescences with 6-8 rays, totally composed of large sterile flowers; flowers on rays of 4th order, not fragrant; sterile flowers white, rotate, up to 4 cm diam. Fl. Apr-May. Distributed in Hubei and Guizhou, also cultivated in other places. Also in Japan.

伞房荚蒾 *Viburnum corymbiflorum*

粉团 *Viburnum plicatum*

蝴蝶戏珠花 *Viburnum plicatum* var. *tomentosum*

蝴蝶戏珠花

Viburnum plicatum Thunb. var. **tomentosum** (Thunb.) Miq.

落叶灌木。叶较狭，两端有时渐尖，下面常带绿白色，侧脉10-17对。花序直径4-10厘米，外围有4-6朵白色、大型的不孕花；中央可孕花直径约3毫米，黄白色。果实最初为红色，后为黑色。花期4-5月，果期8-9月。生海拔240-1800米的山坡混交林或沟谷灌丛，常有栽培。产中国西南、华南、东南、华中、华西和华东。日本亦有。

Deciduous shrubs. Leaves narrow, sometimes both ends of leaves acuminate, abaxially greenish-white, lateral veins 10-17 pairs. Inflorescences 4-10 cm diam, with 4-6 white and large sterile flowers; central breeding flowers yellowish-white, ca. 3 mm diam. Fruits initially turning red, maturing black. Fl. Apr-May. Fr. Aug-Sep. Mixed forests on slopes or thickets in valleys at 200-1800 m, also cultivated. Distributed in SW, S, SE, C, W and E China. Also in Japan.

蝶花荚蒾

Viburnum hanceanum Maxim.

落叶灌木。叶两面被棕黄色星状毛。复伞形式聚伞花序第1辐射枝5条及具2-5个大的不育辐射状花；花无香气，生于第2和第3辐射枝上。果实成熟时红色，卵状圆形；果核具1宽纵沟。花期3-5月，果期8-9月。生海拔200-800米的灌丛。产中国西南、华南、华中和华东。

Deciduous shrubs. Leaves both surfaces yellow-brown stellate-pubescent. Inflorescences a compound umbellike cyme, 1st node of inflorescences usually with 5 rays, with 2-5 large sterile radiant flowers; flowers on rays of 2nd and 3rd orders, not fragrant. Fruits maturing red, ovoid-orbicular; pyrenes with a broad ventral groove. Fl. Mar-May. Fr. Aug-Sep. Thickets at 200-800 m. Distributed in SW, S, C and E China.

蝶花荚蒾 *Viburnum hanceanum*

水红木

Viburnum cylindricum Buch.-Ham. ex D. Don

常绿灌木或小乔木。叶革质，椭圆形至长圆形，近全缘，侧脉3-5对。复伞状聚伞花序顶生，直径4-10厘米；花冠白色或有红晕，钟状，裂片圆卵形；花药紫色。核果先红色，后变蓝黑色，卵球形。花期6-10月，果期8-12月。生海拔500-3300米的阳坡、疏林或灌丛中。产中国西南、华西、华中和东南。南亚和东南亚亦有。

Evergreen shrubs or small trees. Leaves coriaceous, elliptic to oblong, nearly entire, lateral veins 3-5 pairs. Compound umbellike cymes terminal, 4-10 cm diam; corolla white or tinged with reddish, campanulate, lobes ovate; anthers purple. Drupes red and turned bluish-black, ovoid. Fl. Jun-Oct. Fr. Aug-Dec. Sunny slopes, sparse forests or thickets at 500-3300 m. Distributed in SW, W, C and SE China. Also in S and SE Asia.

鳞斑荚蒾

Viburnum punctatum Buch.-Ham. ex D. Don

常绿灌木或小乔木，全株均密被铁锈色、圆形小鳞片。冬芽裸露。叶硬革质，常长圆状椭圆形或长圆状卵形，全缘或有时上部具少数不整齐浅齿，边内卷。聚伞花序复伞形式；花冠辐状，直径6-8毫米。果实长8-15毫米；核扁，有2条背沟和3条浅腹沟。花期3-4月，果期5-10月。生海拔200-1900米的密林中或林缘。产四川和华南。东亚和东南亚亦有。

Evergreen shrubs or small trees, densely rusty and rounded scales. Winter buds naked. Leaves hard

leathery, usually oblong-elliptic or oblong-ovate, margin entire or sometimes with few irregular serration in upper part, margin involute. Inflorescences a compound umbellike cyme; corolla rotate, 6-8 mm diam. Fruits 8-15 mm long; pyrenes compressed, with 2 dorsal grooves and 3 shallow ventral grooves. Fl. Mar-Apr. Fr. May-Oct. Dense forests or forest margins at 200-1900 m. Distributed in Sichuan and S China. Also in E and SE Asia.

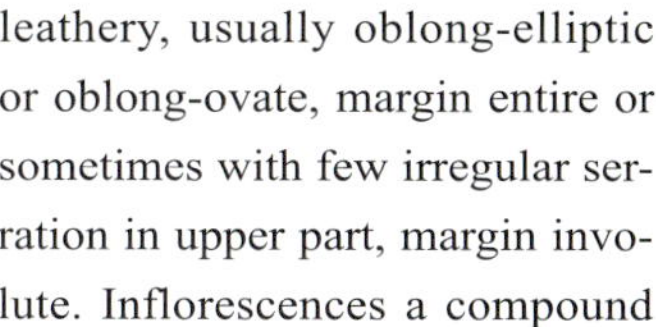

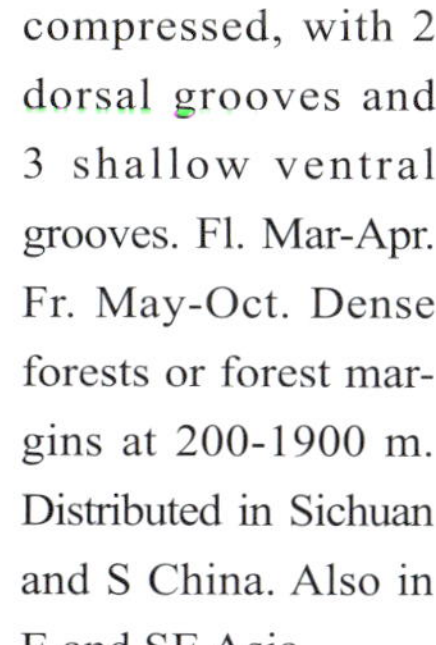

水红木 *Viburnum cylindricum*

鳞斑荚蒾 *Viburnum punctatum*

大果鳞斑荚蒾

Viburnum punctatum Buch.-Ham. ex D. Don var. **lepidotulum** (Merr. et Chun) P. S. Hsu

常绿灌木或小乔木。叶革质，上面光亮，下面覆有鳞片。复伞形状聚伞花序第一辐射枝4或5条；花无香气，生于第三和第四个辐射枝上。果实最初红色，成熟时黑色，1.4-1.5(-1.8) × 约1厘米；果核具2条背沟和3条浅腹沟。花期3-4月，果期5-10月。生海拔200-900米的杂木林中。产广西、广东和海南。

Evergreen shrubs or small trees. Leaves coriaceous, adaxially lustrous, abaxially with lepidote scales. Inflorescences a compound umbellike cyme, 1st node of inflorescences with 4 or 5 rays; flowers on rays of 3rd and 4th orders, not fragrant. Fruits initially turning red, maturing black, 1.4-1.5(-1.8) × ca. 1 cm; pyrenes with 2 dorsal grooves and 3 shallow ventral grooves. Fl. Mar-Apr. Fr. May-Oct. Mixed forests at 200-900 m. Distributed in Guangxi, Guangdong and Hainan.

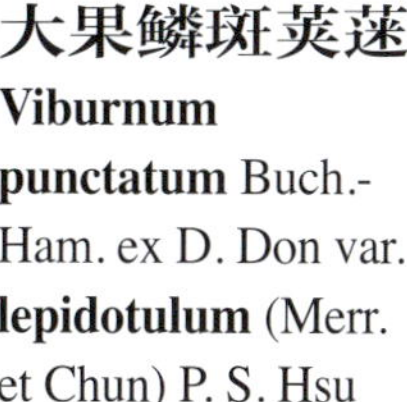

大果鳞斑荚蒾 *Viburnum punctatum* var. *lepidotulum*

三叶荚蒾 *Viburnum ternatum*

三叶荚蒾

Viburnum ternatum Rehd.

落叶灌木或小乔木。叶3枚轮生或对生，下面具星状毛及于脉上杂生丁字毛或单毛。复伞形式聚伞花序第一辐射枝具5-7(-10)分枝；花无香气，生于第二至第六辐射枝上。果实成熟时红色；果核具2条浅背沟和1条腹沟。花期6-7月，果期9月。生海拔600-1400米的林中。产云南、四川、贵州、湖北和湖南。

Deciduous shrubs or small trees. Leaves in whorls of 3 or opposite, abaxially stellate-pubescent and mixed with forklike or simple hairs only on midvein and lateral veins. Inflorescences a compound umbellike cyme, 1st node of inflorescences with 5-7 (-10) rays; flowers on rays from 2nd to 6th orders, not fragrant. Fruits maturing red; pyrenes with 2 shallow dorsal grooves and 1 ventral groove. Fl. Jun-Jul. Fr. Sep. Forests at 600-1400 m. Distributed in Yunnan, Sichuan, Guizhou, Hubei and Hunan.

锥序荚蒾

Viburnum pyramidatum Rehd.

灌木或小乔木，被绒毛。叶卵状长圆形、长圆形或宽椭圆形，边缘具牙齿状锯齿或小锯齿，上面有光泽。圆锥状花序尖塔形，长5-10厘米；花冠白色，辐状，裂片卵形；花药黄色。核果深红色，长圆形或椭圆体形。花期11-12月，果期翌年3-10月。生海拔100-1400米的疏林或灌丛中。产云南和广西。越南北部亦有。

Shrubs or small trees, tomentose. Leaves ovate-oblong, oblong or broadly elliptic, margin dentate-serrate or serrulate, upper surfaces lustrous. Paniculate inflorescences pyramidal, 5-10 cm long; corolla white, rotate, lobes ovate; anthers yellow. Drupes dark red, oblong or ellipsoid. Fl. Nov-Dec. Fr. Mar-Oct next year. Sparse forests or thickets at 100-1400 m. Distributed in Yunnan and Guangxi. Also in N Vietnam.

锥序荚蒾 *Viburnum pyramidatum*

常绿荚蒾
Viburnum sempervirens

常绿荚蒾

Viburnum sempervirens K. Koch

常绿灌木，幼枝、叶柄和花序无毛或疏具星状柔毛。叶革质，椭圆形至椭圆状卵形。复伞状聚伞花序顶生；花冠白色，辐状，裂片近圆形；花药黄色。核果红色，卵球形；分核背面凹陷。花期5月，果期10-12月。生海拔100-1800米的林下、溪边或灌丛中。产广西、广东和江西。

Evergreen shrubs, young branchlets, petioles and inflorescences glabrous or sparsely stellate-pubescent. Leaves coriaceous, elliptic to elliptic-ovate. Compound umbellike cymes terminal; corolla white, rotate, lobes suborbicular, long; anthers yellow. Drupes red, ovoid; pyrenes convex dorsally. Fl. May. Fr. Oct-Dec. Forests, streamsides or thickets at 100-1800 m. Distributed in Guangxi, Guangdong and Jiangxi.

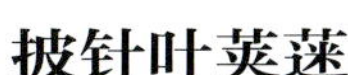

具毛常绿荚蒾

Viburnum sempervirens K. Koch var. **trichophorum** Hand.-Mazz.

常绿灌木，幼枝、叶柄和花序均密被簇状短毛。叶顶端具较明显的锯齿，侧脉5-6对。核果较大，核长约7毫米，直径约6毫米，背面略凸起，腹面不明显凹陷。花期4-5月，果期7-12月。生海拔100-1800米的林下或灌丛中。产中国西南、华南和华东。

Evergreen shrubs, young branchlets, petioles and inflorescences densely stellate-pubescent. Leaves conspicuously serrate nearly the apex, lateral veins 5-6 pairs. Drupes larger, pyrenes ca. 7 mm long, ca. 6 mm diam, slightly convex dorsally, not obviously concave ventrally. Fl. Apr-May. Fr. Jul-Dec. Forests or thickets at 100-1800 m. Distributed in SW, S and E China.

具毛常绿荚蒾 *Viburnum sempervirens* var. *trichophorum*

披针叶荚蒾

Viburnum lancifolium P. S. Hsu

常绿灌木，被红褐色微细腺点和黄褐色簇状毛。叶纸质，长圆状披针形至披针形。复伞状聚伞花序顶生，直径4-6厘米，总花梗纤细；花冠白色，辐状，裂片圆卵形。核果红色，球形。花期4-5月，果期10-11月。生海拔200-600米的疏林、林缘、灌丛或竹林。产广东、福建、浙江和江西。

Evergreen shrubs, minutely reddish-brown glandular spotted and yellowish-brown tufted hairy. Leaves chartaceous, oblong-lanceolate to lanceolate. Compound umbellike cymes terminal, 4-6 cm diam, peduncles slender; corolla white, rotate, lobes orbicular-ovate. Drupes red, globose. Fl. Apr-May. Fr. Oct-Nov. Sparse forests, forest edges, thickets or bamboo forests at 200-600 m. Distributed in Guangdong, Fujian, Zhejiang and Jiangxi.

披针叶荚蒾 *Viburnum lancifolium*

珍珠荚蒾 *Viburnum foetidum* var. *ceanothoides*

直角荚蒾 *Viburnum foetidum* var. *rectangulatum*

珍珠荚蒾

Viburnum foetidum Wall. var. **ceanothoides** (C. H. Wright) Hand.-Mazz.

落叶灌木。枝稍伸长。叶纸质，倒卵状椭圆形至倒卵形，边缘近顶疏生粗牙齿或缺刻，侧脉2-3对。复伞状聚伞花序生于侧生小枝之顶，直径5-8厘米；花冠白色，辐状。核果红色，卵状椭圆体形。花期4-8月，果期8-12月。生海拔900-2600米的山坡密林或灌丛中。产云南、四川和贵州。

Deciduous shrubs. Branches ± elongated. Leaves chartaceous, obovate-elliptic to obovate, margin remotely gross-teethed or incised nearly apex, lateral veins 2-3 pairs. Compound umbellike cymes terminal at the top of the lateral branchlets, 5-8 cm diam; corolla white, rotate. Drupes red, ovoid-ellipsoid. Fl. Apr-Aug. Fr. Aug-Dec. Dense forests on slopes or thickets at 900-2600 m. Distributed in Yunnan, Sichuan and Guizhou.

茶荚蒾(汤饭子)

Viburnum setigerum Hance

落叶灌木。小枝具棱角，无毛。叶纸质，卵状矩圆形至卵状披针形，边缘疏生尖锯齿，近基部两侧有少数腺体。复伞形式聚伞花序顶生，常弯垂；萼齿卵形；花冠白色，芳香，辐状，裂片卵形。果序弯垂，果实红色，卵圆形。花期4-5月，果期9-10月。生海拔800-1650米的山谷或山坡灌丛。产中国西南、华南、华中和华东。

Deciduous shrubs. Branchlets ridged, glabrous. Leaves chartaceous, ovate-oblong to ovate-lanceolate, margin sparsely serrate, with several glandula near base. Compound umbellike cymes terminal, usually drooped; calyx lobes ovate; corolla white, fragrant, rotate, lobes ovate. Infructescences drooped, fruits maturing red, ovoid. Fl. Apr-May. Fr. Sep-Oct. Valleys or slope thickets at 800-1650 m. Distributed in SW, S, C and E China.

茶荚蒾(汤饭子) *Viburnum setigerum*

直角荚蒾

Viburnum foetidum Wall. var. **rectangulatum** (Graebn.) Rehd.

落叶灌木。小枝伸长。叶卵形或椭圆形至长圆状菱形，边缘具浅牙齿或近全缘。花梗极短，最长达2厘米，或近无梗；花冠白色，辐射状。果实成熟时红色，卵球状椭圆体形，压扁。花期5-7月，果期10-12月。生海拔600-2400米的山坡林中或灌丛中。产中国西南、东南、华中、华西和华东。

黑果荚蒾 *Viburnum melanocarpum*

Deciduous shrubs. Branchlets elongated. Leaves ovate or elliptic to oblong-rhombic, margin shallowly dentate or subentire. Peduncles very short to 2 cm at most, or absent; corolla white, rotate. Fruits maturing red, ovoid-ellipsoid, compressed. Fl. May-Jul. Fr. Oct-Dec. Slope forests or thickets at 600-2400 m. Distributed in SW, SE, C, W and E China.

黑果荚蒾
Viburnum melanocarpum P. S. Hsu

落叶灌木。叶下脉上疏被平伏长毛，脉腋疏被星状毛，上面常于中脉被硬毛，后近无毛。复伞形式聚伞花序第一辐射枝5条；花无香气，生第二和第三个辐射枝上。果实最初为深紫红色，成熟后黑色；果核腹面纵向凸起。花期4-5月，果期9-10月。生海拔约1000米的林中或灌丛。产浙江、河南、安徽、江苏和江西。

Deciduous shrubs. Leaves abaxially sparsely adpressed long hairy on midvein and lateral veins and sparsely stellate-pubescent in vein axils, adaxially often stiffly hairy on midvein, later subglabrous. Inflorescences a compound umbellike cyme, 1st node of inflorescences usually with 5 rays; flowers on rays of 2nd and 3rd orders, not fragrant. Fruits initially turning dark purple-red, maturing black; pyrenes longitudinally raised on ventral side. Fl. Apr-May. Fr. Sep-Oct. Forests or thickets at ca. 1000 m. Distributed in Zhejiang, Henan, Anhui, Jiangsu and Jiangxi.

桦叶荚蒾
Viburnum betulifolium Batal.

落叶灌木或小乔木。小枝褐色。叶厚纸质，卵形，边缘具不规则浅波状牙齿。复伞状聚伞花序顶生；萼筒具黄褐色腺点，萼齿小，宽卵状三角形；花冠白色，辐状，裂片圆卵

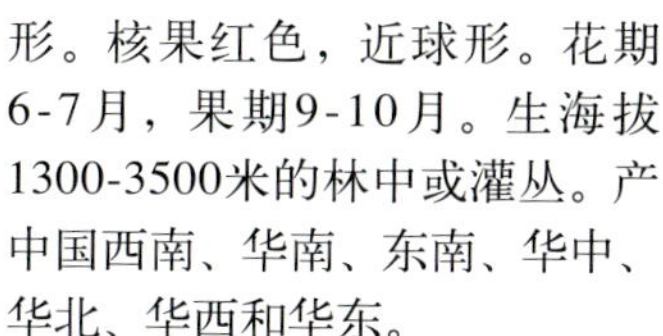

形。核果红色，近球形。花期6-7月，果期9-10月。生海拔1300-3500米的林中或灌丛。产中国西南、华南、东南、华中、华北、华西和华东。

Deciduous shrubs or small trees. Branchlets brown. Leaves thickly chartaceous, ovate, margin shallowly and irregularly undulate-dentate. Compound umbel-like cymes terminal; calyx tubes yellowish-brown glandular punctate, lobes small, broadly ovate-triangular; corolla white, rotate, lobes ovate. Drupes red, subglobose. Fl. Jun-Jul. Fr. Sep-Oct. Forests or thickets at 1300-3500 m. Distributed in SW, S, SE, C, N, W and E China.

桦叶荚蒾 *Viburnum betulifolium*

荚蒾 *Viburnum dilatatum*

荚蒾

Viburnum dilatatum Thunb.

落叶灌木。叶下面被黄色丁字毛和星状毛，常散生腺点，上面具平伏毛。复伞形式聚伞花序，第一辐射枝5条；花芳香，生于第三或第四辐射枝上。果实成熟时红色，椭圆状卵形；果核具2条浅背沟和3条浅腹沟。花期5-7月，果期9-11月。生海拔100-1000米的疏林中、林缘或灌丛。产中国西南、华南、东南、华中、华北、华西和华东。朝鲜半岛和日本亦有。

Deciduous shrubs. Leaves abaxially yellowish forklike pubescent and stellate-pubescent, usually with dispersed glandular dots, adaxially adpressed hairy. Inflorescences a compound umbellike cyme, 1st node of inflorescences with 5 rays; flowers on rays of 3rd and 4th orders, fragrant. Fruits maturing red, ellipsoid-ovoid; pyrenes with 2 shallow dorsal grooves and 3 shallow ventral grooves. Fl. May-Jul. Fr. Sep-Nov. Sparse forests, forest edges or thickets at 100-1000 m. Distributed in SW, S, SE, C, N, W and E China. Also in Korean Peninsula and Japan.

南方荚蒾

Viburnum fordiae Hance

落叶灌木或小乔木。叶下密被星状毛，上部散生棕红色腺点，被星状毛或丁字毛。复伞形式聚伞花序第一辐射枝5条；花无香气，生于第三和第四辐射枝上。果实成熟时红色，卵形；果核具1条背沟和2条腹沟。花期4-5月，果期10-11月。生海拔100-1000米的疏林中或灌丛。产中国西南、华南、东南、华中和华东。

Deciduous shrubs or small trees. Leaves abaxially densely stellate-pubescent, adaxially with dispersed red-brown glandular dots, stellate-pubescent or forklike pubescent. Inflorescences a compound umbellike cyme, 1st node of inflorescences usually with 5 rays; flowers on rays of 3rd and 4th orders, not fragrant. Fruits maturing red, ovoid; pyrenes with 1 dorsal groove and 2 ventral grooves. Fl. Apr-May. Fr. Oct-Nov. Sparse forests or thickets at 100-1000 m. Distributed in SW, S, SE, C and E China.

南方荚蒾 *Viburnum fordiae*

吕宋荚蒾 *Viburnum luzonicum*

吕宋荚蒾
Viburnum luzonicum Rolfe

灌木，被黄褐色簇状毛。叶纸质，卵形、卵状披针形至长圆形，边缘具深波状锯齿，上面具无柄透明圆形微小腺点。复伞状聚伞花序顶生，直径3-5厘米，总花梗极短或几无；花冠白色，辐状，裂片卵形。核果红色，卵球形。花期4-6月，果期8-12月。生海拔100-700米的疏林、灌丛或路边。产云南、广西、广东、台湾、福建、浙江和江西。马来西亚、印度尼西亚和菲律宾亦有。

Shrubs, yellowish-brown tufted hairy. Leaves chartaceous, ovate, ovate-lanceolate to oblong, margin deeply undulate-serrate, upper surfaces with minute sessile transparent orbicular glands. Compound umbellike cymes terminal, 3-5 cm diam, peduncles very short or nearly sessile; corolla white, rotate, lobes ovate. Drupes red, ovoid. Fl. Apr-Jun. Fr. Aug-Dec. Sparse forests, thickets or roadsides at 100-700 m. Distributed in Yunnan, Guangxi, Guangdong, Taiwan, Fujian, Zhejiang and Jiangxi. Also in Malaysia, Indonesia and the Philippines.

宜昌荚蒾
Viburnum erosum Thunb.

落叶灌木。叶纸质，狭卵形、椭圆形至倒卵形，边缘具波状小尖齿，下面密被绒毛。复伞状聚伞花序生于侧生短枝之顶，直径2-4厘米，花具长梗；花冠白色，辐状，裂片圆卵形；花药黄白色。核果红色，宽卵球形。花期4-5月，果期8-10月。生海拔300-2300米的山坡林下或灌丛中。产中国西南、华南、东南、华中、华西和华东。朝鲜半岛和日本亦有。

Deciduous shrubs. Leaves chartaceous, narrowly ovate, elliptic to obovate, margin undulate-serrulate, abaxially densely tomentose. Compound umbellike cymes terminal on the lateral short branchlets, 2-4 cm diam, flowers long pedicellate; corolla white, rotate, lobes ovate; anthers yellowish-white. Drupes red, broadly ovoid. Fl. Apr-May. Fr. Aug-Oct. Slope forests or thickets at 300-2300 m. Distributed in SW, S, SE, C, W and E China. Also in Korean Peninsula and Japan.

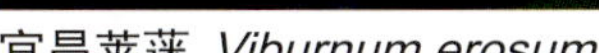

宜昌荚蒾 *Viburnum erosum*

甘肃荚蒾 *Viburnum kansuense*

甘肃荚蒾

Viburnum kansuense Batal.

落叶灌木。叶纸质，3中裂至3深裂，或侧裂片再2裂，掌状3-5出脉，边缘具不规则粗牙齿。复伞状聚伞花序顶生，无大型不孕花；萼筒紫红色；花冠淡红色，辐状；花药红褐色。核果红色，椭圆体形或近球形。花期6-7月，果期9-10月。生海拔2400-3600米的冷杉林或杂木林中。产云南、四川、西藏、陕西和甘肃。

Deciduous shrubs. Leaves chartaceous, 3-lobed to half to 3-parted, or lateral lobes 2-lobed again, palmately 3-5-veined, margin irregularly gross-dentate. Compound umbellike cymes terminal, without large sterile flowers; calyx tube purplish-red; corolla reddish, rotate; anthers reddish-brown. Drupes red, ellipsoid or subglobose. Fl. Jun-Jul. Fr. Sep-Oct. *Abies* forests or mixed forests at 2400-3600 m. Distributed in Yunnan, Sichuan, Xizang, Shaanxi and Gansu.

鸡树条

Viburnum opulus L. subsp. **calvescens** (Rehd.) Sugim.

落叶灌木，无毛。树皮厚，鸡冠状。叶圆卵形至阔卵形，通常3裂，掌状3出脉，边缘具粗牙齿，叶柄粗壮，具腺体。复伞状聚伞花序顶生，周围具5-10个大型不孕花；花冠白色，辐状；花药紫红色。核果红色，近球形。花期5-6月，果期9-10月。生海拔1000-2200米的林下或灌丛中。产华中、华东、华北、东北和西北。俄罗斯、蒙古、朝鲜半岛和日本亦有。

Deciduous shrubs, glabrous. Bark thick, corky. Leaves ovate to broadly ovate, usually 3-lobed, palmately 3-veined, margin gross-dentate, petioles stout, with glands. Compound umbellile cymes terminal, margin with 5-10 large sterile flowers; corolla white, rotate; anthers purplish-red. Drupes red, subglobose. Fl. May-Jun. Fr. Sep-Oct. Forests or thickets at 1000-2200 m. Distributed in C, E, N, NE and NW China. Also in Russia, Mongolia, Korean Peninsula and Japan.

穿心莛子藨

Triosteum himalayanum Wall.

多年生草本，密生长刚毛和腺毛。叶常9-10对，基部连合，连成一体而茎贯穿其中。聚伞花序2-5轮，排成穗状花序状；萼筒与萼裂片间缢缩；花冠黄绿色，筒内紫褐色，筒基部弯曲，一侧膨大成囊。核果红色，近球形。花期5-7月，果期7-9月。生海拔1800-4100米的山坡、山谷、溪边、针叶林下、林缘、灌丛或草地。产中国西南和华中。印度、尼泊尔和不丹亦有。

Perennial herbs, densely long hirsute and glandular hairy. Leaves usually 9-10 pairs, base connate, perforate. Cymes 2-5 whorls, arranged in spike at apex of stem;

鸡树条 *Viburnum opulus* subsp. *calvescens*

穿心莛子藨 *Triosteum himalayanum*

constricted between calyx tube and lobes; corolla yellowish-green, tube purplish-brown inside, base curved at base and shallowly spurred. Drupes red, subglobose. Fl. May-Jul. Fr. Jul-Sep. Mountain slopes, valleys, streamsides, coniferous forests, forest edges, thickets or grasslands at 1800-4100 m. Distributed in SW and C China. Also in India, Nepal and Bhutan.

莛子藨

Triosteum pinnatifidum Maxim.

多年生草本，被白色刚毛和腺毛。叶羽状深裂，裂片1-3对，近全缘。聚伞花序对生，具3花，近无梗，排成顶生短穗状花序；花冠黄绿色，狭钟状，筒基部弯曲，一侧膨大成浅囊，内面具紫色斑点。核果白色，卵球形。花期5-6月，果期8-9月。生海拔1800-2900米的林下、阳坡或溪边中。产华中、华北和西北。日本亦有。

Perennial herbs, hirsute and glandular hairy. Leaves pinnatiparted, lobes 1-3 pairs, margin nearly entire. Cymes opposite, 3-flowered, subsessile, arranged in terminal short spike; corolla yellowish-green, narrowly campanulate, tube curved at base and shallowly spurred, inside purple-brown with paler flecks. Drupes white, subglobose. Fl. May-Jun. Fr. Aug-Sep. Forests, sunny places or streamsides at 1800-2900 m. Distributed in C, N and NW China. Also in Japan.

腋花莛子藨

Triosteum sinuatum Maxim.

多年生草本，密被刚毛及腺毛。叶卵圆形或长圆形，基部抱茎至全贯穿。花序每轮2或3或6朵花生于茎顶；花冠棕紫色，具白斑。核果绿色，梨形，具宿存萼片，密被腺毛；分核3，具5或6条明显突出的棱。花期6月，果期9-10月。生海拔800-900米的林中或溪边。产新疆、吉林和辽宁。俄罗斯和日本亦有。

Perennial herbs, densely hirsute and glandular hairy. Leaves orbicular-ovate or oblong, base amplexicaul to perfoliate. Inflorescences of 2 or 3 or 6-flowered whorls at apex of stems; corolla purple-brown with paler flecks. Drupes greenish, pyriform with persistent calyx, densely glandular hairy; pyrenes 3, with 5 or 6 prominent ribs. Fl. Jun. Fr. Sep-Oct. Forests or streamsides at 800-900 m. Distributed in Xinjiang, Jilin and Liaoning. Also in Russia and Japan.

腋花莛子藨 *Triosteum sinuatum*

莛子藨 *Triosteum pinnatifidum*

七子花 *Heptacodium miconioides*

七子花

Heptacodium miconioides Rehd.

落叶灌木或小乔木。叶对生，全缘，近基部三出脉。聚伞花序每轮含7花，组成顶生圆锥花序；花无梗；总苞片大，宿存；萼筒5裂；花冠白色。果实长椭圆形，萼裂片增大。花期6-7月，果期9-11月。生海拔600-1000米的悬崖峭壁、山坡灌丛和林下。产湖北、浙江和安徽。

Deciduous shrubs or small trees. Leaves opposite, entire, trinerved near base. Each cyme with 7 flowers, formed in terminal panicles; flowers sessile; bracts large, persistent; calyx 5-lobed; corolla white. Fruits oblong, calyx lobes enlarged. Fl. Jun-Jul. Fr. Sep-Nov. Cliffs, thickets and forests at 600-1000 m. Distributed in Hubei, Zhejiang and Anhui.

毛核木

Symphoricarpos sinensis Rehd.

落叶灌木。小枝纤细。叶小，对生，全缘，薄，菱形至卵形。穗状花序顶生，短小；双花3-6，交互对生；花小，无梗；萼齿5枚，先端尖；花冠白色，钟形。核果卵圆形，蓝黑色，具白霜，顶端有1小喙；分核2枚，密生长柔毛。花期7-9月，果期9-11月。生海拔610-2300米的山坡灌木林中。产广西、云南、四川、湖北、甘肃和陕西。

Deciduous shrubs. Branchlets slender. Leaves small, opposite, entire, thin, rhombic to ovate. Spikes terminal, small; Paired flowers 3-6, decussate; flowers small, sessile; sepals 5, acute; corolla white, campanulate. Drupes ovoid, bluish black, pruinose, crowned with a short beak; pyrenes 2, densely pilose. Fl. Jul-Sep. Fr. Sep-Nov. Scrub at 610-2300 m. Distributed in Guangxi, Yunnan, Sichuan, Hubei, Gansu and Shaanxi.

北极花

Linnaea borealis L.

常绿匍匐小灌木，高5-10厘米。茎细长。叶对生，圆形至倒卵形，边缘具1-3对圆齿。花芳香，下垂；主花序梗长6-7厘米；萼片被短柔毛，狭尖；花冠钟状，白色至粉色。瘦果近球形，熟时黄色。花果期7-8月。生海拔700-2300米的林下或岩石上。产内蒙古、新疆、黑龙江、吉林和辽宁。北

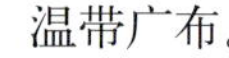

温带广布。

Semishrubs, creeping, evergreen, 5-10 cm tall. Stem long, slender. Leaves opposite, orbicular to obovate, margin with 1-3 pairs of rounded teeth. Flowers fragrant, nodding; main peduncles 6-7 cm long; sepals shortly pubescent, narrowly acute; corolla campanulate, white to pink. Achenes subglobose, yellow at maturity. Fl. and fr. Jul-Aug. Forests or rocks at 700-2300 m. Distributed in Neimenggu, Xinjiang, Heilongjiang, Jilin and Liaoning. Also widely in the north temperate zone.

猬实

Kolkwitzia amabilis Graebn.

直立灌木，多分枝，高达3米。老茎光滑，茎皮剥落。叶椭圆形至卵状椭圆形，全缘，两面被

毛核木 *Symphoricarpos sinensis*

猬实 *Kolkwitzia amabilis*

毛。伞房状聚伞花序；苞片披针形；萼筒被刚毛；花冠粉红色，内面具黄色斑纹。瘦果状核果密被黄色刺刚毛，顶端伸长如角。花期5-6月，果期8-9月。生海拔300-1300米的山坡、路边或灌丛中。产湖北、安徽、河南、山西、陕西和甘肃。

北极花 *Linnaea borealis*

Erect shrubs, strongly branched, to 3 m tall. Old branches glabrous, bark exfoliate. Leaves elliptic to ovate-elliptic, margin entire, both surfaces pubescent. Corymbose-cymes; bracts lanceolate; calyx setaceous; corolla pink, inside yellow striate. Achenial drupes densely with yellow setae, apex elongated as hornlike. Fl. May-Jun. Fr. Aug-Sep. Slopes, roadsides or thickets at 300-1300 m. Distributed in Hubei, Anhui, Henan, Shanxi, Shaanxi and Gansu.

糯米条

Abelia chinensis R. Br.

落叶灌木，多分枝，高达2米。叶卵圆形至椭圆状卵形，边缘有稀疏圆锯齿。聚伞花序生小枝上部叶腋；花芳香，具3对小苞片；花冠白色至红色，漏斗状，花丝细长，伸出花冠筒外。瘦果具宿存而略增大的萼裂片。花期8-9月，果期10-11月。生海拔200-1500米的山地。产中国西南、华南、华中和华东。

Deciduous shrubs, strongly branched, to 2 m tall. Leaves ovate to elliptic-ovate, margin with a few crenatures. Cymes axillary at upper part of branchlets; flowers fragrant, with 3 pairs of bracteoles; corolla white to red, infundibular, filaments slender, extend out of corolla tube. Achenes crowned with persistent and slightly increscent sepals. Fl. Aug-Sep. Fr. Oct-Nov. Mountains at 200-1500 m. Distributed in SW, S, C and E China.

糯米条 *Abelia chinensis*

二翅糯米条 *Abelia macrotera*

二翅糯米条

Abelia macrotera (Graebn. et Buchw.) Rehd.

落叶灌木。叶长渐尖，具不对称顶端。单花腋生，有时稍圆锥状；花冠白色至紫的粉红色，2唇形；上唇2裂；下唇3裂；花萼具锐尖；雄蕊2强。瘦果具柔毛，由2枚宿存萼片包被。花期4-6月，果期8-10月。生海拔200-2000米的灌丛或林中。产中国西南、华中、华

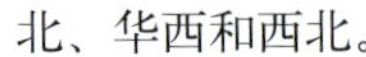

北、华西和西北。

Deciduous shrubs. Leaves long acuminate with asymmetrical apex. Flowers single and axillary, sometimes slightly paniculate; corolla white to purplish pink, bilabiate; upper lips 2-lobed; lower lips 3-lobed; sepals with acute apex; stamens didynamous. Achenes pubescent, crowned with 2 persistent sepals. Fl. Apr-Jun. Fr. Aug-Oct. Thickets or forests at 200-2000 m. Distributed in SW, C, N, W and NW China.

蓪梗花

Abelia uniflora R. Br.

落叶灌木，高1-2米。叶卵形至椭圆状卵形，边缘具疏锯齿。花大，1-2朵生于侧枝顶部叶腋，具3枚小苞片；花冠粉红色至浅紫红色，漏斗状。瘦果冠有2枚宿存而略增大的萼裂片。花期4-6月，果期8-10月。生海拔200-2000米的灌丛、林缘或路边。产中国西南、华中、华东和西北。

Deciduous shrubs, 1-2 m tall. Leaves ovate to elliptic-ovate, margin serrate. Flowers large, 1-2 axillary at the top of branches, with 3 bracteoles; corolla pink to purplish-red, infundibular. Achenes crowned with 2 persistent and slightly increscent sepals. Fl. Apr-Jun. Fr. Aug-Oct. Thickets, forest edges or roadsides at 200-2000 m. Distributed in SW, C, E and NW China.

六道木

Zabelia biflora (Turcz.) Makino

落叶灌木，高1-3米。枝具纵棱。叶长圆形至长圆状披针形，全缘或中部以上羽状浅裂。花单生小枝叶腋；花冠白色、淡黄色或带浅红色，小苞片宿存。瘦果革质，长圆形，具硬毛，冠有4枚宿存增大的萼裂片。花期4-6月，果期8-9月。生海拔1000-2000米的灌丛、林下或沟边。产河北、山西和辽宁。

Deciduous shrubs, 1-3 m tall. Branches with longitudinal ridges. Leaves oblong to oblong-lanceolate, margin entire or pinnately lobed at upper parts. Flowers axillary, solitary; corolla white, yellowish or tinged with pink,

六道木 *Zabelia biflora*

蓪梗花 *Abelia uniflora*

南方六道木 *Zabelia dielsii*

bracteoles persistent. Achenes coriaceous, oblong, strigose, crowned with 4 persistent and increscent sepals. Fl. Apr-Jun. Fr. Aug-Sep. Thickets, forests or ditch sides at 1000-2000 m. Distributed in Hebei, Shanxi and Liaoning.

南方六道木

Zabelia dielsii (Graebn.) Makino

落叶灌木，高2-3米。叶椭圆形、长圆状卵形至披针形，变化幅度很大。花2朵生侧枝顶部叶腋，具明显的总花梗；花冠白色至淡黄色，漏斗状，雄蕊4，内藏。瘦果长1-1.5厘米。花期4-6月，果期8-9月。生海拔800-3700米的灌丛、路边或林下。产中国西南、华南、华东、华北和西北。

Deciduous shrubs, 2-3 m tall. Leaves elliptic, oblong-ovate to lanceolate, very variable. Flowers 2 axillary at upper part of lateral branches, with distinct peduncles; corolla white to yellowish, infundibular, stamens 4, included. Achenes 1-1.5 cm long. Fl. Apr-Jun. Fr. Aug-Sep. Thickets, roadsides or forests at 800-3700 m. Distributed in SW, S, E, N and NW China.

双盾木

Dipelta floribunda Maxim.

落叶灌木或小乔木，高达6米。叶卵状披针形或卵形，全缘，下面灰白色。聚伞花序簇生于侧生短枝顶端叶腋；2对小苞片形状、大小不等，其中1对盾状，圆形至长圆形，宿存，增大，干膜质；萼齿条形；花冠粉红色。肉质核果具棱角。花期4-7月，果期8-9月。生海拔600-2200米的杂木林下或灌丛中。产四川、广西、湖南、湖北、甘肃和陕西。

Deciduous shrubs or small trees, to 6 m tall. Leaves ovate-lanceolate or ovate, margin entire, glaucous abaxially. Cymes fasciated at upper leaf axils of lateral short branchlets; bracteoles 2 pairs, different in shape and size, one pairs peltate, orbicular to oblong, persistent, increscent, dry membranous; calyx lobes linear; corolla pink. Fleshy drupes ridged. Fl. Apr-Jul. Fr. Aug-Sep. Shaws or thickets at 600-2200 m. Distributed in Sichuan, Guangxi, Hunan, Hubei, Gansu and Shaanxi.

双盾木 *Dipelta floribunda*

云南双盾木
Dipelta yunnanensis

云南双盾木

Dipelta yunnanensis Franch.

落叶灌木，高达4米。叶椭圆形至宽披针形。伞房状聚伞花序，有1-4花；小苞片2对，宿存，增大，1对较小，不相等，另1对较大，耳状肾形；萼檐5裂至中部，裂片披针形；花冠白色至粉红色，钟形，长2-4厘米。肉质核果卵球形，被柔毛。花期5-7月，果期7-11月。生海拔800-2400米的山坡疏林或灌丛。产云南、四川、贵州、湖北、陕西和甘肃。

Deciduous shrubs, to 4 m tall. Leaves elliptic to broad lanceolate. Corymbose-cymes, 1-4-flowered; bractlets 2 pairs, persistent, increscent, one pair smaller, unequal, another one pair larger, auriculate-reniform; calyx limb 5-lobed to medium, lobes lanceolate; corolla white to pink, 2-4 cm long. Fleshy drupes ovoid, pubescent. Fl. May-Jul. Fr. Jul-Nov. Sparse forests or thickets on slopes at 800-2400 m. Distributed in Yunnan, Sichuan, Guizhou, Hubei, Shaanxi and Gansu.

半边月 (水马桑)

Weigela japonica Thunb. var. **sinica** (Rehd.) Bailey

落叶灌木。叶长卵形至卵状椭圆形，边缘具锯齿。花单生或3朵成聚伞花序；萼筒长圆柱形；花冠白色或淡红色，后变红色，漏斗状钟形，长2.5-3.5厘米，裂片开展，近整齐。蒴果圆柱形，顶端具喙。种子多少带翅。花期4-5月，果期8-9月。生海拔400-1800米的林下、灌丛或沟谷。产中国西南、华南、华中和华东。朝鲜半岛和日本亦有。

半边月 (水马桑)
Weigela japonica var. *sinica*

Deciduous shrubs. Leaves narrowly ovate to ovate-elliptic, margin serrate. Flowers solitary or 3 arranged in cymes; calyx narrowly cylindric; corolla white or reddish, turned red later, infundibular-campanulate, 2.5-3.5 cm long, lobes spread, nearly equal. Capsules cylindric, apex rostrate. Seeds somewhat winged. Fl. Apr-May. Fr. Aug-Sep. Forests, thickets or valleys at 400-1800 m. Distributed in SW, S, C and E China. Also in Korean Peninsula and Japan.

锦带花

Weigela florida (Bunge) DC.

落叶灌木，被短柔毛。幼枝稍四方形。叶长圆

锦带花 *Weigela florida*

海仙花 *Weigela coraeensis*

形或椭圆形，边缘具锯齿。花单生或成聚伞花序；萼筒长圆柱形，萼齿披针形，不等长；花冠紫红色或玫瑰红色，裂片不整齐。蒴果圆柱形，先端具喙。种子无翅。花期4-6月，果期10月。生海拔100-1500米的林下或灌丛。产华北和东北。俄罗斯、朝鲜半岛和日本亦有。

Deciduous shrubs, shortly pubescent. Twigs somewhat tetragonous. Leaves oblong or elliptic, margin serrate. Flowers solitary or arranged in cymes; calyx narrowly cylindric, lobes lanceolate, unequal; corolla purplish-red or roseate-red, lobes unequal. Capsules cylindric, apex rostrate. Seeds wingless. Fl. Apr-Jun. Fr. Oct. Forests or thickets at 100-1500 m. Distributed in N and NE China. Also in Russia, Korean Peninsula and Japan.

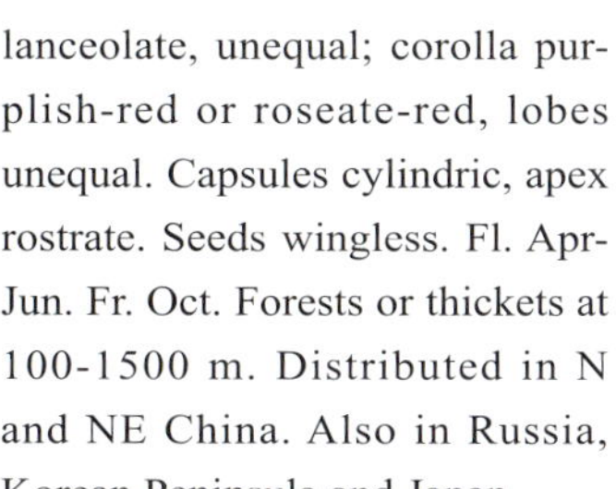

海仙花

Weigela coraeensis Thunb.

落叶灌木。小枝粗壮。叶对生，大而宽，阔椭圆形、近圆形至倒卵形，边缘具锯齿。花单生或3个排成聚伞状；花冠白色或淡红色，后变红色，漏斗状钟形，裂片开展，近整齐。花期5-6月。浙江、江苏、江西、山东、上海和北京有栽培。

Deciduous shrubs. Twigs stout and thick. Leaves opposite, large and broad, broadly elliptic, suborbicular to obovate, margin serrate. Flowers solitary or in threes in cymes; corolla white or reddish, turned red later, infundibular-campanulate, lobes spread, nearly equal. Fl. May-Jun. Cultivated in Zhejiang, Jiangsu, Jiangxi, Shandong, Shanghai and Beijing.

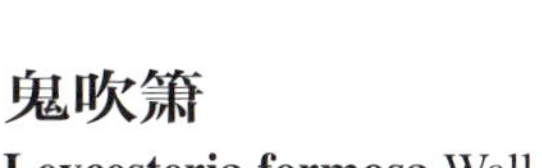

鬼吹箫

Leycesteria formosa Wall.

灌木。叶纸质，披针形、长圆形至卵形，基部圆形至阔楔形，全缘。穗状聚伞花序顶生或腋生，下垂，每节具6花；苞片绿色或紫色；萼筒长圆形，被腺毛；花冠白色或粉红色，漏斗状。浆果幼时红色，熟时黑紫色，卵球形或近球形。花期(5-)6-9(-10)月，果期(8-)9-10月。生海拔1100-3500米的山坡、山谷、溪沟边、河边的林下、林缘或灌丛中。产云南、四川、西藏和贵州。南亚和缅甸亦有。澳大利亚、欧洲、北美洲和太平洋岛屿广泛栽培及归化。

Shrubs. Leaves chartaceous, lanceolate, oblong to ovate, base round to broadly cuneate, margin enire. Spicate cymes terminal or axillary, pendulous, 6-flowered on each node; bracts green or purple; calyx-tube oblong, with glandular hairs; corolla white or pink, infundibular. Berries red at young, dark purple at maturity, ovoid or subglobose. Fl. (May-)Jun-Sep(-Oct). Fr. (Aug-)Sep-Oct. Forests, forest edges or thickets on slopes, in valleys, by streams or on river banks at 1100-3500 m. Distributed in Yunnan, Sichuan, Xizang and Guizhou. Also in S Asia and Myanmar. Widely cultivated and naturalized in Australia, Europe, North America and Pacific Islands.

鬼吹箫 *Leycesteria formosa*

绵毛鬼吹箫 *Leycesteria stipulata*

纤细鬼吹箫
Leycesteria gracilis (Kurz) Airy Shaw

灌木。叶厚纸质，长圆状披针形或长圆状卵形，具疏腺齿和疏缘毛。穗状花序顶生或腋生，每节具2花；苞片和小苞片披针形，短于萼筒；萼檐下部合生成浅杯状；花冠白色，漏斗状。浆果由红色变蓝紫色，长圆形或椭圆体形。花期(9-)10-11(-12)月，果期翌年(3-)4-5月。生海拔2000-3800米的山坡、山谷或溪沟边的林下或灌丛中。产云南和西藏。印度、不丹和缅甸亦有。

Shrubs. Leaves thickly chartaceous, oblong-lanceolate or oblong-ovate, sparsely glandular serrate and ciliate. Spikes terminal or axillary, 2-flowered on each node; bracts and bracteoles lanceolate, shorter than calyx-tube; calyx limb connate as a shallow cup; corolla white, infundibular. Berries red, became blue-purple, oblong or ellipsoid. Fl. (Sep-)Oct-Nov(-Dec). Fr. next (Mar-)Apr-May. Forests or thickets on slopes, by streams or valleys at 2000-3800 m. Distributed in Yunnan and Xizang. Also in India, Bhutan and Myanmar.

绵毛鬼吹箫
Leycesteria stipulata (Hook. f. et Thoms.) Fritsch

灌木。叶背被交织绵毛，两叶柄间具明显的半圆形托叶，托叶长6-8毫米，上面有短柔毛，下面被绵毛，宿存，或有时脱落。穗状花序近球形，腋生，被绵毛；花冠白色。花期3-4月，果期10-11月。生海拔1300-2000米的蕨叶草地或阔叶林中。产云南西北部。印度北部亦有。

Shrubs. Leaves abaxially covered with malted lanate hairs; stipules between two leaves conspicuously semiorcular, 6-8 mm long, adaxially puberulous, abaxially lanate, persistent or sometimes deciduous. Spikes subglobose, axillary, lanate; corolla white. Fl. Mar-Apr. Fr. Oct-Nov. Pteridium-grassland or in broad-leaved forests at 1300-2000 m. Distributed in NW Yunnan. Also in N India.

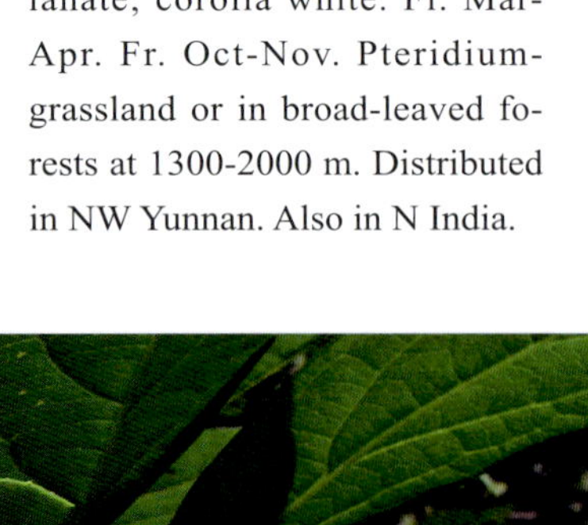

纤细鬼吹箫 *Leycesteria gracilis*

越桔叶忍冬(越橘忍冬)

Lonicera angustifolia Hook. f. et Thoms. var. **myrtillus** (Hook. f. et Thoms.) Q. E. Yang

落叶多枝灌木。叶纸质，卵形、长圆形、椭圆形或披针形。苞片叶状；小苞片合生成杯状，与子房近等长；相邻两萼筒合生至近顶部；花冠白色、淡紫色或紫红色，筒状钟形，裂片圆卵形或近圆形。浆果橘红色，近球形。花期5-7月，果期8-9月。生海拔2400-4500(-4700)米的山坡灌丛、溪旁疏林或河谷石滩地。产云南、四川和西藏。阿富汗至缅甸北部亦有。

Deciduous multi-branched shrubs. Leaves chartaceous, ovate, oblong, elliptic or lanceolate. Bracts foliar; bracteoles connate, cotyloid, subequal to ovary at length; two adjacent calyx tubes connate to nearly top; corolla white, purplish or purplish-red, tubular-campanulate, lobes orbicular-ovate or suborbicular. Berries orange red, subglobose. Fl. May-Jul. Fr. Aug-Sep. Slopes thickets, sparse forests by streams or stony flood plains at 2400-4500(-4700) m. Distributed in Yunnan, Sichuan and Xizang. Also in Afghanistan to N Myanmar.

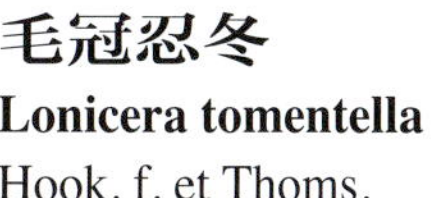

毛冠忍冬

Lonicera tomentella Hook. f. et Thoms.

越桔叶忍冬(越橘忍冬) *Lonicera angustifolia* var. *myrtillus*

落叶灌木。小枝密被短柔毛。叶卵状矩圆形至卵状披针形，边缘背卷。总花梗极短；苞片大，叶状；相邻两萼筒连合过半；萼裂片卵形，先端钝；花冠淡紫红色，筒状漏斗形，内有短伏毛。复果蓝黑色，圆形。花期6-8月，果期9月。生海拔2900-3000米的山坡林中或河边灌丛中。产云南西北部和西藏南部。不丹、印度、缅甸和尼泊尔亦有。

Deciduous shrubs. Branchlets densely lanate. Leaves ovate-oblong to ovate-lanceolate, margin revolute. Peduncles very short; bracts large, leaflike; two adjacent calyxes tubes connate to more than half; sepals ovate, apex pbtuse; corolla purplish-red, tubular-infundibular, pubescent inside. Compound berries bluish-black, globose. Fl. Jun-Aug. Fr. Sep. Forests or scrub at 2000-3200 m. Distributed in NW Yunnan and S Xizang. Also in Bhutan, India, Myanmar and Nepal.

毛冠忍冬 *Lonicera tomentella*

岩生忍冬 *Lonicera rupicola*

岩生忍冬

Lonicera rupicola Hook. f. et Thoms.

落叶灌木。小枝纤细，常呈针刺状。叶3枚轮生，纸质，条形、披针形至长圆形，基部两侧不等。花生于幼枝基部叶腋，芳香；苞片叶状，小苞片合生成杯状；相邻两萼筒分离；花冠淡紫色或紫红色，筒状钟形。浆果红色，椭圆体形。花期5-8月，果期8-10月。生海拔2000-5000米的高山灌丛、岩石缝隙、林缘、荒原、流石滩边缘或河滩草地。产云南、四川、西藏、甘肃、青海和宁夏。印度和尼泊尔亦有。

Deciduous shrubs. Twigs slender, thornlike. Leaves in whorl of 3, chartaceous, linear, lanceolate to oblong, base oblique. Flowers axillary at base of shoots, fragrant; bracts foliar, bracteoles connate, cotyloid; two adjacent calyx tubes separate; corolla purplish or purplish-red, tubular-campanulate. Berries red, ellipsoid. Fl. May-Aug. Fr. Aug-Oct. Alpine thickets, rock crevices, forest edges, deserts, scree margins or flood grasslands at 2000-5000 m. Distributed in Yunnan, Sichuan, Xizang, Gansu, Qinghai and Ningxia. Also in India and Nepal.

唐古特忍冬

Lonicera tangutica Maxim.

落叶灌木。叶纸质，长圆形、倒卵形至椭圆形。总花梗生嫩枝下部叶腋，纤细，下垂；小苞片分离或连合；相邻两萼筒合生至近顶部；花冠白色、黄白色或有淡红晕，筒状漏斗形，筒基部具浅囊；花药内藏。复果红色，球形。花期5-8

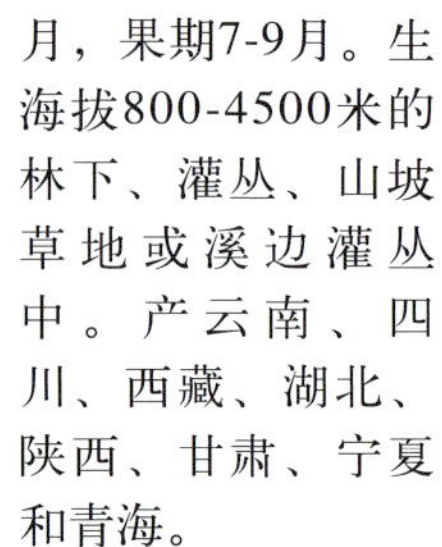

月，果期7-9月。生海拔800-4500米的林下、灌丛、山坡草地或溪边灌丛中。产云南、四川、西藏、湖北、陕西、甘肃、宁夏和青海。

Deciduous shrubs. Leaves chartaceous, oblong, obovate to elliptic. Peduncles axillary at lower parts of shoots, slender, pendulous; bracteoles separate or connate; adjacent calyx lobes connate to near top; corolla white, yellowish-white or tinged with reddish, tube with shallow sac at base; anthers included. Compound berries red, globose.

唐古特忍冬 *Lonicera tangutica*

Fl. May-Aug. Fr. Jul-Sep. Forests, thickets, grasslands on mountain slopes, scrubs at streamsides at 800-4500 m. Distributed in Yunnan, Sichuan, Xizang, Hubei, Shaanxi, Gansu, Ningxia and Qinghai.

理塘忍冬 *Lonicera litangensis*

理塘忍冬

Lonicera litangensis Batal.

落叶多枝矮灌木，全体无毛。小枝节间极度短缩。叶密集，近簇生，椭圆形至倒卵形。总花梗极短或几无；苞片大，叶状；相邻两萼筒全部连合；花冠黄色，筒状或狭漏斗状，筒基部具浅囊。复果红色，后变灰蓝色，球形。花期5-6月，果期8-9月。生海拔3000-4500米的山坡、灌丛、草地、林下或林缘。产云南、四川和西藏。印度、尼泊尔和不丹亦有。

Deciduous multi-branched dwarf shrubs, wholly glabrous. Twigs internodes extremely shortened. Leaves dense, nearly clustered, elliptic to obovate. Peduncles very short or nearly absent; bracts large, foliate; two adjacent calyx tubes wholly connate; corolla yellow, tubular or narrowly infundibular, tube with sac at base. Compound berries red and turned glaucous-blue, globose. Fl. May-Jun. Fr. Aug-Sep. Slopes, thickets, grasslands, forests or forest edges at 3000-4500 m. Distributed in Yunnan, Sichuan and Xizang. Also in India, Nepal and Bhutan.

小叶忍冬

Lonicera microphylla Willd. ex Roem. et Schult.

落叶灌木。叶下常灰白色，两面密柔毛至近无毛。双花腋生；相邻2胚珠完全合生或至少合生至中部；花冠唇形，白色至黄色，有时稍具粉色。浆果红色或橘黄色，球形。花期4-6(-7月)，果期7-8(-9)月。生海拔1100-3600(-4100)米的多石山坡、草地、灌丛、疏林或林缘。产中国西南、华南、华北、华西和西北。南亚、西南亚、中亚和东北亚亦有。

Deciduous shrubs. Leaves abaxially often gray-white, both surfaces densely puberulent to subglabrous. Inflorescences axillary, paired flowers; neighboring 2 ovaries fused completely or at least to middle; corolla labiate, white to yellow, sometimes tinged pink. Berries red or orange-yellow, globose. Fl. Apr-Jun(-Jul). Fr. Jul-Aug(-Sep). Rocky slopes, grasslands, thickets, sparse forests or forest edges at 1100-3600(-4100) m. Distributed in SW, S, N, W and NW China. Also in S, SW, C and NE Asia.

小叶忍冬 *Lonicera microphylla*

杈枝忍冬 *Lonicera simulatrix*

杈枝忍冬

Lonicera simulatrix Pojark.

落叶多分枝灌木。树冠圆球形。叶长圆状卵形或倒披针形，边缘具睫毛。花生嫩枝下部叶腋；总花梗长于叶，无毛；苞片线形，小苞片合生；花冠管状漏斗形，近整齐，淡黄色或白色，筒基部具囊。浆果黄色后变红色最后成黑色，球形。花期5-6月，果期7-8月。生海拔2400-3300米的阳坡灌丛。产新疆。伊朗和中亚亦有。

Deciduous multi-branched shrubs. Crowns globose. Leaves oblong-ovate or oblanceolate, margin ciliate. Flowers axillary at lower parts of shoots; peduncles longer than leaves, glabrous; bracts filiform, bracteoles connate; corolla tubular-infundibular, nearly regular, yellowish or white, tube with sac at base. Berries yellow, turned red later, and turned black finally, globose. Fl. May-Jun. Fr. Jul-Aug. Sunny thicket slopes at 2400-3300 m. Distributed in Xinjiang. Also in Iran and C Asia.

华西忍冬 *Lonicera webbiana*

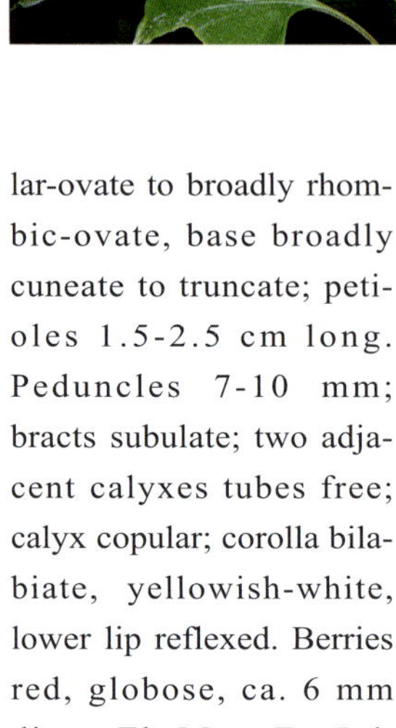

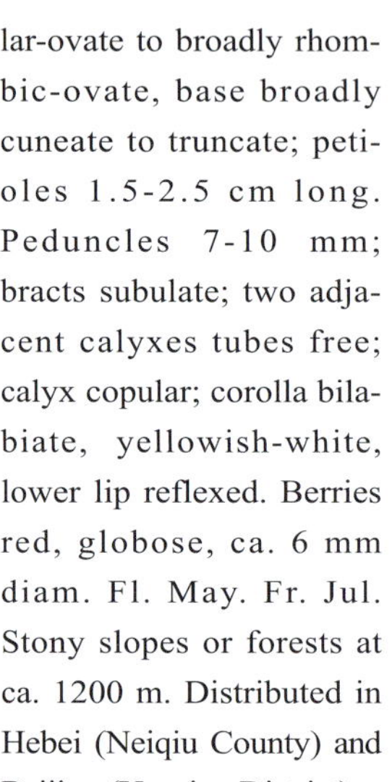

华西忍冬

Lonicera webbiana Wall.

落叶灌木。叶倒卵形或卵状椭圆形至卵状披针形。双花腋生；相邻2子房离生；花冠紫红色，外部疏被柔毛和腺毛或无毛；花冠管内部具柔毛，基部较细，具浅囊肿。浆果红色或黑色，球形。花期5-6月，果期8-9月。生海拔1800-4000米的针阔混交林中、灌丛或草坡。产中国西南、华中、华北、华西、华东和西北。印度、不丹和阿富汗亦有。

Deciduous shrubs. Leaves obovate or ovate-elliptic to ovate-lanceolate. Inflorescences axillary, paired flowers; neighboring 2 ovaries free; corolla purple-red, outside sparsely pubescent and glandular hairy or glabrous; tubes inside puberulent, deeply gibbous above slender base. Berries red or black, globose. Fl. May-Jun. Fr. Aug-Sep. Needle-leaved and broad-leaved mixed forests, thickets or grassy slopes at 1800-4000 m. Distributed in SW, C, N, W, E and NW China. Also in India, Bhutan and Afghanistan.

丁香叶忍冬

Lonicera oblata Hao ex Hsu et H. J. Wang

落叶灌木。叶厚纸质，三角状宽卵形至菱状宽卵形，基部宽楔形至截形；叶柄长1.5-2.5厘米。总花梗长7-10毫米；苞片钻形；相邻两萼筒分离；花萼杯状；花冠二唇形，黄白色，下唇反折。浆果红色，圆形，直径约6毫米。花期5月，果期7月。生海拔约1200米的多石山坡上或山坡林下。产河北(内丘县)和北京(延庆区)。

Deciduous shrubs. Leaves thickly chartaceous, broadly triangular-ovate to broadly rhombic-ovate, base broadly cuneate to truncate; petioles 1.5-2.5 cm long. Peduncles 7-10 mm; bracts subulate; two adjacent calyxes tubes free; calyx copular; corolla bilabiate, yellowish-white, lower lip reflexed. Berries red, globose, ca. 6 mm diam. Fl. May. Fr. Jul. Stony slopes or forests at ca. 1200 m. Distributed in Hebei (Neiqiu County) and Beijing (Yanqing District).

丁香叶忍冬 *Lonicera oblata*

黏毛忍冬

Lonicera fargesii Franch.

落叶灌木。小枝密被腺糙毛。叶倒卵状椭圆形至椭圆状矩圆形，边缘有睫毛，下面脉上密生伏毛及散生短腺毛。总花梗长3-5厘米；苞片叶状；相邻两萼筒全部合生；花冠二唇形，红色或白色，外被柔毛。浆果红色，卵圆形。花期5-6月，果期9-10月。生海拔1600-2900米的林中或灌丛中。产四川、甘肃、陕西、山西和河南。

Deciduous shrubs. Branchlets densely glandular-hairy. Leaves obovate-elliptic to elliptic-oblong, margin ciliate, lower surface densely appressed-hairy and sparsely glandular-hairy. Peduncles 3-5 cm long; bracts leaflike; two adjacent calyxes wholly connate; corolla bilabiate, red or white, outside pubescent. Berries red, ovoid. Fl. May-Jun. Fr. Sep-Oct. Forests or thickets at 1600-2900 m. Distributed in Sichuan, Gansu, Shaanxi, Shanxi and Henan.

黏毛忍冬 *Lonicera fargesii*

下江忍冬 *Lonicera modesta*

黑果忍冬 *Lonicera nigra*

下江忍冬
Lonicera modesta Rehd.

落叶灌木。叶两面具柔毛，无毛或仅于叶脉上疏被柔毛。双花腋生；相邻2个胚珠一半或全部合生；花冠白色，基部红色，后为黄色，外部疏被柔毛或近无毛；花冠管基部具浅囊肿，内部密被毛。浆果从橘红色，后转为红色。花期5-6月，果期7-10月。生海拔500-1700米的杂木林中或灌丛中。产中国东南、华中、华北、华西和华东。

Deciduous shrubs. Leaves pubescent, glabrous or sparsely pubescent only on veins on both sides. Inflorescences axillary, paired flowers; neighboring 2 ovaries half to completely fused; corolla white, base reddish, later yellow, outside sparsely pubescent or subglabrous; tubes shallowly gibbous at base, inside densely hairy. Berries turning from orange-red to red. Fl. May-Jun. Fr. Jul-Oct. Mixed forests or thickets at 500-1700 m. Distributed in SE, C, N, W and E China.

黑果忍冬
Lonicera nigra L.

落叶灌木。叶对生，薄纸质，长圆形、椭圆形、倒卵形或披针形。总花梗细，长1.5-3厘米；苞片小，披针形，小苞片合生成杯状；相邻两萼筒分离，萼齿宽披针形；花冠红色，二唇形，筒基部有囊肿。浆果蓝黑色，球形。花期4-7月，果期8-10月。生海拔1500-3900米的灌丛、针叶林下或林缘。产云南、四川、西藏、贵州、湖北、安徽和吉林(长白山)。印度、尼泊尔、不丹、朝鲜半岛和欧洲中部亦有。

Deciduous shrubs. Leaves opposite, thin chartaceous, oblong, elliptic, obovate or lanceolate. Peduncles slender, 1.5-3 cm long; bracts small, lanceolate, bracteoles connate, cotyloid; two adjacent calyx tubes separate, lobes broadly lanceolate; corolla red, bilabiate, tube with sac at base. Berries bluish-black, globose. Fl. Apr-Jul. Fr. Aug-Oct. Coniferous thickets, forests or forest edges at 1500-3900 m. Distributed in Yunnan, Sichuan, Xizang, Guizhou, Hubei, Anhui and Jilin (Changbai Mountain). Also in India, Nepal, Bhutan, Korean Peninsula and C Europe.

柳叶忍冬
Lonicera lanceolata Wall.

落叶灌木，各部分被短腺毛。叶卵形、卵状披针形或菱状矩圆形。总花梗长0.5-2.5厘米；苞片小；相邻两萼筒分离或下半部合生；花冠二唇形，淡紫色或紫红色，基部有囊，内面有柔毛。浆果黑色，圆形。花期6-7月，果期8-9月。生海拔2000-3900米的针阔混交林、冷杉林或林缘灌丛中。产云南、四川和西藏。尼泊尔和不丹亦有。

Deciduous shrubs, all parts shortly glandular-hairy. Leaves ovate-lanceolate or rhombic-oblong. Peduncles 0.5-2.5 cm long; bracts small; two adjacent calyxes free or connate at lower half; corolla bilabiate, purplish or purplish-red, with sac at base, inside pubescent. Berries black, globose. Fl. Jun-Jul. Fr. Aug-Sep. Mixed broadleaf-conifer forests, fir forests or margin scrub at 2000-3900 m. Distributed in Yunnan, Sichuan and Xizang. Also in Nepal and Bhutan.

柳叶忍冬 *Lonicera lanceolata*

红脉忍冬 *Lonicera nervosa*

红脉忍冬

Lonicera nervosa Maxim.

落叶灌木。幼枝与花梗常被柔毛和腺毛。叶长圆形至椭圆状披针形，除下面中脉上具白毛外无毛。双花腋生；花冠白色，后为黄色；花冠管基部具浅囊肿，外部具柔毛或无毛，内部具柔毛。浆果黑色，常具白粉，球形。花期6-7月，果期8-9月。生海拔2100-4000米的林下或草地灌丛。产中国西南、华北、华西和西北。

Deciduous shrubs. Young branches and peduncles often puberulent and glandular hairy. Leaves oblong to elliptic-lanceolate, glabrous throughout but abaxially often white hairy on midvein. Inflorescences axillary, paired flowers; corolla white becoming yellow; tubes shallowly gibbous at base, outside puberulent or glabrous, inside puberulent. Berries black, often pruinose, globose. Fl. Jun-Jul. Fr. Aug-Sep. Thickets of forests understories or grasslands at 2100-4000 m. Distributed in SW, N, W and NW China.

华北忍冬

Lonicera tatarinowii Maxim.

落叶灌木。叶矩圆状披针形或矩圆形，上面无毛。总花梗纤细，长1-2厘米；苞片三角状披针形；相邻两萼筒合生至中部以上；花冠二唇形，黑紫色，筒长为唇瓣的1/2，基部一侧稍肿大，内面有柔毛。复果红色，近圆形，直径5-6毫米。花期5-7月，果期8-9月。生海拔400-1750米的山坡杂木林或灌丛中。产山东、河北和辽宁。

Deciduous shrubs. Leaves oblong-lanceolate or oblong, upper surface glabrous. Peduncles slender, 1-2 cm long; bracts triangular-lanceolate; two adjacent calyxes connate to more than half; corolla bilabiate, dark purple, tube as long as half of the lower lip, slightly swollen at one side near base, inside pubescent. Compound berries red, subglobose, 5-6 mm diam. Fl. May-Jul. Fr. Aug-Sep. Mixed forests or scrub on slopes at 400-1750 m. Distributed in Shandong, Hebei and Liaoning.

华北忍冬 *Lonicera tatarinowii*

紫花忍冬

Lonicera maximowiczii (Rupr.) Regel

落叶灌木。叶两面疏被糙伏毛或无毛。双花腋生；相邻2个子房超过一半完全合生；花冠紫红色，外部无毛；花冠管基部稍具囊肿，内部密被毛。浆果红色，卵状球形。花期6-7月，果期8-9月。生海拔800-1800米的林中或林缘。产华北、华东和东北。俄罗斯、朝鲜半岛和日本亦有。

Deciduous shrubs. Leaves sparsely strigose or glabrous on both sides. Inflorescences axillary, paired flowers; neighboring 2

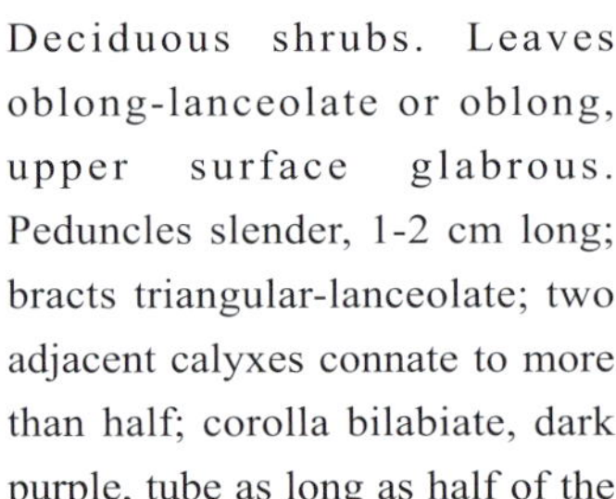

紫花忍冬 *Lonicera maximowiczii*

蕊被忍冬 *Lonicera gynochlamydea*

女贞叶忍冬 *Lonicera ligustrina*

ovaries more than half to completely fused; corolla purple-red, outside glabrous; tubes slightly gibbous toward base, inside densely hairy. Berries red, ovoid-orbicular. Fl. Jun-Jul. Fr. Aug-Sep. Forests or forest edges at 800-1800 m. Distributed in N, E and NE China. Also in Russia, Korean Peninsula and Japan.

蕊被忍冬

Lonicera gynochlamydea Hemsl.

落叶灌木。小枝具实心髓。叶卵形至披针形，下面于中脉基部密具白色长柔毛。花序腋生，双生；花冠白色、粉色或紫红色，内外均具短毛；花冠管基部具明显囊肿。浆果紫红色或白色。花期5月，果期8-9月。生海拔1200-1900(-3000)米的灌丛或林中。产中国西南、华中和华西。

Deciduous shrubs. Branches with solid pith. Leaves ovate to lanceolate, abaxially often densely white villous at base of midvein. Inflorescences axillary, paired flowers; corolla white, pink or purple-red, both sides with short hairs; tubes deeply gibbous at base; cupules not accrescent. Berries purple-red or white. Fl. May. Fr. Aug-Sep. Thickets or forests at 1200-1900(-3000) m. Distributed in SW, C and W China.

女贞叶忍冬

Lonicera ligustrina Wall.

灌木。小枝具向上的硬曲毛。叶卵形至披针形，下部有时稍具黑色腺点，中脉密具硬毛。双花腋生；花冠黄绿色至白色，漏斗形；花冠管基部具浅囊肿，内部具长柔毛，外部具腺点。果实紫色或红色，稍透明，球形。花期4-7月，果期8-12月。生海拔(600-)1000-2000(-3000)米的灌丛或林中。产中国西南、华南、华中和华西。印度、尼泊尔和不丹亦有。

Shrubs. Young branches with stiff and upwardly curved hairs. Leaves ovate to lanceolate, abaxially sometimes with minute black glands, densely stiffly hairy on midvein. Inflorescences axillary, paired flowers; corolla greenish yellow to white, funnelform; tubes shallowly gibbous at base, inside villous, outside glandular. Fruits purple or red, ± translucent, globose. Fl. Apr-Jul. Fr. Aug-Dec. Thickets or forests at (600-)1000-2000(-3000) m. Distributed in SW, S, C and W China. Also in India, Nepal and Bhutan.

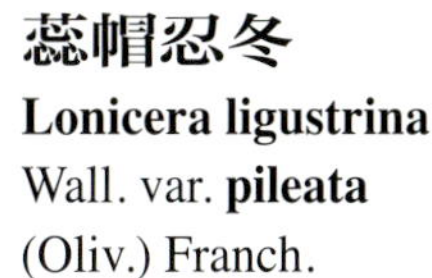

蕊帽忍冬

Lonicera ligustrina Wall. var. **pileata** (Oliv.) Franch.

灌木。叶卵形至披针形，中脉下面凸起，无毛或疏被。双花腋生；花萼杯状，包被小苞片顶端；花冠黄绿色至白色；花冠管基部具浅囊肿，内部具长柔毛，外部具腺点。浆果紫色或白色，稍透明，球形。花期4-7月，果期8-12月。生海拔(300-)600-1700(-2200)米的沙质山坡、疏林中湿润处或山坡灌丛。产中国西南、华南、华中和华西。

Shrubs. Leaves ovate to lanceolate, with midvein raised adaxially, glabrous or sparsely hairy. Inflorescences axillary, paired flowers; calyx cupular, forming a collarlike, overlapping apex of bracteoles; corolla greenish yellow to white; tubes shallowly gibbous at base, inside villous, outside glandular. Berries purple or white, ± translucent, globose. Fl. Apr-Jul. Fr. Aug-Dec. Sandy slopes, moist places in sparse forests or thickets on mountain slopes at (300-)600-1700(-2200) m. Distributed in SW, S, C and W China.

蕊帽忍冬 *Lonicera ligustrina* var. *pileata*

蓝果忍冬 *Lonicera caerulea*

蓝果忍冬

Lonicera caerulea L.

落叶灌木。叶卵形至长圆形或倒卵形。花序腋生，具双花；苞片条形，长为子房的2-3倍；花冠管状漏斗形；裂片整齐。托杯紧包被2浆果，浆果蓝黑色，具白粉。花期5-6月，果期8-9月。生海拔2600-3500米的落叶林或灌丛。产中国西南、华北、华西、西北和东北。东北亚、欧洲和北美洲亦有。

Deciduous shrubs. Leaves ovate to oblong or obovate. Inflorescences axillary, paired flowers; bracts linear, 2-3 × as long as ovaries; corolla tubular-funnelform; lobes regular. Cupules tightly fused around 2 berries, berries blue-black, pruinose. Fl. May-Jun. Fr. Aug-Sep. Deciduous forests or scrubs at 2600-3500 m. Distributed in SW, N, W, NW and NE China. Also in NE Asia, Europe and North America.

葱皮忍冬(秦岭忍冬) *Lonicera ferdinandii*

葱皮忍冬(秦岭忍冬)

Lonicera ferdinandii Franch.

落叶灌木。幼枝被刚毛。叶卵形至披针形，边缘有睫毛，下面常被刚伏毛，有时具腺点。花序顶生；苞片大，叶状；小苞片合生成坛状壳斗，完全包被相邻两萼筒；花冠二唇形，白色，后变淡黄色。壳斗疏松包围浆果；浆果红色，卵圆形。花期4-7月，果期8-10月。生海拔(200-)1000-2700米的向阳山坡林中或林缘灌丛中。产四川、西北、华北和辽宁。朝鲜半岛亦有。

Deciduous shrubs. Branchlets with stiff hairs. Leaves ovate to lanceolate, margin ciliate, abaxially usually hirsute, sometimes with minute glands. Inflorescences terminal; bracts large, leaflike; bracteoles fused into an urceolate cupule, completely surrounding two adjacent calyxes; corolla bilabiate, white, later yellowish. Cupule loosely enclosing paired berries; berries red, ovoid. Fl. Apr-Jul. Fr. Aug-Oct. Forests or margin scrub on sunny slopes at(200-)1000-2700 m. Distributed in Sichuan, NW and N China, and Liaoning. Also in Korean Peninsula.

微毛忍冬

Lonicera cyanocarpa Franch.

落叶灌木，各部密被微糙毛。叶近革质，通常矩圆形，有时椭圆形，两端近圆形。总花梗粗壮；苞片宽卵形，长1-1.5厘米；相邻两萼筒分离，无毛；花冠黄绿色，漏斗状。浆果蓝黑色，卵圆形。花期6-8月，果期10-11月。生海拔3500-4300米的石灰岩山脊、山坡林缘灌丛或多石草原。产四川、云南和西藏。

Deciduous shrubs, all parts densely covered with minute setae. Leaves subcoriaceous, usually oblong, sometimes elliptic,

微毛忍冬 *Lonicera cyanocarpa*

刚毛忍冬 *Lonicera hispida*

both ends subround. Peduncles thick; bracts broadly ovate, 1-1.5 cm long; two adjacent calyxes free, glabrous; corolla greenish-yellow, infundibular. Berries bluish-black, ovoid. Fl. Jun-Aug. Fr. Oct-Nov. Limestone ridges, margin scrub on slopes or rocky grasslands at 3500-4300 m. Distributed in Sichuan, Yunnan and Xizang.

刚毛忍冬

Lonicera hispida Pall. ex Roem. et Schult.

落叶灌木，被刚毛。叶厚纸质，卵状椭圆形至长圆形，基部微心形，边缘具刚睫毛。花序下垂，具双花；相邻两萼筒分离，具刺刚毛和腺毛；花冠白色或淡黄色，漏斗状，基部具囊。浆果橘黄色后变红色，卵球形至长圆筒形。花期5-6月，果期7-9月。生海拔1700-4200(-4800)米的林中、林缘或高山草地上。产中国西南、华北和西北。印度、土耳其、中亚和蒙古亦有。

Deciduous shrubs, hispid. Leaves thick chartaceous, ovate-elliptic to oblong, base subcordate, margin hispid-ciliate. Inflorescences pendulous, biflorate; two adjacent calyx tubes separate, hispid and glandular hairy; corolla white or yellowish, infundibular, base with sac. Berries orange-yellow and turned to red later, ovoid to narrowly cylindric. Fl. May-Jun. Fr. Jul-Sep. Forests, forest edges or alpine grasslands at 1700-4200(-4800) m. Distributed in SW, N and NW China. Also in India, Turkey, C Asia and Mongolia.

矮生忍冬

Lonicera minuta Batal.

落叶矮灌木，高达30厘米，多分枝，各部被微糙毛。小枝落叶后针刺状。叶条形或条状倒披针形，长5-12毫米，边缘背卷。花序腋生，几无总花梗；苞片叶状，条状披针形；相邻两萼筒分离；花冠筒状漏斗形，淡紫红色，芳香，内面有短柔毛。浆果卵圆形或近圆形。花期5-6月，果期9-10月。生海拔3200-3800米的山麓溪流旁石隙中或沙丘。产甘肃西部和青海东部。

Dwarf and deciduous shrubs, to 30 cm tall, multi-branched, all parts covered with minute setae. Twigs thorn-like after leaf fall. Leaves linear or linear-oblanceolate, 5-12 mm long, margin revolute. Inflorescences axillary, subsessile; bracts leaflike, linear-lanceolate; two adjacent calyxes free; corolla tubular-infundibular, purpurish-red, fragrant, inside shortly pubescent. Fl. May-Jun. Fr. Sep-Oct. Scree by mountain stream or sand dunes at 3200-3800 m. Distributed in W Gansu and E Qinghai.

矮生忍冬 *Lonicera minuta*

棘枝忍冬 *Lonicera spinosa*

棘枝忍冬

Lonicera spinosa Jacq. ex Walp.

落叶矮灌木，高达60厘米，常具坚硬、刺状、无叶的小枝。叶条形至条状矩圆形，长4-12毫米，边缘背卷。花生于短枝上叶腋，总花梗极短；苞片叶状，长于萼齿；相邻两萼筒分离；花冠筒状漏斗形，初时淡紫红色，后变白色；柱头伸出筒外。浆果椭圆形。花期6-7月，果期9-10月。生海拔3700-4600米的冰碛灌丛或石砾堆上。产西藏。克什米尔地区亦有。

Dwarf and deciduous shrubs, to 60 cm tall, usually with old twigs rigid, thornlike and afoliate. Leaves linear to linear-oblong, 4-12 mm long, margin revolute. Flowers browned at axils of short branchlets; peduncles very short; bracts leaflike, longer than calyx lobes; two adjacent calyxes free; corolla tubular-infundibular, purpurish-red, become white later; style exserted. Berries ellipsoid. Fl. Jun-Jul. Fr. Sep-Oct. Moraine scrubs or scree at 3700-4600 m. Distributed in Xizang. Also in Kashmir.

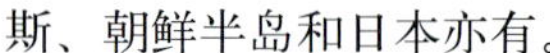

冠果忍冬 *Lonicera stephanocarpa*

冠果忍冬

Lonicera stephanocarpa Franch.

落叶灌木，各部常具倒生刚毛。叶厚纸质，披针形至矩圆形，两面均密被刚伏毛，边缘有刚睫毛。总花梗粗壮；苞片极大，宽卵形，长3-4厘米，下半部连合；相邻两萼筒分离，卵圆形，萼檐宽大，果时可增大至5毫米；花冠黄白色，宽漏斗形。浆果黑褐色，椭圆形。花期7-8月，果期9-10月。生海拔2000-3200米的山地林中或多石灌丛。产四川、甘肃、陕西和宁夏。

Deciduous shrubs, all parts often covered with reverted setae. Leaves thick chartaceous, lanceolate to oblong, both surfaces densely covered with appressed setae, margin setose-ciliate. Peduncles thick; bracts large, broadly ovate, 3-4 cm long, connate at lower half part; two adjacent calyxes free, ovoid, limb broad, enlarged to 5 mm at fruit; corolla yellowish-white, broadly infundibular. Berries blackish-brown, ellipsoid. Fl. Jul-Aug. Fr. Sep-Oct. Mountain forests or rocky scrub at 2000-3200 m. Distributed in Sichuan, Gansu, Shaanxi and Ningxia.

早花忍冬

Lonicera praeflorens Batal.

落叶灌木。叶阔卵形，两面密具平伏柔毛。花序为腋生成对花，生新枝基部，花开放较早；花冠黄白色至粉红色，近整齐，漏斗形，深裂；花药深粉色。浆果红色，球形。花期4月，果期5-6月。生海拔200-600米的林中或灌丛。产黑龙江、吉林和辽宁。俄罗斯、朝鲜半岛和日本亦有。

Deciduous shrubs. Leaves broadly ovate, both surfaces densely adpressed pubescent. Inflorescences axillary paired flowers and at base of new shoots, flowers opening early in season; corolla yellowish white to pink, subregular, funnelform, deeply lobed; anthers dark pink. Berries red, globose. Fl. Apr. Fr. May-Jun. Forests or scrubs at 200-600 m. Distributed in Heilongjiang, Jilin and Liaoning. Also in Russia, Korean Peninsula and Japan.

早花忍冬 *Lonicera praeflorens*

北京忍冬 *Lonicera elisae*

北京忍冬
Lonicera elisae Franch.

落叶灌木。叶纸质，椭圆形、披针形或长圆形，两面被短硬伏毛。苞片宽卵形至披针形；相邻两萼筒分离，萼齿钝，不整齐；花冠白色或带粉红色，长漏斗状，筒细长，裂片卵形或卵状长圆形。浆果红色，椭圆体形。花期4-5月，果期5-6月。生海拔500-1600(-2300)米的沟谷、林下或灌丛中。产华中、华东、华北和西北。

Deciduous shrubs. Leaves chartaceous, elliptic, lanceolate or oblong, both surfaces shortly strigillose. Bracts broadly ovate to lanceolate; two adjacent calyx tubes separate, calyx lobes obtuse, irregular; corolla white or tinged with pink, long infundibular, tube slender, lobes ovate to ovate-oblong. Berries red, elliptic. Fl. Apr-May. Fr. May-Jun. Valleys, forests or thickets at 500-1600(-2300) m. Distributed in C, E, N and NW China.

单花忍冬
Lonicera subhispida Nakai

落叶灌木。叶长圆形、卵形至近圆形；总花梗腋生，长1-2.5厘米。双花之一因退化而不存在；苞片1枚，披针形；萼筒无毛，萼檐极短，口缘截形或浅波状；花冠黄色，漏斗状，筒基部有囊，裂片整齐。浆果红色，纺锤形或椭圆体形。花期5月，果期6月。生海拔约800米的林中。产吉林(临江县)。俄罗斯和朝鲜半岛亦有。

Deciduous shrubs. Leaves oblong, ovate to suborbicular; peduncles axillary, 1-2.5 cm long. 1 flower of bifoliate inflorbescences degenerate and absent; bracts only 1, lanceolate; calyx tube glabrous, calyx limb very short, margin truncate or shallowly undulate; corolla yellow, infundibular, tube with sac at base, lobes regular. Berries red, fusiform or ellipsoid. Fl. May. Fr. Jun. Forests at ca. 800 m. Distributed in Jilin (Linjiang County). Also in Russia and Korean Peninsula.

单花忍冬 *Lonicera subhispida*

郁香忍冬
Lonicera fragrantissima Lindl. et Paxt.

半常绿或落叶灌木。叶厚纸质或革质，椭圆形、卵形至长圆形。花芳香；总花梗长2-10毫米；苞片披针形至近条形；相邻两萼筒合生至中部；花冠白色或淡红色，二唇形，基部具浅囊。复果下部连合，鲜红色，长圆形。花期1-4月，果期4-6月。生海拔100-2700米的山坡灌丛中。产浙江、湖北、河南、河北、安徽和江西。

Semievergreen or deciduous shrubs. Leaves thickly chartaceous or coriaceous, elliptic, ovate to oblong. Flowers fragrant; peduncles 2-10 mm long; bracts lanceolate to sublinear; two adjacent calyx tubes connate to middle part; corolla white or reddish, bilabilate, base with shallow sac. Compound berries connate at lower part, vivid red, oblong. Fl. Jan-Apr. Fr. Apr-Jun. Slope thickets at 100-2700 m. Distributed in Zhejiang, Hubei, Henan, Hebei, Anhui and Jiangxi.

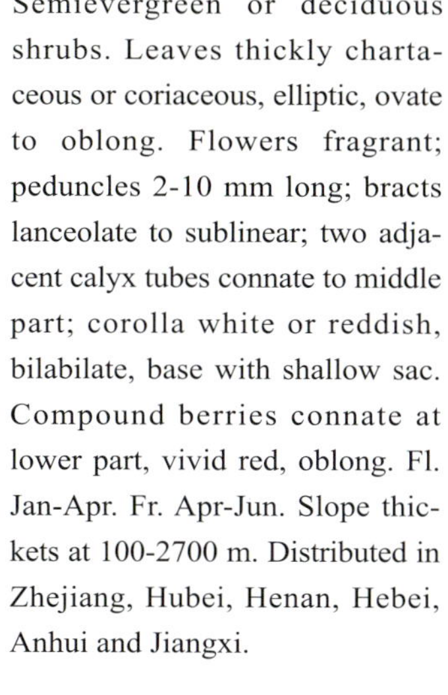

郁香忍冬 *Lonicera fragrantissima*

苦糖果 *Lonicera fragrantissima* subsp. *standishii*

苦糖果

Lonicera fragrantissima Lindl. et Paxt. subsp. **standishii** (Carr.) Hsu et H. J. Wang

落叶灌木。小枝和叶柄有时具短糙毛。叶卵状矩圆形或卵状披针形，两面被刚伏毛。花柱下部疏生糙毛。花期1-4月，果期5-6月。生海拔100-2000米的山坡灌丛中。产华西、华中、华东和华北。

Deciduous shrubs. Branchlets and petioles sometimes shortly strigillose. Leaves ovate-oblong or ovate-lanceolate, both surfaces adpressed strigillose. Lower part of styles sparsely strigillose. Fl. Jan-Apr. Fr. May-Jun. Slope thickets at 100-2000 m. Distributed in W, C, E and N China.

新疆忍冬

Lonicera tatarica L.

落叶灌木，全体近无毛。叶纸质，卵形或卵状长圆形，边缘具短糙毛。总花梗纤细；苞片披针形，小苞片分离，近圆形至卵状长圆形；相邻两萼筒分离；花冠粉红色或白色，二唇形，基部具浅囊。果实红色，球形，双果之一常不发育。花期5-6月，果期6-9月。生海拔700-1600米的石质山坡、沟谷、林缘或灌丛。产新疆北部。俄罗斯亦有。

长白忍冬 *Lonicera ruprechtiana*

新疆忍冬 *Lonicera tatarica*

Deciduous shrubs, whole plant almost glabrous. Leaves chartaceous, ovate or ovate-oblong, margin shortly strigose. Peduncles slender; bracts lanceolate, bracteoles separate, suborbicular to ovate-oblong; two adjacent calyx lobes separate; corolla pink or white, bilabiate, tube with sac at base. Berries red, globose, one of the twin berries usually undeveloped. Fl. May-Jun. Fr. Jun-Sep. Stony slopes, valleys, forest edges or thickets at 700-1600 m. Distributed in N Xinjiang. Also in Russia.

长白忍冬

Lonicera ruprechtiana Regel

落叶灌木。叶纸质，长圆状倒卵形或长圆状披针形，边缘略波状，有时具不规则浅波状大牙齿，具缘毛。苞片条形，小苞片分离，卵形；相邻两萼筒分离；花冠白色，后变黄色，二唇形，筒基部具1深囊。浆果橘红色，球形。花期5-6月，果期7-8月。生海拔300-1100米的阔叶林下或林缘。产黑龙江、吉林和辽宁。俄罗斯和朝鲜半岛亦有。

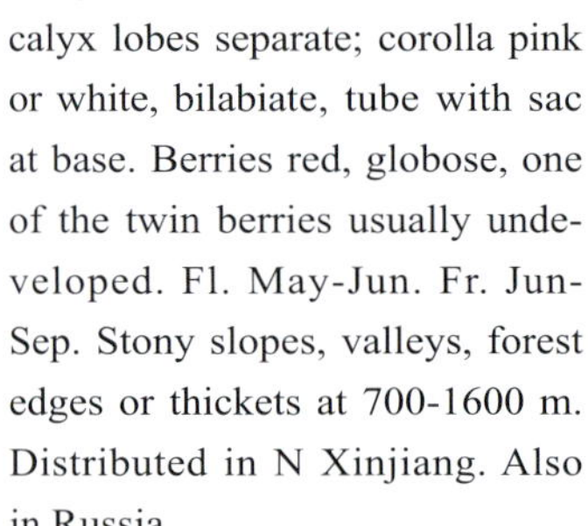

Deciduous shrubs. Leaves chartaceous, oblong-obovate or oblong-lanceolate, margin somewhat undulate, sometimes irregularly and shallowly gross-dentate, ciliate. Bracts linear, bracteoles separate, ovate; two adjacent calyx lobes separate; corolla white and turned yellow later, bilabiate, tube with deep sac at base. Berries orange red, globose. Fl. May-Jun. Fr. Jul-Aug. Broadleaf forests or forest edges at 300-1100 m. Distributed in Heilongjiang, Jilin and Liaoning. Also in Russia and Korean Peninsula.

金花忍冬 *Lonicera chrysantha*

金花忍冬

Lonicera chrysantha Turcz.

落叶灌木，被糙毛。叶纸质，菱形、倒卵形或披针形。总花梗细，长1.5-4厘米；苞片条形，小苞片分离，卵形；相邻两萼筒分离，萼齿卵形或半圆形；花冠白色后变黄色，二唇形。浆果红色，球形。花期5-6月，果期7-9月。生海拔200-3000(-3800)米的沟谷、林下或灌丛中。产华中、华北、东北和西北。俄罗斯和朝鲜半岛亦有。

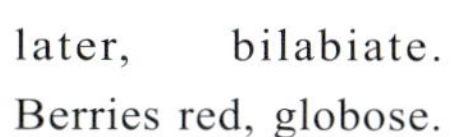

Deciduous shrubs, strigose. Leaves chartaceous, rhombic, obovate or lanceolate. Peduncles slender, 1.5-4 cm long; bracts linear, bracteoles separate, ovate; two adjacent calyx tubes separate, calyx lobes ovate or semiorbicular; corolla white and turned to yellow later, bilabiate. Berries red, globose. Fl. May-Jun. Fr. Jul-Sep. Valleys, forests or thickets at 200-3000 (-3800) m. Distributed in C, N, NE and NW China. Also in Russia and Korean Peninsula.

金银忍冬

Lonicera maackii (Rupr.) Maxim.

落叶灌木，被短柔毛和微腺毛。叶卵形、椭圆形至披针形。花芳香，生幼枝叶腋，总花梗短于叶柄；苞片条形，小苞片合生成对；相邻两萼筒分离，萼齿三角形，不相等；花冠先白色，后变黄色，二唇形。浆果球形，暗红色。花期5-6月，果期8-10月。生海拔100-1800(-3000)米的林中、林缘或溪边。产中国大部分地区；中国各地常见栽培。俄罗斯、朝鲜半岛和日本亦有。

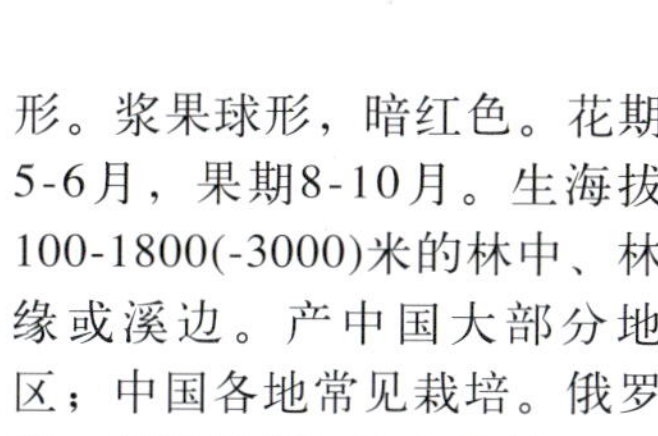

Deciduous shrubs, shortly pubescent and slightly glandular hairy. Leaves ovate, elliptic to lanceolate. Flowers fragrant, axillary at young twigs, peduncles shorter than petioles; bracts linear, bracteoles connate in pairs; two adjacent calyx tubes separate, lobes triangular, unequal; corolla white and turned yellow later, bilabiate. Berries globose, dark red. Fl. May-Jun. Fr. Aug-Oct. Forests, forest edges or streamsides at 100-1800(-3000) m. Distributed in most parts of China; commonly cultivated in China. Also in Russia, Korean Peninsula and Japan.

金银忍冬 *Lonicera maackii*

毛花忍冬 *Lonicera trichosantha*

毛花忍冬

Lonicera trichosantha Bur. et Franch.

落叶灌木。叶常倒卵形、卵形、长圆形或椭圆形，先端钝且具凸尖。花序为成对腋生的花；花冠二唇形，黄色，外面具糙毛和腺毛；冠筒内部密具柔毛。浆果由橘黄色变为橙红色和红色，球形。花期5-7月，果期8-9月。生海拔2400-4600米的林中、林缘或灌丛。产青藏高原和秦岭地区。

Deciduous shrubs. Leaves usually obovate, ovate, oblong or elliptic, apex obtuse and mucronate. Inflorescences axillary paired erect flowers; corolla bilabiate, yellow, outside strigose and glandular hairy; tubes inside densely puberulent. Berries turning from orange-yellow to yellow-red and red, globose. Fl.

May-Jul. Fr. Aug-Sep. Forests, forest edges or scrubs at 2400-4600 m. Distributed in Tibetan Plateau and Qinling Mountain.

长叶毛花忍冬

Lonicera trichosantha Bur. et Franch. var. **deflexicalyx** (Batal.) P. S. Hsu et H. J. Wang

本变种与毛花忍冬的区别在于本变种的叶常卵状披针形至披针形，先端渐尖。生海拔2400-4600米的沟谷、林下、林缘、灌丛或阳坡草地。产云南、四川、西藏、陕西和甘肃。

This variety differs from the typical variety in its leaves usually ovate-lanceolate to lanceolate, apex acuminate. Valleys, forests, forest edges, thickets or sunny grasslands at 2400-4600 m. Distributed in Yunnan, Sichuan, Xizang, Shaanxi and Gansu.

长距忍冬

Lonicera calcarata Hemsl.

木质藤本，长达5米以上，全体无毛。叶革质，卵形至宽披针形，先端尖，基部近圆形或阔楔形。总花梗直而扁；苞片2，宿存，叶状，长2-2.5厘米；相邻两萼筒合生；花冠二唇形，先黄白色后变橙红色，长约3厘米，基部有一长约12毫米的弯距。复果红色，扁卵圆形。花期4-5月，果期6-7月。生海拔1200-2500米的林下、林缘或溪沟旁灌丛。产广西、贵州、

四川、云南和西藏。

Woody vines, to 5 m or more at length, wholly glabrous. Leaves coriaceous, ovate to broadly lanceolate, apex acute, base subround or broadly cuneate. Peduncles erect, somewhat compressed; bracts 2, persistent, leaf-like, 2-2.5 cm long; two adjacent calyxes connate; corolla bilabiate, yellowish-white at first, then turned orange-red later, ca. 3 cm long, base with a curved spur ca. 12 mm long. Compound berries red, compressed ovoid. Fl. Apr-May. Fr. Jun-Jul. Forests, forest margin or scrub by streams at 1200-2500 m. Distributed in Guangxi, Guizhou, Sichuan, Yunnan and Xizang.

长叶毛花忍冬 *Lonicera trichosantha* var. *deflexicalyx*

长距忍冬 *Lonicera calcarata*

匍匐忍冬 *Lonicera crassifolia*

匍匐忍冬
Lonicera crassifolia Batal.

常绿灌木，具匍匐枝，有时亦生根。枝实心，具金色曲毛。叶宽椭圆形，革质，两面无毛，中脉上面具粗毛。腋生双花，生于短枝顶；花冠白色，后为黄色；花冠管红色。浆果黑色，球形。花期6-7月，果期10-11月。生海拔900-1700(-2300)米的溪边或湿润林缘石崖或石缝中。产云南、四川、贵州、湖北和湖南。

Evergreen shrubs, with long prostrate branches sometimes rooting. Branches solid, with crisped golden hairs. Leaves broadly elliptic, coriaceous, both surfaces glabrous but adaxially hirsute on midvein. Inflorescences axillary paired flowers at apices of short branches; corolla white, becoming yellow with reddish tubes. Berries black, globose. Fl. Jun-Jul. Fr. Oct-Nov. Streamsides, rocky cliffs or crevices of moist forest edges at 900-1700(-2300) m. Distributed in Yunnan, Sichuan, Guizhou, Hubei and Hunan.

忍冬(金银花)
Lonicera japonica Thunb.

半常绿木质藤本，被黄褐色糙毛、腺毛和短柔毛。叶卵形至长圆状卵形，具糙缘毛。花序腋生，具双花；苞片大，叶状，卵形至椭圆形；相邻两萼筒分离，萼齿三角形；花冠白色后变黄色，二唇形。浆果球形，熟时蓝黑色，有光泽。花期4-6月，果期10-11月。生海拔约1500米的山坡、灌丛、疏林、乱石堆或路旁。野生或栽培于中国大部分地区。朝鲜半岛和日本亦有。北美洲归化。

Semievergreen woody vines, yellowish-brown strigose, glandular hairy and puberulent. Leaves ovate to oblong-ovate, margin strigose-ciliate. Inflorescences axillary, biflorate; bracts large, foliate, ovate to elliptic; two adjacent calyx tubes separate, calyx lobes triangular; corolla white and turned to yellow later, bilabiate. Berries globose, bluish-black at maturity, lustrous. Fl. Apr-Jun. Fr. Oct-Nov. Slopes, thickets, sparse forests, random ripraps or roadsides at ca. 1500 m. Wild or cultivated in most parts of China. Also in Korean Peninsula and Japan. Naturalized in North America.

红白忍冬
Lonicera japonica Thunb. var. **chinensis** (P. Watson) Baker

半常绿木质藤本。幼枝紫黑色。幼叶带紫红色。小苞片比萼筒狭；花冠外面紫红色，内面白色，上唇裂片较长，4深裂超过唇瓣之半。浆果球形，熟时蓝黑色，有光泽。花期4-6月，果期10-11月。生海拔800米以下的山坡、村旁或庭园。原产安微，栽培于云南、浙江、江苏和江西。

Semievergreen woody vines. Young branchlets purplish-black. Young leaves tinged with purplish-red. Bracteoles narrower than calyx-tubes; corolla purplish-red outside, white inside, upper labium longer, 4-parted more than half of its length. Berries globose, bluish-black at maturity, lustrous. Fl. Apr-Jun. Fr. Oct-Nov. Slopes, villages nearby or gardens below 800 m. Native to Anhui, and cultivated in Yunnan, Zhejiang, Jiangsu and Jiangxi.

忍冬(金银花) *Lonicera japonica*

红白忍冬 *Lonicera japonica* var. *chinensis*

淡红忍冬 *Lonicera acuminata*

淡红忍冬

Lonicera acuminata Wall.

半常绿藤本。小枝常变为中空。叶对生，薄革质至革质，长圆形至披针形，两面至少于中脉具棕黄色柔毛，后变无毛。双花腋生于小枝顶部，有时呈圆锥状；相邻胚珠离生；花冠淡红色。浆果蓝黑色，被白粉，卵形。花期5-7月，果期10-11月。生海拔100-3200米的林中或灌丛。产中国西南、华南、东南、华中、华西和华东。南亚和东南亚亦有。

Climbers, semievergreen. Branches usually becoming hollow. Leaves opposite, thinly coriaceous to coriaceous, oblong to lanceolate, both surfaces brown-yellow hirsute at least on midvein or glabrescent. Flowers paired, axillary at apices of branchlets, sometimes paniculate; neighboring ovaries free; corolla white tinged red. Berries blue-black and pruinose, ovoid. Fl. May-Jul. Fr. Oct-Nov. Forests or thickets at 100-3200 m. Distributed in SW, S, SE, C, W and E China. Also in S and SE Asia.

锈毛忍冬

Lonicera ferruginea Rehd.

木质藤本，密生黄褐色糙毛。叶厚纸质，长圆状卵形或卵状长圆形。花序腋生，具双花，1-3对组成小总状花序，4-5个小花序再组成小圆锥花序；相邻两萼筒分离；花冠先白色后变黄色。浆果卵球形，黑色。花期5-7月，果期8-10月。生海拔600-2000米的林下或灌丛中。产云南、四川、贵州、广西、广东、福建和江西。

Woody vines, densely yellowish-brown strigose. Leaves thickly chartaceous, oblong-ovate or ovate-oblong. Inflorescences axillary, biflorate, 1-3 pairs arranged in small raceme, 4-5 small racemes arranged in small panicle again; two adjacent calyx tubes separate; corolla white and turned to yellow later. Berries ovoid, black. Fl. May-Jul. Fr. Aug-Oct. Forests or thickets at 600-2000 m. Distributed in Yunnan, Sichuan, Guizhou, Guangxi, Guangdong, Fujian and Jiangxi.

菰腺忍冬

Lonicera hypoglauca Miq.

落叶木质藤本，被淡黄褐色短柔毛。叶卵形至卵状长圆形，下面具腺体，腺体蘑菇形，黄色至橘红色。花序腋生，具双花；苞片条状披针形，小苞片卵形；花冠白色，有时有淡红晕，后变黄色，二唇形。浆果熟时黑色，近球形。花期4-5(-6)月，果期10-11月。生海拔200-700(-1800)米的灌丛或疏林中。产中国西南、华南、华中和华东。日本亦有。

Deciduous woody vines, shortly yellowish-brown pubescent. Leaves ovate to ovate-oblong, abaxially glandular punctate, glandular organ mushroom-shaped, yellow to orange-red. Inflorescences axillary, biflorate; bracts linear-lanceolate, bracteoles ovate; corolla white, sometimes tinged with reddish, turned to yellow later, bilabiate. Berries black at maturity, subglobose. Fl. Apr-May(-Jun). Fr. Oct-Nov. Thickets or sparse forests at 200-700(-1800) m. Distributed in SW, S, C and E China. Also in Japan.

锈毛忍冬 *Lonicera ferruginea*

菰腺忍冬 *Lonicera hypoglauca*

大花忍冬（灰毡毛忍冬）*Lonicera macrantha*

大花忍冬(灰毡毛忍冬)

Lonicera macrantha (D. Don) Spreng.

藤本。叶卵形或卵状披针形，下面被灰白色糙伏毛，杂生深橘黄色腺毛。花芳香，在小枝顶端成对着生形成总状花序；花冠2裂，白色，后为黄色，外部具糙伏毛及腺毛；内部密被柔毛，基部不凸出。浆果黑色，球形。花期4-5月，果期7-8月。生海拔300-1800米的山坡、溪边沟谷或杂木林中。产中国西南、华南、东南、华中和华东。印度、尼泊尔、不丹和缅甸亦有。

Climbers. Leaves ovate or ovate-lanceolate, abaxially pale white strigose, mixed with short dark orange glandular hairs. Flowers fragrant, paired and in racemes toward apex of branchlets; corolla bilabiate, white, later yellow, outside with strigose and glandular hairs; inside densely puberulent, not gibbous at base. Berries black, globose. Fl. Apr-May. Fr. Jul-Aug. Mountain slopes, valley by rivesides or mixed forests at 300-1800 m. Distributed in SW, S, SE, C and E China. Also in India, Nepal, Bhutan and Myanmar.

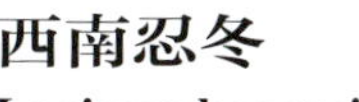

西南忍冬

Lonicera bournei Hemsl.

木质藤本。幼枝、叶柄和总花梗均密被黄色短柔毛。叶薄革质，卵形、椭圆形至披针形。花序具双花，腋生，密集，排成短总状花序；花有香味；苞片披针形，小苞片极小；花冠白色，后变黄色，二唇形。浆果红色。花期2-4月，果期5月。生海拔800-2000米的林中。产云南和广西。缅甸和老挝亦有。

Woody vines. Shoots, petioles and peduncles all densely and shortly yellow pubescent. Leaves thinly coriaceous, ovate, elliptic to lanceolate. Inflorescences biflorate, axillary, close, arranged in short raceme; flowers fragrant; bracts lanceolate, bracteoles minute; corolla white and turned to yellow later, bilabiate. Berries red. Fl. Feb-Apr. Fr. May. Forests at 800-2000 m. Distributed in Yunnan and Guangxi. Also in Myanmar and Laos.

皱叶忍冬

Lonicera reticulata Champ. ex Benth.

常绿木质藤本，植株几乎都为黄褐色或黄白色毡毛所覆盖。叶革质，卵形、椭圆形至长圆形，上面叶脉显著凹陷而呈皱纹状。双花腋生，组成小伞房花序，或在枝端组成圆锥花序；花冠白色后变黄色，二唇形。浆果蓝黑色，椭圆体形。花期6-7月，果期10-11月。生海拔400-1100米的山地灌丛或林中。产广西、广东、福建、湖南和江西。

Evergreen woody vines, whole plant almost covered by yellowish-brown or yellowish-white tomentose hairs. Leaves coriaceous, ovate, elliptic to oblong, adaxial veins conspicuously depressed, furrowed. Biflorate inflorescences axillary, arranged in small corymb, or arranged in panicle on branch apex; corolla white and turned yellow later, bilabiate. Berries bluish-black, ellipsoid. Fl. Jun-Jul. Fr. Oct-Nov. Montane thickets or forests at 400-1100 m. Distributed in Guangxi, Guangdong, Fujian, Hunan and Jiangxi.

西南忍冬 *Lonicera bournei*

皱叶忍冬 *Lonicera reticulata*

细毡毛忍冬 *Lonicera similis*

贯月忍冬 *Lonicera sempervirens*

细毡毛忍冬

Lonicera similis Hemsl.

落叶木质藤本。叶对生，纸质，卵形、长圆形至披针形，顶端尖，基部圆形或截形，下面被由细短柔毛组成的灰白色或灰黄色细毡毛。双花腋生；总花梗长可达4厘米；苞片披针形，小苞片极小，卵形至圆形；萼筒椭圆形至长圆形；花冠白色后变淡黄色，二唇形。浆果蓝黑色，卵球形。花期5-6(-7)月，果期9-10月。生海拔400-1600(-2200)米的山谷、溪旁、阳坡、灌丛或林中。产中国西南、华中、华东和西北。缅甸亦有。

Deciduous woody shrubs. Leaves opposite, chartaceous, ovate, oblong to lanceolate, apex acute, base rotundate or truncate, abaxially finely gray-white or grayish-yellow pannose by covered with fine and short pubescent hairs. Biflorate inflorescences axillary; peduncles up to 4 cm long; bracts lanceolate, bracteoles minute, ovate to orbicular; calyx tube ellipsoid to oblong; corolla white and turned yellowish, bilabiate. Berries bluish-black, ovoid. Fl. May-Jun(-Jul). Fr. Sep-Oct. Valleys, streamsides, sunny slopes, thickets or forests at 400-1600(-2200) m. Distributed in SW, C, E and NW China. Also in Myanmar.

贯月忍冬

Lonicera sempervirens L.

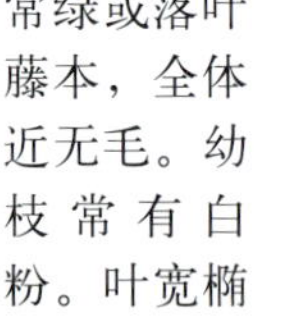

常绿或落叶藤本，全体近无毛。幼枝常有白粉。叶宽椭圆形至矩圆形，顶端钝或圆，基部楔形，下面粉白色，小枝顶端的1-2对叶基部相连成盘状。穗状花序顶生；花轮生，每轮6朵；花冠细长漏斗形，橘红色，长3.5-5厘米，筒细，裂片很短。浆果红色，近球形。花果期4-8月。中国各大城市常有栽植。原产北美洲。

Evergreen or deciduous vines, wholly subglabrous. Branchlets usually glaucous. Leaves broadly elliptic to oblong, apex obtuse or round, base cuneate, abaxially glaucous, 1-2 pair leaves on distal branchlet usually connate, discoid. Inflorescences terminal, spikelike; flowers verticillate, 6 in each whorl; corolla slender infundibular, orange-red, 3.5-5 cm long, tube slender, lobes very short. Berries red, subglobose. Fl. and fr. Apr-Aug. Usually cultivated in many cities of China. Native to North America.

川黔忍冬 *Lonicera subaequalis*

川黔忍冬

Lonicera subaequalis Rehd.

木质藤本，全体无毛。叶椭圆形至矩圆形，顶端钝，基部渐狭，下面通常有白粉，小枝顶端的1对叶合生成盘状。花6朵轮生于小枝顶，无总花梗；小苞片近圆形；花冠黄色，漏斗状，长2.5-3.5厘米。果实红色，近圆形。花期5-6月。生海拔1500-2450米的山坡林下阴湿处。产贵州、四川和云南。

Woody vines, wholly glabrous.

盘叶忍冬 *Lonicera tragophylla*

Leaves elliptic to oblong, apex obtuse, base attenuate, abaxially often glabrous, one pair leaves on distal branchlet connate, discoid. Inflorescences terminal, peduncle absent; flowers verticillate, 6 in a whorl; bracteoles suborbicular; corolla yellow, infundibular, 2.5-3.5 cm long. Berries red, subglobose. Fl. May-Jun. Damp places under slope forests at 1500-2450 m. Distributed in Guizhou, Sichuan and Yunnan.

盘叶忍冬

Lonicera tragophylla Hemsl. ex Forb. et Hemsl.

落叶藤本。叶椭圆形至披针形或卵形，具短硬毛，基部楔形，下延至叶柄。花序生小枝顶端，6朵轮生；每轮2-4朵簇生成头状；花冠黄色至橘黄色；花冠管基部不具囊肿。浆果黄色，成熟时渐变黄红色，近球形。花期5-7月，果期7-10月。生海拔（700-)1000-2000(-3000)米的林中、灌丛或河边石缝。产中国西南、东南、华中、华北、华西和华东。

Deciduous climbers. Leaves elliptic to lanceolate or ovate, shortly white hispid, base attenuate and decurrent to petioles. Inflorescences at apices of branchlets; flowers verticillate, 6 per whorl; whorls 2-4 clustered into a capitulum; corolla yellow to orange-yellow; tubes not gibbous at base. Berries turning from yellow to red-yellow when mature, subglobose. Fl. May-Jul. Fr. Jul-Oct. Forests, thickets or rocky crevices at riversides at (700-)1000-2000(-3000) m. Distributed in SW, SE, C, N, W and E China.

五福花科 Adoxaceae

五福花

Adoxa moschatellina L.

多年生矮小草本。茎高8-15厘米，无毛。基生叶1-3枚；小叶宽卵形或圆形，3裂；茎生叶2枚，对生，3深裂，裂片再3裂。5-9朵花组成聚伞性的紧密头状花序；花黄绿色；子房半下位至下位，4或5室。花期4-7月，果期7-8月。生海拔4000米以下的林中、肥沃土壤、林缘或草地。产中国西南、华北、西北和东北。南亚、东北亚、非洲西北部、欧洲和北美洲亦有。

Perennial dwarf herbs. Stems 8-15 cm tall, glabrous. Basal leaves 1-3; leaflets broadly ovate or orbicular, 3-cleft; cauline leaves 2, opposite, 3-cleft, segments 3-lobed. Inflorescences compact headlike cymes of 5-9 flowers; flowers yellowish green; ovary semi-inferior to inferior, 4- or 5-loculed. Fl. Apr-Jul. Fr. Jul-Aug. Forests, Rich soils, forest edges or meadows below 4000 m. Distributed in SW, N, NW and NE China. Also in S and NE Asia, NW Africa, Europe and North America.

五福花 *Adoxa moschatellina*

败酱科 Valerianaceae

败酱

Patrinia scabiosifolia Fisch. ex Trevir.

多年生草本。根茎和根具陈腐臭味。基生叶丛生，茎生叶对生。聚伞花序多数，组成顶生大型伞房花序；花序梗仅上方一侧被开展白色粗糙毛；苞片条形，甚小；花萼小，萼齿不明显；花冠钟形，黄色。连萼瘦果长圆形，无翅状苞片。花期7-9月，果期9-10月。生海拔50-4000米的林下、林缘、灌丛、路边、田埂或草丛。产中国大部分地区。俄罗斯、蒙古、朝鲜半岛和日本亦有。

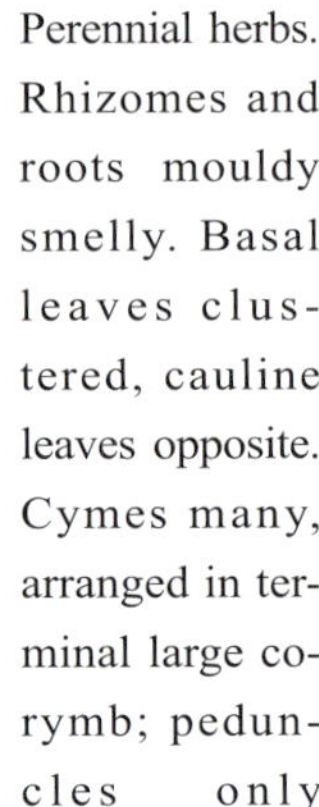

Perennial herbs. Rhizomes and roots mouldy smelly. Basal leaves clustered, cauline leaves opposite. Cymes many, arranged in terminal large corymb; peduncles only spreadly white strigose on upper side; bracts linear, minute; calyx small, lobes unconspicuous; corolla campanulate, yellow. Cypselae oblong, without winged bracts. Fl. Jul-Sep. Fr. Sep-Oct. Forests, forest edges, thickets, roadsides, field ridges or grasslands at 50-4000 m. Distributed in most parts of China. Also in Russia, Mongolia, Korean Peninsula and Japan.

墓头回 *Patrinia heterophylla*

中败酱

Patrinia intermedia (Horn.) Roem. et Schult.

多年生草本。茎粗壮，多分枝。基部叶莲座状；茎生叶2-4对，羽状全裂或二回羽状全裂。花序伞房状或聚伞状；花冠黄色，钟形。瘦果长圆形；翅状果苞具3脉。花期6-8月，果期7-9月。生海拔1000-3000米的林缘、草地、草原或灌丛。产新疆(阿尔泰山、帕米尔、天山)。哈萨克斯坦、吉尔吉斯斯坦、俄罗斯(西伯利亚)和蒙古亦有。

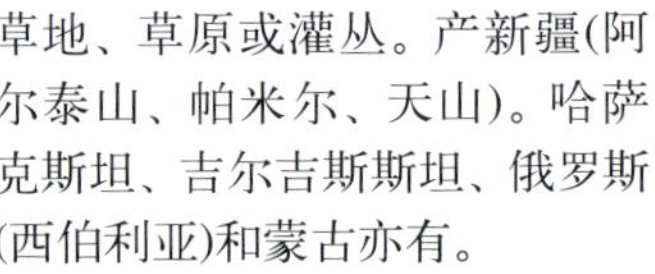

Perennial herbs. Caudex stout, multiple branched. Basal leaves rosulate; cauline leaves in 2-4 pairs, pinnatisect or bipinnatisect. Inflorescence corymbiform or paniculiform; corolla yellow, campanulate. Achenes oblong; bracteoles 3-veined. Fl. Jun-Aug. Fr. Jul-Sep. Forest edges, meadows, steppes or thickets at 1000-3000 m. Distributed in Xinjiang (Altay Mountain, Pamir, Tianshan Mountain). Also in Kazakhstan, Kyrgyzstan, Russia (Siberia) and Mongolia.

墓头回

Patrinia heterophylla Bunge

多年生草本。基生叶莲座状，羽状深裂至全裂；裂片1-5对；下部叶羽状全裂，具2-6对裂片；中上部叶常不裂，具1或2对裂片。花序伞房状；花冠钟形，黄色；雄蕊4。瘦果长圆形或倒卵形；翅状果苞具2(或3)脉。花期7-9月，果期8-10月。

败酱 *Patrinia scabiosifolia*

中败酱 *Patrinia intermedia*

岩败酱 *Patrinia rupestris*

生海拔100-2600米的草坡、开阔林中或路边。产中国西南、东南、华中、华北、华西、华东和西北。

Perennial herbs. Basal leaves rosulate, pinnatifid to pinnatisect; segments 1-5 pairs; lower leaves pinnatisect, with 2-6 pairs of segments; middle and upper leaves often undivided, or with 1 or 2 pairs of segments. Inflorescence corymbiform; corolla campanulate, yellow; stamens 4. Achenes oblong or obovoid; bracteoles 2(or 3)-veined. Fl. Jul-Sep. Fr. Aug-Oct. Grassy slopes, open woods or roadsides at 100-2600 m. Distributed in SW, SE, C, N, W, E and NW China.

岩败酱

Patrinia rupestris (Pall.) Dufr.

多年生草本。基生叶羽状浅裂、深裂至全裂或不分裂；茎生叶羽状深裂至全裂，侧裂片常3-6对。花序伞房状；花冠黄色，漏斗状钟形。瘦果倒卵球状筒形，具硬毛至渐无毛。花期7 9月，果期8-10月。生海拔200-2500米的石质山坡、草甸、桦木或杨属林缘。产华北、华西和东北。俄罗斯(远东地区西伯利亚)和蒙古亦有。

Perennial herbs. Basal leaves pinnatilobate, pinnatifid to pinnatisect or undivided; cauline leaves pinnatifid to pinnatisect, lateral segments usually in 3-6 pairs. Inflorescences corymbiform; corolla yellow, funnelform-campanulate. Achenes obovoid-columnar, hispidulous to glabrescent. Fl. Jul-Sep. Fr. Aug-Oct. Stony slopes, meadows, margins of *Betula* or *Populus* forests at 200-2500 m. Distributed in N, W and NE China. Also in Russia (Far East, Siberia) and Mongolia.

糙叶败酱 *Patrinia rupestris* subsp. *scabra*

糙叶败酱

Patrinia rupestris Pallas subsp. **scabra** (Bunge) H. J. Wang.

多年生草本。块根圆柱状、肉质。基生叶倒披针形、花期枯萎；茎生叶卵状披针形、粗糙、羽状半裂至羽状全裂、具1-5对侧裂片。花冠黄色、漏斗形；雄蕊4。瘦果圆柱状。花期7-8月，果期8-9月。生海拔300-1700米的向阳草坡或林缘。产华北、华西和东北。

Perennial herbs. Taproot columnar, fleshy. Basal leaves oblanceolate, wilted at anthesis; cauline leaves ovate-lanceolate, scaberulose, pinnatifid to pinnatisect, with 1-5 pairs of lateral segments. Corolla yellow, funnelform; stamens 4. Achenes columnar. Fl. Jul-Aug. Fr. Aug-Sep. Sunny grassy slopes or forest edges at 300-1700 m. Distributed in N, W and NE China.

少蕊败酱 *Patrinia monandra*

少蕊败酱

Patrinia monandra C. B. Clarke

二年生或多年生草本。基生叶花期枯萎。叶对生，长圆形，不裂或大头羽状深裂，边缘具粗圆齿或钝齿。花序伞房状或圆锥状；花冠黄色或灰黄色，钟形。瘦果卵球形；翅状果苞宽卵形至近圆形，具2(或3)脉，顶部常极浅3裂。花期8-9月，果期9-10月。生海拔100-3100米的草坡、灌丛、林中或路边。产中国西南、华南、东南、华中、华北、华西、华东和东北。印度北部、尼泊尔和不丹亦有。

Perennial biennial or herbs. Basal leaves wilted at anthesis. Leaves opposite, oblong, unlobed or lyrate-pinnatiparted, margin gross-crenate or obtuse-serrate. Inflorescence corymbiform or paniculiform; corolla yellow or pale yellow, funnelform. Achenes ovoid; bracteoles broadly ovate to suborbicular, 2(or 3)-veined, apex usually shallowly 3-lobed. Fl. Aug-Sep. Fr. Sep-Oct. Grassy slopes, thickets, forests or roadsides at 100-3100 m. Distributed in SW, S, SE, C, N, W, E and NE China. Also in N India, Nepal and Bhutan.

攀倒甑

Patrinia villosa (Thunb.) Dufr.

多年生草本。根茎和根具陈腐臭味。基生叶丛生，茎生叶对生，卵形或披针形，边缘具粗齿。聚伞花序多数，组成顶生圆锥花序或伞房花序；苞叶披针形或条形；花冠钟形，白色。瘦果倒卵球形，与宿存增大苞片贴生；果苞卵形至椭圆形。花期8-10月，果期9-11月。生海拔100-2000米的林下、林缘、路边、灌丛或草丛。产中国西南、华南、华中和华东。日本亦有。

Perennial herbs. Rhizomes and roots mouldy smelly. Basal leaves clustered, cauline leaves opposite, ovate or lanceolate, margin gross-serrate. Cymes many, arranged in terminal panicle or corymb; bract leaves lanceolate or linear; corolla campanulate, white. Cypselae obovoid, adnate with persistent accrescent bracts; fruit bracts ovate to elliptic. Fl. Aug-Oct. Fr. Sep-Nov. Forests, forest edges, roadsides, thickets or grasslands at 100-2000 m. Distributed in SW, S, C and E China. Also in Japan.

蜘蛛香

Valeriana jatamansi W. Jones

植株高20-70厘米。根状茎短而粗壮，节间紧簇。茎生叶2或3对，单叶或具3小叶。花序伞

攀倒甑 *Patrinia villosa*

蜘蛛香 *Valeriana jatamansi*

房状；苞片近钻形；花冠白色或淡粉色，钟形。瘦果狭卵球形。花期(4-)5-7月，果期6-9月。生海拔2500(-3100)以下的草坡、森林或溪边。产中国西南、华中和西北。印度东部和北部、尼泊尔、不丹、泰国北部和越南亦有。

Plants 20-70 cm tall. Rhizomes short, robust, nodes crowded. Cauline leaves in 2 or 3 pairs, simple or trifoliolate. Inflorescences corymbiform; bracts subulate; corolla white or pinkish, funnelform. Achenes narrowly ovoid. Fl. (Apr-)May-Jul. Fr. Jun-Sep. Grassy slopes, forests or by streams below 2500(-3100) m. Distributed in SW, C and NW China. Also in E and N India, Nepal, Bhutan, N Thailand and Vietnam.

长序缬草

Valeriana hardwickii Wall.

多年生大草本。茎粗壮，中空。叶具柄，3-7羽状分裂，基部近圆形，边缘具齿或全缘。大型圆锥状聚伞花序顶生或腋生；苞片条状钻形，小苞片三角状卵形；花小，白色，漏斗状；果序极度伸长，可长达70厘米。连萼瘦果卵形。花期6-8月，果期7-10月。生海拔900-3800米的草坡、林缘、林下或溪边。产中国西南、华南和华中。南亚和东南亚亦有。

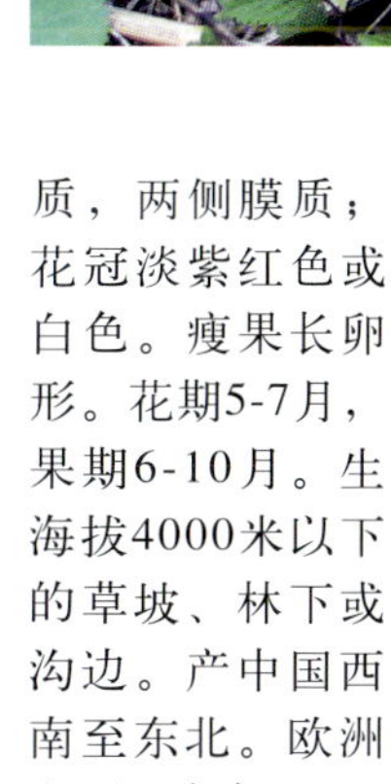

Perennial large herbs. Stems stout, canulate. Leaves petiolate, 3-7 pinnatifid, base subrotundate, margin serrate or entire. Large paniculate-cymes terminal or axillary; bracts linear-subulate, bracteoles triangular-ovate; flowers small, white, infundibular; fruit inflorescences extremely elongated, up to 70 cm long. Cypselae ovoid. Fl. Jun-Aug. Fr. Jul-Oct. Grassy slopes, forest edges, forests or streamsides at 900-3800 m. Distributed in SW, S and C China. Also in S and SE Asia.

缬草

Valeriana officinalis L.

多年生高大草本。根状茎粗短呈头状，须根簇生。茎中空，有纵棱，被粗毛。叶卵形，羽状深裂，裂片披针形或条形，全缘或有疏锯齿。伞房状三出聚伞圆锥花序顶生；小苞片中央纸质，两侧膜质；花冠淡紫红色或白色。瘦果长卵形。花期5-7月，果期6-10月。生海拔4000米以下的草坡、林下或沟边。产中国西南至东北。欧洲和西亚广布。

缬草 *Valeriana officinalis*

Perennial large herbs. Rhizomes thick and short, head-shaped, fibrous roots clustered. Stems canulate, ridged and strigose. Leaves ovate, pinnatiparted, lobes lanceolate or linear, margin entire or sparsely serrate. Ternate corymbose-thyrse terminal; bracteoles centrally papery, laterally membranous; corolla purplish-red or white. Cypselae narrowly obovoid. Fl. May-Jul. Fr. Jun-Oct. Grassy slopes, forests or ditchsides below 4000 m. Widely distributed from SW to NE China. Also widely in Europe and W Asia.

长序缬草 *Valeriana hardwickii*

川续断科
Dipsacaceae

黑水缬草 *Valeriana amurensis*

瑞香缬草 *Valeriana daphniflora*

黑水缬草

Valeriana amurensis P. A. Smirn. ex Kom.

多年生草本。根茎短缩，不明显。茎不分枝，被粗毛，上部被腺毛。叶5-11对，羽状全裂，裂片卵形，具粗牙齿。多歧聚伞花序顶生；花梗被腺毛和粗毛；小苞片草质，边缘膜质；花冠淡红色，漏斗状。瘦果狭三角状卵形，被粗毛。花期6-7月，果期7-8月。生山坡草甸或林下。产吉林和黑龙江。俄罗斯和朝鲜半岛亦有。

Perennial herbs. Rhizomes shortened, unconspicuous. Stems unbranched, strigose, upper parts glandular hairy. Leaves 5-11 pairs, pinnatisected, lobes ovate, margin gross-dentate. Pleiochasial cymes terminal; peduncles glandular hairy and strigose; bracteoles herbaceous, margin membranous; corolla reddish, infundibular. Cypselae narrowly triangular-ovate, strigose. Fl. Jun-Jul. Fr. Jul-Aug. Slope meadows or forests. Distributed in Jilin and Heilongjiang. Also in Russia and Korean Peninsula.

瑞香缬草

Valeriana daphniflora Hand.-Mazz.

纤细草本，高10-40厘米。茎基部叶不分裂或仅基部有1对小裂片，茎中上部叶羽状分裂，有不等疏齿。花冠筒狭长，向上几不扩展，长度为花冠裂片的4-5倍。瘦果卵球状椭圆体形。花期7-8月，果期9月。生海拔2600-3000(-4500)米的山坡草丛中。产中国西南。

Slender herbs, 10-40 cm tall. Basal leaves unlobed or only with a pair of lobes at base, middle and upper cauline leaves pinnately lobed, sparsely and unequally crenate. Corolla narrowly tubular, nearly not dilated upward, 4-5 times longer than corolla-lobes. Achenes ovoid-ellipsoid. Fl. Jul-Aug. Fr. Sep. Grassy slopes at 2600-3000 (-4500) m. Distributed in SW China.

双参

Triplostegia glandulifera Wall. ex DC.

多年生直立草本。茎具棱。叶对生，羽状半裂。花序圆锥状；外苞片4浅裂，下面密被紫色腺毛；内苞片坛状，具8条纵棱；花冠白色或粉红色，钟形。瘦果包于囊苞中；苞片具腺状毛，4裂，具曲钩。花果期7-10月。生海拔1500-4000米的林中、草坡、草地或溪边。产中国西南、华南、华中和华西。南亚和东南亚亦有。

Perennial herbs, erect. Stems angular. Leaves opposite, pinnatifid. Inflorescences paniculiform; outer involucels 4-lobed, abaxially densely covered with purple glandular hairs; inner involucels urceolate, 8-ribbed; corolla white or rose, funnelform. Achenes enveloped by involucels; involucels glandular hairy, 4-lobed, hooked. Fl. and fr. Jul-Oct. Forests, grassy slopes, meadows or by streams at 1500-4000 m. Distributed in SW, S, C and W China. Also in S and SE Asia.

双参 *Triplostegia glandulifera*

刺续断 *Acanthocalyx nepalensis*

刺续断

Acanthocalyx nepalensis (D. Don) M. J. Cannon

多年生直立草本。莲座状叶边缘常具糙毛或刺齿；茎生叶2-4对，椭圆形至线状披针形。花序头状；总苞卵形，具刺；花萼管状，下部绿色，上部紫色，或全为紫色；花冠粉色或紫色；雄蕊4，内藏。瘦果圆柱形，无毛。花期6-8月，果期7-9月。生海拔2800-4200米的草坡。产西藏南部。印度北部、尼泊尔和不丹亦有。

Perennial herbs, erect. Rosulate leaves margins usually setose or spinose; cauline leaves 2-4-paired, elliptic to linear-lanceolate. Inflorescence capitate; involucral bracts ovate, spinose; calyx tubular, green below, purple above, or entirely purple; corolla pink or purple; stamens 4, included. Achenes columnar, glabrous. Fl. Jun-Aug. Fr. Jul-Sep. Grassy slopes at 2800-4200 m. Distributed in S Xizang. Also in N India, Nepal and Bhutan.

大花刺参

Acanthocalyx nepalensis (D. Don) M. J. Cannon subsp. **delavayi** (Franch.) D. Y. Hong

多年生直立草本。莲座状叶边缘常具糙毛或刺齿；茎生叶2-4对，椭圆形至线状披针形。花序头状；总苞卵形，具刺；花萼管状，下部绿色，上部紫色，或全为紫色；花冠粉色或紫色；雄蕊4，内藏。瘦果圆柱形，常具柔毛。花期6-8月，果期7-9月。生海拔3000-4200米的草坡。产云南西北部、四川西部和西藏东南部。印度(锡金)亦有。

Perennial herbs, erect. Rosulate leaves margins usually setose or spinose; cauline leaves 2-4-paired, elliptic to linear-lanceolate. Inflorescence capitate; involucral bracts ovate, spinose; calyx tubular, green below, purple above, or entirely purple; corolla pink or purple; stamens 4, included. Achenes columnar, usually puberulent. Fl. Jun-Aug. Fr. Jul-Sep. Grassy slopes at 3000-4200 m. Distributed in NW Yunnan, W Sichuan and SE Xizang. Also in India (Sikkim).

白花刺续断

Acanthocalyx alba (Hand.-Mazz.) M. J. Cannon

多年生直立草本。莲座状叶线形或线状披针形，全缘；花茎上叶2-4对。花序头状；总苞卵形，具刺；花萼绿色，管状；花冠白色、淡黄色或黄绿色；柱头长于雄蕊。瘦果圆柱状，无毛至密具柔毛。花期6-8月，果期7-9月。生海拔2500-4100米的近高山或高山草甸或森林。产青藏高原。印度北部亦有。

Perennial herbs, erect. Rosulate leaves linear or linear-lanceolate, margins entire; leaves on flowering stems 2-4-paired. Inflorescences capitellate; involucral bracts ovate, spinose; calyx green, tubular; corolla white, yellowish or yellowish green; styles longer than stamens. Achenes columnar, glabrous to densely puberulent. Fl. Jun-Aug. Fr. Jul-Sep. Subalpine or alpine meadows or forests at 2500-4100 m. Distributed in Tibetan Plateau. Also in N India.

大花刺参 *Acanthocalyx nepalensis* subsp. *delavayi*

白花刺续断 *Acanthocalyx alba*

川续断 *Dipsacus asper*

匙叶翼首花 *Pterocephalus hookeri*

川续断

Dipsacus asper Wall.

多年生草本。主根圆柱形，黄褐色，稍肉质。茎具纵棱和皮刺。叶被刺毛，琴状羽裂、羽状深裂或不裂，边缘具疏粗锯齿。头状花序球形，具长总梗；总苞片叶状；花冠淡黄色或白色，管状，裂片4，不相等。瘦果长倒卵柱状。花期7-9月，果期9-11月。生海拔1500-3700米的沟边、路旁、草丛或林缘。产中国西南、华南和华中。

Perennial herbs. Taproots cylindric, yellowish-brown, somewhat succulent. Stems ridged and aculeate. Leaves setose, lyrate pinnatifid, pinnatifid or unlobed, margin sparsely gross-serrate. Capitula globose, long pedunculate; involucral bracts foliar; corolla yellowish or white, tubular, lobes 4, unequal. Cypselae narrowly ovoid-cylindric. Fl. Jul-Sep. Fr. Sep-Nov. Ditchsides, roadsides, grasslands or forest edges at 1500-3700 m. Distributed in SW, S and C China.

大头续断

Dipsacus chinensis Batal.

多年生草本。主根粗壮，红褐色。茎具纵棱和皮刺。茎生叶宽披针形，被黄白色粗毛，3-8琴裂，顶端裂片大。头状花序圆球形，顶生；总花梗粗壮，长达23厘米；总苞片条

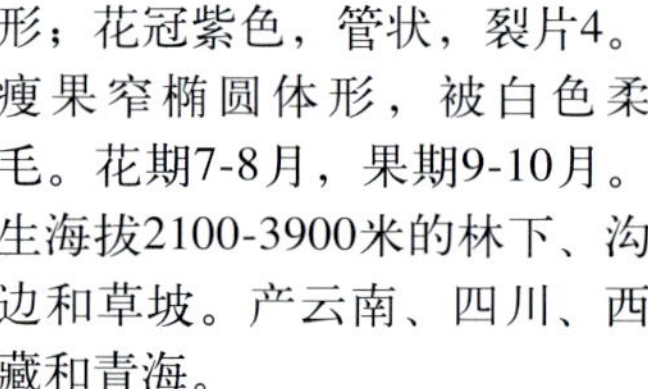

形；花冠紫色，管状，裂片4。瘦果窄椭圆体形，被白色柔毛。花期7-8月，果期9-10月。生海拔2100-3900米的林下、沟边和草坡。产云南、四川、西藏和青海。

Perennial herbs. Taproots stout, reddish-brown. Stems ridged and aculeate. Cauline leaves broadly lanceolate, yellowish-white strigolse, 3-8-lyrate-lobed, terminal lobes large. Capitula globose, terminal; peduncles stout, to 23 cm long; involucral bracts linear; corolla purple, tubular, lobes 4. Cypselae narrowly ellipsoid, white pubescent. Fl. Jul-Aug. Fr. Sep-Oct. Forests, ditchsides or grassy slopes at 2100-3900 m. Distributed in Yunnan, Sichuan, Xizang and Qinghai.

大头续断 *Dipsacus chinensis*

匙叶翼首花

Pterocephalus hookeri (C. B. Clarke) Diels

多年生无茎草本，被白色柔毛。叶基生，莲座状，倒披针形，全缘或羽状深裂。花葶高10-40厘米，无叶；头状花序单个顶生，直径3-4厘米，球形；

蓝盆花 *Scabiosa comosa*

花萼裂片约20，羽毛状；花冠筒状漏斗形，5裂，黄白色至淡紫色。瘦果倒卵形。花果期7-10月。生海拔1800-4800米的草地、高山草甸或耕地附近。产云南、四川、西藏和青海。印度、尼泊尔和不丹亦有。

Perennial acauline herbs, white pubescent. Leaves all basal, rosette, oblanceolate, margin entire or pinnately parted. Scapes 10-40 cm tall, afoliate; capitula solitary, terminal, 3-4 cm diam, globose; calyx segmented to base into ca. 20 soft pinnate hairs; corolla tubular-infundibular, 5-lobed, yellowish-white to purplish. Cypselae obovoid. Fl. and fr. Jul-Oct. Grasslands, alpine meadows or fieldsides at 1800-4800 m. Distributed in Yunnan, Sichuan, Xizang and Qinghai. Also in India, Nepal and Bhutan.

蓝盆花

Scabiosa comosa Fisch. ex Roem. et Schult.

多年生草本。基生叶丛生，羽状全裂，茎生叶基部连接成短鞘，抱茎。头状花序顶生，直径3-3.5厘米，半球形；总苞片6-10片，披针形；花萼5裂，细长针状；花冠蓝紫色；边花二唇形，中央花筒状。瘦果长圆形。花期7-8月，果期9月。生海拔500-1600(-3000)米的干燥沙质地、沙丘、干山坡或草原上。产河南、华北、东北和西北。俄罗斯、蒙古和朝鲜半岛亦有。

Perennial herbs. Basal leaves clustered, pinnatisected, cauline leaves base connate as short sheath, amplexicaul. Capitula terminal, 3-3.5 cm long, semiglobose; involucral bracts 6-10, lanceolate; calyx 5-lobed, slenderly aristate; corolla bluish-purple; margin flowers bilabiate, central flowers tubular. Cypselae oblong. Fl. Jul-Aug. Fr. Sep. Dry sandy places, dunes, dry slopes or grasslands at 500-1600(-3000) m. Distributed in Henan, N, NE and NW China. Also in Russia, Mongolia and Korean Peninsula.

台湾蓝盆花

Scabiosa lacerifolia Hayata

多年生草本。叶无柄，条状披针形，边缘羽状全裂或撕裂，裂片具不整齐锯齿；基生叶长5-12厘米；茎生叶长4-5厘米，先端急尖。头状花序顶生，扁球形；总苞片3轮，披针形；小总苞具4棱，顶端具8窝孔；萼刺刚毛5条，辐状；花冠蓝紫色，边缘辐射花通常较大，花冠裂片5，成二唇形。瘦果倒卵形。生海拔2000-3600米的草坡或山脊。产台湾。

Perennial herbs. Leaves sessile, linear-lanceolate, pinnatisect or lacerate, segments irregularly dentate; basal leaves 5-12 cm long; cauline leaves opposite, 4-5 cm long, apex acute. Capitula terminal, oblate-globose; involucral bracts 3-whorled, lanceolate; involucels 4-ribbed, with 8 pits at apex; calyx 5-setose, radial; corolla blue-purple; marginal ray flowers usually larger, corolla 5-lobed, 2-lipped. Achenes obovoid. Grassy slopes or rocky ridges at 2000-3600 m. Distributed in Taiwan.

台湾蓝盆花 *Scabiosa lacerifolia*

葫芦科 Cucurbitaceae

盒子草

Actinostemma tenerum Griff.

柔弱草本。叶极多变，卵形或箭头状卵形，3浅裂，基部具深凹缺。卷须2歧。雄花数朵，雄花丝顶端稍膨大；雌花单生。果实卵球形或长圆状椭圆体形，具皱。花期7-9月，果期9-11月。生水边草丛中。产中国西南、东南、华中、华北和华东。南亚、东南亚和东北亚亦有。

Slender herbs. Leaf blade very variable, ovate or sagittate-ovate, 3-lobed, base deeply emarginated. Tendrils 2-fid . Staminate many flowers, filaments slightly dilated at apex; pistillate flowers solitary. Fruits ovoid or oblong-ellipsoid, verrucose. Fl. Jul-Sep. Fr. Sep-Nov. Grasslands by rivers. Distributed in SW, SE, C, N and E China. Also in S, SE and NE Asia.

盒子草 *Actinostemma tenerum*

假贝母

Bolbostemma paniculatum (Maxim.) Franch.

攀援草本。茎草质，无毛。叶互生，卵状圆形，掌状5深裂。花序圆锥状；花萼和花冠相似，黄绿色；雄蕊离生；子房卵球形，3室。果实圆柱状，盖裂，顶端圆锥形。花期6-8月，果期8-9月。生山坡或灌丛。产中国西南、华北和华西。

Climbing herbs. Stems herbaceous, glabrous. Leaves alternate, ovate-orbicular, palmately 5-partite. Panicles paniculate; calyx and corolla similar, yellow-green; stamens free; ovary ovoid, 3-locular. Fruits cylindric, circumscissile-dehiscent, conical at apex. Fl. Jun-Aug. Fr. Aug-Sep. Montane slopes or thickets. Distributed in SW, N and W China.

假贝母 *Bolbostemma paniculatum*

棒锤瓜

Neoalsomitra clavigera (Wall.) Hutch.

多年生攀援草本，雌雄异株。茎细长，被短柔毛或近无毛。叶片鸟足状，具5小叶。雄花排列成腋生圆锥花序；花萼筒短；花冠辐状，白色，外面密被短柔毛。雌花花萼与花冠同雄花。蒴果圆柱形，顶端截形，基部钝。种子顶端具1膜质的翅。花期9-11月，果期11月至翌年4月。生海拔550-840(-1600)米的沟谷雨林或次生林中、灌丛中。产云南、西藏东南部、广东、广西和台湾区。南亚、东南亚和大洋洲亦有。

Plants perennial, scandent or creeping, dioecious. Stem slender, much branched, pubescent or subglabrous. Leaf blade pedately 3-5-foliolate. Male inflorescences paniculate, much branched; calyx tube short; corolla white; segments abaxially densely pubescent. Female flowers paniculate; calyx and corolla as in male flowers. Fruit capsular, cylindric, base obtuse, apex truncate. Seeds apex with membranous wing. Fl. Sep-Nov. Fr. Nov to next Apr. Rain forests, broad-leaved forests, valleys, secondary forests at 550-840(-1600) m. Distributed in Yunnan, SE Xizang, Guangdong, Guangxi and Taiwan. Also in S and SE Asia, and Oceania.

棒锤瓜 *Neoalsomitra clavigera*

马铜铃

Hemsleya graciliflora (Harms) Cogn.

多年生攀援草本。纤维根；无块茎。趾状复叶多为7小叶。雌雄异株；雄聚伞花序大型；花冠浅黄绿色；雄蕊5；雌花花柱3，柱头2裂。果实圆锥形。种子长圆形，压扁，具膜质翅。花期6-9月，果期8-11月。生海拔500-2400米的山林中。产广东、广西、四川、湖北、江西和浙江。越南亦有。

Perennial climbing herbs. Root

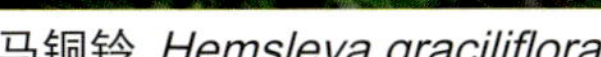
马铜铃 *Hemsleya graciliflora*

fibrous; tuber absent. Leaves pedately 7-foliolate. Flowers dioecious; male inflorescence largely cymose; corolla pale yellow-green; stamens 5; female flowers styles 3, stigma bilobed. Fruits conical. Seeds oblong, compressed, with membranous wing. Fl. Jun-Sep. Fr. Aug-Nov. Montane forests at 500-2400 m. Distributed in Guangdong, Guangxi, Sichuan, Hubei, Jiangxi and Zhejiang. Also in Vietnam.

雪胆

Hemsleya chinensis Cogn. ex Forbes et Hemsl.

多年生攀援草本。鸟足状复叶具5-9小叶。花雌雄异株；花冠橙红色；雌蕊5；子房筒状；花柱3。果长圆状椭圆体形，单生，长3-5(-7)厘米，具有9-10条纵棱；种子褐色，近圆形。花期7-9月，果期9-11月。生海拔400-2800米的杂木林中或林缘沟边。产云南、四川、贵州和湖北。越南亦有。

Perennial climbing herbs. Leaves pedately 5-9-foliolate. Flowers dioecious; corolla orange-red; stamens 5; ovary cylindric; styles 3. Fruits oblong-ellipsoid, solitary, 3-5(-7) cm long, 9-10-ribbed; seeds brown, suborbicular. Fl. Jul-Sep. Fr. Sep-Nov. Mixed forests or forest edges by streams at 400-2800 m. Distributed in Yunnan, Sichuan, Guizhou and Hubei. Also in Vietnam.

翼蛇莲

Hemsleya dipterygia Kuang et A. M. Lu

多年生攀援草本。茎在节部增厚。鸟足状复叶具5-7小叶。聚伞圆锥花序花序梗及分枝纤细；花冠浅黄色至浅黄绿色，辐射状至碗形；花冠裂片开展。果实长圆状高脚碟形或高脚碟形，幼时密具柔毛，后无毛，具明显纵棱。花期6-10月，果期8-11月。生海拔100-1500米的林中。产云南、贵州和广西。越南亦有。

Perennial climbing herbs. Stems usually thickened at nodes. Leaves pedately 5-7-foliolate. Panicles peduncles and branches slender; corolla pale yellow to pale yellow-green, rotate to bowl-shaped; corolla segments spreading. Fruits oblong-claviform or claviform, densely pubescent when immature, glabrous when mature, with indistinct ribs. Fl. Jun-Oct. Fr. Aug-Nov. Forests at 100-1500 m. Distributed in Yunnan, Guizhou and Guangxi. Also in Vietnam.

雪胆 *Hemsleya chinensis*

翼蛇莲 *Hemsleya dipterygia*

曲莲 *Hemsleya amabilis*

曲莲
Hemsleya amabilis Diels

多年生攀援草本。茎和小枝极细弱。鸟足状复叶具5-9小叶。花雌雄异株；花冠黄绿色，直径约1厘米，裂片开展；子房近圆形，密被瘤突；花柱3，柱头马蹄状。果球形。花期6-10月，果期8-11月。生海拔1800-3000米的杂木林或灌丛中。产云南和四川。

Perennial climbing herbs. Stems and branchlets very slender. Leaves pedately 5-9-foliolate. Flowers dioecious; corolla yellowish-green, ca. 1 cm diam, segments spreading; ovary suborbicular, densely verrucose; styles 3; stigmas hoof-shaped. Fruits globose. Fl. Jun-Oct. Fr Aug-Nov. Mixed forests or thickets at 1800-3000 m. Distributed in Yunnan and Sichuan.

肉花雪胆
Hemsleya carnosiflora C. Y. Wu et C. L. Chen

多年生草本。茎草质，攀援，密被短柔毛。卷须条形，长7-16厘米，顶端2歧。小叶7-9，膜质，倒卵形或菱形。花雌雄异株；花冠浅黄绿色，碗形。果实椭圆体形至卵球形。花期6-10月，果期8-11月。生海拔1800-1900米的山谷灌丛中。产云南西北部。

Perennial climbing herbs. Stems herbaceous,densely pubescent. Tendrils linear, 7-16 cm long, 2-fid at apex. Leaves pedately 7-9-foliolate, membranous, obovate or rhombic. Flowers dioecious; corolla pale yellow-green, bowl-shaped. Fruits ellipsoid to ovoid. Fl. Jun-Oct. Fr. Aug-Nov. Thickets in valleys at 1800-1900 m. Distributed in NW Yunnan.

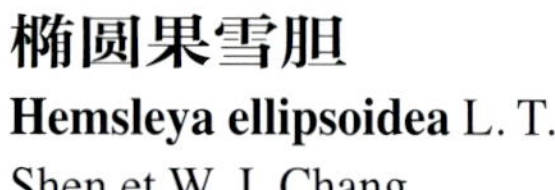

椭圆果雪胆
Hemsleya ellipsoidea L. T. Shen et W. J. Chang

多年生攀援草本。鸟足状复叶具3-5小叶。花雌雄异株；花冠浅黄绿色，灯笼形，裂片先端反折；子房卵球形；花柱3；柱头2裂。果椭圆形，顶端平截，具10条纵棱。果皮革质。花期7-9月，果期9-11月。生海拔约2000米的林缘或灌丛中。产四川。

Perennial climbing herbs. Leaves pedately 3-5-foliolate. Flowers dioecious; corolla yellow-green, lantern-shaped, segments apex reflexed; ovary ovoid; styles 3; stigmas 2-lobed. Fruits ellipsoid, apically truncate, 10-ribbed. Pericarps leathery. Fl. Jul-Sep. Fr. Sep-Nov. Thickets or forest edges at ca. 2000 m. Distributed in Sichuan.

肉花雪胆 *Hemsleya carnosiflora*

椭圆果雪胆 *Hemsleya ellipsoidea*

峨眉雪胆 *Hemsleya emeiensis*

峨眉雪胆

Hemsleya emeiensis L. T. Shen et W. J. Chang

多年生攀援草本。乌足状复叶具7-9小叶。花雌雄异株；聚伞总状花序，长3-8厘米；花冠扁球形，黄绿色；雄蕊5；花柱3；柱头2裂。果近球形，顶端平截。花期7-9月，果期9-11月。生海拔1800-2000米的林缘或山谷灌丛中。产四川中部。

Perennial climbing herbs. Leaves pedately 7-9-foliolate. Flowers dioecious; cymes raceme-like, 3-8 cm long; corolla flat-globose, yellow-green; stamens 5; styles 3; stigmas 2-lobed. Fruits subglobose, flat-truncate at apex. Fl. Jul-Aug. Fr. Sep-Nov. Forest edges or thickets in valleys at 1800-2000 m. Distributed in C Sichuan.

罗锅底

Hemsleya macrosperma C. Y. Wu

多年生攀援草本。乌足状复叶具5-7小叶。花雌雄异株；雄花序二歧聚伞状；花冠橙红色，碗形，裂片长圆形，疏具白色长柔毛。果卵圆形或宽卵形，直径3.5-4厘米。花期7-9月，果期9-11月。生海拔1800-3200米的疏林或灌丛。产云南和四川。

Perennial climbing herbs. Leaves pedately 5-7-foliolate. Flowers dioecious; male panicles dichotomously cymose; corolla orange-red, bowllike, segments oblong, sparsely white villous. Fruits ovoid or broadly ovoid, 3.5-4 cm diam. Fl. Jul-Sep. Fr. Sep-Nov. Open forests or thickets at 1800-3200 m. Distributed in Yunnan and Sichuan.

罗锅底 *Hemsleya macrosperma*

大萼赤爮

Thladiantha grandisepala A. M. Lu et Z. Y. Zhang

草质藤本。叶膜质，卵状心形，卷须纤细，单一，被稀疏短柔毛。花雌雄异株；雄花5-9朵成总状花序；苞片扇形，边缘有不规则三角形小齿；花冠黄色；雌花单生；花萼裂片阔披针形或狭卵形；柱头膨大，圆肾形。果实椭圆体形。花期6-8月，果期8-10月。生海拔2100-2400米的林中或山坡。产云南。

Herbaceous vines. Leaves membranous, ovate-cordate; tendrils slender, simple, sparsely pubescent. Flowers dioecious; staminate flowers 5-9 in pedunculate raceme; bracts flabelliform, with irregular triangular teeth at margin; corolla yellow; pistillate flowers solitary; calyx segments broadly lanceolate or narrowly ovate; stigmas inflated, globose-reniform. Fruits ellipsoid. Fl. Jun-Aug. Fr. Aug-Oct. Forests or montane slopes at 2100-2400 m. Distributed in Yunnan.

大萼赤爮 *Thladiantha grandisepala*

大苞赤爬 *Thladiantha cordifolia*

头花赤爬 *Thladiantha capitata*

大苞赤爬

Thladiantha cordifolia (Bl.) Cogn.

草质藤本。叶片膜质或纸质，卵状心形，两面生淡黄色长柔毛。花雌雄异株；雄花：3至数花密生成短总状，花梗粗壮；苞片覆瓦状，扇形，锐裂，长1.5-2厘米；花冠黄色。果实长圆形，外皮粗糙，稍具柔毛，具10条纵棱。花果期5-11月。生海拔800-2600米的林中或灌丛。产云南、四川、广西和广东。南亚和东南亚亦有。

Herbaceous vines. Leaves membranous or papery, ovate-cordate, both surfaces densely pale yellow pubescent. Flowers dioecious; staminate flowers: 3 to several in dense short raceme, peduncles robust; bracts imbricate, incised at margins, flabelliform, 1.5-2 cm long; corolla yellow. Fruits oblong, with scabrous rind, slightly pubescent, with 10 longitudinal striate. Fl. and fr. May-Nov. Forests or thickets at 800-2600 m. Distributed in Yunnan, Sichuan, Guangxi and Guangdong. Also in S and SE Asia.

头花赤爬

Thladiantha capitata Cogn.

草质藤本。叶阔卵形或阔卵状三角形，膜质。花雌雄异株；雄花成伞形总状花序或近头状花序，具8-15花；雄蕊5；雌花单生或2-3朵聚生；柱头膨大，肾形。果实长圆形，长约4厘米。花果期5-11月。生海拔1000-2700米的林缘、山坡或灌丛中。产四川西部。

Herbaceous vines. Leaves broadly ovate or broadly ovate-triangular, membranous. Flowers dioecious; staminate flowers at apex of peduncle in congested subumbel, umbelliform raceme, or pseudocapitulum, 8-15-flowered; stamens 5; pistillate flowers solitary, or 2-3 at apex of peduncle; stigmas inflated, reniform. Fruits oblong, ca. 4 cm long. Fl. and fr. May-Nov. Forest edges, slopes or thickets at 1000-2700 m. Distributed in W Sichuan.

赤爬 *Thladiantha dubia*

赤爬

Thladiantha dubia Bunge

攀援草本。叶阔卵状心形，两面粗糙。雄花单生或聚集在短枝的上端呈假总状花序，雄蕊4枚合生，1枚离生；雌花单生。果橘黄色，卵状矩圆形，具柔毛及10条纵棱。花期6-8月，果期8-10月。生海拔300-1800米的林缘或沟谷。产中国西南、华北、华西和东北。朝鲜半岛亦有。

Climbing herbs. Leaves broadly ovate-cordate, both surfaces sca-

异叶赤瓟 *Thladiantha hookeri*

brous. Staminate flowers solitary or several congested on short rachises into pseudoracemes, stamens 4 connected in pairs and 1 free; pistillate flowers solitary. Fruits orange-yellow, ovate-oblong, pubescent, with 10 longitudinal striate. Fl. Jun-Aug. Fr. Aug-Oct. Forest edges or valleys at 300-1800 m. Distributed in SW, N, W and NE China. Also in Korean Peninsula.

异叶赤瓟

Thladiantha hookeri C. B. Clarke

攀援草本。叶片多型，单叶或成3-7小叶的鸟足状裂或不分裂，膜质，下面近无毛，上面粗糙。卷须线形，单生。雄花呈总状或单生；雌花单生。果实长圆形，无毛，两端圆。花果期4-10月。生海拔1200-1800米的林下或林缘。产云南。印度、不丹、老挝、越南和柬埔寨亦有。

Climbing herbs. Leaves pedately 3-7-foliolate or undivided, membranous, abaxially subglabrous, adaxially scabrous. Tendrils filiform, simple. Staminate flowers in raceme or solitary; pistillate flowers solitary. Fruits oblong, smooth, rounded at both ends. Fl. and fr. Apr-Oct. Forests or forest edges at 1200-1800 m. Distributed in Yunnan. Also in India, Bhutan, Laos, Vietnam and Cambodia.

南赤瓟 *Thladiantha nudiflora*

鄂赤瓟

Thladiantha oliveri Cogn. ex Mottet

攀援草本。叶阔卵状心形，膜质。花雌雄异株，雄花多数；花冠黄色；雄蕊 5；雌花通常花单生或双生，稀3-4；退化雄蕊条形。果实卵形，无毛，顶端有喙状小尖头。花果期5-10月。生海拔 600-2100米的山坡路旁、灌丛或山谷湿地。产四川、贵州、湖北、陕西和甘肃。

Climbing herbs. Leaves broadly ovate-cordate, membranous. Flowers dioecious; staminate flowers numerous at apex of peduncle; corolla yellow; stamens 5; pistillate flowers usually solitary, rarely 2 or 3-4; staminodes linear. Fruits ovoid, glabrous, apex acute with a beak. Fl. and fr. May-Oct. Roadsides on slopes, thickets or wet places in valleys at 600-2100 m. Distributed in Sichuan, Guizhou, Hubei, Shaanxi and Gansu.

鄂赤瓟 *Thladiantha oliveri*

南赤瓟

Thladiantha nudiflora Hemsl.

攀援草本。植株密生柔毛状硬毛。花雌雄异株；雄花总状花序；花冠黄色；雄蕊5；雌花单生；柱头膨大，圆肾形。果实成熟后红色或红褐色，长圆形，渐无毛，先端钝。花期5-8月，果期8-11月。生海拔900-1700米的沟边、林缘或山坡灌丛中。产华北、华中、华东、西南和东南。越南和菲律宾亦有。

Climbing herbs. Plants densely pubescent-hispid. Flowers dioecious; staminate flowers in racemes; corolla yellow; stamens 5; pistillate flowers solitary; stigmas inflated, globose-reniform. Fruits red or red-brown when mature, oblong, glabrescent, apex obtuse. Fl. May-Aug. Fr. Aug-Nov. Stream edges, forest edges or thickets on slopes at 900-1700 m. Distributed in N, C, E, SW and SE China. Also in Vietnam and the Philippines.

罗汉果 *Siraitia grosvenorii*

罗汉果

Siraitia grosvenorii (Swingle) C. Jeffrey ex A. M. Lu et Z. Y. Zhang

攀援草本。所有部位均具黄褐色柔毛和黑色腺鳞。卷曲二歧分枝。叶卵状心形，膜质。花雌雄异株；雄花：花序总状，具6-10花；雌花单生或2-5。果实球形或长圆形，密具黄褐色绒毛和黑色腺鳞，终无毛。花期5-7月，果期7-9月。生海拔400-1400米的山坡林下、河边湿地或灌丛中。产贵州、广西、广东、湖南和江西。

Herbs climbing. All parts with yellow-brown pubescence and black glandular scales. Tendrils 2-fid. Leaves ovate-cordate, membranous. Flowers dioecious; staminate flowers: panicles racemose, 6-10-flowered; pistillate flowers solitary or 2-5. Fruits globose or oblong, densely yellow-brown velvety and black glandular-scaly, ultimately glabrous. Fl. May-Jul. Fr. Jul-Sep. Mountain slopes, wet places by rivers or thickets at 400-1400 m. Distributed in Guizhou, Guangxi, Guangdong, Hunan and Jiangxi.

翅子罗汉果

Siraitia siamensis (Craib) C. Jeffrey ex Zhong et D. Fang

多年生草质攀援藤本，长达20余米；全体密被黄褐色柔毛和混生红色(干后变黑色)疣状腺鳞。叶片卵状心形。雌雄异株。雄花花萼筒短钟状，裂片扁三角形，密被柔毛和黑色疣状腺鳞；花冠浅黄色。雌花单生或双生；花萼和花冠通常比雄花稍小；子房卵球形，密被短茸毛及黑色腺鳞。果实近球形。种子具3层翅。花期4-6月，果期7-9月。生海拔300-700米的山坡林中。产广西西部和云南东南部。印度尼西亚、马来西亚、泰国和越南北部亦有。

Plants annual, scandent or creeping, up to 20 m long; all parts densely yellow-brown pubescent and red (black when dry) glandular-scaly. Leaf blade ovate-cordate. Male flowers: calyx tube broadly campanulate, segments ovate-triangular; corolla pale yellow, margin with glandular hairs. Female flowers solitary or paired; calyx and corolla as in male flowers but smaller; ovary ovoid, densely velvety and glandular-scaly. Fruit subglobose. Seeds with 3-layered wings. Fl. Apr-Jun. Fr. Jul-Sep. Forests on mountain slopes at 300-700 m. Distributed W Guangxi and SE Yunnan. Also in Indonesia, Malaysia, Thailand and N Vietnam.

马瓟儿

Zehneria japonica (Thunb.) H. Y. Liu

攀援或平卧草本。卷须单一。叶多形，三角状卵形、卵状心形或戟形，膜质。雄花单生或2-3排成总状花序；雌花单生；花冠淡黄色。果橙色或红色，长圆形或渐狭卵球形，1-1.5 × 0.5-0.8(-1)厘米。花期4-7月，果期7-10月。生海拔500-1600米的林下阴湿处、路旁、田边或灌丛中。产中国西南、华中、东南、华南和华东。南亚、东南亚和东北亚亦有。

Plants scandent. Tendrils simple. Leaves polymorphic, triangular-ovate, ovate-cordate or hastate, membranous. Staminate flowers solitary or 2-3 in racemes; pistillate flowers solitary; corolla yellowish. Fruits orange or red, oblong or attenuately ovoid,

翅子罗汉果 *Siraitia siamensis*

马瓟儿 *Zehneria japonica*

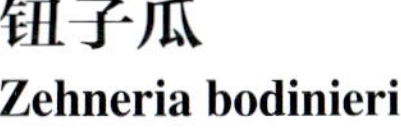

佛手瓜 *Sechium edule*

1-1.5 × 0.5-0.8(-1) cm. Fl. Apr-Jul. Fr. Jul-Oct. Shady and wet forests, roadsides, by fields or thickets at 500-1600 m. Distributed in SW, C, SE, S and E China. Also in S, SE and NE Asia.

钮子瓜

Zehneria bodinieri (H. Lévl.) W. J. de Wilde et Duyfjes

草质藤本。叶阔卵形，膜质，不裂或有时浅3-5裂。卷须线形，单一。雄花3-9成伞房状；雌花单生；花冠白色。果实成熟后红色，囊状，球形，直径1-1.4厘米。种子边缘加厚。花期4-8月，果期8-11月。生海拔500-1000米的疏林或灌丛。产中国西南、华南和东南。南亚和东南亚亦有。

Scandent herbs. Leaves broadly ovate, membranous, undivided or sometimes shortly 3-5-lobed. Tendrils filiform, simple. Staminate flowers 3-9 in corymbs; pistillate flowers solitary; corolla white. Fruits red when ripe, baccate, globose, 1-1.4 cm diam. Seeds thickened at margin. Fl. Apr-Aug. Fr. Aug- Nov. Open forests or thickets at 500-1000 m. Distributed in SW, S and SE China. Also in S and SE Asia.

佛手瓜

Sechium edule (Jacq.) Swartz

多年生草质藤本，雌雄同株。叶近圆形，膜质。雄花10-30朵成总状花序；花冠辐射状；雄蕊3；花丝合生；雌花单生。果实淡绿色，倒卵形，有疏短硬毛，长8-12厘米，直径6-8厘米。花期7-9月，果期8-10月。中国南部有栽培，或逸为野生。在世界温暖地区广泛栽培。原产南美洲。

Perennial herbaceous vines. Leaves suborbicular, membranous. Flowers monoecious; staminate flowers in racemes, 10-30-flowered; corolla rotate; stamens 3; filaments connate; pistillate flowers solitary. Fruits pale green, obovoid, sparsely hirsute, 8-12 cm long, 6-8 cm diam. Fl. Jul-Sep. Fr. Aug-Oct. Cultivated or naturalized in south of China. Commonly cultivated in warm parts of the world. Native to South America.

钮子瓜 *Zehneria bodinieri*

帽儿瓜 *Mukia maderaspatana*

帽儿瓜

Mukia maderaspatana (L.) M. Roem.

一年生平卧或攀援草本，雌雄同株。叶稍坚硬，卵形或卵状心形，常3-5浅裂。雄花束生；花冠黄色；雌花单生或3-5成簇。果实深红色，球形。种子表面具显著的蜂窝状皱纹。花期4-8月，果期8-12月。生海拔400-1700米的山坡岩石或灌丛中。产云南、贵州、广西、广东和台湾。其他热带地区亦有。

Annual scandent or prostrate herbs. Leaves somewhat rigid, ovate or ovate-cordate, usually 3-5-lobed. Flowers monoecious; staminate flowers fascicled; corolla yellow; pistillate flowers solitary or 3-5-fasciculate. Fruits dark red, globose. Seed surfaces conspicuously foveolate. Fl. Apr-Aug. Fr. Aug-Dec. Rocky slopes or thickets at 400-1700 m. Distributed in Yunnan, Guizhou, Guangxi, Guangdong and Taiwan. Also in other tropical areas.

茅瓜

Solena heterophylla Lour.

多年生匍匐草本。叶形极多变，卵形、长圆形、卵状三角形或戟形，不开裂或3-5浅裂，革质。雄花序常具10-20花，花序梗2-5毫米；花黄色或黄白色；花药室弯曲或对折，具柔毛。果实红褐色，阔卵球形、长圆形或近球形。花期5-8月，果期6-11月。生海拔600-2600米的杂木林、灌木、草地、路边或山坡。产中国西南、华南和东南。南亚和东南亚亦有。

Perennial scandent herbs. Leaves very variable, ovate, oblong, ovate-triangular or hastate, undivided or 3-5-lobed, leathery. Male panicles 10-20-flowered, peduncles very short; corolla yellow or yellow-white; anther cells curved or conduplicate, puberulent. Fruits red-brown, broadly ovoid, oblong or subglobose. Fl. May-Aug. Fr. Jun-Nov. Mixed forests, thickets, grasslands, roadsides or mountain slopes at 600-2600 m. Distributed in SW, S and SE China. Also in S and SE Asia.

爪哇帽儿瓜

Mukia javanica (Miq.) C. Jeffrey

一年生攀援草本。叶卵状心形，常3-5裂。卷须细弱，具硬毛。雄花2至数朵簇生在叶腋；雌花簇生在雄花的叶腋；花冠黄色。果矩圆形，红色。种子卵形，边缘凸起。花期4-7月，果期7-10月。生海拔500-1200米的林中、草地或山坡。产云南、广西、广东和台湾。印度、泰国、越南、印度尼西亚和菲律宾亦有。

Annual scandent herbs. Leaves ovate-cordate, usually 3-5-lobed. Tendrils slender, hispid. Staminate flowers 2 to several, clustered in leaf axils; pistillate flowers clustered in same leaf axils as staminate flowers; corolla yellow. Fruits oblong, red. Seeds ovate, margin prominent. Fl. Apr-Jul. Fr. Jul-Oct. Forests, grasslands or slopes at 500-1200 m. Distributed in Yunnan, Guangxi, Guangdong and Taiwan. Also in India, Thailand, Vietnam, Indonesia and the Philippines.

茅瓜 *Solena heterophylla*

爪哇帽儿瓜 *Mukia javanica*

裂瓜 *Schizopepon bryoniifolius*

裂瓜

Schizopepon bryoniifolius Maxim.

一年生攀援草本，长达2-3米；枝细弱。卷须丝状。叶片卵状圆形或阔卵状心形。花极小，两性；花萼裂片披针形；花冠辐状，白色；雄蕊插生于花萼筒的基部；子房卵形。果实成熟后由顶端向基部3瓣裂，有1-3粒种子。种子压扁状。花果期6-11月。生海拔500-1500米的山沟林下或水沟旁。产黑龙江、吉林、辽宁和河北。朝鲜半岛、日本和俄罗斯(远东地区)亦有。

Plants annual, scandent or creeping. Stem 2-3 m. Tendrils filiform. Leaf blade ovate or broadly ovate-cordate. Flowers very small, bisexual; calyx segments lanceolate; corolla rotate, white; stamens inserted at base of calyx tub; ovary ovoid. Fruit broadly ovoid, 3-valved, 1-3-seeded. Seeds compressed. Fl. and fl. Jun-Nov. River valleys, forests at 500-1500 m. Distributed in Heilongjiang, Jilin, Liaoning and Hebei. Also in Korean Peninsula, Japan and Russia (Far East).

湖北裂瓜

Schizopepon dioicus Cogn. ex Oliv.

一年生攀援草本。茎、枝细弱。卷须纤细。叶片宽卵状心形或阔卵形。雌雄异株。雄花生于总状花序上；花冠辐状，白色，裂片披针形或长圆状披针形；雄蕊花丝分离或在基部合生。雌花在叶腋内单生或少数花聚生在短缩的总梗上端；子房卵形。果实通常具2粒种子。种子卵形。花果期6-10月。常生海拔1000-2400米的林下、山沟草丛及山坡路旁。产贵州、湖北西部、湖南西北部、陕西南部、重庆和四川。

Plants annual, scandent or creeping. Stem and branches slender. Tendrils filiform. Leaf blade broadly ovate-cordate. Plants dioecious. Male flowers in a raceme; corolla rotate, white; segments lanceolate or oblong-lanceolate; filaments connate; anthers free or united only at base. Female flowers solitary or a few aggregated at apex of a short peduncle; ovary ovoid. Fruiting pedicel filiform; fruit usually with 2 seeds. Fl. and fr. Jun-Oct. Grasslands, thickets and roadsides at on mountain slopes at 1000-2400 m. Distributed in Guizhou, W Hubei, NW Hunan, S Shaanxi, Chongqing and Sichuan.

苦瓜

Momordica charantia L.

一年生纤细攀援草本，雌雄同株。叶卵状肾形或近圆形，膜质，5-7深裂。雄花单生叶腋；花黄色；雌花单生。果实成熟后橘黄色，纺锤形或圆柱形，外面多瘤皱，自顶端3瓣裂。花果期5-10月。中国普遍栽培。热带和亚热带地区广泛栽培。

Annual climbing vines, slender. Leaves ovate-reniform or suborbicular, membranous, 5-7-partite. Flowers monoecious; staminate flowers solitary in axils of leaves; corolla yellow; pistillate flowers solitary. Fruits orange when mature, fusiform or cylindric, verrucose, 3-valved from apex. Fl. and fr. May-Oct. Widely cultivated in China. Also widely cultivated in tropical and subtropical areas of the world.

湖北裂瓜 *Schizopepon dioicus*

苦瓜 *Momordica charantia*

木鳖子
Momordica cochinchinensis

木鳖子

Momordica cochinchinensis (Lour.) Spreng.

粗壮大藤本。叶柄在基部或中部具2-4个腺体。叶心形或卵状圆形，3-5浅裂。雌雄异株；雄花单生或成短总状；花冠黄色，基部具黄色腺点；雌花单生。果实红色，卵球形，肉质，密具刺，先端具喙。花期6-8月，果期8-10月。生海拔400-1100米的疏林或灌丛。产中国西南、华南、东南和华

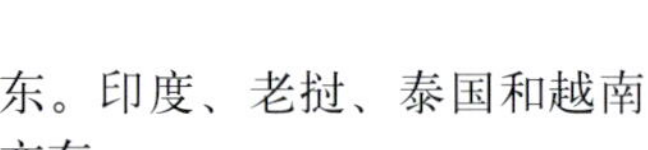

东。印度、老挝、泰国和越南亦有。

Climbers, strong. Petioles with 2-4 glands at base or middle. Leaves cordate or broadly ovate-orbicular, 3-5-lobed. Flowers dioecious; staminate flowers solitary or in a short raceme; corolla yellow, yellow glandular at base; pistillate flowers solitary. Fruits red, ovoid, fleshy, densely spinescent, apex rostellate. Fl. Jun-Aug. Fr. Aug-Oct. Open forests or thickets at 400-1100 m. Distributed in SW, S, SE and E China. Also in India, Laos, Thailand and Vietnam.

丝瓜

Luffa aegyptiaca Mill.

一年生攀援草本，雌雄同株。卷须极强壮，常2-4分叉。叶三角形或近圆形，常掌状5-7裂。花冠黄色，直径5-9厘米；雄花有花15-20朵，总状花序；雌花单生。果实圆柱形，光滑，成熟后内部强烈纤维状。花果期夏秋季。中国南北各地普遍栽培。也广泛栽培在温带和热带地区。云南有野生，果较小。

Annual climbing vines. Tendrils rather robust, usually 2-4-fid. Leaves triangular or suborbicular, often palmately 5-7-lobed. Flowers monoecious; corolla yellow, 5-9 cm diam; staminate flowers in racemes, 15-20-flowered; pistillate flowers solitary. Fruits cylindric, smooth, inside strongly fibrous when mature. Fl. and fr. summer to autumn. Widely cultivated throughout China. Also cultivated in temperate and tropical regions of the world. Wild in Yunnan with smaller fruits.

广东丝瓜

Luffa acutangula (L.) Roxb.

一年生草质藤本。叶近圆形，膜质，常5-7浅裂。雄花17-20朵呈总状花序着生于总花梗顶端；花冠黄色，辐射状；雌花单生。果圆柱状或棍棒状矩圆

丝瓜 *Luffa aegyptiaca*

广东丝瓜 *Luffa acutangula*

喷瓜 *Ecballium elaterium*

形，具8-10棱，无毛。种子黑色，卵球形，边缘无翅。花果期夏季和秋季。栽培于华南。南亚和西南亚亦有。

Annual herbs. Leaves suborbicular, membranous, often 5-7-lobed. Staminate flowers 17-20-flowered in racemes at peduncle apex; corolla yellow, rotate; pistillate flowers solitary. Fruits cylindric or clavate-oblong, 8-10-angled, glabrous. Seeds black, ovoid, margin without wing. Fl. and fr. summer and autumn. Cultivated in S China. Also in S and SW Asia.

喷瓜

Ecballium elaterium (L.) A. Rich.

一年生蔓生草本。茎有短刚毛。叶片卵状长圆形或戟形。雄花生于总状花序，密生黄褐色的长柔毛和短刚毛；花萼裂片披针形，外面密被柔毛和短刚毛；花冠黄色；雄蕊花丝基部有长柔毛。果实长圆形或卵状长圆形，粗糙，有黄褐色短刚毛。常生于干旱的山坡或草地。产新疆。地中海地区和安纳托利亚亦有。

Plants annual, scandent or creeping. Stem scabrous. Leaf blade ovate-oblong or hastate. Male flowers in a raceme; peduncle densely yellow-brown pubescent and setose; calyx segments lanceolate, densely pubescent and hispid; corolla yellow; filaments base villous. Fruit oblong or ovoid-oblong, scabrous, yellow-brown hispid, both ends obtuse. Dry mountain slopes or grasslands. Distributed in Xinjiang. Also in Mediterranean regions and Anatolia.

冬瓜

Benincasa hispida (Thunb.) Cogn.

一年生匍匐或攀援草本，雌雄同株。卷须2-3歧。叶肾状圆形，具5-7浅裂，下面密具柔毛。花单生；花冠黄色；子房卵球形或圆柱形，密具硬毛。种子白色或浅黄色。花期6-9月，果期7-11月。中国广泛栽培。亚洲和其他热带和亚热带地区亦有栽培。

Plants creeping or scandent. Tendrils 2-3-fid. Leaves reniform-orbicular, 5-7-lobed, abaxially densely pubescent. Flowers monoecious, solitary; corolla yellow; ovary ovoid or cylindric, densely hispid. Seeds white or pale yellow. Fl. Jun-Sep. Fr. Jul-Nov. Widely cultivated throughout China. Also cultivated in Asia and other tropical, and subtropical regions.

冬瓜 *Benincasa hispida*

西瓜 *Citrullus lanatus*

西瓜
Citrullus lanatus (Thunb.) Matsum. et Nakai

一年生草本，雌雄同株。茎匍匐。卷须较粗壮，2歧。叶白绿色，三角状卵形，3深裂。花单生叶腋；花冠浅黄色。果大型，球形或长圆形，光滑。种子多数，颜色多变，卵形。花果期4-10月。中国各地栽培。亦广泛栽培于世界热带和温带。

Annual herbs. Stems prostrate. Tendrils stout, 2-fid. Leaves white-green, triangular-ovate, 3-partite. Flowers monoecious, solitary, axillary; corolla pale yellow. Fruits large, globose or oblong, smooth. Seeds numerous, color various, ovate. Fl. and fr. Apr-Oct. Widely cultivated throughout China. Also widely cultivated in tropical and temperate regions of the world.

甜瓜 *Cucumis melo*

甜瓜
Cucumis melo L.

一年生匍匐或攀援草本，雌雄同株。茎、枝有黄褐色或白色的糙硬毛和疣状凸起。卷须纤细。叶片近圆形或肾形。花单生。雄花：数朵簇生于叶腋；花萼筒狭钟形，密被白色长柔毛，裂片近钻形；花冠黄色；雄蕊药室折曲，药隔顶端引长；退化雌蕊长约1毫米。雌花：单生，被柔毛；子房长椭圆形，密被长柔毛和长糙硬毛，柱头靠合。花果期5-9月。全国各地广泛栽培。原产旧世界热带和亚热带；世界温带至热带也广泛栽培。

Plants annual, scandent or creeping. Stems and branches puberulent, or hispid. Tendrils filiform. Leaf blade suborbicular or reniform, abaxially hispid, adaxially scabrous. Plants monoecious or flowers bisexual. Male flowers 1 or 2 to several, fasciculate; calyx tube narrowly campanulate, densely hispidulous to white villous; segments subulate or linear, corolla yellow; anther cells flexuous. Female flowers solitary; ovary ellipsoid, densely white lanate, villous or puberulent. Fl. and fr. May-Sep. Cultivated throughout China. Native to the Old World tropics and subtropics; widely cultivated in tropical and temperate regions.

菜瓜
Cucumis melo subsp. **agrestis** (Naudin) Pangalo

植株纤细。花较小。子房和幼果有光泽丝状短伏毛；果实小，长圆形、球形或陀螺状，栽培状况下无香味和甜味。产华中和华东，广泛栽培。朝鲜半岛亦有。

Plants slender. Flowers small. Ovary and young fruit sericeous with short appressed hairs; fruit in cultivated forms mostly fragrant and not sweet. Farmlands, roadsides. Distributed in C and E China, commonly cultivated in China. Also in Korean Peninsula.

黄瓜
Cucumis sativus L.

一年生蔓生或攀援草本，雌雄同株。茎、枝被白色的糙硬毛。卷须软。叶片宽卵状心形。雄花：常数朵在叶腋簇生；花萼筒狭钟状或近圆筒状，密被白色的长柔毛；花冠黄白色，花冠裂片长圆状披针形。雌花：单生或稀簇生；子房有小刺状凸起。果实表面粗糙，有具刺尖的瘤状凸起。花果期夏季。常生海拔700-

菜瓜 *Cucumis melo* subsp. *agrestis*

和泰国亦有。

Annual climbing herbs. Tendrils filiform, simple. Leaves broadly ovate or triangular-ovate, irregularly 3-5-lobed or not lobed. Staminate flowers solitary or subfasciculate; corolla yellow. Fruits small, oblong, muricate, verrucose. Seeds narrowly ovate. Fl. Jun-Aug. Fr. Aug- Sep. Forests, valleys, thickets or by rivers at 800-1500 m. Distributed in Yunnan. Also in NE India, Myanmar and Thailand.

2000米的山坡、林下、路旁及灌丛中。中国和世界广泛栽培。

Plants annual, scandent or creeping. Stem and branches white hispid. Tendrils slender. Leaf blade broadly ovate-cordate. Plants monoecious. Male flowers fasciculate, puberulent; calyx tube narrowly campanulate, densely white pubescent; corolla yellow-white, segments oblong-lanceolate. Female flowers solitary or fascicled; pedicels pubescent, ovary muricate. Fruit usually verrucose. Fl. and fr. summer. Forests, thickets, mountain slopes at 700-2000 m. Commonly cultivated in China and the world.

野黄瓜

Cucumis hystrix Chakr.

一年生攀援草本。卷须线形，单一。叶片宽卵形或三角状卵形，不规则3-5浅裂或稀不分裂。雄花花单生或近簇生；花冠黄色。果实小，长圆形，多刺，具疣。种子狭卵形。花期6-8月，果期8-9月。生海拔800-1500米的林下、山谷、灌丛或河边。产云南。印度东北部、缅甸

野黄瓜 *Cucumis hystrix*

黄瓜 *Cucumis sativus*

毒瓜 *Diplocyclos palmatus*

毒瓜

Diplocyclos palmatus (L.) C. Jeffrey

攀援草本。根块状。叶阔卵形，掌状5浅裂。雌花和雄花簇生于同一叶腋；花冠黄绿色。果黄绿色至红色，具狭白色条纹，球形，不开裂。种子少数，褐色，卵形。花期3-8月，果期9-12月。生海拔约1000米的林中或灌丛中。产广西、广东、海南和台湾。日本、南亚、东南亚、大洋洲和非洲亦有。

Climbing herbs. Roots tuberous. Leaves broadly ovate, palmately 5-parted. Male and pistillate flowers usually fasciculate at same leaf axil; corolla green-yellow. Fruits yellow-green to red, narrowly white striped, globose, indehiscent. Seeds few, brown, ovate. Fl. Mar-Aug. Fr. Sep-Dec. Forests or thickets at ca. 1000 m. Distributed in Guangxi, Guangdong, Hainan and Taiwan. Also in Japan, S and SE Asia, Oceania and Africa.

波棱瓜

Herpetospermum pedunculosum (Ser.) C. B. Clarke

一年生攀援草本。叶柄长4-8(-10)厘米。叶片膜质，卵状心形。花雌雄异株；雄花排成总状或单生，具5-10花；花冠黄色；萼筒漏斗状；雌花单生；子房密具黄色硬毛。果3浅裂，纤维状。花果期6-10月。生海拔2300-2500米的山坡灌丛、林缘或路旁。产云南和西藏。印度、尼泊尔和不丹亦有。

Annual climbing herbs. Petioles 4-8(-10) cm long. Leaves membranous, ovate-cordate. Flowers dioecious; staminate flowers usually in a raceme or solitary, 5-10-flowered; coralla yellow; calyx tubes funnel-shaped; staminate flowers solitary; ovary densely yellow hispid. Fruits 3-parted, fibrous. Fl. and fr. Jun-Oct. Thickets on montane slopes, forest edges or roadsides at 2300-2500 m. Distributed in Yunnan and Xizang. Also in India, Nepal and Bhutan.

金瓜 *Gymnopetalum chinense*

金瓜

Gymnopetalum chinense (Lour.) Merr.

多年生草本。叶卵状心形，膜质，5棱或3-5浅裂。植株雌雄异株；雄花单生或3-8形成总状；苞片叶状；花冠白色；雌花单生。果实橘红色，卵球状长圆形，具10条纵肋。花期7-9月，果期9-12月。生海拔400-900米的疏林、山坡或灌丛。产云南、广西、广东和海南。南亚和东南亚亦有。

Perennial herbs. Leaves ovate-cordate, membranous, 5-angular or 3-5-lobed. Flowers monoe-

波棱瓜 *Herpetospermum pedunculosum*

凤瓜 *Gymnopetalum scabrum*

cious; staminate flowers solitary or 3-8 in a raceme; bracts leaf-like; corolla white; pistillate flowers solitary. Fruits orange, ovoid-oblong, longitudinally 10-ribbed. Fl. Jul-Sep. Fr. Sep-Dec. Open forests, montane slopes or thickets at 400-900 m. Distributed in Yunnan, Guangxi, Guangdong and Hainan. Also in S and SE Asia.

凤瓜

Gymnopetalum scabrum (Lour.) W. J. de Wilde et Duyfjes

一年生攀援草本。叶肾形或卵状心形，纸质，不裂或波状3-5浅裂。雄花单生或生于总状花序；苞片叶状，开裂，密具黄褐色长柔毛；雌花单生。果近球形，成熟后橙色至红色，光滑，不具棱。花期6-9月，果期9-11月。生海拔400-800米的山坡或草丛中。产云南、贵州、广西、广东和海南。南亚和东南亚亦有。

Annual climbing herbs. Leaves reniform or ovate-cordate, papery, undivided or undulate 3-5-lobed. Staminate flowers solitary or racemes; bracts leaflike, divided, densely yellow-brown villous; pistillate flowers solitary. Fruits subglobose, orange or red when mature, smooth, not ribbed. Fl. Jun-Sep. Fr. Sep-Nov. Slopes or grasses at 400-800 m. Distributed in Yunnan, Guizhou, Guangxi, Guangdong and Hainan. Also in S and SE Asia.

葫芦

Lagenaria siceraria (Molina) Standl.

一年生攀援草本，雌雄同株。叶柄纤细，顶端具2腺体。叶卵状心形或肾状卵形，不裂或3-5浅裂。花单生；花冠黄色。果实尺寸与形状多变，成熟后木质。种子白色。花期夏季，果期秋季。栽培于中国大部分地区。全世界均有栽培。

Annual climbing herbs. Petioles slender, with 2 glands at apex. Leaves ovate-cordate or reniform-ovate, undivided or 3-5-lobed. Flowers monoecious, solitary; coralla yellow. Fruits various in size and shape, woody when mature. Seeds white. Fl. summer. Fr. autumn. Cultivated in most parts of China. Also cultivated worldwide.

葫芦 *Lagenaria siceraria*

五角栝楼 *Trichosanthes quinquangulata*

马干铃栝楼 *Trichosanthes lepiniana*

五角栝楼

Trichosanthes quinquangulata A. Gray

攀援草本。叶片膜质，五角形或阔卵形。花雌雄异株；雄性总状花序具8-10花；花序梗无毛，具条纹；苞片卵形，沿主脉具2裂棕色腺点；花冠白色。果球形，光滑无毛，成熟时红色。花期7-10月，果期9-10月。生海拔500-900米的灌丛、路边或林中。产云南和台湾。东南亚亦有。

Climbing herbs. Leaves membranous, pentagonal or broadly ovate. Flowers monoecious; male racemes 8-10-flowered; peduncles glabrous, striate; bracts ovate, with 2 rows of brownish puncta beside main vein; corolla white. Fruits globose, glabrous, red when mature. Fl. Jul-Oct. Fr. Sep-Oct. Thickets, roadsides or forests at 500-900 m. Distributed in Yunnan and Taiwan. Also in SE Asia.

马干铃栝楼

Trichosanthes lepiniana (Naud.) Cogn.

草质藤本。叶近圆形，具3-5浅裂至中部。花雌雄异株；花冠白色；雄花序总状；苞片近圆形，边缘具锐裂齿；雌花单生。果实红色，卵球形，光滑。种子褐色，阔卵形。花期5-7月，果期8-11月。生海拔700-1200米的林下或灌丛。产云南和西藏。印度和不丹亦有。

Herbaceous vines. Leaves suborbicular, shortly 3-5-lobed up to middle. Flowers dioecious; corolla white; male flowers in racemes; bracts suborbicular, cucullate, margin lacerate; pistillate flowers solitary. Fruits red, ovoid, smooth. Seeds brown, broadly ovate. Fl. May-Jul. Fr. Aug-Nov. Forests or thickets at 700-1200 m. Distributed in Yunnan and Xizang. Also in India and Bhutan.

糙点栝楼

Trichosanthes dunniana Lévl.

一年生攀援藤本。叶片近圆形，纸质，掌状5-7条脉，腹面有白色圆糙点。花雌雄异株；雄花总状花序腋生；萼筒漏斗状，密被短柔毛，裂片浅红色；花冠红色；雄蕊3。果实长圆形，光滑，先端锐尖。种子卵球形，肿胀。花期7-9月，果期10-11月。生海拔900-1900米的山谷密林、山坡疏林或灌丛中，多攀于灌木上。产云南、四川、贵州和广西。缅甸、老挝和泰国亦有。

Annual climbingherbs. Leaves suborbicular, papery, palmately 5-7-veined, adaxially glandular punctate. Flowers dioecious; staminate flowers in racemes; calyx tube narrowly funnelform, densely pubescent, segments pale red; corolla red; stamens 3. Fruits oblong, smooth, apex acute. Seeds ovoid, turgid. Fl. Jul-Sep. Fr. Oct-Nov. Dense forests in valleys, sparse forests on slopes or thickets, usually climbing on shrubs at 900-1900 m. Distributed in Yunnan, Sichuan, Guizhou and Guangxi. Also in Myanmar, Laos and Thailand.

糙点栝楼 *Trichosanthes dunniana*

红花栝楼

Trichosanthes rubriflos Thorel ex Cayla

大型草质攀援藤本。叶阔卵形或近圆形，深掌状3-7浅裂。雄花序为总状；苞片红色，阔卵

红花栝楼 *Trichosanthes rubriflos*

形或倒卵状菱形；花冠粉红色至红色。果实红色，阔卵球形或球形，光滑。种子长圆状椭圆形，棕褐色，平滑。花期5-11月，果期8-12月。生海拔100-1600米的林下或灌丛。产云南、西藏、贵州、广西和广东。南亚和东南亚亦有。

Large climbingherbs. Leaves broadly ovate or suborbicular, deeply palmately 3-7-lobed. Male flowers in racemes; bracts red, broadly ovate or obovate-rhombic; corolla pink to red. Fruits red, broadly ovoid or globose, smooth. Seeds oblong-ellipsoid, brown, glabrous. Fl. May-Nov. Fr. Aug-Dec. Forests or thickets at 100-1600 m. Distributed in Yunnan, Xizang, Guizhou, Guangxi and Guangdong. Also in S and SE Asia.

趾叶栝楼

Trichosanthes pedata Merr. et Chun

草质藤本。茎纤细。3-5小叶复叶，小叶披针形或长圆状倒披针形，膜质，下面无毛，上面具硬毛或白色腺点，远端具锯齿，渐尖。雄花组成总状花序；雌花单生。果实橘黄色，球形。花期6-8月，果期7-12月。生海拔200-1500米的林中或灌丛中。产云南、广西、广东、海南、湖南和江西。越南亦有。

Herbaceous vines. Stems slender. Leaves 3-5-foliolate; leaflets lanceolate or oblong-oblanceolate, membranous, abaxially glabrous, adaxially hispidulous or white punctate, remotely denticulate, acuminate. Staminate flowers in racemes; pistillate flowers solitary. Fruits orange-yellow, globose. Fl. Jun-Aug. Fr. Jul-Dec. Forests or bushes at 200-1500 m. Distributed in Yunnan, Guangxi, Guangdong, Hainan, Hunan and Jiangxi. Also in Vietnam.

趾叶栝楼 *Trichosanthes pedata*

两广栝楼

Trichosanthes reticulinervis C. Y. Wu ex S. K. Chen

多年生攀援草本。茎具纵棱及槽，节上被短柔毛。叶片革质，卵形至阔卵状心形，不分裂，脉上有毛。雌雄异株；雄花排成总状花序或圆锥花序，密被锈色柔毛；花萼筒钟状；花冠白色；雌花单生；花萼裂片线形。果实卵圆形，被毛。种子淡褐色，卵形，压扁。花期5-6月，果期7-8月。生海拔200-400米的山地疏林中。产广东和广西。

Perennial climbing herbs. Stems longituginally ribed and grooved, pubescent at nodes. Leaves leathery, ovate to broadly ovate-cordate, unlobed, pubescent on veins. Flowers dioecious; male flowers in racemes or panicles, densely rust pubescent; calyx tube campanulate; corolla white; female flower solitary; calyx segments linear. Fruits ovoid, densely villous. Seeds pale brown, ovate, compressed. Fl. May-Jun. Fr. Jul-Aug. Mountain forests at 200-400 m. Distributed in Guangdong and Guangxi.

两广栝楼 *Trichosanthes reticulinervis*

密毛栝楼 *Trichosanthes villosa*

栝楼 *Trichosanthes kirilowii*

密毛栝楼

Trichosanthes villosa Blume

攀援藤本。茎密被黄褐色柔毛。叶阔卵形，纸质，不裂或具3尖头，下面密具长柔毛状粗毛，上面密具短褐色长柔毛，边缘具牙齿。雄花序总状，具15-20花；雌花单生。果实红褐色，近球形。花期12月至翌年7月，果期翌年9-11月。生海拔300-1000米的林下阴处。产云南和广西。东南亚亦有。

Climbing herbs. Stems densely yellow-brown villose. Leaves broadly ovate, papery, unlobed or tricuspidate, abaxially densely villous-hirsute, adaxially densely and shortly brownish villous, margin denticulate. Male racemes 15-20-flowered; pistillate flowers solitary. Fruits brown-red, subglobose. Fl. Dec to next Jul. Fr. next Sep-Nov. Shady places in forests at 300-1000 m. Distributed in Yunnan and Guangxi. Also in SE Asia.

大子栝楼 *Trichosanthes truncata*

大子栝楼

Trichosanthes truncata C. B. Clarke

多年生攀援草本。茎有皮孔，无毛；卷须2歧。叶片革质，卵形、狭卵形或宽卵形，不分裂或3浅裂至深裂。花雌雄异株；雄花总状花序，具8-20花；萼筒先端膨大；雌花单生。果实椭圆体形。种子长圆形。花期4-5月，果期7-8月。生海拔300-1600米的密林或灌丛中。产广东、广西和云南南部。孟加拉国和印度亦有。

Perennial climbing herbs. Stems punctate, glabrous; tendrils bifid. Leaves leathery, ovate, narrowly ovate or broadly ovate, unlobed or 3-lobed to 3-parted. Flowers dioecious; male raceme 8-20-flowered; calyx tube dilated at apex; female flowers solitary. Fruits ellipsoid. Seeds oblong. Fl. Apr-May. Fr. Jul-Aug. Forests or thickets at 300-1600 m. Distributed in Guangdong, Guangxi and S Yunnan. Also in Bangladesh and India.

栝楼

Trichosanthes kirilowii Maxim.

攀援藤本。块根肥厚，圆柱形。叶近圆形，纸质，常3-5(-7)裂至中部。雄总状花序单生或在腋处成对生，顶端具5-8花；雌花单生。果实黄褐色或橘黄色，长圆形或球形。花期5-8月，果期8-10月。生海拔200-1800米的山坡林下、灌丛、草地、村边或田边。产华北、华西和华东。老挝、越南、朝鲜半岛和日本亦有。

Climbing herbs. Root tubers plump, cylindric. Leaves suborbicular, papery, usually 3-5(-7)-lobed up to middle. Male raceme solitary or with solitary flowers in axillary pairs, 5-8-flowered at apex; pistillate flowers solitary. Fruits brown-yellow or orange-yellow, oblong or globose. Fl. May-Aug. Fr. Aug-Oct. Forests on montane slopes, thickets, grasslands, village edges or by fields at 200-1800 m. Distributed in N, W and E China. Also in Laos, Vietnam, Korean Peninsula and Japan.

瓜叶栝楼

Trichosanthes cucumerina L.

一年生攀援藤本。叶肾形或阔卵形，膜质，较深5-7裂。雄花序梗成对，早期具1花，后形成总状；总状花序少花；雄花无苞片；雌花单生或有时替换早期雄花。果实卵球状长圆形。花果期秋季。生海拔400-1600米的疏林或灌丛。产云南和广西。南亚、东南亚和澳大利亚亦有。

Annual climbing vines. Leaves reniform or broadly ovate, membranous, ± deeply 5-7-lobed. Male peduncles in pairs, earlier 1-flowered, later bearing a raceme; racemes few flowered; staminate flowers ebracteate; pis-

瓜叶栝楼 *Trichosanthes cucumerina*

全缘栝楼 *Trichosanthes pilosa*

tillate flowers solitary or sometimes replacing earlier staminate flower. Fruits ovoid-oblong. Fl. and fr. autumn. Open forests or thickets at 400-1600 m. Distributed in Yunnan and Guangxi. Also in S and SE Asia, and Australia.

短序栝楼

Trichosanthes baviensis Gagnep.

攀援草本。叶纸质，不分裂，背面密被短绒毛，具掌状脉5条。花雌雄异株；雄花伞房花序，长约2厘米；雄蕊3；雌花单生。果实卵形，长3.5-5厘米，光滑，无毛，顶端有短喙。花期4-5月，果期5-9月。生海拔600-1500米的常绿阔叶林下或灌丛中。产云南、贵州和广西。越南亦有。

Climbing herbs. Leaves papery, unlobed, abaxially densely velutinous, palmately 5-veined. Flowers dioecious; staminate flowers in corymb, ca. 2 cm long; stamens 3; pistillate flowers solitary. Fruits ovoid, 3.5-5 cm long, smooth, glabrous, apex short-beaked. Fl. Apr-May. Fr. May-Sep. Evergreen broad-leaved forests or thickets at 600-1500 m. Distributed in Yunnan, Guizhou and Guangxi. Also in Vietnam.

全缘栝楼

Trichosanthes pilosa Lour.

草质藤本。卷须2-3歧，被短柔毛。叶片纸质，卵状心形至圆心形，不分裂或3-5浅至深裂，上部边缘具小齿或牙齿。花雌雄异株；花冠白色。果实卵球形或长圆形，光滑，先端渐尖，具喙。花期5-9月，果期9-12月。生海拔700-2500米的山谷丛林、山坡疏林、灌丛或林缘。产云南、贵州、广西和广东。南亚、东南亚和日本亦有。

Climbing herbs. Tendrils 2-3-fid, pubescent. Leaves papery, o vate-cordate or orbicular-cordate, not lobed or 3-5-lobed to parted, margin distantly denticulate or dentate. Flowers dioecious; corolla white. Fruits ovoid or oblong, smooth, apex acuminate, beaked. Fl. May-Sep. Fr. Sep-Dec. Forests in valleys, open forests on montane slopes, thickets or forest edges at 700-2500 m. Distributed in Yunnan, Guizhou, Guangxi and Guangdong. Also in S and SE Asia, and Japan.

油渣果

Hodgsonia heteroclita (Roxb.) Hook. f. et Thomson

木质藤本。叶革质，3-5裂。花萼管浅黄色，狭管状，最顶端膨大；花冠外面黄色，内部白色，流苏状边缘长达15厘米。果实红褐色，大型，扁球形，有12条槽沟，含6粒能育种子。花果期6-10月。生海拔300-1500米的灌丛或林下树上。产云南、西藏和广西。南亚和东南亚亦有。

Woody vines. Leaves leathery, 3-5-lobed. Calyx tubes yellowish, narrowly tubular, dilated only at very apex; corolla yellow outside, white inside, fimbriate fringes up to 15 cm long. Fruits reddish brown, large, compressed globose, 12-furrowed, with 6 fertile seeds. Fl. and fr. Jun-Oct. Thickets or forests at 300-1500 m. Distributed in Yunnan, Xizang and Guangxi. Also in S and SE Asia.

短序栝楼 *Trichosanthes baviensis*

油渣果 *Hodgsonia heteroclita*

腺点油瓜
Hodgsonia macrocarpa var. *capinocarpa*

笋瓜 *Cucurbita maxima*

腺点油瓜

Hodgsonia macrocarpa var. **capinocarpa** (Ridl.) Tsai

木质藤本。叶革质，多为5裂。花萼管浅黄色，狭管状，最顶端膨大；花冠外面黄色，内部白色，流苏状边缘长达15厘米。果实红褐色，大型，扁球形，不具棱，腺点明显且多，含6粒能育种子。花果期6-10月。生海拔300-1500米林下树上或灌丛。产云南南部。马来西亚亦有。

Woody vines. Leaves leathery, mostly 5-lobed. Calyx tubes yellowish, narrowly tubular, dilated only at very apex; corolla yellow outside, white inside, fimbriate fringes up to 15 cm long. Fruits reddish brown, large, compressed globose, fruits not ribbed, abundantly and conspicuously glandular., with 6 fertile seeds. Fl. and fr. Jun-Oct. Trees in forests or thickets at 300-1500 m. Distributed in S Yunnan. Also in Malaysia.

笋瓜

Cucurbita maxima Duchesne

攀援藤本。叶片圆形或肾形，近全缘或仅具细锯齿。卷须常多叉，稍具糙毛。萼向钟形；花冠管状，裂片反折。果梗短，圆柱形，不具棱槽，先端不膨大。果实形状、大小和颜色多变。花果期4-11月。栽培于中国大部分地区。全世界有栽培。原产南美洲。

Climbing herbs. Leaves orbicular or reniform, margin subentire or serrulate. Tendrils usually many fid, slightly setose. Calyx tubes campanulate; corolla tubular; segments reflexed. Fruiting pedicels short, cylindric, not angular-sulcate, apex not enlarged. Fruits variable in shape, size, and color. Fl. and fr. Apr-Nov. Cultivated in most parts of China. Also cultivated worldwide. Native to South America.

西葫芦

Cucurbita pepo L.

一年生蔓性草本，雌雄同株。叶片三角形或卵状三角形，不规则5-7浅裂。卷须多裂。花单生；花萼裂片5棱；花冠钟形。果梗粗壮，明显具棱槽，先端稍加厚。果实形状和大小多变。花果期5-11月。栽培于中国大部分地区。世界各地亦广泛栽培。原产北美洲南部。

Annual climbing herbs. Leaves triangular or ovate-triangular, irregularly 5-7-lobed. Tendrils many fid. Plants monoecious; flowers solitary; calyx tubes 5-angled; corolla campanulate. Fruiting pedicels robust, conspicuously angular-sulcate, apex slightly thickened. Fruits variable in shape and size. Fl. and fr. May-Nov. Cultivated in most

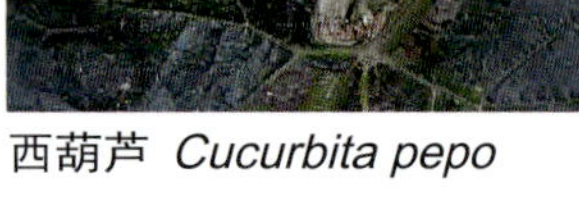

西葫芦 *Cucurbita pepo*

南瓜 *Cucurbita moschata*

parts of China. Also cultivated worldwide. Native to S North America.

南瓜

Cucurbita moschata Duchesne

一年生蔓性草本。叶阔卵形或卵圆形，5角形或5裂。雄花萼筒钟形，裂片条形，先端膨大或叶状；花冠钟形。果梗粗壮，具棱槽，先端强烈膨大。果实形状、大小和颜色多变。花果期4-11月。中国南北各地广泛栽培。世界各地亦普遍栽培。原产墨西哥至中美洲。

Annual prostrate herbs. Leaves broadly ovate or ovate-orbicular, 5-angled or 5-lobed. Male calyx tubes campanulate, segments linear, enlarged or leaflike at apex; corolla campanulate. Fruiting pedicels robust, angular-sulcate, strongly enlarged at apex. Fruits variable in shape, size and color. Fl. and fr. Apr-Nov. Widely cultivated in China. Also cultivated worldwide. Native to Mexico to Central America.

红瓜

Coccinia grandis (L.) Voigt

攀援草本。叶宽心形，常为5角，稀5浅裂，基部具数个腺点。卷须纤细，无毛，不分枝。花雌雄异株，单生；萼筒阔钟形；花冠白色或淡黄色。果熟时深红色，纺锤形。花果期夏季。生海拔100-1100米的山坡灌丛或林中。产云南、广西和广东。热带亚洲和热带非洲亦有。

Herbs climbing. Leaves broadly cordate, often 5-angular or rarely 5-lobed, base with several glands. Tendrils slender, glabrous, not branched. Flowers dioecious, solitary; calyx tubes broadly campanulate; corolla white or yellowish. Fruits deeply red when mature, fusiform. Fl. and fr. summer. Thickets on montane slopes or forests at 100-1100 m. Distributed in Yunnan, Guangxi and Guangdong. Also in tropical Asia and tropical Africa.

红瓜 *Coccinia grandis*

缅甸绞股蓝

Gynostemma burmanicum King ex Chakr.

攀援藤本。卷须2歧。叶纸质，小叶3，下面被短硬毛。花雌雄异株；雄花圆锥状，花冠绿色；花萼裂片长圆形。果实绿色，囊状，球形，无毛，具3粒种子。生海拔600-1300米的疏林或灌丛中。产云南南部。缅甸和泰国亦有。

Herbs climbing. Tendrils 2-fid. Leaves papery, leaflets 3, abaxially hispidulous. Flowers dioecious; staminate flowers in panicle, corolla green; calyx segments oblong; corolla green. Fruits green, baccate, globose, glabrous, 3-seeded. Open forests or thickets at 600-1300 m. Distributed in S Yunnan. Also in Myanmar and Thailand.

缅甸绞股蓝 *Gynostemma burmanicum*

扁果绞股蓝
Gynostemma compressum

扁果绞股蓝

Gynostemma compressum X. X. Chen et D. R. Liang

多年生攀援草本。茎有槽，无毛；卷须二歧。趾状复叶具7小叶。雌雄异株；雄花花序腋生，花绿色；雌花单生或成对，腋生；子房2室，柱头2裂。果实倒三角形，压扁，无毛。花期3-4月，果期3-5月。生海拔400米以下的林下、潮湿的石灰石山麓下。产广西。

Perennial climbing herbs. Stems sulcate, glabrous; tendrils bifid. Leaves pedate, with 7 leaflets. Flowers dioecious; male inflorescence axiallary, flowers greenish; female flowers solitary or paired, axially; ovary 2-loculed, stigma bifid. Fruits ob-triangular, compressed, glabrous. Fl. Mar-Apr. Fr. Mar-May. Humid limestone forests below 400 m. Distributed in Guangxi.

广西绞股蓝

Gynostemma guangxiense X. X. Chen et D. H. Qin

多年生攀援草本。茎无毛，具条纹和沟。叶鸟足状，具(3-)5-7小叶；小叶卵状椭圆形或倒卵形，边缘具锯齿，两面无毛或沿脉疏被微柔毛。雄花淡绿色；花药卵球形。雌花聚伞状；花冠裂片披针形，长约2.5毫米；柱头稍2裂。果淡绿色，长4-5毫米，无毛；果梗丝状，长7-18毫米。花期10月，果期11-12月。生石灰岩山上林中。产广西。

Perennial and climbing herbs. Stems glabrous, striate and sulcate. Leavespedate, with (3-)5-7 leaflets; leaflet ovate-elliptic or obovate, margin serrate, glabrous or with sparse pubescence along midrib and lateral veins on both surfaces. Male flowers greenish; anthers ovoid. Female flowers cymose; corolla segments lanceolate, ca. 2.5 mm long; stigmas slightly bifid. Fruit greenish, 4-5 mm long, glabrous; fruit stalk filiform, 7-18 mm long. Fl. Oct. Fr. Nov-Dec. Forests on limestone mountains. Distributed in Guangxi.

绞股蓝

Gynostemma pentaphyllum (Thunb.) Makino

攀援植物。茎细弱，分枝。叶为鸟足状3-9小叶，常具5-7小叶，膜质或纸质。花雌雄异株；花冠

广西绞股蓝 *Gynostemma guangxiense*

绞股蓝 *Gynostemma pentaphyllum*

浅绿色或白色；雄蕊5；花柱3。果肉质，不开裂，球形，成熟后黑色。花期3-11月，果期4-12月。生海拔 300-3200米山谷密林中、山坡疏林中或路边草丛中。产中国西南、华南、东南、华中和华东。南亚、东南亚和东北亚亦有。

Climbing herbs. Stems slender, branched. Leaves pedately 3-9-foliolate, usually 5-7-foliolate, membranous or papery. Flowers dioecious; corolla pale green or white; stamens 5; styles 3. Fruits fleshy, indehiscent, globose, black when ripe. Fl. Mar-Nov. Fr. Apr-Dec. Dense forests in valleys, sparse forests on slopes or grasslands by roads at 300-3200 m. Distributed in SW, S, SE, C and E China. Also in S, SE and NE Asia.

小雀瓜

Cyclanthera pedata (L.) Schrad.

一年生攀援草本。茎多分枝，无毛。叶片鸟足状5全裂。雄花：生于圆锥花序；花梗丝状；花萼筒杯状，裂片线形；花冠黄色，裂片卵状三角形，两面被疏而短的微柔毛或细腺点；雄蕊花丝极短，连合成一中央轴柱。雌花：花萼和花冠同雄花。果实狭长圆形至狭长椭圆形，具囊状凸起及刺刚毛，具种子8-10粒。花果期5-10月。西藏和云南有栽培。南美洲和中美洲亦有栽培。

Plants annual, scandent or creeping. Stem multibranched, glabrous. Leaf blade 5-pedatisect. Male flowers in a panicle; pedicels filiform; calyx tube cupular; segments linear; corolla yellow; segments ovate-triangular, puberulent; filaments connate; anthers circular. Female flowers: calyx and corolla as in male flowers. Fruit narrowly oblong to narrowly elliptic, setose, apex acuminate, 8-10-seeded. Fl. and fr. May-Oct. Cultivated in Xizang and Yunnan. Also cultivated in South and Central America.

小雀瓜 *Cyclanthera pedata*

桔梗科
Campanulaceae

裂叶蓝钟花
Cyananthus lobatus Wall. ex Benth.

多年生草本。根萝卜形。叶互生；叶片倒卵形、倒披针形、匙形或菱形，近革质，先端3-9深裂或浅裂。花单个顶生于主茎和分枝上；花冠淡蓝色至蓝紫色，管状钟形，外面无毛，内面喉部密被长柔毛；柱头膨人，5裂。花期8-9月。生海拔2800-4500米的草坡或林下。产云南西北部和西藏东南部。印度北部、尼泊尔、不丹和缅甸北部亦有。

Perennial herbs. Roots carrot-shaped. Leaves alternate; leaves obovate, oblanceolate, spatulate or rhombic, subleathery, 3-9-parted or -lobed toward apex. Flowers solitary, terminal on main stems and branches; corolla pale blue to blue-purple, tubular-campanulate, outside glabrous, inside densely villous at throat; stigma large, 5-fid. Fl. Aug-Sep. Grassy slopes or forests at 2800-4500 m. Distributed in NW Yunnan and SE Xizang. Also in N India, Nepal, Bhutan and N Myanmar.

裂叶蓝钟花 *Cyananthus lobatus*

美丽蓝钟花
Cyananthus formosus Diels

多年生草本。根萝卜状。茎丛生，匍匐至斜升，浅紫色。花大，单个顶生于主茎和分枝上，由4或5片叶组成的轮状所托；花冠深蓝色或蓝紫色，外部无毛，喉内部密具柔毛；柱头伸出至花冠喉部；柱头5裂。花期8-9月。生海拔2800-4600米的草坡、林间空地、林缘或高山流石滩上。产云南西北部和东北部、四川西南部。

Perennial herbs. Roots carrotlike. Stems caespitose, prostrate to ascending, pale purple. Flowers large, solitary and terminal on main stems and branches, subtended by a whorllike aggregation of 4 or 5 leaves; corolla dark blue or purple-blue, outside glabrous, inside densely villous at throat; styles extended up to corolla throat; stigma 5-fid. Fl. Aug-Sep. Grassy slopes, forest glades, forest edges or screes at 2800-4600 m. Distributed in NW and NE Yunnan, and SW Sichuan.

美丽蓝钟花 *Cyananthus formosus*

细叶蓝钟花
Cyananthus delavayi Franch.

多年生草本。茎丛生，匍匐或斜升。叶互生，近圆形、卵形或三角形。花单朵顶生；花萼筒形，果时下部膨胀，被褐色硬毛或渐无毛；花冠蓝色，筒状钟形，外面无毛，内面喉部密生长柔毛。花期9月，生海拔1900-4000米的钙质草坡、林下或林缘。产云南西北部和中部、四川西南部。

Perennial herbs. Stems caespitose, prostrate or ascending. Leaves alternate, suborbicular, ovate or deltoid. Flowers solitary, terminal; calyx tubular, inflated at base in fruit, brown hirsute or glabrescent; corolla blue, tubular-campanulate, glabrous outside, densely villose at throat inside. Fl. Sep. Grassy calcareous slopes, forests or forest edges at 1900-4000 m. Distributed in NW and C Yunnan, and SW Sichuan.

细叶蓝钟花 *Cyananthus delavayi*

大萼蓝钟花 *Cyananthus macrocalyx*

大萼蓝钟花

Cyananthus macrocalyx Franch.

多年生草本。叶片卵形至近圆形，边缘全缘。单花顶生，由4或5片聚集成轮状的叶所托；花萼黄绿色或淡紫色，具棕色柔毛，长7-13毫米，果期基部近球形；花冠黄色，有时紫色或具红色脉，管状钟形，外部无毛，内部喉部密具毛。蒴果在花萼上部伸出。花期7-8月。生海拔2500-5000米的高山草甸或草坡。产云南、四川、西藏、甘肃和青海。印度、尼泊尔、不丹和缅甸亦有。

Perennial herbs. Leaves ovate to suborbicular, margin entire. Flowers solitary, terminal, subtended by a whorllike aggregation of 4 or 5 leaves; calyx yellow-green or purplish, brown hirsute, 7-13 mm long, basally subglobose in fruit; corolla yellow, sometimes purple or red-veined, tubular-campanulate, outside glabrous, inside densely villous at throat. Capsules exserted above calyx. Fl. Jul-Aug. Alpine meadows or grassy slopes at 2500-5000 m. Distributed in Yunnan, Sichuan, Xizang, Gansu and Qinghai. Also in India, Nepal, Bhutan and Myanmar.

灰毛蓝钟花

Cyananthus incanus Hook. et Thoms.

黄钟花 *Cyananthus flavus*

多年生草本。茎丛生，单生或下部分枝，具白色绒毛。叶两面具白色柔毛。花单生，顶生于主茎和侧枝上，由4或5片聚成轮状的叶所托；花冠深蓝色或蓝紫色，外面无毛，喉部里面密具柔毛。蒴果超出花萼。花期7-9月，果期9-10月。生海拔2700-5300米的草坡、林间空地、林下、灌丛边缘草地或高山草甸。产云南西北部、四川西部、西藏东部和青海南部。印度(锡金)、尼泊尔和不丹亦有。

Perennial herbs. Stems caespitose, simple or branched below, white villous. Leaves both surfaces white hirsute. Flowers solitary, terminal on main stems and branches, subtended by a whorl-like aggregation of 4 or 5 leaves; corolla dark blue or blue-purple, outside glabrous, inside densely villous at throat. Capsules exserted beyond calyx. Fl. Jul-Sep. Fr. Sep-Oct. Grassy slopes, forest glades, grassy places at forests or thickets edges or alpine meadows at 2700-5300 m. Distributed in NW Yunnan, W Sichuan, E Xizang and S Qinghai. Also in India (Sikkim), Nepal and Bhutan.

黄钟花

Cyananthus flavus Marq.

多年生草本。根萝卜形。茎丛生，斜升。花单生及顶生，由4或5片叶组成的轮状所托；花冠白色或具各种黄色的斑块，外部无毛，喉部内侧密具白色或黄色的柔毛。蒴果与花萼近等长。花期7-8月。生海拔2700-3600米的草坡。产云南西北部和四川西南部。

Perennial herbs. Roots carrot-shaped. Stems caespitose, ascending. Flowers solitary and terminal, subtended by a whorl-like aggregation of 4 or 5 leaves; corolla white or various shades of yellow, outside glabrous, inside densely white or yellow villous at throat. Capsules subequal to calyx in length. Fl. Jul-Aug. Grassy slopes at 2700-3600 m. Distributed in NW Yunnan and SW Sichuan.

灰毛蓝钟花 *Cyananthus incanus*

长花蓝钟花 *Cyananthus longiflorus*

丽江蓝钟花 *Cyananthus lichiangensis*

长花蓝钟花

Cyananthus longiflorus Franch.

多年生草本。叶互生；叶片椭圆形或卵形，下面密被银色丝状毛。花单个顶生于主茎和侧枝上，由3-5片排列成轮状的叶所托；花冠蓝紫色，管状钟形，外面无毛，内面喉部密生长柔毛。蒴果成熟后稍长于花萼。花期7-9月。生海拔2800-4300米的针叶林、干燥山坡或沙丘。产云南西北部。

Perennial herbs. Leaves alternate; leaves elliptic or ovate, abaxially densely silvery sericeous. Flowers solitary, terminal on main stems and branches, subtended by a whorllike aggregation of 3-5 leaves; corolla blue-purple, tubular-campanulate, outside glabrous, inside densely villous at throat. Capsules slightly longer than calyx at maturity. Fl. Jul-Sep. *Pinus* forests, dry slopes or sand dunes at 2800-4300 m. Distributed in NW Yunnan.

丽江蓝钟花

Cyananthus lichiangensis W. W. Sm.

多年生草本。茎丛生，直立。叶两面均具硬毛，边缘反卷。花单生，顶生于主茎和侧枝的顶部，由4或5片排成轮状的叶片所托；花冠淡黄色或绿黄色，圆柱状钟形，外面无毛，内面喉部密被长柔毛。蒴果成熟时超出花萼。花期8-9月。生海拔3000-4400米的草坡或林缘草地。产云南西北部、四川西部和西藏东南部。

Perennial herbs. Stem caespitose, erect. Leaves both surfaces hispidulous, margin revolute. Flowers solitary, terminal on both main stems and branches, subtended by a whorllike aggregation of 4 or 5 leaves; corolla pale yellow or greenish yellow, cylindrical-campanulate, outside glabrous, inside densely villous at throat. Capsules exserted beyond calyx at maturity. Fl. Aug-Sep. Grassy slopes or grassy places at forest edges at 3000-4400 m. Distributed in NW Yunnan, W Sichuan and SE Xizang.

胀萼蓝钟花

Cyananthus inflatus Hook. f. et Thoms.

多年生草本。叶圆形、卵形或三角形，下面具糙毛，上面具微糙毛。花常单生，顶生于主茎和枝条；萼筒花期坛状，花后明显膨大，三角形；花冠浅蓝色，圆柱状钟形。蒴果卵球形。花期8-10月。生海拔1900-4900米的高山草甸或多草多灌山坡。产云南北部、四川西部、西藏南部和贵州西部。印度(锡金)、尼泊尔和不丹亦有。

Perennial herbs. Leaves orbicular, ovate or triangular, abaxially hirsute, adaxially hispidulous. Flowers usually solitary, terminal on main stems and branches; calyx tubes urceolate at anthesis, becoming conspicuously inflated after flowering, deltoid; corolla pale blue, cylindrical-campanulate. Capsules ovoid. Fl. Aug-Oct. Alpine meadows or grassy and shrubby slopes at 1900-4900 m. Distributed in N Yunnan, W Sichuan, S Xizang and W Guizhou. Also in India (Sikkim), Nepal and Bhutan.

蓝钟花

Cyananthus hookeri C. B. Clarke

多年生草本。叶互生；叶片菱形或卵形。花小，单生茎顶或侧枝，由3-5片聚起的叶片组成的轮状所托；花冠蓝紫色，(3-)4(-5)瓣，圆柱形，外面无毛，里面喉部密被长柔毛。蒴果卵球形。花期8-9月。生海拔2700-4500米的灌丛或草地。产云南北部、四川西部、西藏东部、甘肃东南部和青海

胀萼蓝钟花 *Cyananthus inflatus*

蓝钟花 *Cyananthus hookeri*

南部。印度东北部、尼泊尔和不丹亦有。

Perennial herbs. Leaves alternate; leaves rhombic or ovate. Flowers small, solitary, terminal on main stems and branches, subtended by a whorllike aggregation of 3-5 leaves; corolla purple-blue, (3-)4(-5) lobes, cylindrical, outside glabrous, inside densely villous at throat. Capsules ovoid. Fl. Aug-Sep. Thickets or grasslands at 2700-4500 m. Distributed in N Yunnan, W Sichuan, E Xizang, SE Gansu and S Qinghai. Also in NE India, Nepal and Bhutan.

蓝花参
Wahlenbergia marginata
(Thunb.) A. DC.

多年生草本。根为伸长的萝卜形。茎直立或斜升。叶互生，多生于茎下部，匙形、倒披针形、椭圆形或条形；花冠蓝色，宽钟形。蒴果倒圆锥形或近球形，3室。种子黄褐色或深褐色，光滑。花果期2-5月。生低海拔(云南可达2800米)的荒地、田中、山坡或溪中。产中国西南、华南、华中和华东。南亚、东南亚和东北亚亦有，归化于太平洋岛屿(夏威夷群岛)和北美洲。

Perennial herbs. Roots elongate, carrotlike. Stems erect or ascending. Leaves alternate, mostly on lower part of stem, spatulate, oblanceolate, elliptic or linear; corolla blue, broadly campanulate. Capsules obconic, obovoid or subglobose, 3-locular. Seeds yellow-brown or dark brown, smooth. Fl. and fr. Feb-May. Wastelands, fields, slopes or streams at lower elevations (to 2800 m in Yunnan). Distributed in SW, S, C and E China. Also in S, SE and NE Asia, naturalized in the Pacific Islands (Hawaniian Islands) and North America.

羊乳
Codonopsis lanceolata
(Siebold et Zucc.) Trautv.

多年生草本。根常纺锤状加厚。茎缠绕。叶常2-4枚簇生于侧枝顶端。花于小枝顶端单生或成对着生；花冠宽钟形，浅裂；裂片黄绿色或乳白色，具紫色斑点，三角形，反卷。蒴果基部半球形，顶端具喙。花果期7-8月。生海拔200-1500米的灌丛或阔叶林内。产华中、华东和华北。俄罗斯东部、朝鲜半岛和日本亦有。

Perennial herbs. Roots usually fusiform-thickened. Stems twining. Leaves usually 2-4-fascicled on top of branchlets. Flowers solitary or paired on top of branchlets; corolla broadly campanulate, shallowly lobed; lobes yellow-green or milk-white, with purple spots, deltoid, revolute. Capsules hemispherical at base, rostrate toward apex. Fl. and fr. Jul-Aug. Shrublands or broad-leaved forests at 200-1500 m. Distributed in C, E and N China. Also in E Russia, Korean Peninsula and Japan.

蓝花参 *Wahlenbergia marginata*

羊乳 *Codonopsis lanceolata*

雀斑党参
Codonopsis ussuriensis

雀斑党参

Codonopsis ussuriensis (Rupr. et Maxim.) Hemsl.

缠绕藤本。根灰黄色，块状或长圆形。叶在主茎上互生，枝顶常3-5片一簇假轮生。花单生；萼筒贴生到子房中部，半球形；花冠钟形，浅裂；裂片深紫色，三角形，内部具明显的黑紫色条纹或黑点。蒴果基部半球形。种子多数，无翅。花期7-8月。生海拔约800米的沟谷、湿草地或沙质土壤中。产黑龙江和吉林。俄罗斯(远东地区)、朝鲜半岛和日本亦有。

Climbing vines. Roots gray-yellow, tuberous or oblong. Leaves on main stems alternate, those on top of branches usually 3-5-fascicled, pseudoverticillate. Flowers solitary; calyx tubes adnate to ovary up to its middle, semiglobose; corolla campanulate, shallowly lobed; lobes dark purple, deltoid, inside with remarkable black-purple stripes or black spots. Capsules hemispherical at base. Seeds numerous, wingless. Fl. Jul-Aug. Ravines, moist meadows or in sandy soil at ca. 800 m. Distributed in Heilongjiang and Jilin. Also in Russia (Far East), Korean Peninsula and Japan.

党参

Codonopsis pilosula (Franch.) Nannf.

多年生草本。根常肉质，萝卜形或纺锤状圆柱形。茎缠绕，无毛，具分枝。花单生于枝顶，具柄；花冠黄绿色，里面具紫色斑点，宽钟形，浅裂；裂片三角形，先端锐。蒴果基部半球形，顶端圆锥形。花果期7-10月。生海拔900-3900米的林下或灌丛。产中国西南、华中、华东、华北、西北和东北。俄罗斯、蒙古和朝鲜半岛亦有。

Perennial herbs. Roots usually succulent, carrot-shaped or fusiform-cylindrical. Stems twining, glabrous, branched. Flowers solitary and terminal on branches, pedicellate; corolla yellow-green, with purple spots inside, broadly campanulate, shallowly lobed; lobes deltoid, apex acute. Capsules hemispheric at base, conical toward apex. Fl. and fr. Jul-Oct. Forests or thickets at 900-3900 m. Distributed in SW, C, E, N, NW and NE China. Also in Russia, Mongolia and Korean Peninsula.

党参 *Codonopsis pilosula*

球花党参

Codonopsis subglobosa W. W. Sm.

多年生草本。根肥大。茎缠绕，有多数分枝。花顶生或与叶对生；花冠浅黄绿色，裂片或有时在基部具深紫红色斑点，球状钟形。蒴果下面半球形，上部圆锥形，具喙。花果期7-10月。生海拔2500-3700米的多石山地草甸、多树木的石灰石悬崖或溪边灌丛。产云南西北部、四川西部和西藏东部。

Perennial herbs. Roots hypertrophy. Stems twinning, many-branched. Flowers terminal or opposite to leaves; corolla pale yellow-green with dark red-purple on lobes and sometimes at base,

球花党参 *Codonopsis subglobosa*

globose-campanulate. Capsules semiglobose below, conical and rostrate above. Fl. and fr. Jul-Oct. Stony mountain meadows, wooded limestone cliffs or scrubs by streams at 2500-3700 m. Distributed in NW Yunnan, W Sichuan and E Xizang.

大萼党参 *Codonopsis benthamii*

管钟党参 *Codonopsis bulleyana*

大萼党参

Codonopsis benthamii Hook. f. et Thoms.

多年生攀援或匍匐草本。根萝卜状或圆柱状。叶互生或近对生，叶片宽卵形、三角状卵形或卵状披针形，边缘有不规则分裂或浅裂、或有锯齿。花顶生，具长梗；萼筒贴生到子房近中部，具10棱；裂片间的褶窄，具斑点；花冠黄绿色，基部浅红褐色，浅裂。种子黄褐色，卵球形，无翅。花果期7-10月。生海拔2800-3700米的草坡、林缘、灌丛或溪边。产云南西北部、四川西部和西藏。印度东北部、尼泊尔、不丹和缅甸北部亦有。

Perennial herbs, climbing or ascending. Roots carrot-shaped or cylindrical. Leaves alternate or subopposite, broadly ovate, triangular-ovate or ovate-lanceolate, margin irregularly pinnatifid to shallowly lobed, or dentate. Flowers terminal, long pedicellate; calyx tubes adnate to ovary up to middle, 10-ribbed; sinus between lobes narrow and pointed; corolla yellow-green, pale brown-red at base, shallowly lobed. Seeds brown-yellow, ovoid, wingless. Fl. and fr. Jul-Oct. Grassy slopes, forest edges, scrubs or by streams at 2800-3700 m. Distributed in NW Yunnan, W Sichuan and Xizang. Also in NE India, Nepal, Bhutan and N Myanmar.

新疆党参

Codonopsis clematidea C. B. Clarke

多年生草本。主干上的叶较小，互生，分枝上的叶对生，叶片卵形、长圆形、宽披针形或披针形，边缘全缘或偶有小锯齿。花单生，顶生于主茎和分枝上，具长柄；花冠淡蓝色，里面具深蓝色脉纹和紫色斑点。蒴果下部半球形，上部圆锥形，先端具喙尖。花果期7-10月。生海拔1700-2500米的山林、沟壑或溪边。产西藏西部和新疆。南亚和中亚亦有。

Perennial herbs. Leaves on main stems smaller and alternate, those on branches opposite, ovate, oblong, broadly lanceolate or lanceolate, margin entire or occasionally crenulate or sinuate. Flowers solitary, terminal on main stems and branches, long pedicellate; corolla pale blue with dark blue veins and purple spots inside. Capsules semiglobose below, conical above, rostrum acute. Fl. and fr. Jul-Oct. Mountain forests, ravines or by streams at 1700-2500 m. Distributed in W Xizang and Xinjiang. Also in S and C Asia.

管钟党参

Codonopsis bulleyana Forr. ex Diels

多年生攀援草本。根萝卜形。主茎上叶片互生，分枝的叶片近对生，叶片心形、宽卵形或卵形，边缘波状、有小齿或近全缘。花冠浅蓝色，花冠管淡紫色，管状钟形，浅裂；花单一，顶生，微下垂。蒴果基部半球形，先端具喙，圆锥形。种子多数，黄褐色，椭圆体形。花期7-9月，果期9-10月。生海拔3300-4200米的草坡或灌丛。产云南北部、四川西南部和西藏东南部。

Perennial herbs, climbing. Roots carrot-shaped. Leaves on main stems alternate, those on branches subopposite; cordate, broadly ovate or ovate, margin sinuous or obscurely crenate or subentire. Corolla pale blue but tubes purplish, tubular-campanulate, shallowly lobed; flowers solitary, terminal, slightly pendulous. Capsules hemispheric at base, conical toward apex, rostrate. Seeds numerous, brown-yellow, ellipsoid. Fl. Jul-Sep. Fr. Sep-Oct. Grassy slopes or scrubs at 3300-4200 m. Distributed in N Yunnan, SW Sichuan and SE Xizang.

新疆党参 *Codonopsis clematidea*

灰毛党参 *Codonopsis canescens*

脉花党参
Codonopsis foetens subsp. *nervosa*

脉花党参

Codonopsis foetens Hook. f. et Thoms. subsp. **nervosa** (Chipp) Hong

多年生草本。主茎上叶互生，侧枝上叶近对生；叶片宽心形、心形或卵形。花单生，顶生于主茎和侧枝上，稍下垂；花冠淡蓝色，里面基部通常具紫红色斑点，近球状钟形，具浅裂。花期7-9月，果期9-10月。生海拔3300-4500米的草坡、灌丛、林下或山坡向北林缘草甸。产云南西北部、四川西部、西藏东部、甘肃东南部和青海南部。

Perennial herbs. Leaves on main stems alternate, those on branches subopposite; leaves broadly cordate, cordate or ovate. Flowers solitary, terminal on main stems and branches, slightly pendent; corolla pale blue, often with red-purple spots at base inside, subglobose-campanulate, shallowly lobed. Fl. Jul-Sep. Fr. Sep-Oct. Grassy slopes, thickets, forests or meadows at forest edges on N-facing slopes at 3300-4500 m. Distributed in NW Yunnan, W Sichuan, E Xizang, SE Gansu and S Qinghai.

灰毛党参

Codonopsis canescens Nannf.

多年生攀援草本。根萝卜形。主茎密具白色柔毛。花冠淡蓝色或蓝白色，里面基部具蓝色脉纹，宽钟形。蒴果基部半球形，先端圆锥形。种子多数，黄褐色，椭圆体形或长圆形。花期6-9月，果期9-10月。生海拔3000-4200米的草坡、向阳或多石的河边梯田。产四川西北部、西藏东部和青海南部。

Perennial herbs, climbing. Roots carrot-shaped. Main stems densely white villous. Corolla pale blue or blue-white, with blue veins at base inside, broadly campanulate. Capsules hemispherical at base, conical toward apex. Seeds numerous, brown-yellow, ellipsoid or oblong. Fl. Jun-Sep. Fr. Sep-Oct. Grassy slopes, sunny or stony river terraces at 3000-4200 m. Distributed in NW Sichuan, E Xizang and S Qinghai.

绿钟党参

Codonopsis chlorocodon C. Y. Wu

多年生草本。根萝卜形。主茎上的叶互生，侧枝上叶对生，三角形、卵形或披针形。花单生，顶生于主茎；花冠管状钟形，淡黄绿色。蒴果基部钝或稍楔形。种子黄褐色，椭圆体形。花期6-8月，果期9月。生海拔2700-3700米的向阳草坡或疏灌丛。产云南西北部和四川西部。

Perennial herbs. Roots carrot-shaped. Leaves on main stems alternate, those on branches opposite, triangular, ovate or lanceolate. Flowers solitary, ter-

绿钟党参 *Codonopsis chlorocodon*

鸡蛋参 *Codonopsis convolvulacea*

毛叶鸡蛋参 *Codonopsis hirsuta*

minal on main stems; corolla tubular-campanulate, pale yellow-green. Capsules obtuse or slightly cuneate at base. Seeds brown-yellow, ellipsoid. Fl. Jun-Aug. Fr. Sep. Sunny grassy slopes or sparse thickets at 2700-3700 m. Distributed in NW Yunnan and W Sichuan.

毛叶鸡蛋参

Codonopsis hirsuta (Hand.-Mazz.) Hong et L. M. Ma

多年生草本。根卵球形。茎和叶的下表面密被长硬毛。叶片集生于茎下部。单花顶生；花冠浅蓝色或蓝紫色，5裂至近基部。蒴果上部宽圆锥形，下部倒圆锥形。花期9月。生海拔2400-3100米的向阳草坡或开阔灌丛。产云南西北部和四川西南部。

Perennial herbs. Roots ovoid. Stems and leaves abaxially densely hirsute. Leaves mainly on lower part of stems. Flowers solitary, terminal; corolla pale blue or blue-purple, 5-divided to near base. Capsules broadly conical at superior part, obconical at inferior part. Fl. Sep. Sunny grassy slopes or open thickets at 2400-3100 m. Distributed in NW Yunnan and SW Sichuan.

鸡蛋参

Codonopsis convolvulacea Kurz

缠绕草本。叶互生或有时对生，平均的分布于茎上；叶披针形至条状披针形，纸质。花单生，顶生于主茎和侧枝上；花冠淡蓝色或蓝紫色，5中裂至近基部。种子多数，黄褐色，无翅。花果期7-10月。生海拔1000-1800米的草坡或灌丛。产云南南部。缅甸北部亦有。

Twining herbs. Leaves alternate or sometimes opposite, evenly distributed along stems; leaves lanceolate to linear-lanceolate, papery. Flowers solitary, terminal on main stems and branches; corolla pale blue or blue-purple, rotate, 5-fid to near base. Seeds numerous, brown-yellow, wingless. Fl. and fr. Jul-Oct. Grassy slopes or thickets at 1000-1800 m. Distributed in S Yunnan. Also in N Myanmar.

珠子参

Codonopsis convolvulacea Kurz subsp. **forrestii** (Diels) Hong et L. M. Ma

多年生草本。叶片具柄，柄长3-22毫米；叶片三角形或卵形至条状披针形，纸质，基部心形或截形至楔形。花冠辐状，近于5全裂，裂片长1-3.5厘米。种子多数，黄褐色，无翅。花果期7-10月。生海拔1800-3900米的疏林、林缘、灌丛或草甸。产云南中部和北部、四川西南部和贵州西部。缅甸北部亦有。

Perennial herbs. Leaves petiolate, petioles 3-22 mm long; leaves deltoid or ovate to linear-lanceolate, papery, base cordate or truncate to cuneate. Corolla rotate, nearly 5-divided; lobes 1-3.5 cm long. Seeds numerous, brown-yellow, wingless. Fl. and fr. Jul-Oct. Open woods, forest edges, thickets or meadows at 1800-3900 m. Distributed in C and N Yunnan, and W Guizhou. Also in N Myanmar.

珠子参 *Codonopsis convolvulacea* subsp. *forrestii*

金钱豹 *Campanumoea javanica*

小叶轮钟草 *Cyclocodon celebicus*

金钱豹

Campanumoea javanica Blume

缠绕草本。叶对生，具长柄；叶心形或心状卵形，有时3裂。花单生叶腋；花冠钟形，白色或黄绿色，里面紫色或淡红色，开裂至中部。浆果紫色或浅绿色，表面着有红色，球形。花果期5-11月。生海拔400-1800(-2200)米的草坡或灌丛。产云南、贵州西南部、华南和台湾。南亚、东南亚和东北亚亦有。

Twining herbs. Leaves opposite, long petiolate; leaves cordate or cordate-ovate, sometimes becoming trilobate. Flowers solitary, axillary; corolla campanulate, white or yellow-green, purple or reddish inside, cleft to middle. Berries violet or greenish white suffused with red, globose. Fl. and fr. May-Nov. Grassy slopes or shrubberies at 400-1800(-2200) m. Distributed in Yunnan, SW Guizhou, S China and Taiwan. Also in S, SE and NE Asia.

小叶轮钟草

Cyclocodon celebicus (Blume) D. Y. Hong

一年生或两年生草本。叶卵状披针形至披针形，茎上叶稍小。花顶生，常为3个二歧聚伞花序；花冠浅蓝色，钟形，5或6裂至中部；雄蕊5或6；子房球形；花柱5或6浅裂。浆果白色，扁球形。种子极多。花果期7月至翌年1月。生海拔800-2600米的草坡、灌丛或林地。产西藏(波密县、墨脱县、察隅县)和云南。南亚、东南亚和澳大利亚亦有。

Annual herbs or perennial. Leaves ovate-lanceolate to lanceolate, those on branches smaller. Flowers terminal, usually in a dichasium of 3; corolla pale blue, campanulate, 5- or 6-lobed to middle; stamens 5 or 6; ovary globose; stigmas 5- or 6-fid. Berries white, oblate. Seeds extremely numerous. Fl. and fr. Jul to next Jan. Grassy slopes, thickets or woodlands at 800-2600 m. Distributed in Xizang (Bomê County, Mêdog County, Zayü County) and Yunnan. Also in S and SE Asia, and Australia.

轮钟花

Cyclocodon lancifolius (Roxb.) Kurz

草本，直立或斜升，常全株无

轮钟花 *Cyclocodon lancifolius*

细钟花 *Leptocodon gracilis*

毛。叶对生，具短柄。花常单个顶生，或顶生及腋生，有时3朵形成聚伞状；花萼贴生于子房下部；花冠白色或淡红色。浆果成熟时紫黑色，球形。花果期7-11月。生海拔1500米以下的林中、灌丛或草地。产中国西南、华南、华中和华东。南亚、东南亚和东北亚亦有。

Herbs, erect or ascending, usually glabrous throughout. Leaves opposite, shortly petiolate. Flowers usually solitary, terminal or both terminal and axillary, sometimes in a cyme of 3; calyx adnate to lower part of ovary; corolla white or pale red. Berries purple-black when mature, globose. Fl. and fr. Jul-Nov. Forests, thickets or grasslands below 1500 m. Distributed in SW, S, C and E China. Also S, SE and NE Asia.

细钟花

Leptocodon gracilis (Hook.) Hook. et Thoms.

缠绕草本。茎长且细弱，具分枝。叶疏被柔毛，芳香；叶互生，具长柄，卵状圆形。花下垂；花梗细弱，长1-5厘米；花冠蓝色或紫色。蒴果内部半球形或倒圆锥形。种子棕红色。花期8-10月。生海拔2000-2500米的林下或灌丛。产云南西北部和四川西南部。印度北部、尼泊尔、不丹和缅甸北部亦有。

Twining herbs. Stems long, slender, branched. Leaves sparsely villous, odorous; leaves alternate, long petiolate, ovate-orbicular. Flowers pendent; pedicels slender, 1-5 cm long; corolla blue or purple. Inferior part of capsules semiglobose or obconic. Seeds red-brown. Fl. Aug-Oct. Forests or thickets at 2000-2500 m. Distributed in NW Yunnan and SW Sichuan. Also in N India, Nepal, Bhutan and N Myanmar.

桔梗

Platycodon grandiflorus (Jacq.) A. Cand.

多年生草本，高达1.2米，常无毛。根萝卜形，黄褐色。叶上面绿色，卵形、椭圆形或披针形。花冠蓝色或紫色，稀白色或粉红色。蒴果球形、倒圆锥形或倒卵球形。花期7-9月，果期8-10月。生海拔2000米以下的向阳草丛、灌丛或偶生于林下。产中国大部分地区。俄罗斯、朝鲜半岛和日本亦有。

Perennial herbs to 1.2 m tall, usually glabrous. Roots carrotlike, yellow-brown. Leaves adaxially green, ovate, elliptic or lanceolate. Corolla blue or purple, rarely white or pink. Capsules globose, obconic or obovoid. Fl. Jul-Sep. Fr. Aug-Oct. Sunny herbs communities, thickets or rarely in forests below 2000 m. Distributed in most parts of China. Also in Russia, Korean Peninsula and Japan.

桔梗 *Platycodon grandiflorus*

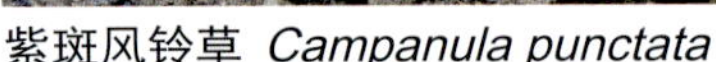
紫斑风铃草 *Campanula punctata*

聚花风铃草 *Campanula glomerata* subsp. *speciosa*

紫斑风铃草
Campanula punctata Lam.

多年生草本，全体被刚毛或少近无毛。花冠长3-6.5厘米，白色、黄色或粉红色，常具紫色或红色斑点。花顶生于主茎和分枝的顶部，下垂。蒴果宽钟形或倒圆锥形。花期6-9月，果期9-10月。生海拔2300米以下的林下、灌丛或草甸。产中国西南、华中、华北、西北和东北。俄罗斯、朝鲜半岛和日本亦有。

Perennial herbs, plants setose throughout or rarely subglabrous. Corolla 3-6.5 cm long, white, yellow or pink, commonly purple or red punctate. Flowers terminal on top of main stems and branches, pendent. Capsules broadly campanulate or obconic. Fl. Jun-Sep. Fr. Sep-Oct. Forests, thickets or meadows below 2300 m. Distributed in SW, C, N, NW and NE China. Also in Russia, Korean Peninsula and Japan.

聚花风铃草
Campanula glomerata L. subsp. **speciosa** (Spreng.) Domin

多年生草本。茎有时上部分枝。茎、叶近无毛至密被白色长柔毛。花冠紫色至蓝色，稀白色；小头状花序多数，多数单生头状花序生于顶生复头状花序两边。蒴果近球形或倒卵球状圆锥形。花期7-9月，果期9-10月。生海拔1300-2600米的草甸或灌丛。产华北和东北。俄罗斯、蒙古、朝鲜半岛和日本广泛栽培。

Perennial herbs. Stems sometimes branched above. Stems and leaves subglabrous to densely white villous. Corolla purple to blue, rarely white; capitula numerous, many simple capitula present besides terminal compound capitula. Capsules subglobose or obovoid-conical. Fl. Jul-Sep. Fr. Sep-Oct. Meadows or thickets at 1300-2600 m. Distributed in N and NE China. Widely cultivated in Russia, Mongolia, Korean Peninsula and Japan.

西南风铃草 *Campanula pallida*

西南风铃草
Campanula pallida Wall.

多年生草本。根萝卜形。茎下部的叶有带翅的柄。花下垂，顶生，有时组成聚伞花序；花冠紫色、蓝紫色或蓝色，管状钟形；柱头短于花冠的2/3，内藏。蒴果倒圆锥形、倒卵球形或球形。花期5-9月。生海拔1000-1400米的草坡或疏林。产中国西南。南亚和西南亚亦有。

灰毛风铃草 *Campanula cana*

Perennial herbs. Roots carrot-shaped. Leaves on lower part of stems with winged petioles. Flowers pendulous, terminal, sometimes in cymes; corolla purple, blue-purple or blue, tubular-campanulate; styles less than 2/3 as long as corolla, included. Capsules obconic, obovoid or globose. Fl. May-Sep. Grassy slopes or open woods at 1000-1400 m. Distributed in SW China. Also in S and SW Asia.

灰毛风铃草

Campanula cana Wall.

多年生草本，植物体除花冠外密被白色长柔毛。茎丛生，常铺散。叶互生。花数朵组成顶生的聚伞状；花冠蓝色、蓝紫色或紫罗蓝色，管状钟形，外部具柔毛；雄蕊内藏；花柱稍伸出。蒴果球形或倒卵球形，具10棱。花果期5-11月。生海拔1000-3200米的开阔石山坡、草坡或灌丛。产中国西南。印度北部、尼泊尔、不丹和缅甸北部亦有。

Perennial herbs, densely white villous on all parts except corolla. Stems caespitose, usually diffuse. Leaves alternate. Flowers several, in a terminal cyme; corolla blue, blue-purple or violet, tubular-campanulate, outside villous; stamens included; styles slightly exserted. Capsules globose or obovoid, 10-ribbed. Fl. and fr. May-Nov. Open rocky slopes, grassy slopes or thickets at 1000-3200 m. Distributed in SW China. Also in N India, Nepal, Bhutan and N Myanmar.

澜沧风铃草

Campanula mekongensis Diels ex C. Y. Wu

多年生草本。根粗大，木质。叶无柄；叶片椭圆形或匙形。花顶生，单生或形成聚伞状；花冠蓝色或白色，钟形。蒴果倒卵球形或倒卵状倒圆锥形。种子黄色，长圆形。花期1-3月。生低海拔的河边沙地草丛、灌丛或低海拔溪边阴处石上。产云南(西双版纳)和广西(天峨县)。

Perennial herbs. Roots thickened, woody. Leaves sessile; leaves elliptic or spatulate. Flowers terminal, solitary or in cymes; corolla blue or white, campanulate. Capsules obovoid or obovoid-obconic. Seeds yellow, oblong. Fl. Jan-Mar. Sandy grass-tufts, thickets by rivers or shaded rocks by streams at lower elevations. Distributed in Yunnan (Sipsongpanna) and Guangxi (Tian'e County).

穿叶异檐花

Triodanis perfoliata (L.) Nieuwl.

一年生草本。茎直立或斜升。叶卵形、近圆形或椭圆形，基部抱茎。花1-3朵生于叶腋，无柄；花冠蓝紫色至玫瑰紫色，旋转。蒴果长圆形，侧瓣开裂。种子浅褐色至棕色，光滑。花果期4-7月。归化于海拔100-1000米的山坡或溪边或混凝土缝中。产台湾、福建和安徽。原产北美洲。

Annual herbs. Stems erect or ascending. Leaves ovate, suborbicular or elliptic, base clasping. Flowers 1-3 in axil, sessile; corolla bluish purple or rose-purple, rotate. Capsules oblong, opening by lateral valves. Seeds light brown to brown, smooth. Fl. and fr. Apr-Jul. Naturalized on mountain slopes, by streams or in cracks of concrete at 100-1000 m. Distributed in Taiwan, Fujian and Anhui. Native to North America.

澜沧风铃草 *Campanula mekongensis*

穿叶异檐花 *Triodanis perfoliata*

甘孜沙参 *Adenophora jasionifolia*

天蓝沙参 *Adenophora coelestis*

甘孜沙参
Adenophora jasionifolia Franch.

多年生草本。茎生叶多集中在茎下半部。花单朵顶生，或少数几朵集成假总状花序；花冠蓝色或蓝紫色，碗形；柱头稍短或有时稍长于花冠。蒴果椭圆状。种子黄褐色。花期7-9月，果期9月。生海拔3000-4700米的草甸或林缘草丛。产云南、四川和西藏。

Perennial herbs. Cauline leaves mainly on lower part of stems. Flowers solitary, terminal or several in a pseudoraceme; corolla blue or purple-blue, bowl-shaped; styles shorter or sometimes longer than corolla. Capsules ellipsoid. Seeds yellow-brown. Fl. Jul-Sep. Fr. Sep. Meadows or grasses of forest edges at 3000-4700 m. Distributed in Yunnan, Sichuan and Xizang.

天蓝沙参
Adenophora coelestis Diels

多年生草本。花排成假总状，或有时在上部具1至数个分枝；萼片狭三角形，边缘具1至数对小齿；花冠蓝色或蓝紫色，钟形；裂片卵状三角形，长为花冠筒的1/2；柱头稍短于花冠。花期8-10月。生海拔1200-4000米的林下、林缘或林间空地。产云南和四川西南部。

Perennial herbs. Flowers in a pseudoraceme or sometimes inflorescences with branches of 1 to several flowers at top; calyx lobes narrowly triangular, margin with 1 to several pairs of denticles; corolla blue or blue-purple, campanulate; lobes ovate-deltoid, ca. 1/2 as long as tube; styles shorter than corolla. Fl. Aug-Oct. Forests, forest edges or glades at 1200-4000 m. Distributed in Yunnan and SW Sichuan.

玉山沙参
Adenophora morrisonensis subsp. **uehatae** (Yamam.) Lammers

多年生草本，被硬毛。茎基具横走的分枝，其上有互生的膜质鳞片。基生叶卵状三角形，基部近截形；茎生叶条状披针形或椭圆形，先端渐尖，边缘具圆齿或锯齿。花1-2朵顶生；花萼裂片边缘具多对细齿；花冠蓝紫色或淡紫色，钟形；花盘环状，高不到1毫米；花柱短于花冠。花期7-11月，果期9-11月。生海拔1000-3700米的松林或竹林。特产台湾。

Perennial herbs, hirsute. Caudexes with horizontal branches bearing alternate membranous scales. Basal leaves ovate-deltoid, base subtruncate; cauline leaves linear-lanceolate or elliptic, apex acuminate, margin crenate or dentate. Flowers 1 or 2, terminal; calyx lobes margin with several pairs of denticles; corolla blue-purple or light purple, campanulate; disk annular, less than 1 mm tall; style shorter than corolla. Fl. Jul-Nov. Fr. Sep-Nov. Conifer forests or bamboo thickets at 1000-3700 m. Endemic to Taiwan.

沙参
Adenophora stricta Miq.

多年生草本。基生叶心形，大而具长柄；茎生叶无柄，或仅下部的叶有极短而带翅的柄。花常排成假总状；花冠蓝色或紫色，宽钟形；柱头常稍长于花冠。蒴果椭圆状球形。花期8-10月。生低至中海拔的草丛或石缝。产华中和华东。朝鲜半岛亦有，日本有归化。

玉山沙参 *Adenophora morrisonensis* subsp. *uehatae*

沙参 *Adenophora stricta*

Perennial herbs. Basal leaves cordate, large and long stipitate; cauline leaves sessile or lower ones with extremely short and winged petioles. Flowers often in a pseudoraceme; corolla blue or purple, broadly campanulate; styles usually slightly longer than corolla. Capsules ellipsoid-globose. Fl. Aug-Oct. Among grasses or rock crevices at lower to middle elevations. Distributed C and E China. Also in Korean Peninsula, naturalized in Japan.

无柄沙参

Adenophora stricta Miq. subsp. **sessilifolia** Hong

多年生草本。茎叶被短硬毛。花萼多被短硬毛或微糙，稀无毛；花常排成假总状；花冠蓝色或紫色，宽钟形；花冠通常无毛或仅外面脉上有少数毛。蒴果椭圆状球形。花期6-8月。生海拔600-2000米的草甸或林缘草地。产中国西南、华南、华中和西北。

Perennial herbs. Stems and leaves hispidulous. Calyx mostly hispidulous or scaberulose, less frequently glabrous; flowers often in a pseudoraceme; corolla blue or purple, broadly campanulate; corolla usually glabrous or only with a few hairs along veins outside. Capsules ellipsoid-globose. Fl. Jun-Aug. Meadows or grasslands at forest edges at 600-2000 m. Distributed in SW, S, C and NW China.

川西沙参

Adenophora stricta Miq. subsp. **aurita** (Franch.) D. Y. Hong et S. Ge

本亚种与沙参的区别在于本亚种的花萼裂片宽1-1.8毫米；花冠长2-2.5厘米；花盘长1.8-2.5毫米。花期7-9月，果期9月。生海拔2100-3300米的草地、林缘或灌丛中。产四川西南部。

This subspecies differs from the typical subspecies in its calyx lobes 1-1.8 mm wide; corolla 2-2.5 cm long; disk 1.8-2.5 mm long. Fl. Jul-Sep. Fr. Sep. Meadows, forest edges or scrubs at 2100-3300m. Distributed in SW Sichuan.

无柄沙参
Adenophora stricta subsp. *sessilifolia*

川西沙参 *Adenophora stricta* subsp. *aurita*

云南沙参 *Adenophora khasiana*

云南沙参

Adenophora khasiana (Hook. et Thoms.) Oliv. ex Coll. et Hemsl.

多年生草本。茎常单生或有时从同一根部发出两枝，高达1米；茎生叶无柄或具短柄，卵状圆形、卵形、倒卵形或椭圆形。花冠漏斗状钟形；花柱稍长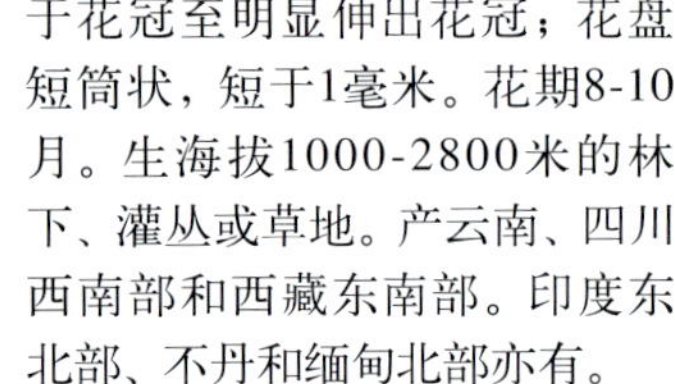于花冠至明显伸出花冠；花盘短筒状，短于1毫米。花期8-10月。生海拔1000-2800米的林下、灌丛或草地。产云南、四川西南部和西藏东南部。印度东北部、不丹和缅甸北部亦有。

Perennial herbs. Stems often single or sometimes 2 from one root, up to 1 m tall; cauline leaves sessile or shortly petiolate, ovate-orbicular, ovate, obovate or elliptic. Corolla funnelform-campanulate; styles slightly longer than corolla to conspicuously exserted; disk shortly tubular, less than 1 mm long. Fl. Aug-Oct. Forests, shrubbery or grasslands at 1000-2800 m. Distributed in Yunnan, SW Sichuan and SE Xizang. Also in NE India, Bhutan and N Myanmar.

石沙参

Adenophora polyantha Nakai

多年生草本。茎生叶无柄，叶片披针形、卵形或椭圆形，稀条状披针形。聚伞花序具单生花形成假总状，或具短花序分枝形成狭圆锥状；花冠紫色或深蓝色，钟形，喉部稍收缩；花柱常稍伸出或有时与花冠等长。蒴果卵状椭圆体形。花果期8-10月。生海拔2000米以下的向阳草坡。产华北、东南和西北。

Perennial herbs. Cauline leaves sessile, leaves lanceolate, ovate or elliptic, rarely linear-lanceolate. Cymes with solitary flowers forming a pseudoraceme or in a narrow panicle with short inflorescences branches; corolla purple or dark blue, campanulate, slightly constricted at throat; styles usually slightly exserted or sometimes as long as corolla. Capsules ovoid-ellipsoid. Fl. and fr. Aug-Oct. Sunny grassy slopes below 2000 m. Distributed in N, SE and NW China.

石沙参 *Adenophora polyantha*

狭长花沙参 *Adenophora elata*

狭长花沙参

Adenophora elata Nannf.

多年生草本。根萝卜形。茎单一；茎生叶极少对生，无柄，或基生叶有时具短的具翅叶柄。花常数个形成假总状，有时单生及顶生；花冠蓝色或蓝紫色；花盘管状。蒴果椭圆体形或长圆形。花期7-9月，果期9月。生海拔1700-3000米的草坡。产河北西部、山西(五台山)和内蒙古(阴山)。

Perennial herbs. Roots carrotlike. Stems single; cauline leaves very rarely opposite, sessile or lower leaves sometimes with short winged petioles. Flowers usually several in a pseudoraceme, sometimes solitary and terminal; corolla blue or purple-blue; disk tubular. Capsules ellipsoid or oblong. Fl. Jul-Sep. Fr. Sep. Grassy slopes at 1700-3000 m. Distributed in W Hebei, Shanxi (Wutai Mountain) and Neimenggu (Yinshan Mountain).

薄叶荠苨
Adenophora remotiflora

薄叶荠苨

Adenophora remotiflora (Siebold et Zucc.) Miq.

多年生草本。茎生叶具长叶柄。叶片膜质，基部平截、圆钝。花常在梗上单生，少生于数花形成的聚伞花序，形成于假总状或狭圆锥状；花冠蓝色，钟形。蒴果倒卵球形。生海拔1700米以下的林缘、林下或草甸。花期7-8月，果期8-9月。产中国东北。俄罗斯、朝鲜半岛和日本亦有。

Perennial herbs. Cauline leaves long petiolate. Leaves membranous, base truncate, obtuse. Flowers often solitary on pedicels, less frequently in several-flowered cymes, forming a pseudoraceme or narrow panicle; corolla blue, campanulate. Capsules obovoid. Fl. Jul-Aug. Fr. Aug-Sep. Forest edges, forests or meadows below 1700 m. Distributed in NE China. Also in Russia, Korean Peninsula and Japan.

荠苨

Adenophora trachelioides Maxim.

多年生草本。茎单生，常之字型扭曲。茎生叶具明显叶柄，基部心形、平截、钝或楔形。花冠蓝色、蓝紫色或白色，钟形；裂片宽椭圆形；柱头与花冠等长。蒴果卵状圆锥形。花期7-8月，果期8-9月。生海拔2400米以下的山坡、草地或林缘。产华东、华北和东北。

Perennial herbs. Stems single, often ± zigzag-twisted. Cauline leaves distinctly petioled, base cordate, truncate, obtuse or cuneate. Corolla blue, blue-purple or white, campanulate; lobes broadly elliptic; styles as long as corolla. Capsules ovoid-conical. Fl. Jul-Aug. Fr. Aug-Sep. Mountain slopes, grasslands or forest edges below 2400 m. Distributed in E, N and NE China.

荠苨 *Adenophora trachelioides*

杏叶沙参

Adenophora petiolata subsp. **hunanensis** (Nannf.) Hong et S. Ge

多年生草本。茎生叶多具柄或者至少基部叶具柄，少近无柄。花序轴和花的各部通常微糙；花萼裂片宽2-4毫米，基部稍叠生；花冠长1.5-2厘米；裂片4-7毫米。花期7-9月，果期10月。生海拔2000米以下的草地、山坡或林缘。产中国西南、华南、华中、华北和华西。

Perennial herbs. Cauline leaves mostly petiolate or at least lower ones petiolate, rarely subsessile. Inflorescences rachises and floral parts often scaberulose; calyx lobes 2-4 mm wide, ± overlapping at base; corolla 1.5-2 cm long; lobes 4-7 mm long. Fl. Jul-Sep. Fr. Oct. Grasslands, slopes or forest edges below 2000 m. Distributed in SW, S, C, N and W China.

杏叶沙参 *Adenophora petiolata* subsp. *hunanensis*

泡沙参 *Adenophora potaninii*

泡沙参
Adenophora potaninii Korsh.

多年生草本。茎生叶无柄，下部叶稀具短柄。花序常自基部具分枝，形成圆锥状，有时仅数花聚合成假总状花序；花冠紫色、蓝色或蓝紫色，稀白色。蒴果球状椭圆体形或椭圆体形。花期7-10月，果期10-11月。生海拔1000-3100米的向阳草坡、灌丛或林下。产中国西北、四川西部和山西西南部。

Perennial herbs. Cauline leaves sessile, lower ones rarely shortly petiolate. Inflorescences usually with branches at base, forming a panicle, sometimes with only several flowers glomerate into a pseudoraceme; corolla purple, blue or blue-purple, rarely white. Capsules globose-ellipsoid or ellipsoid. Fl. Jul-Oct. Fr. Oct-Nov. Sunny grassy slopes, scrubs or forests at 1000-3100 m. Distributed in NW China, W Sichuan and SW Shanxi.

多歧沙参
Adenophora potaninii Korsh. subsp. **wawreana** (Zahlbr.) S. Ge et Hong

多年生草本。茎生叶通常具柄，有时柄很短。叶形多变，从条形至卵形。花序常自基部具分枝，形成圆锥状，有时仅数花聚合成假总状花序；花冠紫色、蓝色或蓝紫色，稀白色。蒴果球状椭圆体形或椭圆体形。花期7-10月，果期10-11月。生海拔2000米以下的向阴山坡草丛、灌丛或疏林，常生于石地或石缝。产华中、华北和东北。

Perennial herbs. Cauline leaves usually petiolate, though sometimes petioles very short. Blades varying greatly, from linear to ovate. Inflorescences usually with branches at base, forming a panicle, sometimes with only several flowers glomerate into a pseudoraceme; corolla purple, blue or blue-purple, rarely white. Capsules globose-ellipsoid or ellipsoid. Fl. Jul-Oct. Fr. Oct-Nov. Grasses on shaded slopes, scrubs or sparse woods, mostly in stony places or rock crevices below 2000 m. Distributed in C, N and NE China.

聚叶沙参
Adenophora wilsonii Nannf.

多年生草本。茎直立，常2至多数支发自同一茎基。花期时下部无叶，中部聚生许多叶；叶两面无毛。圆锥花序具分枝；花萼具1或2对瘤状小齿；花冠紫色或蓝紫色，漏斗状钟形。蒴果球状椭圆体形。花期8-10月，果期9-10月。生海拔1600米以下的灌丛或溪边石缝中。产中国西南、华中和华西。

Perennial herbs. Stems often 2 to several from one root, erect. Densely leafy at middle during anthesis but basal leaves withered; leaves both surfaces glabrous. Inflorescences a panicle with branches; calyx with 1 or 2 pairs of verrucose denticles; corolla purple or blue-purple, funnelform-campanulate. Capsules globose-ellipsoid. Fl. Aug-Oct. Fr. Sep-Oct. Thickets, rocks by streams below 1600 m. Distributed in SW, C and W China.

多歧沙参 *Adenophora potaninii* subsp. *wawreana*

聚叶沙参 *Adenophora wilsonii*

天山沙参 *Adenophora lamarckii*

展枝沙参
Adenophora divaricata Franch. et Sav.

多年生草本。茎单生，高达1米。茎生叶3-5枚轮生，无柄。花序分枝(每个具1至数个聚伞花序)常轮生，或有些互生，形成大型圆锥状；萼片披针形、椭圆状披针形或条状披针形；花冠蓝色或浅紫色，钟形；裂片三角状圆形，长为冠筒的1/3-2/3；柱头与花冠等长或稍突出。蒴果倒卵球形或宽椭圆体形。种子金棕色，椭圆体形。花期7-9月，果期9-10月。生海拔400-1800米的林下、灌丛或草坡。产河北、山西(五台山)、山东(昆嵛山)和东北。俄罗斯(远东地区)、朝鲜半岛和日本亦有。

Perennial herbs. Stems single, up to 1 m tall, simple. Cauline leaves 3-5-verticillate, sessile. Inflorescences branches (each with 1 to several cymes) usually verticillate or sometimes some alternate, forming a large panicle; calyx lobes lanceolate, elliptic-lanceolate or linear-lanceolate; corolla blue or light purple, campanulate; lobes deltoid-orbicular, 1/3-2/3 as long as tube; styles as long as corolla or slightly exserted. Capsules obovoid or broadly ellipsoid. Seeds golden brown, ellipsoid. Fl. Jul-Sep. Fr. Sep-Oct. Forests, shrublands or grassy slopes at 400-1800 m. Distributed in Hebei, Shanxi (Wutai Mountain), Shandong (Kunyu Mountain) and NE China. Also in Russia (Far East), Korean Peninsula and Japan.

展枝沙参 *Adenophora divaricata*

天山沙参
Adenophora lamarckii Fisch.

多年生草本。根萝卜状。茎单生。茎生叶无柄，两面无毛，有时下面具硬毛。数花形成假总状或为具2-4花形成的具短枝的狭圆锥花序；托杯倒卵球形或倒圆锥形，无毛；花冠蓝色，漏斗状钟形。花期7-8月。生林中或林缘。产新疆。哈萨克斯坦、俄罗斯(南西伯利亚)、蒙古北部和朝鲜半岛亦有。

Perennial herbs. Roots carrot-shaped. Stems simple. Cauline leaves sessile, both surfaces glabrous, sometimes hispidulous abaxially. Flowers several in a pseudoraceme or inflorescences with short branches of 2-4 flowers, forming a narrow panicle; hypanthium obovoid or obconic, glabrous; corolla blue, funnelform-campanulate. Fl. Jul-Aug. Forests or forest edges. Distributed in Xinjiang. Also in Kazakhstan, Russia (S Siberia), N Mongolia and Korean Peninsula.

喜马拉雅沙参
Adenophora himalayana Feer.

多年生草本。叶通常宽条形。花萼裂片全缘，偶尔边缘具瘤状小齿；花单生，或数朵形成假总状，从不形成圆锥状；花冠蓝色或蓝紫色，钟形；花柱通常稍外露。花期7-9月，果期8-9月。生海拔(1200-)2500-4700米的林缘高山草甸或灌丛、高山草甸或灌丛。产中国西南和西北。南亚和中亚亦有。

Perennial herbs. Leaves usually broadly linear. Calyx lobes entire, very occasionally with verrucose denticles on margins; flowers solitary or several in a pseudoraceme, never in a panicle; corolla blue or blue-purple, campanulate; styles usually slightly exserted. Fl. Jul-Sep. Fr. Aug-Sep. Alpine meadows or scrubs at forest edges, alpine meadows or scrubs at (1200-) 2500-4700 m. Distributed in SW and NW China. Also in S and C Asia.

喜马拉雅沙参 *Adenophora himalayana*

长柱沙参 *Adenophora stenanthina*

长柱沙参

Adenophora stenanthina (Ledeb.) Kitag.

多年生草本。根萝卜状；茎多数簇生。茎生叶无柄，两面粗糙或近无毛。花序为假总状或圆锥状；托杯无毛；花冠浅蓝色至深蓝色、蓝紫色或紫色，近管状或管状钟形。蒴果椭圆体形。花期7-9月，果期8-9月。生海拔4000米以下的针叶林中、灌丛、草地或沙地。产华北、华西、西北和东北。俄罗斯(远东地区、南西伯利亚)和蒙古亦有。

Perennial herbs. Roots carrotlike; stems several, caespitose. Cauline leaves sessile, both surfaces scaberulose or subglabrous. Inflorescences a pseudoraceme or a panicle; hypanthium glabrous; corolla pale to dark blue, blue-purple, or purple, subtubular or tubular-campanulate. Capsules ellipsoid. Fl. Jul-Sep. Fr. Aug-Sep. Conifer forests, scrubs, grasslands or sandy places at below 4000 m. Distributed in N, W, NW and NE China. Also in Russia (Far East, S Siberia) and Mongolia.

丝裂沙参

Adenophora capillaris Hemsl.

多年生草本。茎单生；茎生叶常无柄。花序具长枝，常形成大而疏松的圆锥花序；托杯椭圆体形或有时卵球形；花萼裂片开展或反折，线形；花冠浅蓝色、浅紫色或白色。蒴果球形、椭圆体形或卵球形。花期7-10月，果期8-10月。生海拔1400-2800米的林中、林缘或草地。产四川西部、贵州西部、重庆、湖北和陕西(秦岭南部)。

Perennial herbs. Stems single; cauline leaves usually sessile. Inflorescences with long branches, often forming a large and lax panicle; hypanthium ellipsoid or sometimes ovoid; calyx lobes spreading or reflexed, filiform; corolla pale blue, pale purple or white. Capsules globose, ellipsoid or ovoid. Fl. Jul-Oct. Fr. Aug-Oct. Forests, forest edges or grasslands at 1400-2800 m. Distributed in W Sichuan, W Guizhou, Chongqing, Hubei and Shaanxi (S Qinling Mountain).

细萼沙参

Adenophora capillaris Hemsl. subsp. **leptosepala** (Diels) Hong

多年生草本。茎和叶常具柔毛。花序具长枝，常形成大而疏松的圆锥花序；托杯椭圆体形或有时卵球形；萼裂片条形，边缘常具牙齿；花冠浅蓝色，狭钟形。蒴果球形或卵球形。花期7-10月，果期9-10月。生海拔2000-3600米的林下、林缘或草地。产云南西北部和四川西南部(木里县)。

丝裂沙参 *Adenophora capillaris* subsp. *capillaris*

细萼沙参 *Adenophora capillaris* subsp. *leptosepala*

Perennial herbs. Stems and leaves usually hirsute. Inflorescences with long branches, often forming a large and lax panicle; hypanthium ellipsoid or sometimes ovoid; calyx lobes linear, margin usually toothed; corolla light blue, narrowly campanulate. Capsules globose or ovoid. Fl. Jul-Oct. Fr. Sep-Oct. Forests, forest edges or grasslands at 2000-3600 m. Distributed in NW Yunnan and SW Sichuan (Muli County).

细叶沙参

Adenophora capillaris Hemsl. subsp. **paniculata** (Nannf.) D. Y. Hong et S. Ge

本亚种与丝裂沙参的区别在于本亚种的花萼裂片长(2-)3-5(-7)毫米，全缘。蒴果卵球形至卵球状长圆形。花期6-9月，果期8-10月。生海拔1100-2800米的草坡。产河北、河南西部、山西、内蒙古东南部和陕西。

This subspecies differs from the typical subspecies in its calyx lobes (2-)3-5(-7) mm long, margin entire. Capsules ovoid to ovoid-oblong. Fl. Jun-Sep. Fr. Aug-Oct. Grassy slopes at 1100-2800 m. Distributed in Hebei, W Henan, Shanxi, SE Neimenggu and Shaanxi.

细叶沙参
Adenophora capillaris subsp. *paniculata*

轮叶沙参 *Adenophora tetraphylla*

轮叶沙参

Adenophora tetraphylla (Thunb.) Fisch.

多年生草本。根萝卜状。茎单生；茎生叶3-6枚轮生。花序为狭圆锥状，伞状分枝多轮生；托杯无毛；花冠蓝色至紫色，管状或狭漏斗形，有时喉部稍收缩。蒴果倒卵球形或宽倒卵球形。花期(2-)3-11月，果期5-11月。生低海拔草原地带或灌丛中。产中国西南、华南、东南、华中、华北、华东和东北。老挝、越南北部、俄罗斯、朝鲜半岛和日本亦有。

Perennial herbs. Roots carrotlike. Stems simple; cauline leaves 3-6-verticillate. Inflorescences a narrow panicle with branches (cymes) mostly verticillate; hypanthium glabrous; corolla blue to purple, tubular or narrowly funnelform, sometimes slightly constricted at throat. Capsules obovoid or broadly obovoid. Fl. (Feb-)Mar-Nov. Fr. May-Nov. Grassy places or scrubs at low elevations. Distributed in SW, S, SE, C, N, E and NE China. Also in Laos, N Vietnam, Russia, Korean Peninsula and Japan.

牧根草 *Asyneuma japonicum*

牧根草

Asyneuma japonicum (Miq.) Briq.

多年生草本。根肉质。茎下部叶有长达4厘米的柄，茎上部叶近无柄。花除在花丝和柱头外均光滑；花萼管球形或钟形；裂片条形；花冠蓝紫色或蓝紫色。蒴果球形。花期7-8月，果期9月。生林下或稀生草甸。产中国东北。俄罗斯、朝鲜半岛和日本亦有。

Perennial herbs. Roots fleshy. Lower cauline leaves with petioles up to 4 cm long, upper cauline leaves almost sessile. Flowers glabrous except on filaments and style; calyx tubes globose or campanulate; lobes linear; corolla purple-blue or blue-purple. Capsules globose. Fl. Jul-Aug. Fr. Sep. Forests or rarely meadows. Distributed in NE China. Also in Russia, Korean Peninsula and Japan.

球果牧根草 *Asyneuma chinense*

球果牧根草

Asyneuma chinense D. Y. Hong

多年生草本。根肉质。茎单生，直立，稍具毛。叶卵形、披针形或椭圆形，边缘有锯齿或圆齿。穗状花序每个总苞片腋内有1-4花；苞片管球形或钟形，花后反折；花冠紫色或蓝色。蒴果球形，具3条宽沟槽，基部截形或凹陷。种子黄褐色，卵球形至长圆形。花果期6-9月。生海拔3000米以下的草坡、林缘或林中。产中国西南和华中。

Perennial herbs. Roots fleshy. Stems single, erect, ± hirsute. Leaves ovate, lanceolate or elliptic, marfin serrate or crenate. Spikes with bract subtending 1-4 flowers; bract tubes globose or campanulate, recurved after anthesis; corolla purple or blue. Capsules globose, with 3 broad pores, base truncate or concave. Seeds brown-yellow, ovoid to oblong. Fl. and fr. Jun-Sep. Grassy slopes, forest edges or forests below 3000 m. Distributed in SW and C China.

卵叶半边莲

Lobelia zeylanica L.

匍匐草本。茎和分枝四棱形，肉质。叶螺旋状互生；叶片宽三角状卵形或卵形。花单生叶腋；花萼钟形，具柔毛；花冠二唇形，紫色、淡紫色或白色。蒴果宽椭圆体形、倒卵球形或长圆形。花果期全年。生海拔1500(-2000)米以下的沟壑水边或溪边。产华南、云南、台湾和福建。南亚和东南亚亦有。

Creeping herbs. Stems and branches 4-angular, succulent. Leaves spirally alternate; leaves broadly deltoid-ovate or ovate. Flowers solitary and axillary; calyx campanulate, puberulent; corolla 2-lipped, purple, pale purple or white. Capsules broadly

卵叶半边莲 *Lobelia zeylanica*

半边莲 *Lobelia chinensis*

ellipsoid, obovoid or oblong. Fl. and fr. whole year round. By water or streams in ravines below 1500(-2000) m. Distributed in S China, Yunnan, Taiwan and Fujian. Also in S and SE Asia.

半边莲

Lobelia chinensis Lour.

多年生草本。茎匍匐，细弱，下部节间生根。叶互生，椭圆形或披针形，通常全缘。花常单生于枝上部叶腋；花冠玫瑰色、白色或淡蓝色，后部深裂至基部，喉部以下具白色长柔毛；裂片全部平展于前方。蒴果倒圆锥形。种子宽椭圆体形，压扁。花果期5-10月。生稻田边、溪边或潮湿草丛。产中国西南、华南、华中和华东。南亚、东南亚和东北亚亦有。

Perennial herbs. Stems decumbent, slender, lower nodes rooted. Leaves alternate, elliptic or lanceolate, usually entire. Flowers usually solitary, axillary at upper leaves of branches; corolla rose, white or bluish, divided to base at back, white villous below throat; lobes all spreading in a plane on anterior side. Capsules obconic. Seeds broadly ellipsoid, compressed. Fl. and fr. May-Oct. By paddy fields, streams or among wet grasses. Distributed in SW, S, C and E China. Also in S, SE and NE Asia.

山梗菜

Lobelia sessilifolia Lamb.

多年生草本。茎单生。叶互生，茎中部叶稍大。总状花序顶生，具8-35花；苞片叶状；花冠蓝紫色或紫色，二唇形；上部2个裂片匙形，下部裂片椭圆形；雄蕊于基部以上贴生。种子红褐色，一侧具翼。花果期7-10月。生海拔3400米以下的湿草地。产中国西南、东南、华中、华东和东北。俄罗斯(远东地区和西伯利亚)、朝鲜半岛和日本亦有。

Perennial herbs. Stems simple. Leaves alternate, larger at middle parts of stems. Racemes terminal, 8-35-flowered; bracts leaflike; corolla blue-purple or violet, 2-lipped; upper 2 lobes spatulate, lower lobes elliptic; stamens connate above base. Seeds brown-red, winged on one side. Fl. and fr. Jul-Oct. Wet meadows below 3400 m. Distributed in SW, SE, C, E and NE China. Also in Russia (Far East and Siberia), Korean Peninsula and Japan.

山梗菜 *Lobelia sessilifolia*

线萼山梗菜

Lobelia melliana F. E. Wimm.

多年生草本。叶互生，稍镰状卵形至镰状披针形，薄纸质。总状花序顶生，花稀疏；花萼裂片线形，果期开展；花冠淡红色或白色，冠檐二唇形。蒴果直立，近球形。花果期8-10月。生海拔1000米以下的沟谷、路边、溪边或林中潮湿处。产华南、华中和华东。

Perennial herbs. Leaves alternate, ± falcate-ovate to falcate-lanceolate, thinly papery. Racemes terminal, flowers sparse; calyx lobes filiform, spreading at fruiting; corolla reddish or white, limbs 2-lipped. Capsules erect, subglobose. Fl. and fr. Aug-Oct. Valleys, roadsides, by streams or damp places in forests below 1000 m. Distributed in S, C and E China.

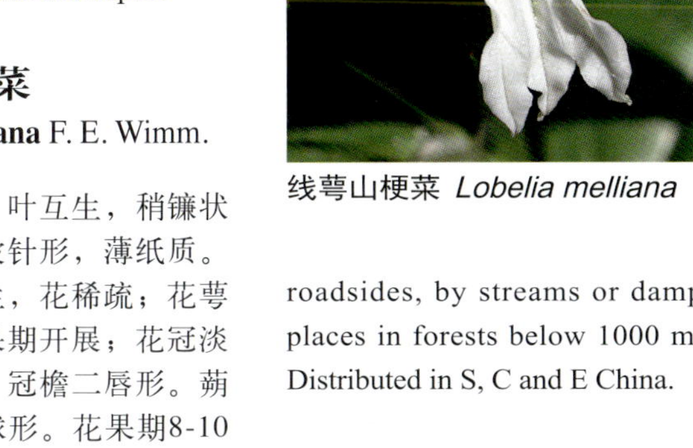

线萼山梗菜 *Lobelia melliana*

西南山梗菜 *Lobelia seguinii*

西南山梗菜
Lobelia seguinii
Lévl. et Van.

亚灌木状草本，高达5米。叶互生，厚纸质，具长柄，叶狭长圆形。总状花序单面向；花冠紫红色、蓝紫色、淡蓝色或粉红色。蒴果下垂，长圆形或椭圆体形。花果期8-10月。生海拔500-3000米的草坡、林缘或路旁。产中国西南、华南、华中和华东。泰国北部和越南北部亦有。

Subshruby herbs, to 5 m tall. Leaves alternate, thickly papery, long petiolate, leaves narrowly oblong. Racemes secund; corolla purple-red, purple-blue, pale blue or pinkish. Capsules pendent, oblong or ellipsoid. Fl. and fr. Aug-Oct. Grassy slopes, forest edges or roadsides at 500-3000 m. Distributed in SW, S, C and E China. Also in N Thailand and N Vietnam.

塔花山梗菜
Lobelia pyramidalis Wall.

灌木状草本。叶近革质，基部渐狭，边缘具细齿，两面无毛。花萼筒钟形或宽椭圆形(稀倒圆锥形)，无毛；花冠白色、粉红色或淡蓝色，外面无毛，近二唇形。蒴果近球状、扁球形或宽卵形，直径3.5-6.5毫米，无毛。花果期1-5月。生海拔1200-2500米的山坡草地、灌丛或路旁。产云南、贵州、广西和西藏。尼泊尔、不丹、印度、缅甸和泰国亦有。

Shrubby herbs. Leaves subleathery, base attenuate, margin serrulate, both surfaces glabrous. Hypanthium campanulate or broadly ellipsoid (rarely obconic), glabrous; corolla white, rose or bluish, outside glabrous, nearly 2-lipped. Capsules subglobose, oblate or broadly ovoid, 3.5-6.5 mm diam, glabrous. Fl. and fr. Jan-May. Grassy slopes, scrub or roadsides at 1200-2500 m. Distributed in Yunnan, Guizhou, Guangxi and Xizang. Also in Nepal, Bhutan, India, Myanmar and Thailand.

密毛山梗菜
Lobelia clavata F. E. Wimm.

草本或亚灌木，可达3.7米。叶互生，无柄，倒披针形或长圆形，厚纸质，两面具硬毛。总状花序近单面向，顶生；花冠近2裂，白色，外面具短毡毛，里面具柔毛。蒴果下垂，密具柔毛。花果期11月至翌年4月。生海拔700-1800米的草坡、林下或路边。产云南南部和贵州西南部。印度东北部、缅甸北部、老挝、泰国北部和越南亦有。

Herbs or subshrubs, up to 3.7 m tall. Leaves alternate, sessile, oblanceolate or oblong, thickly papery, both surfaces hispidulous. Racemes subsecund, terminal; corolla sub-bilabiate, white, outside shortly pannose, inside villous. Capsules pendulous, densely puberulent. Fl. and fr. Nov to next Apr. Grassy slopes, forests or roadsides at 700-1800 m. Distributed in S Yunnan and SW Guizhou. Also in NE India, N Myanmar, Laos, N Thailand and Vietnam.

塔花山梗菜 *Lobelia pyramidalis*

密毛山梗菜 *Lobelia clavata*

大理山梗菜 *Lobelia taliensis*

大理山梗菜

Lobelia taliensis Diels

多年生草本，高50-120厘米。茎直立，常淡紫色。叶倒卵状长圆形至倒卵状披针形或椭圆形，纸质，两面无毛。总状花序顶生，疏松，偏向一侧；花冠淡蓝色或玫瑰色。蒴果长圆形。花果期8-10月。生海拔1600-2600米的草坡。产云南西北部和湖南。

Perennial herbs, 50-120 cm tall. Stems erect, often purplish. Leaves obovate-oblong to obovate-lanceolate or elliptic, papery, both surfaces glabrous. Racemes terminal, lax, one-side-oriented; corolla pale blue or rose. Capsules oblong. Fl. and fr. Aug-Oct. Grassy slopes at 1600-2600 m. Distributed in NW Yunnan and Hunan.

狭叶山梗菜

Lobelia colorata Wall.

多年生草本，高30-100厘米。叶互生，倒卵状长圆形至条状披针形。总状花序顶生，疏松；花偏向一侧；花冠近二唇形，蓝紫色或蓝色，稀白色。蒴果卵球形。花果期9-10月。生海拔1000-3000米的沟壑灌丛或湿草甸。产云南和贵州。印度东北部和泰国北部亦有。

Perennial herbs, 30-100 cm tall. Leaves alternate, obovate-oblong to linear-lanceolate. Racemes terminal, lax; flowers oriented toward one side; corolla nearly 2-lipped, purple-blue or blue, rarely white. Capsules ovoid. Fl. and fr. Sep-Oct. Thickets in ravines or moist meadows at 1000-3000 m. Distributed in Yunnan and Guizhou. Also in NE India and N Thailand.

狭叶山梗菜 *Lobelia colorata*

毛萼山梗菜

Lobelia pleotricha Diels

多年生草本，高60-80厘米。茎深红色，疏具柔毛。叶互生；叶具柄，椭圆状披针形，两面密具白色柔毛。总状花序顶生；苞片叶状；花萼裂片条状披针形，密具白色柔毛；花冠紫红色至蓝紫色。蒴果短圆柱形。花果期8-10月。生海拔2000-3600米的草坡、灌丛或竹林边。产云南西部和西藏(墨脱县)。缅甸北部亦有。

Perennial herbs, 60-80 cm tall. Stems dark red, sparsely pubescent. Leaves alternate; lower leaves petiolate, elliptic-lanceolate, both surfaces densely white hirsute. Racemes terminal; bracts leaf-like; calyx lobes linear-lanceolate, densely white hirsute; corolla purple-red to blue-purple. Capsules shortly columnar. Fl. and fr. Aug-Oct. Grassy slopes, thickets or bamboo forest edges at 2000-3600 m. Distributed in W Yunnan and Xizang (Mêdog County). Also in N Myanmar.

毛萼山梗菜 *Lobelia pleotricha*

江南山梗菜 *Lobelia davidii*

江南山梗菜
Lobelia davidii Franch.

多年生草本，高达180厘米。叶互生，常具柄，叶柄具翅，长达4厘米；叶片卵状椭圆形至条状披针形。总状花序顶生，长20-50厘米；花冠近二唇形，紫红色。蒴果球形。花果期8-10月。生海拔4000米以下的林缘或溪边。产中国西南、华南、华中和华东。印度北部、尼泊尔、不丹和缅甸亦有。

Perennial herbs, up to 180 cm tall. Leaves alternate, usually petiolate, petioles winged, up to 4 cm long; leaves ovate-elliptic to linear-lanceolate. Racemes terminal, 20-50 cm long; corolla nearly 2-lipped, purple-red. Capsules globose. Fl. and fr. Aug-Oct. Forest edges or by streams below 4000 m. Distributed in SW, S, C and E China. Also in N India, Nepal, Bhutan and Myanmar.

马醉草 *Hippobroma longiflora*

铜锤玉带草 *Lobelia nummularia*

铜锤玉带草
Lobelia nummularia Lam.

多年生草本。茎平卧，细长。叶互生，具柄；叶圆形、肾形或卵形。单花腋生；花冠紫红色、浅紫色、粉红色、绿色或浅黄色，冠檐二唇形。浆果紫红色，椭圆体形或球形。花果期全年。生低海拔的田边、路旁、丘陵潮湿地或低海拔疏林。产西藏、广西、台湾、湖北和湖南。南亚、东南亚和新几内亚岛亦有。

Perennial herbs. Stems prostrate, gracile. Leaves alternate, petiolate; leaves orbicular, reniform or ovate. Flowers solitary and axillary; corolla purple-red, pale purple, pink, green or light yellow; limbs 2-lipped. Berries purple-red, ellipsoid or globose. Fl. and fr. all year round. Fieldsides, roadsides, wet places on hills or sparse forests at lower elevations. Distributed in Xizang, Guangxi, Taiwan, Hubei and Hunan. Also in S and SE Asia, and New Guinea.

马醉草
Hippobroma longiflora (L.) G. Don

多年生草本。叶椭圆形或倒披针形。花冠高脚碟状，白色，裂片椭圆形、狭椭圆形或条形；花萼裂片条形，具柔毛，边缘具齿。蒴果倒圆锥形、钟形、宽椭圆体形或倒卵球形，密被柔毛。种子亮褐色至红褐色，宽椭圆体形。生河岸。产广东和台湾。原产牙买加，广

五膜草 *Pentaphragma sinense*

泛归化和引种于热带和亚热带地区。

Perennial herbs. Leaves elliptic or oblanceolate. Corolla salverform, white; lobes elliptic, narrowly elliptic or linear; calyx lobes linear, villous, margin denticulate. Capsules obconic, campanulate, broadly ellipsoid or obovoid, densely villous. Seeds light brown to red-brown, broadly ellipsoid. River banks. Distributed in Guangdong and Taiwan. Native to Jamaica, widely introduced and naturalized in tropics and subtropics.

五膜草
Pentaphragma sinense Hemsl. et Wils.

多年生草本。植株被腺毛及混生星状毛。叶柄长3-10厘米；叶卵形，不对称。花序单生或2个合生，具柄，强烈弯曲，具苞片；花冠白色，裂至中部或至更深。浆果倒卵球形。花果期5-11月。生海拔200-1500米的森林、溪边或阴湿沟岸。产云南南部和广西西南部。越南北部亦有。

Perennial herbs. Plants covered with glandular hairs and these mixed with stellate hairs. Petioles 3-10 cm long; leaves ovate, asymmetric. Inflorescences solitary or 2 together, pedunculate, strongly curved, bracteate; corolla white, cleft for more than halfway. Berries obovoid. Fl. and fr. May-Nov. Forests, by streams or moist shady banks of ravines at 200-1500 m. Distributed in S Yunnan and SW Guangxi. Also in N Vietnam.

尖瓣花
Sphenoclea zeylanica Gaertn.

一年生草本，全株无毛。茎直立，常分枝。叶具柄。穗状花序长1-4厘米；苞片卵形；小苞片阔条形；萼裂片卵状圆形；花冠白色，浅裂；裂片明显。种子黄褐色。花果期终年。生稻田边或湿地。产中国西南、华南和东南。南亚、东南亚、西南亚、非洲东北部和热带、马达加斯加亦有，新大陆热带有引种。

Annual herbs, plants glabrous throughout. Stems erect, usually branched. Leaves petiolate. Spikes 1-4 cm long; bracts ovate; bracteoles broadly linear; calyx lobes ovate-orbicular; corolla white, shallowly lobed; lobes patent. Seeds brown-yellow. Fl. and fr. all year. Paddy fields or wet places. Distributed in SW, S and SE China. Also in S, SE and SW Asia, NE and tropical Africa and Madagascar, introduced in the New World tropics.

尖瓣花 *Sphenoclea zeylanica*

草海桐科 Goodeniaceae

草海桐 *Scaevola taccada*

草海桐

Scaevola taccada (Gaertn.) Roxb.

直立灌木或小乔木，有时下部平卧。叶螺旋状排列，多聚生于枝顶；叶匙形至倒卵形。聚伞花序腋生；花冠白色至浅黄色或紫色；花药在基部毛环处贴生成筒状，花期后分离。核果白色，卵球形，由侧面纵向开裂为两部分，每部分具4棱。花果期4-12月。生近海平面的开阔海岸沙地或岩石处。产华南、华东、南沙群岛、西沙群岛和东沙群岛。南亚、东南亚、东北亚、大洋洲、东非和热带澳大利亚亦有。

Erect shrubs or small trees, sometimes prostrating at base. Leaves spirally arranged, mostly aggregated at apex of branches; leaves spatulate to obovate. Cymes axillary; corolla white to pale yellow or purple; anthers connivent into a tubes with basal part of indusium, becoming free after anthesis. Drupes white, ovoid, divided longitudinally by furrows into 2 parts each 4-ribbed. Fl. and fr. Apr-Dec. Open coastal sands or rocks near sea level. Distributed in S and E China, Nansha Islands, Xisha Islands and Dongsha Islands. Also in S, SE and NE Asia, Oceania, E Africa and tropical Australia.

花柱草科 Stylidiaceae

花柱草

Stylidium uliginosum Swartz

一年生小草本。叶基生莲座状，卵状圆形、卵形或倒卵形，长5-8毫米，边缘全缘，无毛；叶柄极短。花序疏穗状；无花梗；苞片卵形，短于1毫米；花萼筒部狭长，疏生腺毛；花冠白色，长约2毫米，无毛；合蕊柱长约3.5毫米。蒴果细柱状，长约8毫米。花期10-11月。生丘陵溪边湿草地中。产广东和海南。斯里兰卡、柬埔寨、泰国和越南亦有。

Annual and small herbs. Leaves in a basal rosette, ovate-orbicular, ovate or obovate, 5-8 mm long, margin entire, glabrous; petioles very short. Inflorescences lax spikes; flowers sessile; bracts ovate, less than 1 mm; calyx tube linear, with sparse glandular trichomes; corolla white, ca. 2 mm long, glabrous; gynostemium ca. 3.5 mm long. Capsules columnar, ca. 8 mm long. Fl. Oct-Nov. Moist grassy places by streams in coastal hilly areas. Distributed in Guangdong and Hainan. Also in Sri Lanka, Cambodia, Thailand and Vietnam.

花柱草 *Stylidium uliginosum*

菊科 Compositae

栌菊木
Nouelia insignis Franch.

灌木或小乔木。叶互生，厚纸质，长圆形或近椭圆形，下面薄被灰白色绒毛。头状花序直立，单生，无梗，直径达5厘米；总苞钟形；总苞片被黄褐色绒毛；小花两性，白色；缘花二唇形，外唇舌状，内唇2裂，条形，外卷。瘦果圆柱形；冠毛微白色或黄白色。花期2-6月。生海拔1000-2900米的干热河谷。产云南西北部和四川西部。

Shrubs or small trees. Leaves alternate, thick chartaceous, oblong or subelliptic, thinly off-white tomentose abaxially. Capitula erect, solitary, sessile, to 5 cm diam; involucres campanulate; phyllaries yellowish-brown tomentose; florets bisexual, white; margin flowers bilabiate, outer labia ligulate, inner labia bilobed, linear, revolate. Achenes cylindric; pappus whitish or yellowish-white. Fl. Feb-Jun. Dry hot valley at 1000-2900 m. Distributed in NW Yunnan and W Sichuan.

白菊木
Leucomeris decora Kurz

落叶小乔木。幼枝白色，被绒毛。叶互生，纸质，椭圆形或长圆状披针形，下面被绒毛，网脉明显。头状花序直径近1厘米，8-12个或更多聚成复头状花序；总苞倒锥形；花先叶开放，白色，两性；花冠管状。瘦果圆柱形；冠毛淡红色。花期3-4月。生海拔1000-1900米的干热河谷。产云南。东南亚亦有。

Deciduous small trees. Young branchlets white, tomentose. Leaves alternate, chartaceous, elliptic or oblong-lanceolate, tomentose abaxially, reticulate veins conspicuous. Capitula ca. 1 cm diam, 8-12 or more congested in compound head; involucres obconical; flowers hysteranthous, white, bisexual; corolla tubular. Achenes cylindric; pappus reddish. Fl. Mar-Apr. Dry hot valley at 1000-1900 m. Distributed in Yunnan. Also in SE Asia.

和尚菜 *Adenocaulon himalaicum*

和尚菜
Adenocaulon himalaicum Edgew.

多年生草本。茎被蛛丝毛，上部具短柄腺点。头状花序排成松散圆锥花序；苞片5-7；小花白色；边花雌性；花冠管状，顶部4-5裂；两性花雄性，顶部5裂。瘦果棍状倒卵形，具短柄腺点。花期6-8月，果期9-11月。生海拔3400米以下的林中、灌丛、草坡或溪边。产中国西南、东南、华中、华北、华西、华东和东北。印度、俄罗斯、朝鲜半岛和日本亦有。

Perennial herbs. Stems cobwebby, with stipitate glands on upper portion. Heads in loose panicles; phyllaries 5-7; flowers white; marginal florets female; corolla tubular, 4 or 5-lobed at apex; disc florets male; corolla 5-lobed at apex. Achenes clavate-obovate, stipitate glandular. Fl. Jun-Aug. Fr. Sep-Nov. Forests, thickets, grassy slopes or streamsides below 3400 m. Distributed in SW, SE, C, N, W, E and NE China. Also in India, Russia, Korean Peninsula and Japan.

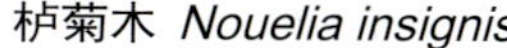
栌菊木 *Nouelia insignis*

白菊木 *Leucomeris decora*

大丁草 *Leibnitzia anandria*

大丁草

Leibnitzia anandria (L.) Turcz.

多年生草本，具春秋二型。春型植株矮小，高8-15厘米。叶基生，呈莲座状，提琴状羽状分裂，下面被白色绵毛。花葶短，头状花序小，舌状花粉红色，后变白色。秋型植株粗壮，高达30厘米，叶大；花葶长，头状花序大，无舌状花。瘦果灰色。花期3-7月和8-11月。生海拔600-2600米的山坡、阴地、林下或灌丛。产中国大部分地区。俄罗斯、朝鲜半岛和日本亦有。

Perennial herbs, with different spring and autumn forms. Spring plants dwarf, 8-15 cm tall. Leaves basal, rosette, lyrate-pinnate-lobed, adaxially white villose. Scapes short, capitula small, ray florets pink to white. Autumn plants stout, to 30 cm tall, leaves large, similar in form to the spring leaves; scapes long, capitula large, ray florets absent. Achenes gray. Fl. Mar-Jul and Aug-Nov. Slopes, shady places, forests or thickets at 600-2600 m. Distributed in most parts of China. Also in Russia, Korean Peninsula and Japan.

箭叶火石花

Gerbera maxima B. L. Rob.

多年生草本。叶莲座状基生；叶柄被绵毛；叶片卵状披针形，有时基部有小裂片，背面密被白色绵毛，基部心形，边缘有不规则锯齿。花葶1-2，被绵毛。头状花序单生顶端，辐射状；总苞宽钟形；总苞片约4层；冠毛白色。花期8-10月。生海拔2300-2800米的林缘草坡。产西藏(吉隆县)。喜马拉雅和泰国亦有。

Perennial herbs. Leaves in basal rosette; petioles lanuginous; leaf blade ovate-lanceolate, sometimes with few small lobes at base, abaxially densely white lanuginous, base cordate-sagittate, margin irregularly denticulate. Scapes 1 or 2, lanuginous. Capitula solitary, terminal, radiate; involucre broadly campanulate; phyllaries ca. 4-seriate; pappus whitish tawny. Fl. Aug-Oct. Grassy slopes at forest margins at 2300-2800 m. Distributed in Xizang (Gyirong County). Also in Himalaya and Thailand.

白背火石花

Gerbera nivea (DC.) Sch. Bip.

多年生草本。叶莲座状；叶柄被绵毛；叶片倒卵状匙形，叶背被灰色绵毛，基部渐狭，边

箭叶火石花 *Gerbera maxima*

白背火石花 *Gerbera nivea*

火石花 *Gerbera delavayi*

缘羽状浅裂或深裂；苞片无或1-2，钻形。头状花序单生，下垂，辐射状；总苞钟形；苞片4层。边缘雌花浅红色或白色。瘦果无毛；冠毛长8-10毫米。生海拔3300-4100米的高山草甸或林缘。产四川西部、西藏南部和云南西北部。不丹、印度和尼泊尔亦有。

Perennial herbs. Leaves in basal rosette; petioles lanuginous; leaf blade obovate-spatulate, abaxially gray lanuginous, base attenuate, margin pinnatilobate to pinnatisect; scapes arachnoid lanate; bracts absent or 1-2, subulate. Capitula solitary, nodding, terminal, radiate; involucre campanulate; phyllaries 4-seriate. Marginal female florets light red to white. Achenes glabrous; pappus 8-10 mm. Fl. Aug-Sep. Alpine meadows or forest margins at 3300-4100 m. Distributed in W Sichuan, S Xizang and NW Yunnan. Also in Bhutan, India and Nepal.

火石花
Gerbera delavayi Franch.

多年生草本。叶基生，披针形或长圆状披针形，下面厚被白色绵毛；花葶质硬，长10-30厘米；苞叶多数，条状钻形。头状花序单生花葶之顶；总苞陀螺状钟形；总苞片带紫红色；雌花舌状，淡红色；两性花管状二唇形。瘦果圆柱形，密被白色柔毛；冠毛黄白色。花期10月至翌年4月。生海拔1800-3200米的旷地、荒坡、松林或林缘。产云南和四川。越南北部亦有。

Perennial herbs. Leaves basal, lanceolate or oblong-lanceolate, densely white villose abaxially; scapes rigid, 10-30 cm long; bracteal leaves numerous, linear-subulate. Capitula solitary, terminal; involucres turbinate-campanulate; phyllaries purplish; pistillate flower ligulate, reddish; bisexual flowers tubular-bilabiate. Achenes cylindric, densely white pubescent; pappus yellowish-white. Fl. Oct to next Apr. Open places, wastelands, pine forests or forest edges at 1800-3200 m. Distributed in Yunnan and Sichuan. Also in N Vietnam.

非洲菊
Gerbera jamesonii Bolus

多年生草本。叶基生，莲座状，长椭圆形至长圆形，边缘不规则羽状分裂，下面被短柔毛；花葶长25-60厘米。头状花序单生花葶之顶，直径6-10厘米；总苞钟形；舌状花淡红色、紫红色、白色或黄色；管状花黄色。瘦果圆柱形，密被白色短柔毛；冠毛污白色。花期11月至翌年4月。中国各地庭园常见栽培。原产非洲。

Perennial herbs. Leaves basal, rosette, narrowly elliptic to oblong, margin irregularly pinnatifid, abaxially shortly pubescent; scapes 25-60 cm. Capitula solitary, terminal, 6-10 cm diam; involucres campanulate; ray florets reddish, purple, white or yellow; tubular florets yellow. Achenes cylindric, densely white puberulent; pappus dirty white. Fl. Nov to next Apr. Commonly cultivated in gardens of China. Native to Africa.

非洲菊 *Gerbera jamesonii*

兔耳一枝箭 *Piloselloides hirsuta*

兔耳一枝箭

Piloselloides hirsuta (Forssk.) C. Jeffrey ex Cufod.

多年生草本。叶莲座状；叶倒卵形、倒卵状长圆形或长圆形，纸质，下面密具白色蛛丝状绵毛。小花序单个顶生；总苞碟形；苞片2层；边花2层、雌性；中央小花多数、两性。瘦果纺锤形、具刚毛及长喙；冠毛正红色或棕色。花期2-5月，果期8-12月。生海拔900-2400米的开阔草坡、林缘或人为干扰区域。产中国西南、华南、东南、华中和华东。南亚、东南亚、东北亚、澳大利亚和非洲亦有。

Perennial herbs. Leaves in basal rosette; leaves obovate, obovate-oblong or oblong, papery, abaxially densely white arachnoid lanuginous. Capitula solitary, terminal; involucres disciform; phyllaries 2-seriate; marginal florets 2-seriate, female; central florets many, bisexual. Achenes fusiform, finely bristly, long beaked; pappus mandarin-red or brownish. Fl. Feb-May. Fr. Aug-Dec. Grassy open areas, forest edges, disturbed sites at 900-2400 m. Distributed in SW, S, SE, C and E China. Also in S, SE and NE Asia, Australia and Africa.

杏香兔儿风 *Ainsliaea fragrans*

杏香兔儿风

Ainsliaea fragrans Champ. ex Benth.

多年生草本。叶柄无翅；叶近革质，基部心形，全缘或疏被胼胝体状小齿。花序轴单生，苞片不明显；侧生分枝穗状；头状花序具3花；苞片约5层；花雌雄同体，开放花与闭锁花生同一植株。瘦果圆柱形或近纺锤形，具纵棱。花期9-12月。生海拔1300米以下的灌丛、路边或溪边草地。产中国西南、华南、东南、华中和华东。日本亦有。

Perennial herbs. Petioles wingless; leaves subleathery, base cordate, margin entire or sparsely callous-denticulate. Inflorescences axises single, obscurely bracteate; paraclades in spike; capitula 3-florous; phyllaries ca. 5-seriate; florets hermaphrodite, chasmogamous and cleistogamous in single plant; chasmogamous florets white; corolla tubular, deeply 5-lobed. Achenes terete or subfusiform, ribbed. Fl. Sep-Dec. Thickets, roadsides or grassy areas by stream banks below sea 1300 m. Distributed in SW, S, SE, C and E China. Also in Japan.

光叶兔儿风

Ainsliaea glabra Hemsl.

多年生草本。叶卵状披针形、披针形、长圆状披针形或椭圆形。头状花序排成圆锥花序，3花；总苞圆筒状。冠毛白色、污白色或浅红色。花期4-10月。生海拔600-2400米的溪边、林下草丛或林缘。产中国西南、华中和华东。

Perennial herbs. Leaves ovate-lanceolate, lanceolate, oblong-lanceolate or elliptic. Capitula arranged in panicles, 3-flowered; involucre cylindric. Pappus white, dark white or light red. Fl. Apr-Oct. Stream banks or moist grassy areas in forests, forest margins at 600-2400 m. Distributed in SW, C and E China.

光叶兔儿风 *Ainsliaea glabra*

大头兔儿风
Ainsliaea macrocephala

大头兔儿风

Ainsliaea macrocephala (Mattf.) Y. C. Tseng

多年生草本。叶基生，莲座状，厚纸质，卵形或卵状披针形。头状花序具花3朵，无梗，3-5或2-6密集成束，穗状花序式排列；总苞圆筒形；花两性，开花受精，紫红色；冠毛褐色至深褐色。花期8-9月。生海拔2600-3100米的山坡林缘、针叶林下或山脊灌丛中。产云南和四川。

Perennial herbs. Leaves in basal rosette; leaf blade ovate or ovate-lanceolate, subleathery. Capitula sessile, (2 or)3-5 (or 6) clustered, arranged in spikes, 3-flowered; involucre cylindric. Florets bisexual, chasmogamous, corollas purplish red; pappus brownish or dark brownish, 8-9 mm. Fl. Aug-Sep. Forests, forest margins, scrubs or grassy areas of mountain slopes at 2300-3600 m. Distributed in Yunnan and Sichuan.

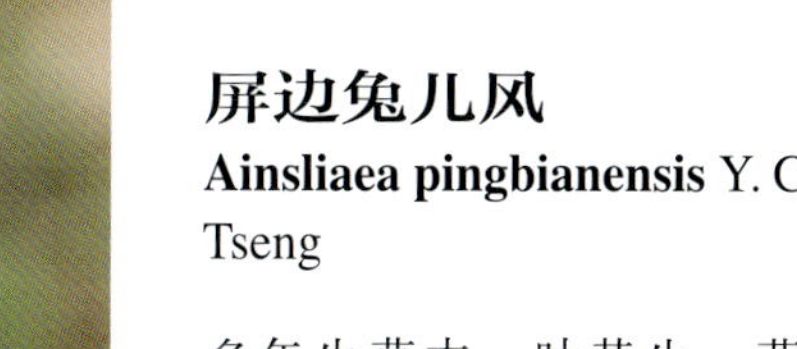

屏边兔儿风

Ainsliaea pingbianensis Y. C. Tseng

多年生草本。叶基生，莲座状；叶长圆形或长圆状披针形，两面被长柔毛。头状花序排成圆锥花序，3-5花；总苞圆筒形；花两性。花期5-8月。生海拔1300-1900米的溪旁或湿润密林中。产云南、四川和广东。

Perennial herbs. Leaves in basal rosette; leaf blade narrowly oblong or oblong-lanceolate, both surfaces villous. Capitula arranged in panicles, 3-5-flowered; involucre cylindric. Florets bisexual, chasmogamous and cleistogamous (at different seasons). Fl. Aug-Dec. Stream banks, moist areas in forests at 1300-1900 m. Distributed in Yunnan, Sichuan and Guangdong.

屏边兔儿风 *Ainsliaea pingbianensis*

长柄兔儿风

Ainsliaea reflexa Merr.

多年生草本。叶柄长1-13厘米，上部具翅，下部无翅；叶缘稍波状，具胼胝状小齿。花序轴单生，苞片不明显；侧生分枝穗状；头状花序具(1-)3花；苞片约5层；花雌雄同株，开放花与闭锁花在同一植株上并存。瘦果圆柱形，具纵棱；冠毛缺无。花期12月至翌年7月。生海拔500-3500米的开阔林中或灌丛。产云南、西藏、海南和台湾。东南亚亦有。

Perennial herbs. Petioles 1-13 cm long, upper parts winged, lower parts wingless; leaves margin slightly repand, callous-denticulate. Inflorescences axises single, obscurely bracteate; paraclades in spike; capitula (1-)3-florous; phyllaries ca. 5-seriate; florets hermaphrodite, chasmogamous and cleistogamous in same plant. Achenes terete, ribbed; pappus absent. Fl. Dec to next Jul. Open forests or thickets at 500-3500 m. Distributed in Yunnan, Xizang, Hainan and Taiwan. Also in SE Asia.

长柄兔儿风 *Ainsliaea reflexa*

云南兔儿风 *Ainsliaea yunnanensis*

云南兔儿风
Ainsliaea yunnanensis Franch.

多年生草本。根状茎圆柱形。基生叶莲座状，大小极不等，近革质，卵形至披针形，基部圆或截平。头状花序具3花，3-6个密集，偏向一侧，排列成间断的穗状花序；总苞圆筒形，总苞片5-6层，带紫红色；花淡红色。瘦果近纺锤形，冠毛黄白色，羽毛状。花期9月至翌年1-5月。生海拔1700-3700米的林下、林缘或草坡。产云南、四川和贵州。

Perennials herbs. Rhizomes terete. Basal leaves rosette, very unequal in size, subcoriaceous, ovate to lanceolate, base round or truncate. Capitula 3-flowered, 3-6-fasciculate, secund, arranged in incontinuous spike on top of stem; involucres cylindric, phyllaries 5-6 rows, tinged with purplish; flowers reddish. Achenes subfusiform, pappus yellowish-white, feathery. Fl. Sep to next Jan-May. Forests, forest edges or grassy slopes at 1700-3700 m. Distributed in Yunnan, Sichuan and Guizhou.

黄毛兔儿风
Ainsliaea fulvipes Jeffrey et W. W. Sm.

多年生草本，被黄褐色硬毛。根状茎圆柱形。基生叶莲座状，具柄，椭圆形或近卵形。头状花序具2-3花，单生或2-3簇生，于花葶之顶排成间断的穗状花序；总苞圆筒形；总苞片约5层；花淡红色；花冠管状。未成熟瘦果近长圆形，冠毛褐色，羽毛状。花期9-10月。生海拔1300-2700米的山坡、路旁或林下。产云南和广东。

Perennials herbs, yellowish-brown hispid. Rhizomes terete. Basal leaves rosette, petiolate, elliptic or subovate. Capitula 2-3-flowered, solitary or 2-3-fasciculate, arranged in incontinuous spike on top of scape; involucres cylindric, phyllaries ca. 5 rows; flowers reddish; corolla tubular. Immature achenes suboblong, pappus bro-

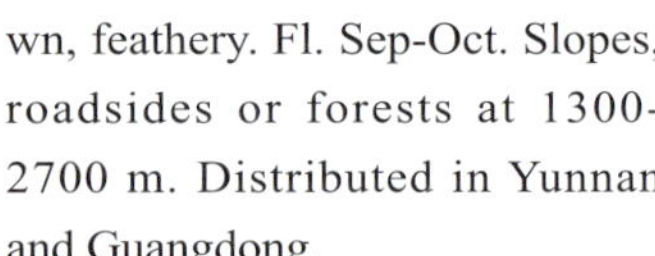
wn, feathery. Fl. Sep-Oct. Slopes, roadsides or forests at 1300-2700 m. Distributed in Yunnan and Guangdong.

黄毛兔儿风 *Ainsliaea fulvipes*

马边兔儿风
Ainsliaea angustata Chang

多年生草本。根状茎粗短。茎直立，单一。基生叶莲座状，具柄，狭椭圆形至披针形。头状花序具3花，单生，具短梗，排成狭圆锥花序；总苞圆筒形，总苞片4-5层，带紫红色；花冠管状。瘦果近纺锤形，冠毛微红色，羽毛状。花期3-5月和10月。生海拔600-1300米的水边、路旁、草地或石上。产四川、陕西和甘肃。

Perennial herbs. Rhizomes thick and short. Stems erect, simple. Basal leaves rosette, petiolate, narrowly elliptic to lanceolate. Capitula 3-flowered, solitary, shortly pedicellate, arranged in narrow panicle; involucres cylindric, phyllaries 4-5 rows, purplish; corolla tubular. Achenes subfusiform, pappus reddish, feathery. Fl. Mar-May and Oct. Watersides, roadsides, grassy areas or rocks at 600-1300 m. Distributed in Sichuan, Shaanxi and Gansu.

马边兔儿风 *Ainsliaea angustata*

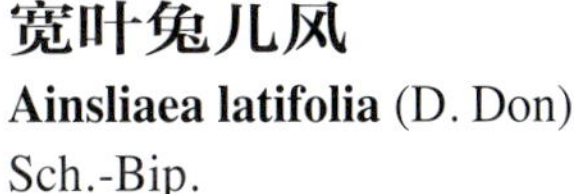
宽叶兔儿风 *Ainsliaea latifolia*

纤枝兔儿风 *Ainsliaea gracilis*

宽叶兔儿风

Ainsliaea latifolia (D. Don) Sch.-Bip.

多年生草本。叶柄宽翅状；叶纸质，卵圆形或狭卵圆形，边缘有胼胝体状细齿。花序轴单生，明显具苞片；侧生分枝穗状；头状花序具3花；苞片约5层；花雌雄同株，开放花与闭锁花在同一植株上并存。瘦果近纺锤形，具纵棱；冠毛棕褐色，羽毛状。花期全年。生海拔800-3600米的疏林下、干燥向阳地带或路边。产中国西南和华南。南亚和东南亚亦有。

Perennial herbs. Petioles broadly winged; leaves papery, ovate or narrowly ovate, margin callous-denticulate. Inflorescences axises single, obviously bracteate; paraclades in spike; capitula 3-florous; phyllaries ca. 5-seriate; florets hermaphrodite, chasmogamous and cleistogamous. Achenes subfusiform, ribbed; pappus brown, feathery. Fl. all year round. Open forests or roadsides at 800-3600 m. Distributed in SW and S China. Also in S and SE Asia.

纤枝兔儿风

Ainsliaea gracilis Franch.

多年生草本，被长柔毛。根状茎极短。叶聚生于茎的中下部，呈轮生状，卵形或卵状披针形，顶端刺芒状，基部心形。头状花序具3花，于茎顶排列成总状花序；总苞圆筒形，总苞片近7层；花两性；花冠管状。瘦果近纺锤形，冠毛淡红色，羽毛状。花期9-10月。生海拔400-1600米的山地丛林或涧旁石缝中。产四川、贵州、广西、广东、湖北、湖南和江西。

Perennials herbs, villose. Rhizomes very short. Leaves clustered at lower and middle part of stem, whorl-like, ovate or ovate-lanceolate, apex aristate, base cordate. Capitula 3-flowered, arranged in raceme on top of stem; involucres cylindric, phyllaries near 7 rows; flowers bisexual; corolla tubular. Achenes subfusiform, pappus reddish, feathery. Fl. Sep-Oct. Montane forests or cracks of stone by streams at 400-1600 m. Distributed in Sichuan, Guizhou, Guangxi, Guangdong, Hubei, Hunan and Jiangxi.

华南兔儿风 *Ainsliaea walkeri*

华南兔儿风

Ainsliaea walkeri Hook. f.

多年生矮小草本。根状茎短。叶聚生于茎的中下部，狭长圆形或条形，顶端突尖，边缘每侧各具1-3刺状尖齿，基出脉1。头状花序具短梗，具3花，于茎顶排列成狭圆锥花序；总苞片约5层，带紫红色；花冠白色。瘦果圆柱形，冠毛污白色，羽毛状。花期10-12月。生海拔700米以下的溪旁石上或密林下湿润处。产广西、广东和福建。

Dwarf perennials herbs. Rhizomes short. Leaves clustered at lower and middle part of stem, narrowly oblong or linear, apex aristate, each side with 1-3 aristate teeth on margin, basal veins 1. Capitula shortly pedicellate, 3-flowered, arranged in narrow panicle on top of stem; phyllaries ca. 5 rows, tinged with purplish; corolla white. Achenes cylindric, pappus dirty white, feathery. Fl. Oct-Dec. Rocks by streams or damp places in dense forests below 700 m. Distributed in Guangxi, Guangdong and Fujian.

针叶帚菊 *Pertya phylicoides*

华帚菊 *Pertya sinensis*

针叶帚菊

Pertya phylicoides Jeffrey

灌木。小枝多而质硬，帚状，被红褐色短柔毛。长枝上的叶互生；短枝上的叶4-6片簇生，针刺状。头状花序多数，无梗，单生；总苞圆筒形；总苞片被长柔毛，先端刺状锐尖；小花4-6，两性，管状，淡紫色。瘦果圆柱形，密被白色长柔毛；冠毛污白色。花期6-9月。生海拔2400-3100米的山坡或干旱河谷。产云南和西藏。

Shrubs. Twigs many, rigid, virgate, brownish-red puberulent. Leaves on long shoot alternate; leaves on short shoot 4-6-rosette, aciculate. Capitula numerous, sessile, solitary; involucres cylindric; phyllaries villose, apex aristate-acute; florets 4-6, bisexual, tubular, purplish. Achenes cylindric, densely white villose; pappus dirty white. Fl. Jun-Sep. Slopes or dry valleys at 2400-3100 m. Distributed in Yunnan and Xizang.

瓜叶帚菊

Pertya henanensis Y. Q. Tseng

多年生直立草本，高约1米。叶卵形至阔卵形，边缘具角状疏粗齿，有缘毛，先端渐尖。头状花序在枝顶腋生，单生或双生，每1头状花序具花7-9朵；总苞圆筒形；总苞片6-7层。花全部两性，花冠管状。花期9月。生海拔950-1110米的林下。产河南西部和重庆。

Herbs, perennial, ca. 1 m tall. Leaf blade ovate to broadly ovate, margin hornlike dentate, ciliate, apex acuminate. Capitula solitary or double, axillary on upper part of branches, 7-9-flowered; involucre cylindric; phyllaries 6-7-seriate. Florets bisexual, corolla tubula. Fl. Sep. Forests at 950-1100 m. Distributed in W Henan and Chongqing.

华帚菊

Pertya sinensis Oliv.

落叶灌木。长枝上的叶互生，长圆状披针形至披针形，全缘；短枝上的叶簇生，叶片长圆状披针形或狭椭圆形。头状花序；总苞狭钟形或近圆筒状。瘦果纺锤形，密被粗毛；冠毛干时黄白色。花期6-8月。生海拔2100-2500米的山坡或溪边灌丛或针叶林中。产甘肃、河南、湖北、宁夏、青海、陕西、山西和四川。

Shrubs, deciduous. Leaves on long shoots alternate, petioles, leaf blade oblong-lanceolate to lanceolate, margin entire; leaves on branchlets, tufted, leaf blade oblong-lanceolate to narrowly elliptic. Capitula; involucre narrowly campanulate or subcylindric. Achenes fusiform, densely hairy; pappus yellow-white. Fl. Jun-Aug. Scrub, coniferous forests at 2100-2500 m Distributed in Gansu, Henan, Hubei, Ningxia, Qinghai, Shaanxi, Shanxi and Sichuan.

瓜叶帚菊 *Pertya henanensis*

蚂蚱腿子 *Myripnois dioica*

蚂蚱腿子
Myripnois dioica Bunge

落叶小灌木。短枝叶椭圆形或近长圆形，长枝叶阔披针形或卵状披针形。头状花序近无梗，单生长枝叶腋；总苞近圆筒形；总苞片长圆形；花雌性和两性异株，先叶开放；雌花紫红色，舌状；两性花白色，管状二唇形。瘦果纺锤形，密被毛；冠毛白色。花期4-5月。生海拔100-600米的山坡、灌丛、林缘或路旁。产湖北、陕西、华北和东北。

Deciduous shrubs.Twigs fastigiated, shortly pubescent. Leaves on short shoots elliptic or nearly oblong, leaves on long shoots broadly lanceolate or ovate-lanceolate. Capitula subsessile, solitary at the axils of long shoots; involucres subcylindric; phyllaries oblong; flowers gynodioecious, hysteranthous; pistillate flowers purplish-red, ligulate; bisexual flowers white, tubular-bilabiate. Achenes fusiform, densely pubescent; pappus white. Fl. Apr-May. Slopes, thickets, forest edges or roadsides at 100-600 m. Distributed in Hubei, Shaanxi, N and NE China.

硬叶蓝刺头
Echinops ritro L.

多年生草本。茎单生或少数茎成簇生。叶革质，两面异色。复头状花序排成伞房花序，稀单生。外层总苞片倒披针形。小花蓝色。冠毛中部以下结合。花果期6-8月。生海拔400-2400米的山坡砾石地。产新疆。蒙古、俄罗斯、中亚、西南亚和欧洲亦有。

Perennial herbs. Stems solitary or tufted. Leaves leathery, discolorous. Pseudocephalia corymbose, rarely solitary. Outermost phyllaries oblanceolate. Corolla blue. Pappus connate in proximal half. Fl. and fr. Jun-Aug. Gravelly places on mountain slopes at 400-2400 m. Distributed in Xinjiang. Also in Mongolia, Russia, C and SW Asia, and Europe.

华东蓝刺头
Echinops grijsii Hance

多年生草本。茎密被蛛丝状绵毛。叶薄纸质，具柄，羽状深裂，边缘具刺状缘毛，上面绿色，下面密被白色蛛丝状绵毛。复头状花序单生枝端或茎顶，直径约4厘米。头状花序多数，各具1小花；花冠紫蓝色。瘦果倒圆锥形，密被长毛；冠毛淡褐色。花果期7-10月。生海拔100-800米的山坡草地。产中国东南、华北和华东。

Perennial herbs. Stems densely arachnoid sericeous. Leaves thin chartaceous, petiolate, pinnatiparted, margin aristate-ciliate, green adaxially, densely white arachnoid sericeous-tomentose abaxially. Compound heads solitary, terminal, ca. 4 cm diam. Capitula numerous, each with 1 floret; corolla purplish-blue. Achenes obconical, densely villose; pappus brownish. Fl. and fr. Jul-Oct. Grassy slopes at 100-800 m. Distributed in SE, N and E China.

硬叶蓝刺头 *Echinops ritro*

华东蓝刺头 *Echinops grijsii*

全缘叶蓝刺头 *Echinops integrifolius*

驴欺口 *Echinops davuricus*

全缘叶蓝刺头

Echinops integrifolius Kar. et Kir.

多年生草本。全部叶无柄，线形或线形披针形，边缘全缘，两面异色有柄。复头状花序单生茎顶。外层苞片线状倒披针形或线形，外面有短糙毛。小花白色。瘦果被淡黄色的长直毛；冠毛量杯状。花果期7-9月。生海拔400-2400米的石质旱燥山坡、沙丘、林间开阔地。产新疆。哈萨克斯坦、蒙古和俄罗斯亦有。

Perennial herbs. Leaves sessile, linear to linear-lanceolate, undivided, discolorous. Pseudocephalium solitary. Outermost phyllaries oblanceolate to linear, abaxially strigose. Corolla white. Achene hairs yellowish; pappus scales subulate. Fl. and fr. Jul-Sep. Dry rocky slopes, sandy slopes, open areas in forests, fields at 400-2400 m. Distributed in Xinjiang. Also in Kazakhstan, Mongolia and Russia.

砂蓝刺头

Echinops gmelinii Turcz.

一年生草本。茎单生，淡黄色。叶薄纸质，条形或条状披针形，基部扩大抱茎，边缘具刺齿，疏被蛛丝状毛和腺点；复头状花序单生茎顶或枝端，直径2-3厘米。小头状花序多数，各具小花1朵；花冠蓝色或白色。瘦果倒圆锥形，密被长毛；冠毛淡褐色。花果期6-9月。生海拔500-3200米的砾石地、荒漠、黄土丘陵或沙地。产华北、东北和西北。俄罗斯(西伯利亚)和蒙古亦有。

Annual herbs. Stems solitary, yellowish. Leaves thin chartaceous, linear or linear-lanceolate, base dilated, amplexicaul, margin aristate-serrate, sparsely arachnoid hairy and glandular punctuate; compound heads solitary, terminal, 2-3 cm diam. Capitula numerous, each with 1 floret; corolla blue or white. Achenes obconical, densely villose; pappus brownish. Fl. and fr. Jun-Sep. Gravelly places, deserts, loess hills or sandy places at 500-3200 m. Distributed in N, NE and NW China. Also in Russia (Siberia) and Mongolia.

驴欺口

Echinops davuricus Trevir.

多年生草本。茎直立，密被蛛丝状绵毛。叶具柄，薄纸质，椭圆形，二回羽状分裂，上面绿色，下面密被灰白色蛛丝状绵毛。复头状花序1-3个，顶生，直径3-3.5厘米。头状花序多数，各具1小花；花冠紫蓝色。瘦果密被淡黄色长毛；冠毛淡褐色。花果期6-9月。生海拔100-2200米的草原、草甸或林中。产华北、东北和西北。俄罗斯和蒙古亦有。

Perennial herbs. Stems erect, densely arachnoid sericeous. Leaves petiolate, thin chartaceous, elliptic, bipinnatiparted, green adaxially, densely white arachnoid sericeous-tomentose abaxially. Pseudocephalium 1-3, terminal, 3-3.5 cm diam. Capitula numerous, each with 1 floret; corolla purplish-blue. Achenes densely yellowish villose; pappus brownish. Fl. and fr. Jun-Sep. Grasslands, meadows or forests at 100-2200 m. Distributed in N, NE and NW China. Also in Russia and Mongolia.

砂蓝刺头 *Echinops gmelinii*

苍术 *Atractylodes lancea*

苍术

Atractylodes lancea (Thunb.) DC.

多年生草本。根状茎粗壮，疙瘩状，坚硬。叶倒卵形、长圆形或倒披针形，质硬，绿色，无毛，边缘具三角形刺齿或重刺齿。头状花序单生茎枝顶端；总苞钟状，直径1-1.5厘米；总苞片5-7层；花冠白色。瘦果倒卵球形；冠毛污白色。花果期6-10月。生海拔200-2500米的草地、林中、灌丛或岩缝。产中国西南、东南、华中、华北、华西、华东和东北。俄罗斯、朝鲜半岛和日本亦有。

Perennial herbs. Rhizomes stout, knobby, rigid. Leaves obovate, oblong or oblanceolate, rigid, green, glabrous, margin serrate or dupli-serrate. Capitula solitary on tops of stems or branches; involucres campanulate, 1-1.5 cm diam; phyllaries 5-7 rows; corolla white. Achenes obovoid; pappus dirty white. Fl. and fr. Jun-Oct. Grasslands, forests, thickets or rock crevices at 200-2500 m. Distributed in SW, SE, C, N, W, E and NE China. Also in Russia, Korean Peninsula and Japan.

白术

Atractylodes macrocephala Koidz.

多年生草本。叶片三至五回羽状全裂，薄纸质，绿色，无毛，边缘具针刺状缘毛或细刺齿，叶柄长3-6厘米，稀无柄。头状花序6-10个，单生茎枝顶端；总苞宽钟状，直径3-4厘米；总苞片9-10层；花冠紫红色。瘦果倒卵球形；冠毛污白色。花果期8-10月。生海拔600-2800米的山坡草地或林下。产四川、浙江、湖南和江西。

Perennial herbs. Leaves 3-5-pinnatisected, thinly chartaceous, green, glabrous, margin aciculate-ciliate or aciculate-serrate, petioles 3-6 cm long, rarely sessile. Capitula 6-10, solitary on tops of stems or branches; involucres broadly campanulate, 3-4 cm diam; phyllaries 9-10 rows; corolla purple. Achenes obovoid; pappus dirty white. Fl. and fr. Aug-Oct. Grassy slopes or forests at 600-2800m. Distributed in Sichuan, Zhejiang, Hunan and Jiangxi.

革苞菊

Tugarinovia mongolica Iljin

多年生草本，全株被黑白色蛛丝状毛。叶莲座状，椭圆形、长圆形或卵形，羽状分裂；裂片有不规则的刺齿；刺坚硬，2-4毫米。花葶单生或几个，匍匐或上升，不分枝，无叶。苞叶革质，有浅刺齿。总苞片线状披针形。花冠白色，瘦果密被绢毛；冠毛污白色。花果期5-6月。生海拔800-1500米的荒漠石质山坡、沙坡或砾石坡。产内蒙古西部。蒙古亦有。

Perennial herbs, whole plant with grayish white arachnoid lanate. Leaves rosulate, elliptic, oblong or ovate, pinnatipartite to pinnately lobed; segments with irregular shallow spine-tipped teeth; spines rigid, 2-4 mm. Flowering scapes solitary or few, decumbent or ascending, unbranched, leafless. Bracts leathery, with shallow spiny lobes. Phyllaries linear-lanceolate, apex spiny. Corolla whitish. Achene densely sericeous-villous; pappus dirty white. Fl. and fr. May-Jun. Stony hillsides, sandy sloping fields, gravelly slopes at 800-1500 m. Distributed in W Neimenggu. Also in Mongolia.

白术 *Atractylodes macrocephala*

革苞菊 *Tugarinovia mongolica*

尚武菊 *Shangwua denticulata*

绒毛苓菊 *Jurinea lanipes*

尚武菊

Shangwua denticulata (DC.) Raab-Straube et Yu J. Wang

多年生草本。基生叶及下部茎叶花期脱落。中部茎叶有短柄，叶片厚纸质，狭卵状椭圆形，两面异色。头状花序，单生茎端，成伞房花序状排列；总苞钟状；总苞片淡黄色，顶端圆，边缘栗色宽膜质。瘦果浅褐色，有棱。冠毛白色，1层。花果期8-10月。生海拔2400-4000米的林下、林缘、灌丛边缘、草地多石处。产云南和西藏。喜马拉雅亦有。

Perennial herbs. Basal and lower stem leaves withered at anthesis. Middle stem leaves petiolate, narrowly ovate-elliptic, discolorous. Capitula solitary, in a lax corymbiform synflorescence, involucre campanulate; phyllaries straw-colored, margin membranous, dark brown, and erose-lacerate, apex rounded. Achene pale brown, ribbed. Pappus 1 row, straw-colored. Fl. and fr. Aug-Oct. Forests, forest or thicket margins, grasslands, moist meadows at 2400-4000 m. Distributed in Yunnan and Xizang. Also in Himalaya.

绒毛苓菊

Jurinea lanipes Rupr.

多年生草本。基生叶狭倒披针形至椭圆形，有柄；中下部茎生叶有短柄至无柄；最上部叶线形，不分裂。头状花序单生茎端。总苞片碗状。小花红色。瘦果有稀疏的刺瘤。冠毛白色，基部不连合成环。花期5-8月，果期6-9月。生海拔1200-2900米的森林草地、草原、荒地和路边。产新疆。中亚亦有。

Perennial herbs. Basal leaves narrowly obovate to elliptic, petiolate; lower and middle stem leaves shortly petiolate to sessile; uppermost stem leaves linear, undivided. Capitula solitary at end of branches. Involucre bowl-shaped. Corolla reddish purple. Achene spinulose-tuberculate. Pappus white, not connate into ring. Fl. May-Aug. Fr. Jun-Sep. Alpine and subalpine forest meadows, steppes, wastelands, roadsides at 1200-2900 m. Distributed in Xinjiang. Also in C Asia.

多花苓菊

Jurinea multiflora (L.) B. Fedtsch.

多年生草本。基生叶和下部茎叶有短柄，两面异色。中部叶线形或宽线形，无柄。苞叶线状钻形。头状花序多数，排成伞房花序；苞片膜质；小花红色或紫色。瘦果褐色或浅红色，4肋。花期6-8月，果期7-9月。生海拔1800-2000米的森林、盐性草地、荒地、田野。产新疆。哈萨克斯坦、蒙古、俄罗斯和欧洲亦有。

Perennial herbs. Basal and lower stem leaves shortly petiolate, discolorous. Middle stem leaves linear to broadly linear, sessile. Bracts linear-subulate. Capitula

多花苓菊 *Jurinea multiflora*

三角叶须弥菊 *Himalaiella deltoidea*

many, in corymbose synflorescence; phyllaries scarious; corolla pink to purple. Achene brown to pale red, 4-ribbed. Fl. Jun-Aug. Fr. Jul-Sep. Forests, saline steppes, meadows, wastelands, fields at 1800-2000 m. Distributed in Xinjiang. Also in Kazakhstan, Mongolia, Russia and Europe.

三角叶须弥菊

Himalaiella deltoidea (DC.) Raab-Straube

二年生高大草本。茎生叶三角形至披针形，下面密被灰白色绒毛。头状花序大，下垂或歪斜，具长花梗；总苞半球形或宽钟状，直径3-4厘米；总苞片披针形，先端草质，边缘具锯齿；小花淡紫红色或白色。瘦果倒圆锥状，黑色；冠毛白色。花果期5-11月。生海拔700-3400米的山坡、草地、林下、林缘、灌丛、荒地、牧场或河谷。产中国西南、华南、华东和华中。尼泊尔和东南亚亦有。

Biennial large herbs. Cauline leaves triangular to lanceolate, densely offwhite tomentose abaxially. Capitula large, drooping or oblique, long pedunculate; involucres semiglobose or broadly campanulate, 3-4 cm diam.; phyllaries lanceolate, apex foliaceous, margin serrate; florets purplish-red or white. Achenes obconical, black; pappus white. Fl. and fr. May-Nov. Slopes, grasslands, forests, forest edges, thickets, wastelands, pastures or valleys at 700-3400 m. Distributed in SW, S, E and C China. Also in Nepal and SE Asia.

白背须弥菊

Himalaiella auriculata (DC.) Raab-Straube

多年生草本。基部和中部茎叶无柄，倒卵形，大头羽状分裂；上部茎叶狭卵形至椭圆形，有短柄。头状花序单生或少数，花期下垂。总苞宽钟状；总苞片5层，边缘被纤毛；花冠深紫色。瘦果4-5棱；冠毛褐色。花果期8-9月。生海拔2700-4000米的混交林、草坡。产西藏。喜马拉雅亦有。

Perennial herbs. Basal to middle stem leaves sessile, obovate, lyrately pinnately parted; upper stem leaves narrowly ovate to elliptic, shortly petiolate. Capitula solitary or few, nodding at anthesis. Involucre broadly campanulate; phyllaries in 5 rows, margin ciliate; corolla dark purple. Achene 4- or 5-ribbed; pappus brown. Fl. and fr. Aug-Sep. Mixed forests, grassy slopes at 2700-4000 m. Distributed in Xizang. Also in Himalaya.

白背须弥菊 *Himalaiella auriculata*

亚东须弥菊

Himalaiella yakla (C. B. Clarke) Fujikawa et H. Ohba

多年生草本，通常无茎。叶莲座状着生，叶片卵形、椭圆形或长圆形，长5-40厘米，宽3-8厘米，羽状分裂，边缘有齿。头状花序着生于莲座的中间；总苞直径0.8-3厘米，总苞片4-6层。花果期8-11月。生海拔3600-4100米的高山草地、灌丛。产西藏南部。喜马拉雅亦有。

Perennial herbs, usually stemless. Rosette leaves petiolate; leaves ovate, elliptic or oblong in outline, 5-40 × 3-8 cm, bipinnately lobed to bipinnately divided, margin dentate and mucronate. Capitula (1-)3-8(-10), in a condensed corymbiform synflorescence in center of leaf rosette; involucres campanulate, 0.8-3 cm diam; phyllaries in 4-6 rows. Fl. and fr. Aug-Nov. Alpine meadows, grasslands, thickets at 3600-4100 m. Distributed in S Xizang. Also in Himalaya.

亚东须弥菊 *Himalaiella yakla*

怒江川木香 *Dolomiaea salwinensis*

怒江川木香
Dolomiaea salwinensis (Hand.-Mazz.) C. Shih

多年生草本，无茎或有短茎。莲座叶椭圆形、窄椭圆形、窄卵形或匙形，无毛，边缘全缘，有钝齿或羽状分裂；上部叶密生于头状花序下，呈苞叶状。头状花序单生；总苞片3-4层；花冠紫红色；冠毛多层，黄褐色。花果期7-9月。生海拔2900-4000米的林缘或高山草甸。产云南西北部。缅甸东北部亦有。

Perennial herbs, stemless or shortly stemmed. Rosette leaves elliptic, narrowly elliptic, narrowly ovate or spatulate, both surfaces glabrous, margin entire, obtusely dentate or pinnately lobed; upper stem leaves subtending and overtopping capitula; capitula solitary; phyllaries in 3-4 rows; corolla purplish red; pappus bristles in several rows, yellowish brown. Fl. and fr. Jul-Sep. Forest margins or alpine meadows at 2900-4000 m. Distributed in NW Yunnan. Also in NE Myanmar.

膜缘川木香
Dolomiaea forrestii (Diels) C. Shih

多年生草本。叶基生，莲座状，具宽柄，质厚，椭圆形、卵形、近三角形或披针形，羽状浅裂或近半裂，疏被糙毛，边缘具锯齿。头状花序3-6个集生；总苞钟状，直径3厘米；总苞片坚硬，近革质，边缘褐色，膜质；小花紫红色。瘦果圆柱状；冠毛黄褐色。花果期7-11月。生海拔3000-4200米的山谷、山坡、草甸、灌丛、林缘或林下。产云南和四川。

Perennial herbs. Leaves basal, rosette, broadly petiolate, thick, elliptic, ovate, subtriangular or lanceolate, pinnately lobed or nearly parted to half, sparsely strigillose, margin serrate. Capitula 3-6, congested; involucres campanulate, ca. 3 cm diam; phyllaries rigid, subcoriaceous, margin brown, membranous; florets purple. Achenes cylindric; pappus yellowish-brown. Fl. and fr. Jul-Nov. Valleys, slopes, meadows, thickets, forest edges or forests at 3000-4200 m. Distributed in Yunnan and Sichuan.

川木香
Dolomiaea souliei (Franch.) C. Shih

多年生草本。叶基生，莲座状，质厚，具宽扁叶柄，椭圆形或披针形，羽状半裂，疏被糙伏毛和黄色小腺点。头状花序6-8个集生；总苞宽钟状；总苞片卵形、椭圆形至披针形，坚硬，先端尾状渐尖成针刺状；小花红色。瘦果圆柱状，稍扁；冠毛黄褐色。花果期7-10月。生海拔3500-4800米的高山草地或灌丛中。产四川和西藏。

Perennial herbs. Leaves basal, rosette, thick, broadly and flattened petiolate, elliptic or lanceolate, pinnatiparted to half, sparsely strigose and yellow glandular punctuate. Capitula 6-8, congested; phyllaries broadly campanulate; phyllaries ovate, elliptic to lanceolate, rigid, apex caudate-acuminate, aristate; florets red. Achenes cylindric, somewhate compressed; pappus yellowish-brown. Fl. and fr. Jul-Oct. Alpine grasslands or thickets at

膜缘川木香 *Dolomiaea forrestii*

川木香 *Dolomiaea souliei*

3500-4800 m. Distributed in Sichuan and Xizang.

厚叶川木香
Dolomiaea berardioidea (Franch.) C. Shih

多年生草本。叶基生，莲座状，质地厚，卵形、椭圆形或长圆形，边缘浅波状凹缺或锯齿，两面绿色，密被短糙毛和黄色小腺点。头状花序单生茎顶；总苞钟状，直径约5.5厘米；总苞片绿色，边缘疏被短缘毛；小花紫红色。瘦果扁三棱形；冠毛黄褐色。花果期6-10月。生海拔2800-3300(-5200)米的山坡、草地或灌丛中。产云南西北部和四川西南部。

Perennial herbs. Leaves basal, rosette, thick, ovate, elliptic or oblong, margin shallowly undulate-incised or serrate, both surfaces green, densely and shortly strigillose and yellow glandular punctuate. Capitula solitary, terminal; involucres campanulate, ca. 5.5 cm diam; phyllaries green, margin sparsely and shortly ciliate; florets purple. Achenes compressed trigonous; pappus yellowish-brown. Fl. and fr. Jun-Oct. Slopes, grasslands or thickets at 2800-3300(-5200) m. Distributed in NW Yunnan and SW Sichuan.

菜川木香
Dolomiaea edulis (Franch.) C. Shih

多年生草本，有或无茎。叶倒披针形或卵形，有长达6厘米的宽扁叶柄，通常羽状浅裂至深裂，被糙伏毛。头状花序单生于莲座叶中或茎顶；总苞宽钟状；苞片5层；小花紫红色。瘦果扁三棱形；冠毛多层，长3厘米。花果期6-10月。生海拔2600-4700米的山坡林缘或草地。产云南西北部、四川西部和西藏东南部。缅甸北部亦有。

Perennial herbs, acaulescent or not. Leaves oblanceolate or ovate, with broadly

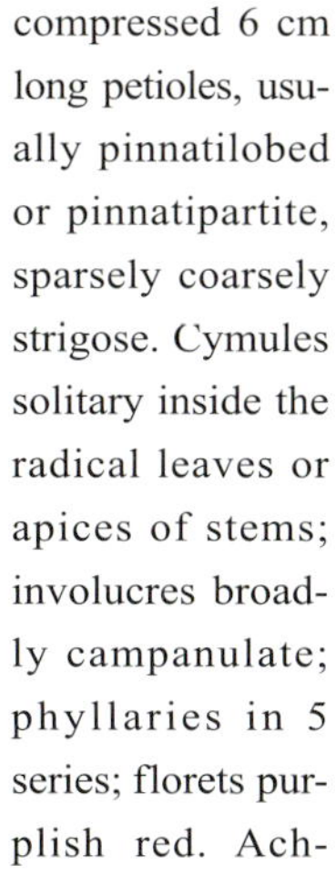
compressed 6 cm long petioles, usually pinnatilobed or pinnatipartite, sparsely coarsely strigose. Cymules solitary inside the radical leaves or apices of stems; involucres broadly campanulate; phyllaries in 5 series; florets purplish red. Achenes compressed prismatic; pappus in many series, up to 3 cm long. Fl. and fr. Jun-Oct. Forest edges, grasslands, fallow open areas at 2600-4700 m. Distributed in NW Yunnan, W Sichuan and SE Xizang. Also in N Myanmar.

厚叶川木香 *Dolomiaea berardioidea*

菜川木香 *Dolomiaea edulis*

蒲公英叶川木香 *Dolomiaea taraxacifolia*

蒲公英叶川木香

Dolomiaea taraxacifolia (J. Anthony) Y. S. Chen et Raab-Straube

多年生草本。叶莲座状，长圆形至椭圆形，边缘不规则羽状深裂。头状花序单生，总苞宽钟状；小花黑紫色。冠毛2-3层。花果期8-9月。生海拔3800-4300米的高山草甸、岩石坡。产云南西北部。缅甸东北部亦有。

Perennial herbs. Leaves in a rosette, oblong to elliptic, deeply and irregularly pinnatilobed to pinnatifid. Capitula solitary; involucre broadly campanulate; florets blackish purple. Pappus in 2-3 row. Fl. and fr. Aug-Sep. Open alpine meadows and on rocky slopes among boulders on marble at 3800-4300 m. Distributed in NW Yunnan. Also in NW Myanmar.

球菊

Bolocephalus saussureoides Hand.-Mazz.

多年生草本，高10-30厘米。茎不分枝，中空。叶片椭圆形和宽批针形，长10-20厘米，宽1-1.5 厘米，倒羽状分裂或羽状分裂，背面被灰白色密毛，正面绿色被疏毛。头状花序单生，球形，下垂，直径5-6厘米；总苞5-6层。花果期7-9月。生海拔3800-4500米的高山碎石坡、陡坡、杜鹃灌丛。产西藏南部。

Perennial herbs 10-30 cm tall. Stem simple, hollow. Leaf blade elliptic to broadly linear, 10-20 cm long, and 1-1.5 cm wide, runcinate, pinnately lobed or pinnatifid, abaxially grayish white and densely tomentose, adaxially green and sparsely arachnoid tomentose but glabrescent. Capitula solitary, drooping; involucres globose, 5-6 cm diam; phyllaries in 5-6 rows. Fl. and fr. Jul-Sep. Alpine scree slopes, cliffs, rocks, Rhododendron thickets at 3800-4500 m. Distributed in S Xizang.

大序齿冠菊

Frolovia frolowii (Ledeb.) Raab-Straube

多年生草本。叶下面密被蛛丝状柔毛，基生叶和茎下部叶有长柄；叶片大头羽状全裂。头状花序大，单一，生于茎顶，俯垂；总苞碗状；苞片约6层。瘦果圆柱形，有纵棱；冠毛淡黄色。花果期7-8月。生海拔约2000米的高山及亚高山草甸或山谷林缘。产新疆西北部。哈萨克斯坦和俄罗斯亦有。

Perennial herbs. Leaves densely arachnoid-pubescent abaxially, basal and lower stem leaves long petiolate; leaves lyrately pinnatipartite. Capitula large, solitary, nodding, on apices of stems; involucres bowl-shaped; phyllaries in ca. 6 series. Achenes cylindrical, ribbed; pappus pale yellow. Fl. and fr. Jul-Aug. Alpine and subalpine meadows or forest edges in valleys at ca. 2000 m. Distributed in NW Xinjiang. Also in Kazakhstan and Russia.

球菊 *Bolocephalus saussureoides*

大序齿冠菊 *Frolovia frolowii*

云木香 *Aucklandia costus*

云木香

Aucklandia costus Falc.

多年生高大草本，高1.5-2米，被短柔毛。主根直径达5厘米。基生叶具长翼柄，心形、卵形或三角形，边缘有大锯齿。头状花序1-5个，顶生；总苞半球形，直径3-4厘米，黑色；总苞片先端针刺状；小花暗紫色。瘦果三棱状；冠毛浅褐色。花果期5-8月。栽培于海拔1800-3000米的药圃中。云南、四川、贵州、广西、福建、陕西和安徽有栽培。原产喜马拉雅西北部。

Perennial large herbs, 1.5-2 m tall, puberulent. Taproots to 5 cm diam. Basal leaves long winged-petiolate, cordate, ovate or triangular, margin gross-serrate. Capitula 1-5, terminal; involucres semiglobose, 3-4 cm diam, black; phyllaries apiculate at apex; florets dark purple. Achenes trigonous; pappus brownish. Fl. and fr. May-Aug. Cultivated in medicinal gardens at 1800-3000 m. Cultivated in Yunnan, Sichuan, Guizhou, Guangxi, Fujian, Shaanxi and Anhui. Native to NW Himalaya.

泥胡菜

Hemisteptia lyrata (Bunge) Fisch. et C. A. Mey.

一年生或二年生草本，被白色蛛丝状毛。基生叶莲座状，倒披针形，大头羽状分裂，下面被白色绒毛；中部叶羽状分裂；上部叶条状披针形至条形。头状花序多数；总苞球形；总苞片顶端具紫红色鸡冠状附片；花冠紫色。瘦果圆柱形；冠毛白色。花果期3-8月。生海拔

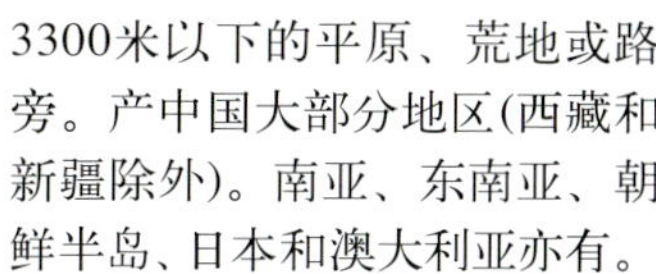

3300米以下的平原、荒地或路旁。产中国大部分地区(西藏和新疆除外)。南亚、东南亚、朝鲜半岛、日本和澳大利亚亦有。

Annual or biennial herbs, white arachnoid hairy. Basal leaves rosette, oblanceolate, lyrate-pinnatifid, white tomentose abaxially; middle cauline leaves pinnatifid; upper leaves linear-lanceolate to linear. Capitula numerous; involucres globose; phyllaries apex with purple cristate appendages; corolla purple. Achenes cylindric; pappus white. Fl. and fr. Mar-Aug. Plains, wastelands or roadsides at below 3300 m. Distributed in most parts of China (except Xizang and Xinjiang). Also in S and SE Asia, Korean Peninsula, Japan and Australia.

泥胡菜 *Hemisteptia lyrata*

洛扎雪兔子 *Saussurea lhozhagensis*

重羽菊 *Saussurea picridifolia*

重羽菊

Saussurea picridifolia (Hand.-Mazz.) Y. S. Chen et Qian Yuan

多年生草本，无茎或有短茎。叶基生长椭圆形或披针形；全缘或具稀疏锯齿。头状花序单生于莲座状叶中；总苞钟状；苞片4-5层；小花紫红色。瘦果倒圆锥形、压扁或三棱形；冠毛2层，白色，均为羽毛状。花果期8-9月。生海拔3600-4000米的山坡草地。产云南西北部与西藏东南部。

Perennial herbs, acaulescent or shortly stemed. Leaves long elliptic or lanceolate; entire or sparsely serrulate. Cymules solitary inside the radical leaves; involucres campanulate; phyllaries in 4-5 rows; florets purplish red. Achenes obconiform, compressed or prismatic; pappus in 2 series, white, both series plumose. Fl. and fr. Aug-Sep. Grassland on slopes at 3600-4000 m. Distributed in NW Yunnan and SE Xizang.

洛扎雪兔子

Saussurea lhozhagensis Y. S. Chen

多年生多次结实草本，高2-5厘米。无茎或有短茎。顶部叶包围头花，卵三角形，密被羊毛状毛。头状花序集生茎顶成伞房花序，花期隐藏在密集的蛛丝状毛中；总苞圆柱状，总苞片膜质；小花白色。瘦果黑色，圆锥形，有棱。冠毛深棕色。花果期8-9月。生海拔4330-4570米的石坡或沙质草甸。产西藏（洛扎县）。

Perennial herbs, 2-5 cm tall, polycarpic. Stemless or shortly stemed. Upper leaves surrounding capitula, ovate-triangular, with dense woollike hairs. Capitula densely clustered on flat top of stem in a synflorescence, hidden by dense arachnoid hairs at anthesis; involucre cylindric; phyllaries membranous; corolla white. Achene blackish, conic, ribbed. Pappus dark brown. Fl. and fr. Aug-Sep. Scree slope or sandy meadows at 4330-4570 m. Distributed in Xizang (Lhozhag County).

拟羽裂雪兔子

Saussurea pseudoleucoma Y. S. Chen

多年生草本，子房多数。地下茎分枝。不育叶莲座状，有柄，卵形；开花茎叶披针形或线状披针形，紫红色。头状花序多数，开花期暴露；总苞圆柱状；总苞片膜质；小花紫色；花药紫黑色。瘦果圆柱状，有棱，顶部有小冠；冠毛深褐色。花果期8-9月。生海拔4600-4800米的冰缘碎石带。产西藏南部。

Perennial herbs, polycarpic. Rhizome usually branched. Sterile leaves rosulate, petiolate, leaf blade ovate; flowering stem leaves lanceolate or linear-lanceolate, purplish red. Capitula many, exposed at anthesis; involucre cylindrical; phyllaries membranous; corolla purple; anthers purplish black. Achene cylindrical, ribbed, apex with a short truncate crown; pappus dark brown. Fl. and fr. Aug-Sep. Alpine periglacial scree zone at 4600-4800 m. Distributed in S Xizang.

拟羽裂雪兔子 *Saussurea pseudoleucoma*

冰川雪兔子

Saussurea glacialis Herder

多年生多次结实草本。基生叶及下部茎叶有短柄，叶片长椭圆形，背面被白色棉毛，正面被浅黄色或白色棉毛。头状花序3-15个，排列成头状伞房花序；总苞圆柱状至钟状。瘦果长椭圆状圆柱形；冠毛污白色。花果期7-8月。生海拔3800-5200米的高山流石滩。产西藏、青海和新疆。南亚、中

冰川雪兔子 *Saussurea glacialis*

左贡雪兔子 *Saussurea zogangensis*

亚、蒙古和俄罗斯亦有。

Perennial herbs, polycarpic. Rosette and stem leaves shortly petiolate, narrowly elliptic to obovate, abaxially white lanate, adaxially yellowish or white lanate. Capitula 3-15, in a hemispheric synflorescence; involucre cylindric to narrowly campanulate. Achene cylindric to obconic; pappus dirty white. Fl. and fr. Jul-Aug. Alpine scree slopes at 3800-5200 m. Distributied in Xizang, Qinghai and Xinjiang. Also in S and C Asia, Mongolia and Russia.

左贡雪兔子

Saussurea zogangensis Y. S. Chen

多年生多次结实草本，高3-10厘米，有时簇生。基部叶有短柄，长圆形或竹片形；顶部叶密集无柄，卵三角形，常反折。茎顶伞房花序半球形，花期裸露；总苞圆柱状总包配膜质；小花粉红色。瘦果长圆柱形，有棱，顶部有小冠；冠毛黑色。花果期8-9月。生海拔4400-5000米的高山流石滩。产西藏东部。

Perennial herbs, 3-10 cm tall, polycarpic, sometimes caespitose. Basal leaves narrowed into short petioles; leaf blade oblong or spatulate; uppermost leaves crowded, sessile, triangular-ovate, usually reflexed. Synflorescence at top of stem, hemispherical, exposed at anthesis; involucre cylindrical; phyllaries membranous; corolla pink. Achenes oblong-cylindrical, ribbed, apex with a short rim; pappus black. Fl. and fr. Aug-Sep. Alpine scree slopes at 4400-5000 m. Distributed in E Xizang.

星状雪兔子

Saussurea stella Maxim.

多年生无茎草本，全株光滑无毛。叶莲座状，星状排列，条状披针形，紫红色，先端长渐尖，基部无柄，扩大，全缘。头状花序多数，密集成半球形总花序；总苞圆柱形；总苞片长圆形至条形；小花紫色。瘦果圆柱状；冠毛白色。花果期7-9月。生海拔2000-5400米的高山草地、灌丛、河边或沼泽地。产青藏高原。印度和不丹亦有。

Perennial acauline herbs, wholly glabrous. Leaves rosette, stellately arranged, linear-lanceolate, purplish-red, apex long acuminate, base sessile, dilated, margin entire. Capitula numerous, congested in semiglobose inflorescences; involucres cylindric; invlocral bracts oblong to linear; florets purple. Achenes cylindric; pappus white. Fl. and fr. Jul-Sep. Alpine grasslands, thickets, riversides or swamps at 2000-5400 m. Distributed in Tibetan Plateau. Also in India and Bhutan.

云状雪兔子

Saussurea aster Hemsl.

一次性结实草本植物，无茎或有短茎。基生叶莲座状，叶片线状匙形、椭圆形或线形。头状花序5-25，无花序梗，在莲座状叶丛中密集成半球形的复合花序；小花紫红色；冠毛鼠灰色。花果期6-8月。生海拔3900-5400米高山流石滩。产青海、西藏和新疆南部。

Monocarpic herbs, stemless or shortly stemmed. Basal leaves rosette, leaves linear-spathulate, elliptic or linear, abaxially white and densely sericeous to lanate, adaxially grayish green or yellowish green, base attenuate, margin entire, apex obtuse. Capitula 5-25, in center of leaf rosette, in a hemispheric synflorescence 2.5-5 cm diam, sessile. Corolla rose-purple. Pappus in 2 rows; outer bristles dirty white, 3-4 mm; inner bristles brown. Fl. and fr. Jun-Aug. Alpine screes at 3900-5400 m. Distributed in Qinghai, Xizang and S Xinjiang.

星状雪兔子 *Saussurea stella*

云状雪兔子 *Saussurea aster*

羌塘雪兔子 *Saussurea wellbyi*

错那雪兔子 *Saussurea conaensis*

羌塘雪兔子

Saussurea wellbyi Hemsl.

多年生无茎草本。叶莲座状，无柄，条状披针形，被白色绒毛，顶端长渐尖，基部扩大，全缘。头状花序多数，密集成半球形总花序；总苞圆柱状；总苞片长圆形，紫红色，密被白色长柔毛；小花紫红色。瘦果圆柱状，黑褐色；冠毛淡褐色。花果期7-9月。生海拔4300-5500米的高山流石滩、山坡沙地或山坡草地。产青藏高原北部。

Perennial acauline herbs. Leaves rosette, sessile, linear-lanceolate, white tomentose, apex long acuminate, base dilated, margin entire. Capitula numerous, densely congested in semiglobose inflorescences; involucres cylindric; phyllaries oblong, purple-red, densely white villose; florets purple. Achenes cylindric, black brown; pappus brownish. Fl. and fr. Jul-Sep. Alpine screes, sandy or grassy slopes at 4300-5500 m. Distributed in N Tibetan Plateau.

槲叶雪兔子 *Saussurea quercifolia*

槲叶雪兔子

Saussurea quercifolia W. W. Sm.

多年生簇生草本，被白色绒毛。基生叶椭圆形，下面密被白色绒毛；茎生叶反折，披针形。头状花序多数，密集成半球形总花序；总苞长圆形；总苞片椭圆形或披针形，紫红色，被长柔毛；小花蓝紫色。瘦果褐色，圆柱状；冠毛鼠灰色。花果期7-10月。生海拔3300-5300米的高山灌丛、草地、流石滩或岩坡。产云南、四川和青海。

Perennial clustered herbs, white tomentose. Basal leaves elliptic, densely white tomentose abaxially; cauline leaves reflexed, lanceolate. Capitula numerous, densely congested in semiglobose inflorescences; involucres oblong; phyllaries elliptic or lanceolate, purple, villose; floret bluish-purple. Achenes brown, cylindric; pappus rat-gray. Fl. and fr. Jul-Oct. Alpine thickets, grasslands, screes or rocky slopes at 3300-5300 m. Distributed in Yunnan, Sichuan and Qinghai.

错那雪兔子

Saussurea conaensis (S. W. Liu) Fujikawa et H. Ohba

多年生草本，高12-20(-30)厘米。上端的茎、叶和花序密被白色长毛；基部叶无毛。叶片线形、窄倒披针形或窄长圆形；顶端的叶片包围头状花序，线形。头状花序单生茎顶；花紫色；冠毛白色或乳白色。花果期8-9月。生海拔(3680-)4000-5500米的高山流石滩、草甸。产西藏南部。不丹亦有。

Perennial herbs 12-20(-30) cm tall. Upper half of stem, stem leaves and capitula covered with dense white woollike multicellular hairs; basal leaves glabrous. Stem leaves linear, narrowly oblanceolate or narrowly oblong; uppermost leaves surrounding capitula, linear. Capitula solitary, on flat top of stem; corolla bright purple or purple; pappus white or creamy. Fl. and fr. Aug-Sep. Alpine scree slopes, sandy gravelly slopes, alpine meadows at (3680-)4000-5500 m. Distributed in S Xizang. Also in Bhutan.

拟三指雪兔子 *Saussurea pseudotridactyla*

拟三指雪兔子

Saussurea pseudotridactyla Y. S. Chen

多年生丛生草本，高5-12厘米。叶片匙形，密被蛛丝状白毛。头状花序多数，在茎端形成半球形直径2.5-3.5厘米的复合花序，花期裸露。总苞圆柱状，直径4-6毫米。花紫色。冠毛褐色，外层冠毛羽毛状。花果期8-9月。生海拔3700-4530米的高山流石滩或草甸。产西藏南部(错那县)。

Perennial herbs, 5-12 cm tall, polycarpic, usually forming a large clump. Leaves spathulate, densely white arachnoid. Capitula many, in terminal of the stem, in a hemispherical synflorescence 2.5-3.5 cm diam, exposed an anthesis. Involucres cylindrical, 4-6 mm diam. Corolla purple. Pappus dark brownish, outer pappus plumose. Fl. and fr. Aug-Sep. Alpine scree zone or meadows at 3700-4530 m. Distributied in S Xizang (Cona County).

羽裂雪兔子

Saussurea leucoma Diels

多年生草本，高14-18厘米，覆盖有白色长棉毛。茎直立，上部膨大。下部叶具宽扁叶柄，叶片长椭圆形，羽状分裂，裂片椭圆形，全缘；上部叶反折，条形，无柄，全缘。头状花序多数，密集成顶生圆锥状或球状总花序；总苞长圆状；总苞片披针形至椭圆形；小花紫黑色。瘦果倒圆锥状，紫黑色；冠毛褐色。花果期8-10月。生海拔3200-5300米的高山草坡、多石地或流石滩。产云南、四川和西藏。

Perennial herbs, 14-18 cm tall, densely white long sericeous-villose. Stem erect, upper part dilated. Lower cauline leaves with broad flattened petioles, blades narrowly elliptic, pinnatifid, lobes elliptic, margin entire; upper ones reflexed, linear, sessile, entire. Capitula numerous, densely congested in terminal conical or globose inflorescences; involucres oblong; phyllaries lanceolate to elliptic; florets dark purple. Achenes obconical, dark purple; pappus brown. Fl. and fr. Aug-Oct. Alpine grassy slopes, stony places or screes at 3200-5300 m. Distributed in Yunnan, Sichuan and Xizang.

三指雪兔子

Saussurea tridactyla Sch.-Bip. ex Hook. f.

多年生草本，密被长棉毛。叶密集，具宽柄，条形、匙形或长圆形，两面密被白色棉毛。头状花序多数，密集成半球形总花序，总花序为白色棉毛所覆盖；总苞长圆状；总苞片紫红色，长圆形；小花紫红色。瘦果倒圆锥状，褐色；冠毛1层，长羽毛状。花果期8-9月。生海拔4300-5300米的高山流石滩、山顶碎石间或山坡草地。产云南和西藏。印度、尼泊尔和不丹亦有。

Perennial herbs, densely long villose. Leaves dense, broadly petiolate, linear, spathulate or oblong, both surfaces densely white villose. Capitula numerous, congested in semiglobose inflorescences which covered by white plumule; involucres oblong; phyllaries purple, oblong; florets purple-red. Achenes obconical, brown; pappus 1 layer, long feathery. Fl. and fr. Aug-Sep. Alpine screes, summit rubbles or grassy slopes at 4300-5300 m. Distributed in Yunnan and Xizang. Also in India, Nepal and Bhutan.

羽裂雪兔子 *Saussurea leucoma*

三指雪兔子 *Saussurea tridactyla*

绵头雪兔子 *Saussurea laniceps*

绵头雪兔子

Saussurea laniceps Hand.-Mazz.

多年生草本，上部覆盖有稠密棉毛。叶极密集，倒披针形、狭匙形或长椭圆形，下面密被褐色绒毛。头状花序多数，密集成顶生圆锥状穗状花序；苞叶条状披针形；总苞宽钟状；总苞片披针形；小花白色。瘦果圆柱状；冠毛鼠灰色。花果期8-10月。生海拔3200-5500米的高山岩石坡或流石滩。产云南、四川和西藏。

Perennial herbs, densely long sericeous-villose. Leaves dense, oblanceolate, narrowly spathulate or elliptic, densely brown tomentose abaxially. Capitula numerous, densely congested in terminal conical spike; bracteal leaves linear-lanceolate; involucres broadly campanulate; phyllaries lanceolate; florets white. Achenes cylindric; pappus rat-gray. Fl. and fr. Aug-Oct. Alpine rocky slopes or screes at 3200-5500 m. Distributed in Yunnan, Sichuan and Xizang.

水母雪兔子

Saussurea medusa Maxim.

多年生草本，覆盖有白色长棉毛。茎直立。叶多数，密集，莲座状，叶形多变，边缘具8-12个粗齿，叶柄紫色。头状花序多数，密集成半球形总花序；苞叶条状披针形；总苞狭圆柱状；外层总苞片长椭圆形，紫色；小花蓝紫色。瘦果纺锤形；冠毛白色。花果期7-9月。生海拔3000-5600米的高山流石滩或多砾石山坡。产青藏高原。

Perennial herbs, wholly densely long sericeous-villose. Stems erect. Leaves numerous, dense, rosette, varied in shape, margin 8-12-gross-dentate, petioles purple. Capitula numerous, densely congested in semiglobose inflorescences; bracteal leaves linear-lanceolate; involucres narrowly cylindric; outer phyllaries oblong, purple; florets bluish-purple. Achenes fusiform; pappus white. Fl. and fr. Jul-Sep. Alpine screes or stony slopes at 3000-5600 m. Distributed in Tibetan Plateau.

雪兔子

Saussurea gossipiphora D. Don

多年生草本，上部覆盖以白色或黄褐色厚棉毛。叶条状长圆形或长椭圆形；苞叶条状披针形，反折。头状花序多数，密集成顶生半球状总花序；总苞宽圆柱状；总苞片卵状披针形或条状长圆形；小花紫红色。瘦果黑色；冠毛淡褐色。花果期8-10月。生海拔4200-5000米的高山流石滩、岩缝中或山顶多石地。产西藏。印度和尼泊尔亦有。

Perennial herbs, upper part densely white or yellowish-brown sericeous-villose. Leaves linear-oblong or narrowly elliptic; bracteal leaves linear-lanceolate, reflexed. Capitula numerous, congested in terminal semiglobose inflorescences; involucres broadly cylindric; phyllaries ovate-lancolate or linear-oblong; florets purplish-red. Achenes black; pappus brownish. Fl. and fr. Aug-Oct. Alpine screes, rocky crevices, or summit stony places at 4200-5000 m. Distributed in Xizang. Also in India and Nepal.

苞叶雪莲

Saussurea obvallata (DC.) Sch.-Bip.

多年生草本，具腺毛。叶具柄，长圆形或卵形，边缘具细齿；苞叶大，膜质，黄色，长圆形。头状花序6-15个，密集成球形总花序；总苞半球形；总苞片边缘黑紫色；小花蓝紫色。瘦果长圆形；冠毛淡褐色。花果期7-9月。生海拔3200-5200米的高山草地、山坡多石处、流石滩或溪边石隙处。产云南、四川、西藏和青海。印度西北部和尼泊尔亦有。

Perennial herbs, glandular hairy. Leaves petiolate, oblong or ovate, margin serrulate; bracteal leaves large, membranous, yellow, oblong. Capitula 6-15, densely congested in globose inflorescences; involucres semiglobse; phyllaries dark purple at margin;

水母雪兔子 *Saussurea medusa*

雪兔子 *Saussurea gossipiphora*

苞叶雪莲 *Saussurea obvallata*

florets bluish-purple. Achenes oblong, pappus brownish. Fl. and fr. Jul-Sep. Alpine grasslands, stony slopes, screes or rocky crevices by rivers at 3200-5200 m. Distributed in Yunnan, Sichuan, Xizang and Qinghai. Also in NW India and Nepal.

垂头雪莲 *Saussurea wettsteiniana*

垂头雪莲

Saussurea wettsteiniana Hand.-Mazz.

多年生草本，被白色长柔毛。基生叶条状披针形或条状长圆形，两面被腺毛；苞叶卵形或椭圆形，膜质，黄色，全缘，被腺毛。头状花序1-3，下垂；总苞半球形；总苞片披针形，黑紫色，被黄白色长柔毛；小花蓝紫色。瘦果长圆形；冠毛淡褐色。花果期7-9月。生海拔3200-4300米的山坡、林下、林缘、湿润草地或草甸。产云南、四川。

Perennial herbs, white villose. Basal leaves linear-lanceolate or linear-oblong, both surfaces glandular hairy; bracteal leaves ovate or elliptic, membranous, yellow, entire, glandular hairy. Capitula 1-3, drooping; involucres semiglobose; phyllaries lanceolate, dark purple, yellowish-white villose; florets bluish-purple. Achenes oblong; pappus brownish. Fl. and fr. Jul-Sep. Slopes, forests, forest edges, damp grasslands or meadows at 3200-4300 m. Distributed in Yunnan and Sichuan.

宝璐雪莲

Saussurea luae Raab-Straube

多年生簇生草本，高30-70厘米。根状茎粗壮，多分枝。叶片狭椭圆形至狭卵形，边缘有波状齿，茎生叶无柄。头状花序2-8个形成伞形的花序，偶尔单生；总苞钟状，总苞片黑色或紫褐色，有长柔毛，先端锐尖；冠毛棕色。花期8-9月，果期9-10月。生海拔4000-5000米砾石坡、流石滩。产四川西北部和西藏东部。

Perennial herbs, 30-70 cm tall, caespitose, forming large clumps. Caudex stout, much branched. Leaf blade narrowly elliptic to narrowly ovate, margin sinuate-dentate, usually sessile. Capitula 2-8 in a corymbiform synflorescence or rarely solitary; involucres campanulate, phyllaries blackish or purplish brown, villous, apex acute; pappus rose-purple. Fl. Aug-Sep. Fr. Sep-Oct. Open gravelly and rocky slopes, stabilized boulder scree slopes, ravine beds at 4000-5000 m. Distributed in NW Sichuan and E Xizang.

宝璐雪莲 *Saussurea luae*

多鞘雪莲

Saussurea polycolea Hand.-Mazz.

无茎莲座状草本，被粗毛和腺毛。叶椭圆形，边缘具小齿；苞叶卵形，紫红色。头状花序单生于莲座状叶丛中；总苞钟状；总苞片卵形，先端长渐尖，边缘褐色；小花多数；冠毛白色。花果期7-10月。生海拔3200-4700米的开阔多石牧场、山坡草地或灌丛。产云南、四川和西藏。

Acauline rosette herbs, hirsute and glandular hairy. Leaves elliptic, margin serrulate; bracteal leaves ovate, purple. Capitula solitary, surrounded by rosette leaves; involucres campanulate; phyllaries ovate, apex acuminate, margin brown; florets numerous; pappus white. Fl. and fr. Jul-Oct. Open stony pastures, grasslands on mountain slopes or thickets at 3200-4700 m. Distributed in Yunnan, Sichuan and Xizang.

多鞘雪莲 *Saussurea polycolea*

毡毛雪莲 *Saussurea velutina*

唐古特雪莲

Saussurea tangutica Maxim.

多年生草本，高6-70厘米。基生叶有叶柄，叶片宽椭圆形，边缘有波状锯齿；顶端叶片卵形或舟形，膜质，紫红色或黄白色，包围头状花序。头状花序1-5个，在茎端密集或单生茎顶；总苞片4层、紫黑色、密被长柔毛；冠毛2层，淡褐色。花果期7-9月。生海拔3600-5300米高山流石滩、高山草甸。产甘肃、青海、四川西北部和西藏。

Perennial herbs, 6-70 cm tall. Rosette and lower stem leaves narrowly elliptic, margin sinuatedentate; uppermost stem leaves enclosing synflorescence, ovate or boat-haped, membranous, both surfaces yellowish rose to purple. Capitula 1-5, subsessile. Phyllaries in ca. 4 rows, blackish purple, densely villous; pappus dirty white. Fl. and fr. Jul-Sep. Alpine scree slopes, alpine meadows at 3600-5300 m. Distributed in Gansu, Qinghai, NW Sichuan and Xizang.

唐古特雪莲 *Saussurea tangutica*

毡毛雪莲

Saussurea velutina W. W. Sm.

多年生簇生草本，高17-40厘米。茎直立，不分枝。下部茎叶有叶柄，叶片线状披针形或披针形，两面密被灰绿色长绒毛，边缘有齿；最上部茎叶苞叶状，倒卵形，紫红色。头状花序单生茎顶；总苞片4-5层，黑紫色或边缘黑紫色，外面被黄褐色长柔毛。冠毛污白色。花果期7-9月。生海拔(3300-)4100-5500米高山草地、灌丛及流石滩。产四川西部和云南西北部。

Perennial herbs, 17-40 cm tall, caespitose. Rosette and lower stem leaves petiolate; leaves narrowly elliptic to linear, both surfaces grayish green and densely lanate-tomentose, margin shallowly sinuate-denticulate to subentire; uppermost stem leaves ovate or narrowly ovate and boat-shaped, membranous, enclosing involucre, both surfaces purplish red. Capitula solitary; phyllaries in 4-5 rows, black or blackish purple, yellowish white villous. Pappus dirty white. Fl. and fr. Jul-Sep. Alpine scree slopes, mat and pastures at (3300-)4100-5500 m. Distributed in W Sichuan and NW Yunnan.

雪莲花

Saussurea involucrata (Kar. et Kir.) Sch.-Bip.

多年生草本。叶密集，无毛，椭圆形或卵状椭圆形；苞叶膜质，淡黄色，宽卵形，包围总花序。头状花序10-20个，密集成顶生球形总花序；总苞半球形；总苞片紫褐色，疏被长柔毛；小花紫色。瘦果长圆形；冠毛污白色。花期7-8月，果期

雪莲花 *Saussurea involucrata*

8-10月。生海拔2400-4100米的山坡、山谷、石缝、水边或草甸。产新疆。俄罗斯和哈萨克斯坦亦有。

Perennial herbs. Leaves dense, glabrous, elliptic or ovate-elliptic; bracteal leaves membranous, yellowish, broadly ovate, surround inflorescences. Capitula 10-20, congested in terminal globose inflorescences; involucres semiglobose; phyllaries purplish-brown, sparsely villose; florets purple. Achenes oblong; pappus dirty white. Fl. Jul-Aug. Fr. Aug-Oct. Slopes, valleys, rocky crevices, watersides or meadows at 2400-4100 m. Distributed in Xinjiang. Also in Russia and Kazakhstan.

红柄雪莲

Saussurea erubescens Lipsch.

多年生草本。基生叶及下部茎叶有叶柄，宽披针形或长椭圆形；最上部茎叶无柄，舟状椭圆形，紫色。头状花序排成伞房花序，极少单生；总苞倒圆锥状；小花紫色。瘦果有棱纹；冠毛污白色。花期7-8月，果期8-10月。生海拔2400-4900米的沼泽草地、河边、山谷、山顶、草甸。产青藏高原。

Perennial herbs. Basal and lower stem leaves petiolate; leaf blade narrowly elliptic; uppermost stem leaves sessile, elliptic and boat-shaped, purple. Capitula in a corymbiform synflorescence or rarely solitary; involucre obconic; corolla purple. Achene indistinctly ribbed; pappus dirty white. Fl. Jul-Aug. Fr. Aug-Oct. Grassy areas in marshes, by rivers, mountain valleys, meadows at 2400-4900 m. Distributed in Tibetan Plateau.

球花雪莲

Saussurea globosa F. H. Chen

多年生草本。茎直立。基生叶有叶柄，长椭圆形、披针形或长圆状披针形；上部苞叶卵状、舟形，紫红色，膜质，半包围伞房状总花序。头状花序排成伞房状总花序，希单生；总苞球形；总苞片全部或边缘紫红色；冠毛淡黄色。花果期7-10月。生海拔3000-4800米的草原、灌木丛、草甸。产四川和云南。

Perennial herbs. Stems usually solitary. Rosette leaves petiolate; leaf blade narrowly ovate, elliptic or linear; uppermost stem leaves ovate and boat-shaped, red to purple, membranous, half-enclosing synflorescence. Capitula in a corymbiform synflorescence or rarely solitary; involucre globose; phyllaries blackish purple or yellowish with dark margin; pappus yellowish. Fl. and fr. Jul-Oct. Alpine grasslands, thickets, meadows at 3000-4800 m. Distributed in Sichuan and Yunnan.

红柄雪莲 *Saussurea erubescens*

球花雪莲 *Saussurea globosa*

紫苞雪莲 *Saussurea iodostegia*

紫苞雪莲

Saussurea iodostegia Hance

多年生草本。基生叶条状长圆形，边缘具疏锐细齿；苞叶膜质，紫色，椭圆形，包围总花序。头状花序4-7个，密集成顶生伞房状总花序；总苞宽钟状；总苞片带紫色，先端钝，被白色长柔毛；小花紫色。瘦果长圆形，淡褐色；冠毛淡褐色。花果期7-9月。生海拔1300-3500米的山坡草地、山地草甸、林缘或盐沼泽地。产华北、东北和西北。

Perennial herbs. Basal leaves linear-oblong, margin sparsely acute-serrulate; bracteal leaves membranous, purple, elliptic, surround inflorescences. Capitula 4-7, congested in terminal corymbose-inflorescences; involucres broadly campanulate; phyllaries purple, apex obtuse, white villose; florets purple. Achenes oblong, brownish; pappus brownish. Fl. and fr. Jul-Sep. Grassy slopes, montane meadows, forest edges or saline swamps at 1300-3500 m. Distributed in N, NE and NW China.

太白山雪莲

Saussurea taipaiensis Y. Ling

多年生草本，簇生。莲座叶和底部叶有柄，椭圆形至长椭圆形；中上部叶无柄，狭卵形至线形和舟形；最顶部叶并入总苞中，被绒毛。头状花序1或2，几无柄；总苞倒圆锥形；总苞片黑紫色，密被绒毛。瘦果圆筒状；冠毛淡黄色。花期8月，果期9-10月。生海拔3200-3900米的高山草甸。产陕西(秦岭)、甘肃和山西。

Perennial herbs, caespitose. Rosette and lower stem leaves petiolate; leaf blade narrowly elliptic to oblong; middle and upper stem leaves sessile, narrowly ovate to linear and boat-shaped; uppermost stem leaves merging into phyllaries, abaxially villous. Capitula 1 (or 2), subsessile; involucre broadly obconic; phyllaries blackish purple, densely villous. Achene cylindric; pappus straw-colored. Fl. Aug. Fr. Sep-Oct. Alpine meadows at 3200-3900 m. Distributed in Shaanxi (Qinling mountain), Gansu and Shanxi.

华中雪莲

Saussurea veitchiana J. R. Drumm. et Hutch.

多年生草本。茎基部和中部叶条状披针形，疏被黄白色柔毛；上部叶包围花序，紫色，膜质。头状花序多数排成伞房状；总苞4层；冠毛棕灰色，2层；小花紫红色。瘦果棕红色。花果期7-9月。生海拔1600-3000米的沼泽地、山坡草地或山顶。产重庆和湖北。

Perennial herbs. Basal and lower stem leaves linear-lanceolate, sparsely yellowish white villous; uppermost stem leaves surrounding inflorescences, purple, membranous. Inflorescences corymbose; capitula numerous; phyllaries in 4 rows; pappus pale brown, in 2 rows; florets purplish red. Achenes reddish brown. Fl. and fr. Jul-Sep. Marshes, grasslands on mountain slopes or mountain peaks at 1600-3000 m. Distributed in Chongqing and Hubei.

太白山雪莲 *Saussurea taipaiensis*

华中雪莲 *Saussurea veitchiana*

膜苞雪莲 *Saussurea bracteata*

无梗风毛菊 *Saussurea apus*

膜苞雪莲

Saussurea bracteata Decne.

多年生草本。莲座叶和基部叶有柄，狭长圆形；中部叶无柄，卵状椭圆形；最上部叶膜质，黄绿色至紫红色，包围头状花序。头状花序单生茎顶；总苞狭钟状或倒圆锥状；小花紫红色；冠毛淡褐色。花果期7-9月。生海拔4000-5400米的高山草甸、高山流石滩。产新疆和西藏。印度西北部和克什米尔地区亦有。

Perennial herbs. Rosette and lower stem leaves petiolate; leaf blade narrowly elliptic; middle and upper stem leaves sessile, ovate-elliptic; uppermost stem leaves membranous, half-enclosing capitula, yellowish green tinged purple to purplish red. Capitula solitary; involucre campanulate to obconic; corolla purplish red; pappus dirty white. Fl. and fr. Jul-Sep. Alpine meadows and scree slopes at 4000-5400 m. Distributed in Xinjiang and Xizang. Also in NW India and Kashmir.

无梗风毛菊

Saussurea apus Maxim.

多年生簇生草本，茎极短。根多分枝，发出多数莲座状叶丛。叶椭圆形或线形，无柄。总苞钟状；总苞片3-4层。花果期7-9月。生海拔4000-4500米的高山草地或山坡。产甘肃、青海和西藏。

Perennial herbs, caespitose, shortly stemmed. Caudex branched, with many sterile and flowering leaf rosettes. Basal leaves rosette, narrowly elliptic to linear, sessile. Involucres campanulate; phyllaries 3-4-seritate. Fl. and fr. Jul-Sep. Alpine steppes or pastures at 4000-5400 m. Distributed in Gansu, Qinghai and Xizang.

草地风毛菊

Saussurea amara (L.) DC.

多年生草本，被白色短柔毛或无毛。叶椭圆形或披针形，边缘全缘或具少数齿。头状花序多数，排成伞房花序或伞房圆锥花序；总苞钟状或圆柱形；总苞片披针形至条形，先端淡紫红色，具扩大圆形附片；小花淡紫色。瘦果长圆形；冠毛白色。花果期7-10月。生海拔500-3200米的荒地、路边、草地、山坡、盐碱地、河堤、沙丘、湖边或水边。产华北、东北和西北。蒙古、俄罗斯、中亚和欧洲亦有。

Perennial herbs, white puberulent or glabrous. Leaves elliptic or lanceolate, margin entire or fewly dentate. Capitula numerous, arranged in corymb or corymbose-panicle; involucres campanulate or cylindric; phyllaries lanceolate to linear, apex purplish-red, attahced with dilated orbicular appendages; florets purplish. Achenes oblong; pappus white. Fl. and fr. Jul-Oct. Wastelands, roadsides, grasslands, slopes, saline or kaline places, river banks, dunes, lakesides or watersides at 500-3200 m. Distributed in N, NE and NW China. Also in Mongolia, Russia, C Asia and Europe.

草地风毛菊 *Saussurea amara*

美花风毛菊 *Saussurea pulchella*

美花风毛菊

Saussurea pulchella (Fisch.) Fisch.

多年生草本。叶具柄，椭圆形、长圆形、披针形至条形，羽状分裂。头状花序多数，排成伞房花序或伞房圆锥花序；总苞球形；总苞片卵形至披针形，先端具扩大的圆形红色膜质附片；小花淡紫色。瘦果倒圆锥状，黄褐色；冠毛淡褐色。花果期8-10月。生海拔300-2200米的草原、草甸、林缘、灌丛、沟谷或河岸。产华北和东北。东北亚亦有。

Perennial herbs. Leaves petiolate, elliptic, oblong, lanceolate to linear, pinnatiparted. Capitula numerous, arranged in corymb or corymbose-panicle; involucres globose; phyllaries ovate to lanceolate, apex with dilated orbicular red membranous appendages; florets purplish. Achenes obconical, yellowish-brown; pappus brownish. Fl. and fr. Aug-Oct. Grasslands, meadows, forest edges, thickets, valleys or river banks at 300-2200 m. Distributed in N and NE China. Also in NE Asia.

风毛菊

Saussurea japonica (Thunb.) DC.

二年生草本，被短柔毛和淡黄色小腺点。叶椭圆形或披针形，羽状深裂。头状花序多数，排成伞房状或伞房圆锥花序；总苞圆柱状；总苞片长卵形至条形，先端紫红色，具扁圆形膜质附片；小花紫色。瘦果深褐色，圆柱形；冠毛白色。花果期6-11月。生海拔200-2900米的山坡、山谷、林下、路旁、灌丛、荒地、水旁或田中。广布中国大部分地区。朝鲜半岛和日本亦有。

Biennial herbs, shortly pubescent and minutely yellowish glandular punctuate. Leaves elliptic or lanceolate, pinnatiparted. Capitula numerous, arranged in corymb or corymbose-panicle; involucres cylindric; phyllaries narrowly ovate to linear, apex purple-red, attached with compressed orbicular membranous appendages; florets purple. Achenes dark brown, cylindric; pappus white. Fl. and fr. Jun-Nov. Slopes, valleys, forests, roadsides, thickets, wastelands, watersides or fields at 200-2900 m. Widely distributed throughout most parts of China. Also in Korean Peninsula and Japan.

风毛菊 *Saussurea japonica*

类尖头风毛菊 *Saussurea pseudomalitiosa*

重齿风毛菊 *Saussurea katochaete*

黑苞风毛菊

Saussurea melanotricha Hand.-Mazz.

多年生草本。叶基生，莲座状，椭圆形或匙状椭圆形，近全缘，下面灰白色，密被贴伏白色绒毛。头状花序单生于莲座状叶丛中；总苞钟状，黑色；总苞片卵形至披针形，被贴伏黑色长柔毛；小花紫色。瘦果圆柱状；冠毛白色。花果期9月。生海拔3500-4700米的高山流石滩或开阔石质山坡。产云南西北部。

Perennial herbs. Leaves basal, rosette, elliptic or spathulate-elliptic, margin nearly entire, offwhite and appressed white tomentose abaxially. Capitula solitary, surrounded by rosette leaves; involucres campanulate, black; phyllaries ovate to lanceolate, appressed black villose; florets purple. Achenes cylindric; pappus white. Fl. and fr. Sep. Alpine screes or open stony slopes at 3500-4700 m. Distributed in NW Yunnan.

类尖头风毛菊

Saussurea pseudomalitiosa Lipsch.

二年生草本。茎单生或少数，不分枝或分枝，有翅，翅具齿。叶片狭椭圆形或线形，羽状半裂或羽状全裂；中部和上部叶片无柄，基部下延。头状花序每3-6个集成伞房状花序，多数伞房花序组成圆锥状花序；小花紫红色；冠毛禾草色。花果期8月。生海拔3300-4200米的戈壁边缘的石砾山坡。产青海。

Biennial herbs. Stems solitary or few, simple or branched from base, narrowly winged; wings dentate. Leaf blade narrowly ovate-elliptic to linear, runcinate-pinnately lobed or pinnatisect; middle and upper stem leaves sessile, base decurrent. Capitula very numerous, 3-6 clustered at end of branches in a corymbiform or paniculiform synflorescence, subsessile to pedunculate; corolla purplish red; pappus straw-colored. Fl. and fr. Aug. Open stonely mountain slopes along Gobi desert at 3300-4200 m. Distributed in Qinghai.

重齿风毛菊

Saussurea katochaete Maxim.

多年生无茎草本。叶基生，莲座状，具宽叶柄，边缘具细密尖锯齿或重锯齿，下面密被白色绒毛。头状花序单生于莲座状叶丛中；总苞宽钟状，直径达4厘米；总苞片无毛，边缘紫色或黑色，狭膜质；小花紫色。瘦果褐色；冠毛浅褐色。花果期7-10月。生海拔2200-4800米的山坡草地、山谷沼泽地、河滩草甸或林缘。产青藏高原。

Perennial acauline herbs. Leaves basal, rosette, broadly petiolate, margin minutely and densely acute-serrulate or doubly serrate, densely white tomentose abaxially. Capitula solitary, surrounded by rosette leaves; involucres broadly campanulate, to 4 cm diam.; phyllaries glabrous, margin purple or black, narrowly membranous; florets purple. Achenes brown; pappus brownish. Fl. and fr. Jul-Oct. Grassy slopes, valley swamps, flood meadows or forest edges at 2200-4800 m. Distributed in Tibetan plateau.

黑苞风毛菊 *Saussurea melanotricha*

柱茎风毛菊 *Saussurea columnaris*

柱茎风毛菊

Saussurea columnaris Hand.-Mazz.

多年生丛生草本，高4-10厘米。叶线形，无柄。头状花序单生，顶生；总苞钟状，直径2-2.5厘米；小花紫红色。冠毛2层，外层白色，内层淡褐色。花果期8-10月。生海拔3200-5300米高山草甸、多石山坡。产云南西北部、四川西南部和西藏东南部。

Herbs perennial, stemless or shortly stemmed, 4-10 cm tall. Leaves linear, sessile. Capitula solitary, terminal; involucres campanulate, 2-2.5 cm diam; flowers purplish. Pappus in 2 rows; outer bristles white; inner bristles pale brown. Fl. and fr. Aug-Oct. Alpine grasslands, rocky slopes at 3200-5300 m. Distributed in NW Yunnan, SW Sichuan and SE Xizang.

不丹风毛菊 *Saussurea bhutanensis*

不丹风毛菊

Saussurea bhutanensis Y. S. Chen

多年生草本，高1-4厘米，无茎。莲座叶无柄，线性；顶部也并入总苞片。头状花序单生，无柄，影藏于绵毛中；总苞钟状。瘦果深棕色，圆锥形；外层冠毛白色，内层冠毛棕色。花果期8-10月。生海拔4500-4900米的高山沙质草甸、流石滩。产西藏(亚东县)。不丹亦有。

Perennial herbs, 1-4 cm tall, stemless. Rosette leaves sessile, linear; uppermost leaves merging into phyllaries. Capitula solitary, sessile, concealed by villous hiars; involucre campanulate. Achene dark brown, conic; outer pappus bristles white; inner pappus brown. Fl. and fr. Aug-Oct. Alpine sandy meadows, alpine scree or limestones at 4500-4900 m. Distributed in Xizang (Yadong County). Also in Bhutan.

钻叶风毛菊 *Saussurea subulata*

钻叶风毛菊

Saussurea subulata C. B. Clarke

多年生垫状草本，高1.5-10厘米。根状茎多分枝，具有多数不育或有花的莲座桩叶丛。叶无柄，钻状线性，长5-12毫米，宽0.5-1毫米，革质。头状花序单生；总苞钟形，直径5-7毫米；花紫红色；冠毛褐色。花果期7-9月。生海拔4100-5300米的河边沙地、高山草地、盐碱地。产青藏高原。印度西北部和克什米尔地区亦有。

Perennial herbs, 1.5-10 cm tall, caespitose. Caudex much branched, with numerous sterile and flowering leaf rosettes forming dense cushions. Leaves sessile, blade subulate-linear, 5-12 mm long, 0.5-1 mm wide, leathery. Capitula solitary; involucre campanulate, 5-7 mm diam. Corolla purplish red; pappus brown. Fl. and fr. Jul-Sep. Gravelly and moist places near rivers, alpine grasslands and meadows, saline or alkaline sandy soils at 4100-5300 m. Distributed in Tibetan Plateau. Also in NW India and Kashmir.

达乌里风毛菊

Saussurea daurica Adams

多年生草本，高4-20(-30)厘米。叶片卵形、披针形或长椭圆形，上部茎叶小；全部叶两面灰绿色，近肉质。头状花序在茎枝顶端排成较紧密的伞房花序；小花粉红色；冠毛白色。花果期8-9月。生海拔1000-3600米河岸碱地、湿河滩、河床林下、盐渍化低湿地、盐化草甸。产甘肃、宁夏、黑龙江、内蒙古、青海和新疆。蒙古和俄罗斯亦有。

Perennial herbs, 4-20(-30) cm tall. Leaf blade narrowly ovate, narrowly elliptic to linear, fleshy, both surfaces grayish green, base cuneate, margin entire, sinuate-dentate, runcinate-pinnately lobed, or pinnately parted, apex acute to subobtuse. Capitula usually numerous, in a globose condensed corymbiform synflorescence; corolla pink; pappus white. Fl. and fr. Aug-Sep. Salt meadows and marshes, alkaline moist soils near rivers and lakes, riverbeds at 1000-3600 m. Distributed in Gansu, Ningxia, Heilongjiang, Neimenggu, Qinghai and Xinjiang. Also in Mongolia and Russia.

盐地风毛菊

Saussurea salsa (Pall.) Spreng.

多年生草本，高15-50厘米。基生叶与下部茎叶宽椭圆形，近肉质，琴状羽状分裂或羽状分裂；中部和上部叶片无柄，长圆形、线状长圆形或狭卵形。头状花序多数，在茎枝顶端排成伞房花序；小花粉紫色；冠毛白色。花果期7-9月。生海拔100-3300米的盐土草地、戈壁滩和湖边。产甘肃、内蒙古、宁夏、青海和新疆。阿富汗、蒙古、俄罗斯、中亚、西南亚和欧洲东部亦有。

Perennial herbs, 15-50 cm tall. Basal leaves ovate to broadly elliptic, fleshy, lyrate-pinnately parted or pinnately lobed; middle and upper stem leaves sessile, oblong, linear-oblong or narrowly ovate. Capitula in a corymbiform synflorescence; corolla pale purple; pappus white. Fl. and fr. Jul-Sep. Saline grasslands, alkaline steppes and meadows at 100-3300 m. Distributed in Gansu, Neimenggu, Ningxia, Qinghai and Xinjiang. Also in Afghanistan, Mongolia, Russia, C and SW Asia, and E Europe.

达乌里风毛菊 *Saussurea daurica*

盐地风毛菊 *Saussurea salsa*

甘青风毛菊 *Saussurea pulvinata*

甘青风毛菊

Saussurea pulvinata Maxim.

多年生草本，丛生。根状茎多分枝。茎多数，不分枝。叶窄椭圆形至线形，边缘全缘或有浅圆齿。头状花序5-12，排列成紧密的伞房状；总苞狭钟形，直径8-10毫米；总苞片4-5层，先端被密绒毛，渐尖或锐尖；小花浅红色。花期7-8月，果期8-9月。生海拔2900-4300米的高山砾石坡。产甘肃、青海和新疆东南部。

Perennial herbs, densely caespitose. Caudex much branched. Stems numerous, simple. Leaves narrowly elliptic to linear, margin entire or slightly crenulate. Capitula 5-12, in densely congested corymbiform; involucres narrowly campanulate, 8-10 mm diam; phyllries in 4-5 series, straw-colored, apically densely villous, apex acuminate to acute; corolla pale red. Fl. Jul-Aug. Fr. Aug-Sep. Alpine scree slopes, rocky dry mountain slopes at 2900-4300 m. Distributed in Gansu, Qinghai and SE Xinjiang.

毓泉风毛菊

Saussurea mae H. C. Fu

多年生草本，高4-15厘米。根状茎粗壮，多分枝。叶片羽状分裂(上部叶片常全缘)，裂片3-5，线形或卵状线形，边缘全缘或有小锯齿。头状花序单生或2-3个形状伞房状花序；总苞钟状；总苞片5或6层；小花粉红色；冠毛白色。花果期7-8月。生海拔约2400米的干旱荒漠山地。产内蒙古西部。

Perennial herbs, 4-15 cm tall. Caudex stout, much branched. Leaves pinnately divided (upper leaves usually entire), segments 3-5, linear to narrowly ovate-linear, margin entire or denticulate. Capitula solitary or 2-3 in a corymbiform synflorescence; involucres campanulate; phyllaries in 5 or 6 rows; corolla rose; pappus white. Fl. and fr. Jul-Aug. Stony mountain slopes at ca. 2400 m. Distributed in W Neimenggu.

中华风毛菊

Saussurea chinensis (Maxim.) Lipsch.

多年生草本。叶无柄，长椭圆形、长椭圆状披针形至条状披针形，边缘具细锯齿或全缘，下面密被白色绒毛。头状花序少数，排成顶生伞房花序；总苞圆柱状；总苞片疏被白色短柔毛，上部黑紫色；小花淡紫色。瘦果褐色；冠毛淡黄色。花果期8-9月。生海拔1900-2300米的山地草甸。产河北和北京。

Perennial herbs. Leaves sessile, narrowly elliptic, narrowly ellip-

毓泉风毛菊 *Saussurea mae*

中华风毛菊 *Saussurea chinensis*

长梗风毛菊 *Saussurea dolichopoda*

Shaanxi, Guizhou and Hubei.

优雅风毛菊
Saussurea elegans Ledeb.

多年生草本。根状茎纤维状撕裂。基生叶和底部叶有柄，长圆形或长圆状卵形，羽状浅裂或几大头羽状浅裂；最上部茎叶线状披针形。头状花序多数，排成顶生伞房花序或圆锥花序；总苞圆柱状。瘦果圆柱状；冠毛白色。花果期7-9月。生海拔1100-3200米的山坡、田间及草坡。产新疆。中亚和蒙古和俄罗斯(西伯利亚)亦有。

tic-lanceolate to linear-lanceolate, margin serrulate or entire, densely white tomentose abaxially. Capitula several, arranged in terminal corymb; involucres cylindric; phyllaries sparsely white puberulent, upper part dark purple; florets purplish. Achenes brown; pappus yellowish. Fl. and fr. Aug-Sep. Montane meadows at 1900-2300 m. Distributed in Hebei and Beijing.

长梗风毛菊
Saussurea dolichopoda Diels

多年生草本，高0.6-1.4米。叶绿色无毛。头状花序多数或少量集成伞房状至伞房状圆锥花序式，具长而粗壮的花梗；总苞钟形至卵形；苞片4-6排；冠毛2层。瘦果棕色。花果期7-10月。生海拔1400-3700米的山谷林中或山坡。产中国西南、华中和华西。

Perennial herbs, 0.6-1.4 m tall. Leaves green, glabrous. Capitula numerous or few corymbose to corymbosely paniculate, with long and stout peduncles; involucres campanulate to ovoid; phyllaries in 4-6 rows; pappus in 2 rows. Achenes brown. Fl. and fr. Jul-Oct. Forests in mountain valleys or mountain slopes at 1400-3700 m. Distributed in SW, C and W China.

假蓬风毛菊
Saussurea conyzoides Hemsl.

多年生草本，高95-120厘米。叶具柄，长椭圆形、长圆形或披针形，边缘具细齿，下面密被灰白色短绒毛。头状花序多数，排成伞房状花序；总苞圆柱状；总苞片长圆形，先端钝；小花紫色。花期7-8月。生海拔1000-2300米的林下。产重庆、河南、陕西、贵州和湖北。

Perennial herbs, 95-120 cm tall. Leaves petiolate, narrowly elliptic, oblong or lanceolate, margin serrulate, densely and shortly off-white tomentose. Capitula numerous, arranged in corymb; involucres cylindric; phyllaries oblong, apex obtuse; florets purple. Fl. Jul-Aug. Forests at 1000-2300 m. Distributed in Chongqing, Henan,

Perennial herbs. Roots fibrous. Basal and lower stem leaves petiolate oblong or ovate-oblong, pinnately lobed or lyrate-pinnately lobed; upper stem leaves narrowly ovate-elliptic to linear. Capitula numerous, in a corymbiform or paniculiform synflorescence; involucre cylindric; pchene cylindric; pappus white. Fl. and fr. Jul-Sep. Grassy or stony mountain slopes, forest meadows and fields at 1100-3200 m. Distributed in Xinjiang. Also in C Asia, Mongolia and Russia (Siberia).

假蓬风毛菊 *Saussurea conyzoides*

优雅风毛菊 *Saussurea elegans*

秦岭风毛菊 *Saussurea megaphylla*

鸢尾叶风毛菊 *Saussurea romuleifolia*

秦岭风毛菊

Saussurea megaphylla (X. Y. Wu) Y. S. Chen

多年生草本。叶长圆形，羽状深裂，侧裂片13-17，长圆形、卵状椭圆形或线形，通常向下弯曲，边缘全缘。头状花序成伞房状排列；总苞近球形；花紫色；冠毛黄白色。花果期8-10月。生海拔1800-2000米的山谷林缘草坡。产陕西(秦岭)。

Perennial herbs. Leaves oblong, pinnatipartite, lateral lobes 13-17 pairs, oblong, narrowly ovate-elliptic or linear, usually curved downward, margin entire. Capitula in loose corymbs; involucre subglobose; florets pink; pappus yellowish white. Fl. and fr. Aug-Oct. Grassy slopes at forest margins in valleys at 1800-2000 m. Distributed in Shaanxi (Qinling Mountain).

拟显鞘风毛菊

Saussurea pseudorockii Y. S. Chen

多年生草本，几无茎。根状茎分枝。叶莲座状或茎生，几无柄；叶线形。头状花序单生；总苞片几乎等长，覆瓦状排列，纸质；花托平，具线形微黄色托片；小花多数。瘦果圆柱状，顶端有具齿小冠；冠毛棕色。花果期8-9月。生海拔约3900米的高山草甸、岩石中。产云南(贡山县)。

Perennial herbs, nearly stemless. Rhizome usually branched. Leaves rosulate or cauline, subsessile; leaf blade linear. Capitula solitary; phyllaries subequal, imbricate, chartaceous; receptacle flat, densely covered with persistent yellowish bristles, bristles linear; florets numerous. Achenes cylindroid, apex with a conspicuous denticulate crown; pappus brown. Fl. and fr. Aug-Sep. Alpine meadows, among rocks at ca. 3900 m. Distributed in Yunnan (Gongshan County).

鸢尾叶风毛菊

Saussurea romuleifolia Franch.

多年生草本，被长柔毛、腺毛或绢状长棉毛。根状茎纺锤状。叶狭条形，坚硬，下面疏被灰白色短柔毛，边缘全缘，内卷。头状花序单生茎端；总苞钟状；总苞片卵形至披针形，紫色，先端具硬刺尖；小花紫色。瘦果顶端有小冠；冠毛污白色。花果期7-8月。生海拔2200-4000米的山坡草地、林下或林缘。产云南和四川。

Perennial herbs, villose, glandular hairy or sericeous-villose. Rhizomes fusiform. Leaves narrowly linear, rigid, sparsely off-white puberulent abaxially, margin entire, involute. Capitula solitary, terminal; involucres campanulate; phyllaries ovate to lanceolate, purple, apex rigidly apiculate; florets purple. Achenes minutely crowned at apex; pappus dirty white. Fl. and fr. Jul-Aug. Grassy slopes, forests or forest edges at 2200-4000 m. Distributed in Yunnan and Sichuan.

密毛风毛菊

Saussurea graminifolia Wall. ex Hook. f.

多年生草本。茎单生或多个，密被白色长棉毛。叶片狭线形，禾草状。头状花序单生茎端；总苞近球形；总苞片4-5层，全部苞片外面被白色长棉毛；小花紫色；冠毛淡黄褐色。花果期7-9月。生海拔4500-4700米山坡草地上、砾石滩边缘草地上。产西藏南部。喜马拉雅亦有。

Perennial herbs. Stem erect, simple or many, densely white lanate. Leaves sessile, narrowly linear, grasslike. Capitula solitary, terminal on stem; involucres subglobose; phyllaries in 4 or 5 rows, densely white lanate; corolla purple; pappus pale yellowish brown. Fl. and fr. Jul-Sep. Grasslands on mountain slopes, grasslands near gravel beaches at 4500-4700 m. Distributed in S Xizang. Also in Himalaya.

拟显鞘风毛菊 *Saussurea pseudorockii*

密毛风毛菊 *Saussurea graminifolia*

禾叶风毛菊
Saussurea graminea Dunn

多年生草本，高3-25厘米。根状茎多分枝。基生叶狭条形，全缘，内卷，背面密被绒毛；花茎多数，丛生，密被白色绢状柔毛。头状花序单生茎端；总苞钟状；总苞片披针形至条形，被绢状长柔毛；小花紫色。瘦果圆柱状；冠毛淡黄褐色。花果期7-9月。生海拔3000-5400米的山坡草地、高山草甸、河滩草甸或杜鹃灌丛中。产青藏高原。

Perennial herbs, 3-25 cm tall. Rhizomes multi-branched. Basal leaves narrowly linear, entire, involute, densely tomentose abaxially; flower stems numerous, clustered, densely white sericeous-villose. Capitula solitary, terminal; involucres campanulate; phyllaries lanceolate to lienar, sericeous-villose; florets purple. Achenes cylindric; pappus yellowish-brown. Fl. and fr. Jul-Sep. Grassy slopes, alpine meadows, flood meadows or *Rhododendron* thickets at 3000-5400 m. Distributed in Tibetan Plateau.

白毛禾叶风毛菊
Saussurea durgae C. Jeffrey et R. C. Srivast.

多年生草本，无茎或茎短，2-5厘米，簇生。叶线性，密被灰白色绵毛。头状花序单生茎顶，几乎被叶掩盖；总苞卵形；花托密被短刚毛；总苞片棕色，边缘黑色，密被毛；小花粉红色或紫罗兰色；冠毛暗白色。花果期7-9月。生海拔4550-5400米的高山草甸或碎石坡。产西藏和四川。喜马拉雅亦有。

Perennial herbs, stemless or short-stemmed, 2-5 cm tall, caespitose. Leaves linear, densely grayish white woolly. Capitula solitary, terminal, almost hidden among leaves; involucre ovoid; receptacle sparsely short setose; phyllaries brown with black near margin, densely hairy; corolla pink or violet; pappus dull white. Fl. and fr. Jul-Sep. Alpine meadows or gravel slope at 4550-5400 m. Distributed in Xizang and Sichuan. Also in Himalaya.

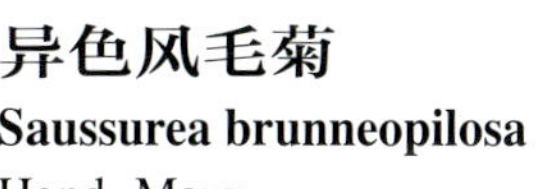

禾叶风毛菊 *Saussurea graminea*

异色风毛菊
Saussurea brunneopilosa Hand.-Mazz.

多年生草本。茎单生、直立。基生与莲座叶狭条形，下面具白毛和密绢毛，上面绿色无毛。小头状花序单生，生于茎顶；总苞宽钟形；苞片约4层，先端紫色；花冠紫色至粉红色。瘦果倒圆锥形；冠毛黄褐色。花果期7-9月。生海拔2900-4900米的高山牧场和草甸或岩石附近草坡。产四川、甘肃、青海和内蒙古。

Perennial herbs. Stem solitary, erect. Rosette and basal leaves narrowly linear, abaxially white and densely sericeous, adaxially green and glabrous. Capitula solitary, terminal on stem; involucres broadly campanulate; phyllaries in ca. 4 rows, apically purple; corolla purple to pink. Achenes obconic; pappus yellowish brown. Fl. and fr. Jul-Sep. Alpine pastures and meadows or grassy slopes among rocks at 2900-4900 m. Distributed in Sichuan, Gansu, Qinghai and Neimenggu.

白毛禾叶风毛菊 *Saussurea durgae*

异色风毛菊 *Saussurea brunneopilosa*

大头风毛菊 *Saussurea baicalensis*

巴东风毛菊 *Saussurea henryi*

大头风毛菊

Saussurea baicalensis (Adams) Robins.

多年生草本。茎单生，直立，被长硬毛。叶多数，密集，椭圆状披针形，边缘具尖锯齿，两面绿色，粗糙，被长柔毛。头状花序大，无梗，多数，沿茎排成紧密的总状花序；总苞钟状；总苞片披针形，被长硬毛；小花紫色。瘦果长圆形；冠毛白色。花果期6-7月。生海拔2000-3200米的山顶草甸或岩石缝。产河北和北京。俄罗斯(西伯利亚)亦有。

Perennial herbs. Stems solitary, erect, long hirsute. Leaves numerous, dense, elliptic-lanceolate, margin acute-serrate, both surfaces green, coarse, villose. Capitula large, sessile, numerous, arranged in closed raceme; involucres campanulate; phyllaries lanceolate, long hirsute; florets purple. Achenes oblong; pappus white. Fl. and fr. Jun-Jul. Meadows or rocky crevies on montane summits at 2000-3200 m. Distributed in Hebei and Beijing. Also in Russia (Siberia).

巴东风毛菊

Saussurea henryi Hemsl.

多年生草本。叶绿色，两面同色，无毛或疏被伏毛，羽状深裂；基部叶基部扩大半抱茎。头状花序1-3，生于茎顶；总苞钟形；苞片5层；冠毛灰黄白色，2层；小花紫色。瘦果黑色，倒圆锥形，无毛。花果期7-8月。生海拔2000-2800米林中或多石山坡。产重庆、湖北和陕西。

Perennial herbs. Leaves green, concolorous, glabrous or sparsely strigose, pinnatisect; basal leaves base enlarged and semiamplexicaul. Capitula 1-3, terminal on stems; involucres campanulate; phyllaries in 5 rows; pappus pale yellow-white, in 2 rows; florets purple. Achenes black, obconic, glabrous. Fl. and fr. Jul-Aug. Forests or rocky slopes at 2000-2800 m. Distributed in Chongqing, Hubei and Shaanxi.

中甸风毛菊

Saussurea dschungdienensis Hand.-Mazz.

多年生小草本，高3-4厘米。茎极短，或几无茎。叶莲座状，有柄，椭圆形或倒卵形，两面异色。头状花序单生，希2个；总苞钟状；总苞片3-4层，顶部淡紫色。瘦果圆柱状，有横皱纹；冠毛浅褐色。花果期8-10月。生海拔3000-4000米的林缘、溪边草地、砾石山坡。产四川和云南。

Perennial herbs, 2-4 cm tall. Stemless or shortly stemmed. Rosette leaves petiolate, narrowly elliptic to obovate, discolorous. Capitula solitary or rarely 2;

中甸风毛菊 *Saussurea dschungdienensis*

involucre campanulate; phyllaries in 3-4 rows, apically pale purple. Achene cylindric, transversely rugose; Pappus pale brown. Fl. and fr. Aug-Oct. Forest margins, grasslands by small streams, gravelly mountain slopes at 3000-4000 m. Distributed in Sichuan and Yunnan.

百裂风毛菊

Saussurea centiloba Hand.-Mazz.

多年生草本。茎单生，不分枝，直立。基生叶有柄，长线形，二回羽状深裂或全裂；茎生叶少数，与基生叶相似。头状花序单生茎端；总苞宽钟状；总苞片紫色，被黄色绒毛。瘦果圆柱状，顶端有小冠；冠毛淡黄色。花期7-8月。生海拔3200-4200米的林缘、灌丛、草地。产四川和云南。

Perennial herbs. Stem solitary, erect, simple. Basal leaves petiolate, narrowly elliptic, bipinnately lobed or bipinnatisect; stem

leaves few, similar to basal leaves. Capitula solitary; involucre broadly campanulate; phyllaries purple, yellowish tomentose. Achene cylindric, apex with short crenulate crown; pappus straw-colored. Fl. Jul-Aug. Forest margins, thickets, grasslands at 3200-4200 m. Distributed in Sichuan and Yunnan.

药山风毛菊

Saussurea bodinieri Lévl.

多年生草本。基部叶莲座状，有柄，柄紫红色，叶椭圆形至狭长圆形。头状花序单生无柄；总苞钟状；总苞片顶部反折；冠毛白色或淡褐色。花果期8-10月。生海拔(2700-)3600-4700米的高山草甸、流石滩、灌木丛、林缘。产中国西南。东喜马拉雅亦有。

Perennial herbs. Basal leaves spreading horizontally, petiolate; petioles purplish red, leaf blade elliptic to narrowly obovate in outline. Capitula solitary, sessile; involucre campanulate; phyllaries apically reflexed; pappus white or brownish. Fl. and fr. Aug-Oct. Alpine meadows, scree slopes, thickets, forest margins at (2700-)3600-4700 m. Distributed in SW China. Also in E Himalaya.

东俄洛风毛菊

Saussurea pachyneura Franch.

多年生草本。基生叶莲座状，长椭圆形或倒披针形，羽状全裂，边缘具粗锯齿，下面密被白色绒毛；叶柄紫红色。头状花序单生茎端；总苞钟状；总苞片坚硬，边缘紫色，披针形；小花紫色。瘦果长圆形，褐色；冠毛白色。花果期8-9月。生海拔3000-4700米的山坡、灌丛、草甸或流石滩。产云南、四川和西藏。

Perennial herbs. Basal leaves rosette, narrowly elliptic or oblanceolate, pinnatisected, margin gross-serrate, densely white tomentose abaxially; petioles purple. Capitula solitary, terminal; involucres campanulate; phyllaries rigid, margin purple, lanceolate; florets purple. Achenes oblong, brown; pappus white. Fl. and fr. Aug-Sep. Slopes, thickets, meadows or screes at 3000-4700 m. Distributed in Yunnan, Sichuan and Xizang.

百裂风毛菊 *Saussurea centiloba*

药山风毛菊 *Saussurea bodinieri*

东俄洛风毛菊 *Saussurea pachyneura*

隆子风毛菊 *Saussurea lhunzensis*

甘肃风毛菊 *Saussurea kansuensis*

隆子风毛菊

Saussurea lhunzensis Y. S. Chen

多年生草本，簇生。莲座叶有柄，叶片长圆形至披针形，羽状浅裂；茎叶有短柄。头状花序单生茎顶；总苞钟状；外层和中层总苞片反则；小花淡蓝色；花药尾部被白色绒毛。瘦果有棱；冠毛淡棕色。花果期7-9月。生海拔3800-4450米的高山沙质草甸或山谷混交林缘灌木丛中。产西藏南部。

Perennial herbs, usually caespitose. Rosette leaves petiolate, leaf blade oblong to lanceolate, pinnately lobed; stem leaves shortly petiolate. Capitula solitary, terminal on stem; involucre campanulate; outer and middle phyllaries reflexed; corolla light blue; anther tails white lanate. Achene ribbed; pappus pale brown. Fl. and fr. Jul-Sep. Alpine sandy grassy meadows or thickets beside mixed forests in valleys at 3800-4450 m. Distributed in S Xizang.

沙生风毛菊

Saussurea arenaria Maxim.

多年生草本。无茎或有短茎；根状茎有分枝；茎极短，或无茎。叶莲座状，有柄，狭椭圆形，边缘全缘或微波状或尖锯齿，两面异色。头状花序单生；总苞宽钟状或宽卵形；总苞片5-6层，顶端紫色。小花紫红色。瘦果圆柱状；冠毛污白色。花果期6-9月。生海拔2800-4500米的山坡、山顶及草甸或沙地、干河床。产甘肃、青海和西藏。

Perennial herbs. Stemless or shortly stemmed; caudex simple or branched. Rosette leaves petiolate; leaf blade narrowly elliptic, leaves discolorous. abaxially grayish white, adaxially green, margin entire or sinuate-dentate. Capitula solitary; involucre broadly campanulate to ovoid; phyllaries in 5-6 rows, apically purplish; corolla purple. Achene cylindric; pappus dirty white. Fl. and fr. Jun-Sep. Mountain slopes, mountaintops, meadows, sandy areas or dry riverbeds at 2800-4500 m. Distributed in Gansu, Qinghai and Xizang.

甘肃风毛菊

Saussurea kansuensis Hand.-Mazz.

多年生无茎莲座状草本。叶莲座状，有柄，长椭圆形，羽状全裂，两面异色。头状花序单生于莲座状叶丛中；总苞钟状。瘦果圆柱状，顶端有小冠；冠毛污白色。花果期8-10月。生海拔3400-4300米的草坡及沙坡。产青藏高原。

Perennial herbs, stemless or shortly stemmed. Rosette leaves petiolate, narrowly elliptic-oblong, pinnatisect, discolorous. Capitula solitary, in center of leaf rosette; involucre campanulate. Achene cylindric, apex with a short crown; pappus dirty white. Fl. and fr. Aug-Oct. Alpine grasslands and steppes in sandy soils at 3400-4300 m. Distributed in Tibetan Plateau.

沙生风毛菊 *Saussurea arenaria*

贡日风毛菊 *Saussurea gongriensis*

帕里风毛菊 *Saussurea pagriensis*

贡日风毛菊

Saussurea gongriensis Y. S. Chen

多年生草本，高4-10厘米。基部和下部叶有柄，椭圆形或长椭圆形，羽状浅裂；中上部茎叶有短柄。头状花序单生茎顶；总苞钟状；总苞片约5层，被白色蛛丝状绒毛。瘦果圆柱状，有横纹，顶端有小冠；冠毛浅棕色。花果期8-9月。生海拔约3420米的山谷森林石缝中。产西藏(错那县)。

Perennial herbs, 4-10 cm tall. Basal and lower stem leaves petiolate, oblong to elliptic, pinnately lobed; middle and upper stem leaves shortly petiolate. Capitula solitary, terminal; involucre broadly campanulate; phyllaries in ca. 5 series, white arachnoid pubescent. Achenes cylindrical, ransversely rugose, apex with a short crown; pappus light brown. Fl. and fr. Aug-Sep. Rocks beside forests in valley at ca. 3420 m. Distribution in Xizang (Cona County).

多裂风毛菊

Saussurea filicifolia (Hook. f.) Y. S. Chen

异名: *S. kunthiana* var. *filicifolia* Hook. f., Fl. Brit. India 3: 369.1881.

多年生草本，高3-8厘米，无茎或几无茎。叶有柄；椭圆形至狭椭圆形，羽状深裂至叶脉。头状花序单生茎顶；总苞钟状。瘦果圆柱状，有棱；冠毛棕色。花果期8-11月。生海拔3000-4700米的石坡、草坡、灌木丛中。产西藏和四川。印度北部和尼泊尔亦有。

多裂风毛菊 *Saussurea filicifolia*

Perennial herbs, 3-8 cm tall, stemless or shortly stemmed. Leaves petiolate; leaf blade oblong to narrowly elliptic, pinnatifid nearly to the midrib. Capitula solitary, terminal; involucre campanulate. Achene cylindrical, ribbed; pappus brown. Fl. and fr. Aug-Nov. Rocky slopes, grassy slopes, thickets at 3000-4700 m. Distributed in Xizang and Sichuan. Also in N India and Nepal.

帕里风毛菊

Saussurea pagriensis Y. S. Chen

多年生草本，簇生，无茎或有短茎。莲座叶无柄，线状披针形。头状花序单生；总苞钟状；花药尾部被绵毛。瘦果深棕色，圆锥形，有棱，顶部有小冠；冠毛灰白色。花果期8-10月。生海拔4600-4870米的高山石坡或沙质草甸。产西藏(亚东县)。不丹亦有。

Perennial herbs, caespitose, usually stemless or shortly stemed. Rosette leaves sessile, linear-lanceolate. Capitula solitary; involucre campanulate; anther tails lanate. Achene dark brown, conic, ribbed, apex shallowly crowned; pappus greyish white. Fl. and fr. Aug-Oct. Alpine scree slope or sandy meadows at 4600-4870 m. Distributed in Xizang (Yadong County). Also in Bhutan.

藏南风毛菊 *Saussurea austrotibetica*

藏南风毛菊

Saussurea austrotibetica Y. S. Chen

多年生草本，簇生。茎常多数或单一，不分枝，被白色稀疏直立绒毛。莲座叶有柄，两面异色，纸质；下部叶披针形。头状花序单生；总苞钟状；总苞片5层，棕紫色，有毛，顶端反折；小花蓝紫色。瘦果圆柱状；冠毛浅棕色。花果期7-9月。生海拔3670-4020米的高山沙质草坡。产西藏南部。

Perennial herbs, caespitose. Stem usually numerous or simple, whitish, erect, sparsely tomentose. Rosette leaves petiolate, discolorous, chartaceous; lower leaves lanceolate. Capitula solitary; involucre campanulate; phyllaries in 5 rows, purplish brown, pubescent, apex usually free and reflexed; corolla blue purplish. Achene cylindric; pappus pale brown. Fl. and fr. Jul-Sep. Alpine sandy grassy slope at 3670-4020 m. Distributed in S Xizang.

泡叶风毛菊

Saussurea bullata W. W. Sm.

多年生矮小草本，高5-16厘米。茎单生，紫色，密被褐色卷毛。莲座叶和底部茎叶有柄，密被褐色卷毛，圆形或宽卵形。头状花序排列成伞房花序；总苞狭钟状；总苞片黑紫色，被毛。瘦果圆柱状，有横皱纹；冠毛基部棕色，顶部污白色。花果期8-9月。生海拔3600-4300米的多石山坡。产云南。

Perennial herbs, 5-16 cm tall. Stem solitary, purple, densely covered with brown articulate hairs. Rosette and lower stem leaves petiolate, densely covered with brown articulate hairs; leaf blade broadly ovate or elliptic to suborbicula. Capitula clustered in a corymbiform synflorescence; involucre narrowly campanulate; phyllaries blackish purple, villous. Achene cylindric, transversely rugose; pappus basally brown, apically dirty white. Fl. and fr. Aug-Sep. Alpine grasslands, consolidated scree slopes at 3600-4300 m. Distributed in Yunnan.

泡叶风毛菊 *Saussurea bullata*

宽翅风毛菊

Saussurea candolleana Wall. ex C. B. Clarke

多年生草本。茎单生，直立，顶部分枝，有窄翼。底部茎叶无茎卵椭圆形、倒卵形或狭长圆形。总状花序排成伞房花序；总苞狭钟状。瘦果倒圆锥形至圆柱形；冠毛浅棕色。花果期8-9月。生海拔2800-

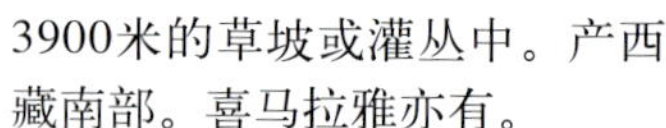
3900米的草坡或灌丛中。产西藏南部。喜马拉雅亦有。

Perennial herbs. Stem solitary, erect, apically branched, narrowly winged. Lower stem leaves sessile, ovate-elliptic, obovate or narrowly elliptic-oblong. Capitula in a clustered corymbiform synflorescence; involucre narrowly campanulate. Achene obconic to cylindric; pappus light brown. Fl. and fr. Aug-Sep. Graasy slopes ot thickets at 2800-3900 m. Distributed in S Xizang. Also in Himalaya.

宽翅风毛菊 *Saussurea candolleana*

尾叶风毛菊

Saussurea caudata Franch.

多年生草本。茎单生或多数。基生叶有长柄，长圆形；下部茎叶宽线形；中上部茎叶长圆形或狭卵状椭圆体形。头状花序单生或2-4个不成明显的伞房花序排列；总苞钟状。瘦果圆柱状，有棱；冠毛污白色。花果期7-9月。生海拔3000-4000米的林缘、高山草甸、开阔多石草场。产四川和云南。

Qinghai and Sichuan.

尾叶风毛菊 *Saussurea caudata*

Perennial herbs. Stem solitary or numerous. Basal leaves petiolate, narrowly ovate-elliptic to oblong; lower stem leaves broadly linear; middle and upper stem leaves sessile, narrowly ovate-elliptic. Capitula 2-4, in a lax corymbiform synflorescence or sometimes solitary, long pedunculate; involucre campanulate. Achene cylindric, ribbed; pappus dirty white; outer bristles ca. 3 mm; inner bristles ca. 1.3 cm. Fl. and fr. Jul-Sep. Forest margins, meadows, open rocky pastures at 3000-4000 m. Distributed in Sichuan and Yunnan.

川西风毛菊

Saussurea dzeurensis Franch.

多年生草本。茎直立，有翼，上部有伞房花序状分枝。基生叶和下部叶有长叶柄，全部叶倒向羽状分裂；中上部茎叶无柄。头状花序多，排成伞房花序；总苞卵形；总苞片革质，被绢状柔毛，边缘黑色；冠毛淡黄褐色。花果期7-10月。生海拔2600-4000米的山坡草地。产甘肃、青海和四川。

Perennial herbs. Stem solitary, apically branched, winged. Basal and lower stem leaves long petiolate, narrowly ovate-elliptic, runcinate-pinnately lobed or dentate; middle and upper stem leaves sessile, narrowly ovate-elliptic. Capitula numerous, in corymbiform synflorescence; involucre ovoid; phyllaries brown but black near margin, leathery, sericeous; pappus pale yellowish brown. Fl. and fr. Jul-Oct. Alpine steppes and grasslands at 2600-4000 m. Distributed in Gansu, Qinghai and Sichuan.

锐齿风毛菊

Saussurea euodonta Diels

多年生草本。中下部茎叶有叶柄，卵形、椭圆形或狭卵状椭圆形；上部叶有短柄或几无柄，宽卵形或椭圆形。头状花序少数，单生或成松散的伞房花序；总苞钟状；总苞片顶端紫色，反折。瘦果圆柱状；外层冠毛污白色。花果期8-10月。生海拔2300-3700米的山坡草地、松林林缘。产四川和云南。

Perennial herbs. Lower and middle stem leaves petiolate, ovate, elliptic or narrowly ovate-elliptic; upper stem leaves shortly petiolate to subsessile, narrowly ovate or elliptic. Capitula few, in a lax corymbiform synflorescence or sometimes solitary; involucre campanulate; phyllaries dark purple apically, and patent to reflexed. Achene cylindric; pappus dirty white. Fl. and fr. Aug-Oct. Montane grasslands, *Pinus* forest margins at 2300-3700 m. Distributed in Sichuan and Yunnan.

川西风毛菊 *Saussurea dzeurensis*

锐齿风毛菊 *Saussurea euodonta*

萎软风毛菊 *Saussurea flaccida*

萎软风毛菊

Saussurea flaccida Ling

多年生草本。叶两面异色，基生叶及下部茎叶有叶柄，卵形或宽卵形；上部茎叶有短柄或几无柄，狭卵状椭圆形或长圆形。头状花序2-4个，成伞房花序状排列或单生；总苞倒圆锥状或长圆状；小花淡黄褐色。瘦果有棱，圆柱状；冠毛淡黄色。花果期8-9月。生海拔2700-2800米的高山草地及灌丛。产陕西(秦岭)和湖北(神农架)。

Perennial herbs. Leaves discolorous; basal and lower stem leaves petiolate, ovate or broadly ovate; upper stem leaves shortly petiolate to subsessile, narrowly ovate-elliptic to oblong. Capitula 2-4, in a corymbiform synflorescence or sometimes solitar; involucre obconic or oblong; corolla pale yellowish brown. Achene cylindric, ribbed; pappus straw-colored. Fl. and fr. Aug-Sep. Montane grasslands and thickets at 2700-2800 m. Distributed in Shaanxi (Qinling Mountain) and Hubei (Shennongjia).

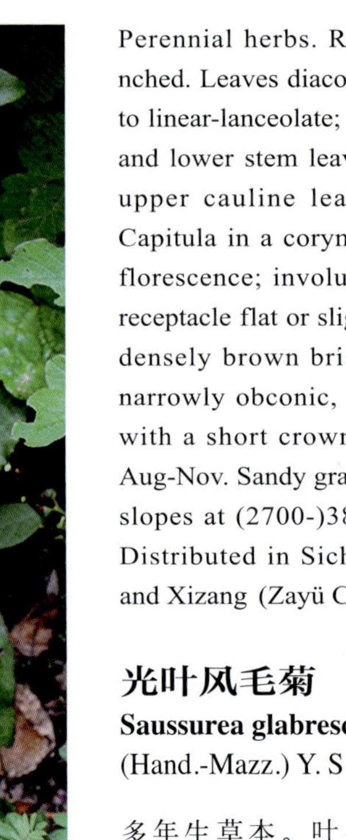

光叶风毛菊 *Saussurea glabrescens*

褐冠风毛菊

Saussurea fuscipappa Y. S. Chen

多年生草本。地下茎分枝。叶两面异色，线形至线状披针形；莲座叶，基部和下部茎叶有柄；上部叶无柄。头状花序排成伞房花序；总苞倒圆锥形；花托平或微隆，密被棕色刚毛。瘦果狭倒圆锥形，有棱，顶部有小冠。花果期8-11月。生海拔(2700-)3800-4600米的沙质草甸和碎石滩。产四川、云南和西藏(察隅县)。

Perennial herbs. Rhizome branched. Leaves diacolorous, linear to linear-lanceolate; rosette, basal and lower stem leaves petiolate; upper cauline leaves sessile. Capitula in a corymbiform synflorescence; involucre obconic; receptacle flat or slightly convex, densely brown bristly. Achene narrowly obconic, ribbed, apex with a short crown. Fl. and fr. Aug-Nov. Sandy grassy and scree slopes at (2700-)3800-4600 m. Distributed in Sichuan,Yunnan and Xizang (Zayü County).

褐冠风毛菊 *Saussurea fuscipappa*

光叶风毛菊

Saussurea glabrescens (Hand.-Mazz.) Y. S. Chen

多年生草本。叶多基生，有柄，卵形或近心形；上部茎叶披针形至线性，无柄，基部下绦到茎，成翼。头状花序再顶部成伞房花序；总苞钟状倒圆锥形；总苞片密被绒毛，深褐色；小花深玫瑰紫；冠毛污白色。花果期8-10月。生海拔(2400-)2900-3850米的山地灌丛、草地或森林。产四川。

Perennial herbs. Leaves mostly basal, petiolate, ovate or subcordate; upper stem leaves lanceolate to linear, sessile, base decurrent along stem, forming narrow and entire wings. Synflorescence flat-topped; involucre campanulate-obconic; phyllaries densely

长毛风毛菊 *Saussurea hieracioides*

昌都风毛菊 *Saussurea qamdoensis*

昌都风毛菊

Saussurea qamdoensis Y. S. Chen

多年生草本。地下茎分枝。莲座叶、基部和下部叶有柄，叶线形；中上部叶少，小，线形至线状披针形。头状花序2-6簇生或排成松散的伞房花序；总苞倒圆锥形；花药尾部被绵毛。瘦果狭倒圆锥形，有棱，顶部有小冠。花果期8-9月。生海拔3650-4000米的云杉林缘或草坡。产西藏(昌都)。

Perennial herbs. Rhizome usually branched. Rosette, basal and lower stem leaves petiolate, leaf blade linear; middle and upper cauline leaves few, small, linear to linear-lanceolate. Capitula 2-6, in a clustered or lax, corymbiform synflorescence; involucre obconic; anthers tails lanate; Achene narrowly obconic, ribbed, apex with a short crown. Fl. and fr. Aug-Sep. *Picea* forest margins or grassy slopes at 3650-4000 m. Distributed in Xizang (Qamdo).

villous, dark brown; corolla deeply rose purple; pappus dirty white. Fl. and fr. Aug-Oct. Montane thickets, grasslands or forests at (2400-)2900-3850 m. Distributed in Sichuan.

Achene cylindric; pappus pale brown. Fl. and fr. Jun-Sep. Alpine scree slopes, grasslands, rocky slopes at 4400-5200 m. Distributed in SW China. Also in Himalaya.

长毛风毛菊

Saussurea hieracioides Hook. f.

多年生草本。基生叶有柄，叶片椭圆形或长椭圆状倒披针形；茎生叶少，狭卵形或线形，无柄。头状花序单生茎顶；总苞宽钟状，直径2-3.5厘米；总苞片黑紫色，密被长柔毛。瘦果圆柱状；冠毛淡褐色。花果期6-9月。生海拔4440-5200米的山碎石土坡、高山草坡。产中国西南。喜马拉雅亦有。

Perennial herbs. Basal leaves distinctly petiolate, elliptic or narrowly elliptic-oblanceolate; stem leaves few, sessile, narrowly ovate-elliptic or linear. Capitula solitary, terminal on stem; involucre campanulate; phyllaries blackish purple, densely villous.

椭圆风毛菊

Saussurea hookeri C. B. Clarke

多年生草本，4-30厘米。茎不分枝。莲座叶和底部茎叶狭线形；中上部茎叶无柄，线状锥形至狭卵形；顶部茎叶三角卵形、卵形或椭圆形，黑紫色。头状花序单生；总苞片黑色。瘦果淡黄色带黑斑，圆筒状，有棱，顶部有小冠；冠毛污白色。花果期8-9月。生海拔4300-5300米的高山湿草甸。产青藏高原。喜马拉雅亦有。

Perennial herbs, 4-30 cm tall. Stems simple. Rosette and lower stem leaves narrowly linear; middle and upper stem leaves sessile, linear-subulate to narrowly ovate; uppermost stem leaves triangular-ovate, ovate or elliptic, blackish purple. Capitula solitary; phyllaries black. Achene straw-colored with black spots, cylindric, ribbed, apex with a short crown; pappus dirty white. Fl. and fr. Aug-Sep. Alpine grasslands at 4300-5300 m. Distributed in Tibetan Plateau. Also in Himalaya.

椭圆风毛菊 *Saussurea hookeri*

小金风毛菊 *Saussurea xiaojinensis*

小金风毛菊
Saussurea xiaojinensis Y. S. Chen

多年生草本。地下茎纤维状撕裂，有时分枝。叶无柄，叶片长圆形至披针形。头状花序成松散的伞房花序；总苞钟状至近球形；总苞片绿色至紫色或黑紫色；花托平，密被黄色托片；小花20-23。瘦果圆筒状，有纵棱；冠毛黄白色。花果期7-8月。生海拔2840-2900米的灌丛或草坡。产四川(小金县)。

Perennial herbs. Rhizome sometimes apically branched, with fibrous lacerate. Leaves sessile, blade oblong to lanceolate. Capitula in sparse corymbs; involucre campanulate to subglobose; phyllaries green to purple or blackish purple; receptacle flat, densely yellowish bristles; florets 20-23. Achenes cylindroid, longitudinally striate; pappus yellowish white. Fl. and fr. Jul-Aug. Thickets or grassy slopes at 2840-2900 m. Distributed in Sichuan (Xiaojin County).

盐源风毛菊
Saussurea yanyuanensis Y. S. Chen

多年生草本。叶禾草状，无柄，线形。头状花序单生茎顶；总苞半球形；总苞片革质，深紫色，覆瓦状排列；花药黑色。瘦果圆柱形，10个明显的棱，顶部有小冠；冠毛灰褐色。花果期7-9月。生海拔3900-4200米的高山草坡或流石滩。产四川(盐源县)。

Perennial herbs. Leaves grass-like, sessile, blade linear. Capitula solitary, terminal; involucre subglobose; phyllaries imbricate, dark brown, leathery; florets many, dark purple; anthers blackish. Achenes cylindrical, with 10 distinct ribs; apex with conspicuous denticulate crown; pappus grayish brown. Fl. and fr. Jul-Sep. Alpine scree slopes or grassy slope at 3900-4200 m. Distributed in Sichuan (Yanyuan County).

阿拉善风毛菊
Saussurea alaschanica Maxim.

多年生草本，高20-30厘米。茎单生，直立。基生叶及下部茎叶椭圆形或卵状椭圆形；中部茎叶逐渐变小；上部茎叶披针形或椭圆状披针形。头状花序1-5个，在茎顶密集排列成伞房花序；总苞钟状；总苞片4-5层，被长柔毛。花果期7-9月。生海拔2500-3500米的山坡灌丛或岩石缝隙中。产内蒙古西部和宁夏(贺兰山)。

Perennial herbs, 20-30 cm tall. Stem solitary, erect. Basal and lower stem leaves elliptic or ovate-elliptic; upper leaves smaller, lanceolate or elliptic-lanceolate. Capitula 1-5, densely arranged in corymb; involucres campanulate; phyllaries 4-5-seritate, villose. Fl. and fr. Jul-Sep. Thickets, mountain slopes or rock fissures at 2500-3500 m. Distributed in W Neimenggu and Ningxia (Helan Mountain).

盐源风毛菊 *Saussurea yanyuanensis*

阿拉善风毛菊 *Saussurea alaschanica*

拟柳叶风毛菊 *Saussurea leucota*

齿苞风毛菊 *Saussurea odontolepis*

拟柳叶风毛菊

Saussurea leucota Hand.-Mazz.

多年生草本，高20-50厘米。叶无柄，叶片线状披针形至狭卵状椭圆形，基部下延并稍有耳，划过伞房状、顶生。头状花序4-15(-30)。总苞钟状，直径7-8.5毫米。花冠紫色。冠毛污白色。花果期8-9月。生海拔3100-4000米的林缘或草甸。产四川西北部和甘肃南部。

Perennial herbs, 20-50 cm tall. Leaves sessile; leaf blade linear-lanceolate to narrowly ovate-elliptic, base narrowly decurrent and slightly auriculate Synflorescence corymbiform, ample, at-topped to rounded. Capitula 4-15(-30). Involucres campanulate, 7-8.5 mm diam. Corolla purple. Pappus dirty white. Fl. and fr. Aug-Sep. Forest margins or meadows at 3100-4000 m. Distributed in NW Sichuan and S Gansu.

齿苞风毛菊

Saussurea odontolepis Sch.-Bip. ex Herder

多年生草本。叶绿色，下面无毛，上面粗糙，密被糙毛；茎生叶羽状浅裂或近羽状深裂。头状花序小，排成伞房状；总苞卵形至卵状钟形；苞片4-6层，外层边缘具篦齿；冠毛白色，2层；小花紫色。瘦果圆柱形，无毛。花果期8-10月。生海拔100-700米的林缘或草地。产山西、内蒙古、黑龙江、吉林和辽宁。俄罗斯(远东地区)、蒙古和朝鲜半岛亦有。

Perennial herbs. Leaves green, abaxially glabrous, adaxially scabrous and densely strigose; middle stem leaves pinnately parted or subpinnatisect. Capitula small, corymbose; involucres ovate to ovate-campanulate; phyllaries in 4-6 rows, outer phyllaries margin pectinate; pappus white, in 2 rows; florets purple. Achenes cylindric, glabrous. Fl. and fr. Aug-Oct. Forest edges or grasslands at 100-700 m. Distributed in Shanxi, Neimenggu, Heilongjiang, Jilin and Liaoning. Also in Russia (Far East), Mongolia and Korean Peninsula.

篦苞风毛菊

Saussurea pectinata Bunge ex DC.

多年生草本，被短糙毛。叶具柄，卵形、披针形或椭圆形，羽状深裂。头状花序数个，排成伞房花序；总苞钟状；总苞片披针形，被蛛丝毛，先端反折，栉齿状，草质，绿色或紫色；小花紫色。瘦果圆柱状；冠毛污白色。花果期8-10月。生海拔300-1900米的山坡、林下、林缘、路旁、草原或沟谷。产华北、东北和西北。

Perennial herbs, shortly hirsute. Leaves petiolate, ovate, lanceolate or elliptic, pinnatiparted. Capitula several, arranged in corymb; involucres campanulate; phyllaries lanceolate, arachnoid hariy, apex revolute, pectinate, foliaceous, green or purple; florets purple. Achenes cylindric; pappus dirty white. Fl. and fr. Aug-Oct. Slopes, forests, forest edges, roadsides, grasslands or valleys at 300-1900 m. Distributed in N, NE and NW China.

篦苞风毛菊 *Saussurea pectinata*

蒙古风毛菊 *Saussurea mongolica*

杨叶风毛菊
Saussurea populifolia

杨叶风毛菊
Saussurea populifolia Hemsl.

多年生草本。茎中部及下部叶绿色，心形至卵状心形，下面无毛，上面密被糙毛。头状花序1或2个生于茎顶；总苞宽钟形；苞片带紫色，5-7层；冠毛棕灰色，2层；小花紫色。瘦果棕色，近圆柱形，具棱，无毛。花果期7-10月。生海拔1700-3600米的山坡草地或沼泽地。产四川、重庆、湖北、河南、陕西和甘肃。

Perennial herbs. Lower and middle stem leaves green, cordate to ovate-cordate, abaxially glabrous, adaxially densely strigose. Capitula 1 or 2, terminal on stems; involucres broadly campanulate; phyllaries with a purple flush, in 5-7 rows; pappus pale brown, in 2 rows; florets purple. Achenes brown, subcylindric, ribbed, glabrous. Fl. and fr. Jul-Oct. Grasslands on mountain slopes or marshes at 1700-3600 m. Distributed in Sichuan, Chongqing, Hubei, Henan, Shaanxi and Gansu.

蒙古风毛菊
Saussurea mongolica (Franch.) Franch.

多年生草本。叶卵状三角形、卵形或披针形，羽状分裂。头状花序多数，排成伞房花序或伞房圆锥花序；总苞长圆状；总苞片卵形至长圆形，先端反折，具长渐尖马刀形附属物；小花紫红色。瘦果圆柱状，褐色；冠毛上部白色，下部淡褐色。花果期7-10月。生海拔500-2900米的山坡、林下、灌丛、路旁或草地。产华北、东北和西北。朝鲜半岛亦有。

Perennial herbs. Leaves ovate-triangular, ovate or lanceolate, pinnatifid. Capitula numerous, arranged in corymb or corymbose-panicle; involucres oblong; phyllaries ovate to oblong, apex revolute, attached with acuminate saber-shape appendages; florets purple. Achenes cylindric, brown; pappus white on upper part, brownish on lower part. Fl. and fr. Jul-Oct. Slopes, forests, thickets, roadsides or grasslands at 500-2900 m. Distributed in N, NE and NW China. Also in Korean Peninsula.

利马川风毛菊
Saussurea leclerei Lévl.

多年生草本。茎单生、直立、具狭翅。叶卵形至椭圆形，上面浅绿色和蛛丝状绒毛，上面深绿色无毛。头状花序多数、簇生为复合伞房状；总苞钟形；苞片约5层，棕色；花冠紫色。瘦果圆柱状；冠毛浅棕色。花果期8-10月。生海拔2000-3300米的山地草原。产云南、四川、重庆和湖北。

Perennial herbs. Stem solitary, erect, narrowly winged. Leaves ovate to elliptic, abaxially grayish green and arachnoid tomentose, adaxially dark green and

利马川风毛菊 *Saussurea leclerei*

大叶风毛菊 *Saussurea grandifolia*

glabrous. Capitula numerous, in a clustered corymbiform synflorescence; involucres campanulate; phyllaries in ca. 5 rows, brown; corolla purple. Achenes cylindric; pappus pale brown. Fl. and fr. Aug-Oct. Montane grasslands at 2000-3300 m. Distributed in Yunnan, Sichuan, Chongqing and Hubei.

大叶风毛菊

Saussurea grandifolia Maxim.

多年生直立草本。叶坚硬，7-20×4-13厘米，边缘具粗锯齿，两面疏被短糙毛。头状花序3-18个，排成伞房花序或圆锥花序；总苞钟状；总苞片质薄，椭圆形至条形，被白色蛛丝毛；小花暗红色。瘦果稍弯曲；冠毛白色。花果期8-10月。生海拔200-1100米的林缘、山谷或草地。产黑龙江和吉林。俄罗斯和朝鲜半岛亦有。

Perennial erect herbs. Leaves rigid, 7-20 × 4-13 cm, margin gross-serrate, both surface shortly hirsute. Capitula 3-18, arranged in corymb or panicle; involucres campanulate; phyllaries thin, elliptic to linear, white arachnoid hairy; florets dark purple. Achenes somewhat bent; pappus white. Fl. and fr. Aug-Oct. Forest edges, valleys or grasslands at 200-1100 m. Distributed in Heilongjiang and Jilin. Also in Russia and Korean Peninsula.

乌苏里风毛菊

Saussurea ussuriensis Maxim.

多年生草本。茎中部和上部叶长圆状卵形、披针形或条形。头状花序伞房状，具条形托叶；总苞狭钟形；苞片5-7层，边缘及顶部常稍紫红色；冠毛白色，2层；小花紫红色。瘦果棕灰色，无毛。花果期7-9月。生海拔1100-2900米的山坡草地、林中、河边或沟谷。产华北、华西、华东、西北和东北。俄罗斯(远东地区)、蒙古、朝鲜半岛和日本亦有。

Perennial herbs. Middle and upper stem leaves oblong-ovate, lanceolate or linear. Capitula corymbose, with linear subtending leaves; involucres narrowly campanulate; phyllaries in 5-7 rows, margin and apically usually slightly purplish red; pappus white, in 2 rows; florets purplish red. Achenes pale brown, glabrous. Fl. and fr. Jul-Sep. Grasslands on mountain slopes, forests, by rivers or ravines at 1100-2900 m. Distributed in N, W, E, NW and NE China. Also in Russia (Far East), Mongolia, Korean Peninsula and Japan.

乌苏里风毛菊 *Saussurea ussuriensis*

银背风毛菊

Saussurea nivea Turcz.

多年生草本。茎单生、直立。叶狭三角状卵形，下面具白色和密蛛丝状毡毛，上面绿色无毛；叶三角状卵形至狭卵状椭圆形。头状花序形成伞房状复合花序，具梗；总苞钟形；苞片6或7层，具白色毡毛；花冠紫色。瘦果棕色，圆柱形；冠毛白色。花果期7-9月。生海拔400-2200米的林中、林缘或灌丛中。产华北、华西和东北。朝鲜半岛亦有。

Perennial herbs. Stems solitary, erect. Leaves narrowly triangular-ovate, abaxially white and densely arachnoid lanate, adaxially green and glabrous; leaves triangular-ovate to narrowly ovate-elliptic. Capitula in a corymbiform synflorescence, pedunculate; involucres campanulate; phyllaries in 6 or 7 rows, white lanate; corolla purple. Achenes brown, cylindric; pappus white. Fl. and fr. Jul-Sep. Forests, forest edges or thickets at 400-2200 m. Distributed in N, W and NE China. Also in Korean Peninsula.

银背风毛菊 *Saussurea nivea*

大耳叶风毛菊 *Saussurea macrota*

大耳叶风毛菊
Saussurea macrota Franch.

多年生草本。叶下面疏被棕色腺毛，上面疏被伏毛；茎中部及下部叶基部深心形，大耳状抱茎。头状花序2-10，伞房状；总苞卵球形至长圆形；苞片5或6层，革质；冠毛棕灰色，2层；小花深紫色。瘦果圆柱形，具纵棱，无毛。花果期7-8月。生海拔2200-3300米的山坡、林中或灌丛。产四川、重庆、湖北、陕西、甘肃和宁夏。

Perennial herbs. Leaves abaxially sparsely brown glandular punctate, adaxially sparsely strigose; lower and middle stem leaves base deeply cordate with amplexicaul large auricles. Capitula 2-10, corymbose; involucres ovoid to oblong; phyllaries in 5 or 6 rows, leathery; pappus pale brown, in 2 rows; florets dark purple. Achenes cylindric, ribbed, glabrous. Fl. and fr. Jul-Aug. Mountain slopes, forests or thickets at 2200-3300 m. Distributed in Sichuan, Chongqing, Hubei, Shaanxi, Gansu and Ningxia.

柳叶菜风毛菊
Saussurea epilobioides Maxim.

多年生草本。茎单生或数个生。茎中部及下部叶无柄，狭椭圆形至条形，上部灰绿色，常具腺点。头状花序多数，簇生成复合伞房状，具短梗；总苞狭钟形；苞片4或5层；花冠紫色。瘦果圆柱形；冠毛污白色。花果期7-9月。生海拔2600-4200米的山坡或高山草甸。产四川、甘肃、宁夏和青海。

Perennial herbs. Stems solitary or few. Lower and middle stem leaves sessile, narrowly elliptic to linear, abaxially grayish green, usually gland-dotted. Capitula numerous, in a clustered corymbiform synflorescence, shortly pedunculate; involucres narrowly campanulate; phyllaries in 4 or 5 rows; corolla purple. Achenes cylindric; pappus dirty white. Fl. and fr. Jul-Sep. Mountain slopes or alpine meadows at 2600-4200 m. Distributed in Sichuan, Gansu, Ningxia and Qinghai.

小花风毛菊
Saussurea parviflora (Poiret) DC.

多年生草本。茎下部叶叶柄具翅，椭圆形至长圆状椭圆形，基部下延至叶柄，边缘具锯齿。头状花序多数，呈伞房状排列；总苞钟形；苞片5层；冠毛白色，2层；小花紫色。花果期7-9月。生海拔1600-3800米的山坡湿润处、山地沟谷灌丛、林中或岩石裂缝中。产华北和西北。哈萨克斯坦、俄罗斯和蒙古亦有。

Perennial herbs. Lower stem leaves with winged petioles, elliptic to oblong-elliptic, base decurrent on stems, margin toothed. Capitula numerous, corymbosely arranged; involucres campanulate; phyllaries in 5 rows; pappus white, in 2 rows; florets purple. Fl. and fr. Jul-Sep. Moist places on mountain slopes, thickets in mountain valleys, forests or among fractured rock at 1600-3800 m. Distributed in N and NW China. Also in Kazakhstan, Russia and Mongolia.

柳叶菜风毛菊 *Saussurea epilobioides*

小花风毛菊 *Saussurea parviflora*

牛耳风毛菊
Saussurea woodiana

牛耳风毛菊
Saussurea woodiana Hemsl.

多年生草本。茎单生。叶椭圆形至倒卵形，下面具浅绿色或棕色密绒毛，上面绿色，具腺毛。头状花序单生于莲座叶丛中间或顶生于茎；总苞钟形；苞片5或6层，禾秆色，但是顶端和边缘紫色；花冠紫色。瘦果圆柱形；冠毛浅褐色。花果期7-9月。生海拔3000-4200米的山地草原。产四川和青海。

Perennial herbs. Stems solitary. Leaves elliptic to obovate, abaxially grayish green or brownish and densely tomentose, adaxially green and glandular hairy. Capitula solitary, in center of leaf rosette or terminal on stem; involucres campanulate; phyllaries in 5 or 6 rows, straw-colored but purplish apically and near margin; corolla purple. Achenes cylindric; pappus pale brown. Fl. and fr. Jul-Sep. Montane grasslands at 3000-4200 m. Distributed in Sichuan and Qinghai.

硬苞刺头菊
Cousinia sclerolepis C. Shih

二年生草本。叶纸质，疏被蛛丝毛；基生叶椭圆形，羽状半裂或深裂，边缘具3-7个不等齿及侧生针刺。头状花序1个；苞片6或7排；外部及中部苞片硬革质，顶部窄成一个硬三角形针刺；花冠紫红色。瘦果灰色，斜倒卵球形；无冠毛。花果期7月。生海拔约3200米的山坡。产新疆(乌恰县)。

Biennial herbs. Leaves papery, sparsely cobwebby; basal leaves elliptic, pinnatifid or pinnatipartite, margin with 3-7 unequal teeth and lateral spines. Capitula solitary; phyllaries in 6 or 7 rows; outer and middle phyllaries, rigid, leathery, apex narrowed into a rigid triquetrous spine; corolla purplish red. Achenes pale, obliquely obovoid; pappus absent. Fl. and fr. Jul. Mountain slopes at ca. 3200 m. Distributed in Xinjiang(Wuqia County).

毛苞刺头菊 *Cousinia thomsonii*

硬苞刺头菊
Cousinia sclerolepis

毛苞刺头菊
Cousinia thomsonii C. B. Clarke

多年生草本，高30-80厘米，上部分枝。茎直立。基生叶长椭圆形或倒披针形，羽状全裂；中部茎叶与基生叶及下部茎叶同形并等样分裂。头状花序单生枝端；总苞近球形；总苞片9层，质地坚硬，革质；小花紫红色、粉红色或紫红色。花果期7-9月。生海拔3700-4300米山坡草地、河滩砾石地。产西藏西南部。印度、尼泊尔和巴基斯坦亦有。

Perennial herbs, 30-80 cm tall. Basal cauline leaves elliptic to oblanceolate, middle cauline leaves similar to lower ones. Capitula solitary, terminal; involucres subglobose; phyllaries in ca. 9 rows, rigid, leathery; corolla purple to pink. Fl. and fr. Jul-Sep. Grasslands, gravelly places in flooded lands at 3700-4300 m. Distributed in SW Xizang. Also in India, Nepal and Pakistan.

牛蒡 *Arctium lappa*

牛蒡
Arctium lappa L.

二年生粗壮草本。肉质直根粗大。叶宽卵形，长达30厘米，边缘波状，基部心形，具长叶柄。总苞卵球形，绿色，无毛；总苞片多层，三角形至条形，顶端具倒钩刺；花冠紫红色。瘦果倒长卵形，冠毛浅褐色。花果期6-9月。生海拔700-3500米的山坡、山谷、村庄附近、路边、林缘、林中、灌丛、湿地或荒地。中国除西藏、海南和台湾外都有生长。印度、尼泊尔、不丹、巴基斯坦、阿富汗、日本、亚洲西南部和欧洲亦有。

Biennial stout herbs. Tap roots fleshy, stout. Leaves broadly ovate, to 30 cm long, margin undulate, base cordate, long petiolate. Involucres ovoid, green, glabrous; phyllaries many rows, deltoid to linear, apex hooked spiny; corolla purple. Achenes oblong-obovoid, pappus pale brown. Fl. and fr. Jun-Sep. Slopes, valleys, near villages, roadsides, forest edges, forests, thickets, wetlands or wastelands at 700-3500 m. Distributed throughout China except for Xizang, Hainan and Taiwan. Also in India, Nepal, Bhutan, Pakistan, Afghanistan, Japan, SW Asia and Europe.

毛头牛蒡
Arctium tomentosum Mill.

二年生草本。根肉质，粗壮，肉红色。基生叶卵形，基部心形。头状花序在茎枝顶端作伞房花序或圆锥状伞房花序式排列；苞片多层，被蛛丝毛；小花紫红色。瘦果浅褐色；冠毛多层，基部不连合成环。花果期7-9月。生海拔1200-2100米的山坡草地。产新疆。俄罗斯、中亚和欧洲亦有。

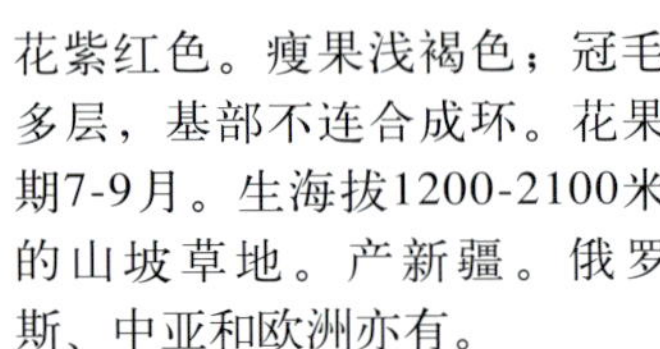

Biennial herbs. Roots fleshy, stout, flesh red. Basal leave ovate, base cordate. Capitula arranges into corymb or umbel-like panicles at apices of branches; phyllaries in many series, arachnoid; florets purple-red. Achenes light-brown; pappus in many series, base not connate into a loop. Fl. and fr. Jul-Sep. Moutain slopes at 1200-2100 m. Distributed in Xinjiang. Also in Russia, C Asia and Europe.

毛头牛蒡 *Arctium tomentosum*

山牛蒡
Synurus deltoides (Ait.) Nakai

多年生草本，被密厚绒毛。叶具长叶柄，心形、卵形、卵状三角形或戟形，下面密被灰白色绒毛，边缘具粗大锯齿。头状花序顶生，大，下垂；总苞球形，直径3-6厘米；总苞片多数，披针形；小花管状，紫红色。瘦果长圆形；冠毛褐色。花果期6-10月。生海拔500-2200米的山坡、林缘、林下或草甸。产华中、华东、华北和东北。俄罗斯、蒙古、朝鲜半岛和日本亦有。

Perennial herbs, densely tomentose. Leaves long petiolate, cordate, ovate, ovate-triangular or

hastate, densely offwhite tomentose abaxially, margin gross-dentate. Capitula terminal, large, drooping; involucres globose, 3-6 cm diam; phyllaries numerous, lanceolate; florets tubular, purple. Achenes oblong; pappus brown. Fl. and fr. Jun-Oct. Slopes, forest edges, forests or meadows at 500-2200 m. Distributed in C, E, N and NE China. Also in Russia, Mongolia, Korean Peninsula and Japan.

山牛蒡 *Synurus deltoides*

薄叶翅膜菊 *Alfredia acantholepis*

薄叶翅膜菊

Alfredia acantholepis Kar. et Kir.

多年生草本。茎粗壮，不分枝或有单一分枝，紫红色。叶纸质，两面异色。头状花序1或2；中外层总苞片中部边缘膜质附属流苏状撕裂，顶端具骨针状针刺；内层总苞片两侧边缘有流苏状撕裂；小花黄色；冠毛易脆折。花果期7-10月。生海拔1600-3200米的草甸、草原或疏林中或阴湿处。产新疆。哈萨克斯坦亦有。

Perennial herbs. Stem, stout, unbranched or rarely with 1 branch purplish red. Leaves papery, discolorous; capitula 1 or 2, erect; outer and middle phyllaries middle portion laterally expanded into scarious fimbriate-lacerate wings, apex narrowed into spine; inner phyllaries margin sometimes expanded into small fimbriate-lacerate scarious wings; corolla yellowm; pappus fragile. Fl. and fr. Jul-Oct. Meadows, steppes, open forests, moist places at 1600-3300 m. Distributed in Xinjiang. Also in Kazakhstan.

糙毛翅膜菊

Alfredia aspera C. Shih

多年生草本。茎紫红色，有棱，上部有单一分枝。全部叶纸质，两面异色，边缘有淡黄色的缘毛状针刺。头状花序2，歪斜；中外层苞片中部边缘有膜质撕裂的附片，顶端具骨刺；内层苞片顶端有针刺，近顶端有膜质附片；小花黄色；冠毛褐色。花期7-9月。生海拔1700-3100米的林间空旷地。产新疆。

Perennial herbs. Stem purplish red, apically once branched, ribbed. Leaves papery, discolorous, margin ciliate with yellowish spinules. Capitula 2, nodding; outer and middle phyllaries middle portion laterally expanded into scarious fimbriate-lacerate wings, apex narrowed into spine; inner phyllaries distally with a small scarious lacerate appendage narrowed into a short spine; corolla yellow; pappus brown. Fl. Jul-Sep. Open places in forests at 1700-3100 m. Distributed in Xinjiang.

糙毛翅膜菊 *Alfredia aspera*

翅膜菊 *Alfredia cernua*

厚叶翅膜菊 *Alfredia nivea*

翅膜菊
Alfredia cernua (L.) Cass.

多年生草本，高1-3米。茎紫红色，上部有长分枝。叶纸质，边缘有刺齿，两面异色。头状花序多数，下垂；外层和中层苞片边缘具齿状膜质附属物，顶端具骨刺；最内层顶端附片边缘全缘或撕裂；小花黄色。花果期7-9月。生海拔1400-2000米的阳坡森林、潮湿草地、岩缝。产新疆。哈萨克斯坦和俄罗斯亦有。

Perennial herbs, 1-3 m tall. Stem purplish red, apically long branched. Leaves papery, discolorous, margin spinulose denticulate. Capitula several, nodding; outer and middle phyllaries distally with a roundish scarious denticulate appendage apically narrowed into spine; inner phyllarieswith a small scarious entire to lacerate apical appendage; corolla yellow. Fl. and fr. Jul-Sep. Sunny slopes in forests, wet grasslands, rocky crevices at 1400-2000 m. Distributed in Xinjiang. Also in Kazakhstan and Russia.

厚叶翅膜菊
Alfredia nivea Kar. et Kir.

多年生草本，高35-60厘米。茎红紫色，不分枝或上部有1个长分枝，有多数条棱。叶革质，两面异色，边缘有齿状淡黄色针刺。头状花序1或2，下垂；中外层苞片中部边缘有膜质撕裂的附片，顶端具骨刺；内层苞片干膜质，全缘。花果期7-9月。生海拔1400-2400米的河边草地、路边、云杉林。产新疆。哈萨克斯坦亦有。

Perennial herbs, 35-60 cm tall. Stem purplish red, unbranched or with 1 branch, many ribbed. Leaves leathery, discolorous, margin with teeth ending in straw-colored needle-shaped spines. Capitula 1 or 2, nodding; outer and middle phyllaries middle portion laterally expanded into scarious fimbriate-lacerate wings, apex narrowed into spine; inner most phyllaries scarious, margin entire. Fl. and fr. Jul-Sep. Grasslands in river valleys, trailsides, *Picea* forests at 1400-2400 m. Distributed in Xinjiang. Also in Kazakhstan.

火媒草
Olgaea leucophylla (Turcz.) Iljin

多年生草本，密被灰白色蛛丝状绒毛。叶厚纸质，长椭圆形至披针形，边缘具针刺，羽状浅裂或具三角形大刺齿。头状花序单生茎枝顶端；总苞钟状，直径3-4厘米；总苞片长三角形至披针形，先端渐尖成针刺；小花紫色或白色。瘦果长椭圆形；冠毛浅褐色。花果期5-10月。生海拔700-1800米的草地、田边或沟边。产华北、东北和西北。蒙古亦有。

Perennial herbs, densely offwhite arachnoid tomentose. Leaves thick chartaceous, narrowly elliptic to lanceolate, margin aristate, pinnately lobed or triangularly gross-serrate. Capitula solitary, terminal; involucres campanulate, 3-4 cm diam; phyllaries narrowly triangular to lanceolate, apex acuminate, aristate; florets purple or white. Achenes narrowly elliptic; pappus brownish. Fl. and fr. May-Oct. Grasslands, fields or ditchsides at 700-1800 m. Distributed in N, NE and NW China. Also in Mongolia.

火媒草 *Olgaea leucophylla*

刺疙瘩 *Olgaea tangutica*

黄缨菊 *Xanthopappus subacaulis*

刺疙瘩
Olgaea tangutica Iljin

多年生草本。茎翼羽状浅裂，裂片具刺齿，上面绿色，无毛，下面被灰白色绒毛。叶近革质，宽条形，边缘具刺齿。头状花序单生枝顶；总苞宽卵形，直径2-3厘米；总苞片条状披针形，革质，顶端针刺状，被微柔毛；花冠紫色。瘦果稍扁；冠毛污黄色。花果期6-9月。生海拔1200-2000米的山坡、沙滩、山顶、荒地或农田。产华北和西北。

Perennial herbs. Stem wings pinnately lobed, lobes aristate, green and glabrous adaxially, offwhite tomentose abaxially. Leaves subcoriaceous, broadly linear, margin aristate-serrate. Capitula solitary, terminal; involucres broadly ovoid, 2-3 cm diam; phyllaries linear-lanceolate, coriaceous, apex aristate, puberulent; corolla purple. Achenes somewhat compressed; pappus dirty yellow. Fl. and fr. Jun-Sep. Slopes, sandy plains, loess hilltops, wastelands or fields at 1200-2000 m. Distributed in N and NW China.

黄缨菊
Xanthopappus subacaulis C. Winkl.

多年生无茎草本。基生叶莲座状，坚硬，革质，长圆形，羽状深裂，下面密被灰白色蛛丝状绒毛，边缘和顶端具针刺。头状花序多达20，密集成团球状；总苞宽钟状；总苞片披针形至条形；花冠黄色。瘦果偏斜倒长卵形；冠毛淡黄色或棕黄色。花果期7-9月。生海拔2400-4000米的草甸、草原或干燥山坡。产甘肃、内蒙古西部、宁夏、青海、四川和云南西北部。

Perennial stemless herbs. Basal leaves rosette, rigid, coriaceous, oblong or linear-oblong, pinnatipartite, densely offwhite arachnoid-tomentose, margin and apex aciculate. Capitula up to 20, glomerate; involucres broadly campanulate; phyllaries lanceolate to linear; corolla yellow. Achenes obliquely obovoid; pappus yellowish or brownish-yellow. Fl. and fr. Jul-Sep. Meadows, grasslands or dry slopes at 2400-4000 m. Distributed in Gansu, W Neimenggu, Ningxia, Qinghai, Sichuan and NW Yunnan.

大翅蓟
Onopordum acanthium L.

二年生大草本。叶长椭圆形、宽卵形或倒披针形，边缘具刺齿或羽状浅裂；茎翅宽2-5厘米。头状花序多数或少数，顶生；总苞卵形或球形，直径达5厘米；总苞片坚硬，革质，钻形，先端针刺状长渐尖；小花紫红色或粉红色。瘦果倒卵球形，3棱状；冠毛土红色。花果期6-9月。生海拔400-1200米的山坡、荒地或沟边。产新疆。伊朗、俄罗斯和欧洲亦有。

Biennial large herbs. Leaves narrowly elliptic, broadly ovate or oblanceolate, margin aristate-dentate or pinnately lobed; cauline wings 2-5 cm wide. Capitula many or few, terminal; involucres ovoid or globose, to 5 cm diam; phyllaries rigid, coriaceous, subulate, apex aristate-acuminate; florets purple or pink. Achenes obovoid, trigonous; pappus latericeous. Fl. and fr. Jun-Sep. Slopes, wastelands or ditch sides at 400-1200 m. Distributed in Xinjiang. Also in Iran, Russia and Europe.

大翅蓟 *Onopordum acanthium*

肋果蓟 *Ancathia igniaria*

肋果蓟

Ancathia igniaria (Spreng.) DC.

多年生草本。茎生叶厚革质，条状披针形或条形，边缘具黄白色针刺。头状花序单生茎顶；苞片多层，外层顶端针刺状渐尖；小花紫色或紫红色。瘦果长椭圆体形，有多数纵肋；冠毛多层，基部连生成环，长达2.7厘米。花果期6-9月。生海拔1100-1500米的山坡或多石贫瘠滩。产新疆。哈萨克斯坦、俄罗斯和蒙古亦有。

Perennial herbs. Caudline leaves thick leathery, linear-lanceolate or linear, margin with yellow white spins. Capitula solitary at apices of stems; phyllaries in many series, outside one apex spinescent; florets purple or purple-red. Achenes long ellipsoid, many-ribbed; pappus many series, base connate into a loop, 2.7 cm long. Fl. and fr. Jun-Sep. Slopes or stony barren beach at 1100-1500 m. Distributed in Xinjiang. Also in Kazakhstan, Russia and Mongolia.

南蓟

Cirsium argyracanthum DC.

多年生草本。叶两面同色，基部扩大耳状抱茎，中部茎叶长椭圆状披针形或长椭圆形，羽状分裂。头状花序多数或少数，排成总状花序或穗状花序或总状圆锥花序；总苞卵状或卵球形；小花紫色或白色；冠毛浅褐色。花果期6-10月。生海拔2100-3700米的山坡林缘、林下、草地、河边灌丛中或田边。产西藏和云南西北部。喜马拉雅亦有。

Perennial herbs. Leaves concolorous, green; middle cauline leaves sessile, elliptic-lanceolate to elliptic, pinnately divided, auriculate semiamplexicaul. Capitula many or few, spicate, racemose or racemose-paniculate; involucre ovoid; corolla purple or white; pappus bristles brownish. Fl. and fr. Jun-Oct. Forests, forest margins, grasslands, thickets by rivers or by farmlands at 2100-3700 m. Distributed in Xizang and NW Yunnan. Also in Himalaya.

丝路蓟

Cirsium arvense (L.) Scop.

多年生草本。叶椭圆形或椭圆状披针形，羽状浅裂或半裂。总苞有极稀疏的蛛丝毛，外层苞片顶端有反折或开展的短针刺。花果期6-9月。生海拔700-4300米的沟边水湿地、田间或湖滨地区。产甘肃、新疆和西藏。南亚、西南亚、中亚和欧洲亦有。

Perennial herbs. Leaves elliptic to elliptic-lanceolate, pinnately lobed or pinnatifid. Involucre very sparsely cobwebby, outer and middle phyllaries with a ca. 0.5 mm patent to reflexed apical spinule. Fl. and fr. Jun-Sep. Moist places by ditches, farmlands or lakesides at 700-4300 m. Distributed in Gansu, Xinjiang and Xizang. Also in S, SW and C Asia, and Europe.

刺儿菜

Cirsium arvense var. **integrifolium** Wimm. et Grabowski

多年生草本。基生叶和中部茎叶不分裂或羽状半裂，常无柄，叶缘具细密的针刺，叶两面同色，两面无毛。头状花序单生或少数作伞房花序式排列；苞片6层，具短针刺；花冠紫红色。瘦果椭圆体形；冠毛白色。花果期5-9月。生海拔100-2700米的山坡、河边或荒地。中国广泛分布。俄罗斯、蒙古、朝鲜半岛、日本和欧洲亦广泛分布。

Perennial herbs. Basal and middle cauline leaves undivided or pinnatifid, usually sessile, margin

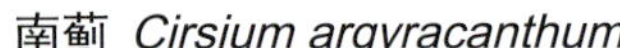
南蓟 *Cirsium argyracanthum*

丝路蓟 *Cirsium arvense*

刺儿菜 *Cirsium arvense* var. *integrifolium*

with fine spins, concolor and glabrous on both sides. Heads solitary or arrange into corymb at apices of branches; involucres in ca. 6 series, with short spins; corolla purple. Achenes ellipsoid; pappus white. Fl. and fr. May-Sep. Moutain slopes, riversides or waste lands at 100-2700 m. Widely distributed in China. Also widely distributed in Russia, Mongolia, Korean Peninsula, Japan and Europe.

绿蓟

Cirsium chinense Gardn. et Champ.

多年生草本。茎被多细胞长节毛。叶长圆形、披针形或条形，羽状分裂或不裂，绿色，无毛，坚硬，边缘具针刺。头状花序少数顶生，排成不规则伞房花序；总苞卵球形，直径约2厘米；总苞片无毛，绿色，顶端针刺状；小花紫红色。瘦果楔状倒卵球形，压扁，顶端截形；冠毛污白色。花果期6-10月。生海拔100-1600米的山坡草丛中。产中国西南、东南、华中和华北。

Perennial herbs. Stems septate-setose. Leaves oblong, lanceolate or linear, pinnatifid or undivided, green, glabrous, rigid, margin aristate. Capitula several, terminal, arranged in irregular corymb; involucres ovoid, ca. 2 cm diam; phyllaries glabrous, green, apex aristate; florets purple. Achenes cuneate-obovoid, compressed, apex truncate; pappus dirty white. Fl. and fr. Jun-Oct. Grassy slopes at 100-1600 m. Distributed in SW, SE, C and N China.

贡山蓟

Cirsium eriophoroides (Hook. f.) Petrak

多年生高大草本。叶羽状分裂，薄纸质，边缘具刺齿或针刺。头状花序数个，排成伞房状花序；苞叶边缘有长针刺；总苞球形，被稠密而蓬松的棉毛，直径达5厘米；总苞片具刺毛；小花紫色。瘦果倒披针状长圆形，黑褐色，顶端截形；冠毛污白色或浅褐色。花果期7-10月。生海拔2000-4100米的山坡、灌丛、林缘、草地、草甸、河滩地或水边。产云南、四川和西藏。不丹、印度北部亦有。

绿蓟 *Cirsium chinense*

Perennial stout herbs. Leaves pinnately lobed, thin chartaceous, margin aristate-dentate or aristate. Capitula several, arranged in corymb; bracteal leaves margin long aristate; involucres globose, densely and laxly cottony; phyllaries aristate; florets purple. Achenes oblanceolate-oblong, black-brown, apex truncate; pappus dirty white or brownish. Fl. and fr. Jul-Oct. Slopes, thickets, forest edges, grasslands, meadows, flood plains or watersides at 2000-4100 m. Distributed in Yunnan, Sichuan and Xizang. Also in Bhutan and N India.

贡山蓟 *Cirsium eriophoroides*

丽江蓟 *Cirsium lidjiangense*

刺苞蓟 *Cirsium henryi*

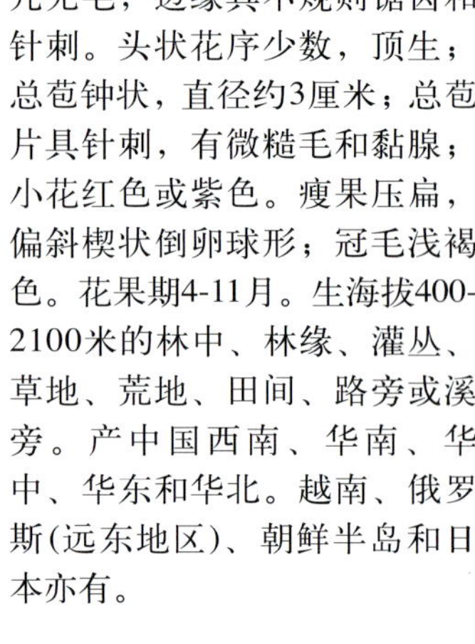

丽江蓟

Cirsium lidjiangense Petr. et Hand.-Mazz.

多年生草本。叶两面异色；下部茎叶半抱茎，无柄，椭圆形，二回羽状分裂。头状花序下垂，排成总状或总状圆锥花序；总苞球形，密被绵毛；小花红紫色。瘦果褐色，长约5毫米；冠毛浅褐色。花果期6-8月。生海拔1800-3200米的草坡。产四川西南部和云南西北部。

Perennial herbs. Leaves discolorous; lower cauline leaves sessile, elliptic, bipinnatipartite, semiamplexicaul. Capitula racemose to racemose-paniculate, nodding; involucre globose, densely lanate; corolla reddish purple. Achene brown, ca. 5 mm; pappus bristles brownish. Fl. and fr. Jun-Aug. Grassy slopes at 1800-3200 m. Distributed in SW Sichuan and NW Yunnan.

刺苞蓟

Cirsium henryi (Franch.) Diels

多年生草本，无明显主根。叶羽状半裂或全裂，基部渐狭成柄，半抱茎，两面沿脉具多细胞长节毛，边缘有三角形状刺齿。头状花序通常3-5个集生于枝端作伞房花序式排列，或多至18个成圆锥花序式排列；苞片7层，具短针刺。花果期6-10月。生海拔2700-3500米的草甸。产云南、四川和湖北。

Perennial herbs, without obvious taproots. Leaves pinnatifid or pinnatisect, attenuate into a petiolelike basal portion, semiamplexicaul, both sides with multicellular hairs, margin with triquetrous spins. Heads often 3-5 in corymb at apices of branches, or several to 18 range in paniculate; involucres in 7 series, with short spins. Fl. and fr. Jun-Oct. Meadows at 2700-3500 m. Distributed in Yunnan, Sichuan and Hubei.

蓟

Cirsium japonicum DC.

多年生草本。叶卵形、椭圆形或长圆形，羽状深裂，被多细胞节毛或几无毛，边缘具不规则锯齿和针刺。头状花序少数，顶生；总苞钟状，直径约3厘米；总苞片具针刺，有微糙毛和黏腺；小花红色或紫色。瘦果压扁，偏斜楔状倒卵球形；冠毛浅褐色。花果期4-11月。生海拔400-2100米的林中、林缘、灌丛、草地、荒地、田间、路旁或溪旁。产中国西南、华南、华中、华东和华北。越南、俄罗斯(远东地区)、朝鲜半岛和日本亦有。

Perennial herbs. Leaves ovate, elliptic or oblong, pinnatisected, septate-setose or nearly glabrous, margin irregularly serrate and aristate. Capitula several, terminal; involucres campanulate, ca. 3 cm diam; phyllaries ca. 6 rows, apex aristate, outside puberulent and viscid glandular; florets red or purple. Achenes compressed, obliquely cuneate-obovoid; pappus brownish. Fl. and fr. Apr-Nov. Forests, forest edges, thickets, grasslands, wastelands, fields, roadsides or streamsides at 400-2100 m. Distributed in SW, S, C, E and N China. Also in Vietnam, Russia (Far East), Korean Peninsula and Japan.

蓟 *Cirsium japonicum*

魁蓟 *Cirsium leo*

魁蓟

Cirsium leo Nakai et Kitag.

多年生草本。叶长圆形或长披针形，羽状深裂，被多细胞长节毛，边缘具针刺。头状花序少数，顶生，排成伞房花序；总苞钟状，直径达4厘米；总苞片披针形，边缘具针刺；小花紫色或红色。瘦果压扁，斜椭圆体形；冠毛污白色。花果期5-9月。生海拔700-3400米的山谷、山坡、河滩、溪旁、路边、林缘、草地、石砾地、岩石缝、潮湿地或田间。产四川、河南、河北、山西、陕西、甘肃和宁夏。

Perennial herbs. Leaves oblong or long lanceolate, pinnatifid, long septate-setose, margin aristate. Capitula several, terminal, arranged in corymb; involucres campanulate, to 4 cm diam; phyllaries lanceolate, margin aristate; florets purple or red. Achenes compressed, obliquely ellipsoid; pappus dirty brown. Fl. and fr. May-Sep. Valleys, slopes, flood plains, streamsides, roadsides, forest edges, grasslands, stony places, rock cracks, damp places or fields at 700-3400 m. Distributed in Sichuan, Henan, Hebei, Shanxi, Shaanxi, Gansu and Ningxia.

线叶蓟

Cirsium lineare Sch.-Bip.

多年生草本。叶长圆形、长圆状披针形、披针形或宽条形，不分裂，质厚，下面被密厚的绒毛。头状花序顶生，排成伞房花序；总苞卵球形，直径2-2.5厘米，无毛；总苞片具黑色黏腺；小花紫红色或粉红色。瘦果偏斜楔状倒卵球形，压扁；冠毛浅褐色。花果期8-10月。生海拔500-2500米的山坡、灌丛、林缘、草丛、荒地或田间。产中国西南、华中、华东和华北。

Perennial herbs. Leaves oblong, oblong-lanceolate, lanceolate or broadly linear, undivided, thick, densely and thickly tomentose abaxially. Capitula terminal, arranged in corymb; involucres ovoid, 2-2.5 cm diam, glabrous; phyllaries blackly viscid glandular; florets purple or pink. Achenes obliquely cuneateobovoid, compressed; pappus brownish. Fl. and fr. Aug-Oct. Slopes, thickets, forest edges, grasslands, wastelands or fields at 500-2500 m. Distributed in SW, C, E and N China.

线叶蓟 *Cirsium lineare*

马刺蓟

Cirsium monocephalum (Vaniot) H. Lévl.

多年生草本。叶两面同色，绿色；中部茎叶全形椭圆形至长椭圆形，羽状深裂，无柄。总苞宽钟状至半球形；小花白色或淡黄色。花果期7-10月。生海拔700-2000米的山谷或山坡林缘、林下或灌丛中或荒地、农田或潮湿地。产四川、重庆、贵州、湖北、甘肃、陕西和山西南部。

Perennial herbs. Leaves concolorous, green; middle cauline leaves sessile, narrowly elliptic to lanceolate, pinnatipartite. Capitula few, corymbose to paniculate; involucre broadly campanulate to hemispheric; corolla white or yellowish. Fl. and fr. Jul-Oct. Mountain valleys, forest margins, forests, thickets, wastelands at 700-2000 m. Distributed in Sichuan, Chongqing, Guizhou, Hubei, Gansu, Shaanxi and S Shanxi.

马刺蓟 *Cirsium monocephalum*

烟管蓟 *Cirsium pendulum*

赛里木蓟 *Cirsium sairamense*

烟管蓟

Cirsium pendulum Fisch. ex DC.

多年生草本。茎被白色蛛丝状毛或多细胞长节毛。叶二回羽状深裂，边缘齿状，齿端具针刺。头状花序多数，顶生，下垂，排成总状圆锥花序；总苞钟状，直径3.5-5厘米；总苞片披针形；小花紫色或红色。瘦果稍压扁，偏斜楔状倒披针形；冠毛污白色。花果期6-9月。生海拔300-2300米的山谷、山坡草地、林缘、林下、岩石缝、溪旁或路边。产华北、东北和西北。俄罗斯、朝鲜半岛和日本亦有。

Perennial herbs. Stems white arachnoid hairy or long septate-setose. Leaves bipinnatisected, margin dentate, tooth top aristate. Capitula numerous, terminal, drooping, arranged in racemose-panicle; involucres campanulate, 3.5-5 cm diam; phyllaries lanceolate; florets purple or red. Achenes compressed, obliquely and cuneately oblanceolate; pappus dirty white. Fl. and fr. Jun-Sep. Valleys, grassy slopes, forest edges, forest, rock cracks, riversides or roadsides at 300-2300 m. Distributed in N, NE and NW China. Also in Russia, Korean Peninsula and Japan.

赛里木蓟

Cirsium sairamense (C. Winkl.) O. Fedtsch. et B. Fedtsch.

多年生草本。叶两面异色；中下部茎叶长椭圆形、披针形或长披针形。头状花序多数，在茎枝顶端排成伞房状圆锥花序、伞房花序或短总状花序；总苞卵球形，直径2.5厘米；总苞钟状或卵形；小花紫色；冠毛污白色。花果期7-9月。生海拔1700-2300米的山坡、山谷、水边或湿地。产新疆。中亚亦有。

Perennial herbs. Leaves discolorous; lower and middle cauline leaves sessile, elliptic or narrowly lanceolate. Capitula many, terminal, corymbose-paniculate; involucre campanulate to ovoid; corolla purple; pappus bristles dirty white. Fl. and fr. Jul-Sep. Mountain slopes, mountain valleys, by water or moist places at 1700-2300 m. Distributed in Xinjiang. Also in C Asia.

天山蓟

Cirsium alberti Regel et Schmalh.

多年生草本。叶椭圆状披针形或披针形，羽状深裂，边缘有3-5个刺齿及缘毛状针刺，长5-11毫米；向上的叶渐小，披针形或长披针形，基部耳状扩大半抱茎。头状花序排成伞房花序或伞房圆锥花序；总苞卵球形或卵形；小花黄色或白色。花果期7-9月。生海拔1000-2400米的山坡、山谷林缘、草滩、河滩地或溪旁。产新疆。哈萨克斯坦亦有。

Perennial herbs. Leaves elliptic-lanceolate or lanceolate, pinnatipartite, with 3-5 teeth tipped with a 5-11 mm spine; middle and upper cauline leaves sessile, lanceolate or narrowly lanceolate. Capitula corymbose to corymbose-paniculate; involucre subovoid to ovoid; corolla yellow or white. Fl. and fr. Jul-Sep. Forest margins in mountain valleys, mountain slopes, flooded lands, by streams at 1000-2400 m. Distributed in Xinjiang. Also in Kazakhstan.

天山蓟 *Cirsium alberti*

麻花头蓟 *Cirsium serratuloides*

薄叶蓟 *Cirsium shihianum*

麻花头蓟
Cirsium serratuloides (L.) Hill

多年生草本。叶不分裂，两面同色；中部茎叶披针形，基部耳状扩大半抱茎。头状花序直立，排成疏松伞房花序；总苞卵形；小花紫红色；冠毛刚毛白色。花果期7-10月。生海拔1200-2600米的林下、河边或水边。产新疆。蒙古和俄罗斯亦有。

Perennial herbs. Leaves undivided, concolorous; middle cauline leaves sessile, lanceolate, auriculate semiamplexicaul. Capitula erect, laxly corymbose; involucre ovoid; corolla purplish red; pappus bristles white. Fl. and fr. Jul-Oct. Forests on mountain slopes, by rivers or by water at 1200-2600 m. Distributed in Xinjiang. Also in Mongolia and Russia.

牛口刺
Cirsium shansiense Petr.

多年生草本，被多细胞长节毛或绒毛。叶羽状分裂，下面灰白色，密被绒毛，侧裂片不等大2齿裂，边缘和齿裂顶端具针刺。头状花序数个，顶生；总苞卵球形；总苞片长三角形至披针形，顶端具短针刺；小花粉红色或紫色。瘦果偏斜椭圆状倒卵球形；冠毛浅褐色。花果期5-11月。生海拔1300-3400米的山地、林下、草地、河边、湿地或路边。产中国西南、华南、华中、华北和西北。南亚亦有。

Perennial herbs, long septate-setose or tomentose. Leaves pinnatifid, offwithe and densely tomentose abaxially, lateral lobes unequally 2 dentate-lobed, margin and tooth top aristate. Capitula several, terminal; involucres ovoid; phyllaries narrowly triangular to lanceolate, apex shortly aristate; florets pink or purple. Achenes obliquely ellipsoido-bovoid; pappus brownish. Fl. and fr. May-Nov. Montanes, forests, grasslands, riversides, wetlands or roadsides at 1300-3400 m. Distributed in SW, S, C, N and NW China. Also in S Asia.

薄叶蓟
Cirsium shihianum Greuter

一年生草本。中部茎生叶基部半抱茎，质地薄或稍厚，边缘有针刺，两面同色，上面具疏节毛，下面无毛或疏被蛛丝毛。头状花序单生茎顶或植株生2个头状花序；苞片约7层，外层苞片三角形，具针刺。花果期7-8月。生海拔1400-1600米的山谷林下或杂草丛中。产新疆。

Annual herbs. Middle cauline leaves semiamplexicaul, thin or slightly thick, concolor on both sides, margin with spins, with multicellular hairs adaxially, glabrous or sparsely arachnoid hairy abaxially. Heads solitary at apices of branches or plants with 2 heads; phyllaries in ca. 7 series, outer ones with triquetrous spins. Fl. and fr. Jul- Aug. Forests in valleys or weeds at 1400-1600 m. Distributed in Xinjiang.

牛口刺 *Cirsium shansiense*

附片蓟 *Cirsium sieversii*

葵花大蓟 *Cirsium souliei*

附片蓟

Cirsium sieversii (Fisch. et C. A. Mey.) Petr.

多年生草本。叶两面同色，绿色；上部茎叶无柄，长椭圆形或披针形，羽状分裂；接头状花序下部的叶边缘锯齿针刺化。头状花序3-5个集生于顶端或多数排成圆锥状花序；总苞卵球形；小花紫红色；冠毛刚毛浅褐色。花果期7-10月。生海拔1600-2900米的山坡林中草地或近水旁。产新疆北部。中亚和俄罗斯亦有。

Perennial herbs. Leaves concolorous, green; upper cauline leaves sessile, elliptic or lanceolate, pinnately lobed or pinnatifid. Capitula 3-6 in terminal cluster or many, terminal, and paniculate, surrounded by pectinately spiny bracts with long pungent spines; involucre narrowly ovoid; corolla purplish red; pappus bristles brownish. Fl. and fr. Jul-Oct. Forests on mountain slopes or by water at 1600-2900 m. Distributed in N Xinjiang. Also in C Asia and Russia.

葵花大蓟

Cirsium souliei (Franch.) Mattf.

多年生无茎草本。叶莲座状，羽状分裂，具柄，被多细胞长节毛，边缘具长针刺或刺齿。头状花序簇生，近无梗；总苞宽钟状，无毛；总苞片披针形，顶端和边缘具长针刺；小花紫红色。瘦果长圆状倒圆锥形；冠毛白色或浅褐色。花果期7-9月。生海拔1900-4800米的山坡、路旁、林缘、荒地、河滩地、田间或潮湿地。产青藏高原。

Perennial herbs, acauline. Leaves rosette, pinnatifid, petiolate, long septate-setose, margin long aristate or aristate-dentate. Capitula fasciate, subsessile; involucres broadly campanulate, glabrous; phyllaries lanceolate, apex and margin long aristate; florets purple. Achenes oblong-obovoid; pappus white or brownish. Fl. and fr. Jul-Sep. Slopes, roadsides, forst margins, wastelands, flood plains, fields or damp places at 1900-4800 m. Distributed in Tibetan Plateau.

莲座蓟

Cirsium esculentum (Siev.) C. A. Mey.

多年生草本。叶通常基生，莲座状，稀有明显的地上茎而多数叶生于茎顶端；叶片倒披针形或椭圆形或长椭圆形。头状花序5-9个簇生；总苞钟状，无毛；小花淡紫色或白色。瘦果淡黄色；冠毛白色或污白色。花果期8-9月。生海拔500-3200米的平原或山地潮湿地或水边。产中国东北、内蒙古和新疆。中亚、蒙古和俄罗斯亦有。

莲座蓟 *Cirsium esculentum*

峨眉蓟 *Cirsium fangii*

Perennial herbs. All leaves usually basal, rosulate, rarely with a distinct stem and most leaves clusted on top of the stem; leaf blade oblanceolate or narrowly elliptic. Capitula 5-9, in a cluster; involucre campanulate, glabrous; corolla light purple to white. Achene yellowish; pappus bristles white or dirty white. Fl. and fr. Aug-Sep. Moist places, by water in plains or hilly areas at 500-3200 m. Distributed in NE China, Neimenggu and Xinjiang. Also in C Asia, Mongolia and Russia.

峨眉蓟

Cirsium fangii Petr.

多年生草本，高达1.2米。叶片两面同色，绿色；下部叶片有叶柄，叶片披针形至线状披针形，近羽裂，裂片6-7对，椭圆形，有刺；中部和上部叶无柄。头状花序少数，生于枝顶，下垂；总苞宽钟形，直径约4.5厘米；总苞片约7层，三角状披针形、披针形或线形，先端有刺；花冠红色；冠毛褐色，长约2厘米。花期7-8月。生海拔2300-2400米的山地草坡。产四川。

Perennial herbs, to 1.2 m tall. Leaves concolorous, green; lower cauline leaves petiolate; leaf blade lanceolate to linear-lanceolate, subpinnatisect; segments 6 or 7 pairs, elliptic, fringed with spinules; middle and upper cauline leaves similar but sessile. Capitula few, terminal on long branches, nodding; involucre broadly campanulate, ca. 4.5 cm diam; phyllaries in ca. 7 rows, triangular-lanceolate, lanceolate or linear, apex tipped with a spinule; corolla red; pappus bristles brown, ca. 2 cm. Fl. Jul.-Aug. Grasslands on mountain slopes at 2300-2400 m. Distributed in Sichuan.

钻苞蓟

Cirsium subulariforme C. Shih

多年生草本。叶纸质，两面异色；下部茎叶椭圆形，有柄，长14厘米，有翼；中上部茎无柄；苞叶针刺化。头状花序排成伞房花序或总状花序；总苞钟状；小花紫红色。花果期7-8月。生海拔1500-2500米的山坡草地、河谷灌丛或松林下。产云南西北部和西藏东南部。

Perennial herbs. Leaves papery, discolorous; lower cauline leaves with winged petioles to 14 cm; leaf blade elliptic; middle and upper cauline leaves sessile; bracts reduced to spines. Capitula corymbose to racemose; involucre campanulate; corolla purple. Fl. and fr. Jul-Aug. Grasslands on mountain slopes, thickets in river valleys, forests at 1500-2500 m. Distributed in NW Yunnan and SE Xizang.

绒背蓟

Cirsium vlassovianum Fisch. ex DC.

多年生草本。叶两面异色；茎叶披针形或椭圆状披针形，不分裂，边缘有长约1毫米的针刺状缘毛。头状花序1或少数，或直立的伞房花序；总苞长卵形；小花紫色。花果期5-9月。生海拔300-1500米的山坡林中、林缘、河边或潮湿地。产华北和东北。朝鲜半岛、蒙古和俄罗斯亦有。

Perennial herbs. Leaves discolorous; leaf blade lanceolate or elliptic-lanceolate undivided, fringed with ca. 1 mm spinules. Capitula 1 to few, corymbose, erect; involucre narrowly ovoid; corolla purple. Fl. and fr. May-Sep. Forests, forest margins, by rivers, moist places at 300-1500 m. Distributed in N and NE China. Also in Korean Peninsula, Mongolia and Russia.

钻苞蓟 *Cirsium subulariforme*

绒背蓟 *Cirsium vlassovianum*

苞叶蓟

Cirsium verutum (D. Don) Spreng.

多年生草本。叶两面同色，绿色；中部茎叶倒披针形，无柄或基部扩大耳状抱茎；苞叶卵圆形，边缘多刺或具刺状锯齿。头状花序排成伞房状花序；总苞钟状；小花红紫色；冠毛浅褐色。花果期7-9月。生海拔2900-3900米的林下、林缘或山坡草丛中。产西藏。喜马拉雅和越南亦有。

Perennial herbs. Leaves concolorous, green; middle cauline leaves sessile, oblanceolate, semiamplexicaul; bracts ovate, margin spiny or with spiny teeth. Capitula corymbose; involucre campanulate; corolla reddish purple; pappus bristles brownish. Fl. and fr. Jul-Sep. Slopes, forests, forest margins at 2900-3900 m. Distributed in Xizang. Also in Himalaya and Vietnam.

苞叶蓟 *Cirsium verutum*

块蓟 *Cirsium viridifolium*

块蓟

Cirsium viridifolium (Hand.-Mazz.) C. Shih

多年生草本。块根纺锤状。茎有纵棱。下部叶椭圆形或披针形，无柄，基部半抱茎，边缘及顶端具针刺，叶两面同色，无毛或有多细胞节毛。头状花序常单生茎顶；总苞片约7层，有黑色黏腺。花果期8-9月。生海拔200-2000米的湿地、溪旁、路边或山坡。产河北、内蒙古和吉林。

Perennial herbs. Rootstocks fusiform. Stems ribbed. Basal leaves elliptic or lanceolate, sessile, semiamplexicaul, margin and apex with fine spins, concolor on both sides, glabrous or with multicellular hairs. Capitula solitary at apices of branches; phyllaries in ca. 7 series, with black cement glands. Fl. and fr. Aug-Sep. Wetland, roadsides, riversides or mountain slopes at 200-2000 m. Distributed in Hebei, Neimenggu and Jilin.

翼蓟 *Cirsium vulgare*

翼蓟

Cirsium vulgare (Savi) Ten.

多年生草本。叶两面异色；中部茎叶全形披针形、倒披针形或线状披针形，羽状深裂。头状花序多数或少数排成圆锥状伞房花序或总状花序；总苞卵球形；小花红色；冠毛白色。花果期7-8月。生海拔400-1800米的田间及湿润草地。产新疆北部。中亚、西南亚、俄罗斯、北非和欧洲亦有。

Perennial herbs. Leaves discolorous; middle cauline leaves sessile, linear-lanceolate, lanceolate or oblanceolate, bipinnatipartite. Bracts linear, margin with long spines. Capitula few to many, paniculate-corymbose to racemose, erect; involucre ovoid; corolla red; pappus bristles white. Fl. and fr. Jul-Aug. Farmlands and wet grasslands at 400-1800 m. Distributed in N Xinjiang. Also in C and SW Asia, Russia, N Africa and Europe.

丝毛飞廉

Carduus crispus L.

二年生或多年生植物，被多细胞长节毛。茎单生，具叶质狭翅，上部被蛛丝状毛。叶羽状深裂或半裂，边缘有针刺。头状花序1-5个顶生，近无梗；总苞卵圆形，直径1.5-2.5厘米；总苞片具针刺；花冠红色或紫色。瘦果楔状长圆形；冠毛白色。花果期2-10月。生海拔400-3600米的山坡、草地、田间、荒地、河旁或林下。广布中国各地。中亚、东北亚、欧洲和北美洲亦有。

Biennial or perennial herbs, septate-setose. Stems solitary, with narrow foliaceous wings, upper part arachnoid hairy. Leaves pinnately parted or sected, margin aristate. Capitula 1-5, terminal,

丝毛飞廉 *Carduus crispus*

subsessile; involucres ovoid, 1.5-2.5 cm diam; phyllaries aristate; corolla red or purple. Achenes cuneately oblong; pappus white. Fl. and fr. Feb-Oct. Slopes, grasslands, fields, wastelands, riversides or forests at 400-3600 m. Distributed in most parts of China. Also in C and NE Asia, Europe and North America.

节毛飞廉

Carduus acanthoides L.

二年生或多年生植物，被多细胞长节毛。茎单生，具叶质狭翅。叶长圆形或长倒披针形，羽状浅裂、半裂或深裂，边缘有三角形刺齿，齿端有黄白色针刺。头状花序3-5个顶生，近无梗；总苞卵球形；总苞片具针刺；花冠红紫色。瘦果长椭圆体形；冠毛白色。花果期5-10月。生海拔200-3500米的山坡、山谷、水边、草地、林缘、灌丛或田间。广布中国各地。中亚、东北亚和欧洲亦有。

Biennial or perennial herbs, septate-setose. Stems solitary, with narrow foliaceous wings. Leaves oblong or narrowly oblanceolate, pinnately lobed, parted or divided, margin triangular-setose-serrate, teeth top yellowish-white aristate. Capitula 3-5, terminal, subsessile; involucres ovoid; phyllaries aristate; corolla purple. Achenes narrowly ellipsoid; pappus white. Fl. and fr. May-Oct. Slopes, valleys, watersides, grasslands, forest edges, thickets or fields at 200-3500 m. Distributed in most parts of China. Also in C and NE Asia, and Europe.

节毛飞廉 *Carduus acanthoides*

飞廉

Carduus nutans L.

二年生或多年生草本，高30-100厘米。茎直立，多分枝，有大小不等的三角形刺齿裂翼。中下部茎叶长卵圆形或披针形；向上叶与下部叶相似，但渐小。头状花序下垂或下倾，单生茎顶或长分枝的顶端，总苞钟状或宽钟状；总苞片多层；花冠紫色，裂片狭线形；冠毛白色，多层，不等长。花果期6-9月。生海拔540-2300米山谷、田边或草地。产新疆北部。俄罗斯、欧洲、北非和西南亚亦有。

Biennial or perennial herbs, 30-100 cm tall. Stems erect, usually branched, teeth triangular, margin and apex with spines. Lower and middle cauline leaves sessile, ovate to lanceolate; upper cauline leaves similar but gradually smaller upward. Capitula solitary at end of stem and branches, nodding; involucres broadly campanulate; corolla purplish red; pappus bristles white, unequal. Fl. and fr. Jun-Sep. By croplands, grasslands, valleys at 540-2300 m. Distributed in N Xinjiang. Also in Russia, Europe, N Africa and SW Asia.

飞廉 *Carduus nutans*

顶羽菊 *Rhaponticum repens*

顶羽菊

Rhaponticum repens (L.) Hidalgo

多年生草本，高25-70厘米。茎单生，自基部分枝，分枝斜升。叶长椭圆形或线形。头状花序多数在茎枝顶端排成伞房花序或伞房圆锥花序；总苞卵形或椭圆状卵形；总苞片约8层；花冠粉红色或淡紫色。瘦果倒长卵形，顶端圆形；冠毛白色，多层，短羽毛状。花果期5-9月。生海拔600-2600米的山坡、丘陵、平原、农田或荒地。产华北和西北。俄罗斯、蒙古、巴基斯坦、中亚、西南亚和欧洲亦有。

Perennial herbs, 25-70 cm tall. Stem solitary, branched at base. Leaf blade narrowly elliptic or filiform. Capitula many, arranged in terminal corymb or corymbose-panicle; involucres ovate or elliptic-ovate; phyllaries ca. 8-seritate; flowers pink or light purple; flowers pink or light purple; pappus white, several seriates, short feathery. Fl. and fr. May-Sep. Mountain slopes, hills, waste places, farmlands or plains at 600-2600 m. Distributed in N and NW China. Also in Russia, Mongolia, Pakistan, C and SW Asia, and Europe.

漏芦 *Rhaponticum uniflorum*

漏芦

Rhaponticum uniflorum (L.) DC.

多年生草本，灰白色，密被棉毛。叶具柄，椭圆形，羽状深裂或几全裂，边缘有锯齿。头状花序单生茎顶；总苞半球形，直径3.5-6厘米；总苞片具浅褐色膜质附属物；小花两性，管状，紫红色。瘦果楔状，具3-4纵棱；冠毛褐色。花期4-7月，果期5-8月。生海拔100-2700米的山坡、丘陵、草甸或林下。产华北、东北和西北。东南亚亦有。

Perennial herbs, offwhite, densely villose. Leaves petiolate, elliptic, pinnately parted or nearly sected, margin serrate. Capitula solitary, terminal; involucres semiglobose, 3.5-6 cm diam; phyllaries attached with brownish menbranous appendages; florets bisexual, tubular, purple. Cypsesale cuneate, 3-4-ribbed; pappus brown. Fl. Apr-Jul. Fr. May-Aug. Slopes, hills, meadows or forests at 100-2700 m. Distributed in N, NE and NW China. Also in NE Asia.

华漏芦

Rhaponticum chinense (S. Moore) L. Martins et Hidalgo

多年生草本。叶缘有锯齿，两面粗糙，被多细胞短节毛及棕黄色小腺点。头状花序少数，单生茎枝顶端，不明显的伞房花序式排列；苞片6-7层；小花两性；花冠紫红色。瘦果长椭圆形；冠毛多层。花果期7-10月。生海拔300-1400米的山坡草地、林缘、林下或灌木中。产华南、东南、华中、华北、华西和华东。

华漏芦 *Rhaponticum chinense*

麻花头 *Klasea centauroides*

Perennial herbs. Leaves margin serrate, both sides scabrous, with short multicellular hairs and yellowish brown glandular. Capitula solitary, on apices of stems, obscurely range in corymb; phyllaries in 6-7 series; florets bisexual; corolla pupurish red. Achenes long elliptic; pappus many seriate. Fl. and fr. Jul-Oct. Meadows on slopes, forest edges, forests or shrubs at 300-1400 m. Distributed in S, SE, C, N, W and E China.

麻花头

Klasea centauroides (L.) Cass. ex Kitag.

多年生草本，被多细胞节毛。叶粗糙，长椭圆形，羽状深裂，全缘或具锯齿。头状花序少数，单生茎枝顶端；总苞卵形，直径1.5-2厘米；总苞片先端具刺尖；全部小花两性，红色、红紫色或白色。瘦果楔状长圆形，褐色；冠毛褐色。花果期6-9月。生海拔800-1700米的山坡、林缘、草原、草甸、路旁或田间。产华北、东北和西北。俄罗斯和蒙古亦有。

Perennial herbs, septate-setose. Leaves coarse, narrowly elliptic, pinnatiparted, margin entire or serrate. Capitula several, terminal, solitary; involucres ovoid, 1.5-2 cm diam; apex of invloucral bracts apiculate; all florets bisexual, red, purple or white. Achenes cuneate-oblong, brown; pappus brown. Fl. and fr. Jun-Sep. Slopes, forest edges, grasslands, meadows, roadsides or fields at 800-1700 m. Dsitributed in N, NE and NW China. Also in Russia and Mongolia.

钟苞麻花头

Klasea centauroides (L.) Cass. ex Kitag. subsp. **cupuliformis** (Nakai et Kitag.) L. Martins

多年生草本。茎基部被残存的纤维状撕裂的叶柄。全部叶两面粗糙，被多细胞节毛或渐脱落。头状花序单生茎顶或少数头状花序生茎枝顶端；总苞卵形；苞片约10层；小花紫红色；冠毛刚毛锯齿状。花果期7-10月。生海拔900-2400米的山坡草地或疏林下。产华北、华西与东北。

Perennial herbs. Stem basally with densely fibrous-lacerate leaf sheath. Whole leaves both sides scabrous, with multicellular hairs or glabrescent. Capitula solitary or few on apices of stems; involucres ovate; phyllaries ca. 10 series; florets pupurish red; pappus of serrate bristle. Fl. and fr. Jul-Oct. Meadows on slopes or sparce forests at 900-2400 m. Distributed in N, W and NE China.

钟苞麻花头 *Klasea centauroides*

碗苞麻花头
Klasea centauroides

碗苞麻花头

Klasea centauroides (L.) Cass. ex Kitag. subsp. **chanetii** (H. Lévl.) L. Martins

多年生草本，被多细胞节毛。叶具柄，粗糙，椭圆形或披针形，羽状分裂。头状花序数个，顶生，排成伞房花序，具长花序梗；总苞碗状，直径2-3厘米；总苞片先端具刺尖；全部小花两性，紫色或粉红色。瘦果楔状长圆形，黄白色；冠毛淡黄白色。花果期5-9月。生海拔200-2100米的山坡草地、林下、荒地或田间。产华东、华北和西北。蒙古亦有。

Perennial herbs, septate-setose. Leaves petiolate, coarse, elliptic or lanceolate, pinnatifid. Capitula several, terminal, arranged in corymb, long petiolate; involucres bowl-shaped, 2-3 cm diam; apex of invloucral bracts apiculate; all florets bisexual, purple or pink. Achenes cuneate-oblong, yellowish-white; pappus lightly yellowish-white. Fl. and fr. May-Sep. Grassy slopes, forests, wastelands or fields at 200-2100 m. Distributed in E, N and NW China. Also in Mongolia.

多花麻花头

Klasea centauroides subsp. **polycephala** (Iljin) L. Martins

多年生草本。茎上部伞房状分枝。基部和茎下部叶羽状深裂；全部叶两面沿脉有稀疏的多细胞节毛。头状花序10-20个在茎顶排成伞房状；总苞长卵形；苞片8-9层；小花两性，紫色或粉红色。瘦果楔状椭圆体形；冠毛刚毛锯齿状。花果期7-9月。生海拔600-2000米的山坡、路旁或农田中。产河北、山西、内蒙古和辽宁。

Perennial herbs. Upper stems corymbose branched. Basal and lower stem leaves pinnatifid; whole leaves with sparce multicellular hairs. Capitula 10-20, in corymbose; involucres long ovate; phyllaries 8-9 series; whole florets bisexual, purple or purpulish red. Achenes cuneate-ellipsoid; pappus serrated bristle. Fl. and fr. Jul-Sep. Moutain slopes, roadside or field at 600-2000 m. Distributed in Hebei, Shanxi, Neimenggu and Liaoning.

多花麻花头 *Klasea centauroides*

薄叶麻花头

Klasea marginata (Tausch) Kitag.

多年生草本。茎不分枝，大部分无叶。全部叶薄，两面有稀疏多细胞节毛或渐脱落。头状花序单生于茎端；总苞碗状；苞片7-8层；全部小花两性，紫红色；冠毛长达1.2厘米。花果期6-8月。生海拔1500-2300米的高山草原和丘陵地区。产内蒙古、甘肃、新疆北部和黑龙江。中亚、俄罗斯和蒙古亦有。

Perennial herbs. Stem s unbranched, large portion leafless. Whole leaves thin, both sides sparcely multicellular hairs or glabrescent. Capitula solitary on apices of stems; involucres bowl-like; phyllaries 7-8 series; whole florets bisexual, purpulish red; pappus to 1.2 cm. Fl. and fr. Jun-Aug. Mountain steppes and hilly areas at 1500-2300 m. Distributed in Neimenggu, Gansu, N Xinjiang and Heilongjiang. Also in C Asia, Russia and Mongolia.

薄叶麻花头 *Klasea marginata*

滇麻花头
Archiserratula forrestii

针苞菊 *Tricholepis furcata*

滇麻花头

Archiserratula forrestii (Iljin) L. Martins

多年生草本。茎粗壮，基部常木质化。全部叶无柄，边缘具细锯齿，两面无毛。头状花序多数，单生于茎枝顶端；总苞狭圆柱状；苞片约10层；全部小花两性；冠毛褐色，多层，刚毛锯齿状。花果期8-11月。生海拔1300-2000米的山坡草地或岩石缝中。产云南西北部和四川西南部。

Perennial herbs. Stem robust, basally sometimes woody. Whole leaves sessile, margin finely toothed, both sides glabrous. Capitula solitary on apices of stems; involucres narrowly cylindric; phyllaries ca. 10 series; whole florets bisexual; pappus brown, many seriate, finely serrulate, pinnate toward apex. Fl. and fr. Aug-Nov. Meadows on slopes or among fractured rocks at 1300-2000 m. Distributed in NW Yunnan and SW Sichuan.

针苞菊

Tricholepis furcata DC.

多年生草本，茎高0.6-1.4米。叶椭圆形或披针形，边缘有锯齿。头状花序半球形，下垂；总苞片多层，多数，针芒状；小花黄色。花果期10月。生海拔2600-3000米山谷林缘。产西藏南部。喜马拉雅亦有。

Perennial herbs, 0.6-1.4 m tall. Leaf blade elliptic or lanceolate-base, margin serrate, apex acute to acuminate. Capitula nodding; involucres hemispheric; phyllaries in many rows; appendage usually blackish, awnlike, somewhat recurved; corolla yellow. Fl. and fr. Oct. Forest margins in mountain valleys at 2600-3000 m. Distributed in S Xizang. Also in Himalaya.

伪泥胡菜

Serratula coronata L.

多年生草本。叶羽状全裂，两面有短糙毛或渐脱落。头状花序少数，在茎枝顶端排成伞房花序；苞片约7层；中央花两性，紫色。瘦果倒披针状椭圆形，有多数高起的细条纹；冠毛黄褐色，长达1.2厘米。花果期7-10月。生海拔100-1600米的山坡林下、林缘、草原或河岸。产中国西南、华中、华北、华西、西北和东北。俄罗斯、日本和欧洲亦有。

Perennial herbs. Leaves pinnatisect, both sides short strigillose or glabrescent. Capitula few, range in corymb on apices of stems; phyllaries in ca. 7 series; central disc bisexual, purple. Achenes lanceolate-elliptic, many-ribbed; pappus yellowish brown, to 1.2 cm long. Fl. and fr. Jul-Oct. Forests on slopes, forest edges, grasslands or riverbanks at 100-1600 m. Distributed in SE, C, N, W, NW and NE China. Also in Russia, Japan and Europe.

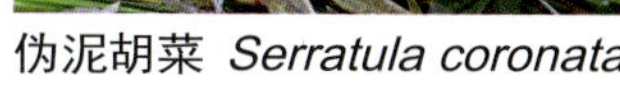

伪泥胡菜 *Serratula coronata*

欧亚矢车菊

Rhaponticoides ruthenica (Lam.) M. V. Agababjan et Greuter

多年生草本。基生叶与下部茎叶全形倒披针形，羽状全裂，有长叶柄；中部及上部茎叶无叶柄。头状花序；总苞卵形；小花黄色。瘦果长椭圆形，有横皱纹；冠毛2列，刚毛或膜片白色或浅褐色。花果期7-9月。生海拔1200-1900米的山坡或山沟近水处。产新疆。西南亚、中亚、北亚、欧洲亦有。

欧亚矢车菊 *Rhaponticoides ruthenica*

Perennial herbs. Basal and lower stem leaves, leaf blade oblanceolate in outline, pinnatisect to pinnate; middle and upper stem leaves sessile. Capitula, involucre ovoid. Achene narrowly ellipsoid, apically rugulose; pappus 2. Fl. and fr. Jul-Sep. Mountain slopes or meadows at 1200-1900 m. Distributed in Xinjiang. Also in SW, C and N Asia, and Europe.

红花 *Carthamus tinctorius*

红花

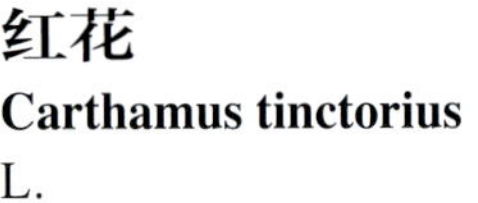

Carthamus tinctorius L.

一年生草本，无毛。茎白色。叶披针形或长圆形，硬革质，有光泽，无柄，半抱茎，边缘具针刺。头状花序顶生；苞叶椭圆形或卵状披针形，边缘有针刺；总苞卵形，直径约2.5厘米；总苞片坚硬；小花管状，红色、橘红色或黄色。瘦果倒卵球形，乳白色；无冠毛。花果期5-8月。栽培于海拔3000米以下的药圃中。中国广泛栽培。原产地不详。

Annual herbs, glabrous. Stems white. Leaves lanceolate or oblong, rigid, coriaceous, lustrous, base sessile, semi-amplexicaul, margin aristate. Capitula terminal; bracteal leaves elliptic or ovate-lanceolate, margin ari-

针刺矢车菊 *Centaurea iberica*

糙叶矢车菊 *Centaurea scabiosa*

state; involucres ovate, ca. 2.5 cm diam; phyllaries rigid; florets tubular, red, orange or yellow. Achenes obovoid, milky; pappus absent. Fl. and fr. May-Aug. Cultivated in medicinal gardens below 3000 m. Widely cultivated in China. Native origin unknown.

针刺矢车菊

Centaurea iberica Trevir. ex Spreng.

二年生草本，高20-100厘米。基生叶大头羽状深裂至大头羽状全裂；中部茎叶羽状深裂至全裂；向上的叶渐小。总苞卵形或卵球形，总苞片4-6层；花冠粉红色或紫色。花果期6-9月。生海拔500-1200米山坡。产新疆西北部。巴基斯坦、俄罗斯、中亚、西南亚和欧洲亦有。

Biennial herbs, 20-100 cm tall. Basal leaves lyrately pinnatipartite to lyrately pinnatisect; middle stem leaves pinnatipartite or pinnatisect; upper leaves decreasing in size upward. Involucres ovate or ovoid; phyllaries in 4-6 rows; corolla pink or purple. Fl. and fr. Jun-Sep. Mountain slopes at 500-1200 m. Distributed in NW Xinjiang. Also in Pakistan, Russia, C and SW Asia, and Europe.

糙叶矢车菊

Centaurea scabiosa L. subsp. **adpressa** (Ledeb.) Gugler

多年生草本。基生叶倒披针形或长椭圆形，羽状全裂，叶柄长5-8厘米。头状花序排成伞房花序或伞房状圆锥花序；总苞7层；苞片顶端有针刺；边花无性；两性花两性，紫红色。花果期6-9月。生海拔500-1400米的草地、河边或荒漠地带。产新疆。俄罗斯和欧洲亦有。

Perennial herbs. Basal leaves oblanceolate or long elliptic, pinnatisect, petioles 5-8 cm long. Capitula in corymb or corymbose-paniculate; involucres in 7 series, apex spinescent; marginal florets asexual; disc florets bisexual, purpule red. Fl. and fr. Jun-Sep. Grasslands, riversides or desert areas at 500-1400 m. Distributed in Xinjiang. Also in Russia and Europe.

小花矢车菊

Centaurea virgata subsp. **squarrosa** (Boiss.) Gugler

多年生草本。基生及下部茎叶椭圆形，一或二回羽状全裂；中部茎叶一回羽状全裂；上部茎叶不分裂。头状花序多数排成圆锥状花序；总苞卵形或长椭圆状卵形或圆柱状；花冠紫色到粉红色。花果期6-9月。

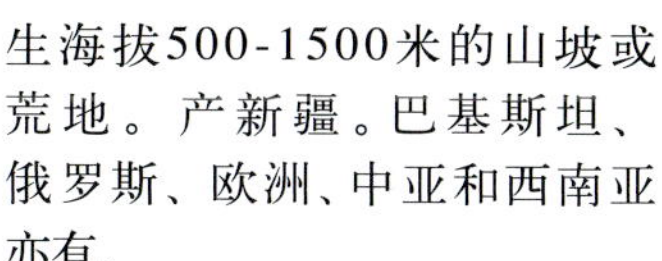
生海拔500-1500米的山坡或荒地。产新疆。巴基斯坦、俄罗斯、欧洲、中亚和西南亚亦有。

Perennial herbs. Basal and lower stem leaves elliptic, pinnatipartite to bipinnatipartite; middle stem leaves pinnatipartite; upper leaves undivided. Capitula many, paniculate; involucre ovoid, ellipsoid-ovoid, or cylindric; corolla purple to pink. Fl. and fr. Jun-Sep. Mountain slopes, wastelands at 500-1500 m. Distributed in Xinjiang. Also in Pakistan, Russia, Europe, C and SW Asia .

小花矢车菊 *Centaurea virgata*

莛菊 *Cavea tanguensis*

莛菊

Cavea tanguensis (J. R. Drumm.) W. W. Sm. et J. Small

多年生草本。叶片匙状长圆形、卵圆状披针形至长圆状匙形。头状花序单生茎端，近球形；总苞半球形；总苞片4-5层；小花紫色；雌花细管状，顶端3-4细裂。瘦果被黄白色绢状密毛。花期7-8月，果期9-10月。生海拔4000-5100米高山近雪线地带的砾石坡地、干燥沙地和河谷或灌丛间。产西藏南部和四川西部。不丹和印度北部亦有。

Perennial herbs. Leaves spatulate-oblong, ovate-lanceolate to oblong-spatulate. Capitula solitary, terminal, subglobose; involucres semiglobose; phyllaries in 4-5 series; florets purple; female florets tubular, apex 3-4 lobed. Achene yellow-white densely sericeous. Fl. Jul-Aug. Fr. Sep-Oct. Gravelly ground near streams and glaciers at 4000-5100 m. Distributed in S Xizang and SW Sichuan. Also in Bhutan and N India.

毛梗鸦葱

Scorzonera radiata Fisch. ex Ledeb.

多年生莲座状草本。茎具膜质叶鞘残余，被蛛丝毛，渐无毛。茎生叶0-3，条形至条状披针形。头状花序单生；总苞宽圆柱状至钟形；苞片顶端常具一红色点；花黄色。瘦果圆柱形，具突出纵棱；冠毛污黄色。花果期5-7月。生海拔900-2600米的林缘、林中、草地或多石河岸。产内蒙古、新疆、黑龙江、吉林和辽宁。哈萨克斯坦、乌兹别克斯坦和俄罗斯亦有。

Perennial herbs, rosulate. Caudex with scarious leaf sheath residues, arachnoid hairy, glabrescent with age. Stem leaves 0-3, linear to linear-lanceolate. Capitula solitary; involucres broadly cylindric to campanulate; phyllaries apex usually with a red spot; florets yellow. Achenes cylindric, with smooth elevated ribs; pappus dirty yellow. Fl. and fr. May-Jul. Forest edges, forests, grasslands or gravelly river banks at 900-2600 m. Distributed in Neimenggu, Xinjiang, Heilongjiang, Jilin and Liaoning. Also in Kazakhstan, Uzbekistan and Russia.

鸦葱

Scorzonera austriaca Willd.

多年生草本，光滑无毛。根垂直直伸，黑褐色。茎多数，簇生，不分枝。基生叶条形、披针形至长圆形，全缘；茎生叶2-3，鳞片状。头状花序单生茎端；总苞圆柱状；总苞片三角形至长圆形；舌状小花黄色。瘦果圆柱状；冠毛淡黄色，羽毛状。花果期4-7月。生海拔400-2000米的山坡、草滩或河滩地。产华东、华北、东北和西北。哈萨克斯坦、俄罗斯、蒙古、地中海地区和欧洲中部亦有。

Perennial herbs, wholly glabrous. Taproots vertically straight, dark brown. Stems numerous, clustered, unbranched. Basal leaves linear, lanceolate to oblong, entire; cauline leaves 2-3, squamiform. Capitula solitary, terminal; involucres cylindric; phyllaries triangular to oblong; ray florets yellow. Achenes terete; pap-

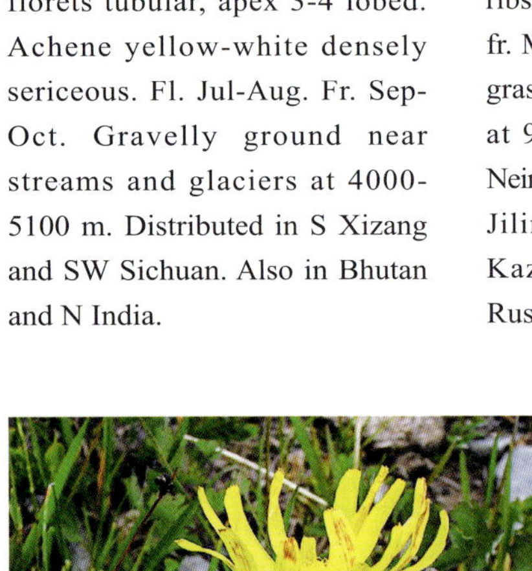

毛梗鸦葱 *Scorzonera radiata*

鸦葱 *Scorzonera austriaca*

桃叶鸦葱 *Scorzonera sinensis*

pus yellowish, feathery. Fl and fr. Apr-Jul. Slopes, meadows or flood plains at 400-2000 m. Distributed in E, N, NE and NW China. Also in Kazakhstan, Russia, Mongolia, Mediterranean regions and C Europe.

桃叶鸦葱

Scorzonera sinensis Lipsch. et Krasch.

多年生草本，全株光滑无毛。茎不分枝。基生叶较宽，卵形、披针形、椭圆形至条形，边缘皱波状；茎生叶少数，鳞片状。头状花序单生茎顶；总苞圆柱状；总苞片三角形至披针形；舌状小花黄色。瘦果圆柱状；冠毛污黄色，羽毛状。花果期2-9月。生海拔200-2500米的山坡、丘陵、沙丘、荒地或灌丛。产华中、华东、华北、东北和西北。

Perennial herbs, wholly glabrous. Stems unbranched. Basal leaves broader, ovate, lanceolate, elliptic to linear, margin rugosely undulate; cauline leaves several, squamiform. Capitula terminal, solitary; involucres cylindric; phyllaries triangular to lanceolate; ray florets yellow. Achenes terete; pappus dirty yellowish, feathery. Fl. and fr. Feb-Sep. Slopes, hills, dunes, wastelands or thickets at 200-2500 m. Distributed in C, E, N, NE and NW China.

东北鸦葱

Scorzonera manshurica Nakai

多年生草本，莲座状。茎密被纤维撕裂状叶鞘残遗。茎生叶1-3，苞片状，边缘和下面具棉毛。头状花序单生；总苞钟形。瘦果污黄色，圆柱形，具光滑的纵棱；冠毛污黄色。花果期4-5月。生旱山坡。产中国东北。

Perennial herbs, rosulate. Caudex densely covered with brown fibrous and lacerate leaf sheath residues. Stem leaves 1-3, scale-like, margin and adaxially lanate. Capitula solitary; involucres campanulate. Achenes dirty yellow, cylindric, with smooth ribs; pappus dirty yellow. Fl. and fr. Apr-May. Dry mountain slopes. Distributed in NE China.

华北鸦葱

Scorzonera albicaulis Bunge

多年生草本，被白色绒毛。茎单生或簇生，上部分枝。叶条形，全缘。头状花序数个，排成伞房花序或聚伞花序；总苞圆柱状；总苞片卵形、披针形至条形，被薄柔毛；舌状小花黄色。瘦果圆柱状；冠毛污黄色，羽毛状。花果期5-9月。生海拔200-2500米的山谷、山坡、林下、林缘、灌丛、荒地或田边。产中国西南、华中、华东、华北和东北。俄罗斯和朝鲜半岛亦有。

Perennial herbs, white tomentose. Stems solitary or clustered, upper part branched. Leaves linear, entire. Capitula several, arranged in corymb or cyme; involucres cylindric; phyllaries ovate, lanceolate to linear, thinly pubescent; ray florets yellow. Achenes terete; pappus dirty yellowish, feathery. Fl. and fr. May-Sep. Valleys, slopes, forests, forest edges, thickets, wastelands or fields at 200-2500 m. Distributed in SW, C, E, N and NE China. Also in Russia and Korean Peninsula.

东北鸦葱 *Scorzonera manshurica*

华北鸦葱 *Scorzonera albicaulis*

蒙古鸦葱 *Scorzonera mongolica*

蒙古鸦葱

Scorzonera mongolica Maxim.

多年生草本，无毛。叶基生和茎生，厚，肉质，披针形、椭圆形、长圆形至条形，全缘。头状花序单生茎端；总苞狭圆柱状；总苞片卵形至披针形；舌状小花19朵，黄色，稀白色。瘦果圆柱状，淡黄色；冠毛白色，羽毛状。花果期4-8月。生海拔3200米以下的盐碱地、沙地或干河滩地。产华北和西北。哈萨克斯坦和蒙古亦有。

Perennial herbs, glabrous. Leaves basal and cauline, thick, succulent, anceolate, elliptic, oblong to linear, margin entire. Capitula terminal, solitary; involucres narrowly cylindric; phyllaries ovate to lanceolate; ray florets 19, yellow, rare white. Achenes terete, yellowish; pappus white, feathery. Fl. and fr. Apr-Aug. Saline or kaline places, sandy places or dry flood plains below 3200 m. Distributed in N and NW China. Also in Kazakhstan and Mongolia.

阿尔泰婆罗门参

Tragopogon altaicus S. A. Nikitin et Schischk.

二年生草本，高40-120米。茎无毛。基部和下部茎叶披针形。头状花序；舌状花橙黄色；瘦果外面浅褐色、弯曲或近乎直的；冠毛污白色。花果期6-8月。生海拔1500-3000米的高山草甸、石质山坡。产新疆。哈萨克斯坦、蒙古和俄罗斯中南部亦有。

Biennial herbs, 40-120 cm tall. Stem glabrous. Basal and lower cauline leaves lanceolate. Capitula; ligules yellowish orange; outer achenes pale brown, ± curviform or almost straight; pappus dirty white. Fl. and fr. Jun-Aug. Mountain meadows, stony slopes in hills at 1500-3000 m. Distributed in Xinjiang. Also in Kazakhstan, Mongolia and SC Russia.

狭锥花佩菊

Faberia faberi (Hemsl. ex Hemsl.) N. Kilian

多年生草本，高1.2-2.5米。根状茎粗壮。茎直立，单生。茎生叶三角卵形至五角形，有柄。复伞房花序，头状花序密集，具小花5；总苞片绿色；小花淡紫色。瘦果棕色。花果期6-8月。生海拔1800-3000米的山坡、林缘。产重庆、贵州、四川和云南。

Perennial herbs, 1.2-2.5 m tall. Rhizomes stout. Stem solitary, erect. Stem leaves triangular-ovate to pentagonal, with petioles. Synflorescences contracted paniculiform; capitula clustered, each with 5 florets; phyllaries green; florets pale purple. Achene brown. Fl. and fr. Jun-Aug. Mountain slopes, forest margins at 1800-3000 m. Distributed in Chongqing, Guizhou, Sichuan and Yunnan.

阿尔泰婆罗门参 *Tragopogon altaicus*

狭锥花佩菊 *Faberia faberi*

假花佩菊

Faberia nanchuanensis C. Shih

多年生草本。叶狭椭圆形，下面稍具紫色；叶柄紫色。复合花序具1至少量短分枝，生于茎上部叶腋，每个分枝具1-3个头状花序；头状花序常具15-20花；总苞绿色，略带紫色；小花蓝紫色。花期6-8月。生海拔600-700米的沟谷湿地或溪边。产重庆(南川区)。

Perennial herbs. leaves narrowly elliptic, abaxially somewhat tinged purplish; Petioles purplish. Synflorescences of 1 to few short branchlets subtended by upper stem leaves, each with 1-3 capitula; capitula with usually 15-20 florets; involucres green and tinged purple; florets bluish purple. Fl. Jun-Aug. Wet places in ravines or creeks at 600-700 m. Distributed in Chongqing (Nanchuan District).

大花毛鳞菊

Melanoseris atropurpurea (Franch.) N. Kilian et Z. H. Wang

多年生草本。茎被多细胞节毛。叶椭圆形、长圆形或披针形，大头羽状、羽状分裂或不裂。头状花序常具25-35花，排成总状花序；总苞宽钟状；总苞片无毛；舌状花约30朵，蓝色或蓝紫色。瘦果黑褐色，倒披针形，压扁；冠毛2层，白色。花果期7-11月。生海拔2800-4000米的林中、林缘、灌丛或高山草甸。产云南、四川和西藏。缅甸亦有。

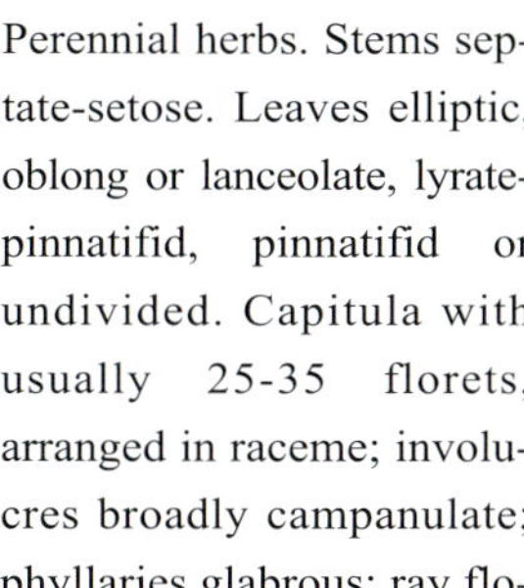

Perennial herbs. Stems septate-setose. Leaves elliptic, oblong or lanceolate, lyrate-pinnatifid, pinnatifid or undivided. Capitula with usually 25-35 florets, arranged in raceme; involucres broadly campanulate; phyllaries glabrous; ray florets ca. 30, blue or bluish-purple. Achenes black-brown, oblanceolate, compressed; pappus double, white. Fl. and fr. Jul-Nov. Forests, forest edges, thickets or alpine meadows at 2800-4000 m. Distributed in Yunnan, Sichuan and Xizang. Also in Myanmar.

康滇毛鳞菊 *Melanoseris souliei*

康滇毛鳞菊

Melanoseris souliei (Franch.) N. Kilian

多年生莲座状草本，无茎。外部叶不分裂；内部叶大头羽状深裂至全裂；次级头状花序密被头状花序及1个椭圆形托叶。头状花序具4-6花；苞片5；花紫红色至蓝色。瘦果黑色至棕黑色；冠毛单生，早落。花果期8月。生海拔2300-4300米的高山草甸、山坡碎石滩、沼泽地或林缘。产云南、四川和西藏。不丹和缅甸亦有。

Perennial herbs, rosulate, acaulescent. Leaves of outer leaves undivided; leaves of more inner leaves lyrately pinnatipartite to pinnatisect; secondary capitula with densely congested capitula with 1 elliptic subtending leaf. Capitula with 4-6 florets; phyllaries 5; florets purplish red to blue. Achenes dark to blackish brown; pappus single, caducous. Fl. and fr. Aug. Alpine meadows, scree slopes, marshes or forest edges at 2300-4300 m. Distributed in Yunnan, Sichuan and Xizang. Also in Bhutan and Myanmar.

假花佩菊 *Faberia nanchuanensis*

大花毛鳞菊 *Melanoseris atropurpurea*

苞叶毛鳞菊 *Melanoseris bracteata*

黑苞毛鳞菊 *Melanoseris lessertiana*

苞叶毛鳞菊

Melanoseris bracteata (Hook. f. et Thoms. ex C. B. Clarke) N. Kilian

多年生草本。茎高20-120厘米。基生和下部叶卵形、椭圆状卵形或倒披针形，基部渐狭或具耳，边缘有波状齿。花序总状或疏圆锥状，具叶；头状花序少数或达20个，下垂；花淡蓝色。花果期9月。生海拔800-3000米的林缘。产西藏。喜马拉雅亦有。

Perennial herbs. Stem solitary, 20-120 cm tall, strongly glandular hairy to glandular hispid. Basal and lower stem leaves ovate, ellipticovate or oblanceolate, base attenuate and auriculately clasping, margin sinuate-dentate. Synflorescence raceiform or sparsely paniculiform, leafy, with several to 20 capitula; capitula nodding; florets pale blue. Fl. and fr. Sep. Forest margins at 800-3000 m. Distributed in Xizang. Also in Himalaya.

黑苞毛鳞菊

Melanoseris lessertiana (DC.) Decne.

多年生草本，高5-40厘米。茎有分枝，无毛或上部有红色长柔毛。下部叶片椭圆形至倒披针形、不分裂或琴状分裂。总花序总状，具有1个或数个头状花序；花紫红色或蓝色。花果期8-9月。生海拔2700-4500米的草坡。产西藏和云南西北部。喜马拉雅亦有。

Perennial herbs, 5-40 cm tall. Stems sparsely branched, glabrous to apically reddish villous. Basal and lower stem leaves elliptic to oblanceolate, undivided to lyrately pinnatisect. Synflorescence racemiform, with 1 to several capitula; florets purplish red to bluish. Fl. and fr. Aug-Sep. Grasslands on mountain slopes at 2700-4500 m. Distributed in Xizang and NW Yunnan. Also in Himalaya.

云南毛鳞菊

Melanoseris yunnanensis (C. Shih) N. Kilian et Z. H. Wang

多年生草本。中下部茎叶有柄，叶片椭圆形至卵形，羽状深裂或羽状全裂至二回羽状浅裂；上部茎叶浅裂；最上部叶片线形披，不分裂。头状花序排成圆锥状花序；总苞圆柱状；小花黄色或白色；冠毛2层。花果期9-10月。生海拔700-3400米的草坡、河谷地或林下。产四川和云南。

Perennial herbs. Lower and middle stem leaves with petioles, leaves elliptic to ovate, pinnatipartite or pinnatisect to bipinnately lobed; upper stem leaves less divided; uppermost leaves linear-lanceolate, undivided.

云南毛鳞菊 *Melanoseris yunnanensis*

细莴苣 *Melanoseris graciliflora*

头嘴菊 *Cephalorrhynchus macrorhizus*

Synflorescence racemiform to narrowly paniculiform, capitula; involucre cylindric; florets yellow to whitish; pappus double. Fl. and fr. Sep-Oct. Grasslands on mountain slopes, river valleys, forests at 700-3400 m. Distributed in Sichuan and Yunnan.

细莴苣

Melanoseris graciliflora (DC.) N. Kilian

多年生草本。中下部茎叶大头羽状全裂，具叶柄；最上部茎叶及接花序下部的叶不裂，有短柄或无柄。头状花序排成圆锥状复合花序；总苞狭圆柱状；小花蓝紫色；冠毛黄白色。花果期6-9月。生海拔2800-3500米的山坡灌丛及林缘。产中国西南。喜马拉雅亦有。

Perennial herbs. Lower and middle stem leaves with petioles, leaves pinnatisect to lyrately pinnatisect; upper stem leaves shortly petiolate to subsessile, less or not divided. Synflorescence paniculiform; involucre narrowly cylindric; florets bluish purple; pappus yellowish-white. Fl. and fr. Jun-Sep. Thickets, forest margins, grasslands at 2800-3500 m. Distributed in SW China. Also in Himalaya.

头嘴菊

Cephalorrhynchus macrorhizus (Royle) Tuisl

多年生草本。中下部茎叶椭圆形至匙形，大头羽状分裂；上部茎叶披针形至宽线形，不分裂。头状花序排成伞房花序或伞房圆锥花序，含8-15朵小花；总苞片黑紫色；小花紫红色、蓝紫色，稀白色。花果期7-10月。生海拔2700-4000米的山谷、山坡林下、灌丛或草地。产西藏和云南。喜马拉雅和西南亚亦有。

Perennial herbs. Lower and middle stem leaves elliptic to spatulate, lyrately pinnatifid to lyrately pinnatisect; upper leaves lanceolate to broadly linear, undivided. Synflorescence loosely to paniculately corymbiform; capitula with usually 10-15 florets; phyllaries dark purplish; florets purple, bluish purple, or rarely white. Fl. and fr. Jul-Oct. Mountain valleys, forests, thickets or grasslands at 2700-4000 m. Distributed in Xizang and Yunnan. Also in Himalaya and SW Asia.

岩生头嘴菊

Cephalorrhynchus saxatilis (Edgew.) C. Shih

本种与头嘴菊相似，但是总苞片外面沿中脉有长刚毛、叶片较小。生海拔3500-4000米的山坡水沟边及岩石缝隙中。产西藏西部。喜马拉雅亦有。

This species is similar to *Cephalorrhynchus macrorrhizus*, but differs by its phyllaries hispid along mid-vein, and leaves smaller. Streamside or stony slopes at 3500-4000 m. Distributed in W Xizang. Also in Himalaya.

岩生头嘴菊 *Cephalorrhynchus saxatilis*

假福王草 *Paraprenanthes sororia*

假福王草

Paraprenanthes sororia (Miq.) C. Shih

一年生草本，光滑无毛。下部叶具翼柄，大头羽状半裂至全裂；上部叶小，不裂。头状花序多数，排成圆锥状花序；总苞圆柱状；总苞片卵形至披针形，无毛，有时淡紫红色；舌状小花粉红色，约10朵。瘦果黑色，纺锤状，淡黄白色；冠毛白色。花果期5-8月。生海拔200-3200米的山坡、山谷、灌丛或林下。产中国西南、华南、华中和华东。中南半岛、朝鲜半岛和日本亦有。

Annual herbs, glabrous. Lower cauline leaves winged petiolate, lyrate-pinnately parted to sected; upper leaves small, unlobed. Capitula numerous, arranged in panicle; involucres cylindric; phyllaries ovate to lanceolate, glabrous, sometimes purplish; ray florets pink, ca. 10. Achenes black, fusiform, yellowish-white; pappus white. Fl. and fr. May-Aug. Slopes, valleys, thickets or forests at 200-3200 m. Distributed in SW, S, C and E China. Also in Indo-China Peninsula, Korean Peninsula and Japan.

云南假福王草

Paraprenanthes yunnanensis (Franch.) C. Shih

一年生草本。基部和下部叶具柄，不裂，卵形三角形；中部茎叶具短柄或无柄，椭圆形至倒披针形，羽状深裂至羽状全裂；上部茎叶无柄，披针形，多不裂；最上部茎叶线形披针形，全缘。头状花序组成复合圆锥花序；外苞片三角形至线形披针形；小花浅紫色。花果期7-8月。生海拔1500-2700米的河谷和森林。产云南。

Perennial herbs. Basal and lower leaves undivided and triangular-ovate, with petioles; middle stem leaves with a short petioles or sessile, leaves elliptic to oblanceolate, pinnatipartite to pinnatisect; upper stem leaves sessile, lanceolate, mostly not divided; uppermost leaves linear-lanceolate, margin entire. Synflorescence paniculiform, capitula; outer phyllaries triangular to linear-lanceolate; florets pale purple. Fl. and fr. Jul-Aug. River valleys and forests at 1500-2700 m. Distributed in Yunnan.

云南假福王草 *Paraprenanthes yunnanensis*

黑花紫菊

Paraprenanthes melanantha (Franch.) Z. H. Wang

多年生草本。叶片卵形，羽状深裂，裂片边缘具锯齿至浅裂，两面粗糙，被短糙毛。头状花序多数，排成圆锥状花序；总苞圆柱状；总苞片长椭圆形，无毛，紫红色；小花5朵，舌状，红色或粉红色。瘦果棕红色，压扁，倒披针形；冠毛白色。花果期6-12月。生海拔1300-2700米的山坡、林缘或林下。产中国西南、华南和华中。

Perennial herbs. Leaves ovate, pinnatiparted, margin of lobes serrate to lobed, both surfaces coarse, shortly strigillose.

黑花紫菊 *Paraprenanthes melanantha*

Capitula numerous, arranged in panicle; involucres cylindric; phyllaries narrowly elliptic, glabrous, purplish-red; florets 5, ligulate, red or pink. Achenes brownish-red, compressed, oblanceolate; pappus white. Fl. and fr. Jun-Dec. Slopes, forest edges or forests at 1300-2700 m. Distributed in SE, S and C China.

峨眉紫菊

Paraprenanthes wilsonii (C. C. Chang) Z. H. Wang

多年生草本。下部和中部茎叶具柄，叶片羽状深裂至羽状全裂，有时匙状羽裂；上部茎叶具短翼柄。头状花序排成圆锥状花序；总苞紫色，无毛，外苞片卵形三角形至线形披针形；小花蓝紫色。花果期5-7月。生海拔1000-2800米的森林和林缘。产四川。

Perennial herbs. Lower and middle stem leaves petiolate, leaves pinnatipartite to pinnatisect and sometimes lyrately; upper stem leaves sessile or with shorter usually broader winged petioles. Synflorescence paniculiform; phyllaries purple, glabrous; outer phyllaries triangular-ovate to linear-lanceolate; florets purplish blue. Fl. and fr. May-Jul. Forests and forest margins at 1000-2800 m. Distributed in Sichuan.

峨眉紫菊 *Paraprenanthes wilsonii*

光苞紫菊

Notoseris macilenta (Vaniot et Lévl.) N. Kilian

多年生草本。基部和中下部茎叶具柄，卵圆形、卵形三角形或近圆形；上部茎叶三角形至狭菱形；最上部茎叶无柄，狭菱形至长椭圆形。头状圆锥花序；苞片紫色，无毛，外苞片卵形三角形至线形披针形；小花紫红色。花果期9-11月。生海拔800-2300米的山谷、森林中。产中国西南、华中、华南和华东。

Perennial herbs. Basal, lower and middle stem leaves petiolate, leaves ovate, triangular-ovate or more rarely suborbicular; upper stem leaves triangular to narrowly rhombic; uppermost leaves sessile, narrowly rhombic to narrowly elliptic. Synflorescence paniculiform; phyllaries purple, glabrous, outer phyllaries triangular-ovate to linear-lanceolate; florets purplish red. Fl. and fr. Sep-Nov. By water in mountain valleys, forests at 800-2300 m. Distributed in SW, C, S and E China.

光苞紫菊 *Notoseris macilenta*

乳苣 *Lactuca tatarica*

乳苣

Lactuca tatarica (L.) C. A. Mey.

多年生草本，光滑无毛。叶质厚，长圆形至条形，边缘浅裂或具锯齿。头状花序多数，排成圆锥花序；总苞圆柱状或楔形；总苞片卵形至披针形，带紫红色；小花20朵，紫色或紫蓝色。瘦果长圆状披针形，灰黑色；冠毛白色。花果期6-9月。生海拔1200-4300米的河滩、湖边、草甸、田边、固定沙丘或砾石地。产华北、东北和西北。亚洲北部和欧洲亦有。

Perennial herbs, glabrous. Leaves thick, oblong to linear, margin lobed or serrate. Capitula numerous, arranged in panicle; involucres cylindric or cuneate; phyllaries ovate to lanceolate, purplish-red; florets ca. 20, purple or purplish-blue. Achenes oblong-lanceolate, grayish-black; pappus white. Fl. and fr. Jun-Sep. Flood plains, lakesides, meadows, fieldsides, fixed dunes or gravelly places at 1200-4300 m. Distributed in N, NE and NW China. Also in N Asia and Europe.

毛脉翅果菊

Lactuca raddeana Maxim.

二年生或多年生草本。茎中部及下部叶基部楔形或具翅状叶柄，不开裂、羽状深裂或大头羽状深裂。头状花序集成狭圆锥状复合花序；总苞片常浅紫红色；小花亮黄色。瘦果红色至深棕色，具宽翅。花果期5-10月。生海拔200-3000米的林中、林缘、灌丛、山地沟谷或路边。产中国西南、华南、东南、华中、华北、华西、华东和东北。越南、俄罗斯东部、朝鲜半岛和日本亦有。

Biennial or perennial herbs. Lower and middle stem leaves with basal portion cuneate or winged petiolelike, either undivided, pinnatipartite or lyrately pinnatipartite. Synflorescences narrowly paniculate; phyllaries often pale purplish red; florets bright yellow. Achenes reddish to dark brown, broadly winged. Fl. and fr. May-Oct. Forests, forest edges, thickets, mountain valleys or trailsides at 200-3000 m. Distributed in SW, S, SE, C, N, W, E and NE China. Also in Vietnam, E Russia, Korean

Peninsula and Japan.

翅果菊

Lactuca indica L.

一年生或二年生高大草本。茎粗壮，单生。叶无柄，草质，条形或长圆形。头状花序多数，排成圆锥花序或总状圆锥花序；总苞卵球形；总苞片卵形至披针形，边缘紫红色；小花黄色。瘦果椭圆体形，黑色，压扁，边缘具宽翅；冠毛白色。花果期4-11月。生海拔200-3000米的山谷、山坡、林缘、林下、灌丛、水边、草地或田间。广布中国大部分地区。南亚、东南亚和东北亚亦有，各地引种。

Annual or biennial large herbs. Stems stout, solitary. Leaves sessile, herbaceous, linear or oblong. Capitula numerous, arranged in panicle or racemose-panicle; involucres ovoid; phyllaries ovate to lanceolate, margin purple; corolla yellow. Achenes

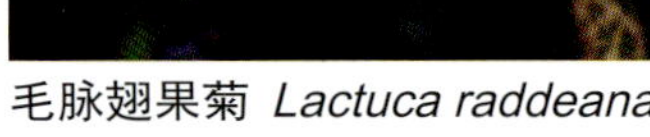

毛脉翅果菊 *Lactuca raddeana*

翅果菊 *Lactuca indica*

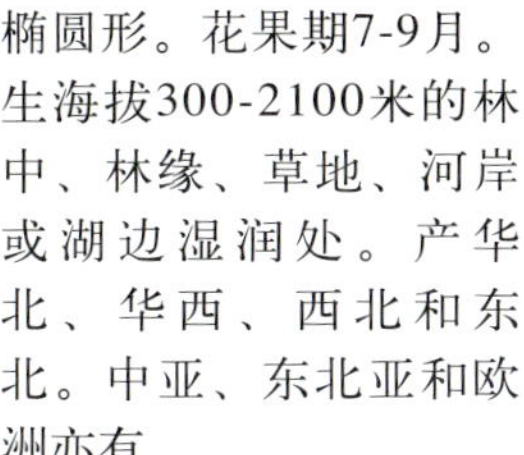

ellipsoid, black, flattened, margin broadly winged; pappus white. Fl. and fr. Apr-Nov. Valleys, slopes, forest edges, forests, thickets, watersides, grasslands or fields at 200-3000 m. Widely distributed in most parts of China. Also in S, SE and NE Asia, introduced elsewhere.

山莴苣

Lactuca sibirica (L.) Benth. ex Maxim.

多年生草本，具根状茎。叶薄，无毛；茎下部及中部叶无柄，常不分裂。多数头状花序排成伞房状至伞房圆锥状；头状花序具约20小花；总苞圆柱形；内轮总苞片约8；小花蓝色。瘦果棕色至橄榄绿色，狭椭圆形。花果期7-9月。生海拔300-2100米的林中、林缘、草地、河岸或湖边湿润处。产华北、华西、西北和东北。中亚、东北亚和欧洲亦有。

Perennial herbs, with a taproot. Leaves thin, glabrous; lower and middle stem leaves sessile, often undivided. Synflorescences corymbiform to corymbosely paniculiform, with many capitula; capitula with ca. 20 florets; involucres cylindric; inner phyllaries ca. 8; florets blue. Achenes brown to olive green, narrowly ellipsoid. Fl. and fr. Jul-Sep. Forests, forest edges, meadows, river banks, moist places or by lakes at 300-2100 m. Distributed in N, W, NW and NE China. Also in C and NE Asia, and Europe.

莴苣

Lactuca sativa L.

一年生或二年生草本。根垂直直伸。叶椭圆形或披针形，边缘波状或有细锯齿。头状花序排成圆锥花序；总苞卵球形；总苞片三角形、披针形或长椭圆形；舌状小花约15朵，黄色。瘦果倒披针形，压扁，浅褐色；冠毛白色。花果期2-9月。中国各地常见栽培，有时逸生。原产地不详。

Annual or biennial herbs. Taproots erectly straight. Leaves elliptic or lanceolate, margin undulate or serrulate. Capitula arranged in panicle; involucres ovoid; phyllaries triangular Lanceolate or narrowly elliptic; ray florets ca. 15, yellow. Achenes oblanceolate, compressed, brownish; pappus white. Fl. and fr. Feb-Sep. Commonly cultivated in China, sometimes escaped. Origin unknown.

山莴苣 *Lactuca sibirica*

莴苣 *Lactuca sativa*

河西菊 *Launaea polydichotoma*

河西菊

Launaea polydichotoma (Ostenf.) Amin ex N. Kilian

多年生草本。茎下部叶条形，基部半抱茎。复合花序二歧状圆锥形，具多数头状花序。头状花序具5-11花；总苞圆柱形；总苞片边缘膜质；内轮苞片4-5。瘦果灰黄色至棕黄色，同形，圆柱形，具5条主肋；冠毛早落。花果期5-9月。生海拔400-2100米的沙质土壤或沙丘之间。产甘肃和新疆。

Perennial herbs. Lower stem leaves linear, base semiamplexicaul. Synflorescences divaricately paniculiform, with numerous capitula. Capitula with 5-11 florets; involucres cylindric; phyllaries with indistinct scarious margin; inner phyllaries 4-5. Achenes pale yellow to yellowish brown, homomorphic, cylindric, with 5 main ribs; pappus deciduous. Fl. and fr. May-Sep. Sandy soils or between sand dunes at 400-2100 m. Distributed in Gansu and Xinjiang.

苣荬菜

Sonchus wightianus DC.

多年生草本，具根状茎。叶倒披针形或长圆形，不裂或羽状分裂，边缘具小锯齿或小尖头，基部渐窄，半抱茎。头状花序排成伞房花序；总苞钟状，被绒毛；总苞片披针形，长渐尖；舌状小花多数，黄色。瘦果长圆形；冠毛白色。花果期1-10月。生海拔300-2300米的草地、潮湿地、路边或河边。产中国西南、华南、华中、华东和西北。几遍全球分布。

Perennial herbs, with rhizomes. Leaves oblanceolate or oblong, undivided or pinnatifid, margin serrulate or apiculate, base attenuate, semi-amplexicaul. Capitula arranged in corymb; involucres campanulate, tomentose; phyllaries lanceolate, acuminate; ligulate florets numerous, yellow. Achenes oblong; pappus white. Fl. and fr. Jan-Oct. Grasslands, damp places, roadsides or riversides at 300-2300 m. Distributed in SW, S, C, E and NW China. Widely distributed all around the world.

花叶滇苦菜

Sonchus asper (L.) Hill

一年生草本。叶长圆形、倒卵形、匙形或披针形，不裂或羽状分裂，顶端渐尖，基部耳状抱茎，边缘有尖刺齿。头状花序排成伞房花序；总苞宽钟状；总苞片绿色，草质，先端急尖；舌状小花黄色。瘦果倒披针状；冠毛白色。花果期5-10月。生海拔1500-3700米的山坡、林缘、路边、田野或水边。产中国西南、华中、华东和西北。西亚、南亚、北亚、东北亚和欧洲亦有。

Annual herbs. Leaves oblong, obovate, spathulate or lanceolate, undivided or pinnatifid, apex acuminate, base auriculately amplexicaul, margin aristately serrate. Capitula arranged in corymb; involucres broadly campanulate; phyllaries green, herbaceous, apex acute; ray florets yellow. Achenes oblanceolate; pappus white. Fl. and fr. May-Oct. Slopes, forest edges, roadsides, fields or watersides at 1500-3700 m. Distributed in SW, C, E and NW China. Also in W, S, N and NE Asia, and Europe.

长裂苦苣菜

Sonchus brachyotus DC.

一年生草本。叶卵形、长圆形或倒披针形，羽状分裂，基部耳状半抱茎，边缘全缘。头状花序少数，排成伞房状花序；总苞钟状；小花黄色。瘦果长圆状；冠毛白色。花果期5-9月。生海拔300-4000米的山地草坡、河边或盐碱地。产华北和东北。泰国、哈萨克斯坦、吉尔吉斯斯坦、俄罗斯、蒙古和日本亦有。

苣荬菜 *Sonchus wightianus*

长裂苦苣菜 *Sonchus brachyotus*

花叶滇苦菜 *Sonchus asper*

Annual herbs. Leaves ovate, oblong or oblanceolate, pinnatifid, base auriculately semi-amplexicaul, margin entire. Capitula several, arranged in corymbs; involucres campanulate; corolla yellow. Achenes oblong; pappus white. Fl. and fr. May-Sep. Grassy slopes, riversides or saline places at 300-4000 m. Distributed in N and NE China. Also in Thailand, Kazakhstan, Kyrgyzstan, Russia, Mongolia and Japan.

暗苞粉苞菊

Chondrilla phaeocephala Rupr.

多年生草本，扫帚状。叶狭披针形至条状披针形。头状花序常具10-12花；总苞具蛛丝毛；苞片深绿色至黑色；内轮8片。瘦果具5个短的不裂至稍不等3裂的直立冠鳞；冠毛长6-7毫米。花果期6-9月。生海拔900-4000米的沙漠砾石地。产新疆。阿富汗和中亚亦有。

Perennial herbs, broomlike. Leaves narrowly elliptic to linear-elliptic. Capitula with usually 10-12 florets; involucres arachnoid hairy; phyllaries dark green to blackish; inner phyllaries 8. Achenes with a corona of 5 short unlobed to ± unequally 3-lobed erect scales; pappus 6-7 mm long. Fl. and fr. Jun-Sep. Gravelly areas in deserts at 900-4000 m. Distributed in Xinjiang. Also in Afghanistan and C Asia.

暗苞粉苞菊 *Chondrilla phaeocephala*

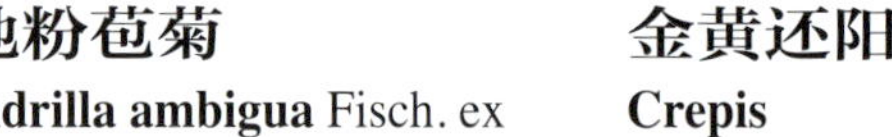
沙地粉苞菊 *Chondrilla ambigua*

沙地粉苞菊

Chondrilla ambigua Fisch. ex Kar. et Kir.

多年生草本，扫帚状，基部常木质化。茎下部叶条状披针形至披针形，早落；中上部叶条状披针形至线形。头状花序具约5花；苞片绿色至黄绿色，内轮5片。瘦果圆柱形，顶部渐狭成1粗喙；冠毛长6-8毫米。花果期5-9月。生海拔300-800米的沙丘或砾石和黄土地。产新疆。俄罗斯、中亚和欧洲南部亦有。

Perennial herbs, broomlike, basally sometimes woody. Lower stem leaves linear-lanceolate to lanceolate, early deciduous; middle and upper stem leaves linear-lanceolate to filiform. Capitula with ca. 5 florets; phyllaries green to yellowish green, inner phyllaries 5. Achenes cylindric, apically shortly attenuate into a stout beak; pappus 6-8 mm long. Fl. and fr. May-Sep. Sand dunes or gravel and loess areas at 300-800 m. Distributed in Xinjiang. Also in Russia, C Asia and S Europe.

金黄还阳参

Crepis chrysantha (Ledeb.) Turcz.

多年生草本。茎绿色或下部淡紫色。基生叶多数，倒披针形、长椭圆状倒披针形或匙形；茎生叶2-3(或4)。头状花序1-2，小花多数；总苞钟状，黑绿色。瘦果纺锤状，红褐色或黑紫色，有12-15条细肋。冠毛白色，长5-7毫米，宿存。花果期7-9月。生海拔500-1500米的河滩砾石地及石质坡地。产新疆。哈萨克斯坦、蒙古和俄罗斯亦有。

Perennial herbs. Stems green or basally tinged with purple. Basal leaves crowded, oblanceolate, narrowly elliptic-oblanceolate or spatulate; stem leaves 2 or 3 (or 4). Capitula solitary or 2, many flowered; involucre dark green to blackish, campanulate. Achene reddish brown to dark purple, fusiform, with 12-15 ribs; pappus white, 5-7 mm, persistent. Fl. and fr. Jul-Sep. Gravelly areas on floodplains, stony slopes at 500-1500m. Distributed in Xinjiang. Also in Kazakhstan, Mongolia and Russia.

金黄还阳参 *Crepis chrysantha*

北方还阳参

Crepis crocea (Lam.) Babc.

多年生草本，莲座状。基生叶多数，倒披针形至椭圆形，羽状浅裂或半裂；无茎生叶或茎生叶1-3，不分裂。头状花序单生茎端或茎生1-4；总苞钟状；总苞片果期绿色。瘦果纺锤状，有10-15条纵肋；冠毛白色。花果期5-8月。生海拔800-2900米的山坡、荒地、黄土丘陵地。产甘肃、河北、内蒙古、青海、陕西和山西。蒙古和俄罗斯亦有。

Perennial herbs, rosulate. Rosette leaves numerous, oblanceolate to elliptic, shallowly pinnatifid to pinnatisect; stem leaves 0-3, undivided. Capitula 1-4 per stem, with many florets; involucre narrowly campanulate; phyllaries green in fruit. Achene fusiform, with 10-15 ribs; pappus white. Fl. and fr. May-Aug. Mountain

藏滇还阳参 *Crepis elongata*

北方还阳参 *Crepis crocea*

slopes, loess hills, wastelands at 800-2900 m. Distributed in Gansu, Hebei, Neimenggu, Qinghai, Shaanxi and Shanxi. Also in Mongolia and E Russia.

藏滇还阳参
Crepis elongata Babc.

多年生草本。茎直立，高25-65厘米。基生叶多数，倒披针形至椭圆形。头状花序(1-)3-12枚排成伞房花序至圆锥状伞房花序，极少单生；头状花序具30-40小花；总苞绿色至墨绿色，圆柱状至钟状。瘦果纺锤形，有10纵肋；冠毛白色。花果期6-8月。生海拔2600-4200米的山坡草地、灌丛、林下、林缘及草甸。产云南、四川、西藏。喜马拉雅亦有。

Perennial herbs. Stem 25-65 cm tall, erect. Leaves mostly basal, rosulate to crowded, oblanceolate to elliptic. Synflorescence corymbiform to paniculately corymbiform, with (1-)3-12 capitula; capitula with 30-40 florets; involucre green to dark green cylindric to chmpanulate. Achene fusiform, with 10 ribs; pappus white. Fl. and fr. Jun-Aug. Mountain slopes, thickets, forests, forest margins and meadows at 2600-4200 m. Distributed in Yunnan, Sichuan and Xizang. Also in Himalaya.

西伯利亚还阳参
Crepis sibirica L.

多年生草本，具粗壮根系。茎生叶与茎下部叶具带翅叶柄。复合花序伞房状，具少量至数个头状花序；头状花序多花；总苞钟形，果期深绿色。瘦果纺锤形，具20条纵棱；冠毛白色或淡黄白色。花果期5-9月。生海拔1000-2700米的山坡、林缘、灌丛或林中草地。产内蒙古、新疆、黑龙江和辽宁。俄罗斯、蒙古、中亚和欧洲东部亦有。

Perennial herbs, with stout rhizomes. Basal leaves and lower cauline leaves with a winged petioles. Synflorescences laxly corymbiform, with few to several capitula; capitula many-flowered; involucres campanulate, dark green in fruit. Achenes fusiform, with 20 ribs; pappus white or pale yellowish white. Fl. and fr. May-Sep. Mountain slopes, forest edges, thickets or grasslands in forests at 1000-2700 m. Distributed in Neimenggu, Xinjiang, Heilongjiang and Liaoning. Also in Russia, Mongolia, C Asia and E Europe.

西伯利亚还阳参 *Crepis sibirica*

宽叶还阳参 *Crepis coreana*

绿茎还阳参 *Crepis lignea*

宽叶还阳参

Crepis coreana (Nakai) Sennikov

多年生草本，25-55厘米高。叶片匙形至椭圆形，边缘有锯齿或大头羽状分裂；叶柄有翅至无柄。花序伞房状；总苞宽柱形或钟形。瘦果圆柱形或纺锤形，长4-6毫米；冠毛白色，长约6毫米。花果期7-9月。生海拔1600-2200米的森林、林缘、草甸或陡坡。产中国东北。朝鲜半岛亦有。

Perennial herbs, 25-55 cm tall. Leaves spatulate to elliptic, margin sharply and pectinately dentate to rarely lyrately pinnatifid; petioles winged to sessile. Synflorescence corymbiform; involucre broadly cylindric to campanulate. Achene cylindric to fusiform, 4-6 mm long; pappus white, ca. 6 mm. Fl. and fr. Jul-Sep. Forests, forest margins, meadows or steppes at 1600-2200 m. Distributed in NE China. Also in Korean Peninsula.

绿茎还阳参

Crepis lignea (Vant.) Babcock

多年生草本。根木质。茎绿色，伞状分枝或帚状分枝，下部木质。叶线形，长达3厘米，全缘。头状花序直立，排成伞房状花序；总苞钟状至圆柱状；舌状小花黄色。瘦果纺锤形，稍弯曲；冠毛白色。花果期2-8月。生海拔1500-2700米的向阳山坡。产云南、四川和贵州。

Perennial herbs. Roots woody. Stems green, corymbosely or fastigiately branched, lower part woody. Leaves filiform, to 3 cm long, entire. Capitula erect, arranged in corymb; involucres campanulate to cylindric; ray florets yellow. Achenes fusiform, somewhat bent; pappus white. Fl. and fr. Feb-Aug. Sunny slopes at 1500-2700 m. Distributed in Yunnan, Sichuan and Guizhou.

无茎黄鹌菜

Youngia simulatrix (Babc.) Babc. et Stebbins

多年生矮小丛生草本，光滑无毛。根直伸。茎极短缩，长约1厘米。叶莲座状，倒披针形，边缘全缘或具波状钝齿。头状花序4-7个，密集簇生；总苞圆柱状钟形；舌状花13-18朵，黄色。瘦果黑褐色，纺锤状；冠毛白色。花果期7-10月。生海拔2700-5000米的山坡草地、河滩砾石地或湿草甸。产四川、西藏、甘肃和青海。印度和尼泊尔亦有。

Perennial dwarf tufty herbs, glabrous. Taproot straight. Stems very short, ca. 1 cm long. Leaves rosette, oblanceolate, margin entire or undulated-crenate. Capitula 4-7, closely fasciate; involucres cylindric-campanulate; ray florets 13-18, yellow. Achenes black-brown, fusiform; pappus white. Fl. and fr. Jul-Oct. Grassy slopes, stony places on flood plain or damp meadows at

无茎黄鹌菜 *Youngia simulatrix*

2700-5000 m. Distributed in Sichuan, Xizang, Gansu and Qinghai. Also in India and Nepal.

总序黄鹌菜 *Youngia racemifera*

总序黄鹌菜

Youngia racemifera (Hook. f.) Babc. et Stebbins

多年生草本。茎基部和下部叶无毛，基部渐狭或缩短为1个长3-9厘米，具翅的叶柄状部分。下垂的头状花序在茎和枝上排成偏向一侧的总状花序式；头状花序具10-20朵小花；内轮苞片8-10。瘦果近圆柱形；冠毛黄色至浅棕色。花果期8-9月。生海拔2800-4200米的山坡草地、林缘、林中、开阔林中或灌丛。产云南、四川和西藏。印度、尼泊尔和不丹亦有。

Perennial herbs. Basal and lower stem leaves glabrous, base attenuate or contracted into a 3-9 cm long, winged petiolelike portion. Synflorescences of stem and branches secundly racemiform, with few to many drooping capitula; capitula with 10-20 florets; inner phyllaries 8-10. Achenes ± cylindric; pappus yellowish to pale brown. Fl. and fr. Aug-Sep. Grasslands on mountain slopes, forest edges, forests, forest openings or thickets at 2800-4200 m. Distributed in Yunnan, Sichuan and Xizang. Also in India, Nepal and Bhutan.

戟叶黄鹌菜 *Youngia longipes*

戟叶黄鹌菜

Youngia longipes (Hemsl.) Babc. et Stebbins

一年生草本。基生叶心状戟形至有时卵形，无毛。多个头状花序组成圆锥伞房状；头状花序具15-20朵小花；总苞圆柱形；内轮苞片8。瘦果浅红色，密具黄色斑点，纺锤形，纵棱顶端有细刺；冠毛白色。花果期6月。生海拔1000-1500米的沙地或沟谷。产浙江、重庆和湖北。

Annual herbs. Basal leaves cordate-hastate to sometimes ovate, glabrous. Synflorescences paniculiform-corymbiform, with many capitula; capitula with 15-20 florets; involucres cylindric; inner phyllaries 8. Achenes pale red, finely mottled with yellow, fusiform, ribs apically finely spiculate; pappus white. Fl. and fr. Jun. Sandy areas or valleys at 1000-1500 m. Distributed in Zhejiang Chongqing and Hubei.

异叶黄鹌菜 *Youngia heterophylla*

异叶黄鹌菜

Youngia heterophylla (Hemsl.) Babc. et Stebbins

一年生或二年生草本。茎被稀疏多细胞节毛。叶异型，羽状分裂或不裂，下面紫红色，上面绿色。头状花序多数，排成伞房花序；总苞圆柱状；舌状花11-25朵，黄色。瘦果黑褐紫色，纺锤形；冠毛白色。花果期4-10月。生海拔400-2300米的山坡、林缘、林下或荒地。产云南、四川、重庆、贵州、湖北、湖南和陕西。

Annual or biennial herbs. Stems sparsely septate-hairy. Leaves heterotypic, lyrate-pinnatipartite or undivided, purple abaxially, green adaxially. Capitula numerous, arranged in corymb; involucres cylindric; ray florets 11-25, yellow. Achenes dark brownish-purple, fusiform; pappus white. Fl. and fr. Apr-Oct. Slopes, forest edges, forests or wastelands at 400-2300 m. Distributed in Yunnan, Sichuan, Chongqing Guizhou, Hubei, Hunan and Shaanxi.

黄鹌菜 *Youngia japonica*

黄鹌菜

Youngia japonica (L.) DC.

一年生草本。茎被疏毛，无叶或具1-2叶。基生叶倒披针形、椭圆形或宽条形，大头羽状分裂，被皱波状柔毛。头状花序少数或多数，排成伞房花序；总苞圆柱状；舌状花10-20朵，黄色。瘦果纺锤形，褐色；冠毛白色。花果期2-12月。生海拔100-4500米以下的山地、沼泽地、田间或荒地上。产中国西南、华南、华中、华东、华北和西北。南亚、东南亚和东亚亦有。

Annual herbs. Stems sparsely villose, afoliate or with 1-2 leaves. Basal leaves oblanceolate, elliptic or broadly linear in outline, lyrate-pinnatipartite, crisped-villose. Capitula several or numerous, arranged in corymb; involucres cylindric; ray florets 10-20, yellow. Achenes fusiform, brown; pappus white. Fl. and fr. Feb-Dec. Montanes, swamps, fields or wastelands below 100-4500 m. Distributed in SW, S, C, E, N and NW China. Also in S, SE and E Asia.

心叶黄鹌菜

Youngia humifusa (Dunn) Y. L. Peng et X. F. Gao

多年生草本。叶大头羽状深裂至全裂，基部心形或截形。由2-7个头状花序组成疏松伞房状复合花序；头状花序具10-14小花；总苞圆柱形；内轮苞片8。瘦果棕色，椭圆形，具约10条纵棱；冠毛白色。花果期8-9月。生海拔900-2500米阴湿山谷悬崖上。产云南东北部、四川、重庆(巫山县)和湖北(巴东县和神农架)。

心叶黄鹌菜 *Youngia humifusa*

Perennial herbs. Leaves lyrately pinnatisect or pinnatipartite, base cordate or truncate. Synflorescences laxly corymbiform with 2-7 capitula; capitula with 10-14 florets; involucres cylindric; inner phyllaries 8. Achenes brown, ellipsoid, with ca. 10 ribs; pappus white. Fl. and fr. Aug-Sep. Cliffs in shaded and damp valleys at 900-2500 m. Distributed in NE Yunnan, Sichuan, Chongqing (Wushan County) and Hubei (Badong County and Shennongjia).

羽裂黄鹌菜

Youngia paleacea (Diels) Babc. et Stebbins

多年生草本，高30-100厘米。基生叶及下部叶片倒披针形或长圆形，羽状半裂、羽状深裂、倒向羽状或大头羽裂；中上部茎叶与基生叶相似但较之更小而窄分裂或不分裂；茎上部叶小，线钻形。头状花序多数，在茎枝顶端排成圆锥状伞房花序；舌状小花9-16朵，黄色。花果期5-9月。生海拔1800-3800米山坡或山谷林下、林缘、灌丛中或草坡。产甘肃、四川、云南和西藏。

Perennial herbs, 30-100 cm tall. Basal and lower stem leaves oblanceolate, elliptic or narrowly elliptic, pinnatifid, pinnatipartite, or runcinately or lyrately. Middle and upper stem leaves similar to lower ones but smaller, narrower, and divided or undivided, upper most leaves reduced subulate. Synflorescence corymbiform, with few to many capitula; capitula with 9-16 yellow florets. Fl. and fr. May-Sep. Forests on mountain slopes, mountain valleys, forest margins, thickets, grassy slopes at 1800-3800 m. Distributed in Gansu, Sichuan, Yunnan and Xizang.

羽裂黄鹌菜 *Youngia paleacea*

川黔黄鹌菜 *Youngia rubida*

川黔黄鹌菜

Youngia rubida Babc. et Stebbins

一年生草本，高20-70厘米。基生叶倒披针形，羽状裂或全裂；中下部茎叶与基生叶相似；上部茎叶线形，小，不分裂或近基部浅裂。头状花序含13-15枚黄色舌状小花，多数在茎枝顶端排成伞房圆锥花序；总苞圆柱状；冠毛白色。花果期4-11月。生海拔600-1000米的林中、林缘或荒地中。产贵州、湖南和四川。

Annual herbs, 20-70 cm tall. Basal leaves oblanceolate, lyrately pinnatipartite to pinnatisect; lower and middle stem leaves similar to basal leaves; upper stem leaves linear, reduced in size, undivided or acutely 2-lobed near base. Synflorescence paniculiform-corymbiform; capitula with 13-15 florets, yellow; involucres cylindric; pappus white. Fl. and fr. Apr-Nov. Forests, forest margins or waster fields at 600-1000 m. Distributed in Guizhou, Hunan and Sichuan.

鼠冠黄鹌菜 *Youngia cineripappa*

长裂黄鹌菜 *Youngia henryi*

鼠冠黄鹌菜

Youngia cineripappa (Babc.) Babc. et Stebbins

多年生草本。基生叶倒卵形至倒披针形，不裂、羽状分裂或匙状羽裂；茎叶椭圆形至披针形；最上面的叶线形至锥形。头状花序；总苞圆筒状。瘦果棕色，近纺锤形或近压扁，肋细针状；冠毛浅灰色。花果期6-10月。生海拔600-3000米的湿润的山谷、山坡、疏林、灌丛。产中国西南和华南。印度东北部、缅甸和越南亦有。

Perennial herbs. Basal leaves obovate to oblanceolate, undivided to pinnatifid or lyrately pinnatifid; stem leaves elliptic to lanceolate; uppermost stem leaves linear to subulate. Synflorescence corymbiform, capitula; involucre cylindric. Achene brown, subfusiform, subcompressed, ribs finely spiculate; pappus grayish. Fl. and fr. Jun-Oct. Moist areas by water in mountain valleys, grasslands on mountain slopes, sparse forests, thickets at 600-3000 m. Distributed in SW and S China. Also in NE India, Myanmar and Vietnam.

长裂黄鹌菜

Youngia henryi (Diels) Babc. et Stebbins

多年生草本，莲座状。基生叶二型，具柄，早期基生叶宽卵形，不分裂，全缘或羽状半裂，晚期基生叶披针形倒披针形或椭圆形，羽状深裂或全裂；中下部茎叶长椭圆形，羽状深裂或几全裂。头状花序排成伞房花序或伞房圆锥花序；总苞狭圆柱状。瘦果近纺锤状或柱状；冠毛白色。花果期7-8月。生海拔1500-2000米的山坡草地。产陕西、湖北和四川。

Perennial herbs, rosulate. Rosette leaves distinctly dimorphic, early leaves broadly ovate, undivided and coarsely sinuate-dentate to pinnatifid, later leaves oblanceolate, elliptic, or lanceolate, pinnatipartite to pinnatisect; lower and middle stem leaves, lanceolate, pinnatisect. Synflorescence laxly corymbiform; involucre narrowly cylindric. Achene subfusiform to columnar; pappus white. Fl. and fr. Jul-Aug. Grasslands on mountain slopes at 1500-2000 m. Distributed in Shaanxi and Hubei, Sichuan.

川西黄鹌菜

Youngia prattii (Babc.) Babc. et Stebbins

多年生草本，莲座状。基生叶及下部茎叶全形倒披针形、长椭圆形，大头羽状或倒向羽状浅裂、半裂或深裂；中部和上部茎叶不分裂。头状花序排成伞房花序或伞房圆锥花序；总苞狭圆柱状；舌状小花黄色。瘦果褐色，近纺锤形或圆柱状；冠毛白色。花果期6-9月。生海拔1500-2700米的山坡灌丛或草地。产河南、湖北和四川。

Perennial herbs, ± rosulate. Rosette leaves and lower stem leaves oblanceolate to narrowly elliptic, lyrately or runcinately pinnatifid to pinnatisect, glabrous; middle and upper stem leaves, undivided. Synflorescence corymbiform, capitula; involucre narrowly cylindric. Florest yellow, Achene brown, subfusiform to columnar; pappus white. Fl. and fr. Jun-Sep. Thickets or grasslands on mountain slopes at 1500-2700 m. Distributed in Henan, Hubei and Sichuan.

川西黄鹌菜 *Youngia prattii*

稻槎菜

Lapsanastrum apogonoides (Maxim.) Pak et K. Bremer

一年生矮小草本。基生叶莲座状，大头羽状全裂，顶裂片大，边缘具极稀疏小尖头。头状花序小，果期下垂或歪斜，花序梗纤细；总苞椭圆形或长圆形；总苞片披针形，草质，无毛；小花舌状，黄色，两性。瘦果淡黄色，长椭圆形，顶端具2枚下垂长钩刺；无冠毛。花果期11月至翌年6月。生田野、荒地或路边。产中国西南、华南、华中和华东。朝鲜半岛和日本亦有。

Annual dwarf herbs. Basal leaves rosette, lyrate-pinnatisected, apex lobes large, margin very sparsely and minutely apiculate. Capitula small, drooping or deflective at fruiting, peduncules slender; involucres elliptic or oblong; phyllaries lanceolate, herbaceous, glabrous; florets

稻槎菜 *Lapsanastrum apogonoides*

叉枝假还阳参 *Crepidiastrum akagii*

ligulate, yellow, bisexual. Achenes yellowish, narrowly elliptic, apex with 2 drooping long hooked spines; pappus absent. Fl. and fr. Nov to next Jun. Fields, wastelands or roadsides. Distributed in SW, S, C and E China. Also in Korean Peninsula and Japan.

叉枝假还阳参

Crepidiastrum akagii (Kitag.) Sennikov

垫状小灌木。主根和主茎木质。莲座叶倒披针形至狭椭圆形，羽状半裂至全裂。茎生叶少数，不分裂，常苞叶状。二岐聚伞花序；头状花序有小花10-12；总苞狭椭圆形；总苞片深绿色。瘦果梭形，10棱。花果期7-9月。生海拔1400-4900米的草坡、砾石地。产甘肃、河北、内蒙古和新疆。蒙古和俄罗斯东部亦有。

Shrublets cushion forming. Taproot and caudex woody. Rosette leaves oblanceolate to narrowly elliptic, pinnatifid to pinnatisect. Stem leaves few, undivided, and often bractlike. Synflorescence divaricately corymbiform; capitula with 10-12 florets; involucre narrowly cylindric; phyllaries dark green. Achene fusiform, with 10 ribs. Fl. and fr. Jul-Sep. Grasslands on mountain slopes, gravelly areas at 1400-4900 m. Distributed in Gansu, Hebei, Neimenggu and Xinjiang. Also in Mongolia and E Russia.

黄瓜假还阳参

Crepidiastrum denticulatum (Houtt.) Pak et Kawano

一年生或二年生草本，光滑无毛。叶全缘、具锯齿、不规则分裂至羽状分裂，基部圆形或耳状抱茎。头状花序多数，排成伞房花序或伞房圆锥花序；总苞圆柱状；总苞片卵形至披针形；舌状小花15朵，黄色。瘦果长椭圆体形，压扁，黑色；冠毛白色。花果期5-11月。生海拔100-2000米的山坡、林缘、林下、田边或岩缝中。广布中国大部分地区。东北亚亦有。

Annual or biennial herbs, glabrous. Leaves margin entire, serrate, irregularly parted to pinnatiparted, base rotundate or auriculate-amplexicual. Capitula numerous, arranged in corymb or corymbose-panicle; involucres cylindrical; phyllaries ovate to lanceolate; ray florets 15, yellow. Achenes narrowly ellipsoid, compressed, black; pappus white. Fl. and fr. May-Nov. Slopes, forest edges, forests, fieldsides or rocky crevices at 100-2000 m. Widely distributed in most parts of China. Also in NE Asia.

细裂假还阳参

Crepidiastrum diversifolium (Ledeb. ex Spreng.) J. W. Zhang et N. Kilian

多年生莲座状草本，高18-40厘米。莲座叶有叶柄，叶片窄椭圆形，羽状分裂；茎生叶较小，分裂程度较低或不分裂。复合头状花序伞房状或圆锥状伞房花序。瘦果具10-14条肋，顶端变薄至略有喙。花果期7-9月。生海拔1800-4700米的山坡、岩石坡、砾石地。产甘肃、新疆和西藏。西喜马拉雅、哈萨克斯坦、蒙古和俄罗斯东部亦有。

Perennial herbs, 18-40 cm tall. rosulate. Rosette leaves petiolate, leaves narrowly elliptic, pinnatipartite to pinnatisect; stem leaves similar to rosette leaves but gradually smaller and less or not divided upward on stem. Synflorescence corymbiform or paniculiform-corymbiform; achene with 10-14 ribs, apex attenuate to weakly beaked. Fl. and fr. Jul-Sep. Mountain slopes, rock slopes, gravelly areas on floodplains at 1800-4700 m. Distributed in Gansu, Xinjiang and Xizang. Also in W Himalaya, Kazakhstan, Mongolia and E Russia.

细裂假还阳参 *Crepidiastrum diversifolium*

黄瓜假还阳参 *Crepidiastrum denticulatum*

尖裂假还阳参

Crepidiastrum sonchifolium (Maxim.) Pak et Kawano

多年生草本，无毛。基生叶莲座状，边缘具锯齿或大头羽状深裂；茎生叶基部扩大，心形或耳状抱茎。头状花序排成伞房花序或伞房圆锥花序；总苞圆柱形；总苞片卵形至披针形；舌状小花约17朵，黄色。瘦果黑色，纺锤形；冠毛白色。花果期3-5月。生海拔100-1900米的路旁、林下、河滩地、岩石或荒地。产中国西南至东北的大部分地区。朝鲜半岛和日本亦有。

Perennial herbs, glabrous. Basal leaves rosette, margin serrate or lyrate-pinnatiparted; cauline leaves base dilated, cordate or auriculate-amplexicaul. Capitula arranged in corymb or corymbose-panicle; involucres cylindric; phyllaries ovate to lanceolate; ligualte flowers ca. 17, yellow. Achenes black, fusiform; pappus white. Fl. and fr. Mar-May. Roadsides, forests, flood plains, rocks or wastelands at 100-1900 m. Distributed in most parts of SW to NE China. Also in Korean Peninsula and Japan.

紫花蒲公英

Taraxacum lilacinum Schischk.

多年生草本，高4-12厘米。叶片匙状倒披针形，不分裂，边缘全缘或羽状分裂，或有齿；花葶通常高于叶片，无毛。头状花序直径2-2.5厘米；小花紫红色和堇紫色；冠毛白色。花期7-8月。生海拔3000-4000米的高山湿草坡、石砾地、高山草甸。产新疆。哈萨克斯坦和吉尔吉斯斯坦亦有。

Perennial herbs, 4-12 cm tall. Leaves spatulate-oblanceolate, undivided, margin usually remotely dentate or less often with remote short triangular lobules; scapes usually overtopping leaves, glabrous. capitula 2-2.5 cm wide; ligules deep pink to light pinkish violet; pappus white. Fl. Jul-Aug. Wet alpine slopes, gravelly sites, alpine meadows at 3000-4000 m. Distributed in Xinjiang. Also in Kazakhstan and Kyrgyzstan.

紫花蒲公英 *Taraxacum lilacinum*

葱岭蒲公英

Taraxacum pseudominutilobum S. Koval.

多年生草本。叶条状披针形，羽状深裂或浅裂，无毛。花葶2-11，高3-6厘米；总苞窄钟状；总苞片暗绿色，被蛛丝状毛；舌状花黄色，背面有紫色条纹。瘦果于顶部有极少量的小刺；冠毛白色。生海拔3000-3700米的高山草甸、沟谷草甸或洼地。产新疆。哈萨克斯坦和乌兹别克斯坦亦有。

Perennial herbs. Leaves linear-lanceolate, pinnately parted or lobed, glabrous. Scapes 2-11, 3-6 cm tall; involucres narrowly campanulate; phyllaries dark green, arachnoid hairy; ray florets yellow, purple striate abaxially. Achenes with a few small prickles at top; pappus white. Alpine meadows, valley meadows or low-lying lands at 3000-3700 m. Distributed in Xinjiang. Also in Kazakhstan and Uzbekistan.

华蒲公英

Taraxacum borealisinense Kitam.

多年生草本。叶倒卵状披针形或狭披针形，全缘、羽状浅裂至深裂，具波状齿；花葶1至数个，长于叶。头状花序直径2-2.5厘米；总苞片卵状披针形

尖裂假还阳参 *Crepidiastrum sonchifolium*

葱岭蒲公英 *Taraxacum pseudominutilobum*

华蒲公英 *Taraxacum borealisinense*

至披针形，先端淡紫色；舌状花黄色。瘦果倒卵球状披针形，淡褐色；冠毛白色。花期春天至夏天。生海拔600-2000米的潮湿盐碱地、原野或砾石中。产中国西南、华北、东北和西北。俄罗斯和蒙古亦有。

Perennial herbs. Leaves obovate-lanceolate or narrowly lanceolate, entire, pinnately lobed to parted, undulate; scapes 1 to several, longer than leaves. Capitula 2-2.5 cm diam; phyllaries ovate-lanceolate to lanceolate, apex purplish; ray florets yellow. Achenes obovoid-lanceolate, brownish; pappus white. Fl. spring to summer. Damp saline or kaline places, open fields or gravelly places at 600-2000 m. Distributed in SW, N, NE and NW China. Also in Russia and Mongolia.

白花蒲公英
Taraxacum leucanthum (Ledeb.) Ledeb.

多年生矮小草本。叶条状披针形，近全缘至浅裂，具很小的小齿，无毛；花葶1至数个，长2-6厘米。头状花序直径2.5-3厘米；总苞片卵状披针形；舌状花白色。瘦果倒卵球状长圆形，上部具小刺；冠毛淡红色。花期春季至夏季。生海拔2500-6000米的山坡、草地、沟谷、河滩或草甸。产西藏、甘肃、青海和新疆。印度、巴基斯坦、伊朗和俄罗斯亦有。

Perennial dwarf herbs. Leaves linear-lanceolate, nearly entire to lobed, minutely dentate, glabrous; scapes solitary to several, 2-6 cm long. Capitula 2.5-3 cm diam; phyllaries ovate-lanceolate; ray florets white. Achenes obovoid-oblong, upper part prickled; pappus reddish. Fl. spring to summer. Slopes, grasslands, valleys, river plains or meadows at 2500-6000 m. Distributed in Xizang, Gansu, Qinghai and Xinjiang. Also in India, Pakistan, Iran and Russia.

白花蒲公英 *Taraxacum leucanthum*

深裂蒲公英 *Taraxacum scariosum*

深裂蒲公英

Taraxacum scariosum

(Tausch) Kirschner et Štepanek

多年生草本。叶被蛛丝毛，长圆形或长圆状条形，羽状深裂。头状花序宽2-3厘米；外轮苞片14-18；花葶1-6；总苞宽钟状；总苞片披针状卵圆形至披针形；舌状花黄色。瘦果浅禾秆色；冠毛白色。花期春天及夏天，无性生殖。生海拔900-3000米的干草原及半干旱草原、路边或旱牧草地。产中国西南、华北、西北和东北。哈萨克斯坦、俄罗斯(阿勒泰)和蒙古亦有。

Perennial herbs. Leaves arachnoid, oblong or oblong-linear, pinnatipartite. Capitula 2-3 cm wide; outer phyllaries 14-18; scapes 1-6; involucres broad campaniform; phyllaries lanceolate-ovate to lanceolate; ligules yellow. Achenes pale grayish straw-colored; pappus white. Fl. spring and summer; agamosperm. Dry steppe and sub-steppe habitats, roadsides or dry pastures at 900-3000 m. Distributed in SW, N, NW and NE China. Also in Kazakhstan, Russia (Altay) and Mongolia.

堆叶蒲公英

Taraxacum compactum

Schischk.

多年生草本。叶腋被疏松褐色皱曲毛；叶狭倒披针形至长圆形，具齿至大头羽状分裂，裂片三角形。花葶2-5，粗壮，顶端密生蛛丝状毛；总苞片绿色，无角；舌状花黄色，疏生短柔毛。瘦果黄褐色，圆柱形，上部有大量小刺；冠毛白色。花果期6-8月。生海拔700-1700米的森林草甸、低山草原或荒漠草原带。产新疆北部。哈萨克斯坦亦有。

Perennial herbs. Base of leaf axils sparsely brown rugose floccose; leaves narrowly oblanceolate to oblong, dentate to lyrate-pinnatiparted, lobes triangular. Scapes 2-5, stout, apex densely arachnoid villose; phyllaries green, hornless; ray florets yellow, sparsely puberulent. Achenes yellowish-brown, cylindric, upper part densely prickled; pappus white. Fl. and fr. Jun-Aug. Forest meadows, hill grasslands or desert grasslands at 700-1700 m. Distributed in N Xinjiang. Also in Kazakhstan.

蒲公英

Taraxacum mongolicum

Hand.-Mazz.

多年生草本。叶披针形，边缘具波状齿或羽状深裂。花葶1至数个，上部紫红色，密被蛛丝状白色长柔毛；总苞钟状，淡绿色；总苞片披针形，先端增厚或具角状凸起；舌状花黄色。瘦果上部具小刺，下部具小瘤；冠毛白色。花期春季，通常到夏末。生海拔800-2000(-2800)米以下的山坡草地、路边、田野或河滩。产中国大部分地区。俄罗斯、蒙古和朝鲜半岛亦有。

Perennial herbs. Leaves lanceolate, margin undulately creanate or pinnatiparted. Scapes 1 to several, upper part purplish-red, densely white arachnoid villose; involucres campanulate, greenish; phyllaries lanceolate, apex thickened or cornuate; ray florets yellow. Achenes densely prickled on upper part and verruculose on lower part; pappus white. Fl. spring, occasionally also to late summer. Grassy slopes, roadsides, fields or flood plains below 800-2000(-2800) m. Distributed in most parts of China. Also in Russia, Mongolia and Korean Peninsula.

大头蒲公英

Taraxacum calanthodium

Dahlst.

多年生草本。叶宽披针形或倒卵状披针形，羽状深裂，背面疏被蛛丝状长柔毛。头状花序大，直径5-6厘米；总苞片卵形、卵状披针形或宽条形，边

堆叶蒲公英 *Taraxacum compactum*

蒲公英 *Taraxacum mongolicum*

缘明显膜质，先端增厚或具短小角；舌状花黄色。瘦果倒披针形，黄褐色；冠毛淡污黄白色。花期夏季。生海拔3000-4000米的高山草地。产四川、西藏、陕西、甘肃和青海。

Perennial herbs. Leaves broadly lanceolate or obovate-lanceolate, pinnatiparted, sparsely arachnoid villose abaxially. Capitula large, 5-6 cm diam; phyllaries ovate, ovate-lanceolate or broadly linear, margin conspicuously membranous, apex thickened or shortly cornuate; ray florets yellow. Achenes oblanceolate, yellowish-brown; pappus lightly dirty yellowish-white. Fl. summer. Alpine grasslands at 3000-4000 m. Distributed in Sichuan, Xizang, Shaanxi, Gansu and Qinghai.

大头蒲公英 *Taraxacum calanthodium*

川甘蒲公英

Taraxacum lugubre Dahlst.

多年生草本。叶条状披针形，羽状深裂，裂片宽三角形，背面疏被蛛丝状柔毛；叶柄长，粉紫色；花葶数个，顶端密被蛛丝状柔毛。头状花序直径3.5-5.5厘米；总苞暗紫色；舌状花黄色。瘦果倒卵球状楔形，上部有短而尖的小瘤；冠毛白色。花期夏季。生海拔4000-4600米的高山草地。产云南、四川、西藏、甘肃和青海。

Perennial herbs. Leaves linear-lanceolate, pinnatiparted, lobes broadly triangular, sparsely arachnoid villose; petioles long, pink-purple; scapes several, apex densely arachnoid villose. Capitula 3.5-5.5 cm diam; involucres dark purple; ray florets yellow. Achenes obovoid-cuneate, upper part with small short and acute papules; pappus white. Fl. summer. Alpine grasslands at 4000-4600 m. Distributed in Yunnan, Sichuan, Xizang, Gansu and Qinghai.

川甘蒲公英 *Taraxacum lugubre*

白缘蒲公英 *Taraxacum platypecidum*

白缘蒲公英

Taraxacum platypecidum Diels

多年生草本。叶倒披针形，羽状分裂。花亭1至数个，高达45厘米；头状花序大型，直径40-45毫米；总苞宽钟状；总苞片宽卵形，边缘为宽白色膜质，上端粉红色；舌状花黄色，背面有紫红色条纹。瘦果上部有刺状小瘤；冠毛白色。花果期夏季。生海拔1900-3000米的山坡草地或路旁。产华中、华北和东北。俄罗斯、朝鲜半岛和日本亦有。

Perennial herbs. Leaves oblanceolate, pinnately lobed. Scapes 1 to several, to 45 cm tall; Capitula large, 40-45 mm diam; involucres broad campanulate; phyllaries broadly ovate, margin broadly white membranous, apex pink; ray florets yellow, purple striate abaxially. Achenes with prickle-like strumae; pappus white. Fl. summer. Grassy slopes or roadsides at 1900-3000 m. Distributed in C, N and NE China. Also in Russia, Korean Peninsula and Japan.

东北蒲公英

Taraxacum ohwianmn Kitam.

多年生草本。叶倒披针形，不规则羽状浅裂至深裂。花亭多数，高10-20厘米；外层总苞片宽卵形，暗紫色，具狭窄的白色膜质边缘；舌状花黄色，背面有紫色条纹。瘦果长圆形，上部有刺状凸起；冠毛污白色。花果期4-6月。生低海拔的荒地、山坡或路旁。产黑龙江、吉林和辽宁。俄罗斯和朝鲜半岛亦有。

Perennial herbs. Leaves oblanceolate, irregularly pinnately lobed to parted. Scapes numerous, 10-20 cm tall; outer phyllaries broadly ovate, dark purple, margin narrowly white membranous; ray florets yellow, purle striate abaxially. Achenes oblong, upper part with prickle-like papules; pappus dirty white. Fl. and fr. Apr-Jun. Wastelands, slopes or roadsides at low altitude. Distributed in Heilongjiang, Jilin and Liaoning. Also in Russia and Korean Peninsula.

药用蒲公英

Taraxacum officinale F. H. Wigg.

多年生草本。叶狭倒卵形或长圆形，大头羽状深裂或羽状浅裂。花亭多数，基部红紫色；总苞宽钟状；总苞片绿色，无角，外层总苞片披针形，反卷；舌状花亮黄色，背面有紫色条纹。瘦果中部以上有大量小尖刺，其余部分具小瘤状凸起；冠毛白色。花果期6-8月。生海拔700-2200米的低山草原、森林草甸、田间或路边。产新疆。中亚、欧洲和北美洲亦有。

Perennial herbs. Leaves narrowly obovate or oblong, lyrate-pinnately parted or pinnately lobed. Scapes numerous, base purple; involucres broad campanulate; phyllaries green, hornless, outer ones lanceolate, revolute; ray florets lightly yellow, purle striate abaxially. Achenes with many prickles above middle part and small papules at other parts; pappus white. Fl. and fr. Jun-Aug. Hill grasslands, forest meadows, fields or roadsides at 700-2200 m. Distributed in Xinjiang. Also in C Asia, Europe and North America.

东北蒲公英 *Taraxacum ohwianmn*

药用蒲公英 *Taraxacum officinale*

灰果蒲公英

Taraxacum maurocarpum Dahlst.

多年生草本。叶狭披针形，羽状深裂，具齿，裂片狭三角形或近条状披针形，全缘；花葶长于叶，高10-25厘米。头状花序直径约3厘米；总苞片披针形至条形，粉红色；舌状花黄色。瘦果倒卵球状长圆形，灰色至深灰褐色，上部具小刺；冠毛淡污黄色。花期夏季。生海拔约4000米的高山草坡或河边。产四川、西藏和青海。巴基斯坦、阿富汗和伊朗亦有。

Perennial herbs. Leaves narrowly lanceolate, pinnatiparted, dentate, lobes narrowly triangular or nearly linear-lanceolate, margin entire; scapes longer than leaves, 10-25 cm tall. Capitula ca. 3 cm diam; phyllaries lanceolate to linear, pink; ray florets yellow. Achenes obovoid-oblong, gray to dark gray-brown, upper part prickled; pappus slightly dirty yellow. Fl. summer. Alpine grassy slopes or riversides at ca. 4000 m. Distributed in Sichuan, Xizang and Qinghai. Also in Pakistan, Afghanistan and Iran.

弯茎假苦菜

Askellia flexuosa (Ledeb.) W. A. Weber

多年生草本。基生叶和底部茎叶羽状浅裂至深裂，稀不裂；中上部叶线状披针形至狭线形。头状花序具9-13朵小花；总苞狭圆柱形；总苞片绿色，无毛；小花黄色。瘦果浅黄色。花果期6-10月。生海拔800-5100米的溪边、湖边、沼泽地、沙地和砾石地。产华西。中亚、西南亚、蒙古和俄罗斯亦有。

Perennial herbs. Basal and lower stem leaves shallowly pinnatifid to pinnatisect or more rarely undivided; middle and upper stem leaves linear-lanceolate to narrowly linear. Capitula with 9-13 florets; involucre narrowly cylindric; phyllaries green, abaxially glabrous; florets yellow. Achene pale yellow. Fl. and fr. Jun-Oct. Stream banks, lake margins, marshes and floodplains, sandy areas, gravel and loess areas at 800-5100 m. Distributed in W China. Also in C and SW Asia, Mongolia and Russia.

灰果蒲公英 *Taraxacum maurocarpum*

弯茎假苦菜 *Askellia flexuosa*

矮小假苦菜 *Askellia pygmaea*

矮小假苦菜
Askellia pygmaea (Ledeb.) Sennikov

多年生草本。根常分枝。基生叶和茎生叶有柄，卵形、圆形或椭圆形。头状花序有9-11朵小花；总苞狭圆柱形；总苞片绿色，无毛，边缘膜质；小花黄色。瘦果浅黄色，圆柱形或梭形。花果期6-9月。生海拔4600-4700米的河滩多石地和山坡底部、岸边。产新疆和西藏。哈萨克斯坦、蒙古、俄罗斯东部和北美洲亦有。

Perennial herbs. Roots often branched. Basal and stem leaves with petioles, ovate, orbicular or elliptic. Capitula with 9-11 florets; involucre narrowly cylindric; phyllaries green, abaxially glabrous, margin scarious; florets yellow. Achene pale yellowish, columnar to fusiform. Fl. and fr. Jun-Sep. Gravelly areas on floodplains and bases of slopes, stream banks at 4600-4700 m. Distributed in Xinjiang and Xizang. Also in Kazakhstan, Mongolia, E Russia and North America.

细叶小苦荬 *Ixeridium gracile*

细叶小苦荬
Ixeridium gracile (DC.) Pak et Kawano

多年生草本，无毛。叶长椭圆形、狭披针形、条形或狭条形，无毛，全缘。头状花序多数，排成伞房花序或伞房圆锥花序，花序梗极纤细；总苞极小，圆柱状；总苞片卵形或条状长椭圆形；舌状小花6朵，黄色。瘦果褐色，长圆锥状；冠毛褐色或淡黄色。花果期5-7月。生海拔1400-2700米的山坡、山谷、林缘、林下、田间、荒地或草甸。产中国西南、华南、华中、华东和西北。印度、尼泊尔、不丹和缅甸亦有。

Perennial herbs, glabrous. Leaves narrowly elliptic, narrowly lanceolate, linear or narrowly linear, glabrous, entire. Capitula numerous, arranged corymb or corymbose-panicles, peduncles very slender; involucres very small, cylindric; phyllaries ovate or narrowly linear-elliptic; ray florets ca. 6, yellow. Achenes brown, narrowly conical; pappus brown or brownish. Fl. and fr. May-Jul. Slopes, valleys, forest edges, forests, fields, wastelands or meadows at 1400-2700 m. Distributed in SW, S, C, E and NW China. Also in India, Nepal, Bhutan and Myanmar.

剪刀股
Ixeris japonica (Burm. f.) Nakai

多年生草本。基生叶匙状倒披针形或舌形，边缘具锯齿至羽

剪刀股 *Ixeris japonica*

中华苦荬菜 *Ixeris chinensis*

状分裂。茎生叶少数，无柄。头状花序1-6，排成伞房花序；总苞钟状；总苞片卵形或长披针形，有时先端具小鸡冠状凸起；舌状小花24朵，黄色。瘦果褐色，近纺锤形；冠毛白色。花果期3-5月。生海拔500米以下的路边潮湿地、弃荒地或田边。产广东、福建、浙江和河南。朝鲜半岛和日本亦有。

Perennial herbs. Basal leaves spathulate-oblanceolate or ligulate, margin serrate to pinnatiparted. Cauline leaves few, sessile. Capitula 1-6, arranged in corymb; involucres campanulate; phyllaries ovate or narrowly lanceolate, apex sometimes minutely cristate; ray florets ca. 24, yellow. Achenes brown, subfusiform; pappus white. Fl. and fr. Mar-May. Damp places by roads, disturbed places or fieldsides below 500 m. Distributed in Guangdong, Fujian, Zhejiang and Henan. Also in Korean Peninsula and Japan.

中华苦荬菜

Ixeris chinensis (Thunb.) Kitag.

多年生草本，无毛。基生叶密集，长椭圆形、倒披针形、条形或舌形；茎生叶2-4枚，披针形。头状花序排成伞房花序，花序梗纤细；总苞圆柱状；总苞片宽卵形至长椭圆状倒披针形；小花舌状，黄色。瘦果褐色，长椭圆形；冠毛白色。花果期6-10月。生海拔100-4000米路旁、山坡、田野、河边、灌丛或岩缝中。产中国西南、华东、华北和东北。俄罗斯、朝鲜半岛和日本亦有。

Perennial herbs, glabrous. Basal leaves dense, narrowly elliptic, oblanceolate, linear or ligulate; cauline leaves 2-4, lancelate. Capitula arranged corymb, puduncules slender; involucres cylindric; phyllaries broadly ovate to narrowly elliptic-oblanceolate; florets ligulate, yellow. Achenes brown, narrowly ellipsoid; pappus white. Fl. and fr. Jun-Oct. Roadsides, slopes, fields, riversides, thickets or rocky crevices at 100-4000 m. Distributed in SW, E, N and NE China. Also in Russia, Korean Peninsula and Japan.

碱小苦苣菜

Sonchella stenoma (Turcz. ex DC.) Sennikov

多年生草本，高10-50厘米。基生叶及下部茎叶线形或线状披针形或线状倒披针形，常肉质；中上部茎叶渐小，线形。头状花序排成总状花序或总状狭圆锥花序；总苞圆柱状；总苞片4层，全部总苞片外面无毛。瘦果纺锤形；冠毛白色，糙毛状。花果期7-9月。生海拔900-1500米的草原沙地及盐渍地。产甘肃和内蒙古。蒙古和俄罗斯(西伯利亚)亦有。

Perennial herbs, 10-50 cm tall. Basal and lower stem leaves narrowly lanceolate, narrowly elliptic or narrowly oblanceolate, often somewhat fleshy; middle and upper stem similar to lower ones but gradually smaller. Synflorescence narrowly racemiform; involucre narrowly cylindric; phyllaries 4 series, glabrous. Achene fusiform; pappus white, scabrid. Fl. and fr. Jul-Sep. Sandy soil in steppes and alkaline areas at 900-1500 m. Distributed in Gansu and Neimenggu. Also in Mongolia and Russia (Siberia).

碱小苦苣菜 *Sonchella stenoma*

矮小厚喙菊 *Dubyaea gombalana*

矮小厚喙菊
Dubyaea gombalana (Hand.-Mazz.) Stebbins

多年生草本，高达10厘米，莲座状，无茎，有直根。莲座叶窄椭圆形或倒披针形，无毛，边缘全缘。总苞窄钟形；总苞片无毛，约3层；花蓝色或蓝紫色；冠毛褐色。花期7-8月。生海拔3200-3900米的山坡、林下或高山草甸。产西藏东南部和云南(贡山县)。

Perennial herbs, to 10 cm tall, rosulate, acaulescent, with a taproot. Rosette leaves narrowly elliptic to oblanceolate, glabrous, margin entire. Capitula solitary; involucre narrowly campanulate; phyllaries glabrous, ca. 3 rows; florets blue to bluish purple; pappus brownish. Fl. Jul-Aug. Ravines on mountain slopes, forests, alpine meadows at 3200-3900 m. Distributed in NE Xizang and Yunnan (Gongshan County).

紫花厚喙菊
Dubyaea atropurpurea Stebbins

多年生草本，高30-80(-120)厘米。基生叶及下部茎叶不裂，僧帽形、箭头状心形或卵状心形；中上部茎叶与基生叶及下部茎叶同形，但较小。头状花序下垂或歪斜，3-7枚在茎枝顶端排成伞房状或聚伞状花序；总苞宽钟状；总苞片3-4层；小花蓝色到蓝紫色。冠毛污黄色，2层，糙毛状。花果期7-10月。生海拔3000-4100米冷杉林缘、高山草甸及灌丛中。产四川和云南。缅甸东北部亦有。

Perennial herbs, 30-80(-120) cm tall. Basal and lower stem leaves hairy or sometimes glabrescent, leaf margin mucronulately sinuate-dentate. Synflorescence corymbiform, with 3-7capitula; involucres broadly campanulate; phyllaries 3-4 seriates; florets blue to bluish purple. Pappus brownish. Fl. and fr. Jul-Oct. *Picea* forest margins, alpine meadows, thickets at 3000-4100 m. Distributed in Sichuan and Yunnan. Also in NE Myanmar.

厚喙菊
Dubyaea hispida DC.

多年生草本，高20-60厘米。茎直立，自中下部分枝。基生叶及下部茎叶椭圆形或长椭圆形；中上部茎叶与基生叶同形，较小。头状花序下垂或歪斜，2-7枚在茎枝顶端排成伞房花序或聚伞花序；总苞宽钟状；总苞片3层；舌状小花黄色；冠毛淡黄色，2层。花果期7-11月。生海拔2700-4500米高山林缘、林下、草甸或灌丛中。产云南西北部、西藏和四川西南部。喜马拉雅亦有。

Perennial herbs, 20-60 cm tall. Stem erect, branched from below middle. Basal and lower stem leaves elliptic or long elliptic, middle and upper stem leaves similar to lower stem leaves but smaller. Synflorescence corymbiform, with 2-7 capitula. Capitula nodding; involucres campanulate; phyllaries 3-seriate; florets yellow; pappus yellowish, 2-seriate. Fl. and fr. Jul-Nov. Forests, forest margins, meadows, thickets at 2700-4500 m. Distributed in NW Yunnan, Xizang and SW Sichuan. Also in Himalaya.

黄花合头菊
Syncalathium chrysocephalum (C. Shih) S. W. Liu

多年生莲座状草本，茎极短或几无茎。叶片卵圆形或卵圆状圆形，边缘有锯齿。头状花序含5朵舌状小花，茎端密集成直径2厘米的团伞花序；总苞狭圆柱状，直径3毫米；总苞片1层，5枚；舌状小花5朵。花期7月。生海拔4100-4700米高山流石滩。产青海、四川西北部和西藏东部。

Perennial herbs, 3-5 cm tall, rosulate, acaulescent. Rosette leaves ovate to ovate-orbicular, margin dentate. Synflorescence flat conical with few to some capitula; capitula with 5 florets; involucres narrowly cylindric; phyllaries 5, lanceolate; florets yellow. Fl. Jul. Scree slopes at

紫花厚喙菊 *Dubyaea atropurpurea*

厚喙菊 *Dubyaea hispida*

黄花合头菊 *Syncalathium chrysocephalum*

Distributed in Qinghai and Xizang.

4100-4700 m. Distributed in Qinghai, NW Sichuan and E Xizang.

合头菊

Syncalathium kawaguchii (Kitam.) Ling

一年生莲座草本。叶有柄，通常能够紫红色，卵形、倒披针形或椭圆形，不分裂或基部琴状羽裂。头状花序少数或多数，在茎端排成直径为2-5厘米的团伞花序；总苞片1层，3枚，舌状小花3枚；冠毛白色。花果期6-10月。生海拔3800-5400米的山坡及河滩砾石地、流石滩。产青海和西藏。

Annual herbs, 1-5 cm tall, rosulate. Rosette leaves petiolate; leaves often dark purple, ovate, oblanceolate or elliptic, undivided to basally lyrately pinnate. Synflorescence flat conical, 2-6 cm diam, with some to numerous capitula; capitula with 3 florets; florets purple; pappus white. Fl. and fr. Jun-Oct. Alpine steppes, scree slopes, gravelly areas in dry river valleys at 3800-5400 m.

合头菊 *Syncalathium kawaguchii*

盘状合头菊

Syncalathium disciforme (Mattfeld) Y. Ling

多年生草本，莲座状。叶狭卵形或倒披针形，无柄，羽状裂。头状花序在茎顶排成直径为2-7厘米的团伞花序；总苞片1层，5枚，长椭圆形；舌状小花黄色，5枚；冠毛灰白色。花果期9-10月。生海拔3900-4800米的高山草甸、小石子山坡、山坡、砾石地区。产甘肃、青海和四川。

Perennial herbs, rosulate, acaulescent. Leaves narrowly obovate to oblanceolate, sinuate-dentate to pinnately lobed with toothlike lobes. Synflorescence flat conical, 2-7 cm diam; capitula with 5 florets; phyllaries in 1 row, 5; florets yellow, ligule ca. 3 mm; pappus greyish white. Fl. and fr. Sep-Oct. Alpine meadows, scree slopes, mountain slopes, gravelly areas at 3900-4800 m. Distributed in Gansu, Qinghai and Sichuan.

盘状合头菊 *Syncalathium disciforme*

brown; pappus lightly lateritious. Fl. and fr. Jul-Oct. Valleys, slopes, forests, grasslands or wetlands at 1100-2700 m. Distributed in Sichuan, Henan, Hebei, Shanxi, Shaanxi and Gansu.

盘果菊 *Nabalus tatarinowii*

盘果菊

Nabalus tatarinowii (Maxim.) Nakai

多年生草本。叶互生，具长柄，心形或卵形，边缘具不整齐细齿，叶柄具卵形耳状小裂片。头状花序多数，排成圆锥状；总苞狭圆柱状；总苞片卵形至条形，疏被卷毛；舌状小花5朵，紫色、粉红色、白色或黄色。瘦果长椭圆状，紫褐色；冠毛淡褐色。花果期8-10月。生海拔500-3000米的山谷、山坡、林缘、林下、水边或潮湿地。产中国西南、华中、华北、东北和西北。俄罗斯和朝鲜半岛亦有。

Perennial herbs. Leaves alternate, long petiolate, cordate or ovate, margin irregularly serrulate, petioles atthached with ovate-auriculate lobes. Capitula numerous, arranged in panicle; involucres narrowly cylindric; phyllaries ovate to linear, sparsely curly pubescent; ray florets 5, purple, pink, white or yellow. Achenes narrowly ellipsoid, purplish-brown; pappus brownish. Fl. and fr. Aug-Oct. Valleys, slopes, forest edges, forests, watersides or wetlands at 500-3000 m. Distributed in SW, C, N, NE and NW China. Also in Russia and Korean Peninsula.

多裂耳菊

Nabalus tatarinowii (Maxim.) Nakai subsp. **macrantha** (Stebbins) N. Kilian

多年生草本。叶互生，具长柄，圆形至长圆形，掌状羽裂，叶柄下部具卵形耳状小裂片。头状花序多数，排成圆锥状；总苞狭圆柱状；总苞片卵形至披针形；舌状小花5朵，淡红紫色。瘦果圆柱状，棕色；冠毛浅土红色。花果期7-10月。生海拔1100-2700米的山谷、山坡、林下、草地或潮湿地。产四川、河南、河北、山西、陕西和甘肃。

Perennial herbs. Leaves alternate, long petiolate, orbicular to oblong, palmately pinnatiparted, lower part of petioles atthached with ovate-auriculate lobes. Capitula numerous, arranged in panicle; involucres narrowly cylindric; phyllaries ovate to lanceolate; ray florets 5, reddish-purple. Achenes cylindric,

绢毛苣

Soroseris glomerata (Decne.) Stebbins

多年生草本。地上基生叶莲座状，匙形、宽椭圆形或倒卵形。头状花序多数，在莲座状叶丛中密集成团伞花序；总苞狭圆柱状；总苞片条形或长圆形，被白色长柔毛；小花4-6朵，舌状，黄色，稀白色或粉红色。瘦果长圆柱状；冠毛灰色或浅黄色。花果期5-9月。生海拔3200-5600米的高山流石滩或高山草甸。产云南、四川和西藏。印度北部和尼泊尔亦有。

Perennial herbs. Overground basal leaves rosette, spathulate, broadly elliptic or obovate. Capitula numerous, glomerate in the center of rosette leaves; invo-

多裂耳菊 *Nabalus tatarinowii* subsp. *macrantha*

lucres narrowly cylindric; phyllaries linear or oblong, white villose; florets 4-6, ligulate, yellow, rare white or pink. Achenes nearly terete; pappus gray or brownish. Fl. and fr. May-Sep. Alpine screes or meadows at 3200-5600 m. Distributed in Yunnan, Sichuan and Xizang. Also in N India and Nepal.

绢毛苣 *Soroseris glomerata*

皱叶绢毛苣 *Soroseris hookeriana*

皱叶绢毛苣

Soroseris hookeriana Stebbins

多年生草本，无毛。茎极短或伸长，直径达3厘米，上部中空。叶螺旋状或莲座状排列，倒披针形至长椭圆形，羽状深裂。头状花序多数，团伞状，生于茎顶或莲座状叶丛中；总苞狭圆柱状；小花4朵，舌状，黄色。瘦果圆柱状；冠毛黄色或灰色。花果期7-9月。生海拔2800-5500米的高山流石滩、岩石坡或草甸。产青藏高原。喜马拉雅亦有。

Perennial herbs, glabrous. Stems very short or elongate, to 3 cm diam, upper part canulate. Leaves spiral or rosette, oblanceolate to oblong, pinnatiparted. Capitula numerous, glomerate, on top of stems or in center of rosette leaves; involucres narrowly cylindric; florets 4, ligulate, yellow. Achenes cylindric; pappus yellow or gray. Fl. and fr. Jul-Sep. Alpine screes, rocky slopes or meadows at 2800-5500 m. Distributed in Tibetan Plateau. Also in Himalaya.

空桶参 *Soroseris erysimoides*

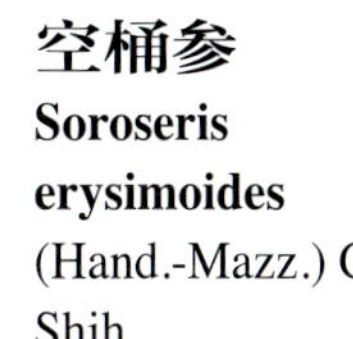

空桶参

Soroseris erysimoides (Hand.-Mazz.) C. Shih

多年生草本，高5-30厘米常合生。茎直立，单生，中空，不分枝。叶线舌形、椭圆形或线形。花序半球形，由数朵紧聚于一起的头状花序组成，每个头状花序由4朵小花组成总苞片2层；舌状小花黄色。冠毛鼠灰色或淡黄色，细锯齿状。花果期6-10月。生海拔3000-3500米高山灌丛、草甸或流石滩。产青藏高原和陕西。喜马拉雅亦有。

Perennial herbs, 5-30 cm tall, usually conspicuously caulescent. Stem solitary, erect, hollow, leafy. Leaves oblanceolate, lanceolate, elliptic, or linear. Synflorescence hemispheric, with numerous closely crowded capitula; capitula with 4 florets; phyllaries in 2 rows florets yellow; pappus whitish to straw-colored and grayish apically. Fl. and fr. Jun-Oct. Alpine thickets, meadows, scree slopes at 3000-3500 m. Distributed in the Tibetan Plateau and Shaanxi. Also in Himalaya.

矮生绢毛苣 *Soroseris depressa*

矮生绢毛苣

Soroseris depressa (Hook. f. et Thoms.) J. W. Zhang, N. Kilian et H. Sun

多年生草本，高2-3厘米，莲座状，无茎。莲座叶有叶柄，圆形、宽卵形或四角形，边缘近全缘或有波状锯齿。复合花序近伞形或半球形。头状花序有15-20朵小花；总苞片2层；小花黄色冠毛白色。花果期8-9月。生海拔3200-4500米的草坡、高山草甸。产西藏。喜马拉雅、不丹、印度和尼泊尔亦有。

Perennial herbs, 2-3 cm tall, rosulate, acaulescent. Rosette leaves orbicular, broadly ovate or deltoid, margin subentire to sinuate-dentate. Synflorescence subumbellate to hemispheric. Capitula with 15-20 florets; phyllaries in 2 rows; florets yellow; pappus white. Fl. and fr. Aug-Sep. Grasslands on mountain slopes, alpine meadows at 3200-4500 m. Distributed in Xizang. Also in Himalaya, Bhutan, India and Nepal.

柱序绢毛苣

Soroseris teres C. Shih

多年生草本。茎单生，中空，不分枝。叶长椭圆形，羽状或倒向羽状浅裂，全部叶两面无毛。头状花序多数，沿茎排列成圆柱状花序长达13厘米，头状花序数朵紧密排列；总苞狭圆柱状；舌状小花黄色，4朵；冠毛白色。花果期7-9月。生海拔3900-4250米高山草甸或灌丛中。产西藏（亚东县）。不丹亦有。

Perennial herbs, Stem solitary, erect, hollow, leafy. Leaves narrowly elliptic, pinnatipartite to pinnatisect, glabrous or sparsely pilose. Synflorescence cylindric, elongate, to 13 cm, with numerous densely crowded capitula; involucres narrowly cylindric; capitula with 4 florets; florets yellow; pappus whitish. Fl. and fr. Jul-Sep. Alpine grasslands or thickets at 3900-4250 m. Distributed in Xizang (Yadong County). Also in Bhutan.

柱序绢毛苣 *Soroseris teres*

肉菊

Soroseris umbrella (Franch.) Stebbins

多年生草本，肉质。无地上茎。叶莲座状，稠密，紫红色，卵形或卵状椭圆形，边缘具稀疏细尖齿；叶柄宽厚。头状花序多数，在茎顶莲座状叶丛中密集成团伞花序；全部小花舌状，黄白色。瘦果近长圆柱状。花果期7-9月。生海拔2600-4600米的高山草甸或流石滩。产云南、四川和西藏。

Perennial herbs, succulent. Acaulescent. Leaves rosette, dense, purplish-red, ovate or ovate-elliptic, margin sparsely and minutely serrulate; petioles wide and thick. Capitula numerous, glomerate in the center of rosette leaves; all florets ligulate, yellowish-white. Achenes nearly terete. Fl. and fr. Jul-Sep. Alpine meadows or screes at 2600-4600 m. Distributed in Yunnan, Sichuan and Xizang.

日本毛连菜

Picris japonica Thunb.

多年生短命草本。茎具柔毛、混生深绿色至黑色分叉钩状硬毛。基部及茎下部叶倒披针形、椭圆状披针形或椭圆状倒披针形。复合花序伞房状至圆锥状伞房状、具多数小头状花序；总苞圆柱状钟形至卵形；小花黄色。瘦果棕红色，纺锤形；冠毛早落。生海拔600-3700米的山坡草地、林中、林中弃荒地、溪边或高山草甸。产中国西南、华北、华西、西北和东北。哈萨克斯坦、俄罗斯东部、蒙古和日本亦有。

Short-lived perennial herbs. Stem hirsute with dark green to blackish rigid 2-hooked hairs. Basal leaves and lower stem leaves oblanceolate, elliptic-lanceolate or elliptic-oblanceolate. Synflorescences corymbiform to paniculately corymbiform, with many capitula; involucres cylindriccampanulate to ovoid; florets yellow. Achenes reddish brown, fusiform; pappus caducous. Grasslands on mountain slopes, forests, waste places in forests, river margins or alpine meadows at 600-3700 m. Distributed in SW, N, W, NW and NE China. Also in Kazakhstan, E Russia, Mongolia and Japan.

日本毛连菜 *Picris japonica*

毛连菜

Picris hieracioides L.

二年生草本，全部密被亮色分叉钩状硬毛。茎单生，直立，细长。叶长圆形、披针形或条形，全缘或具锯齿，基部半抱茎。头状花序多数，排成伞房花序；总苞圆柱状钟形；总苞片条形；舌状花黄色。瘦果纺锤形；冠毛白色。花果期6-10月。生海拔200-3600米的山坡草地、林下、沟边、田间、荒地或沙地。产中国西南、华中、华北、东北和西北。伊朗、哈萨克斯坦、俄罗斯、地中海地区和欧洲亦有。

Biennial herbs, wholly and densely light furcate-hooked-strigillose. Stems solitary, erect, slender. Leaves oblong, lanceolate or linear, margin entire or serrate, base semi-amplexicaul. Capitula numerous, arranged in corymb; involucres cylindric-campanulate; phyllaries linear; ray florets yellow. Achenes fusiform; pappus white. Fl. and fr. Jun-Oct. Grassy slopes, forests, ditchsides, fields, wastelands or sandy places at 200-3600 m. Distributed in SW, C, N, NE and SW China. Also in Iran, Kazakhstan, Russia, Mediterranean regions and Europe.

肉菊 *Soroseris umbrella*

毛连菜 *Picris hieracioides*

滇苦菜 *Picris divaricata*

菊苣 *Cichorium intybus*

滇苦菜
Picris divaricata Vaniot

二年生草本。基生叶长椭圆形，被硬单毛，边缘浅波状或全缘；茎生叶小，极少或近无。头状花序少数至多数，单生枝顶；总苞钟状；总苞片条形至披针形，被短硬毛；舌状小花黄色。瘦果长椭圆形，无喙；冠毛白色。花果期4-11月。生海拔1400-3200米的山坡草地、林缘或灌丛。产云南和西藏。

Biennial herbs. Basal leaves narrowly elliptic, unisetose, margin shallowly undulate or entire; cauline leaves small, very few or nearly absent. Capitula few to many, solitary, terminal; involucres campanulate; phyllaries linear to lanceolate, shortly strigillose; ligulate florets yellow. Achenes narrowly ellipsoid, erostrate; pappus white. Fl. and fr. Apr-Nov. Grassy slopes, forest edges or thickets at 1400-3200 m. Distributed in Yunnan and Xizang.

菊苣
Cichorium intybus L.

多年生草本。基生叶莲座状，渐狭成叶柄状，不开裂至常羽状深裂。主轴与大枝上的复花序穗状至圆锥状；头状花序具15-20花；总苞圆柱形；小花蓝色。花期5-10月。生低海拔河边、海边荒地、山坡或沟边。产华南、华北、华西、华东、西北和东北。中亚、西南亚、北非和欧洲亦有。

Perennial herbs. Basal leaves rosulate, attenuate into a petiole-like basal portion, undivided to usually pinnatipartite. Synflorescences of main axises and larger branches spiciform-paniculiform; capitula with usually 15-20 florets; involucres cylindric; florets blue. Fl. May-Oct. By rivers, wastelands along seashore, slopes or by ditches at low elevations. Distributed in S, N, W, E, NW and NE China. Also in C and SW Asia, N Africa and Europe.

山柳菊
Hieracium umbellatum L.

多年生草本。茎中部及上部叶无柄，披针形，基部狭楔形。复合花序伞房状至伞房圆锥状，具少量至多数头状花序；总苞深绿色，钟形；苞片数层至多层；小花黄色。瘦果深紫色，圆柱形，具10条纵棱；冠毛浅黄色。花果期7-9月。生海拔(200-)1000-3000(-3300)米的林地、草地或冲积平原的沙质土壤上。产中国西南、华中、华北、华西、华东、西北和东北。南亚、中亚、东北亚、西南亚、欧洲和北美洲亦有。

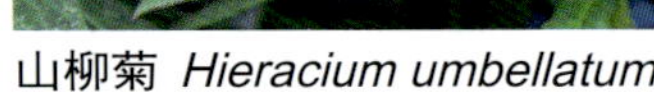
山柳菊 *Hieracium umbellatum*

Perennial herbs. Middle and upper stem leaves sessile, lanceolate, base narrowly cuneate. Synflorescences corymbiform to corymbosely paniciform, with few to numerous capitula; involucres dark green, campanulate; phyllaries in several to many rows; florets yellow. Achenes dark purple, cylindric, with 10 ribs; pappus pale yellow. Fl. and fr. Jul-Sep. Forests, grassland or sandy soils on floodplains at (200-) 1000-3000(-3300) m. Distributed in SW, C, N, W, E, NW and NE China. Also in S, C, NE and SW Asia, Europe and North America.

新疆山柳菊 *Hieracium korshinskyi*

卵叶山柳菊 *Hieracium regelianum*

新疆山柳菊

Hieracium korshinskyi Zahn

多年生草本，有长根状茎。茎直立，紫红色。基生叶花期生存，椭圆形或披针形，有翼柄；下部茎有短柄或无柄。头状花序少数，伞房花序状；总苞钟状；总苞片3层，暗绿色；小花黄色。瘦果有10条纵肋。花果期7-9月。生海拔1600-2200米的林中空地、林间。产新疆。哈萨克斯坦、俄罗斯和蒙古亦有。

Perennial herbs, with long rhizomes. Stem purplish red, erect. Basal leaves present at anthesis; petioles winged; leaves elliptic to lanceolate; lower stem leaves shortly petiolate or sessile. Synflorescence corymbiform, with (1 or)2 to several capitula; involucre campanulate; phyllaries in 3 rows, dark green; florets yellow. Achene with 10 ribs. Fl. and fr. Jul-Sep. Forests, open places in forests at 1600-2200 m. Distributed in Xinjiang. Also in Kazakhstan, Mongolia and Russia.

卵叶山柳菊

Hieracium regelianum Zahn

多年生草本，根状茎粗短。茎直立，单生或少数茎成簇生。基生叶及下部茎叶花期枯萎；中部茎叶无柄。头状花序端排成疏松的伞房圆锥花序；总苞钟状；总苞片3层，暗绿色；小花黄色。瘦果暗褐色，有8-10条细肋；冠毛污白色。花果期7-9月。生海拔1700-2000米的林间开阔地。产新疆。哈萨克斯坦亦有。

Perennial herbs, with short and thick rhizomes. Stem solitary to few fascicled, erect. Basal and lower stem leaves withered in fruit; middle stem leaves sessile. Synflorescence sparsely corymbosely paniculiform; involucre campanulate; phyllaries in 3 rows, dark green; florets yellow. Achene dark brown, with 8-10 ribs; pappus dirty white. Fl. and fr. Jul-Sep. Open places in forests at 1700-2000 m. Distributed in Xinjiang. Also in Kazakhstan.

粗毛山柳菊

Hieracium virosum Pallas

多年生草本。茎中部叶无柄，多为卵形，基部心形抱茎。复合花序短伞房状至长圆锥状，具多个头状花序；总苞钟形或基部螺旋状；苞片具少数几层；小花黄色。瘦果棕黑色，圆柱形，具10条纵棱；冠毛灰黄色。花果期6-10月。生海拔1700-2100米的草地、林中或灌丛。产新疆。西南亚、中亚和欧洲东南部亦有。

Perennial herbs. Middle stem leaves sessile, mainly ovate, base cordate and amplexicaul. Synflorescences shortly corymbiform to long paniciform, with many capitula; involucres campanulate or basally turbinate; phyllaries in few rows; florets yellow. Achenes blackish brown, cylindric, with 10 ribs; pappus pale yellow. Fl. and fr. Jun-Oct. Grasslands, forests or thickets at 1700-2100 m. Distributed in Xinjiang. Also in SW and C Asia, and SE Europe.

粗毛山柳菊 *Hieracium virosum*

树斑鸠菊 *Vernonia arborea*

树斑鸠菊

Vernonia arborea Buch.-Ham.

小乔木或灌木，高10-20米。枝密被黄褐色绒毛。叶近革质，卵形或长圆形，全缘，侧脉8-12对。头状花序多数，排成宽复伞房状圆锥花序；总苞杯状；小花5-6朵，淡紫色或白色。瘦果压扁；冠毛污白色。花果期8-11月。生海拔(100-)800-1200米的山谷、山坡或疏林。产云南东南部和广西。南亚和东南亚亦有。

Small trees or shrubs, 10-20 m tall. Stems densely yellowish-brown tomentose. Leaves subcoriaceous, ovate or oblong, entire, lateral veins 8-12 pairs. Capitula numerous, arranged in broad compound corymbose-panicle; involucres cup-shaped; florets 5-6, purplish or white. Achenes compressed; pappus dirty white. Fl. and fr. Aug-Nov. Valleys, slopes or sparse forests at (100-) 800-1200 m. Distributed in SE Yunnan and Guangxi. Also in S and SE Asia.

毒根斑鸠菊

Vernonia cumingiana Benth.

攀援灌木或藤本。叶卵状长圆形或长圆状披针形。头状花序顶生或腋生；总苞卵状球形或钟状；总苞片5层，被锈色或黄褐色绵毛；花淡红或淡红紫色；冠毛红色或红褐色。花期10月至翌年4月。生海拔300-1500米，攀援于栎树林或山谷灌丛中。产云南、四川、贵州、广西、广东、福建和台湾。东南亚亦有。

Scandent shrubs or vine. Leaves ovate-oblong or oblong-lanceolate. Synflorescences terminal or axillary; involucres ovoid-globose or campanulate; phyllaries 5-seriate, ferruginous or fulvous tomentulose; corolla reddish or reddish purple; pappus red or reddish brown. Fl. Oct to next Apr. Climbing on trees in *Quercus* forests or thickets in valleys at 300-1500 m. Distributed in Yunnan, Sichuan, Guizhou, Guangxi, Guangdong, Fujian and Taiwan. Also in SE Asia.

大叶斑鸠菊

Vernonia volkameriifolia (Wall.) DC.

小乔木，高5-8米。叶倒卵形，边缘深波状或具疏粗齿，侧脉12-17对，下面被疏柔毛。头状花序多数，排成宽圆锥花序；总苞圆柱状狭钟形；小花10-12朵，淡红色或淡红紫色。瘦果

毒根斑鸠菊 *Vernonia cumingiana*

大叶斑鸠菊 *Vernonia volkameriifolia*

长圆状圆柱形；冠毛污白色。花期10月至翌年4月。生海拔800-1600米的灌丛或杂木林中。产云南、西藏、贵州和广西。南亚和越南亦有。

Small trees, 5-8 m tall. Leaves obovate, margin deeply undulate or sparsely gross-dentate, lateral veins 12-17 pairs, sparsely pilose abaxially; capitula numerous, arranged in broad pannicle; florets 10-12, reddish or reddish-purple. Achenes oblong-cylindric; pappus dirty white. Fl. Oct to next Apr. Thickets or mixed forests at 800-1600 m. Distributed in Yunnan, Xizang, Guizhou and Guangxi. Also in S Asia and Vietnam.

斑鸠菊 *Vernonia esculenta*

斑鸠菊

Vernonia esculenta Hemsl.

灌木或小乔木，高2-6米，密被灰色绒毛。叶具柄，硬纸质，披针形，侧脉9-13对。头状花序多数，排成宽圆锥花序；总苞倒圆锥状；总苞片革质，暗绿色；小花5-6朵，淡红紫色。瘦果近圆柱状；冠毛白色或污白色。花期7-12月。生海拔1000-2700米的山坡阳处、灌丛、山谷、疏林或林缘。产云南、四川、贵州和广西。

Shrubs or small trees, 2-6 m tall, densely gray tomentose. Leaves petiolate, rigid chartaceous, lanceolate, lateral veins 9-13 pairs. Capitula numerous, arranged in broad panicle; involucres obconical; phyllaries coriaceous, dark green; florets 5-6, reddish-purple. Achenes subcylindric; pappus white or dirty white. Fl. Jul-Dec. Sunny slopes, thickets, valleys, sparse forests or forest edges at 1000-2700 m. Distributed in Yunnan, Sichuan, Guizhou and Guangxi.

南漳斑鸠菊

Vernonia nantcianensis (Pamp.) Hand.-Mazz.

一年生或多年生草本。头状花序单生；花梗强壮，上部极度膨大，密被柔毛和腺点；苞片5-6层，上部和边缘紫红色；花多数；花冠紫粉色。瘦果圆柱形；冠毛棕黄色。花果期8-10月。生海拔700-2000米的沟谷林缘或山坡。产重庆和湖北西部。

Annual or perennial herbs. Capitula solitary; peduncles robust, rather dilated above, densely puberulent and glandular; phyllaries 5-6-seriate, upper parts and margin purple-red; florets many; corolla pink-purple. Achenes cylindrical; pappus yellowish-brown. Fl. and fr. Aug-Oct. Forest edges in valleys or slopes at 700-2000 m. Distributed in Chongqing and W Hubei.

南漳斑鸠菊 *Vernonia nantcianensis*

夜香牛 *Vernonia cinerea*

夜香牛
Vernonia cinerea (L.) Less.

一年生或多年生草本，高20-100厘米，密被短柔毛。叶菱状卵形、菱状长圆形或卵形。头状花序小，排成伞房状圆锥花序；总苞钟状；总苞片先端渐尖至刺尖；小花19-23朵，淡红紫色。瘦果圆柱形；冠毛白色。花期全年。生海拔1500-1800米的山坡、荒地、田边或路旁。产中国西南、华南、华中和华东。南亚、日本和非洲亦有。

Annual or perennial herbs, 20-100 cm tall, densely puberulent. Leaves rhomboid-ovate, rhomboid-oblong or ovate. Capitula small, arranged in corymbose panicle; involucres campanulate; phyllaries acuminate to spinescent; florets 19-23, purplish. Achenes terete; pappus white. Fl. year-round. Slopes, wastlands, fields or roadsides at 1500-1800 m. Distributed in SW, S, C and E China. Also in S Asia, Japan and Africa.

咸虾花
Vernonia patula (Ait.) Merr.

一年生粗壮草本，被灰色短柔毛。叶具柄，卵形，边缘具圆齿状浅齿，侧脉4-5对。头状花序通常2-3个生于枝顶端；总苞片绿色，披针形；小花淡红紫色。瘦果近圆柱状；冠毛白色。花期7月至翌年5月。生海拔160-800米的山坡开阔地、荒地、田边或路旁。产云南、贵州、广西、广东、福建和台湾。南亚和东南亚亦有。

Annual stout herbs, shortly gray pubescent. Leaves petiolate, ovate, margin shallowly crenate-serrate, lateral veins 4-5 pairs. Capitula generally 2-3 on tops of branches; phyllaries green, lanceolate; florets reddish-purple. Achenes subcylindric; pappus white. Fl. Jul to next May. Open places on slopes, wastelands, fields or roadsides at 160-800 m. Distributed in Yunnan, Guizhou, Guangxi, Guangdong, Fujian and Taiwan. Also in S and SE Asia.

地胆草
Elephantopus scaber L.

多年生草本，被白色贴生硬毛。基生叶莲座状，匙形或倒披针形。头状花序密集成顶生团球状复头状花序，基部被3朵叶状苞片所包围；苞片绿色，草质，卵形；总苞狭；小花4朵，淡紫色或粉红色。瘦果长圆状条形；冠毛污白色。花期7-11月。生海拔1400米以下的山坡、路边、林缘或荒地。产中国西南、华南、华中和东南。世界热带地区广布。

Perennial herb, appressed white strigose. Basal leaves rosette, spathulate or oblanceolate. Capitula congested in terminal glomerate compound head, base surrounded by 3 foliaceous floral bracteal leaves; bracteal leaves green, herbaceous, ovate; involucres slender; florets 4, purplish or pink. Achenes oblong-linear; pappus dirty white. Fl. Jul-Nov. Slopes, roadsides, forest edges or wastelands below 1400 m. Distributed in SW, S, C and SE China. Also in tropical regions of the world.

咸虾花 *Vernonia patula*

地胆草 *Elephantopus scaber*

狭舌多榔菊 *Doronicum stenoglossum*

狭舌多榔菊

Doronicum stenoglossum Maxim.

多年生草本，被白色柔毛。叶椭圆形、长圆形或披针形，基部心形，半抱茎，膜质，边缘具细尖齿或近全缘。头状花序小，2-10个，排成总状花序；总苞半球形或宽钟状；总苞片绿色，超出花盘；舌状花短于总苞，条形。瘦果近圆柱状，褐色；冠毛白色或微红色。花期7-9月。生海拔2100-3900米的高山草坡、林缘、灌丛或林下。产云南、四川、西藏、甘肃和青海。

Perennial herbs, white pubescent. Leaves elliptic, oblong or lanceolate, base cordate, semi-amplexicaul, membranous, margin minutely acute-serrate or nearly entire. Capitula small, 2-10, arranged in raceme; involucres semiglobose or broadly campanulate; phyllaries green, across the floral disc; ray florets shorter than involucre, linear. Achenes subcylindric, brown; pappus white or slightly reddish. Fl. Jul-Sep. Alpine slopes, forest edges, thickets or forests at 2100-3900 m. Distributed in Yunnan, Sichuan, Xizang, Gansu and Qinghai.

甘肃多榔菊 *Doronicum gansuense*

甘肃多榔菊

Doronicum gansuense Y. L. Chen

多年生草本。茎直立，不分枝。基部叶具长柄；下部茎叶倒卵形或倒卵状匙形，中部及上部茎叶卵形至卵状长圆形。头状花序单生茎端；总苞半球形或宽钟形，总苞片2层；舌状花舌片黄色；冠毛黄褐色。花期7-8月。生海拔3100-3700米草坡和砾石坡。产甘肃和陕西。

Perennial herbs. Stems erect, simple. Radical and stoloniferous leaves long petiolate; lower stem leaves obovate or obovate-spatulate, median and upper leaves sessile, ovate to ovate-oblong. Capitula solitary, terminal; involucres hemispheric or broadly campanulate; phyllaries 2-seriate; ray florets yellow; pappus yellow-brown. Fl. Jul-Aug. Grassy slopes and gravelly slopes at 3100-3700 m. Distributed in Gansu and Shaanxi.

大吴风草

Farfugium japonicum (L.) Kitam.

多年生草本。叶基生，莲座状，肾形，边缘全缘至掌状浅裂，近革质；花葶高达70厘米，被柔毛。头状花序辐射状，排成伞房状花序；总苞钟形或宽陀螺形；总苞片长圆形；舌状花8-12朵，黄色，长圆形管状花多数。瘦果纺锤形。花果期8月至翌年3月。生低海拔的林下、山谷或草丛。产中国西南、华南、东南和华中。

Perennial herbs. Leaves basal, rosette, reniform, margin entire to palmately lobed, subcoriaceous; scapes to 70 cm tall, pubescent. Capitula radiate, arranged in corymb; involucres campanulate or broad turbinate; phyllaries oblong; ray florets 8-12, yellow, oblong; tubular florets numerous. Achenes cylindric. Fl. and fr. Aug to next Mar. Forests, valleys or grasslands at low altitude. Distributed in SW, S, SE and C China.

大吴风草 *Farfugium japonicum*

翅柄橐吾 *Ligularia alatipes*

亚东橐吾 *Ligularia atkinsonii*

翅柄橐吾

Ligularia alatipes Hand.-Mazz.

多年生草本，高达150厘米。基生叶卵状心形，具翼柄；茎中部叶具短柄，柄有翅；茎上部叶卵形或肾形，苞叶状，具膨大的鞘状柄；苞片卵状披针形。头状花序组成总状花序总苞宽钟状；苞片10或11个，2层；舌花黄色；冠毛白色，与管状花花冠等长。花期7-8月。生海拔2700-3600米的草地及草丛中。产云南西北部和四川西南部。

Perennial herbs, to 150 cm tall. Basal leaves ovate-cordate, petiolate, winged. Middle stem leaves shortly petiolate, winged; distalmost stem leaves ovate or reniform, bracteal, with enlarged sheathed petioles; leaflike bracts ovate-lanceolate. Capitula in racomose; involucre broadly campanulate; phyllaries 10 or 11, in 2 rows; ray florets yellow; pappus white, as long as tubular corolla. Fl. Jul-Aug. Grassy slopes at 2700-3600 m. Distributed in NW Yunnan and SW Sichuan.

亚东橐吾

Ligularia atkinsonii (C. B. Clarke) S. W. Liu

多年生草本。基生叶肾形或心状卵形，具柄；茎中上部叶与下部叶同形，具柄。总状花序长；苞片线形；头状花序组成总状花序，辐射状；总苞钟形；总苞片8-9，2层；舌状花黄色；冠毛白色比花冠稍短。花期6-7月。生海拔3000-3500米的水边及林下草地。产西藏南部。不丹和印度北部亦有。

Perennial herbs. Basal leaves reniform or cordate-ovate, petiolate; middle to distal stem leaves similar but smaller, petiolate. Synflorescence racemose; leaflike bracts linear; capitula numerous; involucre campanulate; phyllaries 8 or 9, in 2 rows; ray florets yellow; pappus white, slightly shorter than tubular corolla. Fl. Jun-Jul. Stream banks, forests at 3000-3500 m. Distributed in S Xizang. Also in Bhutan and N India.

黑紫橐吾

Ligularia atroviolacea (Franch.) Hand.-Mazz.

多年生草本，高25-60厘米。丛生叶及茎下部叶卵状心形，具长柄，基部具狭鞘，叶脉羽状。头状花序4-10，排列成伞房状花序；苞片及小苞片钻形；总苞狭钟形或狭陀螺形；总苞片10-12，2层；小花全部管状，黄色；冠毛淡黄色，与管状花花冠近等长。花果期8-12月。生海拔3000-4000米的林下、草甸。产云南和四川。

Perennial herbs, 25-60 cm tall. Basal leaves ovate-cordate, petiolate, base narrowly sheathing, pinnately veined. Capitula 4-10, in corymb; leaflike and supplementary bracts subulate; involucre narrowly campanulate or turbinate; phyllaries 10-12, in 2 rows; florets all tubular, yellow; pappus yellowish, as long as tubular corolla. Fl. and fr. Aug-Dec. Forests, grasslands at 3000-4000 m. Distributed in Yunnan and Sichuan.

黑紫橐吾 *Ligularia atroviolacea*

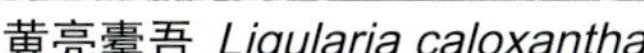
黄亮橐吾 *Ligularia caloxantha*

密花橐吾 *Ligularia confertiflora*

黄亮橐吾

Ligularia caloxantha (Diels) Hand.-Mazz.

多年生草本，高40-115厘米。中下部叶三角状或卵状心形，具翼柄；茎上部叶无柄，卵形或圆形。总状花序疏离；苞片卵状披针形；头状花序10-25；总苞宽钟形；总苞片8-10，2层；舌状花黄色；冠毛淡黄色与管部等长。花期7-10月。生海拔1600-4000米的草坡、水边及山顶草地。产四川西南部和云南西北部。

Perennial herbs, 40-115 cm tall. Basal and middle leaves triangular or ovate-cordate, petiolate, winged; distal stem leaves sessile, ovate or orbicular. Synflorescence racemose, lax; leaflike bracts ovate-lanceolate; capitula 10-25; involucre broadly campanulate; phyllaries 8-10, in 2 rows; ray florets yellow; pappus pale yellow, as long as tube of tubular corolla. Fl. Jul-Oct. Stream banks, grassy slopes, alpine grasslands at montane summits at 1600-4000 m. Distributed in SW Sichuan and NW Yunnan.

密花橐吾

Ligularia confertiflora C. C. Chang

多年生草本，高60-100厘米。基生叶卵状心形或肾状心形，具柄，基部鞘状；茎中上部叶与下部者相似而渐小，柄短。总状花序组成总状花序；苞片狭披针形；总苞狭筒形；总苞状5-6个，2层；小花5-14，全部管状；冠毛淡黄色，与管状花花冠管部近等长。花果期8-10月。生海拔3200-3300米的草坡和林中。产云南西北部。

Perennial herbs, 60-100 cm tall. Basal leaves ovate-cordate or reniform-cordate, petiolate, base sheathed; middle to distal stem leaves similar but smaller, shortly petiolate. Synflorescence racemose; leaflike bracts narrowly lanceolate; involucre narrowly cylindric; phyllaries 5 or 6, in 2 rows; florets 5-14, all tubular; pappus pale yellow. Fl. and fr. Aug-Oct. Grassy slopes and forests at 3200-3300 m. Distributed in NW Yunnan.

垂头橐吾

Ligularia cremanthodioides Hand.-Mazz.

多年生草本，高15-60厘米。茎直立。基生叶肾形，有柄；茎生叶柄基部膨大成鞘状；中上部叶只有膨大的叶鞘。头状花序2-11，呈伞房花序，稀单生；苞片及小苞片线形；总苞半球形或钟形；总苞片10-13，2层；舌花黄色，管花多数。花期7-8月。生海拔3600-4000(-5300)米的山谷林下和岩石下。产西藏和云南。尼泊尔亦有。

Perennial herbs, 15-60 cm tall. Basal leaves reniform, petiolate; stem leaves petiolar base broadly sheathed; middle to distal leaves only base broadly sheathed. Capitula 2-11, in corymb or rarely solitary; leaflike and supplementary bracts linear; involucre hemispheric or campanulate; phyllaries 10-13, in 2 rows; ray florets yellow; tubular florets numerous. Fl. Jul-Aug. Forests of valleys and rocky areas at 3600-4000 (-5300) m. Distributed in Xizang and Yunnan. Also in Nepal.

垂头橐吾 *Ligularia cremanthodioides*

浅苞橐吾 *Ligularia cyathiceps*

网脉橐吾 *Ligularia dictyoneura*

浅苞橐吾

Ligularia cyathiceps Hand.-Mazz.

多年生草本，高57-90厘米。基生叶宽卵状心形或肾形，具柄；茎中部叶肾状心形，具短柄，鞘膨大。总状花序疏散；叶状总苞紫红色，卵状披针形；总苞浅杯状；总苞片9-13，2层；舌状花黄色；冠毛浅黄色，与管状花花冠等长。花期7-8月。生海拔3000-4000米的河边、谷地及草坡。产云南。

Perennial herbs, 57-90 cm tall. Basal leaves broadly ovate or reniform, petiolate; middle stem leaves reniform-cordate, shortly petiolate, sheath enlarged. Synflorescence racemose, lax; leaflike bracts purplish red, ovate-lanceolate; involucre shallowly cupular; phyllaries 9-13, in 2 rows; ray florets yellow; pappus yellowish, as long as tubular corolla. Fl. Jul-Aug. Stream banks, valleys and grassy slopes at 3000-4000 m. Distributed in Yunnan.

网脉橐吾

Ligularia dictyoneura (Franch.) Hand.-Mazz.

多年生草本，全株灰绿色。茎直立，紫红色。丛生叶卵形、长圆形或近圆形，具柄；茎上部叶无柄，卵状披针形至线形。头状花序多数，组成总状花序；苞片及小苞片线形；总苞片6-8，2层；舌状花黄色，4-6；冠毛黄白色，与管状花花冠等长。花期6-8月。生海拔1900-3600米的水边、林下、灌丛及山坡草地。产云南、四川和西藏。

Perennial herbs, plants grayish green. Stem erect, purplish red. Basal leaves oblong, ovate or suborbicular, petiolate; distalmost stem leaves sessile, ovate-lanceolate to linear. Synflorescence racemose; leaflike and supplementary bracts linear; phyllaries 6-8, in 2 rows; ray florets 4-6, yellow; pappus yellowish white, as long as tubular corolla. Fl. Jun-Aug. Stream banks, forests, grassy slopes and scrubs at 1900-3600 m. Distributed in Yunnan, Sichuan and Xizang.

盘状橐吾

Ligularia discoidea S. W. Liu

多年生草本，高35-40厘米。基生叶长圆形或长圆状倒卵形，基部渐狭成翅状的柄，叶脉羽状；茎中上部叶无柄；最上部叶长圆形，筒状抱茎。头状花4-8个排列成伞房状；苞片及小苞片线形；总苞宽陀螺形；总苞片10-12，2层；小花全部管状，黄色；冠毛浅黄色，与管状花花冠等长。花果期8-9月。生海拔约4300米的灌丛中。产西藏。

盘状橐吾 *Ligularia discoidea*

Perennial herbs, 35-40 cm tall. Basal leaves oblong-obovate or oblong, pinnately veined, base narrowed into a winged petioles; middle to distal stem leaves sessile; distalmost stem leaves oblong, tubular-amplexicaul. Capitula 4-8, in corymb; leaflike and supplementary bracts linear; involucre broadly turbinate; phyllaries 10-12, in 2 rows; florets all tubular, yellow; pappus yellowish, as long as tubular corolla. Fl. and fr. Aug-Sep. Scrub at ca. 4300 m. Distributed in Xizang.

太白山橐吾

Ligularia dolichobotrys Diels

多年生草本，高20-80厘米。基生叶心状戟形，有柄，基部具窄鞘；茎中部叶肾形，叶柄短，具膨大的鞘。总状花序；苞片及小苞片钻形；头状花序组成总状花序；总苞狭圆筒形；总苞片4-5，2层；舌状花2-3，黄色；管状花2-3；冠毛褐色，与管状花花冠等长。花果期6-8月。生海拔2000-3300米的河边、山坡、林下及岩石上。产陕西。

太白山橐吾 *Ligularia dolichobotrys*

Perennial herbs, 20-80 cm tall. Basal leaves cordate-hastate, petiolate, base narrowly sheathed; middle stem leaves reniform, shortly petiolate, sheath enlarged. Synflorescence racemose; leaflike and supplementary bracts subulate; involucre narrowly cylindric; phyllaries 4 or 5, in 2 rows; ray florets 2 or 3, yellow; tubular florets 2 or 3; pappus brown, as long as tube of tubular corolla. Fl. and fr. Jun-Aug. Stream banks, grassy slopes, forests and at base of rocks at 2000-3300 m. Distributed in Shaanxi.

大黄橐吾

Ligularia duciformis (C. Winkl.) Hand.-Mazz.

多年生草本，高达170厘米。基生叶与中部茎叶肾形或心形，具柄，基部具极为膨大的鞘；最上部叶仅有叶鞘。头状花序组成复伞房状聚伞花序，分枝开展；苞片与小苞片极小；总苞片5，2层；小花管状，5-7，黄色。花果期7-9月。生海拔1900-4300米的河边、林下、草地及高山草地。产云南、四川、甘肃和宁夏。

Perennial herbs, to 170 cm tall. Basal leaves and middle stem leaves reniform or cordate, petiolate, base enlarged sheathed; distalmost stem leaves only sheathed. Capitula numerous, in compound corymbs; leaflike and supplementary bracts minute; phyllaries 5, in 2 rows; florets all tubular, 5-7, yellow. Fl. and fr. Jul-Sep. Stream banks, forests, grasslands and alpine meadows at 1900-4300 m. Distributed in Yunnan, Sichuan, Gansu and Ningxia.

紫花橐吾

Ligularia dux (C. B. Clarke) Ling

多年生草本，高达90厘米。基生叶圆肾形或肾形，具柄，基部具窄鞘；茎叶肾形，具柄，基部鞘状抱茎。头状花序5-8或更多，组成复伞房状聚伞花序；苞片和小苞片线形；总苞狭筒形；总苞片4-5，2层；小花管状，4或5，紫红色。花果期8月。生海拔3200-4200米的草丛中及林下。产西藏东南部。印度东北部和缅甸北部亦有。

Perennial herbs, to 90 cm tall. Basal leaves orbicular-reniform or reniform, petiolate; stem leaves reniform, petiolate, sheath amplexicaul. Capitula 5-8 or more, in compound corymbs; leaflike and supplementary bracts linear; involucre narrowly cylindric; phyllaries 4 or 5, in 2 rows; florets all tubular, 4 or 5, purplish red. Fl. and fr. Aug. Grassy slopes and forests at 3200-4200 m. Distributed in SE Xizang. Also in NE India and N Myanmar.

大黄橐吾 *Ligularia duciformis*

紫花橐吾 *Ligularia dux*

小紫花橐吾 *Ligularia dux* var. *minima*

异叶橐吾 *Ligularia heterophylla*

小紫花橐吾

Ligularia dux var. **minima** S. W. Liu

多年生草本，高25-40厘米。茎细长，直径1.5-2.5毫米。叶肾形，基部近截形。头状花序5-8个，成伞房花序。生海拔3200-4200米的草坡。产西藏东南部。

Perennial herbs, 25-40 cm tall. Stem slender, 1.5-2.5 mm diam. Leaves reniform, base subtruncate. Capitula 5-8, in corymb. Grassy slopes at 3200-4200 m. Distributed in SE Xizang.

异叶橐吾

Ligularia heterophylla Rupr.

多年生草本，植株灰绿色，高20-200厘米。茎基部叶卵状长圆形或椭圆形，具柄，有宽翅，基部膨大成鞘；茎生叶无柄，向上渐小。头状花序组成圆锥状总状花序；苞片及小苞片线状钻形或丝状。总苞片6-8(10)，2层。舌状花(4)5-7，黄色。花期6-8月。生海拔2200-2500米的山坡草地、溪边和湿地。产新疆。中亚亦有。

Perennial herbs, plants grayish green, 20-200 cm tall. Basal leaves ovate-oblong or oblong, petiolate, narrowly winged, base sheathed. Middle to distal stem leaves sessile, smaller distally. Synflorescence racemose-paniculate; leaflike and supplementary bracts linear-subulate or filiform. Phyllaries 6-8 (-10), in 2 rows. Ray florets (4 or) 5-7, yellow. Fl. Jul-Aug. Slopes, grasslands, stream banks, swamps at 2200-2500 m. Distributed in Xinjiang. Also in C Asia.

细茎橐吾

Ligularia hookeri (C. B. Clarke) Hand.-Mazz.

多年生草本。基生叶具柄，基部鞘状，叶脉掌状；茎中部叶1，具短柄；最上部叶1，苞片状，舟形。头状花序单生或2-7(-16)，排列成总状花序；总苞片8-10，2层；舌状花黄色；冠毛褐色或淡褐色。花果期5-9月。生海拔3000-4500米的山坡、灌丛、林中、水边及高山草甸。产中国西南和陕西。喜马拉雅亦有。

Perennial herbs. Basal leaves petiolate, base sheathed, palmately veined; middle stem leaf solitary, shortly petiolate; distalmost stem leaf solitary, bracteal, cymbiform. Capitula solitary or 2-7(-16), in raceme; phyllaries 8-10, in 2 rows; ray florets yellow; pappus brown or pale brown. Fl. and fr. May-Sep. Grassy slopes, scrub, forests, stream banks, alpine meadows at 3000-4500 m. Distributed in SW China and Shaanxi. Also in Himalaya.

细茎橐吾 *Ligularia hookeri*

沼生橐吾

Ligularia lamarum (Diels ex H. Limpr.) C. C. Chang

多年生草本，高37-52厘米。基生叶具柄，基部鞘状，叶脉掌状；中上部叶具短柄，鞘膨大抱茎。头状花序组成总状花序，近穗状或疏离；苞片线形；总苞钟状陀螺形；总苞片6-8，2层；舌花5-8，黄色；冠毛淡黄色，稍短于花冠。花果期7-9月。生海拔3300-5300米的沼泽地、潮湿草地、灌丛及林下。产中国西南和甘肃。缅甸北部亦有。

Perennial herbs, 37-52 cm tall.

沼生橐吾 *Ligularia lamarum*

上部叶无柄，长圆形至披针形。头状花序组成总状花序；苞片线形；总苞陀螺形，黑灰色；总苞片7-8，2层；舌状花8-12，黄色；冠毛黄白色，与管状花花冠等长。花果期8-9月。生海拔3400-4000米的林缘、林下及草坡。产四川西部和云南西北部。

Perennial herbs, to 100 cm tall, plants grayish green. Basal leaves shortly petiolate, winged, base enlarged sheathed, pinnately veined; middle to distal stem leaves sessile, oblong to lanceolate. Synflorescence in rax racemose; leaflike bracts linear; involucre blackish gray, turbinate; phyllaries 7-8, in 2 rows; ray florets 8-12, yellow; pappus yellowish white, as long as tubular corolla. Fl. and fr. Aug-Sep. Forest margins, forests and grassy slopes at 3400-4000 m. Distributed in W Sichuan and NW Yunnan.

心叶橐吾

Ligularia microcardia
Hand.-Mazz.

多年生草本，高达45厘米。基生叶心状卵形或卵形，具柄，基部具鞘，叶脉羽状；茎中上部叶具短柄，鞘膨大，近舟形。头状花序2-6，排列成伞房状花序；苞片线形，小苞片缺少；总苞宽陀螺形；总苞片约13，2层；小花多数，黄色，全部管状。花果期9-10月。生海拔3300-4000米的沟边、山坡草地。产云南和四川。

Perennial herbs, to 45 cm tall. Basal leaves ovate-cordate or ovate, petiolate, pinnately veined; middle to distal stem leaves shortly petiolate, sheath enlarged, cymbiform. Capitula 2-6, in corymb; leaflike bracts linear; supplementary bracts absent; involucre broadly turbinate; phyllaries ca. 13, in 2 rows; florets numerous, all tubular, yellow. Fl. and fr. Sep-Oct. Stream banks, grassy slopes at 3300-4000 m. Distributed in Yunnan and Sichuan.

Basal leaves petiolate, sheathed, palmately veined; middle to distal stem leaves shortly petiolate, sheath enlarged, amplexicaul. Synflorescence racemose, spicate or lax; leaflike bracts linear; involucre campanulate-turbinate; phyllaries 6-8, in 2 rows; ray florets 5-8, yellow; pappus yellowish, slightly shorter than tubular corolla. Fl. Jul-Aug. Fr. Aug-Sep. Swamps, wet grasslands, scrub or forests at 3300-5300 m. Distributed in SW China and Gansu. Also in N Myanmar.

黑苞橐吾

Ligularia melanocephala
(Franch.) Hand.-Mazz.

多年生草本，高达100厘米，植株灰绿色。基生叶具翼柄，基部具膨大的鞘，叶脉羽状；茎中

黑苞橐吾 *Ligularia melanocephala*

心叶橐吾 *Ligularia microcardia*

木里橐吾 *Ligularia muliensis*

马蹄叶橐吾 *Ligularia odontomanes*

木里橐吾

Ligularia muliensis Hand.-Mazz.

多年生草本，高28-32厘米。茎生叶长圆形或卵状长圆形，向上渐小，叶脉羽状。头状花序4-13，组成总状花序；苞片最下部1枚为卵形，余者均为线形；总苞绿色或紫黑色；总苞片8-10，2层；舌状花8-10，黄色；冠毛白色，比管状花花冠稍短。花果期7-10月。生海拔3800-4200米的草坡、林下及灌丛中。产云南和四川。

Perennial herbs, 28-32 cm tall. Stem leaves oblong or ovate-oblong, smaller distally, pinnately veined. Capitula 4-13, in racemose; leaflike bracts linear (except lower one ovate); involucre green or blackish purple; phyllaries 8-12, in 2 rows; ray florets 8-10, yellow; pappus white, shorter than tubular corolla. Fl. and fr. Jul-Oct. Grassy slopes, forests and scrubs at 3800-4200 m. Distributed in Yunnan and Sichuan.

马蹄叶橐吾

Ligularia odontomanes Hand.-Mazz.

多年生草本，高达60厘米，全株被红黄色长柔毛。基生叶具柄，基部具窄鞘，叶脉掌状；茎中上部叶无柄，鞘膨大。头状花序组成总状花序；苞片卵形；总苞片7-8，2层；舌状花5，黄色；冠毛红黄色，短于花冠管部。花期7-8月。生海拔2500-2800米的林缘及草坡。产四川。

Perennial herbs, to 60 cm tall; plants long yellowish red pilose. Basal leaves petiolate, base narrowly sheathed, palmately veined; middle to distal stem leaves sessile, sheath enlarged. Synflorescence racemose; leaflike bracts ovate; phyllaries 7 or 8, in 2 rows; ray florets 5, yellow; pappus reddish yellow, shorter than tube of tubular corolla. Fl. Jul-Aug. Forest margins and grassy slopes at 2500-2800 m. Distributed in Sichuan.

裸柱橐吾

Ligularia petiolaris Hand.-Mazz.

多年生草本，高约30厘米。叶卵状心形，黑绿色，叶脉羽状，叶柄与叶片等长，无翅，基部有鞘。头状花序组成总状花序；苞片线形；小苞片丝状；总苞钟形；总苞片5，2层；边花雌性，1-2，花冠缺失，花柱裸露；冠毛白色，比花冠短。花期7月，果期8-9月。生海拔3600-4100米的多石草坡或灌丛。产西藏南部。

裸柱橐吾 *Ligularia petiolaris*

Perennial herbs, ca. 30 cm tall. Petioles as long as leaves, unwinged, base sheathed; leaves dark green, ovate-cordate, pinnately veined. Synflorescence racemose; leaflike bracts linear; involucre campanulate; phyllaries 5, in 2 rows; outer florets female, 1 or 2, corolla absent; style naked; pappus white, shorter than tubular corolla. Fl. Jul. Fr. Aug-Sep. Stony grassy slopes or scrub at 3600-4100 m. Distributed in S Xizang.

浅齿橐吾

Ligularia potaninii (C. Winkl.) Ling

多年生草本。基部叶具柄，叶片宽肾形，先端圆形或凹，边缘具波状圆形；中部茎叶肾形，叶柄极度膨大成鞘状；最上部茎叶仅有膨大的鞘，稀披针形。头状排列成伞房状花序；总苞陀螺形；小花黄色，全部管状；冠毛白色。花期7月。生海拔约4000米的沼泽地。产四川和甘肃。

浅齿橐吾 *Ligularia potaninii*

褐毛橐吾 *Ligularia purdomii*

Perennial herbs. Basal leaves petiolate, leaves broadly reniform, margin undulate-crenate and ciliate, apex rounded or retuse; middle stem leaves reniform, petioles enlarged into a sheath; distalmost stem leaves with only enlarged sheath or very rarely lanceolate. Capitula in corymb; involucre turbinate; florets yellow, all tubular; pappus white. Fl. Jul. Grassy areas in swamps at ca. 4000 m. Distributed in Sichuan and Gansu.

褐毛橐吾

Ligularia purdomii (Turrill) Chitt.

多年生草本。基部叶具柄，基部具长而窄的鞘，叶片肾形或圆肾形，边缘具整齐的浅齿；中部茎叶具短柄，先端深凹；最上部茎叶仅有膨大的鞘。头状花序组成复伞房花序；总苞钟状陀螺形；小花黄色，全部管状；冠毛幼时白色，老时褐色。花期6-7月。生海拔3700-4100米的河边、沼泽。产四川、青海和甘肃。

Perennial herbs. Basal leaves petiolate, base narrowly long sheathed, leaves reniform or orbicular-reniform, margin regularly dentate; middle stem leaves shortly petiolate, apex retuse; distalmost stem leaves only broadly sheathed. Capitula in compound corymbs; involucre campanulate-turbinate; florets yellow, all tubular; pappus white or brown when mature. Fl. Jun-Jul. Stream banks, swamps at 3700-4100 m. Distributed in Sichuan, Qinghai and Gansu.

巧家橐吾

Ligularia qiaojiaensis Y. S. Chen et H. J. Dong

多年生草本。基部叶具柄，叶两面异色，叶片圆心形；茎叶具短柄，基部扩大成鞘；上部叶具柄。头状花序组成总状花序；总苞绿色，陀螺形；舌状花黄色，舌片披针形，管状花钟状；冠毛褐色。花期7-8月。生海拔2700-3400米的林缘边的湿润草坡、河流边的高山草甸。产云南东北部。

Perennial herbs. Basal leaves petiolate, leaves discolorous, cordate-orbicular; stem leaves petiolate, base enlarged, sheathing; upper leaves shorter petioles. Synflorescence racemose; involucre green, turbinate; ray florets yellow, lamina lanceolate, tubular florets campanulate; pappus brown. Fl. Jul-Aug. Wet grassy slopes along forest margins, alpine meadows along streams at 2700-3400 m. Distributed in NE Yunnan.

巧家橐吾 *Ligularia qiaojiaensis*

黑毛橐吾 *Ligularia retusa*

藏橐吾 *Ligularia leesicotal*

黑毛橐吾

Ligularia retusa DC.

多年生草本。基部叶具柄，肾形，边缘具整齐的小齿；中部茎叶具短柄，鞘成舟形；最上部茎生叶仅有膨大的鞘。头状花序单生或排列成伞房花序；总苞钟形或半球形；舌状花黄色，管状花黄棕色；冠毛黄色。花果期7-10月。生海拔3800-4500米的水沟边、草坡及山顶坡地。产西藏和云南。喜马拉雅亦有。

Perennial herbs. Basal leaves petiolate, leaves reniform, margin regularly denticulate; middle stem leaves shortly petiolate, base broadly sheathed, sheath cymbiform; distalmost stem leaves only broadly sheathed. Capitula solitary or in corymb; involucre hemispheric or campanulate; ray florets yellow, tubular florets yellowish brown; pappus yellow. Fl. and fr. Jul-Oct. Stream banks, grassland slopes and alpine grasslands at 3800-4500 m. Distributed in Xizang and Yunnan. Also in Himalaya.

藏橐吾

Ligularia leesicotal Kitam.

多年生草本。基部叶具柄，卵状长圆形；中上部茎叶无柄，卵形或卵状披针形；最上部茎叶披针形或线状披针形，近全缘。头状花序组成伞房花序或圆锥花序；总苞陀螺形或钟状陀螺形；舌状花黄色；冠毛白色或浅褐色。花果期7-10月。生海拔3700-4500米的湖边、林下、灌丛及山坡。产四川和西藏。尼泊尔亦有。

Perennial herbs. Basal leaves petiolate, leaves ovate-oblong; middle to distal stem leaves without petioles, leaves ovate or ovate-lanceolate; distalmost stem leaves lanceolate or linear-lanceolate, margin subentire. Capitula in compound corymbs or panicles; involucre turbinate or campanulate-turbinate; ray florets yellow; pappus white or pale brownish. Fl. and fr. Jul-Oct. Lake shores, forests, scrub, slopes at 3700-4500 m. Distributed in Sichuan and Xizang. Also in Nepal.

纤细橐吾

Ligularia tenuicaulis C. C. Chang

多年生草本。基部叶具柄，基部具窄鞘，叶片卵状心形或近肾形，边缘具不整齐的三角状大齿；中部茎叶具柄，叶片肾形，先端圆形，边缘具小齿；最上部茎生叶卵形，不抱茎。头状花序组成复伞房花序；总苞圆柱形；小花黄色，管状；冠毛下部淡褐色，上部褐色。花果期9-10月。生海拔3200-4500米的草地及灌丛中。产云南。

Perennial herbs. Basal leaves petiolate, base narrowly sheathed, leaves ovate-cordate or subreniform, margin coarsely irregularly triangular-dentate; middle stem leaves petiolate, leaf blade reniform, margin denticulate, apex rounded; distalmost stem leaves ovate, base not amplexicaul. Capitula in compound corymbs; involucre cylindric; florets yellow, all tubular; pappus pale brown below, brown above. Fl. and fr. Sep-Oct. Grasslands and scrub and 3200-4500 m. Distri-

纤细橐吾 *Ligularia tenuicaulis*

簇梗橐吾 *Ligularia tenuipes*

buted in Yunnan.

簇梗橐吾

Ligularia tenuipes (Franch.) Diels

多年生草本。基部叶具柄，被白色柔毛，基部鞘状，叶片心形或宽卵状心形，边缘具整齐的齿；中部叶具短柄，柄有翅，鞘略膨大，半抱茎；上部叶无柄，卵状披针形。头状花序组成总状花序；总苞陀螺形；舌状花黄色，舌片线形；冠毛污褐色。花果期7-9月。生海拔2200-3200米的水边、山坡湿地及草坡。产四川、重庆和湖北西部。

Perennial herbs. Basal leaves petiolate, white puberulent, base sheathed, leaf blade cordate or broadly ovate-cordate, margin regularly dentate and ciliate; middle stem leaves shortly petiolate, petioles winged, sheath slightly enlarged, semiamplexicaul; distal stem leaves sessile, ovate-lanceolate. Synflorescence racemose; involucre turbinate; ray florets yellow, lamina linear; pappus dirty brown. Fl. Jul-Sep. Stream banks, wet slopes, grassy slopes at 2200-3200 m. Distributed in Sichuan, Chongqing and W Hubei.

塔序橐吾

Ligularia thyrsoidea (Ledeb.) DC.

多年生草本。基部叶具柄，叶两面异色，卵状三角形、箭形或三角形，基部心形，边缘具不整齐的齿；中上部叶具短柄或无柄，卵状三角形至线状披针形。头状花序组成圆锥状复伞房状花序；总苞杯状；舌状花黄色；冠毛白色，与管状花花冠等长。花果期7-8月。生海拔500-2000的湿地和溪岸。产新疆。中亚、蒙古西部和俄罗斯亦有。

Perennial herbs. Basal leaves petiolate, leaf blade discolorous, triangular, sagittate or triangular-ovate, base cordate, margin irregularly dentate; middle to distal stem leaves shortly petiolate or sessile, triangular-ovate to linear-lanceolate. Capitula in compound corymbs, pyramidal, paniculate; involucres copular; ray florets yellow; pappus white, as long as tubular corolla. Fl. and fr. Jul-Aug. Wet grasslands and stream banks at 500-2000 m. Distributed in Xinjiang. Also in C Asia, W Mongolia and Russia.

天山橐吾

Ligularia tianschanica Chang Y. Yang et S. L. Keng

多年年生草本。丛生叶与茎下部叶具柄，圆心形，基部心形，边缘具波状齿或尖锯齿，先端圆形；末端茎具短柄，卵圆形或三角形，基部耳状抱茎；最末端叶无柄，苞片状。头状花序组成伞房状花序；总苞半球形或杯状，直径15-20毫米；苞片2列；舌状花黄色；冠毛白色，与管状花花冠等长。花果期6-7月。生海拔2400-2700米的亚高山草甸。产新疆。

Perennial herbs. Basal and proximal stem leaves petiolate, leaf blade orbicular-reniform, base cordate, margin irregularly sharply dentate, apex obtuse; distal stem leaves shortly petiolate, leaf blade ovate or triangular, base auriculate-amplexicaul; distalmost leaves sessile, bracteal. Capitula in subcorymb; involucres campanulate or hemispheric, 15-20 mm diam; phyllaries in 2 rows; ray florets yellow; pappus white, as long as tubular corolla. Fl. and fr. Jun-Jul. Subalpine meadows at 2400-2700 m. Distributed in Xinjiang.

塔序橐吾 *Ligularia thyrsoidea*

天山橐吾 *Ligularia tianschanica*

东俄洛橐吾 *Ligularia tongolensis*

棉毛橐吾 *Ligularia vellerea*

东俄洛橐吾

Ligularia tongolensis (Franch.) Hand.-Mazz.

多年生草本。基部叶具柄，卵状心形或卵状长圆形，边缘具细齿，先端圆形；中部至末端茎叶具短柄。头状花序排成伞房花序或单生；总苞钟状，直径5-7毫米；舌状花黄色；瘦果浅褐色，圆柱形；冠毛浅褐色。花果期7-8月。生海拔2100-4000米的林缘、林中落叶层、河谷、灌丛、高山草甸。产中国西南。

Perennial herbs. Basal leaves petiolate, leaf blade ovate-cordate or ovate-oblong, margin denticulate, apex obtuse; middle to distal stem leaves shortly petiolate. Capitula in corymb or solitary; involucre campanulate, 5-7 mm diam; ray florets yellow; achenes pale brown, cylindric; pappus pale brown. Fl. and fr. Jul-Aug. Forest margins, forests, wet valleys, scrub, alpine meadows at 2100-4000 m. Distributed in SW China.

棉毛橐吾

Ligularia vellerea (Franch.) Hand.-Mazz.

多年生草本。叶全部基生，具柄，叶形卵形、椭圆形或近圆形，边缘具整齐的细齿；花葶无苞叶。头状花序组成总状花序；总苞钟形；舌状花黄色。瘦果狭倒披针形，光滑；冠毛淡黄色。瘦果黄色。花果期6-9月。生海拔2100-4600米的水边、林下及草坡。产四川西南部和云南西北部。

Perennial herbs. Leaves all basal, petiolate, leaf blade ovate, elliptic or suborbicular, margin regularly denticulate; scape without bracteal leaf. Synflorescence racemose; involucre campanulate; ray florets yellow. Achenes yellowish, narrowly oblanceolate; pappus pale yellow. Achene yellow Fl. and fr. Jun-Sep. Stream banks, forests and grassy slopes at 2100-4600 m. Distributed in SW Sichuan and NW Yunnan.

齿叶橐吾

Ligularia dentata (A. Gray) H. Hara

多年生草本。基生叶的叶柄长22-60厘米，基部具鞘；叶肾形，边缘具规则锯齿。头状花序组成开展的伞房状或复伞房状；总苞半球形；苞片8-14；舌状花黄色。瘦果棕色，圆柱形，具纵棱；冠毛棕红色。花果期7-10月。生海拔700-3200米的草坡、河边、林缘、林下，或有时栽培于花园中。产中国西南、东南、华中、华北、华西和华东。缅甸、越南和日本亦有；欧洲有栽培。

Perennial herbs. Petioles of basal leaves petiolate 22-60 cm long, base sheathed; leaves reniform, margin regularly dentate. Inflorescences corymbose or compound corymbose, spreading; involucres hemispherical; phyllaries 8-14; ligulate florets yellow. Achenes brown, cylindric, ribbed; pappus reddish brown. Fl. and fr. Jul-Oct. Grassy slopes, riversides, forest edges, forests at 700-3200 m, or sometimes culti-

齿叶橐吾 *Ligularia dentata*

大头橐吾 *Ligularia japonica*

vated in gardens. Distributed in SW, SE, C, N, W and E China. Also in Myanmar, Vietnam and Japan; cultivated in Europe.

大头橐吾

Ligularia japonica (Thunb.) Less.

多年生草本。从生叶和茎下部叶肾形，掌状3-5全裂，再作掌状浅裂，叶脉掌状，灰绿色，具紫斑。头状花序大，2-8，排成伞房状花序；总苞半球形；总苞片宽长圆形，被白色柔毛，背部隆起，先端三角形。瘦果细圆柱形；冠毛红褐色，与管状花花冠管部等长。花果期4-9月。生海拔600-2300米的河岸、草坡或林下。产华南、华中和华东。印度、朝鲜半岛和日本亦有。

Perennial herbs. Rosette and lower cauline leaves reniform, palmately 3-5-sected, lobes palmately lobed again, nerves palmate, glaucous, purple spotted. Capitula large, 2-8, arranged in corymb; involucres semiglobose; phyllaries broadly oblong, white pubescent, keeled adaxially, apex triangular. Achenes narrowly cylindric; pappus reddish-brown, as long as tubular corolla. Fl. and fr. Apr-Sep. River banks, grassy slopes or forests at 600-2300 m. Distributed in S, C and E China. Also in India, Korean Peninsula and Japan.

鹿蹄橐吾

Ligularia hodgsonii Hook. f.

多年生草本，被柔毛。丛生叶和茎下部叶肾形或心状肾形，边缘具三角状齿或圆齿，掌状脉。头状花序单生至多数，排成伞房状花序；总苞宽钟形；总苞片长圆形，紫红色，背部隆起；舌状花黄色，长圆形。瘦果圆柱形；冠毛红褐色，与管状花花冠等长。花果期7-10月。生海拔900-2800米的河边、山坡草地或林下。产云南、四川、贵州、广西、湖北、陕西和甘肃。俄罗斯和日本亦有。

Perennial herbs, pubescent. Rosette and lower cauline leaves reniform or cordate-reniform, margin triangular-toothed or crenate, palmately nerved. Capitula solitary to numerous, arranged in corymb; involucres broadly campanulate; phyllaries oblong, purple, outside keeled; ray florets yellow, oblong. Achenes cylindric; pappus purplish-brown, as long as tubular corolla. Fl. and fr. Jul-Oct. Riversides, grassy slopes or forests at 900-2800 m. Distributed in Yunnan, Sichuan, Guizhou, Guangxi, Hubei, Shaanxi and Gansu. Also in Russia and Japan.

鹿蹄橐吾 *Ligularia hodgsonii*

裂舌橐吾 *Ligularia stenoglossa*

裂舌橐吾

Ligularia stenoglossa (Franch.) Hand.-Mazz.

多年生草本。基生叶叶柄长达70厘米；叶缘具三角状齿。复伞房花序开展；头状花序多数；总苞狭钟形；苞片7-10；舌状花3-5，黄色，3-5裂；两性花多至10朵，棕黄色。瘦果棕色，圆柱形；冠毛红棕色，比管状花花冠稍短。花果期9-11月。生海拔2100-4000米的林下或山坡草地。产云南西部。

Perennial herbs. Petioles of basal leaves to 70 cm long; leaves margin triangular-dentate. Corymbs compound, spreading; capitula numerous; involucres narrowly campanulate; phyllaries 7-10; ligulate florets 3-5, yellow; ligules 3-5-divided; tubular florets more numerous, to 10, yellowish brown. Achenes brown, cylindric; pappus reddish brown, slightly shorter than tubular corolla. Fl. and fr. Sep-Nov. Forests or grasslands slopes at 2100-4000 m. Distributed in W Yunnan.

刚毛橐吾

Ligularia achyrotricha (Diels) Y. Ling

多年生草本。叶片肾形，掌状叶脉，边缘有粗锯齿。伞房花序开展；总苞狭钟形或圆柱形，外面被黄棕色柔毛；总苞片8-10，2层，外面被密而短的黄柔毛；小花10-16，黄色。瘦果狭圆柱形，长7-9毫米；冠毛白色或浅黄色，与管状花花冠管部等长。花果期6-8月。生海拔3300-3700米的草坡或林缘。产陕西(秦岭)。

Perennial herbs. Leaf blade reniform, palmately veined, margin irregularly coarsely dentate. Compound corymbs spreading; involucre narrowly campanulate or cylindric, outside yellowish brown pubescent; phyllaries 8-10, in 2 rows, outside densely shortly yellow pilose; florets 10-16, yellow. Achenes narrowly cylindric, 7-9 mm; pappus white or lower yellow, as long as tube of tubular corolla. Fl. and fr. Jun-Aug. Grassy slopes or forest margins at 3300-3700 m. Distributed in Shaanxi (Qinling Mountain).

莲叶橐吾

Ligularia nelumbifolia (Bureau et Franch.) Hand.-Mazz.

多年生草本。基生叶叶柄具白色蛛丝状柔毛，长10-50厘米；基部扩大成鞘状；叶盾状，肾形，边缘具锐锯齿。复伞房花序开展；头状花序多数；总苞狭圆柱形；苞片5-7，2层；小花管状具6-8花；冠毛白色或黄色。花期7-9月。生海拔2400-3900米的林下、山坡或高山草甸。产云南、四川、湖北和甘肃。

Perennial herbs. Petioles of basal leaves white arachnoid-puberulous, 10-50 cm long, base enlarged sheathed; leaves peltate, reniform, margin sharply dentate. Compound corymbs spreading; capitula numerous; involucres narrowly cylindric; phyllaries 5-7, in 2 rows; florets 6-8; pappus white or yellowish. Fl. Jul-Sep. Forests, slopes or alpine

刚毛橐吾 *Ligularia achyrotricha*

莲叶橐吾 *Ligularia nelumbifolia*

云南橐吾 *Ligularia yunnanensis*

meadows at 2400-3900 m. Distributed in Yunnan, Sichuan, Hubei and Gansu.

云南橐吾

Ligularia yunnanensis (Franch.) C. C. Chang

多年生草本。基生叶2；叶片肾形或圆肾形，掌状叶脉，边缘粗三角状锯齿。头状花序9-14，伞房状排列；总苞狭钟形，直径5-7毫米；总苞片5-8，2层；小花6-20，黄色，全部为管状花；冠毛白色，与管状花花冠等长。花果期5-10月。生海拔3100-4000米的草坡、林下或岩石间。产云南西北部。

Perennial herbs. Basal leaves 2; leaf blade reniform or orbicular-reniform, palmately veined, margin coarsely triangular-dentate. Capitula 9-14, in corymb; involucres narrowly campanulate, 5-7 mm diam; phyllaries 5-8, in 2 rows; florets 6-20, yellow, all tubular; pappus white, as long as tubular corolla. Fl. and fr. May-Oct. Grassy slopes, forests or between rocks at 3100-4000 m. Distributed in NW Yunnan.

舟叶橐吾

Ligularia cymbulifera (W. W. Sm.) Hand.-Mazz.

多年生草本，被白色柔毛。丛生叶和茎下部叶椭圆形或卵状长圆形，叶柄具宽翅；茎中部叶无柄，舟形，鞘状抱茎。头状花序50个以上，组成具多数分枝的大型复伞房状花序；总苞钟形；总苞片披针形，边缘膜质。瘦果狭长圆形；冠毛淡黄色或白色。花果期7-9月。生海拔2900-4800米的荒地、草坡、林缘、灌丛、草甸或河边。产云南和四川。

Perennial herbs, white pubescent. Rosette and lower cauline leaves elliptic or ovate-oblong, petioles broadly winged; middle cauline leaves sessile, cymbiform, sheathlike amplexicaul. Capitula more than 50, arranged in large multi-branched compound corymb; involucres campanulate; phyllaries lanceolate, margin membranous. Achenes narrowly oblong; pappus yellowish or white. Fl. and fr. Jul-Sep. Wastelands, grassy slopes, forest edges, thickets, meadows or riversides at 2900-4800 m. Distributed in Yunnan and Sichuan.

舟叶橐吾 *Ligularia cymbulifera*

牛蒡叶橐吾 *Ligularia lapathifolia*

橐吾 *Ligularia sibirica*

牛蒡叶橐吾

Ligularia lapathifolia (Franch.) Hand.-Mazz.

多年生草本，被白色蛛丝状柔毛。丛生叶和茎下部叶卵形或卵状长圆形，叶脉羽状，叶柄长，基部具鞘；茎中上部叶无柄，鞘状抱茎。头状花序6-23，排成伞房状花序；总苞半球形或宽钟形；总苞片近革质，披针形，缘花黄色。瘦果长圆形；冠毛红褐色，与管状花花冠等长。花果期7-10月。生海拔1800-3300米的草坡、林下或灌丛。产云南和四川。

Perennial herbs, white arachnoid villose. Rosette and lower cauline leaves ovate or ovate-oblong, nerves pinnate, petioles long, base sheathlike; middle cauline leaves sessile, sheathlike amplexicaul. Capitula 6-23, arranged in corymb; involucres semiglobose or broadly campanulate; phyllaries subcoriaceous, lanceolate; ray florets yellow. Achenes oblong; pappus reddish-brown, as long as tubular corolla. Fl. and fr. Jul-Oct. Grassy slopes, forests or thickets at 1800-3300 m. Distributed in Yunnan and Sichuan.

橐吾

Ligularia sibirica (L.) Cass.

多年生草本。基生叶叶柄长14-39厘米，边缘具规则锯齿，卵状心形、三角状心形、肾状心形或阔心形。头状花序组成总状花序，常簇生；头状花序多数；总苞常钟形；苞片7-12，2层；舌状花黄色。瘦果棕色，圆柱形；冠毛黄色。花果期7-10月。生海拔至2200米的沼泽、湿草地或溪边。产内蒙古、黑龙江和吉林。俄罗斯、蒙古和欧洲亦有。

Perennial herbs. Petioles of basal leaves 14-39 cm long, margin regularly dentate, ovate-cordate, triangular-cordate, reniform-cordate or broadly cordate. Inflorescences racemose, often clustered; capitula numerous; involucres often campanulate; phyllaries 7-12, in 2 rows; ray florets yellow. Achenes brown, cylindric; pappus yellowish. Fl. and fr. Jul-Oct. Swamps, wet grasslands or stream banks at up to 2200 m. Distributed in Neimenggu, Heilongjiang and Jilin. Also in Russia, Mongolia and Europe.

阿勒泰橐吾

Ligularia altaica DC.

多年生草本。基生叶的叶柄长13-30厘米，叶无毛，全缘。头状花序5-17(-35)，组成总状花序；总苞钟形或近杯状；苞片7-9，2层；内轮苞片边缘膜质；舌状花4或5，黄色；两性花10-20。瘦果棕黄色，圆柱形；冠毛白色。花果期6-8月。生海拔1100-3000米的草质山坡或林缘。产新疆北部。哈萨克斯坦、俄罗斯和蒙古亦有。

Perennial herbs. Petioles of basal leaves 13-30 cm long; leaves glabrous, margin entire. Inflorescences racemose; capitula 5-17(-35); involucres campanulate or subcupular; phyllaries 7-9, in 2 rows; inner phyllaries membranous at margin; ligulate florets 4 or 5, yellow; tubular florets 10-20. Achenes yellowish brown, cylindric; pappus white. Fl. and fr. Jun-Aug. Grassy slopes or forest edges at 1100-3000 m. Distributed in N Xinjiang. Also in Kazakhstan, Russia and Mongolia.

离舌橐吾 *Ligularia veitchiana*

阿勒泰橐吾 *Ligularia altaica*

离舌橐吾

Ligularia veitchiana (Hemsl.) Greenm.

多年生草本。丛生叶和茎下部叶三角状或卵状心形，叶脉掌状，叶柄光滑，基部具窄鞘。头状花序多数，排成总状花序；苞片大，宽卵形至卵状披针形，近膜质；总苞钟形；总苞片长圆形，被有节短柔毛；冠毛黄白色，短于管状花花冠。花期7-9月，果期8-9月。生海拔1100-3300米的河边、山坡或林下。产云南、四川、贵州、湖北、陕西和甘肃。

Perennial herbs. Rosette and lower cauline leaves triangular-cordate or ovate-cordate, nerves palmate, petioles glabrous, base narrowly sheathlike. Capitula numerous, arranged in raceme; bracts large, broadly ovate to ovate-lanceolate, submembranous; involucre campanulate; phyllaries oblong, septate-puberulent; pappus yellowish-white, shorter than tubular corolla. Fl. Jul-Sep. Fr. Aug-Sep. Riversides, slopes or forests at 1100-3300 m. Distributed in Yunnan, Sichuan, Guizhou, Hubei, Shaanxi and Gansu.

蹄叶橐吾

Ligularia fischeri (Ledeb.) Turcz.

多年生草本。丛生叶和茎下部叶光滑，肾形，边缘具锯齿，掌状脉；茎中部叶具短柄，肾形，鞘膨大。头状花序多数，排成总状花序；总苞钟形；总苞片长圆形，先端急尖。瘦果圆柱形；冠毛红褐色，短于管部。花果期7-10月。生海拔2500米以下的河岸、草甸、草坡、灌丛、林缘或林下。产中国大部分地区。喜马拉雅、日本、朝鲜半岛、蒙古和俄罗斯亦有。

Perennial herbs. Rosette and lower cauline leaves glabrous, reniform, margin serrate, palmately nerved; middle cauline leaves shortly petiolate, reniform, sheathes dilated. Capitula numerous, arranged in raceme; involucres campanulate; phyllaries oblong, apex acute. Achenes cylindric; pappus purplish-brown, shorter than tube of florets. Fl. and fr. Jul-Oct. River banks, meadows, grassy slopes, thickets, forest edges or forests below 2500 m. Distributed in most parts of China. Also in Himalaya, Japan, Korean Peninsula, Mongolia and Russia.

蹄叶橐吾 *Ligularia fischeri*

宽戟橐吾 *Ligularia latihastata*

宽戟橐吾

Ligularia latihastata (W. W. Sm.) Hand.-Mazz.

多年生草本，被柔毛。丛生叶和茎基部叶光滑，宽戟形或三角状戟形，叶脉掌状；茎中部叶有柄或无柄，鞘膨大，边缘具齿；茎上部叶鞘状。头状花序7-24，排成总状花序；总苞宽钟形；总苞片长圆形。瘦果圆柱形；冠毛淡褐色或红褐色，与管部等长。花果期7-10月。生海拔2400-4000米的水边、沼泽草地、林下或草地。产云南西北部。

Perennial herbs, pubescent. Rosette and lower cauline leaves glabrous, broadly hastate or triangular-hastate; middle cauline leaves petiolate or sessile, sheath dilated, margin serrate; upper ones sheathlike. Capitula 7-24, arranged in raceme; involucres broadly campanulate; phyllaries oblong. Achenes cylindric; pappus brownish or reddish-brown, equal to tube of florets. Fl. and fr. Jul-Oct. Watersides, marshes, forests or grasslands at 2400-4000 m. Distributed in NW Yunnan.

狭苞橐吾 *Ligularia intermedia*

窄头橐吾 *Ligularia stenocephala*

窄头橐吾

Ligularia stenocephala (Maxim.) Matsum. et Koidz.

多年生草本。基生叶叶柄长23-75厘米，基部具狭鞘；叶无毛，边缘具规则锐锯齿。花序总状；头状花序多数；总苞圆柱形；苞片5(-7)，2层；舌状花1-5朵，黄色；两性花5-10朵。瘦果纺锤形；冠毛白色或黄白色。花果期7-12月。生海拔900-3300米的河岸、草坡、林下或岩石基部。产中国西南、华南、东南、华中、华北和华东。日本亦有。

Perennial herbs. Petioles of basal leaves 23-75 cm long, base narrowly sheathed; leaves glabrous, margin regularly sharply dentate. Inflorescences racemose; capitula numerous; involucres cylindric; phyllaries 5(-7), in 2 rows; ligulate florets 1-5, yellow; tubular florets 5-10. Achenes fusiform; pappus white or yellowish white. Fl. and fr. Jul-Dec. Stream banks, grassy slopes, forests or base of rocks at 900-3300 m. Distributed in SW, S, SE, C, N and E China. Also in Japan.

狭苞橐吾

Ligularia intermedia Nakai

多年生草本，被白色蛛丝状柔毛。丛生叶和茎下部叶光滑，肾形或心形，先端钝，边缘具三角状齿或小齿，掌状脉。头状花序多数，排成总状花序；总苞钟形；总苞片长圆形，先端三角状。瘦果圆柱形；冠毛紫褐色，短于管状花管部。花果期7-10月。生海拔100-3400米的水边、山坡、林缘、林下或草甸。产中国西南、华中、华

洱源橐吾 *Ligularia lankongensis*

北、西北和东北。朝鲜半岛和日本亦有。

Perennial herbs, white arachnoid pubescent. Rosette and lower cauline leaves glabrous, reniform or cordate, apex obtuse, margin triangular-toothed or serrulate, palmately nerved. Capitula numerous, arranged in raceme; involucres campanulate; phyllaries oblong, apex triangular. Achenes cylindric; pappus purplish-brown, shorter than tube of tubular corolla. Fl. and fr. Jul-Oct. Watersides, slopes, forest edges, forests or meadows at 100-3400 m. Distributed SW, C, N, NW and NE China. Also in Korean Peninsula and Japan.

洱源橐吾

Ligularia lankongensis

(Franch.) Hand.-Mazz.

多年生草本，密被白色蛛丝状柔毛。丛生叶卵形或三角形；茎下部叶鳞片状，卵形；中部叶与丛生叶相似，无鞘，不抱茎；上部叶箭形或卵状披针形。头状花序多数，排成总状花序；总苞宽浅钟形；总苞片条状长圆形。瘦果圆柱形；冠毛白色，与管状花花冠等长。花果期4-9月。生海拔2100-3800米的山坡、灌丛或林下。产云南西北部和四川西南部。

Perennial herbs, densely white arachnoid villose. Rosette leaves ovate or triangular; lower cauline leaves squamiform, ovate; middle ones similar to rosette ones, athecate; upper leaves sagittate or ovate-lanceolate. Capitula numerous, arranged in raceme; involucres broadly and shallowly campanulate; phyllaries linear-lanceolate. Achenes cylindric; pappus white, equal to tubular corolla. Fl. and fr. Apr-Sep. Slopes, thickets or forests at 2100-3800 m. Distributed in NW Yunnan and SW Sichuan.

干崖子橐吾 *Ligularia kanaitzensis*

干崖子橐吾

Ligularia kanaitzensis

(Franch.) Hand.-Mazz.

多年生草本。丛生叶和茎基部叶光滑，卵状长圆形或卵状三角形，边缘具整齐小齿，叶脉羽状，叶柄光滑，具狭翅，膨大成鞘；茎生叶卵状披针形。头状花序多数，排成总状花序；总苞钟状；总苞片长圆形。瘦果圆柱形；冠毛黄色，与管状花花冠等长。花果期7-10月。生海拔2400-4300米的水边、山坡、沼泽地或灌丛中。产云南西北部。

Perennial herbs. Rosette and lower cauline leaves glabrous, ovate-oblong or ovate-triangular, margin regularly serrulate, nerves pinnate, petioles glabrous, narrowly winged, dilated as sheath-like; cauline leaves ovate-lanceolate. Capitula numerous, arranged in raceme; involucres campanulate; phyllaries oblong. Achenes cylindric; pappus yellow, equal to tubular corolla. Fl. and fr. Jul-Oct. Watersides, slopes, swamps or thickets at 2400-4300 m. Distributed in NW Yunnan.

苍山橐吾 *Ligularia tsangchanensis*

苍山橐吾

Ligularia tsangchanensis (Franch.) Hand.-Mazz.

多年生草本。丛生叶和茎下部叶光滑，长圆状卵形或卵形，叶脉羽状，叶柄具翅，基部鞘状；茎中上部叶无柄，长圆形，基部半抱茎。头状花序多数，排成总状花序；总苞钟形；总苞片长圆形或披针形，先端黑褐色。瘦果白色，长圆形；冠毛白色，与管状花花冠等长。花果期6-9月。生海拔2800-4100米的草坡、林下、灌丛或高山草甸。产云南、四川和西藏。

Perennial herbs. Rosette and lower cauline leaves glabrous, oblong-ovate or ovate, nerves pinnate, petioles winged, base sheathlike; middle and upper cauline leaves sessile, oblong, base semi-amplexicaul. Capitula numerous, arranged in raceme; involucres campanulate; phyllaries oblong or lanceolate, apex blackish-brown. Achenes white, oblong; pappus white, equal to tubular corolla. Fl. and fr. Jun-Sep. Grassy slopes, forests, thickets or alpine meadows at 2800-4100 m. Distributed in Yunnan, Sichuan and Xizang.

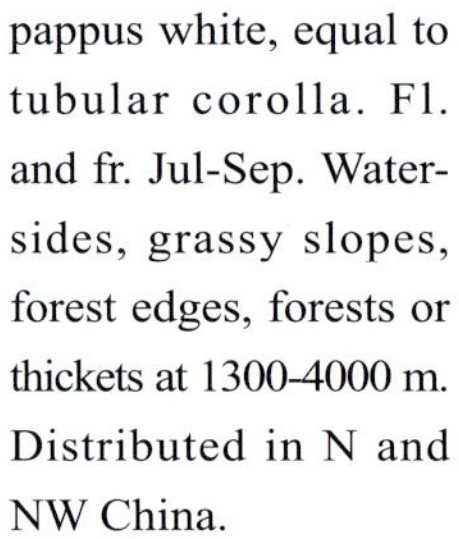

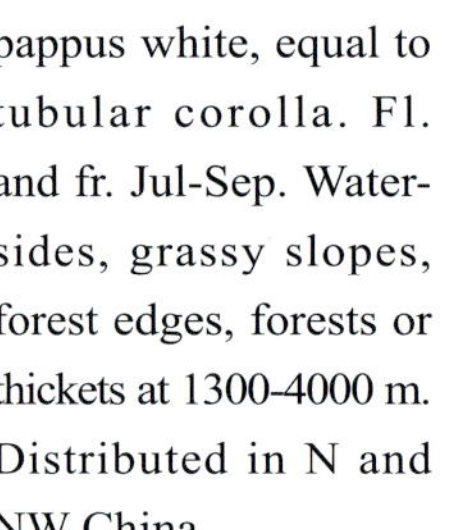

箭叶橐吾

Ligularia sagitta Mattf. ex Rehd. et Kobuski

多年生草本。丛生叶和茎下部叶箭形、戟形或长圆状箭形，叶脉羽状；茎中部叶小，具短柄，鞘状抱茎；上部叶披针形。头状花序多数，排成总状花序；总苞钟形；总苞片长圆形或披针形。瘦果长圆形；冠毛白色，与管状花花冠等长。花果期7-9月。生海拔1300-4000米的水边、草坡、林缘、林下或灌丛。产华北和西北。

Perennial herbs. Rosette and lower cauline leaves sagittate, hastate or oblong-sagittate, nerves pinnate; middle leaves small, shortly petiolate, sheath-like amplexicaul; upper leaves lanceolate. Capitula numerous, arranged in raceme; involucres campanulate; phyllaries oblong or lanceolate. Achenes oblong; pappus white, equal to tubular corolla. Fl. and fr. Jul-Sep. Watersides, grassy slopes, forest edges, forests or thickets at 1300-4000 m. Distributed in N and NW China.

箭叶橐吾 *Ligularia sagitta*

长白山橐吾

Ligularia jamesii (Hemsl.) Kom.

多年生草本。基生叶叶柄长达29厘米，基部具狭鞘；叶缘具锐锯齿。头状花序单生；总苞宽钟形，外被白色蛛丝状柔毛；苞片约13，2层；舌状花13-16朵，黄色。瘦果深棕色，圆柱形；冠毛淡黄色。花果期7-8月。生海拔300-2500米的林下、灌丛或高山草甸。产内蒙古、吉林和辽宁。朝鲜半岛亦有。

Perennial herbs. Petioles of basal leaves petiolate to 29 cm long, base narrowly sheathed; leaves margin sharply dentate. Capitula solitary; involucres broadly campanulate, outside white arachnoid-puberulous; phyllaries ca. 13, in 2 rows; ligulate florets 13-16, yellow. Achenes dark brown, cylindric; pappus pale yellow. Fl. and fr. Jul-Aug. Forests, thickets or alpine meadows at 300-2500 m. Distributed in Neimenggu, Jilin and Liaoning. Also in Korean Peninsula.

全缘橐吾

Ligularia mongolica (Turcz.) DC.

多年生草本。基生叶叶柄长达35厘米，基部具狭鞘；叶无毛，全缘。头状花序组成总状

全缘橐吾 *Ligularia mongolica*

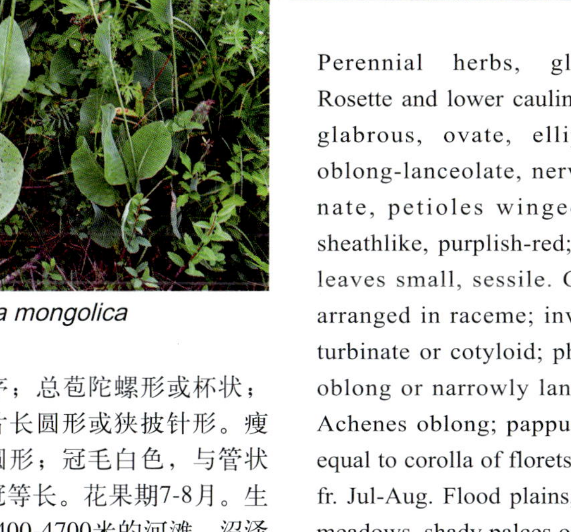

长白山橐吾 *Ligularia jamesii*

花序，簇生，近头状；总苞狭钟形或圆柱形；苞片5或6，2层；舌状花1-4朵，黄色；两性花5-10朵。瘦果棕色，楔状圆柱形；冠毛棕红色。花果期6-8月。生海拔1500米以下的沼泽草甸、山坡、灌丛或林窗下。产河北、内蒙古和黑龙江。俄罗斯、蒙古和朝鲜半岛亦有。

Perennial herbs. Petioles of basal leaves to 35 cm long, base narrowly sheathed; leaves glabrous, margin entire. Inflorescences racemose, clustered, subcapitate; involucres narrowly campanulate or cylindric; phyllaries 5 or 6, in 2 rows; ligulate florets 1-4, yellow; tubular florets 5-10. Achenes brown, cuneate-cylindric; pappus reddish brown. Fl. and fr. Jun-Aug. Swamp meadows, slopes, thickets or canopy gaps in forests below 1500 m. Distributed in Hebei, Neimenggu and Heilongjiang. Also in Russia, Mongolia and Korean Peninsula.

黄帚橐吾

Ligularia virgaurea Mattf. ex Rehd. et Kobuski

多年生草本，灰绿色。丛生叶和茎基部叶光滑，卵形、椭圆形或长圆状披针形，叶脉羽状，叶柄具翅，基部鞘状，紫红色；茎生叶小，无柄。头状花序排成总状花序；总苞陀螺形或杯状；总苞片长圆形或狭披针形。瘦果长圆形；冠毛白色，与管状花花冠等长。花果期7-8月。生海拔2400-4700米的河滩、沼泽草甸、阴湿地或灌丛。产青藏高原。尼泊尔和不丹亦有。

Perennial herbs, glaucous. Rosette and lower cauline leaves glabrous, ovate, elliptic or oblong-lanceolate, nerves pinnate, petioles winged, base sheathlike, purplish-red; cauline leaves small, sessile. Capitula arranged in raceme; involucres turbinate or cotyloid; phyllaries oblong or narrowly lanceolate. Achenes oblong; pappus white, equal to corolla of florets. Fl. and fr. Jul-Aug. Flood plains, swamp meadows, shady palces or thickets at 2400-4700 m. Distributed in Tibetan Plateau. Also in Nepal and Bhutan.

黄帚橐吾 *Ligularia virgaurea*

侧茎橐吾 *Ligularia pleurocaulis*

侧茎橐吾

Ligularia pleurocaulis

(Franch.) Hand.-Mazz.

多年生草本，灰绿色。根肉质，近纺锤形。丛生叶和茎基部叶条状长圆形至宽椭圆形；茎生叶小，椭圆形至条形。头状花序排成圆锥状总状花序或总状花序；总苞陀螺形；总苞片卵形或披针形。瘦果倒披针形；冠毛白色，与管状花花冠等长。花果期7-11月。生海拔3000-4700米的山坡、草甸、溪边或灌丛中。产云南和四川。

Perennial herbs, glaucous. Roots succulent, subfusiform. Rosette and lower cauline leaves linear-oblong to broadly elliptic; cauline leaves small, elliptic to linear. Capitula arranged in paniculate-raceme or raceme; involucres turbinate; phyllaries ovate to lanceolate. Achenes oblanceolate; pappus white, equal to corolla of florets. Fl. and fr. Jul-Nov. Slopes, meadows, riversides or thickets at 3000-4700 m. Distributed in Yunnan and Sichuan.

药山垂头菊

Cremanthodium pirolaefolia

(H. Lévl.) T. L. Ming

多年生草本。基生叶椭圆形或卵形，有长叶柄；茎生叶2-4片，长圆形至披针形，无柄，抱茎。头状花序下垂，通常排成总状花序；总苞宽钟状，舌花16-20，黄色，舌片披针形；冠毛白色，与管状花近等长。花期7-8月，果期8-9月。生海拔3600-4000米的高山草甸。产云南(巧家县)和四川(冕宁县)。

Perennial herbs. Basal leaves elliptic to ovate, long petiolate; stem leaves 2-4, oblong to lanceolate, sessile and amplexicaul. Capitula pendulus, in a racemose synflorescence; involucre broad campanulate, ray florets 16-20, yellow, ligules lanceolate; pappus white, as long as tubular corolla. Fl. Jul-Aug. Fr. Aug-Sep. Alpine meadows at 3600-4000 m. Distributed in Yunnan (Qiaojia County) and Sichuan (Mianning

药山垂头菊 *Cremanthodium pirolaefolia*

硕首垂头菊 *Cremanthodium obovatum*

宽舌垂头菊 *Cremanthodium arnicoides*

County).

硕首垂头菊

Cremanthodium obovatum

Ling et S. W. Liu

多年生草本。基生叶有叶柄，叶倒卵形、长圆形或椭圆形，基部宽楔形、有下延到叶柄的翅，边缘全缘或稀疏的小齿，先端圆形或钝形；中上部叶较小，无柄。头状花序单生，下垂；总苞半球形，直径3-4厘米，外面密被白色和褐色柔毛；花黄色；冠毛白色，和管状花近等长。花期7-8月。生海拔4700-5000米高山流石滩。产西藏东南部。

Perennial herbs. Leaves obovate, oblong or elliptic, base broadly cuneate, attenuate into a winged petioles, margin entire or remotely denticulate, apex rounded or obtuse; middle to distal stem leaves smaller, sessile. Capitula solitary, nodding; involucre hemispheric, 3-4 cm diam, outside densely white puberulent and black pilose; flowers yellow; pappus white, as long as tubular corolla. Fl. Jul-Aug. Alpine screes at 4700-5000 m. Distributed in SE Xizang.

宽舌垂头菊

Cremanthodium arnicoides

(DC. ex Royle) R. D. Good

多年生草本。下部叶具柄，卵形或卵状长圆形；茎中部叶具短柄，卵形；最上部叶苞叶状，披针形或长圆形，两面有毛。头状花序1至4个排列呈疏总状花序，下垂；总苞片2层；舌状花黄色。瘦果棕色，有棱；冠毛白色，与管状花花冠等长。花期8月。生海拔3600-4600米的高山流石滩。产西藏。西喜马拉雅亦有。

Perennial herbs. Basal leaves petiolate, ovate or ovate-oblong; middle stem leaves shortly petiolate, ovate; distalmost stem leaves bracteal, lanceolate or oblong, both surfaces hairy. Capitula 1-4, in lax raceme, nodding; phyllaries in 2 rows; ray florets yellow. Achenes brown, ribbed; pappus white, as long as tubular corolla. Fl. Aug. Gravelly areas on mountains at 3600-4600 m. Distributed in Xizang. Also in W Himalaya.

不丹垂头菊

Cremanthodium bhutanicum

Ludlow

多年生草本，蓝绿色。茎直立，单生，高10-25厘米。茎基部叶具柄，线形、线状倒披针形或线状长圆形；茎中上部叶线形。头状花序单生，下垂；总苞半球形，总苞片2层，边缘具狭膜；小花黄色；冠毛白色，与管状花花冠等长。花期8月。生海拔约4300米的山坡草甸。产西藏东南部。不丹亦有。

Perennial herbs, plants bluish green. Stem solitary, erect, 10-25 cm tall. Basal leaves petiolat, linear, linear-lanceolate, or linear-oblong; middle to distal stem leaves linear. Capitula solitary, nodding; involucre hemispheric; phyllaries in 2 rows, margin narrowly membranous; florets yellow; pappus white, as long as tubular corolla. Fl. Aug. Alpine meadows at ca. 4300 m. Distributed in SE Xizang. Also in Bhutan.

不丹垂头菊 *Cremanthodium bhutanicum*

褐毛垂头菊 *Cremanthodium brunneopilosum*

褐毛垂头菊

Cremanthodium brunneopilosum S. W. Liu

多年生草本，全株灰绿色或蓝绿色。茎单生，直立，高1米。茎下部叶具柄，长椭圆形至披针形；最上部叶苞叶状，披针形。头状花序下垂，1-13，排列成总状花序；总苞片2层；花黄色；冠毛白色，与管状花花冠等长。花果期6-9月。生海拔3000-4300米的高山沼泽草甸、河滩草甸、水边。产青藏高原。

Perennial herbs, plants grayish green or bluish green. Stem solitary, erect, to 1 m tall. Basal leaves petiolate, narrowly elliptic to lanceolate; distalmost stem leaves bracteal, lanceolate. Capitula 1-13, in raceme, nodding; phyllaries in 2 rows; florets yellow; pappus white, as long as tubular corolla. Fl. and fr. Jun-Sep. Alpine swamp meadows, stream banks and associated meadows at 3000-4300 m. Distributed in Tibetan Plateau.

珠芽垂头菊

Cremanthodium bulbilliferum W. W. Sm.

多年生草本，高8-25厘米。基部叶具长柄，肾形或宽肾形；茎生叶肾形；茎生叶3-4，肾形；上部叶较小，仅有叶鞘，里面有黑色的珠芽。头状花序单生，下垂或半下垂；总苞半球形；总苞片2层；小花黄色；冠毛白色，与管状花花冠等长。花果期8-10月。生海拔3000-4000米的山坡草地、石质坡地上。产西藏东南部和云南西北部。

Perennial herbs, 8-25 cm tall. Basal leaves long petiolate, glabrous, reniform or broadly reniform; stem leaves reniform; stem leaves 3 or 4, reniform; smaller distally, distalmost stem leaves with only sheath, inside with bulbil, black. Capitula solitary, nodding or semierect; involucre hemispheric; phyllaries in 2 rows. florets yellow; pappus white, as long as tubular corolla. Fl. and fr. Aug-Oct. Grassy slopes, rocky slopes at 3000-4000 m. Distributed in SE Xizang and NW Yunnan.

长鞘垂头菊

Cremanthodium calcicola W. W. Sm.

多年生草本，高20-50厘米。基部叶具柄，基部呈鞘状，鞘长6-9厘米，近膜质；茎叶1-5；中部叶肾形，具柄，柄基部膨大呈鞘状；上部叶卵状长圆形，苞叶状，无柄，无鞘。头状花序单生，下垂；总苞片20，2层；小花全部管状，黄色。花果期6-7月。生海拔3400-3500米的山顶草坡、石灰石峭壁、水沟边。产云南。

Perennial herbs, 20-50 cm tall. Basal leaves petiolate, sheath 6-9 cm, submembranous; stem leaves 1-5; middle leaves reniform, petiolate, base sheathed;

珠芽垂头菊 *Cremanthodium bulbilliferum*

长鞘垂头菊 *Cremanthodium calcicola*

错那垂头菊 *Cremanthodium conaense*

心叶垂头菊 *Cremanthodium cordatum*

米，紫红色。基生叶具柄，基部有鞘；下部茎叶具短的鞘状柄；上部茎叶叶钻形。头状花序单生，下垂；总苞半球形；总苞片14-16，2层；舌花黄色；管状花紫色；冠毛白色，略长于管状花花冠。花期7-8月，果期8-9月。生海拔4700-5400米的河边、高山沼泽、高山草甸。产四川西南部。

Perennial herbs. Stem solitary, erect, purplish red, to 50 cm tall. Basal leaves petiolate, base sheathed; proximal stem leaves petiolate; petioles short, enlarged into a sheath; distal stem leaves subulate. Capitula solitary, nodding; involucre hemispheric; phyllaries 14-16, in 2 rows; ray florets yellow; tubular florets purple; pappus white, slightly longer than tubular corolla. Fl. Jul-Aug. Fr. Aug-Sep. Stream banks, alpine swamp meadows, alpine meadows at 4700-5400 m. Distributed in SW Sichuan.

distal leaves sessile, bracteal, ovate-oblong, without sheath. Capitula solitary, nodding; phyllaries 20, in 2 rows; florets all tubular, yellow. Fl. and fr. Jun-Jul. Grassy slopes, calcitic cliffs, stream banks at 3400-3500 m. Distributed in Yunnan.

错那垂头菊

Cremanthodium conaense S. W. Liu

多年生草本，高10-15厘米，全株灰绿色或蓝灰绿色。基部叶具柄，基部具鞘，叶片长圆形或倒披针形；中上部叶3-4，长圆形至线形，基部半抱茎。头状花序单生，下垂；总苞半球形，紫红色；总苞片10-12，2层；小花黄色；冠毛白色，与管状花花冠等长。花期7-8月。生海拔4300-4600米的山顶坡地、草甸。产西藏(错那县)。

Perennial herbs, 10-15 cm tall, plants bluish grayish green. Basal leaves petiolate, base sheathed, oblong or oblanceolate; middle to distal stem leaves 3 or 4, oblong to linear, base semiamplexicaul. Capitula solitary, nodding; involucre purplish red, hemispheric; phyllaries 10-12, in 2 rows; florets yellow; pappus white, as long as tubular corolla. Fl. Jul-Aug. Alpine meadows, mountain summits at 4300-4600 m. Distributed in Xizang (Cona County).

心叶垂头菊

Cremanthodium cordatum S. W. Liu

多年生草本，全株灰绿色，高15-25厘米。基部叶具柄，心形或卵状心形，稀长圆状心形。头状花序单生，下垂；总苞半球形，干后黑色；小花黄色；冠毛白色，与管状花花冠等长。花果期7-8月。生海拔4200米的山谷草地。产西藏南部。

Perennial herbs, plants grayish green, 15-25 cm tall. Basal leaves petiolate, cordate or ovate-cordate, rarely oblong-cordate. Capitula solitary, nodding; involucre black when dry, hemispheric; florets yellow; pappus white, as long as tubular corolla. Fl. and fr. Jul-Aug. Grasslands in valleys at 4200 m. Distributed in S Xizang.

稻城垂头菊

Cremanthodium daochengense Ling et S. W. Liu

多年生草本。茎直立，高达50厘

稻城垂头菊 *Cremanthodium daochengense*

向日垂头菊 *Cremanthodium helianthus*

矮垂头菊 *Cremanthodium humile*

向日垂头菊

Cremanthodium helianthus (Franch.) W. W. Sm.

多年生草本，全株灰绿色。茎单生，直立，高7-56厘米。基生叶卵状椭圆形至宽椭圆形，具柄，基部有长鞘。头状花序单生，下垂，被苞叶环绕；总苞干后灰绿色或深灰色；总苞片12-20，2层。瘦果有棱。花果期7-11月。生海拔2800-4500米的林下、灌丛中、草坡、高山草甸。产四川西南部和云南西北部。

Perennial herbs, plants grayish green. Stem solitary, erect, 7-56 cm tall. Basal leaves petiolate, base long sheathed, ovate-elliptic to broadly elliptic. Capitula solitary, nodding; leaf-like bracts several, usually surrounding capitula; involucre grayish green or blackish gray when dry; phyllaries 12-20, in 2 rows. Achenes ribbed. Fl. and fr. Jul-Nov. Forests, scrub, grassy slopes, alpine meadows at 2800-4500 m. Distributed in SW Sichuan and NW Yunnan.

矮垂头菊

Cremanthodium humile Maxim.

多年生草本，高5-20厘米。茎下部叶卵形或卵状长圆形，有时近圆形，具柄；茎中上部叶无柄或有短柄。头状花序单生；总苞半球形；总苞片8-12，1层；小花黄色；冠毛白色，与管状花花冠等长。花果期7-11月。生海拔3500-5300米的高山流石滩。产青藏高原。不丹亦有。

Perennial herbs, 5-20 cm tall. Proximal stem leaves ovate or ovate-oblong, sometimes orbicular, petiolate; middle to distal stem leaves sessile or shortly petiolate. Capitula solitary; involucre hemispheric; phyllaries 8-12, in 1 row; florets yellow; pappus white, as long as tubular corolla. Fl. and fr. Jul-Nov. Gravelly areas on mountains at 3500-5300 m. Distributed in Tibetan Plateau. Also in Bhutan.

墨脱垂头菊

Cremanthodium medogense Y. S. Chen

多年生草本。茎直立，单生，高18-38厘米。基生叶多数，心状肾形，有柄，基部有狭鞘；茎生叶2或3。头状花序单生，下垂。总苞半球形；总苞片8，2层；小花黄色；冠毛棕色，略短于管状花花冠。花期8月。生海拔3700-4200米的高山灌丛和草甸。产西藏东南部。

Perennial herbs. Stems caespitose, erect, 18-38 cm tall. Basal leaves numerous, cordate-reniform, petiolate, base narrowly sheathed; stem leaves 2 or 3. Capitula solitary, nodding; involucre hemispheric; phyllaries 8, in 2 rows; florets yellow; pappus brown, shorter than tubular corolla. Fl. Aug. Alpine thickets and meadows at 3700-4200 m. Distributed in SE Xizang.

小舌垂头菊

Cremanthodium microglossum S. W. Liu

多年生草本。茎深紫色，高4-15厘米。基生叶卵形或宽卵形，有紫棕色柄；茎生叶3，卵形到长卵形。头状花序单生，直立；总苞半球形；总苞片9-12，1层；外层花白色，中央花橙色。瘦果有棱；冠毛白色，与管状花花冠等长。花果期7-9月。生海拔4000-5400米的山谷斜坡、高山草甸、沼泽草甸。产青藏高原。

墨脱垂头菊 *Cremanthodium medogense*

小舌垂头菊 *Cremanthodium microglossum*

尼泊尔垂头菊 *Cremanthodium nepalense*

壮观垂头菊 *Cremanthodium nobile*

Perennial herbs. Stem dark purple, 4-15 cm tall. Basal leaves ovate or broadly ovate, petiolate, petioles purplish brown; stem leaves 3, ovate to oblong-ovate. Capitula solitary, erect; involucre hemispheric; phyllaries 9-12, in 1 row; outer florets white; central florets orange. Achenes ribbed; pappus white, as long as tubular corolla. Fl. and fr. Jul-Sep. Grassy slopes in gravelly areas on mountains, alpine meadows, swamp meadows at 4000-5400 m. Distributed in Tibetan Plateau.

尼泊尔垂头菊

Cremanthodium nepalense Kitam.

多年生草本，高14-30厘米。基生叶卵形至近圆形，具柄；茎生叶2-4；下部叶卵形，有柄，基部鞘状抱茎；中上部叶线状披针形至线形。头状花序单生，下垂；总苞半球形；总苞片10-14，2层；小花黄色；冠毛白色，与管状花花冠等长。花果期8-9月。生海拔4300-4800米的草坡、溪边、岩石中。产西藏南部。尼泊尔亦有。

Perennial herbs, 14-30 cm tall. Basal leaves ovate to suborbicular, petiolate; stem leaves 2-4; proximal leaves ovate, petiolate, base sheathed; middle and distal leaves linear-lanceolate to linear. Capitula solitary, nodding; involucre hemispheric; phyllaries 10-14, in 2 rows; florets yellow; pappus white, as long as tubular corolla. Fl. and fr. Aug-Sep. Grassy slopes, stream banks, rocky places at 4300-4800 m. Distributed in S Xizang. Also in Nepal.

壮观垂头菊

Cremanthodium nobile Kitam.

多年生草本，高15-40厘米。基生叶倒卵形、宽椭圆形或近圆形，无柄或有短柄；茎生叶少，狭长圆形至线形，无柄。头状花序单生，下垂；总苞半球形；总苞片10-14，2层；小花黄色。瘦果有棱；冠毛白色，与管状花花冠等长。花果期7-9月。生海拔3400-5000米的高山灌层、草甸。产四川、西藏和云南。

Perennial herbs, 15-40 cm tall. Basal leaves obovate, broadly elliptic or suborbicular, sessile or shortly petiolate; stem leaves few, sessile, narrowly oblong to linear. Capitula solitary, nodding; involucre hemispheric; phyllaries 10-14, in 2 rows; florets yellow. Achenes ribbed; pappus white, as long as tubular corolla. Fl. and fr. Jul-Sep. Scrub, alpine meadows at 3400-5000 m. Distributed in Sichuan, Xizang and Yunnan.

矩叶垂头菊

Cremanthodium oblongatum C. B. Clarke

多年生草本。茎直立，1-2，高8-20厘米，紫红色。基生叶长圆形、圆形或椭圆形，具柄，紫红色；茎生叶3-4，无柄，长圆形至披针形。头状花序单生，下垂；总苞半球形；总苞片10-14，2层；小花黄色；冠毛白色，与管状花花冠等长。花期7-8月，果期8-9月。生海拔4500-5300米的高山草甸、高山流石滩。产西藏南部。印度北部和尼泊尔亦有。

Perennial herbs. Stems 1 or 2, erect, often purplish red, 8-20 cm tall. Basal leaves oblong, orbicular or elliptic, petiolate; petioles purplish red; stem leaves 3-4, sessile, oblong to lanceolate. Capitula solitary, nodding; involucre hemispheric; phyllaries 10-14, in 2 rows; florets yellow; pappus white, as long as tubular corolla. Fl. Jul-Aug. Fr. Aug-Sep. Gravelly areas on mountains, alpine meadows at 4500-5300 m. Distributed in S Xizang. Also in N India and Nepal.

矩叶垂头菊 *Cremanthodium oblongatum*

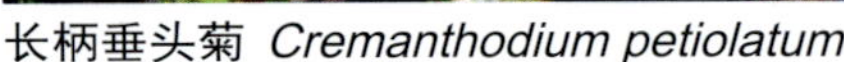
长柄垂头菊 *Cremanthodium petiolatum*

叶状柄垂头菊 *Cremanthodium phyllodineum*

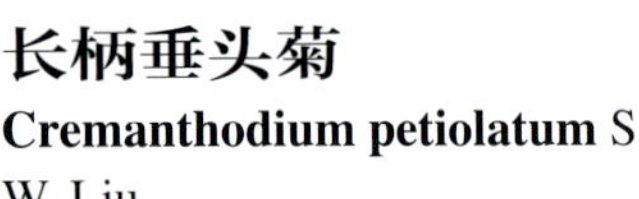

长柄垂头菊
Cremanthodium petiolatum S. W. Liu

多年生草本，高达55厘米。基生叶宽椭圆形，具柄，有棱，下半部紫红色，基部有鞘；茎生叶向上渐小。头状花序3，排列成伞房状总状花序；总苞半球形；总苞片12-14，1层；舌状花黄色；管状花黄褐色；冠毛白色，与管状花花冠等长。花果期7月。生海拔4200-4500米的高山水边或草甸。产西藏西南部。

Perennial herbs, to 55 cm tall. Basal leaves broadly elliptic, petiolate, base sheathed, proximally purplish red; stem leaves smaller distally. Capitula 3, in corymb-raceme; involucre hemispheric; phyllaries 12-14, in 1 row; ray florets yellow; tubular florets yellowish brown; pappus white, as long as tubular corolla. Fl. and fr. Jul. Stream banks or meadows on mountains at 4200-4500 m. Distributed in SW Xizang.

叶状柄垂头菊
Cremanthodium phyllodineum S. W. Liu

多年生草本，高35-60厘米。基生叶肾形或三角状肾形，具长柄，基部鞘状；茎生叶1-3。头状花序单生，下垂；总苞半球形；总苞片约12个，2层；舌状花黄色；管状花先端略带褐色；冠毛淡褐色，短于管状花花冠。花期7-8月。生海拔3700-4200米的山坡润湿草地、高山草地。产西藏东南部和云南西北部。

Perennial herbs, 35-60 cm tall. Basal leaves reniform or triangular-reniform, long petiolate, base sheathed; stem leaves 1-3. Capitula solitary, nodding; involucre hemispheric; phyllaries ca. 12, in 2 rows; ray florets yellow; tubular florets apex brown; pappus pale brown, shorter than tubular corolla. Fl. Jul-Aug. Alpine meadows, wet grassy slopes at 3700-4200 m. Distributed in SE Xizang and NW Yunnan.

羽裂垂头菊
Cremanthodium pinnatifidum Benth.

多年生草本，高10-15 厘米。基生叶长圆形，具短柄，基部膨大成鞘状；茎中上部无叶或仅具1叶，苞片状，基部有鞘。头状花序单生，下垂；总苞半球形；总苞片2层；小花黄色；冠毛白色，与管状花花冠等长。花期8月。生海拔4300-4600米的高山砾石地。产西藏南部。东喜马拉雅亦有。

Perennial herbs, 8-15 cm tall. Basal leaves oblong, shortly petiolate, base enlarged sheathed; middle to distal stem leaf absent or solitary, bracteal, base sheathed. Capitula solitary, nodding; involucre hemispheric; phyllaries in 2 rows; florets yellow; pappus white, as long as tubular corolla. Fl. Aug. Alpine rocky places at 4300-4600 m. Distributed in S Xizang. Also in E Himalaya.

戟叶垂头菊
Cremanthodium potaninii C. Winkl.

多年生草本，高5-30厘米。基生叶具柄，基部有鞘；茎中上部叶线状披针形至线形，无柄。头状花序单生，下垂；总苞宽钟状；总苞片12-14，2层；小花黄色；冠毛褐色，与管状花花冠等长。花果期7-9月。生海拔3600-4500米的灌丛中、山坡湿地及高山草甸。产四川西北部和甘肃西南部。

Perennial herbs, 5-30 cm tall. Basal leaves petiolate, base sheathed; middle to distal stem leaves sessile, linear-lanceolate to linear. Capitula solitary, nodding; involucre broadly campanulate; phyllaries 12-14, in 2 rows; florets yellow; pappus brown, as long as tubular corolla. Fl. and fr. Jul-Sep. Wet grassy slopes, thickets, alpine meadows at 3600-4500 m. Distributed in NW Sichuan and SW Gansu.

羽裂垂头菊 *Cremanthodium pinnatifidum*

戟叶垂头菊 *Cremanthodium potaninii*

方叶垂头菊 *Cremanthodium principis*

alpine scrub, alpine meadows at 3600-4500 m. Distributed in NW Sichuan and SW Gansu.

方叶垂头菊
Cremanthodium principis (Franch.) R. D. Good

多年生草本，高10-30厘米。基生叶长圆形、四方形或近圆形，具柄，基部鞘状；茎中上部叶少，向上渐小，苞叶状，无柄，四方形至线形。头状花序单生，下垂；总苞半球形；总苞片约12，2层；小花黄花；冠毛褐色，与管状花花冠等长。花果期6-7月。生海拔3600-4600米的高山灌丛、草甸和砾石地。产云南西北部和四川西南部。

Perennial herbs, 10-30 cm tall. Basal leaves oblong, square or suborbicular, petiolate, base sheathed; middle to distal stem leaves few, smaller distally, sessile, bracteal, square to linear. Capitula solitary, nodding; involucre hemispheric; phyllaries ca. 12, in 2 rows; florets yellow; pappus brown, as long as tubular corolla. Fl. and fr. Jun-Jul. Alpine scrub, alpine meadows, rocky places at 3600-4600 m. Distributed in NW Yunnan and SW Sichuan.

箭叶垂头菊
Cremanthodium sagittifolium Ling et Y. L. Chen

多年生草本，高10-20厘米。叶革质，亮绿色；基部叶具柄，柄紫红色；茎生叶1-2，中部叶箭形，具柄，基部鞘状，紫红色；上部叶线形，苞叶状。头状花序单生，下垂；总苞半球形；总苞片12-14，2层；小花黄色；冠毛淡褐色，略短于管状花花冠。花果期7-10月。生海拔3400-4400米的高山草甸。产云南东北部。

箭叶垂头菊 *Cremanthodium sagittifolium*

Perennial herbs, 10-20 cm tall. Leaves nitid, leathery; basal leaves petiolate, petioles purplish red; stem leaves 1 or 2; middle stem leaves sagittate, petiolate, petioles purplish red, base sheathed; distal stem leaves bracteal, linear. Capitula solitary, nodding; involucre hemispheric; phyllaries 12-14, in 2 rows; florets yellow; pappus pale brown, shorter than tubular corolla. Fl. and fr. Jul-Oct. Alpine meadows at 3400-4400 m. Distributed in NE Yunnan.

铲叶垂头菊
Cremanthodium sino-oblongatum R. D. Good

多年生草本，高15-25厘米。基生叶铲形，具柄，基部鞘状；茎生叶铲形至线形，无柄或有短柄。头状花序常单生，下垂；总苞半球形；总苞片14-18，2层；小花黄色；冠毛白色，与管状花花冠等长。花果期7-9月。生海拔3900-5000米的山坡、灌丛中。产云南西北部。

Perennial herbs, 15-25 cm tall. Basal leaves spatulate, petiolate, base sheathed; stem leaves sessile or shortly petiolate, spatulate to linear. Capitula usually solitary, nodding; involucre hemispheric; phyllaries 14-18, in 2 rows; florets yellow; pappus white, as long as tubular corolla. Fl. and fr. Jul-Sep. Grassy slopes, scrub at 3900-5000 m. Distributed in NW Yunnan.

铲叶垂头菊 *Cremanthodium sino-oblongatum*

紫茎垂头菊 *Cremanthodium smithianum*

Yunnan and Sichuan. Also in N Myanmar.

膜苞垂头菊
Cremanthodium stenactinium
Diels ex H. Limpr.

多年生草本，全株灰绿色或蓝绿色。茎高40-50厘米。基生叶长圆形或椭圆形，具宽柄，基部具宽鞘；茎生叶5-6；最上部者膜质，苞叶状。头状花序下垂，1-13，排列成总状花序；苞叶浅黄色，卵形，膜质；总苞片12-16，2层；小花黄色；冠毛白色，与管状花花冠等长。花期7-9月。生海拔3600-4400米的灌丛中、草地、水边。产四川西北部和西藏东北部。

Perennial herbs, plants grayish green or bluish green. Stem 40-50 cm tall. Basal leaves oblong or elliptic, broadly petiolate; base broadly sheathed; stem leaves 5-6. Capitula 1-13, in raceme, nodding; leaflike bracts yellowish white, ovate, membranous; phyllaries 12-16, in 2 rows; florets yellow; pappus white, as long as tubular corolla. Fl. Jul-Sep. Grasslands, stream banks at 3600-4400 m. Distributed in NW Sichuan and NE Xizang.

狭舌垂头菊
Cremanthodium stenoglossum
Ling et S. W. Liu

多年生草本，高10-32厘米。基生叶圆肾形或肾形，具柄，基部鞘状；茎下部叶1个，宽肾形，无柄或有短柄，基部鞘状。头状花序单生，下垂；总苞片9-14，紫红色，2层；舌状花黄色；冠毛白色。花果期7-8月。生海拔3700-5000米的灌丛中、水边、沼泽地、高山草甸、岩石隙中、高山流石滩。产四川西北部和青海。

Perennial herbs, 10-32 cm tall. Basal leaves orbicular-reniform or reniform, petiolate, base sheathed; proximal stem leaf solitary, broadly reniform, shortly petiolate o sessile, base sheathed. Capitula solitary, nodding; phy-

紫茎垂头菊
Cremanthodium smithianum
(Hand.-Mazz.) Hand.-Mazz.

多年生草本，高10-25厘米。基生叶肾形，紫红色，具长柄，柄紫红色，基部有鞘；茎中上部叶1-2，肾形至线状披针形。头状花序单生，下垂或近直立；总苞片12-14，2层；小花黄色；冠毛白色，与管状花花冠等长。花果期7-9月。生海拔3000-5200 米的山坡草地、水边、高山草甸、高山流石滩。产西藏、云南和四川。缅甸北部亦有。

Perennial herbs, 10-25 cm tall. Basal leaves purplish red, reniform, long petiolate; petioles purplish red, base sheathed; middle to distal stem leaves 1 or 2, reniform to linear-lanceolate. Capitula solitary, nodding or suberect; phyllaries 12-14, in 2 rows; florets yellow; pappus white, as long as tubular corolla. Fl. and fr. Jul-Sep. Grassy slopes, stream banks, gravelly areas on mountains, alpine meadows at 3000-5200 m. Distributed in Xizang,

膜苞垂头菊 *Cremanthodium stenactinium*

狭舌垂头菊 *Cremanthodium stenoglossum*

木里垂头菊 *Cremanthodium suave*

叉舌垂头菊 *Cremanthodium thomsonii*

llaries 9-14, in 2 rows, purple; ray florets yellow; pappus white. Fl. and fr. Jul-Aug. Swamps, stream banks, scrub, alpine meadows, alpine crevices, gravelly areas on mountains at 3700-5000 m. Distributed in NW Sichuan and Qinghai.

木里垂头菊

Cremanthodium suave W. W. Sm.

多年生草本，全株灰绿色。茎高20-40厘米。基生叶狭椭圆形或狭匙形，具柄，具狭翅，基部有鞘；茎生叶4-6，无柄，披针形，向上渐小。头状花序单生，下垂；总苞半球形；总苞片2层；舌状花黄色；冠毛白色，比管状花花冠短。花期7-8月。生海拔3000-4300米的林缘、山坡、高山草甸。产四川西南部和云南西北部。

Perennial herbs, plants grayish green. Stem 20-40 cm tall. Basal leaves narrowly elliptic or narrowly spatulate, petiolate, with narrow wings, base sheathed; stem leaves 4-6, sessile, lanceolate, smaller distally. Capitula solitary, nodding; involucre hemispheric; phyllaries in 2 rows; ray florets yellow; pappus white, shorter than tubular corolla. Fl. Jul-Aug. Fr. Aug-Sep. Forests, grassy slopes, alpine meadows at 3000-4300 m. Distributed in SW Sichuan and NW Yunnan.

叉舌垂头菊

Cremanthodium thomsonii C. B. Clarke

多年生草本，高25-50厘米。基生叶圆肾形或肾形，具柄，有窄叶鞘；茎中部叶1，具柄，柄有鞘；茎上部叶1-2，无柄，线状披针形。头状花序单生，下垂；总苞片10-14，2层；小花黄色；冠毛褐色。花果期6-8月。生海拔3500-4800米的林下、草坡、高山草甸、高山流石滩。产西藏东南部和云南西北部。喜马拉雅亦有。

Perennial herbs, 25-50 cm tall. Basal leaves long reniform or orbicular-reniform, petiolate, base narrowly sheathed; middle stem leaf solitary, petiolate, base sheathed; distal stem leaves 1 or 2, sessile, linear-lanceolate. Capitula solitary, nodding; phyllaries 10-14, in 2 rows; florets yellow; pappus brown. Fl. and fr. Jun-Aug. Alpine meadows, grassy slopes, forests, gravelly areas on mountains at 3500-4800 m. Distributed in SE Xizang and NW Yunnan. Also in Himalaya.

变叶垂头菊

Cremanthodium variifolium R. D. Good

多年生草本，高8-25厘米。基生叶椭圆形、卵形或倒卵形，具柄；茎生叶多数，密集，无柄，线状长圆形。头状花序单生，下垂；总苞半球形；总苞片12-14，2层；小花黄色；冠毛白色，与管状花花冠等长。花果期7-10月。生海拔3200-4500米的林中草地、竹林边缘、山坡草地及高山草甸。产中国西南。

Perennial herbs, 8-25 cm tall. Basal leaves elliptic, ovate or obovate, petiolate; stem leaves numerous, dense, sessile, linear-oblong. Capitula solitary, nodding; involucre hemispheric; phyllaries 12-14, in 2 rows; florets yellow; pappus white, as long as tubular corolla. Fl. and fr. Jul-Oct. Grasslands beneath bamboo forests, margins of bamboo forests, grassy slopes, alpine meadows at 3200-4500 m. Distributed in SW China.

变叶垂头菊 *Cremanthodium variifolium*

钟花垂头菊 *Cremanthodium campanulatum*

长柱垂头菊 *Cremanthodium rhodocephalum*

钟花垂头菊
Cremanthodium campanulatum (Franch.) Diels

多年生草本，被紫色有节柔毛。叶肾形、卵形或披针形，下面紫色，叶脉掌状。头状花序单生，下垂；总苞钟形；总苞片淡紫红色至紫红色，花瓣状，近全缘；小花多数，全部管状，紫红色。瘦果倒卵球形；冠毛白色。花果期5-9月。生海拔3200-4800米的林中、林缘、灌丛、草地、高山草甸或高山流石滩。产中国西南。缅甸东北部亦有。

Perennial herbs, purple septate-villose. Leaves reniform, ovate or lanceolate, purple abaxially, nerves palmate. Capitula solitary, drooping; florets numerous, all tubular, purple. Achenes obovoid; pappus white. Fl. and fr. May-Sep. Forests, forest edges, thickets, grasslands, alpine meadows or alpine screes at 3200-4800 m. Distributed in SW China. Also in NE Myanmar.

黑垂头菊
Cremanthodium atrocapitatum R. D. Good

多年生草本。基部叶肾形，鞘状抱茎，无毛。单生头状花序下垂；总苞半球形，深紫色或黑色，基部具短黑柔毛；苞片2层；小花全为管状，深棕黄色。瘦果棕黄色，圆柱形；冠毛白色。生海拔约4000米的草质山坡。花期7-8月，果期8-9月。产云南西北部。缅甸东北部亦有。

Perennial herbs. Basal leaves reniform, sheath amplexicaul, glabrous. Capitula solitary, nodding; involucres hemispherical, dark purple or black, base shortly black pilose; phyllaries in 2 rows; florets all tubular, deep yellowish brown. Achenes yellowish brown, cylindric; pappus white. Fl. Jul-Aug. Fr. Aug-Sep. Grassy slopes at ca. 4000 m. Distributed in NW Yunnan. Also in NE Myanmar.

长柱垂头菊
Cremanthodium rhodocephalum Diels

多年生草本。叶肾形，无叶鞘。单生头状花序少量生于茎顶，下垂；总苞半球形，外部具长紫红色短柔毛；苞片10-16，2层；舌状花紫红；两性花紫红色；冠毛白色。花果期6-9月。生海拔3000-5000米

黑垂头菊 *Cremanthodium atrocapitatum*

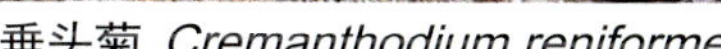
垂头菊 *Cremanthodium reniforme*

喜马拉雅垂头菊 *Cremanthodium decaisnei*

的高山草甸、林缘或山地多石处。产中国西南。

Perennial herbs. Leaves reniform, without sheath. Capitula solitary or few, on apex of stem or branches, nodding; involucres hemispherical, outside long purplish red pilose; phyllaries 10-16, in 2 rows; ligulate florets purplish red; tubular florets purplish red; pappus white. Fl. and fr. Jun-Sep. Alpine meadows, forest edges or gravel zones on mountains at 3000-5000 m. Distributed in SW China.

垂头菊

Cremanthodium reniforme (DC.) Benth.

多年生草本，被紫褐色有节柔毛。叶片肾形、圆肾形或心状肾形，柔软，光滑，叶脉掌状。头状花序单生，下垂，辐射状；总苞半球形；总苞片披针形至长圆形，背部被黑色有节柔毛；舌状花黄色，倒披针形。瘦果倒卵球形；冠毛白色。花果期8-9月。生海拔3300-4500米的林缘或高山草甸。产中国西南。喜马拉雅亦有。

Perennial herbs, purplish-brown septate-villose. Leaves reniform, orbicular-reniform or cordate-reniform, thin, glabrous, nerves palmate. Capitula solitary, drooping, radiate; involucres semiglobose; phyllaries lanceolate to oblong, dark septate-villose abaxially; ray florets yellow, oblanceolate. Achenes obovoid; pappus white. Fl. and fr. Aug-Sep. Forest edges or alpine meadows at 3300-4500 m. Distributed in SW China. Also in Himalaya.

喜马拉雅垂头菊

Cremanthodium decaisnei C. B. Clarke

多年生草本，被褐色有节柔毛。叶具柄，肾形或圆肾形，先端圆形，边缘具圆钝齿，下面密被褐色有节柔毛，叶脉掌状。头状花序单生，下垂，辐射状；总苞半球形；总苞片披针形；舌状花黄色；冠毛白色。花果期7-9月。生海拔3500-5400米的草地、高山草甸或高山流石滩。产青藏高原。喜马拉雅亦有。

Perennial herbs, brown septate-pubescent. Leaves petiolate, reniform or orbicular-reniform, apex rotundate, margin crenate, brown septate-pubescent abaxially, nerves palmate. Capitula solitary, drooping, radiate; involucres semiglobose; phyllaries lanceolate; ray florets yellow; pappus white. Fl. and fr. Jul-Sep. Grasslands, alpine meadows or alpine screes at 3500-5400 m. Distributed in Tibetan Plateau. Also in Himalaya.

车前状垂头菊

Cremanthodium ellisii (Hook. f.) Kitam.

多年生草本。叶近肉质，卵形、椭圆形、长圆形至条形，叶脉羽状。头状花序1(-5)，下垂，辐射状；总苞半球形，近黑色；总苞片披针形，先端急尖，被白色睫毛；舌状花黄色；管状花深黄色；冠毛白色。花果期7-10月。生海拔3400-5600米的高山流石滩、沼泽草地或河滩。产青藏高原。喜马拉雅亦有。

Perennial herbs. Leaves subsucculent, ovate, elliptic, oblong to linear, nerves pinnate. Capitula 1(-5), drooping, radiate; involucres semiglobose, nearly black; phyllaries lanceolate, apex acute, white ciliate; ray florets yellow; tubular florets deep yellow; pappus white. Fl. and fr. Jul-Oct. Alpine screes, marsh meadows or flood plains at 3400-5600 m. Distributed in Tibetan Plateau. Also in Himalaya.

车前状垂头菊 *Cremanthodium ellisii*

条叶垂头菊 *Cremanthodium lineare*

双花华蟹甲 *Sinacalia davidii*

条叶垂头菊

Cremanthodium lineare Maxim.

多年生草本，蓝绿色。叶条形或条状披针形，全缘，光滑。头状花序单生，辐射状，下垂；总苞半球形；总苞片披针形，边缘具白色纤毛；舌状花黄色，条状披针形。瘦果长圆形；冠毛白色。花果期7-10月。生海拔2400-4800米的高山草地、水边、沼泽草地和灌丛中。产四川、西藏、甘肃和青海。

Perennial herbs, aeruginous. Leaves linear or linear-lanceolate, entire, glabrous. Capitula solitary, radiate, drooping; involucres lanceolate, margin ciliate; ray florets yellow, linear-lanceolate. Achenes oblong; pappus white. Fl. and fr. Jul-Oct. Alpine grasslands, watersides, swamps and thickets at 2400-4800 m. Distributed in Sichuan, Xizang, Gansu and Qinghai.

狭叶垂头菊

Cremanthodium angustifolium W. W. Sm.

多年生草本。茎紫红色，被有节长柔毛。叶狭披针形至条形，全缘，光滑，具平行脉。头状花序单生，稀为2个，下垂，盘状；总苞半球形，密被紫褐色有节柔毛；总苞片披针形；小花全部管状。瘦果圆柱形；冠毛白色。花果期7-10月。生海拔3200-4800米的灌丛、草坡、河边或高山沼泽地。产中国西南。

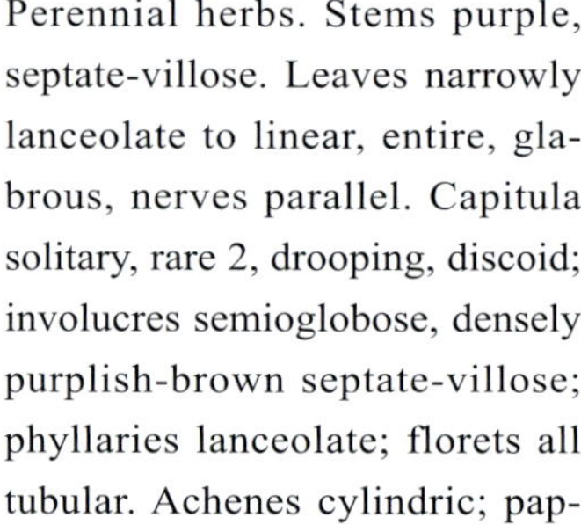

Perennial herbs. Stems purple, septate-villose. Leaves narrowly lanceolate to linear, entire, glabrous, nerves parallel. Capitula solitary, rare 2, drooping, discoid; involucres semioglobose, densely purplish-brown septate-villose; phyllaries lanceolate; florets all tubular. Achenes cylindric; pappus white. Fl. and fr. Jul-Oct. Thickets, grassy slopes, riversides or alpine swamps at 3200-4800 m. Distributed in SW China.

双花华蟹甲

Sinacalia davidii (Franch.) H. Koyama

茎粗壮、中空、无毛。基部及下部茎叶具柄；中部茎叶叶片三角形或五角形，上面浅绿色或

狭叶垂头菊 *Cremanthodium angustifolium*

华蟹甲 *Sinacalia tangutica*

紫色，下面浅绿色；最上部叶卵状三角形，具短柄。头状花排成顶生复圆锥状花序；总苞圆柱形；花冠黄色；冠毛白色。花期7-8月，果期10月。生海拔900-3200米的草坡、悬崖、路边及林缘。产中国西南和陕西。

Stems robust, fistulose, glabrous. Radical and lower stem leaves, petiolate; middle stem leaves with petioles, blade abaxially pale green or purplish, adaxially dark green, triangular or pentagonal; uppermost leaves shortly petiolate, ovate-triangular. Capitula arranged in terminal compound paniculoid thyrses; involucres cylindric; corolla yellow; pappus white. Fl. Jul-Aug. Fr. Oct. Grassy slopes, cliffs, roadsides, forest margins at 900-3200 m. Distributed in SW China and Shaanxi.

华蟹甲

Sinacalia tangutica (Maxim.) B. Nord.

多年生草本，被腺状短柔毛。叶厚纸质，卵形或心形，羽状深裂，侧裂片长圆形，边缘具小尖齿，羽状脉明显。头状花序小，排成多分枝宽塔状复圆锥花序；总苞圆柱状；总苞片5，条状长圆形；舌状花2-3朵，黄色。瘦果圆柱形；冠毛白色。花期7-8月，果期10月。生海拔1200-3500米的草坡、悬崖、沟边、草甸、林缘或路边。产华中、华北和西北。

Perennial herbs, glandular puberulent. Leaves thick chartaceous, ovate or cordate, pinnatiparted, lateral lobes oblong, margin acutely serrulate, nerves conspicuously pinnate. Capitula small, arranged in multi-branched broad pyramidalis compound panicle; involucres cylindric; phyllaries 5, linear-oblong; ray florets 2-3, yellow. Achenes cylindric; pappus white. Fl. Jul-Aug. Fr. Oct. Grassy slopes, cliff, ditchsides, meadows, forest edges or roadsides at 1200-3500 m. Distributed in C, N and NW China.

蟹甲草

Parasenecio forrestii W. W. Sm. et Small

多年生草本。叶具柄，基生叶心形；茎生叶无柄或近无柄，叶片椭圆状长三角形。头状花序排列成窄圆锥状；总苞钟状；花冠淡粉白色。瘦果圆柱形，无毛；冠毛白色。花期8月，果期9-10月。生海拔2300-3700米的林中落叶层。产四川西南部和云南西北部。

Perennial herbs. Leaves petiolate, radical leaf cordate; stem leaves sessile or subsessile, elliptic-triangular. Capitula, narrowly paniculate; involucres campanulate; corolla pale pink-white. Achenes cylindric, glabrous; pappus white. Fl. Aug. Fr. Sep-Oct. Forests at 2300-3700 m. Distributed in SW Sichuan and NW Yunnan.

蟹甲草 *Parasenecio forrestii*

戟状蟹甲草

Parasenecio hastiformis Y. L. Chen

多年生草本。叶具长叶柄，两面异色，中部茎叶叶片宽三角状戟形或卵状三角形；上部叶与中部茎叶同形或卵状三角形，具短柄；最上部叶狭三角形至披针形。头状花序排成偏侧生的总状或圆锥花序；总苞圆柱形或窄钟状；花冠黄色；冠毛褐色。花期9月。生海拔约2400米的山谷溪边。产云南。

Perennial herbs. Leaves long petiolate; median leaf discolorous, broadly triangular-hastate or ovate-triangular; upper leaves similar to median leaves, or ovate-triangular, shortly petiolate; uppermost narrowly triangular to lanceolate. Capitula apically and in upper leaf axils in lateral racemes or panicles; involucres cylindric or narrowly campanulate; corolla yellow; pappus brown. Fl. Sep. Riversides in valleys at ca. 2400 m. Distributed in Yunnan.

戟状蟹甲草 *Parasenecio hastiformis*

秦岭蟹甲草 *Parasenecio tsinlingensis*

黄山蟹甲草

Parasenecio hwangshanicus (Ling) Y. L. Chen

多年生草本。根状茎具束状被绒毛的须根。叶集生于茎中部，叶片宽圆肾形或宽卵圆状心形；上部叶片具短柄，卵状心形；最上部叶苞片状，卵形或卵形披针形。疏圆锥花序；总苞窄钟状圆柱形；花冠黄色；冠毛白色。花期6-8月，果期9月。生海拔1500-1800米的山顶草地或山坡阴湿处。产安徽、浙江、江西和台湾。

Perennial herbs, Rhizomes with fascicled tomentose fibrous roots. Leaves crowded at middle part, broadly reniform or ovate-orbicular; upper leaves shortly petiolate, ovate-cordate; uppermost leaves bractlike, ovate or ovate-lanceolate. Capitula arranged in lax panicles; involucres narrowly campanulate-cylindric; corolla yellow; pappus white. Fl. Jun-Aug. Fr. Sep. Grasslands at mountain summits, shaded wet places on slopes at 1500-1800 m. Distributed in Anhui, Zhejiang, Jiangxi and Taiwan.

黄山蟹甲草 *Parasenecio hwangshanicus*

瓜拉坡蟹甲草 *Parasenecio koualapensis*

瓜拉坡蟹甲草
Parasenecio koualapensis
(Franch.) Y. L. Chen

多年生草本。中部和下部茎叶具柄，叶两面异色，宽卵状三角形或宽卵形，稀菱形或心形；上部叶呈苞叶状，披针形或线状披针形，具短叶柄或近无柄，全缘。头状花序排列成窄圆锥花序；总苞圆柱形；花冠黄色；冠毛白色。花期8-9月，果期10月。生海拔2800-3200米的山坡杂木林下。产云南西北部。

Perennial herbs, Median and lower stem leaves, petioles, blade discolorous, broadly ovate-triangular or broadly ovate, rarely rhombic or cordate; upper leaves shortly petiolate or subsessile, bracteal-leaflike, lanceolate or linear-lanceolate, entire. Capitula arranged in narrow panicles; involucres cylindric; corolla yellow; pappus white. Fl. Aug-Sep. Fr. Oct. Understories of mixed forests on slopes at 2800-3200 m. Distributed in NW Yunnan.

秦岭蟹甲草
Parasenecio tsinlingensis
(Hand.-Mazz.) Y. L. Chen

多年生草本。叶片肾形或卵状心形，基部5-7脉，边缘浅波状齿，叶背有蛛丝状和棕色柔毛。头状花序多数，顶生，排列成总状花序或密圆锥花序；总苞圆柱形，直径4-6毫米；总苞片5；小花5-6，花冠白色，长4-5毫米。冠毛白色，长约5毫米。花期7-8月，果期9月。生海拔1400-1800米的疏林下、山谷阴湿处。产甘肃和陕西。

Perennial herbs. Leaf blade reniform or ovate-cordate, basally 5-7-veined, margin shallowly undulate-lobed, abaxially sordid arachnoid and brown pubescent. Capitula many, terminal, arranged in racemes or dense panicles; involucres cylindric, 4-6 mm diam; phyllaries 5; florets 5-6; corolla white, 4-5 mm; pappus white, ca. 5 mm. Fl. Jul-Aug. Fr. Sep. Understories of lax forests, shaded wet places in valleys at 1400-1800 m. Distributed in Gansu and Shaanxi.

丽江蟹甲草
Parasenecio lidjiangensis
(Hand.-Mazz.) Y. L. Chen

多年生草本。茎具明显的条棱。叶三角形或戟状三角形；上部叶披针形或线状披针形。头状花序排列成总状或复总状花序；总苞圆柱形或狭钟状；花冠黄色。瘦果圆柱形，无毛，具肋；冠毛白色。花期7-8月，果期9-10月。生海拔3400-3500米的山坡松林或冷杉林下。产云南西北部。

Perennial herbs. Stem distinctly striate. Lower leaves triangular or hastate-triangular; upper leaves lanceolate or linear-lanceolate. Capitula arranged in racemes or compound racemes; involucres cylindric or narrowly campanulate; corolla yellow. Achenes cylindric, glabrous, ribbed; pappus white. Fl. Jul-Aug. Fr. Sep-Oct. Understories of *Pinus* or *Picea* forests on slopes at 3400-3500 m. Distributed in NW Yunnan.

深山蟹甲草
Parasenecio profundorum
(Dunn) Y. L. Chen

多年生草本。叶具长叶柄，基部近抱茎，宽卵形或卵状菱形；上部叶有短叶柄。头状花序排列成疏散的圆锥花序；总苞圆柱形；花冠黄色。瘦果圆柱形，无毛而具肋；冠毛白色。花果期8-9月。生海拔1000-2100米的山坡林缘或山谷潮湿处。产重庆和湖北。

Perennial herbs. Leaves long petiolate, basally subamplexicaul, blade broadly ovate or ovate-rhombic; upper leaves shortly petiolate. Capitula arranged in terminal lax panicles; involucres cylindric; corolla yellow. Achenes cylindric, glabrous, ribbed; pappus white. Fl. and fr. Aug-Sep. Forest margins on slopes, wet places in valleys at 1000-2100 m. Distributed in Chongqing and Hubei.

丽江蟹甲草 *Parasenecio lidjiangensis*

深山蟹甲草 *Parasenecio profundorum*

五裂蟹甲草 *Parasenecio quinquelobus*

五裂蟹甲草
Parasenecio quinquelobus (Wall. ex DC.) Y. L. Chen

多年生草本。中部茎叶具柄，两面异色，肾状五角形或宽卵状五角；上部叶具短柄，三角形或长三角形；最上部叶狭披针形或线形，苞叶状。窄圆锥花序；总苞圆柱形；花冠黄色；冠毛白色。花期8月，果期9-10月。生海拔2800-4100米的栎林和冷杉林下及高山草甸。产中国西南。喜马拉雅亦有。

Perennial herbs. Median leaves, petioles, blade discolorous, reniform-pentagonal to triangular or lanceolate; upper leaves shortly petiolate, triangular or narrowly triangular; uppermost leaves narrowly lanceolate or linear, bracteal-leaflike. Capitula arranged in terminal narrow or broad panicles; involucres cylindric; corolla yellow; pappus white. Fl. Aug. Fr. Sep-Oct. Understories of *Quercus* or *Abies* forests, alpine meadows at 2800-4100 m. Distributed in SW China. Also in Himalaya.

川西蟹甲草
Parasenecio souliei (Franch.) Y. L. Chen

多年生草本。茎常紫红色，具纵条纹，无毛。叶常着生于茎中下部及中部，具柄，肾状三角形或宽卵形；上部叶卵形或卵状披针形，基部有缺刻或小裂片。头状花序排列成疏总状狭圆锥花序；总苞圆柱形或窄钟状；花冠黄色；冠毛白色。花果期8-9月。生海拔3100-3700米的森林下层植物或灌层、草甸、树阴或湿山坡。产四川西部。

Perennial herbs. Stem often purple-red, striate, glabrous. Leaves usually at middle and lower part of stem, petioles, blade reniform-triangular or broadly ovate; upper leaves ovate or ovate-lanceolate, incised or lobulate at base. Capitula numerous, arranged in lax racemose panicles; involucres cylindric or narrowly campanulate; corolla yellow; pappus white. Fl. and fr. Aug-Sep. Understories of forests or thickets, grasslands, shaded and wet slopes at 3100-3700 m. Distributed in W Sichuan.

山尖子
Parasenecio hastatus (L.) H. Koyama

多年生草本。中部叶叶柄长4-5厘米，具狭翅。叶三角状戟形，边缘具不规则密锯齿。头状花序多数，下垂，顶部及上部叶腋狭金字塔状圆锥形；苞片7或8；小花8-15(-20)；花冠白色。瘦果棕灰色，圆柱形，具肋纹。花期7-8月，果期9-10月。生林下、山坡、林缘草地或路边。产华北、华西和东北。俄罗斯(远东地区)、蒙古和朝鲜半岛亦有。

Perennial herbs. Median leaf petioles 4-5 cm long, narrowly winged. Leaves triangular-hastate, margin irregularly finely toothed. Capitula numerous, pendulous, apically and in upper leaf axils narrowly pyramidal-paniculate; phyllaries 7 or 8; florets 8-15(-20); corolla whitish. Achenes pale brownish, cylindrical, ribbed. Fl. Jul-Aug. Fr. Sep-Oct. Forests, slopes, grasslands along forest edges or roadsides. Distributed in N, W and NE China. Also in Russia (Far East), Mongolia and Korean Peninsula.

川西蟹甲草 *Parasenecio souliei*

蛛毛蟹甲草
Parasenecio roborowskii (Maxim.) Y. L. Chen

多年生草本。茎高60-100厘米，不分枝。叶卵状三角形或长三角形，边缘有不规则的锯齿。头状花序多数，通常在茎端或上部叶腋排列成塔状疏圆锥状花序偏向一侧着生，下垂；小苞片线形或线状披针形；总苞圆柱形；总苞片3-4，稀2；小花通常3-4，稀1-2，花冠白色；冠毛白色。花期7-8月，果期9-10月。生海拔1700-3400米山坡林下、林缘、灌丛和草地。产陕西、甘肃、青海、四川和云南。

Perennial herbs. Stems 60-100 cm tall, simple. Leaf blade ovate-triangular or narrowly triangular, margin irregularly serrate, teeth mucronulate. Capitula numerous, usually terminal or in upper leaf axils, arranged in pyramidal lax panicles on one side, spreading pendulous; involucres cylindric;

山尖子 *Parasenecio hastatus*

phyllaries 3- 4, rarely 2; florets usually 3-4, rarely 1-2; corolla and pappus white. Fl. Jul-Aug. Fr. Sep-Oct. Understories of forests on slopes, forest margins, thickets and grasslands at 1700-3400 m. Distributed in Shaanxi, Gansu, Qinghai, Sichuan and Yunnan.

耳翼蟹甲草

Parasenecio otopteryx (Hand.-Mazz.) Y. L. Chen

多年生草本。茎生叶叶柄具宽翅，扩大成一耳状抱茎；叶宽卵状心形或宽心形，基部具3出脉。头状花序多数顶生，排成复总状；苞片(3-)5；小花3(-5)；花冠黄白色。瘦果棕色，圆柱形，具肋纹。花果期7-9月。生海拔1400-2800米的山坡林下、林缘或灌丛阴湿处。产四川、湖北、河南和陕西。

Perennial herbs. Stem leaves petioles with wide wings, expanded into an amplexicaul large auricle; leaves broadly ovate-cordate or broadly cordate, basally 3-veined. Capitula numerous, terminal, compound racemose; phyllaries (3-)5; florets 3(-5); corolla yellow-white. Achenes brown, cylindrical, ribbed. Fl. and fr. Jul-Sep. Forests on slopes, forest edges or shaded places in thickets at 1400-2800 m. Distributed in Sichuan, Hubei, Henan and Shaanxi.

蛛毛蟹甲草 *Parasenecio roborowskii*

耳翼蟹甲草 *Parasenecio otopteryx*

兔儿风蟹甲草 *Parasenecio ainsliaeiflorus*

大理蟹甲草 *Parasenecio taliensis*

兔儿风蟹甲草

Parasenecio ainsliaeiflorus (Franch.) Y. L. Chen

多年生草本。中部叶心状肾形或圆肾形，基部5出脉，具长柄，柄无翅；叶缘常5-7三角状裂。头状花序小，多数，顶生或生于上部叶腋，排成总状或复总状；苞片5；小花5；花冠白色；冠毛白色或污白色。花期7-8月，果期9-10月。生海拔1500-2600米的林缘、林下、灌丛或山坡草地。产四川、贵州、湖北和湖南。

Perennial herbs. Median cordate-reniform or orbicular-reniform, basally 5-veined, long petiolate, petioles not winged; leaves margin often 5-7-triangularly lobed. Capitula small, numerous, terminal or in upper leaf axils, racemose or compound racemose; phyllaries 5; florets 5; corolla white; pappus white or sordid white. Fl. Jul-Aug. Fr. Sep-Oct. Forests edges, forests, thickets or grasslands on slopes at 1500-2600 m. Distributed in Sichuan, Guizhou, Hubei and Hunan.

大理蟹甲草

Parasenecio taliensis (Franch.) Y. L. Chen

多年生草本。中部及上部叶疏松，具长梗，宽卵形或卵状扁圆形，膜质或近膜质，基部具5出脉，边缘不明显波状或具不规则三角状粗齿。头状花序5-7，顶生或上部叶叶腋，排成总状或复总状；苞片5；小花(6-)8或9；花冠黄色。花期7-8月，果期9月。生海拔3000-3400米的开阔山坡或冷杉林下。产云南西北部。

Perennial herbs. Leaves lax at middle to upper part, long petiolate, broadly ovate or ovate-oblate, membranous or submembranous, basally 5-veined, margin inconspicuously undulate or irregularly triangular coarsely toothed. Capitula 5-7, terminal or upper leaf axils, arranged in racemes or compound racemes; phyllaries 5; florets (6-)8 or 9; corolla yellow. Fl. Jul-Aug. Fr. Sep. Open slopes or understories of *Abies* forests at 3000-3400 m. Distributed in NW Yunnan.

掌裂蟹甲草

Parasenecio palmatisectus (Jeffrey) Y. L. Chen.

多年生草本。叶具长梗；中部叶宽卵状圆形或五角状心形，羽裂至5-7掌裂。头状花序多数，排成顶生的总状或疏松的圆锥状，花期后开展或下垂；苞片4；小花4或5；花冠黄色。瘦果圆柱形。花期7-8月，果期9-10月。生海拔2400-3800米林下、林缘或山坡灌丛。产中国西南。不丹亦有。

掌裂蟹甲草 *Parasenecio palmatisectus*

兔儿伞 *Syneilesis aconitifolia*

长白蜂斗菜 *Petasites rubellus*

Perennial herbs. Leaves long petiolate; median leaves broadly ovate-orbicular or pentagonous-cordate, pinnate-palmately 5-7-divided. Capitula many, arranged in terminal racemes or lax panicles, spreading or pendulous after anthesis; phyllaries 4; florets 4 or 5; corolla yellow. Achenes cylindrical. Fl. Jul-Aug. Fr. Sep-Oct. Forests, forest edges or thickets on slopes at 2400-3800 m. Distributed in SW China. Also in Bhutan.

兔儿伞

Syneilesis aconitifolia (Bunge) Maxim.

多年生草本。茎直立，紫褐色，无毛，不分枝。叶具长柄，盾状圆形，掌状深裂，被密蛛丝状绒毛，渐变无毛。头状花序多数，在茎端密集成复伞房状；总苞筒状；总苞片淡粉红色；小花8-10，管状，淡粉白色。瘦果圆柱形；冠毛污白色。花期6-7月，果期8-10月。生海拔500-1800米的山坡、荒地、林缘或路旁。产中国东南、华中、华东和东北。东北亚亦有。

Perennial herbs. Stems erect, purplish-brown, glabrous, solitary. Leaves long petiolate, clypeiform, palmately parted, densely arachnoid villose, glabrescent. Capitula numerous, clustered as compound corymb on top of stem; involucres cylindric; phyllaries pinkish; florets 8-10, tubular, pinkish-white. Achenes cylindric; pappus dirty white. Fl. Jun-Jul. Fr. Aug-Oct. Slopes, wastelands, forest edges or roadsides at 500-1800 m. Distributed in SE, C, E and NE China. Also in NE Asia.

款冬

Tussilago farfara L.

多年生草本。茎具鳞片状紫色苞叶。叶圆心形，边缘波状。单个头状花序顶生，稍下垂；舌状花雌性，多层，辐射状。瘦果圆柱形；冠毛白色。生海拔600-3400米的湿地、林下或沟谷。产中国西南、东南、华中、华北、华西、华东、西北和东北。南亚、东北亚、西南亚、北非和欧洲西部亦有。

Perennial herbs. Stems with scale-shaped alternate purple-violet bracteate leaves. Leaves orbicular-cordate, margin undulate. Capitula solitary, terminal, nodding; ray florets pistillate, many seriate, radiate. Achenes cylindrical; pappus white. Wet places, forests or valleys at 600-3400 m. Distributed in SW, SE, C, N, W, E, NW and NE China. Also in S, NE and SW Asia, N Africa and W Europe.

长白蜂斗菜

Petasites rubellus (J. F. Gmel.) Toman

多年生草本。基生叶具长柄，叶两面异色，叶片肾形或肾状心形，边缘有具小尖头波状粗齿；茎生叶鳞片状，抱茎，卵状披针形。头状花序排成伞房状花序；总苞倒锥状；花冠黄色；冠毛白色。花果期5-7月。生海拔1800-2800米的高山地区、林下或林缘。产中国东北。东北亚亦有。

Perennial herbs. Basal leaves long petiolate; blade discolorous, reniform or reniform-cordate, margin acutely toothed, teeth emarginate, apex rounded; stem leaves scale-shaped, ovate-lanceolate, basally amplexicaul. Capitula arranged in corymbs; involucres conical; corolla yellow; pappus white. Fl. and fr. May-Jul. Alpine regions, adjacent forests, forest margins at 1800-2800 m. Distributed in NE China. Also in NE Asia.

款冬 *Tussilago farfara*

毛裂蜂斗菜 *Petasites tricholobus*

毛裂蜂斗菜

Petasites tricholobus Franch.

多年生草本，被白色薄蛛丝状绵毛。基生叶具长柄，宽肾状心形，边缘具细齿。雌花茎高27-60厘米；雌头状花序多数，排成密集聚伞圆锥花序，总苞钟状，总苞片披针形；花冠顶端4-5撕裂；雄头状花序排成伞房状花序。瘦果圆柱形；冠毛白色。花期4-6月。生海拔700-4300米的山谷、路旁或水旁。产中国西南和西北。印度北部、尼泊尔和越南亦有。

Perennial herbs, thinly white arachnoid tomentose. Basal leaves long petiolate, broadly reniform-cordate, margin serrulate. Female scapes 27-60 cm tall; female capitula numerous, arranged in dense thyrse, involucres campanulate, phyllaries lanceolate; corolla 4-5-lacerated; male capitula arranged in corymb. Achenes cylindric; pappus white. Fl. Apr-Jun. Valleys, roadsides or watersides at 700-4300 m. Distributed in SW and NW China. Also in N India, Nepal and Vietnam.

蜂斗菜

Petasites japonicus (Siebold et Zucc.) Maxim.

多年生草本。基生叶具长梗；叶纸质，基部心形，边缘具细锯齿。花莛具20-30朵头状花序，集成顶生伞房状；所有花为管状完全可育花；小花多数；花冠线形，白色；冠毛为多数密硬毛。花期4-5月，果期6月。生溪边、草地或灌丛。产中国西南、东南、华中、华北、华西和华东。东北亚亦有。

Perennial herbs. Basal leaves long petiolate; leaves papery, base cordate, margin finely toothed. Scapes with 20-30 capitula crowded in a terminal corymb; phyllaries biseriate; all florets tubular, perfect, sterile; florets numerous; corolla filiform, white; pappus of many fine bristles. Fl. Apr-May. Fr. Jun. Streamsides, grasslands or thickets. Distributed in SW, SE, C, N, W and E China. Also in NE Asia.

蜂斗菜 *Petasites japonicus*

滇黔蒲儿根

Sinosenecio bodinieri (Vaniot) B. Nord.

多年生草本，具莛、根状茎。叶基生，莲座状，具长柄，两面异色，叶片圆形或近圆形。头状花序排列成顶生伞房状；总苞钟状；花冠黄色。瘦果圆柱形，无毛；冠毛白色。花期3-4月，果期5月。生于海拔900-1000米的山麓、溪流边及林下阴湿处。产贵州和重庆。

Perennial herbs, scapigerous, rhizomes. Leaves rosulate, petioles, discolorous, broadly ovate or broadly elliptic. Capitula arranged in terminal corymbs; involucres campanulate; corolla yellow. Achenes cylindric, glabrous; pappus white. Fl. Mar-Apr. Fr. May. Riverbanks, waterfalls, rocky cliffs at 900-1000 m. Distributed in Guizhou and Chongqing.

滇黔蒲儿根 *Sinosenecio bodinieri*

仙客来蒲儿根
Sinosenecio cyclaminifolius (Franch.) B. Nord.

多年生具匍匐枝葶状草本。叶基生，莲座状，具长柄，叶片卵状心形或稀近圆形。头状花序单生于茎端，花葶具几个小苞片，小苞片膜质，紫色，全缘或具疏齿；总苞倒锥状钟形；花冠黄色；冠毛髯毛状，易脱落。花期4-5月，果期5-6月。生海拔1300-1900米的树木、河边、草地或岩石处。产四川东部和重庆。

Perennial herbs, scapigerous, stoloniferous, rhizomes. Leaves radical, rosulate, long petiolate, ovate-cordate or rarely suborbicular. Capitula solitary, scape with a few bracteoles, bracteoles purplish, ovate-lanceolate, membranous, margin entire or sparsely denticulate; involucres obconic-campanulate; corolla yellow; pappus scant, soon deciduous. Fl. Apr-May. Fr. May-Jun. Woods, streamsides, rocky places, grasslands at 1300-1900 m. Distributed in E Sichuan and Chongqing.

七裂蒲儿根
Sinosenecio septilobus (C. C. Chang) B. Nord.

近葶状多年生草本，具根状茎。基生叶莲座状，具长柄，叶片轮廓圆形；茎叶掌状分裂，具短柄。头状花序排列成近伞形状伞房花序；总苞近半球形；花冠黄色；冠毛白色。花期5-7月，果期6-7月。生海拔400-2300米的灌丛、岩上或路边。产重庆和贵州。

仙客来蒲儿根 *Sinosenecio cyclaminifolius*

Perennial herbs, subscapigerous, rhizomes. Radical leaves rosulate, long petiolate; stem leaves palmately lobed, shortly petiolate. Capitula arranged in terminal subumbelliform corymbs; involucres campanulate; corolla yellow; pappus white. Fl. May-Jul. Fr. Jun-Jul. Thickets, rocks, bushes or roadsides at 400-2300 m. Distributed in Chongqing and Guizhou.

白脉蒲儿根
Sinosenecio albonervius Y. Liu et Q. E. Yang

多年生草本。基生叶肾形至圆肾形，7-9掌状浅裂，叶脉掌状，白色。头状花序伞房状排列。果圆柱状，无毛；冠毛无。花果期4-6月。生海拔800-1200米的溪边阴湿处。产湖北与湖南交界的壶平山。

Perennial herbs. Radical leaves reniform to orbicular-reniform, shallowly palmately 7-9-lobed, palmately white veined. Capitula many in apical compound corymbs. Achenes cylindrical, glabrous; pappus absent. Fl. and fr. Apr-Jun. Shady places along streamside at 800-1200 m. Distributed in Huping Mountain between the border of Hubei and Hunan.

七裂蒲儿根 *Sinosenecio septilobus*

白脉蒲儿根 *Sinosenecio albonervius*

保靖蒲儿根

Sinosenecio baojingensis Y. Liu et Q. E. Yang

多年生草本。基生叶卵状心形，掌状叶脉，基部心形，边缘有细齿。头状花序复伞房状排列；总苞钟状；冠毛白色。花果期3-5月。生海拔约270米的稻田旁、山坡、深草丛。产湖南西北部。

Perennial herbs. Radical leaves ovate-cordate, palmately veined, base cordate, margin denticulate. Capitula in apical compound corymbs; involucres campanulate; pappus white. Fl. and fr. Mar-May. Deep grass near paddy-fields on an open hillside at ca. 270 m. Distributed in NW Hunan.

保靖蒲儿根 *Sinosenecio baojingensis*

壶瓶山蒲儿根

Sinosenecio hupingshanensis Y. Liu et Q. E. Yang

多年生草本。基生叶肾形至圆肾形，掌状叶脉。头状花序伞房状排列；总苞钟状，无副萼。瘦果圆柱形，无毛。花果期4-6月。生海拔1000-1600米的混交林林缘阴处。产湖北与湖南交界的壶平山。

Perennial herbs. Radical leaves reniform to orbicular-reniform, palmately veined. Capitula many in apical compound corymbs; involucres campanulate, ecalyculate. Achenes cylindrical, glabrous; pappus white. Fl. and fr. Apr-Jun. Shady places along the margin of mixed evergreen and deciduous broad-leaved forests at 1000-1600 m. Distributed in Huping Mountain between the border of Hubei and Hunan.

江西蒲儿根

Sinosenecio jiangxiensis Y. Liu et Q. E. Yang

多年生草本。基生叶莲座状，近圆形或肾形，掌状5-7脉，基部心形到深心形。茎生叶无或1，苞片状。头状花序单生，稀

壶瓶山蒲儿根 *Sinosenecio hupingshanensis*

江西蒲儿根 *Sinosenecio jiangxiensis*

吉首蒲儿根 *Sinosenecio jishouensis*

2-3。瘦果被柔毛；冠毛白色。花果期4-6月。生海拔1400-1700米的松林溪边草丛或岩石上，或山顶岩石上。产江西。

Perennial herbs. Radical leaves rosulate, suborbicular or reniform, palmately 5-7-veined, base cordate to deeply cordate. Cauline leaf absent or 1, bract-like. Capitula solitary, rarely 2-3. Achenes pubescent; pappus white. Fl. and fr. Apr-Jun. Rocks or in grass along streamside in pine forests, or dry rocks at mountain summits at 1400-1700 m. Distributed in Jiangxi.

吉首蒲儿根

Sinosenecio jishouensis D. G. Zhang, Y. Liu et Q. E. Yang

多年生草本。叶椭圆状披针形，羽状叶脉3-5。头状花序伞房状排列；总苞钟状，无副萼；冠毛无。花果期3-5月。生海拔约250米的溪边岩石上或瀑布附近。产湖南(吉首)。

Perennial herbs. Leaves elliptic-lanceolate, pinnately 3-5-veined. Capitula in apical corymbs; involucres campanulate, ecalyculate; pappus absent. Fl. and fr. Mar-May. Limestone rocks along streams or under waterfalls at ca. 250 m. Distributed in Hunan (Jishou).

四川蒲儿根

Sinosenecio sichuanicus Y. Liu et Q. E. Yang

多年生草本。叶宽心形至肾形，叶脉掌状。头状花序单生或少数；总苞倒圆锥状钟形，无副萼。冠毛在成熟时期通常缺失。花果期5-6月。生海拔1300-2400米的落叶阔叶林缘的溪边草坡或岩石上。产四川西部。

Perennial herbs. Leaves broadly cordate to reniform, palmately veined. Capitula solitary or several; involucres obconic-campanulate, ecalyculate; pappus usually absent in mature achenes. Fl. and fr. May-Jun. Grassy slopes or rocks along streamside in deciduous broad-leaved forests at 1300-2400 m. Distributed in W Sichuan.

艺林蒲儿根

Sinosenecio yilingii Y. Liu et Q. E. Yang

多年生草本。叶肾形，有时宽卵形，叶脉掌状。头状花序单生，顶生；总苞倒圆锥状，无副萼；冠毛无。花果期5-7月。生海拔约2200米的林缘溪边草丛或岩石上。产四川(宝兴县)。

Perennial herbs. Leaves reniform or sometimes broadly ovate, palmately veined. Capitula solitary, terminal; involucre obconic, ecalyculate; pappus absent. Fl. and fr. May-Jul. Grasses or rocks of streamside along margin of deciduous broad-leaved forests at ca. 2200 m. Distributed in Sichuan (Baoxing County).

四川蒲儿根 *Sinosenecio sichuanicus*

艺林蒲儿根 *Sinosenecio yilingii*

白背蒲儿根 *Sinosenecio latouchei*

广西蒲儿根 *Sinosenecio guangxiensis*

白背蒲儿根
Sinosenecio latouchei (Jeffrey) B. Nord.

多年生草本，被黄褐色长柔毛或绒毛。基生叶莲座状，具长柄，近圆形，基部心形，边缘掌状浅裂或具粗齿，掌状5-7脉。头状花序3-4个排列成顶生伞房状花序；总苞半球状钟形；舌状花11-13，黄色，长圆形；冠毛白色。花期3-5月，果期5-6月。生海拔200-400米的潮湿沟边或山谷。产福建和江西。

Perennial herbs, yellowish-brown villose or tomentose. Basal leaves rosette, long petiolate, suborbicular, base cordate, margin palmately lobed or gross-dentate, nerves 5-7-palmate. Capitula 3-4, arranged in terminal corymb; involucres semiglobose-campanulate; ray florets 11-13, yellow, oblong; pappus white. Fl. Mar-May. Fr. May-Jun. Damp ditchsides or valleys at 200-400 m. Distributed in Fujian and Jiangxi.

广西蒲儿根
Sinosenecio guangxiensis C. Jeffrey et Y. L. Chen

多年生草本。基生叶莲座状，具长柄，厚纸质，近圆形或肾形，基部心形，下面密被白色绒毛或脱毛呈深紫红色，5-7掌状脉。头状花序2-7个排列成顶生伞房花序；总苞半球形，具8-10个条形外层小苞片。瘦果圆柱形，被短柔毛；冠毛白色。花期5-10月，果期6-11月。生海拔800-2300米的林下、溪边或岩石潮湿处。产广西北部和湖南西南部。

Perennial herbs. Basal leaves rosette, long petiolate, thick chartaceous, suborbicular or reniform, base cordate, densely white tomentose or glabrescent and purple abaxially, nerves 5-7-palmate. Capitula 2-7, arranged in terminal corymb; involucres semiglobose, with 8-10 linear bracteoles. Achenes cylindric, puberulent; pappus white. Fl. May-Oct. Fr. Jun-Nov. Forests, streamsides or damp rocks at 800-2300 m. Distributed in N Guangxi and SW Hunan.

耳柄蒲儿根
Sinosenecio euosmus (Hand.-Mazz.) B. Nord.

具匍枝茎草本。中部茎叶具长柄，叶片卵形或宽卵形，边缘浅裂或有时具5-13较深掌状裂；叶柄基部常常扩大成半抱茎的耳。头状花序5-15，或更多排列成顶生近伞形状伞房花序或复伞房花序；总苞近钟形，无外苞片；总苞片草质，1层；舌状花黄色；冠毛白色。花期7-8月，果期7-9月。生海拔1800-4000米的林中、草地或溪边。产中国西南、湖北、甘肃和陕西。缅甸北部亦有。

Herbs, stoloniferous. Median stem leaves long petiolate; blade ovate, broadly ovate, margin shallowly to sometimes rather deeply palmately divided into 5-13 lobes or large teeth; petioles usually gradually expanded into subamplexicaul auricles. Capitula 5-15 or more arranged in terminal subumbelliform corymbs or compound corymbs; involucres subcampanulate, not calyculate; phyllaries herbaceous, in 1 series; ray florets yellow; pappus white. Fl. Jun-Aug. Fr. Jul-Sep. Woods, grasslands or streamsides at 1800-4000 m. Distributed in SW China, Hubei, Gansu and Shaanxi. Also in N Myanmar.

蒲儿根
Sinosenecio oldhamianus (Maxim.) B. Nord.

多年生或二年生草本，被白色蛛丝状毛和疏长柔毛。叶具柄，卵形，顶端尖，基部心形或楔形，边缘具重锯齿，掌状5脉。头状花序多数，排成顶生复伞房状花序；总苞宽钟状；舌状花约13，黄色，长圆形。瘦果圆柱形；冠毛白色。花期1-12月。生海拔400-2100米的林缘、溪边、潮湿岩石边、草坡或田边。产中国西南、华南、华中、华东、华北和西北。东南亚亦有。

Perennial or biennial herbs, white arachnoid hairy and sparsely villose. Leaves petiolate, ovate, apex acute, base cordate or cuneate, margin doubly serrate,

耳柄蒲儿根 *Sinosenecio euosmus*

蒲儿根 *Sinosenecio oldhamianus*

nerves 5-palmate. Capitula numerous, arranged in terminal compound corymb; involucres broadly campanulate; ray florets ca. 13, yellow, oblong. Achenes cylindric; pappus white. Fl. Jan-Dec. Forest edges, streamsides, damp rocksides, grassy slopes or fieldsides at 400-2100 m. Distributed in SW, S, C, E, N and NE China. Also in SE Asia.

莲座狗舌草

Sinosenecio changii (B. Nord.) B. Nord.

多年生草本。茎葶状，单生，不分枝，被丛绵毛。叶片卵状长圆形，羽状脉，背面被密棉状绒毛，基部心形，边缘具不规则浅波状齿。头状花序6-15，排列成顶生简单或复近伞形状伞房花序；总苞宽钟状至半球形；总苞片线状披针形；花黄色。瘦果圆柱形；冠毛白色，长约3毫米。花期5-7月，果期6-8月。生海拔1800-2700米的路边或林缘。产重庆(南川区)。

Perennial herbs. Stems scapiform, simple, floccose-tomentose. Leaf blade ovate-oblong, pinnately veined, abaxially densely floccose-tomentose, base cordate, margin irregularly shallowly repand-dentate. Capitula 6-15, arranged in simple to compound subumbelliform terminal corymbs; involucres broadly campanulate to hemispheric; phyllaries linear-lanceolate; florets yellow. Achenes cylindric; pappus white, ca. 3 mm. Fl. May-Jul. Fr. Jun-Aug. Roadsides, forest margins at 1800-2700 m. Distributed in Chongqing (Nanchuan District).

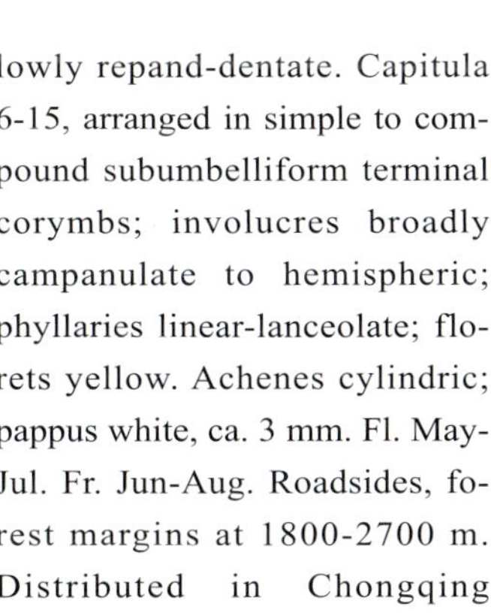

莲座狗舌草 *Sinosenecio changii*

天山狗舌草

Tephroseris turczaninovii (DC.) Holub

多年生草本。茎单生，不分枝，高10-20厘米，被紫色腺毛。基生叶莲座状，具长柄，长圆状卵形，边缘全缘或有波状齿；中部茎叶无柄，长圆状披针形或披针形。头状花序径2-6个排成顶生伞房花序；总苞宽钟状；总苞片18-20，密被深紫色腺毛；舌状花深黄色；冠毛白色。花期7-8月。生海拔3000米以下的岩石坡中草甸。产新疆北部。蒙古和俄罗斯(西伯利亚)亦有。

Perennial herbs. Stems 10-20 cm tall, simple, glandular villous with purplish hairs in upper part. Radical leaves rosulate, blade oblong-ovate, margin entire or sinuate-dentate; median stem leaves sessile, oblong-lanceolate or lanceolate. Capitula 2-6 arranged in terminal corymbs; involucres broadly campanulate; phyllaries 18-20, densely glandular villous with dark purple hairs; ray florets deep yellow; pappus white. Fl. Jul-Aug. Montane meadows among rocky slopes below 3000 m. Distributed in N Xinjiang. Also in Mongolia and Russia (Siberia).

天山狗舌草 *Tephroseris turczaninovii*

橙舌狗舌草 *Tephroseris rufa*

橙舌狗舌草

Tephroseris rufa (Hand.-Mazz.) B. Nord.

多年生草本。茎单生、具绵状绒毛。叶卵形、椭圆形或倒披针形，纸质，两面初疏具蛛丝状绒毛，后无毛。头状花序辐射状，2-20朵排列成顶生近伞形伞房花序；总苞钟形；舌状花黄色；管状花花冠橘黄色至橘红色或黄色，具橘红色裂片；冠毛稍红色。花期6-8月。生海拔2600-4000米的山地草甸、山坡或路边。产中国西南、华北、华西和西北。

Perennial herbs. Stems solitary, white floccose-tomentose. Leaves ovate, elliptic or oblanceolate, papery, both surfaces at first sparsely arachnoid-tomentose, glabrescent. Capitula radiate, 2-20 arranged in terminal subumbelliform corymbs; involucres campanulate; ray florets yellow; disk florets corolla orange to orange-red or yellow with orange lobes; pappus somewhat rubescent. Fl. Jun-Aug. Montane meadows, slopes or roadsides at 2600-4000 m. Distributed in SW, N, W and NW China.

长白狗舌草 *Tephroseris phaeantha*

长白狗舌草

Tephroseris phaeantha (Nakai) C. Jeffrey et Y. L. Chen

多年生草本，被蛛丝状毛。茎单生，近葶状。基生叶莲座状，卵状长圆形或椭圆形，基部心形或截形，边缘具锯齿。头状花序2-8，排成顶生伞形状伞房花序；总苞钟状；总苞片披针形，紫色；舌状花约13朵，黄色。瘦果圆柱形；冠毛白色。花期7-8月。生海拔2000-2500米的多石山坡。产

吉林(长白山)。朝鲜半岛亦有。

Perennial herbs, arachnoid hairy. Stems solitary, scapiform. Basal leaves rosette, ovate-oblong or elliptic, base cordate or truncate, margin serrate. Capitula 2-8, arranged in terminal umbellate-corymb; involucres campanulate; phyllaries lanceolate, purple; ray florets ca. 13, yellow. Achenes terete; pappus white. Fl. Jul-Aug. Stony slopes at 2000-2500 m. Distributed in Jilin (Changbai Mountain). Also in Korean Peninsula.

狗舌草

Tephroseris kirilowii (Turcz. ex DC.) Holub

多年生草本，密被白色蛛丝状绒毛。茎单生，近葶状。基生叶莲座状，长圆形或卵状长圆形，基部楔形。头状花序3-11，排成顶生伞形状伞房花序；总苞近圆柱状钟形；总苞片披针形，绿色；舌状花13-15，黄色。瘦果圆柱形；冠毛白色。花期2-8月。生海拔200-2000米的山坡草地或山顶

狗舌草 *Tephroseris kirilowii*

阳处。中国西南、华中、华东、华北和东北广布。东北亚亦有。

Perennial herbs, densely white arachnoid tomentose. Stems solitary, scapiform. Basal leaves rosette, oblong or ovate-elliptic, base cuneate. Capitula 3-11, arranged in terminal umbellate-corymb; involucres subcylindric-campanulate; phyllaries lanceolate, green; ray florets 13-15, yellow. Achenes terete; pappus white. Fl. Feb-Aug. Grassy slopes or sunny summits at 200-2000 m. Widely distributed in SW, C, E, N and NE China. Also in NE Asia.

红轮狗舌草

Tephroseris flammea (Turcz. ex DC.) Holub

多年生草本。茎下部叶厚纸质，基部渐狭成具翅叶柄，半抱茎，叶柄稍下延。2-9个头状花序排成顶生近聚伞状伞房花序；苞片约25，深紫色；舌状花13-15，深橘黄色；两性花多数，黄色或黄紫色；冠毛米白色。花期7-8月。生海拔1200-2100米的山地草甸或林中。产华中、华北、华西和东北。俄罗斯亦有。

Perennial herbs. Lower stem leaves thickly papery, base cuneately narrowed into winged, subamplexicaul, slightly decurrent petioles. Capitula 2-9 arranged in ± umbelliform terminal corymbs; phyllaries ca. 25, dark purple; ray florets 13-15, dark orange; disc florets many; corolla yellow or purplish yellow; pappus whitish. Fl. Jul-Aug. Montane meadows or forests at 1200-2100 m. Distributed in C, N, W and NE China. Also in Russia.

红轮狗舌草 *Tephroseris flammea*

台湾刘寄奴 *Nemosenecio formosanus*

台湾刘寄奴

Nemosenecio formosanus (Kitam.) B. Nord.

二年生草本，有匍匐茎。叶片卵状长圆形，羽状深裂或近羽状深裂，每边各有侧裂片3-6，裂片长圆状卵形或长圆形，尖至钝，通常具3-6细齿。头状花序少数到多数，排列成近伞形花序，直径8-10毫米；总苞杯状；总苞片1层，狭长圆形；冠毛白色，长4.5-5毫米，宿存。花期8月。生海拔2300-2900米的潮湿山坡。产台湾(高雄)。

Biennial herbs, stoloniferous. Leaves ovate-oblong, pinnatipartite or subbipinnatipartite with 3-6 lateral lobes on each side; lateral lobes ovate-oblong or oblong, usually 3-6-denticulate, apex acute to obtuse. Capitula few to many, pseudoumbellate, pedunculate, 8-10 mm across ray florets; involucres cupuliform; phyllaries uniseriate, narrowly oblong; pappus persistent, white, 4.5-5 mm. Fl. Aug. Mountain slopes at 2300-2900 m. Distributed in Taiwan (Kaohsiung).

缅甸合耳菊

Synotis birmanica C. Jeffrey et Y. L. Chen

多年生草本，具根状茎。叶无柄或近无柄，椭圆形至倒披针形，边缘具较疏尖细齿，纸质。头状花序具舌状花，排列成较密的伞房花序；总苞狭钟状；花冠黄色；冠毛禾秆色。花期9月。生海拔3000-3300米的高山草坡。产云南西部。缅甸亦有。

Perennial herbs, rhizomatous. Leaves sessile or subpetiolate; blade elliptic to oblanceolate, margin rather remotely denticulate with mucronulate teeth, apex shortly acuminate. Capitula radiate, dense terminal corymbs; involucres narrowly campanulate; corolla yellow; pappus straw-colored. Fl. Sep. Grassy slopes at 3000-3300 m. Distributed in W Yunnan. Also in Myanmar.

缅甸合耳菊 *Synotis birmanica*

美头合耳菊 *Synotis calocephala*

丽江合耳菊 *Synotis lucorum*

美头合耳菊

Synotis calocephala C. Jeffrey et Y. L. Chen

灌木状草本或亚灌木。茎被黄褐色柔毛或脱毛。叶具短柄，椭圆形或披针状椭圆形，边缘具尖细齿或细锯齿，先端渐尖。头状花序具舌状花，排成较密的复伞房花序；总苞圆柱形；花冠黄色；冠毛禾秆色。花期8-9月。生海拔2100-2700米的灌丛边。产云南西部。缅甸亦有。

Subshrubs or shrubby herbs. Stems fulvous pubescent or glabrescent. Leaves shortly petiolate, blade elliptic or lanceolate-elliptic, margin apiculate-denticulate or serrulate, apex acuminate. Capitula radiate, arranged in terminal rather dense compound corymbs; involucres cylindric; corolla yellow; pappus straw-colored. Fl. Aug-Sep. Thicket margins at 2100-2700 m. Distributed in W Yunnan. Also in Myanmar.

丽江合耳菊

Synotis lucorum (Franch.) C. Jeffrey et Y. L. Chen

多年生草本，具根状茎。茎被黄褐色柔毛，通常脱毛。叶无柄或稀近有柄，狭长圆状椭圆形至披针形；上部叶披针形。头状花序具同形小花，无舌状花，排成较密伞房花序；总苞狭钟状；花冠黄色；冠毛禾秆色。花期7-8月。生海拔2800-4000米的森林、灌丛及开阔山坡。产云南西北部和四川西南部。

Perennial herbs, rhizomatous. Stems fulvous pubescent, usually ± glabrescent. Leaves sessile or rarely subpetiolate, blade narrowly oblong-elliptic to lanceolate; upper leaves lanceolate. Capitula homogamous, discoid, dense terminal corymbs; involucres narrowly campanulate; corolla yellow; pappus straw-colored. Fl. Jul-Aug. Forests, thickets, open slopes at 2800-4000 m. Distributed in NW Yunnan and SW Sichuan.

腺毛合耳菊

Synotis saluenensis (Diels) C. Jeffrey et Y. L. Chen

灌木状草本或亚灌木。叶具短柄，椭圆形或椭圆状披针形，边缘具细至较粗的不规则具小尖锯齿。头状花序具异形小花，排列成圆形复伞房状；总苞狭钟状；花冠黄色；冠毛白色。花期10月至翌年2月。生海拔1000-3000米的林下、林缘及灌丛边。产西藏东南部和云南西北部。缅甸和越南亦有。

Subshrubs or shrubby herbs. Leaves shortly petiolate, blade elliptic or elliptic-lanceolate, margin finely to rather coarsely and irregularly mucronate-serrate. Capitula heterogamous, disciform, arranged in lax to rather dense rounded compound corymbs; involucres narrowly campanulate; corolla yellow; pappus white. Fl. Oct to next Feb. Forests, forest and thicket margins at 1000-3000 m. Distributed in SE Xizang and NW Yunnan. Also in Myanmar and Vietnam.

四花合耳菊

Synotis tetrantha (DC.) C. Jeffrey et Y. L. Chen

攀援多年生草本。茎被柔短柔毛或近无毛。叶具柄，叶片卵形或卵状披针形，边缘具不规则深波状锯齿，齿端具尖头，膜质。头状花序具异形小花，排列成圆锥状圆锥聚伞花序；总苞狭圆柱形；花冠黄色；冠毛白色或淡红色。花期8-9月。生海拔2300-2700米的林内或混交林中。产西藏南部。喜马拉雅亦有。

Perennial herbs, scandent. Stems sparsely pubescent or subglabrous. Leaves petiolate, blade ovate or ovate-lanceolate, margin mucronately irregularly sinuate-serrate or dentate, apex acutely long acuminate, membranous. Capitula heterogamous, minutely radiate, arranged in terminal and axillary paniculoid thyrse; involucres narrowly cylindric; corolla yellow; pappus white or pale reddish. Fl. Aug-Sep. Mixed woods or forests at

腺毛合耳菊 *Synotis saluenensis*

四花合耳菊 *Synotis tetrantha*

川西合耳菊 *Synotis solidaginea*

2300-2700 m. Distributed in S Xizang. Also in Himalaya.

密花合耳菊

Synotis cappa (Buch.-Ham. ex D. Don) C. Jeffrey et Y. L. Chen

灌木状草本或亚灌木。叶纸质，披针形或长圆形，下面被黄褐色柔毛和白色绒毛，羽状脉。头状花序辐射状，多数，排成复伞房花序或圆锥状聚伞花序；总苞狭钟状；总苞片条状披针形，草质；舌状花8。瘦果圆柱形；冠毛白色。花期10月至翌年1月。生海拔1500-2300米的林缘、灌丛、溪边或草地。产中国西南。南亚和东南亚亦有。

Suffrutescent herbs or subshrubs. Leaves chartaceous, lanceolate or oblong, yellowish-brown pubescent and white tomentose abaxially, nerves pinnate. Capitula radiate, numerous, arranged in compound corymb or paniculate-cyme; involucres narrowly campanulate; phyllaries linear-lanceolate, herbaceous; ray florets 8. Achenes cylindric; pappus white. Fl. Oct to next Jan. Forest margin, thickets, stream-sides or grasslands at 1500-2300 m. Distributed in SW China. Also in S and SE Asia.

川西合耳菊

Synotis solidaginea (Hand.-Mazz.) C. Jeffrey et Y. L. Chen

多年生簇生草本。叶披针形或椭圆状长圆形，基部偏斜，边缘具规则密尖锯齿。头状花序盘状，多数，排成塔状复圆锥聚伞花序；总苞狭圆柱形；总苞片长圆形，近革质；小花3，管状，两性，淡黄色或乳黄色。瘦果圆柱形，被柔毛；冠毛白色。花期7-10月。生海拔2900-3900米的开阔阳坡。产中国西南。

Perennial fasciate herbs. Leaves lanceolate or elliptic-oblong, base oblique, margin regularly and densely acute-serrate. Capitula discoid, numerous, arranged in pyramidalis compound paniculate-cyme. involucres narrowly cylindric; phyllaries oblong, subcoriaceous; florets 3, tubular, bisexual, yellowish or milk-yellow. Achenes cylindric, villose; pappus white. Fl. Jul-Oct. Open sunny slopes at 2900-3900 m. Distributed in SW China.

密花合耳菊 *Synotis cappa*

尾尖合耳菊 *Synotis acuminata*

翅柄合耳菊 *Synotis alata*

尾尖合耳菊

Synotis acuminata (Wall. ex DC.) C. Jeffrey et Y. L. Chen

多年生草本。茎直立，高40-120厘米，上部分枝。叶片狭椭圆形至披针形，边缘有疏而细小的锯齿。头状花序多数，在茎枝顶端排成顶生复伞房花序；总苞狭圆柱形；总苞片3-4；舌状花1，花冠黄色；冠毛禾秆色。花期8-9月。生海拔2600-3400米的山谷林缘。产西藏南部。喜马拉雅亦有。

Perennial herbs. Stems erect, 40-120 cm tall, usually corymbosely branched in upper part. Leaves narrowly elliptic to lanceolate, margin rather remotely mucronate-serrulate. Capitula minutely radiate, numerous, arranged in terminal compound corymbs; involucres narrowly cylindric; phyllaries 3-4; ray floret 1; corolla yellow; pappus straw-colored. Fl. Aug-Sep. Forest margins at 2600-3400 m. Distributed in S Xizang. Also in Himalaya.

术叶合耳菊 *Synotis atractylidifolia*

术叶合耳菊

Synotis atractylidifolia (Ling) C. Jeffrey et Y. L. Chen

亚灌木。茎数个，高20-60厘米。叶柄有或无，叶片披针形。头状花序数个至较多数在茎端及枝端排成顶生复伞房花序；花总苞近钟形；苞总片8，长圆形到线形；舌花3-5，舌片黄色；冠毛白色。花期8月。生海拔1500-2300米的山谷中岩石山坡。产宁夏和内蒙古。

Subshrubs. Aerial stems several, 20-60 cm tall. Leaves very shortly petiolate or sessile; blade lanceolate. Capitula radiate, numerous, arranged in terminal compound corymbs; involucres subcampanulate; phyllaries 8; oblong-linear; ray florets 3-5, yellow; pappus white. Fl. Aug. Rocky places in wet valleys at 1500-2300 m. Distributed in Ningxia and Neimenggu.

翅柄合耳菊

Synotis alata C. Jeffrey et Y. L. Chen

多年生草本，高20-60厘米。花茎近葶状。基部叶片近莲座状或稍疏生，叶柄有翅，基部半抱茎；叶片宽卵形或卵形，基部心形、截形或宽楔形。头状花序盘状或辐射状，排成较密的复伞房花序；总苞圆柱形；总苞片4-5；舌花0-2；冠毛白色。花期8-11月。生海拔1900-4000米的混交林中。产贵州、西藏南部和云南。喜马拉雅亦有。

Perennial herbs, 20-60 cm tall. Flowering stem subscapiform. Leaves at base of flowering stems subrosulate or somewhat distant; petioles winged, base subamplexicaul; blade broadly ovate to lanceolate, base cordate, truncate or broadly cuneate. Capitula disciform or minutely radiate, arranged in narrowly to broadly pyramidal thyrsoid corymbs or branched thyrsoid panicles; involucres cylindric; phyllaries 4 or 5; ray florets 0-2; pappus white. Fl. Aug-Nov. Mixed forests at 1900-4000 m. Distributed in Guizhou, S Xizang and Yunnan. Also in Himalaya.

肇骞合耳菊

Synotis changiana Y. L. Chen

多年生草本。茎直立，高40-50厘米，不分枝。叶在茎中部莲座状或近莲座状，叶片倒卵状匙形或椭圆状倒卵形；上部叶小，线形。头状花序2-5个在茎端或花序枝上排列成伞房状；总苞钟状，总苞片9-11；舌状花5-6，舌片黄色；冠毛白色。花期10-11月。生海拔400-1000米的疏林中。产广西。

肇骞合耳菊 *Synotis changiana*

红缨合耳菊 *Synotis erythropappa*

Perennial herbs, rhizomatous, 40-50 cm tall. Leaves usually densely crowded at middle part, rosulate or subrosulate, shortly petiolate; blade obovate-spatulate or elliptic-obovate; upper leaves linear, small. Capitula usually 2-5 in terminal corymbs; involucres campanulate, phyllaries 9-11; ray florets 5 or 6, yellow; pappus white. Fl. Oct-Nov. Lax forests margin at 400-1000 m. Distributed in Guangxi.

红缨合耳菊

Synotis erythropappa (Bureau et Franch.) C. Jeffrey et Y. L. Chen

多年生草本。茎单生或数个，高达100厘米。叶卵形、卵状披针形或长圆状披针形。头状花序多数，排列顶生或上部腋生的圆锥状聚伞花序；总苞狭圆柱形；总苞片2-3；舌花无；管状花2-3；冠毛污白色至淡红褐色。花期7-10月。生海拔1500-3900米林缘或灌丛边、草坡。产湖北西部、四川、西藏东南部和云南。

Perennial herbs. Stems to 100 cm tall. Leaf blade ovate, ovate-lanceolate or oblong-lanceolate. Capitula arranged in terminal and upper axillary, broadly pyramidal compound thyrses; involucres narrowly cylindric; phyllaries 2 or 3; ray florets absent; disk florets 2-3; pappus dirty white to pale reddish brown. Fl. Jul-Oct. Forests, thicket margins, open grassy places at 1500-3900 m. Distributed in W Hubei, Sichuan, SE Xizang and Yunnan.

木里合耳菊

Synotis muliensis Y. L. Chen

多年生草本，高20-60厘米。叶片三角状卵形或卵状披针形，掌状3-5裂。头状花序多数，在茎端和上部叶腋排列成圆锥花序；总苞圆柱形；总苞片5；小花全部管状，5；冠毛白色。花果期8-9月。生海拔2400-2700米的草坡或岩石坡。产四川西南部。

Perennial herbs, 20-60 cm tall. Leaf blade triangular-ovate or ovate-lanceolate, palmately 3-5-veined. Capitula discoid, in terminal and upper axillary pyramidal thyrses, pedunculate or subsessile; involucres cylindric; phyllaries 5; florets 5, all tubular; pappus white. Fl. and fr. Aug-Sep. Forest margins or rocky places at 2400-2700 m. Distributed in SW Sichuan.

木里合耳菊 *Synotis muliensis*

锯叶合耳菊 *Synotis nagensium*

锯叶合耳菊

Synotis nagensium (C. B. Clarke) C. Jeffrey et Y. L. Chen

多年生草本。茎密被白色绒毛。叶具短柄，倒卵状椭圆形，倒披针状椭圆形或椭圆形，叶被密被白色、褐色或最初为褐色的柔毛，在脉处尤为明显。头状花序具多数小花，排成锥形聚伞花序；总苞倒锥状钟形；总苞片13-15，密被绒毛；边缘小花12-13，黄色。花期8月至翌年3月。生海拔100-2000米的森林、灌丛及草甸。产中国西南、华南、华中和甘肃。印度东北部、缅甸北部和泰国北部亦有。

Perennial herbs. Stem densely whitish tomentose or fulvous tomentose. Leaf blade obovate-elliptic, oblanceolate-elliptic or elliptic, abaxially densely whitish tomentose or fulvous tomentose and brownish setulose especially on veinsnt. Capitula disciform or minutely radiate, arranged in terminal and upper axillary paniculoid thyrses. Phyllaries very densely tomentose. Female florets 12-13. Fl. Aug to next Mar. Woods, thickets, meadows at 100-2000 m. Distributed in SW, S and C China, and Gansu. Also in NE India, N Myanmar and N Thailand.

红脉合耳菊

Synotis rufinervis (DC.) C. Jeffrey et Y. L. Chen

亚灌木，高达1.5米。茎和枝条被灰棕色绒毛。叶片长圆状卵圆形或椭圆形，边缘有锐尖的锯齿。头状花序排列成密集的聚伞圆锥状伞房花序；总苞钟形，被白色绒毛；总苞片线状长圆形；舌花0-4，黄色；管状花3-4。花果期7-11月。生海拔2500-3000米的林缘。产西藏西南部。印度西北部和尼泊尔亦有。

Subshrubs, up to 1.5 m tall. Stems, branches greyish tomentose with brownish pubescence. Leaves oblong-ovate or elliptic, sharply mucronulate toothed. Capitula arranged in rather dense compound thyrsoid corymbs; bracteoles 4-5, linear; involucres campanulate, sparsely white tomentose; phyllaries linear-oblong; ray florets 3-4, yellow; disc florets 3-4. Fl. and fr. Jul-Nov. Forest margins at 2500-3000 m. Distributed in SW Xizang. Also in NW India and Nepal.

合耳菊

Synotis wallichii (DC.) C. Jeffrey et Y. L. Chen

多年生草本，具地下茎。叶密集于花茎基部，叶片宽卵圆形至卵圆形，边缘有疏锯齿。头状花序盘状或不明显辐射状，排列成密集的聚伞圆锥状伞房花序；

红脉合耳菊 *Synotis rufinervis*

合耳菊 *Synotis wallichii*

藤菊 *Cissampelopsis volubilis*

总苞圆柱形；雌花2，两性管状花；冠毛白色。花期8-9月。生海拔2300-3000米的林中。产西藏南部。喜马拉雅亦有。

Perennial herbs, rhizomatous. Leaves subrosulate at base of flowering stems; blade broadly ovate or ovate, margin remotely, obscurely to coarsely sinuate-dentate with mucronulate teeth. Capitula disciform or minutely radiate, arranged in rather dense compound thyrsoid corymbs; involucres cylindric; phyllaries 5; female florets 2; bisexual florets 3; pappus white. Fl. Aug-Sep. Mixed forests at 2300-3000 m. Distributed in S Xizang. Also in Himalaya.

藤菊
Cissampelopsis volubilis (Blume) Miq.

高大草质藤本或亚灌木，疏被白色蛛丝状绒毛。叶互生，卵形或宽卵形，下面被灰白色绒毛，基生5-7掌状脉。头状花序盘状，多数，排成顶生或腋生复伞房花序；总苞圆柱形；总苞片条状长圆形；小花8-10朵，白色、淡黄色或粉色。瘦果圆柱状；冠毛白色。花期10月至翌年1月。生海拔800-3000米的疏林或灌丛，攀援于乔木或灌木上。产云南、贵州、广西、广东和海南。南亚和东南亚亦有。

Large herbaceous vines or sub-shrubs, sparsely white arachnoid-tomentose. Leaves alternate, ovate or broadly ovate, offwhite tomentose abaxially, palmate veins 5-7, basal. Capitula discoid, numerous, arranged in terminal or axillary corymb; involucres cylindric; phyllaries linear-oblong; florets 8-10, white, yellowish or purple. Achenes cylindric; pappus white. Fl. Oct to next Jan. Sparse forests or thickets at 800-3000 m, climbing on trees or shrubs. Distributed in Yunnan, Guizhou, Guangxi, Guangdong and Hainan. Also in S and SE Asia.

双舌千里光
Senecio biligulatus W. W. Sm.

多年生草本，具根状茎。茎单生，高60-90厘米。下部和中部茎叶三角形或三角状披针形；上部叶无柄，三角状披针形至线状披针形。头状花序，有舌状花，复伞房花序；总苞圆柱形；花冠黄色；冠毛白色。花期6-9月。生海拔3000-3900米的山坡开阔处。产西藏南部。东喜马拉雅亦有。

Perennial herbs, rhizomatous. Stem solitary, 60-90 cm tall. Lower and median stem leaf, triangular or triangular-lanceolate; upper leaves sessile, triangular-lanceolate to linear-lanceolate. Capitula, radiate, arranged in dense terminal compound corymbs; involucres cylindric; corolla yellow; pappus white. Fl. Jun-Sep. Open places at 3000-3900 m. Distributed in S Xizang. Also in E Himalaya.

瓜叶千里光
Senecio cinarifolius H. Lévl.

多年生草本，具根状茎。茎单生，高60-120厘米。中部茎叶全形卵状椭圆形或卵状长圆形；上部茎叶具少数浅裂片；最上部叶线形或线状披针形。头状花序具舌状花，伞房花序；总苞圆柱形或狭钟状；花冠黄色；冠毛白色。花期8-9月。生海拔2300-3200米的山坡草甸。产云南。

Perennial herbs, rhizomatous. Stem solitary, 60-120 cm tall. Median stem leaf ovate-elliptic or ovate-oblong; upper leaves few, less deeply lobed; uppermost leaves linear-lanceolate or linear. Capitula radiate, corymbs; involucres cylindric or narrowly campanulate; corolla yellow; pappus white. Fl. Aug-Sep. Montane meadows at 2300-3200 m. Distributed in Yunnan.

双舌千里光 *Senecio biligulatus*

瓜叶千里光 *Senecio cinarifolius*

芥叶千里光 *Senecio desfontainei*

异羽千里光 *Senecio diversipinnus*

芥叶千里光
Senecio desfontainei Druce

一年生草本。叶无柄，全形长圆形，羽状浅裂至羽状全裂；最上部叶长圆状线形或线形，羽状浅裂或具齿。头状花序有舌状花，疏伞房花序；总苞狭钟状；花冠黄色；冠毛白色。花期7-8月。生海拔3100-4600米的溪边。产西藏。喜马拉雅西北部、西南亚、北非和加那利群岛亦有。

Annual herbs. Leaves sessile, oblong, pinnatifid to pinnatisect; uppermost leaves oblong-linear or linear, pinnatifid or dentate. Capitula radiate, laxly corymbose; involucres narrowly campanulate; corolla yellow; pappus white. Fl. Jul-Aug. Gravelly places by streams at 3100-4600 m. Distributed in Xizang. Also in NW Himalaya, SW Asia, N Africa and Canary Islands.

异羽千里光
Senecio diversipinnus Ling

多年生草本，具根状茎。基生叶和下部茎叶全形倒披针状匙形，大头羽状分裂；中部茎叶与下部茎叶具短柄或无柄；上部茎叶无柄。头状花序有舌状花或无舌状花，复伞房花序；总苞狭钟状；花冠黄色；冠毛白色。花期6-8月，果期9-10月。生海拔1900-3800米的开阔草坡、岩石山坡和灌丛。产四川、青海和甘肃。

Perennial herbs, rhizomatous. Basal and lower stem leaves oblanceolate-spatulate, lyrate-pinnatipartite with large; median stem leaves shortly petiolate or sessile; upper stem leaves sessile. Capitula radiate or discoid, corymbs; involucres narrowly campanulate; corolla yellow; pappus white. Fl. Jun-Aug. Fr. Sep-Oct. Open grassy and rocky slopes at 1900-3800 m. Distributed in Sichuan, Qinghai and Gansu.

黑缘千里光
Senecio dodrans C. Winkl.

多年生草本，具根状茎。基生叶具长柄，宽卵形至近圆形；下部叶与基生叶同形；中部茎叶卵形至近圆形；上部叶无柄，披针形，渐尖边缘有不规则的齿或撕裂。头状花序有舌状花，伞房花序；总苞钟状或半球形，花冠黄色，冠毛白色。花果期8-9月。生海拔3800-4400米的高山草甸。产四川北部。

Perennial herbs, rhizomatous. Basal leaves broadly ovate to suborbicular; lower stem leaves similar to basal leaves; median stem leaf ovate to suborbicular; upper leaves sessile, lanceolate, margin irregularly dentate or lacerate, apex acuminate. Capitula radiate, corymb; involucres campanulate or hemispheric; corolla yellow; pappus white. Fl. and fr. Aug-Sep. Alpine meadows at 3800-4400 m. Distributed in N Sichuan.

黑缘千里光 *Senecio dodrans*

裸缨千里光
Senecio echaetus Y. L. Chen et K. Y. Pan

多年生草本，具根状茎。基部叶具柄；下部和中部茎叶无柄，全形长圆状披针形，羽状浅裂或近羽状深裂；上部叶长圆状披针形至披针形，具羽状细裂。头状花序有舌状花，伞房花序；总苞钟状或半球形；花冠黄色；全部小花无冠毛。花期7-10月。生海拔2700-3700米的铁杉林下或草坡。产西藏(吉隆县)。尼泊尔亦有。

Perennial herbs, rhizomatous. Basal leaves petiolate; lower and median stem leaves sessile, blade oblong-lanceolate, pinnatifid or subpinnatisect; upper leaves oblong-lanceolate to lanceolate, pinnately lobulate. Capitula radiate, corymbs; involucres campanulate or hemispheric; corolla yellow; pappus absent from all florets. Fl. Jul-Oct. *Tsuga* forests or grassy slopes at 2700-3700 m. Distributed in Xizang (Gyirong County). Also in Nepal.

裸缨千里光 *Senecio echaetus*

峨眉千里光

Senecio faberi Hemsl. ex Hemsl.

多年生高大草本，具匍匐茎。基生叶具长柄，卵形，大头羽状分裂；下部和中部茎叶大头羽状浅裂；上部叶无柄，卵状披针形至长圆形；最上部叶线状披针形至线形。头状花序有舌状花，复伞房花序；总苞狭钟状；花冠黄色；冠毛白色。花期6-8月。生海拔900-2700米的林下、灌丛及草坡、阴湿处。产四川、贵州和陕西。

Perennial large herbs, stoloniferous. Basal leaves long petiolate, blade ovate, margin lyrate-pinnate with large irregularly coarsely dentate; lower and median stem leaves lyrate-pinnatifid; upper leaves sessile, blade ovate-lanceolate to oblong; uppermost leaves linear-lanceolate to linear. Capitula radiate, corymbs; involucres narrowly campanulate; corolla yellow; pappus white. Fl. Jun-Aug. Forests, thickets, grassy slopes, shaded wet places at 900-2700 m. Distributed in Sichuan, Guizhou and Shaanxi.

匍枝千里光

Senecio filifer Franch.

多年生草本，具匍匐枝和根状茎。基生叶具柄，提琴形或大头羽状；中部茎叶长圆状披针形，无柄或有具宽翅的叶柄；上部叶无柄，披针形或线状披针形；最上部叶较狭，顶端渐狭或尾状。头状花序有舌花，伞房花序；总苞狭钟状；花冠黄色；冠毛白色。花期5-8月。生海拔700-3700米的混交林下潮湿处、灌丛边缘和草坡。产四川西南部、云南。

Perennial herbs, stoloniferous, rhizomatous. Basal leaves, blade pandurate or usually lyrate; median stem leaves with broadly winged petioles or sessile, oblong-lanceolate; upper leaves sessile, lanceolate or linear-lanceolate; uppermost leaves narrower, apically attenuate or caudate. Capitula simple or in compound subumbelliform corymbs; involucres narrowly campanulate; corolla yellow; pappus white. Fl. May-Aug. Wet places in mixed woods, thicket margins and grassy slopes at 700-3700 m. Distributed in SW Sichuan and Yunnan.

峨眉千里光 *Senecio faberi*

匍枝千里光 *Senecio filifer*

纤花千里光 *Senecio graciliflorus*

纤花千里光
Senecio graciliflorus (Wall.) DC.

多年生草本，具根状茎。叶具柄，中部茎叶全形卵形或卵状长圆形，羽状分裂；上部叶浅裂。头状花序具不明显的舌状花，复伞房花序；总苞狭圆柱形；花冠黄色；冠毛白色。花期5-10月。生海拔2000-4100米的草坡、林缘、林中开阔处或溪边。产中国西南。喜马拉雅和马来西亚西部亦有。

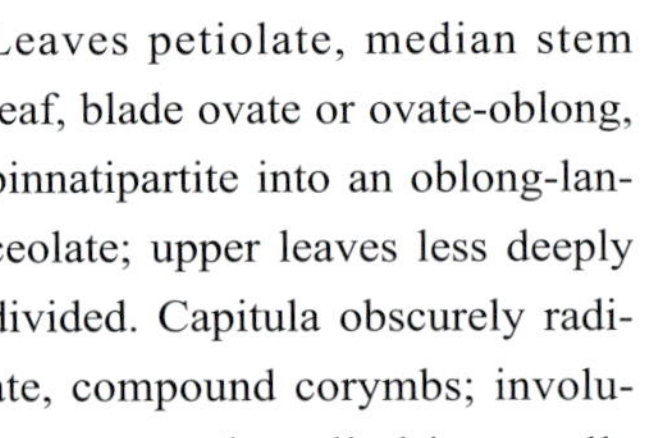

Perennial herbs, rhizomatous. Leaves petiolate, median stem leaf, blade ovate or ovate-oblong, pinnatipartite into an oblong-lanceolate; upper leaves less deeply divided. Capitula obscurely radiate, compound corymbs; involucres narrowly cylindric; corolla yellow; pappus white. Fl. May-Oct. Open places in forests, forest margins, grassy slopes, or streamsides at 2000-4100 m. Distributed in SW China. Also in Himalaya and W Malaysia.

新疆千里光
Senecio jacobaea L.

二年生草本，具根状茎。下部茎叶具柄，全形长圆状倒卵形，具钝齿或大头羽状浅裂；中部茎叶无柄，羽状全裂；上部叶同形，但较小。头状花序有舌状花，复伞房花序；总苞宽钟状或半球形；花冠黄色；冠毛白色。花期5-7月。生海拔500-2000米的疏林或草甸。产新疆。欧洲、中亚、蒙古和俄罗斯(西伯利亚)亦有。

Biennial, herbs, rhizomatous. Lower leaves blade oblong-obovate, obtusely dentate or lyrate-pinnatifid; median stem leaves sessile, pinnatisect; upper leaves similar but smaller. Capitula radiate, compound corymbs; involucres broadly campanulate or hemispheric; corolla yellow; pappus white. Fl. May-Jul. Meadows or lax forests at 500-2000 m. Distributed in Xinjiang. Also in Europe, C Asia, Mongolia and Russia (Siberia).

工布千里光
Senecio kongboensis Ludlow

多年生草本。基生叶莲座状，具柄，倒披针状长圆形或匙形；茎叶近无柄，长圆形或长圆状披针形，具疏生齿至深锯齿，或大头羽状浅裂，尖端钝。头状花序具舌状花，伞房花序；总苞钟状；花冠黄色；冠毛白色。花期8-9月。生海拔3600-3900米的高山草甸和流石滩湿处。产西藏。

Perennial, herbs. Basal leaves rosulate, blade oblanceolate-oblong or spatulate; stem leaves subsessile, blade oblong or oblong-lanceolate, margin remotely dentate to deeply serrate or lyrate-pinnatifid, apex obtuse. Capitula radiate, corymb; involucres campanulate; corolla yellow; pappus white. Fl. Aug-Sep. Alpine grasslands and wet scree slopes at 3600-3900 m. Distributed in Xizang.

须弥千里光
Senecio kumaonensis Duthie ex C. Jeffrey et Y. L. Chen

多年生草本。中部茎叶具长柄，叶片卵形或卵状长圆形；上部

新疆千里光 *Senecio jacobaea*

工布千里光 *Senecio kongboensis*

须弥千里光 *Senecio kumaonensis*

菊状千里光 *Senecio analogus*

叶狭披针形，基部楔形。头状花序盘状，下垂，圆锥花序；总苞圆柱形；花冠黄色；冠毛污白色。花期6-8月。生海拔3600-4500米灌丛或草坡。产西藏南部。喜马拉雅亦有。

Perennial herbs. Median stem leaves long petiolate; blade ovate or ovate-oblong; upper leaves lanceolate, base cuneate. Capitula discoid, pendulous, panicles; involucres cylindric; corolla yellow; pappus dirty white. Fl. Jun-Aug. Thickets or grassy slopes at 3600-4500 m. Distributed in S Xizang. Also in Himalaya.

菊状千里光
Senecio analogus DC.

多年生草本。基生叶和最下部茎叶具柄；中部茎叶椭圆形或倒披针状长圆形；上部叶长圆状披针形或长圆状线形。头状花序有舌状花，伞房花序或复伞房花序；总苞钟状；花冠黄色；冠毛污白色、禾秆色或稀淡红色。花期4月。生海拔1100-3800米的林下、林缘、开阔草坡、田边和路边。产中国西南和华中。喜马拉雅亦有。

Perennial herbs. Basal and lowest stem leaves petiolate; median stem leaves oblong or oblanceolate-oblong; upper leaves oblong-lanceolate to oblong-linear. Capitula radiate, corymbs or compound corymbs; involucres campanulate; corolla yellow; pappus dirty white, straw-colored or rarely rufous. Fl. Apr. Forests, forest and thicket margins, open grassy places, field margins and roadsides at 1100-3800 m. Distributed in SW and C China. Also in Himalaya.

凉山千里光
Senecio liangshanensis C. Jeffrey et Y. L. Chen

多年生草本，具根状茎。中部茎叶具柄，叶片狭三角形；上部叶叶柄较短，具较宽的翅和明显叶耳；最上部叶无柄，披针形，长渐尖，边缘具细锯齿。头状花序盘状，排列成顶生伞房状花束；总苞圆柱状钟形；花冠黄色；冠毛白色。花期8-9月。生海拔2600-3400米的高山草甸和林下。产四川和云南。

Perennial herbs, rhizomatous. Median stem leaves narrowly deltoid; upper leaves with shorter, rather more broadly winged petioles; uppermost leaves sessile, lanceolate-attenuate, base subamplexicaul, margin denticulate. Capitula soon pendulous, discoid, arranged in rather congested terminal corymbiform stalked clusters; involucres cylindric-campanulate; corolla yellow; pappus white. Fl. Aug-Sep. Montane woods and meadows at 2600-3400 m. Distributed in Sichuan and Yunnan.

凉山千里光 *Senecio liangshanensis*

木里千里光 *Senecio muliensis*

风毛菊状千里光 *Senecio saussureoides*

木里千里光

Senecio muliensis C. Jeffrey et Y. L. Chen

多年生草本。基生叶和下部茎叶具长柄；中部茎叶全形狭倒披针状长圆形，大头羽状浅裂；上部茎叶具短柄或近无柄，线状披针形或线形。头状花序无舌状花，下垂，顶生伞房花序；总苞宽钟状；花冠紫色；冠毛白色。花期9-10月。生海拔3800-4300米的高山草地。产四川西南部。

Perennial herbs. Basal and lower leaves, long petiolate; median stem leaf petioles purplish, blade narrowly oblanceolate-oblong; upper stem leaves shortly petiolate or subsessile, linear-lanceolate or linear. Capitula discoid, nodding, terminal corymb; involucres broadly campanulate; corolla purplish; pappus white. Fl. Sep-Oct. Grassy mountain slopes at 3800-4300 m. Distributed in SW Sichuan.

黑苞千里光 *Senecio nigrocinctus*

黑苞千里光

Senecio nigrocinctus Franch.

多年生草本，具根状茎。基生叶和下部茎叶具长柄，叶片卵状三角形；中部茎叶卵状三角形或三角形；上部叶无柄，卵状披针形或长圆状披针形。头状花序盘状，初期直立，后期下垂，密集簇生的圆状伞房状花序；总苞倒锥状钟形；花冠黄色；冠毛白色。花期7-9月。生海拔3200-4000米的高山草甸、开阔山坡和林缘。产西藏东南部和云南西北部。

Perennial herbs, rhizomatous. Basal and lower stem leaves, long petiolate, blade ovate-triangular; median stem leaves, blade ovate-triangular; upper leaves sessile, blade ovate-lanceolate or oblong-lanceolate. Capitula at first erect, then nodding, discoid, dense rounded corymbiform clusters; involucres obconic-campanulate; corolla yellow; pappus white. Fl. Jul-Sep. Alpine meadows, slopes, forest margins at 3200-4000 m. Distributed in SE Xizang and NW Yunnan.

风毛菊状千里光

Senecio saussureoides Hand.-Mazz.

多年生草本。基生叶和下部茎具长柄；中部茎叶全形椭圆形或披针形，羽状深裂；上部叶具短柄或近无柄；最上部叶线形，苞片状。头状花序无舌状花，疏散顶生伞房花序；总苞宽钟状无毛；花冠黄色；冠毛白色。花期9月。生海拔3900-4200米的灌丛草地。产西藏和四川。

Perennial herbs. Basal and lower leaves, long petiolate; median stem leaf, blade elliptic or lanceolate, margin pinnatipartite to (in basal half) pinnatisect; upper leaves, shortly petiolate or subsessile; uppermost leaves linear, bractiform. Capitula discoid, nodding, in lax terminal corymb; involucres broadly campanulate; corolla yellow; pappus white. Fl. Sep. Shrubs or meadows at 3900-4200 m. Distributed in Xizang and Sichuan.

天山千里光

Senecio thianschanicus Regel et Schmalh.

多年生草本。基生叶和下部茎叶近全缘，具浅齿或浅裂；中部茎叶无柄，具浅齿至羽状浅裂，或稀羽状深裂；上部叶全缘或近全缘。头状花序具舌状花，疏伞房花序；总苞钟状；花冠黄色；冠毛白色。花期7-9月。生海拔2400-5000米的草坡、开阔湿处或溪边。产中国西南和西北。俄罗斯、中亚和缅甸亦有。

Perennial herbs, rhizomatous. Basal and lower stem leaves, margin subentire, shallowly den-

天山千里光 *Senecio thianschanicus*

三尖千里光 *Senecio tricuspis*

tate, or pinnatifid; median stem leaves sessile, margin shallowly dentate to pinnatifid or rarely pinnatipartite; upper leaves margin ± entire. Capitula radiate, lax corymbs; involucres campanulate; corolla yellow; pappus white. Fl. Jul-Sep. Grassy slopes, open wet places or streamsides at 2400-5000 m. Distributed in SW and NW China. Also in Russia, C Asia and Myanmar.

三尖千里光
Senecio tricuspis Franch.

多年生草本，具根状茎。中部茎叶具长柄，叶片宽三角形；上部叶与中部叶具短柄。头状花序盘状，下垂，单生排成疏散顶生总状聚伞花序；总苞钟状；花冠黄色；冠毛白色。花期9-10月。生海拔3500-3800米的高山草地。产四川西南部和云南西北部。

Perennial herbs, rhizomatous. Median stem leaves long petiolate, blade broadly triangular; upper leaves shortly petiolate. Capitula nodding, discoid, solitary, terminal and in axils of upper leaves, or forming a lax terminal racemose cyme; involucres campanulate; corolla yellow; pappus white. Fl. Sep-Oct. Montane meadows at 3500-3800 m. Distributed in SW Sichuan and NW Yunnan.

麻叶千里光 *Senecio cannabifolius*

麻叶千里光
Senecio cannabifolius Less.

多年生草本。叶纸质，下面被皱曲柔毛，上面无毛。头状花序辐射状，多数，排成宽顶生复伞房花序；苞片8-10；舌状花8-10，黄色，具3齿。瘦果圆柱形，无毛；冠毛禾秆色。花期7月。生草甸或林中。产河北、内蒙古、黑龙江和吉林。蒙古和东北亚亦有。

Perennial herbs. Leaves papery, abaxially crisped-puberulous, adaxially glabrous. Capitula radiate, numerous, arranged in broad terminal compound corymbs; phyllaries 8-10; ray florets 8-10, yellow, 3-denticulate. Achenes cylindrical, glabrous; pappus stramineous. Fl. Jul. Meadows or forests. Distributed in Hebei, Neimenggu, Heilongjiang and Jilin. Also in Mongolia and NE Asia.

林荫千里光 *Senecio nemorensis*

林荫千里光

Senecio nemorensis L.

多年生直立草本，被疏柔毛或近无毛。叶披针形、长圆状披针形至条形，基部渐狭或半抱茎，羽状脉。头状花序多数，排成复伞房花序；总苞近圆柱形；总苞片长圆形；舌状花8-10，黄色，条状长圆形。瘦果圆柱形；冠毛白色。花期6-12月。生海拔700-3000米的林中开阔处、草地或溪边。广布中国大部分地区。俄罗斯、蒙古、朝鲜半岛、日本和欧洲亦有。

Perennial erect herbs, sparsely pubescent or subglabrous. Leaves lanceolate, oblong-lanceolate to linear, base attenuate or semi-amplexicaul, nerves pinnate. Capitula numerous, arranged in compound corymb; involucres subcylindric; phyllaries oblong; ray florets 8-10, yellow, linear-oblong. Achenes cylindric; pappus white. Fl. Jun-Dec. Open places in forests, grasslands or streamsides at 700-3000 m. Widely distributed throughout most parts of China. Also in Russia, Mongolia, Korean Peninsula, Japan and Europe.

君范千里光

Senecio lingianus C. Jeffrey et Y. L. Chen

多年生草本。叶膜质或薄纸质，下面疏被柔毛，上面无毛；茎中部叶柄具狭翅。头状花序辐射状，多数，排成极密的顶生复伞房状；苞片8；舌状花黄色，顶部深2或3裂。瘦果无毛；冠毛白色。花期8-9月。生海拔3600-4000米的开阔或高山灌丛草甸。产西藏东南部。

Perennial herbs. Leaves membranous or thinly papery, abaxially sparsely puberulous, adaxially glabrous; median stem leaf petioles narrowly winged. Capitula radiate, numerous, arranged in rather crowded terminal compound corymbs; phyllaries 8; rays yellow, apically deeply 2- or 3-lobed. Achenes glabrous; pappus white. Fl. Aug-Sep. Open forests or alpine shrubby meadows at 3600-4000 m. Distributed in SE Xizang.

节花千里光

Senecio nodiflorus C. C. Chang

多年生草本。叶下面白色，密被蛛丝毛，上面疏被蛛丝毛，后渐无毛。头状花序1-15，生花序分枝顶端，极疏松，聚伞圆锥状，稍下垂；苞片约20，脉及上部紫黑色；舌状花18-21，亮黄色。瘦果无毛；冠毛白色。花期8-10月。生海拔3000-4500米的湿生石牧场、多石地或巨砾滩。产云南(德钦县)和西藏东南部。

Perennial herbs. Leaves abaxially white and densely arachnoid, adaxially sparsely arachnoid and ± glabrescent. Capitula 1-15, terminal on inflorescences branches, rather lax, thyrsoid, nodding; phyllaries ca. 20, blackish purple on veins and in upper part; ray florets 18-21, bright yellow. Achenes glabrous; pappus white. Fl. Aug-Oct. Wet stony pastures, rocky places or boulder screes at 3000-4500 m. Distributed in Yunnan (Dêqên County) and SE Xizang.

君范千里光 *Senecio lingianus*

节花千里光 *Senecio nodiflorus*

湖南千里光 *Senecio actinotus*

湖南千里光

Senecio actinotus Hand.-Mazz.

多年生草本。基部和茎生叶三角形，无毛，基部深心形。头状花序辐射状，多数，排成一个充实开展的顶生复伞房花序；总苞6-8，条形；舌状花3朵，顶部3裂；两性花7-9；花冠黄色。瘦果圆柱形，无毛；冠毛白色。花期6-8月。生海拔700-1600米的山地灌丛或荒野地。产广西、贵州和湖南西南部。

Perennial herbs. Basal and lower stem leaves triangular, glabrous, base deeply cordate. Capitula radiate, numerous, arranged in an abundant, spreading terminal compound corymb; phyllaries 6-8, linear; ray florets 3, apically 3-denticulate; disc florets 7-9; corolla yellow. Achenes cylindrical, glabrous; pappus white. Fl. Jun-Aug. Montane thickets or moors at 700-1600 m. Distributed in Guangxi, Guizhou and SW Hunan.

蕨叶千里光

Senecio pteridophyllus Franch.

多年生直立草本，被卷柔毛或变无毛。叶薄纸质，长圆形，羽状分裂。头状花序多数，排成顶生复伞房花序；总苞狭钟状；总苞片条形，上端紫色，被短柔毛；舌状花5，黄色，长圆形；管状花11-13；花冠黄色。瘦果圆柱形，无毛；冠毛白色。花期7-10月。生海拔3000-3800米的高山牧场和草甸。产云南西北部。

蕨叶千里光 *Senecio pteridophyllus*

Perennial erect herbs, curly-pubescent or glabrescent. Leaves thin chartaceous, oblong, pinnatisected. Capitula numerous, arranged in terminal compound corymb; involucres narrowly campanulate; phyllaries linear, apex purple, puberulent; ray florets 5, yellow, oblong; tubular florets 11-13; corolla yellow. Achenes cylindric, glabrous; pappus white. Fl. Jul-Oct. Alpine pastures or meadows at 3000-3800 m. Distributed in NW Yunnan.

额河千里光 *Senecio argunensis*

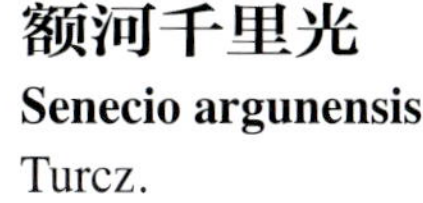

额河千里光

Senecio argunensis Turcz.

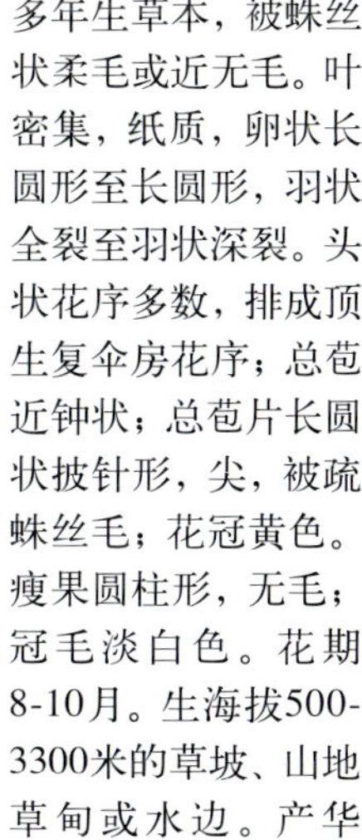

多年生草本，被蛛丝状柔毛或近无毛。叶密集，纸质，卵状长圆形至长圆形，羽状全裂至羽状深裂。头状花序多数，排成顶生复伞房花序；总苞近钟状；总苞片长圆状披针形，尖，被疏蛛丝毛；花冠黄色。瘦果圆柱形，无毛；冠毛淡白色。花期8-10月。生海拔500-3300米的草坡、山地草甸或水边。产华中、华北、东北和西北。俄罗斯、蒙古和朝鲜半岛亦有。

Perennial herbs, arachnoid villose or subglabrous. Leaves dense, chartaceous, ovate-oblong to oblong, pinnatisected to pinnatiparted. Capitula numerous, arranged in terminal compound corymb; involucres subcampanulate; phyllaries oblong-lanceolate, acute, sparsely arachnoid; corolla yellow. Achenes cylindric, glabrous; pappus whitish. Fl. Aug-Oct. Grassy slopes, montane meadows or watersides at 500-3300 m. Distributed in C, N, NE and NW China. Also in Russia, Mongolia and Korean Peninsula.

菊状千里光 *Senecio analogus*

菊状千里光

Senecio analogus DC.

多年生直立草本，被疏蛛丝状毛或变无毛。叶具柄，纸质，椭圆形、长圆形或披针形，羽状分裂。头状花序多数，排成顶生伞房花序；总苞钟状；总苞片长圆状披针形，具柔毛；花冠黄色。瘦果圆柱形，被疏柔毛；冠毛污白色。花期4月。生海拔1100-3800米的林下、林缘、草坡、田边或路边。产云南、西藏、贵州、重庆、湖北和湖南。喜马拉雅亦有。

Perennial erect herbs, arachnoid or glabrescent. Leaves petiolate, chartaceous, elliptic, oblong or lanceolate, pinnatifid. Capitula numerous, arranged in terminal corymb; involucres campanulate; phyllaries oblong-lanceolate, pubescent; corolla yellow. Achenes cylindric, sparsely pubescent; pappus dirty white. Fl. Apr. Forests, forest edges, grassy slopes, fieldsides or roadsides at 1100-3800 m. Distributed in Yunnan, Xizang, Guizhou, Chongqing, Hubei and Hunan. Also in Himalaya.

莱菔叶千里光

Senecio raphanifolius Wall. ex DC.

多年生直立草本，被疏蛛丝状毛或变无毛。叶纸质，长圆形或披针形，羽状分裂，边缘具缺刻状齿或细裂。头状花序多数，排成顶生伞房花序；总苞宽钟状或半球形；舌状花12-16，黄色；冠毛淡红色。花期7-9月。生海拔2700-4400米的山地林下、草甸、草坡或河边。产西藏。喜马拉雅亦有。

Perennial erect herbs, arachnoid or glabrescent. Leaves chartaceous, oblong or lanceolate, pinnatifid, margin incised-dentate or minutely lobed. Capitula numerous, arranged in terminal corymb; involucres campanulate or semiglobose; ray florets 12-16, yellow; pappus reddish. Fl. Jul-Sep. Montane forests, meadows, grassy slopes or riversides at 2700-4400 m. Distributed in Xizang. Also in Himalaya.

莱菔叶千里光 *Senecio raphanifolius*

岩生千里光

Senecio wightii Benth. ex C. B. Clarke

多年生草本。基生叶花期枯萎；叶纸质，椭圆形、披针形至条形，基部楔形或半抱茎，边缘具粗齿或锯齿。头状花序少数，排成顶生疏伞房花序；

岩生千里光 *Senecio wightii*

闽粤千里光 *Senecio stauntonii*

总苞半球形；总苞片具3脉；舌状花11-13，黄色，长圆形。瘦果圆柱形；舌状花无冠毛。花期8-11月。生海拔1100-3000米的溪边、潮湿地或路边。产云南、四川和贵州。印度、不丹和缅甸亦有。

Perennial herbs. Basal leaves withered at anthesis; leaves chartaceous, elliptic, lanceolate to linear, base cuneate or semi-amplexicaul, margin grossly dentate or serrate. Capitula several, arranged in terminal lax corymb; involucres semiglobose; phyllaries triveined; ray florets 11-13, yellow, oblong. Achenes cylindric; pappus absent in ray florets. Fl. Aug-Nov. Riversides, damp places or roadsides at 1100-3000 m. Distributed in Yunnan, Sichuan and Guizhou. Also in India, Bhutan and Myanmar.

闽粤千里光

Senecio stauntonii DC.

多年生草本。茎生叶多数，无柄；叶卵状披针形至狭长圆状披针形，上面散生柔毛至无毛。头状花序辐射状，疏松排列成顶生伞房状；总苞钟形；苞片13，条状披针形；舌状花8-13；冠毛白色。花期10-11月。生海拔约600米的灌丛、开阔林地、石灰山、旱山坡或峡谷。产广西、广东、香港、福建、江西和湖南。

Perennial herbs. Stem leaves numerous, sessile; leaves ovate-lanceolate to narrowly oblong-lanceolate, adaxially scattered pilose to glabrous. Capitula radiate, arranged in lax terminal corymbs; involucres campanulate; phyllaries 13, linear-lanceolate; ray florets 8-13; pappus white. Fl. Oct-Nov. Thickets, open forests, limestone hills, dry slopes or ravines at ca. 600 m. Distributed in Guangxi, Guangdong, Hong Kong, Fujian, Jiangxi and Hunan.

千里光

Senecio scandens Buch.-Ham. ex D. Don

多年生攀援草本。根状茎木质。叶具柄，披针形至狭三角形，边缘具牙齿，羽状脉。头状花序多数，排成顶生大型复聚伞圆锥花序；总苞圆柱状钟形；总苞片条状披针形，具3脉；舌状花8-10，黄色，长圆形，花冠黄色。瘦果圆柱形，被柔毛；冠毛白色。花期8月至翌年4月或10-11月。生海拔4000米以下的森林、灌丛或溪边，常攀援于灌木或岩石上。广布中国南部。喜马拉雅、东南亚和日本亦有。

Perennial climbing herbs. Rhizomes woody. Leaves petiolate, lanceolate to narrowly triangular, margin dentate, nerves pinnate. Capitula numerous, arranged in terminal large compound cymose-panicle; involucres cylindric-campanulate; phyllaries linear-lanceolate, triveined; ray florets 8-10, yellow, oblong; corolla yellow. Achenes cylindric, pubescent; pappus white. Fl. Aug to next Apr or Oct-Dec. Forests, thickets or streamsides, often climbing on shrubs or rocks below 4000 m. Widely distributed throughout southern parts of China. Also in Himalaya, SE Asia and Japan.

千里光 *Senecio scandens*

近全缘千里光 *Senecio subdentatus*

北千里光 *Senecio dubitabilis*

近全缘千里光
Senecio subdentatus Ledeb.

一年生草本。茎单生，纤细。叶无柄，长圆形或宽条形，基部半抱茎，近全缘，具数齿或细裂。头状花序，数个至多数，排成顶生疏散伞房花序；总苞圆柱形；总苞片长圆状披针形；舌状花7-8，黄色，长圆形。瘦果圆柱形，被密柔毛；冠毛白色。花期5-6月。生海拔400-700米多沙砾处。产新疆。俄罗斯、蒙古、中亚和西南亚亦有。

Annual herbs. Stems solitary, slender. Leaves sessile, oblong to broadly linear, base semi-amplexicaul, margin nearly entire, sparsely dentate or minutely lobed. Capitula several to numerous, arranged in terminal lax corymb; involucres cylindric; phyllaries oblong-lanceolate; ray florets 7-8, yellow, oblong. Achenes cylindric, densely pubescent; pappus white. Fl. May-Jun. Sandy or pebbly places at 400-700 m. Distributed in Xinjiang. Also in Russia, Mongolia, C and SW Asia.

北千里光
Senecio dubitabilis C. Jeffrey et Y. L. Chen

一年生草本。下部叶基部渐狭成柄状，中部叶常基部膨大成不规则近抱茎耳状物，羽状浅裂。头状花序盘状，顶部，常排成疏松伞房状；苞片约15；无舌状花；两性花多数；花冠黄色。瘦果圆柱形，密被柔毛；冠毛白色。花期5-9月。生海拔2000-4800米的路边或多沙石地。产中国西南、华北、华西和西北。南亚、中亚和东北亚亦有。

Annual herbs. Lower leaves attenuate into a petioloid base, middle ones often ± expanded at base into irregularly subamplexicaul auricles, pinnately shortly lobulate. Capitula discoid, terminal, usually laxly corymbose; phyllaries ca. 15; ray florets absent; disc florets many; corolla yellow. Achenes cylindrical, densely pubescent; pappus white. Fl. May-Sep. Sandy and rocky places or field margins at 2000-4800 m. Distributed in SW, N, W and NW China. Also in S, C and NE Asia.

欧洲千里光
Senecio vulgaris L.

一年生草本，被蛛丝状毛至无毛。叶倒披针状匙形或长圆形，羽状分裂，边缘具不规则齿。头状花序盘状，排列成顶生密集伞房花序；总苞钟状；总苞片条

细梗千里光 *Senecio krascheninnikovii*

欧洲千里光 *Senecio vulgaris*

形，顶端具长黑色尖头；无舌状花；管状花多数；花冠黄色。瘦果圆柱形；冠毛白色。花期4-10月。生海拔300-2300米的开阔山坡、草地或路旁。产中国西南、华北和东北。广布亚洲、欧洲和北非。

Annual herbs, arachnoid to glabrous. Leaves oblanceolate-spathulate to oblong, pinnatifid, margin irregularly crenate. Capitula discoid, arranged in terminal clustered corymb; involucres campanulate; phyllaries linear, apex with long black acumens; ray florets absent; tubular florets numerous; corolla yellow. Achenes cylindric; pappus white. Fl. Apr-Oct. Open slopes, grasslands or roadsides at 300-2300 m. Distributed in SW, N and NE China. Widely distributed in Asia, Europe and N Africa.

细梗千里光
Senecio krascheninnikovii Schischk.

一年生草本。叶无柄，全形卵状长圆形，羽状浅裂至羽状全裂；上部叶羽状裂至线形，近全缘。头状花序有舌状花，疏伞房花序；总苞狭钟状；花冠黄色；冠毛白色。花期6-9月。生海拔1800-3900米的多沙砾山坡或沙地。产新疆、青海和西藏。中亚、西南亚和俄罗斯亦有。

Annual herbs. Leaves sessile, blade ovate-oblong, pinnatifid to finely and deeply pinnatisect; upper leaves pinnately lobed to linear and subentire. Capitula radiate, laxly corymbose; involucres narrowly campanulate; corolla yellow; pappus white. Fl. Jun-Sep. Gravelly slopes or sandy places at 1800-3900 m. Distributed in Xinjiang, Qinghai and Xizang. Also in C and SW Asia, and Russia.

野茼蒿
Crassocephalum crepidioides (Benth.) S. Moore

一年生草本，光滑无毛。茎肉质。叶膜质，边缘有不规则锯齿或重锯齿。头状花序数个，顶生；总苞钟状，基部截形；总苞片条状披针形；全部小花管状，两性，红褐色或橙红色。瘦果狭圆柱形，赤红色；冠毛白色，绢毛状。花果期7-12月。生海拔300-1800米的山坡、灌丛、路旁、水边或田间。产中国西南、华南、华中和华东。东南亚和非洲亦有。

Annual herbs, glabrous. Stems succulent. Leaves membranous, irregularly serrate or doubly serrate. Capitula several, terminal; involucres campanulate, base truncate; phyllaries linear-lanceolate; all florets tubular, bisexual, reddish-brown or orange. Achenes narrowly cylindric, crimson; pappus white, villose. Fl. Jul-Dec. Slopes, thickets, roadsides, watersides or fields at 300-1800 m. Distributed in SW, S, C and E China. Also in SE Asia and Africa.

野茼蒿 *Crassocephalum crepidioides*

败酱叶菊芹

Erechtites valerianifolius (Link ex Spreng.) DC.

一年生草本，近无毛。叶具柄，长圆形至椭圆形，顶端尖或渐尖，基部斜楔形，边缘具不规则重锯齿或羽状深裂。头状花序多数，排成密集伞房状圆锥花序；总苞圆柱状钟形；总苞片条形；小花全为管状花，淡黄紫色。瘦果圆柱形；冠毛淡红色。花期全年。生海拔1700米以下的田边、路旁。云南、广东、海南和台湾有归化。原产南美洲。

Annual herbs, subglabrous. Leaves petiolate, blade oblong to elliptic, apex acute or acuminate, base obliquely cuneate, margin irregularly double-serrate or pinnately lobed. Capitula numerous, arranged in dense corymbose-panicle; involucres cylindric-campanulate; phyllaries linear; florets all tubular, yellowish-purple. Achenes cylindric; pappus purplish. Fl. year-round. Fieldsides or roadsides below 1700 m. Naturalized in Yunnan, Guangdong, Hainan and Taiwan. Native to South America.

败酱叶菊芹 *Erechtites valerianifolius*

菊三七 *Gynura japonica*

红凤菜

Gynura bicolor DC.

多年生草本，高50-100厘米。叶片倒卵形或倒披针形，稀长圆状披针形；上部和分枝上的叶小，披针形至线状披针形。头状花序多数，在茎、枝端排列成疏伞房状；花序梗细，有1-2(3)丝状苞片；总苞钟状；总苞片1层；小花橙黄色至红色。花果期5-10月。生海拔600-1500米山坡林下、岩石上或河边湿处。产中国西南、华南、东南和华东。缅甸和泰国亦有。

Perennial herbs, 50-100 cm tall. Leaves obovate or oblanceolate, rarely oblong-lanceolate; upper leaves lanceolate to linear-lanceolate, small. Capitula in terminal lax corymbs; peduncles slender, with 1-2(3) filiform bracts; involucres campanulate; bracts of calyculus 1-seriate; florets orange to reddish. Fl. and fr. May-Oct. Forests on slopes, rocky or wet places by rivers at 600-1500 m. Distributed in SW, S, SE and E China. Also in Myanmar and Thailand.

红凤菜 *Gynura bicolor*

菊三七

Gynura japonica (Thunb.) Juel.

多年生高大草本。叶椭圆形或长圆状椭圆形，羽状深裂，下面常紫色，基部有具圆锯齿或羽状裂的叶耳。头状花序多数，排成伞房状圆锥花序；总苞钟状，基部具9-11条形小苞片；小花全部为管状花，花冠黄色或橙黄色。瘦果圆柱形，棕褐色；冠毛白色。花期8-10月。生海拔1200-3000米的山谷、草坡、林下或林缘。产中国西南、华南、华中和华东。尼泊尔、泰国和日本亦有。

Perennial large herbs. Leaves elliptic or oblong-elliptic, pinnatiparted, generally purple aba-

木耳菜 *Gynura cusimbua*

xially, base with crenate or pinnatilobed auricles. Capitula numerous, arranged in corymbose-panicle; involucres campanulate, base with 9-11 linear bracteoles; florets all tubular, corolla yellow or orange. Achenes cylindric, brown; pappus white. Fl. Aug-Oct. Valleys, grassy slopes, forests or forest edges at 1200-3000 m. Distributed in SW, S, C and E China. Also in Nepal, Thailand and Japan.

木耳菜

Gynura cusimbua (D. Don) S. Moore

多年生草本。叶大，无柄或有短柄，叶片长圆状椭圆形、椭圆形或长圆状披针形。头状花序常4-15朵呈顶生聚伞圆锥状；总苞狭钟形或圆柱形；苞片13-15，明显具3条肋纹；小花全部为管状，花冠橘黄色；冠毛白色。花期9-10月。生海拔1300-3400米的林中、山坡、草地或路边。产中国西南。喜马拉雅和东南亚亦有。

Perennial herbs. Leaves large, sessile or shortly petiolate; blade oblong-elliptic, elliptic or oblong-lanceolate. Capitula usually 4-15 in terminal corymbose-panicles; involucres narrowly campanulate or cylindrical; phyllaries 13-15, conspicuously 3-ribbed; florets all tubular, corolla orange; pappus white. Fl. Sep-Oct. Forests, slopes, grasslands or roadsides at 1300-3400 m. Distributed in SW China. Also in Himalaya and SE Asia.

白凤菜

Gynura formosana Kitam.

多年生草本。茎被短糙毛。叶椭圆形或匙形，肉质，边缘具波状小尖齿，叶柄基部有1对耳状假托叶。头状花序2-5，排成疏伞房状；花序梗细长；总苞筒状，基部陀螺形，具数个条形小苞片；总苞片披针形；小花多数；花冠黄色。瘦果圆柱形；冠毛白色。花果期5-7月。生海拔500米以下的沙质海滨。产台湾。

Perennial herbs. Stems shortly strigillose. Leaves elliptic or spathulate, succulent, margin undulate-serrulate with 1 pair of auriculate pseudostipules at base. Capitula 2-5, arranged in lax corymb; peduncles slender; involucres cylindric, base turbinate, with several linear bracteoles; phyllaries lanceolate; florets numerous; corolla yellow. Achenes cylindric; pappus white. Fl. and fr. May-Jul. Sandy shores below 500 m. Distributed in Taiwan.

白凤菜 *Gynura formosana*

平卧菊三七 *Gynura procumbens*

小一点红 *Emilia prenanthoidea*

平卧菊三七

Gynura procumbens (Lour.) Merr.

攀援多年生草本。叶卵形、卵状长圆形或椭圆形，边缘全缘或有齿，先端锐尖或渐尖。头状花序3-5形成顶生或腋生的伞房状花序；总苞钟形或漏斗形；花冠橙色。瘦果圆柱形，无毛，具10肋；冠毛白色，丝状。花期3-4月。生海拔200-900米的溪边林中、沙坡、树上、灌丛或岩石上。产中国西南和华南。东南亚和非洲亦有。

Scandent perennial herbs. Leaves ovate, ovate-oblong, or elliptic, margin entire or repand-dentate, apex acute or acuminate. Capitula 3-5 in terminal or axillary corymbs; involucres campanulate or funnelform; corolla orange. Achenes cylindric, glabrous, 10-ribbed; pappus white, silky. Fl. Mar-Apr. By streams in forests, sandy slopes, climbing on shrubs or trees or rocks at 200-900 m. Distributed in SW and S China. Also in SE Asia and Africa.

一点红 *Emilia sonchifolia*

小一点红

Emilia prenanthoidea DC.

一年生草本。基部叶倒卵形或倒卵状长圆形；中部叶长圆形或线状长圆形，无柄两面异色；上部叶线状披针形。头状花序排列成疏伞房状；总苞片10，长圆形；小花管状，红色或紫红色。瘦果具5肋；冠毛白色。花果期5-10月。生海拔500-2000米的路旁、疏林或林中潮湿处。产中国西南、华南和华东。南亚和东南亚亦有。

Annual herbs. Basal leaves obovate or obovate-oblong; median leaves sessile, discolorous, oblong or linear-oblong; upper leaves linear-lanceolate. Capitula in lax corymbs; phyllaries 10, oblong; florets tubular, red or purple-red. Achenes 5-ribbed; pappus white. Fl. and fr. May-Oct. Roadsides on slopes, lax forests, wet places in forests at 500-2000 m. Distributed in SW, S and E China. Also in S and SE Asia.

一点红

Emilia sonchifolia (L.) DC.

一年生草本，灰绿色，基部分枝。下部叶密集，质厚，大头

海南菊 *Hainanecio hainanensis*

羽状分裂，边缘具不规则齿。头状花序2-5，顶生；总苞圆柱形；总苞片条形，无毛；小花管状，粉红色或紫红色。瘦果圆柱形；冠毛白色。花期7-10月。生海拔2100米以下的山坡、荒地、田边或路旁。产中国西南、华南、华中和华东。泛热带广布。

Annual herbs, glaucous, branched from base. Lower leaves dense, thick, lyrate-pinnatifid, margin irregularly serrate. Capitula 2-5, terminal; involucres cylindric; phyllaries linear, glabrous; florets all tubular, pink or purplish-red. Achenes cylindric; pappus white. Fl. Jul-Oct. Slopes, wastelands, fieldsides or roadsides below 2100 m. Distributed in SW, S, C and E China. Also in pantropical.

海南菊

Hainanecio hainanensis (C. C. Chang et Y. C. Tseng) Y. Liu et Q. E. Yang

多年生葶状草本。叶基生，莲座状，具长柄；叶片卵形、卵状长圆形或倒卵形，顶端近圆形，基部圆形至宽楔形，边缘具线波状齿，齿端具小尖；羽状脉。头状花序单生于茎端；花葶无苞片；总苞钟状半球形；总苞片草质，1层，披针形；舌状花白色。瘦果倒卵状椭圆形；无冠毛。花期7-10月，果期8-11月。生海拔900-1200米的山坡阳处或林中。产海南。

Perennial herbs, scapigerous. Leaves rosulate, long petiolate; blade ovate, ovate-oblong or obovate, pinnately veined, base rounded to broadly cuneate and margin shallowly repand-dentate, with mucronulate teeth. Capitula solitary, terminal; scapes not bracteate; involucres campanulate-hemispheric; phyllaries in 1 row, lanceolate, herbaceous; ray florets white. Achenes obovoid-elliptic; pappus absent. Fl. Jul-Oct. Fr. Aug-Nov. Shaded places in woods at 900-1200 m. Distributed in Hainan.

毛冠菊

Nannoglottis carpesioides Maxim.

多年生草本。茎高60-100厘米。叶片长圆形或卵状披针形。头状花序3-12个在茎端排成总状或伞房状聚伞花序；总苞半球形；舌状花棕褐色；管状花淡黄色。花期6-7月。生海拔2000-3400米山坡林下、草甸或牧场。产云南西北部、陕西、甘肃南部和青海东部。

Perennial herbs. Stems 60-100 cm tall. Leaves oblong or ovate-oblanceolate. Capitula 3-12, in loose racemiform or corymbiform synflorescences; involucres hemispheric; ray florets brownish; disk florets yellowish. Fl. Jun-Jul. Forests on slopes, meadows or pastures at 2000-3400 m. Distributed in NW Yunnan, Shaanxi, S Gansu and E Qinghai.

毛冠菊 *Nannoglottis carpesioides*

狭舌毛冠菊 *Nannoglottis gynura*

虎克毛冠菊

Nannoglottis hookeri (Clarke ex Hook. f.) Kitam.

多年生草本。叶无柄或最下部叶柄有翅，叶片披针形至倒披针形，边缘具小圆齿或锐牙齿，基部下延。头状花序2-7个排成伞房状聚伞花序；总苞半球状，总苞片2-3层；舌状花和管状花黄色。花期6-8月。生海拔3400-4100米的松树林缘、灌丛或草坡。产西藏南部。喜马拉雅亦有。

Perennial herbs. Leaves sessile or lowest winged petiolate, leaves lanceolate to oblanceolate, margin denticulate to sharply toothed, decurrent on stem. Capitula 2-7 in corymbiform-cymose synflorescences; involucres hemispheric; phyllaries 2-3-seriate; ray florets and disc florets yellow. Fl. Jun-Aug. *Pinus* forests, among shrubs or meadows at 3400-4100 m. Distributed in S Xizang. Also in Himalaya.

宽苞毛冠菊

Nannoglottis latisquama Y. Ling et Y. L. Chen

多年生草本。叶片卵形、披针形或长圆形。头状花序2-14，排成伞房状聚伞花序；总苞半球形，径3-3.5厘米；总苞片2-3层，卵状披针形；花黄色。花果期7-9月。生海拔3400-3900米的针叶林或混交林下、灌丛、草坡或高

狭舌毛冠菊

Nannoglottis gynura (C. Winkl.) Y. Ling et Y. L. Chen

多年生草本。叶长圆形或卵状长圆形，边缘具不规则牙齿。头状花序排成较密圆锥状聚伞花序；总苞钟状，总苞片3-4层；舌状花和中央的管状花黄色。花期7-8月。生海拔3500-4000米云杉林下或灌丛草坡。产青藏高原。尼泊尔亦有。

Perennial herbs. Leaves oblong or ovate-oblong, margin irregularly teeth mucronulate. Capitula in corymbiform synflorescences; involucre campanulate, phyllaries in 3-4 rows; ray florets and disk florets yellow. Fl. Jul-Aug. *Picea* forests, thickets, grasslands on slopes at 3500-4000 m. Distributed in Tibetan Plateau. Also in Nepal.

虎克毛冠菊 *Nannoglottis hookeri*

宽苞毛冠菊 *Nannoglottis latisquama*

云南毛冠菊 *Nannoglottis yunnanensis*

山草甸。产云南西北部和四川西南部。

Perennial herbs. Leaves ovate to lanceolate or oblong. Capitula 2-14, in corymbiform synflorescences; involucres hemispheric, 3-3.5 cm diam; phyllaries 2-3-seriate, lanceolate-ovate; flowers yellow. Fl. and fr. Jul-Sep. *Pinus*, *Abies*, or mixed forests, thickets, grasslands or alpine meadows at 3400-3900 m. Distributed in NW Yunnan and SW Sichuan.

云南毛冠菊

Nannoglottis yunnanensis Hand.-Mazz.

多年生草本。基生叶有长叶柄，叶片心形，下部叶近心形，宽卵形、卵形或近椭圆形，基部下延，叶缘有不规则的齿，上部叶渐小，无叶柄，基部有耳。头状花序3-10个组成伞形聚伞花序；舌花淡棕色，管花浅黄色。花期6-7月。生海拔2900-4000米的云杉林中或草甸。产云南西北部和四川西部。

Perennial herbs. Basal leaves long winged petiolate, cordate; lower cauline leaves subcordate, broadly ovate, ovate or subelliptic, decurrent on stem, margin irregularly dentate; upper cauline gradually diminished, uppermost sessile, auriculate; capitula 3-10 in corymbiform-cymose synflorescences; ray florets brownish; disk florets yellowish. Fl. Jun-Jul. *Picea* forests or open pastures at 2900-4000 m. Distributed in NW Yunnan and W Sichuan.

鱼眼草

Dichrocephala integrifolia (L. f.) Kuntze

一年生草本。叶卵形、椭圆形或披针形，大头羽裂，边缘具重粗锯齿或缺刻状。头状花序小，球形，多数，排成伞房花序；外围雌花多层，紫色，条形；中央两性花少数，黄绿色，管状。瘦果压扁，倒披针形；无冠毛。花果期全年。生海拔200-2000米的山坡、山谷、林下、耕地、荒地或沟边。产中国西南、华南、华中和华东。亚洲和非洲的热带亚热带地区亦有。

Annual herbs. Leaves ovate, elliptic or lanceolate, lyrate-pinnatifid, margin doubly gross-serrate or incised. Capitula small, globose, numerous, arranged in corymb; margin pistillate flowers in numerous layers, purple, linear; central bisexual flowers few, grennish-yellow, tubular. Achenes compressed, oblanceolate; pappus absent. Fl. and fr. year-round. Slopes, valleys, forests, fields, wastelands or ditchsides at 200-2000 m. Distributed in SW, S, C and E China. Also in tropical and subtropical regions of Asia and Africa.

鱼眼草 *Dichrocephala integrifolia*

菊叶鱼眼草 *Dichrocephala chrysanthemifolia*

小鱼眼草 *Dichrocephala benthamii*

菊叶鱼眼草

Dichrocephala chrysanthemifolia DC.

一年生草本。叶长圆形或倒卵形，羽状分裂，基部扩大，耳状抱茎，两面被白色柔毛。头状花序较大，球形或长圆状，单生茎上部叶腋，近总状花序式排列，具显著花序梗和条形苞叶；外围雌花多层，紫色；花冠短漏斗形；中央两性花少数，花冠管状；无冠毛。花果期7-9月。生海拔约2900米的山坡路旁草丛中。产云南和西藏。日本、东南亚、南亚、非洲热带和澳大利亚亦有。

Annual herbs. Leaves oblong or obovate, pinnatifid, base dilated and auriculate-amplexicaul, both surfaces white pubescent. Capitula larger, globose or oblong, solitary in axils of upper leaves, nearly arranged in racemes, distinctly pedunculate and linear-bracteate; margin pistillate flowers in numerous layers, purple; corolla shortly funnelform; central bisexual flowers few; corolla tubular; pappus absent. Fl. and fr. Jul-Sep. Grasslands at roadsides of mountain slopes at ca. 2900 m. Distributed in Yunnan and Xizang. Also in Japan, SE and S Asia, tropical Africa, and Australia.

小鱼眼草

Dichrocephala benthamii C. B. Clarke

一年生草本，高15-35厘米。叶倒卵形、匙形或长圆形，边缘羽裂、大头羽裂或具深圆锯齿，基部扩大，耳状抱茎。头状花序小，扁球形，排成伞房花序；外围雌花多层，白色；花冠卵球形或坛形；中央两性花少数，黄绿色；花冠管状；无冠毛。花果期全年。生海拔700-3200米的山坡、山谷、草地、河岸、溪旁、路旁、田边或荒地。产云南、四川、西藏、贵州、广西、湖北和甘肃。南亚和东南亚亦有。

Annual herbs, 15-35 cm tall. Leaves obovate, spathulate or oblong, margin pinnatifid, lyrate-pinnatifid or deeply crenate, base dilated and auriculate-amplexicaul. Capitula small, compressed globose, arranged in corymb; margin pistillate flowers in numerous layers, white; corolla ovoid or urceolate; central bisexual flowers few, greenish-yellow; corolla tubular; pappus absent. Fl. and fr. year-round. Slopes, valleys, grasslands, river banks, streamsides, roadsides, fieldsides or wastelands at 700-3200 m. Distributed in Yunnan, Sichuan, Xizang, Guizhou, Guangxi, Hubei and Gansu. Also in S and SE Asia.

杯菊 *Cyathocline purpurea*

杯菊

Cyathocline purpurea Kuntze

一年生草本，高10-15厘米。茎红紫色，被黏质长柔毛。叶无柄，卵形或倒卵形，二回羽状分裂，基部耳状抱茎。头状花序小，排成伞房状花序或圆锥状伞房花序；总苞半球形，顶端紫色；花冠条形，红紫色，顶端2齿裂。瘦果长圆形；无冠毛。花果期近全年。生海拔100-2000米的林下、山坡草地、路边或田边。产云南、四川、贵州、广西和广东。南亚和东南亚亦有。

Annual herbs, 10-15 cm tall. Stems reddish-purple, viscid-villose. Leaves sessile, ovate or obovate, bipinnatifid, base auriculately amplexicaul. Capitula small, arranged in corymb or panicu-late-corymb; involucres semiglobose, apex purplish; corolla linear, reddish-purple, apex 2-lobed. Achenes oblong; pappus absent. Fl. and fr. nearly year-round. Forests, grassy slopes, roadsides or field margins at 100-2000 m. Distributed in Yunnan, Sichuan, Guizhou, Guangxi and Guangdong. Also in S and SE Asia.

圆舌黏冠草

Myriactis nepalensis Less.

多年生草本。叶卵状长椭圆形，边缘有大锯齿或圆锯齿，有时分裂，侧面的耳状分裂1-2（或3）对。头状花序球形或半球形，单生茎顶或枝端，多数头状花序排列成疏松的伞房状或伞房状圆锥花序。总苞片2-3层，披针形，几等长。瘦果压扁。无冠毛。花果期4-11月。生海拔700-3700米的山坡山谷林缘、林下、灌丛中，或近水潮湿地或荒地上。产中国西南、华南和华中。越南和喜马拉雅亦有。

圆舌黏冠草 *Myriactis nepalensis*

Perennial herbs. Leaves ovate-elliptic, margin coarsely serrate or crenate, sometimes lobed or parted, lateral lobes 1- 2(or 3)-paired. Capitula globose or hemispheric, solitary or numerous in laxly corymbose-paniculiform synflorescences. Phyllaries 2-3-seriate, lanceolate, nearly equal. Achenes compressed. Pappus absent. Fl. and fr. Apr-Nov. Forests, forest margins, thicket margins, wet areas on grassy-shrubby disturbed slopes, steep moist slopes, meadows, open and moist or humid areas along streamsides or ravines at 700-3700 m. Distributed in SW, S and C China. Also in Vietnam and Himalaya.

黏冠草

Myriactis wightii DC.

一年生草本，被白色柔毛。叶卵形。头状花序排成疏散伞房花序或伞房圆锥花序，具长花梗；花序直径达1.2厘米；总苞半球形；总苞片狭长圆形，无毛，先端撕裂状。瘦果压扁，倒披针形，边缘脉状加厚，顶端有短喙，喙顶有黏质分泌物；无冠毛。花果期7-11月。生海拔1900-3600米的山坡、林下、草地、溪旁或沟边。产中国西南。南亚亦有。

Annual herbs, white pubescent. Leaves ovate. Capitula arranged in lax corymb or corybose-panicle, long pedunculated. Capitula to 1.2 cm diam; involucres semiglobose; phyllaries narrowly oblong, glabrous, apex lacerate. Achenes compressed, oblanceolate, margin veined-thickened, apex shortly rostrate, rostra top with viscid secretion; pappus absent. Fl. and fr. Jul-Nov. Slopes, forests, grasslands, riversides or ditchsides at 1900-3600 m. Distributed in SW China. Also in S Asia.

黏冠草 *Myriactis wightii*

狐狸草

Myriactis wallichii Less.

一年生草本。叶具柄，叶表被白色稀疏的糙伏毛，有时无毛；中部茎叶长椭圆状披针形或长椭圆形。头状花序半球形，排成疏松的伞房状花序或近伞房状花序；总苞片长圆形或倒披针状长圆形。花期8-10月。生海拔2600-3600米的山坡草地及林下。产华中和西南。西南亚、南亚和东南亚亦有。

Annual herbs. Petioles, leaf surfaces sparsely white strigillose, sometimes glabrate; mid cauline leaves narrowly elliptic-lanceolate or narrowly elliptic. Capitula hemispheric, laxly corymbiform or subcorymbiform synflorescences; those of lower branches in paniculate-corymbiform or racemiform synflorescences; phyllaries oblong or oblanceolate. Fl. Aug-Oct. Grasslands, forests on slopes at 2600-3600 m. Distributed in C and SW China. Also in SW, S and SE Asia.

狐狸草 *Myriactis wallichii*

羽裂黏冠草 *Myriactis delavayi*

羽裂黏冠草

Myriactis delavayi Gagnep.

多年生草本。叶具柄；基部和下部茎叶倒披针形，大头羽裂或羽状深裂；中部茎叶同形，倒披针形至狭倒披针形；上部茎叶狭倒披针形或长椭圆形至线形，羽状半裂或羽状浅裂至锯齿状或全裂。疏总状花序；总苞披针形或椭圆形至倒披针形。花果期8-9月。生海拔2700-3000米的草地、山坡。产四川和云南。越南亦有。

Perennial herbs. Leaves petioles; basal and lower cauline blade oblanceolate, lyrate or pinnatipartite; mid cauline blade homomorphic, oblanceolate to narrowly oblanceolate; upper blade narrowly oblanceolate or narrowly elliptic to (uppermost) linear, pinnatifid or pinnatilobate, to (uppermost) serrate or entire. Lax racemiform synflorescences (terminal seeming subcorymbiform); phyllaries lanceolate or oblong to oblanceolate (inner). Fl. and fr. Aug-Sep. Grasslands, forests on slopes at 2700-3000 m. Distributed in Sichuan and Yunnan. Also in Vietnam.

白酒草 *Eschenbachia japonica*

白酒草

Eschenbachia japonica (Thunb.) J. Koster

一年生或两年生草本，全株被白色长柔毛或短糙毛。叶通常密集于茎较下部，呈莲座状，基部叶倒卵形或匙形。头状花序通常在茎顶密集成球状或伞房状；总苞半球形；花冠线形；中央的两性花15-16朵。瘦果长圆形；冠毛粉白色或绯红色至淡红色。花期3-9月。生海拔400-2500米的山坡草地或林缘。产中国西南、华南、东南、华中和华东。南亚、东南亚、西南亚和东北亚亦有。

Annual or biennial herbs, whole plants with long pubescence or short strigillose. Leaves usually congested at lower parts of stems, rosulate, basal leaves obovate or spatulate. Capitula in a dense apical globose or corymb; involucres hemispherical; corolla filiform; center pistillate florets 15-16. Achenes oblong; pappus pinkish white or cinnamon to reddish. Fl. Mar-Sep. Grassland on slopes or forest edges at 400-2500 m. Distributed in SW, S, SE, C and E China. Also in S, SE, SW and NE Asia.

小舌菊 *Microglossa pyrifolia*

小舌菊

Microglossa pyrifolia (Lam.) O. Kuntze

半灌木。茎攀援状，叉状分枝。叶互生，纸质，卵形或卵状长圆形。头状花序多数，排成密复伞房花序；总苞钟状；总苞片披针形至条形，干膜质；边缘雌花多数，线形，舌片极小；中央两性花2-3，管状。瘦果长圆形；冠毛浅红色。花果期全年。生海拔1800米以下的灌丛、溪岸、弃荒地或疏林。产云南、贵州、广西、广东和台湾。南亚和东南亚亦有。

Subshrubs. Stems scandent, divaricately branched. Leaves alternate, chartaceous, ovate or ovate-oblong. Capitula numerous, arranged in dense compound corymb; involucres campanulate; phyllaries lanceolate to linear, scarious; margin pistillate flowers numerous, filiform, ligules minute; central bisexual flowers 2-3, tubular. Achenes cylindric; pappus reddish. Fl. and fr. year-round. Thickets, stream banks, wastelands or sparse forests below 1800 m. Distributed in Yunnan, Guizhou, Guangxi, Guangdong and Taiwan. Also in S and SE Asia.

窄叶乳菀 *Galatella angustissima*

乳菀 *Galatella punctata*

碱菀

Tripolium pannonicum (Jacq.) Dobrocz.

一年生肉质草本，光滑无毛。叶无柄，条状或长圆状披针形。头状花序排成伞房状，有长花序梗；总苞片2-3层，披针形至长圆形；舌状花蓝紫色或粉红色。瘦果圆柱形；冠毛多层，不等长，白色或浅红色。花果期8-12月。生海拔2500米以下的海岸、湖滨、沼泽地或盐碱地。产华东、华北、东北和西北。亚洲西南部、欧洲、非洲北部和北美洲广布。

Annual succulent herbs, glabrous. Leaves sessile, lanceolate or oblong-lanceolate. Capitula arranged in corymb, long pedunculate; phyllaries 2-3 rows, lanceolate to oblong; ray florets bluish purple or pinkish. Achenes terete; pappus several layers, unequal in length, white or reddish. Fl. and fr. Aug-Dec. Seashores, lakesides, swamps or saline places below 2500 m. Distributed in E, N, NE and NW China. Widely Distributed in SW Asia, Europe, N Africa and North America.

碱菀 *Tripolium pannonicum*

窄叶乳菀

Galatella angustissima (Tausch) Novopokr.

多年生草本。叶密集，无柄，绿色，具3脉；中上部叶狭线形，具1条脉；最上部叶苞叶状。头状花序排列成疏伞房花序；总苞宽倒锥形；总苞片顶端淡紫红色；舌花紫罗兰色；管状花淡黄色。瘦果3棱；冠毛微黄色。花果期7-9月。生海拔900-2000米的干旱草地、多石地。产新疆。哈萨克斯坦、蒙古和俄罗斯亦有。

Perennial herbs. Leaves dense, sessile, green, 3-veined; mid and upper usually linear, 1-veined; uppermost bractlike. Capitula in lax corymbiform synflorescences; involucre broadly obconic; phyllaries pinkish at tip; ray florets pinkish violet; disk florets pale yellow. Achenes 3-ribbed; pappus yellowish. Fl. Jul-Sep. Dry grasslands, steppes, stony slopes at 900-2000 m. Distributed in Xinjiang. Also in Kazakhstan, Mongolia and Russia.

乳菀

Galatella punctata (Waldst. et Kit.) Nees

多年生草本。叶密集，披针状长圆形或线状披针形至线形，无柄。头状花序多数，排列成伞房花序；总苞宽倒锥形至钟状；总苞片3-4层；舌花5-10，淡紫红色；盘花10-15，淡黄色，有时带紫色；冠毛褐色。花期7-9月。生海拔800-1700米的山坡草地或潮湿草甸。产新疆。哈萨克斯坦、俄罗斯、西南亚和欧洲亦有。

Perennial herbs. Leaves dense, sessile, lanceolate-oblong or linear-lanceolate to linear. Capitula numerous, corymbiform synflorescences; involucre broadly obconic to campanulate; phyllaries 3-4-seriate; ray florets 5-10, pinkish violet; disk florets 10-15, pale yellow, sometimes pinkish violet tinged; pappus brownish. Fl. Jul-Sep. Grasslands on slopes or flood meadows at 800-1700 m. Distributed in Xinjiang. Also in Kazakhstan, Russia, SW Asia and Europe.

新疆麻菀 *Crinitina tatarica*

灌木紫菀木 *Asterothamnus fruticosus*

新疆麻菀

Crinitina tatarica (Less.) Soják

多年生草本。茎数个，直立或斜升。叶线形或线状倒披针形或椭圆形至线形，具1条脉。头状花序多数，排列成伞房花序；总苞短圆柱形或窄倒圆锥形；总苞片3-4层，淡黄绿色；花全部两性花，5-10朵，淡黄色；冠毛褐色。花期7-9月。生海拔700-1200米的盐碱地、草原或干旱砾石坡。产新疆。哈萨克斯坦、俄罗斯和欧洲亦有。

Perennial herbs. Stems several, erect or ascending. Leaves linear to linear-oblanceolate or oblong-linear, 1-veined. Capitula numerous, in corymbiform synflorescences; involucre shortly cylindric or obconic; phyllaries 3-4-seriate, yellowish green; disk florets 5-10, pale yellow; pappus brownish. Fl. Jul-Sep. Salt marshes, steppes or dry stony slopes at 700-1200 m. Distributed in Xinjiang. Also in Kazakhstan, Russia and Europe.

灌木紫菀木

Asterothamnus fruticosus (C. Winkl.) Novopokr.

亚灌木。叶条形，两面被蛛丝绒毛或有时上面近无毛。头状花序大形，作疏松伞房状排列，具膜质小花；苞片3层，革质，具宽膜质边缘；舌状花7-10，紫色；两性花15-18。连萼瘦果长圆形；冠毛白色。花果期6-9月。生海拔1000-1600米的沙石质山脚或石质河床。产新疆。中亚和俄罗斯亦有。

Subshrubs. Leaves linear, arachnoid-tomentulose on both surfaces or sometimes adaxially subglabrous. Capitula large, in loose corymbs, with membranous florets; phyllaries 3-seriate, coriaceous, with broadly membranous margins; ray florets 7-10, purplish; disc florets 15-18. Achenes oblong; pappus white. Fl. and fr. Jun-Sep. Gritty-stony foothills or stony riverbeds at 1000-1600 m. Distributed in Xinjiang. Also in C Asia and Russia.

中亚紫菀木

Asterothamnus centraliasiaticus Novopokr.

多分枝亚灌木。茎多数，簇生，基部木质，被灰白色短绒毛。叶长圆状条形或近条形，边缘反卷，中脉明显。头状花序直径约1厘米；总苞宽倒卵形；总苞片3-4层，顶端紫红色；舌状花7-10个，淡紫色至白色；冠毛白色，糙毛状。瘦

中亚紫菀木 *Asterothamnus centraliasiaticus*

果长圆形，稍扁。花果期7-9月。生海拔1300-3400米的草原、石质河床、开阔沙丘或荒漠。产内蒙古、甘肃、宁夏和青海。蒙古南部亦有。

Subshrubs, multi-branched. Stems many, clustered, base woody, shortly offwhite tomentose. Leaves oblong-linear or sub-linear, margin revolute, midrib conspicuous. Capitula ca. 1 cm diam, form corymb on tops of stems and branches; involucres broadly obovate; phyllaries 3-4 rows, apex purple-red; ray florets 7-10, purplish to white; pappus white, strigillose. Achenes oblong, somewhat compressed. Fl. and fr. Jul-Sep. Grasslands, stony riverbeds, open sand dunes or deserts at 1300-3400 m. Distributed in Neimenggu, Gansu, Ningxia and Qinghai. Also in S Mongolia.

翠菊

Callistephus chinensis (L.) Nees

一年生或二年生草本，被白色糙毛。叶具柄，被短硬毛，边缘具不规则粗锯齿。头状花序单生，直径6-8厘米，有长花序梗；总苞半球形；总苞片叶质；舌状花多变异，红色、淡红色、蓝色、黄色或淡蓝紫色；冠毛白色。花果期5-10月。生海拔300-2700米的山坡、荒地、草地、水边或疏林阴处。产中国东北、西南、西北、华北和华中。朝鲜半岛和日本亦有。

Annual or biennial herbs, white strigillose. Leaves petiolate, shortly strigillose, margin irregularly gross-serrate. Capitula solitary, 6-8 cm diam, long pedunculate; involucres semiglobose; phyllaries foliaceous; ray florets varied, red, reddish, blue, yellow or light purplish blue; pappus white. Fl. and fr. May-Oct. Slopes, wastelands, grasslands, watersides or shady sparse forests at 300-2700 m. Distributed in NE, SW, NW, N and C China. Also in Korean Peninsula and Japan.

翠菊 *Callistephus chinensis*

复芒菊

Formania mekongensis W. W. Sm. et J. Small

小灌木，高30-100厘米。小枝细，浅灰褐色。叶卵形，基部渐狭成宽的短柄，两面无毛。头状花序2-12，圆筒形，长约7毫米，花序梗长1-10毫米；总苞片长5-6毫米；舌花白色，盘花淡黄色。瘦果长2毫米。花果期7-8月。生海拔2500-3000米的干热河谷石砾山坡。产云南西北部及四川西部。

Shrubs, 30-100 cm tall. Branchlets gray-brown. Leaves narrowly winged petiolate, blade ovate, surfaces glabrous. Capitula 2-12, cylindrical, ca. 7 mm; peduncle 1-10 mm; involucres 5-6 mm; ray florets white, disk florets yellow. Achenes ca. 2 mm. Fl. and fr. Jul-Aug. Dry rocky slopes in dry and hot vallies at 2500-3000 m. Distributed in NW Yunnan and W Sichuan.

复芒菊 *Formania mekongensis*

藏寒蓬

Psychrogeton poncinsii (Franch.) Ling et Y. L. Chen

多年生草本，高3-19厘米。茎不分枝，葶状。叶片被疏至密毡毛至绒毛，倒披针形至倒卵形或匙形，边缘上部具粗齿或稍波状，稀全缘。头状花序单生；总苞片2-3层；舌花长于冠毛，白色；管状花黄色；冠毛黄白色。生海拔3000-4600米的高寒荒漠地区。产西藏西部和新疆。西喜马拉雅、中亚和西南亚亦有。

Perennial herbs, 3-19 cm tall. Stems scapiform, simple. Leaf surfaces sparsely or densely lanate to tomentose; leaves oblanceolate to obovate or spatulate, margin remotely coarsely serrate or slightly repand, rarely entire. Capitula solitary phyllaries 2- or 3-seriate; ray florets longer than pappus, white; disk florets yellow; pappus yellowish white. Alpine deserts at 3000-4600 m. Distributed in W Xizang and Xinjiang. Also in W Himalaya, C and SW Asia.

藏寒蓬 *Psychrogeton poncinsii*

香短星菊

Neobrachyactis anomala (DC.) Brouillet

多年生草本，高60-95厘米。基部叶倒卵形或长圆状披针形，上部叶渐小。头状花序单生或3-4个密集于茎或枝顶端；总苞半球状钟形，总苞片2-3层。花果期5-9月。生海拔3300-4000米的高山灌丛边缘或山坡草丛中。产西藏南部。喜马拉雅亦有。

Perennial herbs, 60-95 cm tall. Basal leaves obovate or oblong-lanceolate; upper leaves smaller. Capitula solitary or 3-4 in the top of a stem or branch; involucres semiglobose-campanulate, phyl-

香短星菊 *Neobrachyactis anomala*

西疆短星菊 *Neobrachyactis roylei*

laries 2-3 seriates. Fl. and fr. May-Sep. Alpine shrub edges or grassy slopes at 3300-4000 m. Distributed in S Xizang. Also in Himalaya.

西疆短星菊

Neobrachyactis roylei (DC.) Brouillet

一年生草本。基部叶倒卵形或倒卵状长圆形，上部叶渐小。头状花序多数，在茎或枝端排列成总状或总状圆锥花序；总苞半球形，总苞片3-4层；瘦果倒披针形，扁压，被贴生短微毛；冠毛淡黄色至浅黄褐色，2层。花果期7-9月。生海拔1800-4300米的岩石上或灌丛中。产西藏西部和新疆北部。南亚、西南亚、中亚和俄罗斯亦有。

Annual herbs. Basal leaves obovate or obovate-oblong; upper smaller. Capitula many, arranged in terminal raceme or racemose panicle; involucres hemispheric; phyllaries 3-4-seritate. Achenes lanceolate, compressed strigose. Pappus yellowish to cinnamon, 2-seriate. Fl. and fr. Jul-Sep. Among rocks and shrubs at 1800-4300 m. Distributed in W Xizang and N Xinjiang. Also in S, SW and C Asia, and Russia.

虾须草

Sheareria nana S. Moore

一年生草本。叶条形或倒披针形；上部叶鳞片状。头状花序顶生或腋生；苞片2层，4-5个；雌花舌状，白色或有时淡红色；两性花管状，有5齿。瘦果长椭圆形，褐色；无冠毛。花期4-11月，果期10-11月。生海拔700米以下的山坡、田边、湖边草地或河边沙滩上。产中国西南、华南、东南、华中和华东。

Annual herbs. Leaves linear or oblanceolate; upper leaves scaly. Capitula terminal or axillary; phyllaries in 2 rows, 4-5; ray florets pistillate, white or flush red; bisexual tubular, 5-toothed. Achenes long-elliptic,brown; pappus absent. Fl. Apr-Nov. Fr. Oct-Nov. Moutain slopes, Field, grassland near lake or sand by riversides below 700 m. Distributed in SW, S, SE, C and E China.

虾须草 *Sheareria nana*

小舌紫菀 *Aster albescens*

小舌紫菀

Aster albescens (DC.) Hand.-Mazz.

多分枝灌木。叶近纸质，卵形、椭圆形或矩圆状披针形，全缘或具浅齿。头状花序直径5-7毫米，多数，排成复伞房状；总苞倒圆锥形；总苞片边缘宽膜质；舌状花白色、浅红色或紫红色；筒状花黄色。瘦果椭圆形；冠毛污白色。花期6-9月，果期8-10月。生海拔500-4100米的林下、灌丛、草甸、溪边或丘陵至高山区域。产中国西南和西北。东喜马拉雅亦有。

Multi-branched shrubs. Leaves subchartaceous, ovate, elliptic or oblong-lanceolate, margin entire or shallowly serrate. Capitula 5-7 mm diam, numerous, arranged in compound corymb; involucres obconical; phyllaries broadly membranous at margin; ray florets white, reddish or purplish-red; tubular florets yellow. Achenes oblong; pappus dirty white. Fl. Jun-Sep. Fr. Aug-Oct. Forests, shrubs, meadows, streamsides or hills to alpine regions at 500-4100 m. Distributed in SW and NW China. Also in E Himalaya.

线叶紫菀

Aster lavandulifolius Hand.-Mazz.

灌木。茎多分枝。叶狭线形。头状花序成伞房状或3-5个生于枝端；总苞狭钟状；舌状花5-7个，舌片白色；管状花黄色。瘦果圆柱形，被毛；冠毛浅黄色。花期6-9月，果期8-10月。生海拔2000-2900米的亚高山石砾坡地或溪岸。产四川西部和云南西北部。

Shrubs. Stems multibranched. Leaves narrowly linear. Capitula in densely corymbiform synflorescences, terminal on current-year branches or 3-5 at ends of lateral branches; involucres narrowly campanulate; ray florets 5-7, white; disk florets yellow. Achenes cylindric, moderately strigose; pappus straw-colored. Fl. Jun-Sep. Fr. Aug-Oct. Subalpine stony slopes or riverbanks at 2000-2900 m. Distributed in W Sichuan and NW Yunnan.

东风菜

Aster scaber Thunb.

多年生草本，被微毛。叶具柄，心形、卵状三角形至披针形，边缘具锯齿，具三或五出脉。头状花序圆锥伞房状排列；总苞半球形；总苞片无毛，边缘宽膜质；舌状花约10个，白色；冠毛污黄白色。花期6-10月，果期8-10月。生海拔2000米以下的山坡、山谷、草地或灌丛中。产华南、华中、华东、华北和东北。俄罗斯、朝鲜半岛和日本亦有。

线叶紫菀 *Aster lavandulifolius*

东风菜 *Aster scaber*

Perennial herbs, puberulent. Leaves petiolate, cordate, ovate-triangular to lanceolate, margin serrate, 3-5-nerved. Capitula arranged in paniculate-corymb; involucres semiglobose; phyllaries glabrous, margin broadly membranous; ray florets ca. 10, white; pappus dirty yellowish-white. Fl. Jun-Oct. Fr. Aug-Oct. Slopes, valleys, grasslands or thickets below 2000 m. Distributed in S, C, E, N and NE China. Also in Russia, Korean Peninsula and Japan.

三脉紫菀

Aster trinervius Roxb. ex D. Don subsp. **ageratoides** (Turcz.) Grierson

多年生草本。叶卵形、椭圆形或披针形，基部狭楔形，边缘有3-7对锯齿，具离基三出脉。头状花序多数，直径1.5-2厘米；总苞片条状长圆形；舌状花紫色至白色；冠毛浅红褐色或污白色。花果期7-12月。生海拔100-3400米的林下、林缘、灌丛或山谷湿地。产中国除西北以外各地。喜马拉雅和东北亚亦有。

Perennial herbs. Leaves ovate, elliptic or lanceolate, base narrowly cuneate, margin with 3-7 pairs of teeth, triplicostate. Capitula numerous, 1.5-2 cm diam; phyllaries linear-oblong; ray florets purple to white; pappus red-brown or dirty white. Fl. and fr. Jul-Dec. Forests, forest edges, thickets or valley swamps at 100-3400 m. Distributed in most parts of China, except NW China. Also in Himalaya and NE Asia.

三脉紫菀 *Aster trinervius* subsp. *ageratoides*

Capitula arranged in lax corymb; involucres semiglobose; phyllaries linear or oblong-lanceolate; ray florets more than 20, purplish; pappus short, deciduous. Fl. May-Oct. Fr. Jul-Nov. Slopes, forest edges, thickets or roadsides below 1600 m. Distributed in SW, C, E, N and NE China. Also in NE Asia.

全叶马兰 *Aster pekinensis*

马兰 *Aster indicus*

马兰
Aster indicus L.

多年生草本。根状茎有匍枝。叶质薄，倒披针形或倒卵状矩圆形，基部渐狭成具翅的长柄，边缘具锯齿或羽状裂片。头状花序排成疏伞房状；总苞半球形；总苞片倒披针形；舌状花15-20朵，浅紫色；冠毛短，易脱落。花果期5-11月。生海拔3900米以下的林缘、草丛、溪岸或路边。产中国西南、华南、华中和华东。广布南亚和东亚。

Perennial herbs. Rhizomes with stolons. Leaves thin, oblanceolate or obovoid-oblong, base attenuate to winged long petioles, margin serrate or pinnately lobed. Capitula arranged in lax corymb; involucres semiglobose; phyllaries oblanceolate; ray florets 15-20, purplish; pappus short, deciduous. Fl. and fr. May-Nov. Forest edges, grasslands, riversides or roadsides below 3900 m. Distributed in SW, S, C and E China. Also widely distributed in S and E Asia.

全叶马兰
Aster pekinensis (Hance) F. H. Chen

多年生草本。直根长纺锤状。叶密集，披针形、矩圆形或条形，边缘全缘，稍反卷，两面密被粉状短绒毛。头状花序排成疏伞房状；总苞半球形；总苞片条形或矩圆状披针形；舌状花20余朵，淡紫色；冠毛短，易脱落。花期5-10月，果期7-11月。生海拔1600米以下的山坡、林缘、灌丛或路旁。产中国西南、华中、华东、华北和东北。东北亚亦有。

Perennial herbs. Taproots long fusiform. Leaves dense, lanceolate, oblong or linear, margin entire, slightly revolute, both surfaces densely shortly pulverulent-tomentose.

裂叶马兰 *Aster incisus*

裂叶马兰
Aster incisus Fisch.

多年生草本。叶纸质，长椭圆状披针形或披针形，边缘疏生缺刻状锯齿或间有羽状披针形尖裂片。头状花序直径2.5-3.5厘米，排成伞房状；总苞半球形；总苞片长椭圆状披针形；舌状花淡蓝紫色；管状花黄色。瘦果倒卵球形；冠毛短，长0.5-1.2毫米，淡红色。花果期6-10月。生海拔400-1000米的草坡、灌丛、林缘或湿草地。产内蒙古和东北。东北亚亦有。

Perennial herbs. Leaves chartaceous, narrowly elliptic-lanceolate or lanceolate, margin sparsely incised-serrate or mixed with

山马兰 *Aster lautureanus*

长柄马兰 *Aster longipetiolatus*

pinnate-lanceolate acute lobes. Capitula 2.5-3.5 cm diam, arranged in corymb; involucres semiglobose; phyllaries narrowly elliptic-lanceolate; ray florets bluish-purple; tubular florets yellow. Achenes obovoid; pappus short, 0.5-1.2 mm long, reddish. Fl. and fr. Jun-Oct. Grassy slopes, thickets, forest edges or damp grasslands at 400-1000 m. Distributed in Neimenggu and NE China. Also in NE Asia.

山马兰

Aster lautureanus Franch.

多年生草本，被白色糙毛。叶厚或近革质，披针形或矩圆状披针形，基部渐狭，边缘具疏齿、羽状浅裂或全缘。头状花序排成伞房状；总苞半球形；总苞片长椭圆形至倒披针状长圆形；舌状花淡蓝色；管状花黄色；冠毛淡红色。花果期7-10月。生海拔100-2200米的山坡、溪岸、草原或灌丛。产华中、华东、华北和东北。

Perennial herbs, white strigillose. Leaves thick or subcoriaceous, lanceolate or oblong-lanceolate, base attenuate, margin sparsely serrate, pinnately lobed or entire. Capitula arranged in corymb; involucres semiglobose; phyllaries narrowly elliptic to oblanceolate-oblong; ray florets bluish; tubular florets yellow; pappus reddish. Fl. and fr. Jul-Oct. Slopes, stream banks, grasslands or thickets at 100-2200 m. Distributed in C, E, N and NE China.

长柄马兰

Aster longipetiolatus C. C. Chang

多年生草本，高50-70厘米。全部叶质薄，宽卵形，近掌状基出脉，边缘有粗齿，基部截形或浅心形。头状花序1或2；总苞片2-3层，等长，条形或倒披针形；舌状花约30朵，淡蓝色；管状花黄色；冠毛污白色，有不等长的膜片状毛。花期7月。生海拔约2500米的山谷溪岸。产四川西部。

Perennial herbs, 50-70 cm tall. Leaves thin, broadly ovate, basally palmately veined, margin coarsely serrate, base truncate or shallowly cordate. Capitula 1 or 2; phyllaries 2-3-seriate, equal, linear or oblanceolate; ray florets ca. 30, bluish; disk florets yellow; pappus dirty white, of scale-like bristles, unequal. Fl. Jul. Brooklet banks, valleys at ca. 2500 m. Distributed in W Sichuan.

圆齿狗娃花

Aster crenatifolius Hand.-Mazz.

一年生或二年生草本。下部叶渐尖成细或有翅的长柄；中部叶无柄。头状花序单生枝顶；舌状花35-40朵，淡紫色或蓝色；管状花黄色；舌状花冠毛常较少，或极短，或不存在；管状花的冠毛2层，淡黄色或近褐色，有不等长的微糙毛。花果期5-10月。生海拔2400-3500米的开阔山坡、田野、路旁、河边、冲积平原、高山草甸或林地。产中国西南、西北和华北。尼泊尔亦有。

Annual or biennial herbs. Lower cauline leaves shortly winged petiolate; middle leaves sessile. Capitula solitary, terminal on branches; ray florets 35-40, mauve or blue; disk florets yellow; pappus of ray florets often few and short, sometimes absent or equaling that of disk; pappus of disk florets 2-seriate, straw-colored or brownish, of unequal bristles. Fl. and fr. May-Oct. Open disturbed slopes, pebbly-sandy river floodplains, riverbanks, alpine or montane meadows, Juniperus slopes, montane woodlands, fields, roadsides at 2400-3500 m. Distributed in SW, NW and N China. Also in Nepal.

圆齿狗娃花 *Aster crenatifolius*

拉萨狗娃花 *Aster gouldii*

半卧狗娃花 *Aster semiprostratus*

拉萨狗娃花

Aster gouldii C. E. C. Fisch.

一年生草本，高8-30厘米，披散。茎自基部起分枝，纤细，被毛。基部叶有翼柄，柄基部抱茎。头状花序单生，少数生于枝顶；总苞绿色；舌状花约25-40朵淡紫色或浅蓝色；管状花黄色；舌状花冠毛1层，长0.3-0.5毫米；管状花冠毛2层，微红色，外层长约0.5毫米，内层长3-3.7毫米。花果期6-9月。生海拔2900-5600米的山坡草地、高山草甸、田边、河滩。产西藏和青海。不丹和印度北部亦有。

Annual herbs, 8-30 cm tall, spreading. Stems branched from base, tenuous densely strigose. Basal leaves winged petiolate, base of petioles clasping. Capitula solitary or few at ends of branches; involucres green; ray florets 25-40, purplish or bluish; disk florets yellow; pappus of ray florets 1-seriate, scales 0.3-0.5 mm; disk florets 2-seriate, reddish; outer bristles ca. 0.5 mm; inner bristles 3.3-3.7 mm. Fl. and fr. Jun-Sep. Open grassy slopes, disturbed open or bare slopes in alpine meadows, sometimes sandy soils and riverbanks, field margins at 2900-5600 m. Distributed in Xizang and Qinghai. Also in Bhutan and N India.

半卧狗娃花

Aster semiprostratus (Grierson) H. Ikeda

多年生草本。茎簇生，平卧或斜升，被平贴硬柔毛。叶条形或匙形，全缘，被平贴柔毛。头状花序单生枝端，直径1.5-3厘米；总苞半球形；总苞片披针形，绿色；舌状花20-35朵，蓝色或浅紫色；管状花黄色；冠毛浅棕红色。花果期6-8月。生海拔3200-4600米的干燥多砂石山坡、冲积扇或河滩沙地。产西藏和青海。印度西北部和尼泊尔亦有。

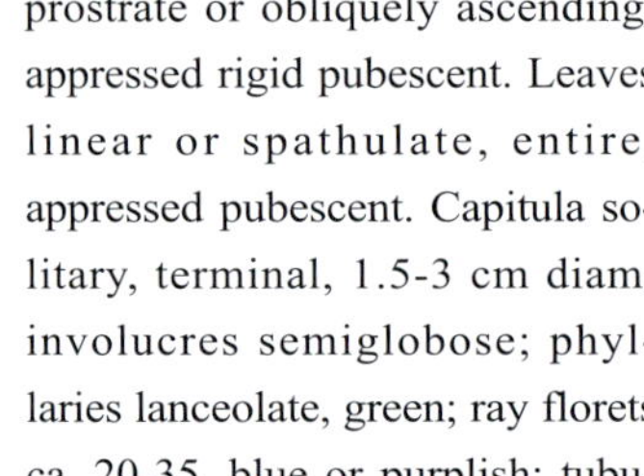

Perennial herbs. Stems clustered, prostrate or obliquely ascending, appressed rigid pubescent. Leaves linear or spathulate, entire, appressed pubescent. Capitula solitary, terminal, 1.5-3 cm diam; involucres semiglobose; phyllaries lanceolate, green; ray florets ca. 20-35, blue or purplish; tubular florets yellow; pappus lightly brownish-red. Fl. and fr. Jun-Aug. Dry gravelly slopes, alluvial fans or flood sandy places at 3200-4600 m. Distributed in Xizang and Qinghai. Also in NW India and Nepal.

狗娃花

Aster hispidus Thunb.

一年生或二年生草本，多少被粗毛。主根纺锤状。叶互生，狭矩圆形或倒披针形，全缘，被疏毛。头状花序直径3-5厘米，排成圆锥伞房状；总苞片草质，被粗毛；舌状花约30朵，浅红色或白色，条状矩圆形；筒状花黄色。瘦果倒卵球形，扁；冠毛白色。花期7-9月，果

狗娃花 *Aster hispidus*

阿尔泰狗娃花 *Aster altaicus*

期8-10月。生海拔2400米以下的荒地、路旁、林缘或草地。产华中、华东、华北、东北和西北。俄罗斯、蒙古、朝鲜半岛和日本亦有。

Annual or biennial herbs, hispid. Taproots fusiform. Leaves alternate, narrowly oblong or oblanceolate, entire, sparsely pubescent. Capitula 3-5 cm diam, arranged in paniculate-corymb; phyllaries herbaceous, strigillose; ray florets ca. 30, reddish or white, linear-oblong; tubular florets yellow. Achenes obovoid, compressed; pappus white. Fl. Jul-Sep. Fr. Aug-Oct. Wastelands, roadsides, forest edges or grasslands below 2400 m. Distributed in C, E, N, NE and NW China. Also in Russia, Mongolia, Korean Peninsula and Japan.

阿尔泰狗娃花

Aster altaicus Willd.

多年生草本。茎直立，被毛，分枝。叶条形或披针形，全缘或有疏浅齿。头状花序直径2-3.5厘米，单生或排成伞房状；总苞半球形；总苞片矩圆状披针形或条形；舌状花浅蓝紫色；冠毛污白色或红褐色。花果期7-9月。生海拔4000米以下的草原、荒漠、沙地或干旱山地。产中国大部分地区。中亚、东亚、北亚和东北亚亦有。

Perennial herbs. Stems erect, pubescent, branched. Leaves linear or lanceolate, margin entire or sparsely and shallowly serrate. Capitula 2-3.5 cm diam, solitary or arranged in corymb; involucres semiglobose; phyllaries oblong-lanceolate or linear; ray florets bluish-purple; pappus dirty white or reddish-brown. Fl. and fr. Jul-Sep. Grasslands, deserts, sandy places or dry montanes below 4000 m. Distributed in most parts of China. Also in C, E, N and NE Asia.

青藏狗娃花

Aster boweri Hemsl.

二年生或多年生草本，低矮，垫状。叶条形，质厚，密生白色长粗毛，全缘，具缘毛。头状花序单生茎端，直径2.5-3厘米；总苞半球形；总苞片条形；舌状花约50朵，蓝紫色；冠毛污白色或稍褐色。花果期7-8月。生海拔2200-5200米的高山砾石地、高山草甸或山地河流冲刷平原。产西藏、甘肃和青海。

Biennial or perennial herbs, dwarf, pulvinate. Leaves linear, thick, densely white long strigillose, margin entire, ciliate. Capitula solitary, terminal, 2.5-3 cm diam; involucres semiglobose; phyllaries linear; ray florets ca. 50, bluish-purple; pappus dirty white or somewhat brown. Fl. and fr. Jul-Aug. Alpine gravelly places, alpine meadows or montane river floodplains at 2200-5200 m. Distributed in Xizang, Gansu and Qinghai.

青藏狗娃花 *Aster boweri*

紫菀 *Aster tataricus*

褐毛紫菀 *Aster fuscescens*

紫菀
Aster tataricus L. f.

多年生草本。基生叶大，长圆状或椭圆状匙形，边缘具圆齿，茎生叶匙状长圆形或长圆形，边缘具密锯齿。头状花序多数；总苞片条形或条状披针形；舌状花蓝紫色；冠毛污白色或淡红色。花期7-9月，果期8-10月。生海拔400-3300米的低山、阴坡、山顶、草地或沼泽地。产华北、东北和西北。俄罗斯、朝鲜半岛和日本亦有。

Perennial herbs. Basal leaves large, oblong or elliptic-spathulate, margin crenate, cauline leaves spathualte-oblong or oblong, margin densely serrate. Capitula numerous; phyllaries linear or linear-lanceolate; ray florets purple; pappus dirty white or reddish. Fl. Jul-Sep. Fr. Aug-Oct. Hills, shady slopes, summits, grasslands or swamps at 400-3300 m. Distributed in N, NE and NW China. Also in Russia, Korean Peninsula and Japan.

褐毛紫菀
Aster fuscescens Bureau et Franch.

多年生草本。茎上部分枝，被柔毛。基部叶多数，叶柄长达15厘米，叶片宽卵圆形，近纸质或革质，基部圆形或心形，顶端钝或急尖，边缘具锯齿。头状花序径约3厘米，排成伞房状；舌状花蓝紫色；冠毛红褐色。花期7-10月，果期8-12月。生海拔2700-4200米的高山草坡、灌丛边缘、石砾地或沟涧旁。产横断山区南部。缅甸亦有。

Perennial herbs. Stems branched at upper part, pubescent. Basal leaves numerous, petioles to 15 cm long, blades broadly ovate, subchartaceous or coriaceous, base rotundate or cordate, apex obtuse or acute, margin serrate. Capitula ca. 3 cm diam, arranged in corymb; ray florets bluish-purple; pappus reddish-brown. Fl. Jul-Oct. Fr. Aug-Dec. Alpine grasslands, thicket margins, rocky places or streamsides at 2700-4200 m. Distributed in S Hengduan Mountain. Also in Myanmar.

耳叶紫菀
Aster auriculatus Franch.

多年生草本。根状茎粗壮，茎被开展的长粗毛。下部叶稍狭，基部扩大成圆形抱茎的耳部。头状花序在茎和枝端排列成圆锥伞房状或伞房状；舌状花18-30朵，白色；管状花黄色。瘦果狭倒卵圆形，压扁。花果期4-8月。生海拔800-3000米的疏林下、混交林、灌丛及草地。产广西、甘肃、湖北和西南。

Perennial herbs. Rhizomes ± woody, stems sparsely to densely villous Lowest leaves broadly winged petiolate, petioles base auriculate. Capitula in terminal corymbiform synflorescences; ray florets 18-30, white; disk florets yellow. Achenes lanceolate, compressed. Fl. and fr. Apr-Aug. Open forests, mixed forests, thickets, grasslands at 800-3000 m. Distributed in Guangxi, Gansu, Hubei and SW China.

耳叶紫菀 *Aster auriculatus*

琴叶紫菀 *Aster panduratus*

琴叶紫菀

Aster panduratus Nees ex Walp.

多年生草本。叶匙状长圆形或卵状长圆形，两面被长平伏毛及短柔毛及腺点。头状花序单生或作疏松伞房状排列；苞片3层；舌状花紫色；两性花密被柔毛；冠毛白色或淡红色。花期2-9月，果期6-10月。生海拔100-1400米灌丛、山坡草地、沟边或路边。产中国西南、华南、东南、华中和华东。

Perennial herbs. Leaves spatulate-oblong or ovate-oblong, covered with appressed long hairs and pilose on both surfaces, glandular. Capitula solitary or laxly corymbose; phyllaries 3-seriate; ray florets purplish; disc florets densely puberulent; pappus white or reddish. Fl. Feb-Sep. Fr. Jun-Oct. Thickets, grasslands on slopes, canal sides or roadsides at 100-1400 m. Distributed in SW, S, SE, C and E China.

密毛紫菀

Aster vestitus Franch.

多年生草本。茎单生，被长密毛。叶密集，长圆状披针形，无柄，全部叶被密腺毛，离基三出脉。头状花序少数至数十个排列成复伞房状，具花序梗；总苞片约3层，顶部常紫红色；舌状花20-30朵，白色或浅紫红色；冠毛1层，污白色或稍红色。花果期9-12月。生海拔2200-3200米的高山或亚高山林缘、草坡、溪边或沙地。产中国西南。印度东北部、不丹和缅甸亦有。

Perennial herbs. Stems solitary, villose. Leaves dense, oblong-lanceolate, sessile, densely glandular hairy, triplicostate. Capitula several to dozens arranged in compound corymb, long pedunclate; phyllaries ca. 3 rows, apex generally purple; ray florets 20-30, white or purplish; pappus in one layer, dirty white or somewhat red. Fl. and fr. Sep-Dec. Alpine or subalpine forest edges, grassy slopes, riversides or sandy places at 2200-3200 m. Distributed in SW China. Also in NE India, Bhutan and Myanmar.

凉山紫菀

Aster taliangshanensis Y. Ling

多年生草本。中部叶卵形至长披针形，边缘锯齿状；上部叶披针形，全缘或细锯齿状。头状花序组成疏散伞房花序；总苞半球形；舌状花蓝紫色，管状花黄色；冠毛污白色或红褐色。花期6-8月，果期8-9月。生海拔2500-3100米的低山及亚高山山坡林下、草地及路旁。产四川。

Perennial herbs. Middle leaves ovate to broadly lanceolate, margin serrate; upper leaves lanceolate, margin entire or serrulate. Capitula in open corymbiform synflorescences; involucres hemispheric; ray florets blue-purple, disk florets yellow; pappus dirty white or reddish brown. Fl. Jun-Aug. Fr. Aug-Sep. Low mountains and alpine forests, grasslands, dry stony shrubby slopes, roadsides at 2500-3100 m. Distributed in Sichuan.

密毛紫菀 *Aster vestitus*

凉山紫菀 *Aster taliangshanensis*

甘川紫菀 *Aster smithianus*

甘川紫菀
Aster smithianus Hand.-Mazz.

木质草本或亚灌木。茎灰褐色，被微柔毛。叶狭卵圆形或披针形，两面密被贴伏微柔毛，具离基三出脉。头状花序多数，直径1.5-2.5厘米，排列成伞房状；总苞半球形；总苞片2-3层；舌状花约30朵，白色或浅紫红色；冠毛白色或稍红色。花期8-10月，果期9-10月。生海拔1300-3400米的低山或亚高山的山坡草地或石砾河岸。产云南、四川和甘肃。

Woody herbs or suffruticoses. Stems grayish-brown, puberulent. Leaves narrowly ovate or lanceolate, densely appressed puberulent, triplicostate. Capitula numerous, 1.5-2.5 cm diam, arranged in corymb; involucres semiglobose; phyllaries 2-3 rows; ray florets ca. 30, white or purplish-red; pappus white or reddish. Fl. Aug-Oct. Fr. Sep-Oct. Grassy slopes or rocky banks of hills or subalpines at 1300-3400 m. Distributed in Yunnan, Sichuan and Gansu.

长梗紫菀
Aster dolichopodus Y. Ling

多年生草本，高38-90厘米。叶基部圆形(下部叶)或有耳部，常抱茎。头状花序1-16个，在枝端单生或排列成伞房状；舌状花19-26朵，舌片浅紫色或紫色；冠毛4层。花果期7-9月。生海拔2400-3500米的草坡、干山坡灌丛、灌层边缘沟边及路旁。产四川、甘肃和陕西。

Perennial herbs, 38-90 cm tall. Leaves base rounded (lower leaves) or auriculate, clasping. Capitula 1-16 in terminal corymbiform synflorescences, sometimes solitary; ray florets 19-26, light purple to purple; pappus 4-seriate. Fl. and fr. Jul-Sep. Grasslands on slopes, scrub on dry slopes, scrub, thicket margins, dry stream banks, canal sides, roadsides at 2400-3500 m. Distributed in Sichuan, Gansu and Shaanxi.

长叶紫菀
Aster dolichophyllus Y. Ling

多年生草本，高约45厘米，具根状茎和长匍枝。叶匙状至条状披针形，上部具疏锯齿或近全缘，基部半抱茎，质厚，无毛，具离基三出脉。头状花序约8个，直径约2.5厘米，排列成疏散伞房状，总苞片条形；舌状花白色，条形；冠毛浅红褐色。花期9-10月。生海拔500-1250米的疏林中水旁岩石上。产广西北部。

长梗紫菀 *Aster dolichopodus*

长叶紫菀 *Aster dolichophyllus*

Perennial herbs, ca. 45 cm tall, with rhizomes and long stolons. Leaves spathulate to linear-lanceolate, upper part sparsely serrate or nearly entire, base semi-amplexicaul, thick, glabrous, tripli-costate. Capitula ca. 8, ca. 2.5 cm diam, arranged in sparse corymb; phyllaries linear; ray florets white, linear; pappus brownish. Fl. Sep-Oct. Rocks by waters in sparse forests at 500-1250 m. Distributed in N Guangxi.

等苞紫菀

Aster homochlamydeus
Hand.-Mazz.

多年生草本。基部叶在花期枯萎，具狭翅柄。头状花序排列成伞房状；总苞倒锥形；舌状花10-14朵，舌片白色或紫色；管状花黄色；冠毛稍红色。花期4-8月，果期7-9月。生海拔3000-3700米的高山及亚高山杂木林中。产云南、四川及甘肃。

Perennial herbs. Lowest leaves withered by anthesis, narrowly winged petiolate. Capitula in terminal corymbiform synflorescences; involucres campanulate; ray florets 10-14, white or purple; disk florets yellow; pappus reddish. Fl. Apr-Aug. Fr. Jul-Sep. Alpine and subalpine mixed forests at 3000-3700 m. Distributed in Yunnan, Sichuan and Gansu.

四川裸菀

Aster pseudosimplex Brouillet, Semple et Y. L. Chen

多年生草本，被白色糙毛状长柔毛。茎斜生，丛生。基生叶匙形，具长柄；茎生叶矩圆状披针形、近椭圆形或条形，全缘。头状花序单生；总苞半球形；总苞片披针形或长圆，绿色；舌状花淡紫红色；无冠毛。花果期7-9月。生海拔2600-3000米的山坡草地或沟边。产四川西北部。

Perennial herbs, white strigillose-villose. Stems obliquely ascending, clustered. Basal leaves spathulate, long petiolate; cauline leaves oblong-lanceolate, subelliptic or linear, entire. Capitula solitary; involucres semiglobose; phyllaries lanceolate or oblong, green; ray florets purplish-red; pappus absent. Fl. and fr. Jul-Sep. Grassy slopes or ditchsides at 2600-3000 m. Distributed in NW Sichuan.

等苞紫菀 *Aster homochlamydeus*

四川裸菀 *Aster pseudosimplex*

秋分草 *Aster verticillatus*

密叶紫菀 *Aster pycnophyllus*

秋分草

Aster verticillatus (Reinwardt) Brouillet, Semple et Y. L. Chen

多年生草本。叶柄具狭翅，边缘具波状锯齿。头状花序顶生或腋生，单生或近总状；总苞宽钟形或半球形；苞片极度不等；舌状花2或3层，管短，具腺体；冠毛为小而早落的硬毛。花果期8-11月。生海拔400-2500米的林缘、灌丛、林中阴处或水边。产中国西南、华南、东南、华中和华东。印度、不丹、缅甸、马来西亚和日本亦有。

Perennial herbs. Petioles narrowly winged, margin repand-serrate. Capitula terminal or axillary, solitary or subracemose; involucres broadly campanulate or hemispherical; phyllaries rather unequal; ray florets 2- or 3-seriate, tubes very short, glandular; pappus of minute, easily deciduous bristles. Fl. and fr. Aug-Nov. Forest edges, thickets, shady places of forests or watersides at 400-2500 m. Distributed in SW, S, SE, C and E China. Also in India, Bhutan, Myanmar, Malaysia and Japan.

密叶紫菀

Aster pycnophyllus Franch. ex Diels

多年生草本。中上部叶卵形至披针形，边缘有具小尖头的锯齿；顶部叶狭卵圆形至披针形或线形。圆锥伞房花序；总苞钟状；舌状花蓝紫色或近白色，管状花黄色或带紫色；冠毛污白色。花期6-10月，果期9-12月。生海拔1000-3800米的亚高山山地、草地及灌丛中。产中国西南。印度和缅甸亦有。

Perennial herbs. Middle and upper leaves ovate to lanceolate, margin coarsely serrate; uppermost leaves narrowly ovate to lanceolate or linear. Capitula in corymbiform synflorescences; involucres campanulate; ray florets purplish blue to nearly white, disk florets yellow or sometimes purplish tinged; pappus dirty white. Fl. Jun-Oct. Fr. Sep-Dec. Subalpine slopes, grasslands and thickets at 1000-3800 m. Distributed in SW China. Also in India and Myanmar.

黑山紫菀

Aster nigromontanus Dunn

多年生草本。叶长圆形，边缘有粗齿，先端渐尖。头状花序多数，排列成复伞房状；总苞钟状；总苞片2-3层，几等长；舌花25-30朵，白色；管状花黄色；冠毛3-4层，污白色或浅红色。花果期10-11月。生海拔1500-3000米的山坡草地。产云南。

黑山紫菀 *Aster nigromontanus*

白舌紫菀 *Aster baccharoides*

Perennial herbs. Leaves oblong, margin coarsely serrate, apex acuminate. Capitula numerous, corymbiform synflorescences; involucres campanulate; phyllaries 2-3-seriate, subequal; ray florets 25-30, white; disk florets yellow; pappus 3-4-seriate, dirty white or slightly reddish. Fl. and fr. Oct-Nov. Grasslands on slopes at 1500-3000 m. Distributed in Yunnan.

白舌紫菀
Aster baccharoides (Benth.) Steetz.

木质草本或亚灌木。叶长圆形，基部渐狭，全缘或上部有疏齿，被短糙毛，侧脉3-4对。头状花序顶生，直径1.5-2厘米，多数排列成圆锥伞房状；舌状花白色；冠毛白色。花期7-10月，果期8-11月。生海拔1000米以下的山坡、路旁、草地和沙地。产广东、广西、福建、浙江、湖南和江西。

Woody herbs or subshrubs. Leaves oblong, base attenuate, margin entire or sparsely serrate on upper part, shortly strigillose, lateral veins 3-4 pairs. Capitula terminal, 1.5-2 cm diam, many arranged in panicuate-corymb; ray florets white; pappus white. Fl. Jul-Oct. Fr. Aug-Nov. Slopes, roadsides, grasslands and sandy places below 1000 m. Distributed in Guangdong, Guangxi, Fujian, Zhejiang, Hunan and Jiangxi.

高山紫菀
Aster alpinus L.

多年生草本，具粗壮的根状茎和基生莲座状叶丛。茎丛生，高10-35厘米，不分枝。基生叶匙状或条状长圆形，全缘，被柔毛。头状花序单生茎顶，直径3-3.5厘米；舌状花紫色、蓝色或浅红色；冠毛白色。瘦果长圆形，褐色。花期6-8月，果期7-9月。生海拔1500米以上的山顶石缝或草甸中。产山西、河北、东北和新疆。亚洲、欧洲和北美洲广布。

Perennial herbs, with stout rhizomes and basal rosette leaves. Stems clustered, 10-35 cm tall, simple. Basal leaves spathulate or linear-oblong, entire, pubescent. Capitula solitary, terminal, 3-3.5 cm diam; ray florets purple, blue or reddish; pappus white. Achenes oblong, brown. Fl. Jun-Aug. Fr. Jul-Sep. Summit rocky cracks or meadows above 1500 m. Distributed in Shanxi, Hebei, NE China and Xinjiang. Widely distributed in Asia, Europe and North America.

高山紫菀 *Aster alpinus*

石生紫菀 *Aster oreophilus*

石生紫菀

Aster oreophilus Franch.

多年生草本，簇生。基生叶和下部茎叶匙形、倒卵形或倒披针形。头状花序2-30，成伞房花序，或单生；总苞片约3层，几等长；舌状花蓝紫色到粉色或粉白色；管状花黄色。瘦果2肋；冠毛4层，污白色。花果期8-10月。生海拔2000-4000米的高山和亚高山林缘、草地、开阔坡地或山坡路旁。产云南、四川和西藏。

Perennial herbs, caespitose. Basal leaves and lower cauline leaves narrowly spatulate to obovate or oblanceolate. Capitula 2-30 in corymbiform synflorescences, sometimes solitary; phyllaries 3-seriate, subequal; ray florets blue-purple to pink or pinkish white; disk florets yellow. Achenes 2-ribbed; pappus 4-seriate, dirty white. Fl. and fr. Aug-Oct. Alpine and subalpine forest margins, grasslands, pastures, open slopes or roadsides at 2000-4000 m. Distributed in Yunnan, Sichuan and Xizang.

舌叶紫菀

Aster lingulatus Franch.

多年生草本，具根状茎和莲座状叶丛。中部叶长圆形或条状长圆形，半抱茎，质厚，两面密被微糙毛。头状花序直径2-3厘米，2-4个成伞房状；总苞半球形；总苞片条状长圆形，顶端常褐色，具缘毛；舌状花蓝紫色；冠毛淡红色。花期8-10月。生海拔2600-3600米的高山或亚高山草地或松林下。产云南和四川。

Perinnial herbs, with rhizomes and basal rosette leaves. Middle cauline ones oblong or linear-oblong, semi-amplexicaul, thick, densely strigillose on both surfaces. Capitula 2-3 cm diam, 2-4 arranged in corymb; involucres semiglobose; phyllaries linear-oblong, apex generally brown, ciliate; ray florets bluish-purple; pappus reddish. Fl. Aug-Oct. Alpine or subalpine grasslands or pine forests at 2600-3600 m. Distributed in Yunnan and Sichuan.

舌叶紫菀 *Aster lingulatus*

红冠紫菀

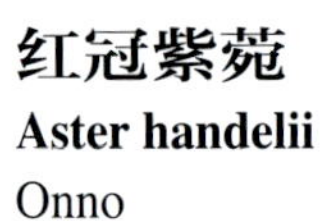

Aster handelii Onno

多年生草本，常有丛生花茎。叶片向上渐减；下部叶在花期生存。头状花序在茎端单生；舌状花28-40朵，浅蓝紫色；管状花橙色。瘦果宽卵圆形；冠毛淡紫黄色，有极多数细糙毛。花果期7-9月。生海拔3000-3500米的高山或亚高山草坝和干旱草地、牧场、湿草甸、混合灌木地、橡木林的空地。产云南和四川。

Perennial herbs, often with stolons bearing rosettes. Leaves gradually reduced upward; basal rosette present at anthesis. Capitula terminal, solitary; ray florets 28-40, bluish purple to lavender blue; disk florets orange. Achenes broadly obovoid; pappus purplish straw-colored, bristles barbellate throughout. Fl. and fr. Jul-Sep. Alpine or subalpine grasslands and dry grasslands, pastures, wet meadows, mixed shrublands, clearings in *Quercus* forests at 3000-3500 m. Distributed in Yunnan and Sichuan.

红冠紫菀 *Aster handelii*

东俄洛紫菀

Aster tongolensis Franch.

多年生草本，具细根状茎和匍匐枝。茎纤细，被长毛，不分枝。叶匙形、长圆状或条状披针形，被长粗毛，侧脉和离基三出脉显明。头状花序单生，

东俄洛紫菀 *Aster tongolensis*

直径3-6.5厘米；总苞片长圆状披针形，被密毛；舌状花蓝色或浅红色；冠毛紫褐色。花期6-8月，果期7-9月。生海拔2500-4000米的高山、亚高山林下、水边或草地。产青藏高原。

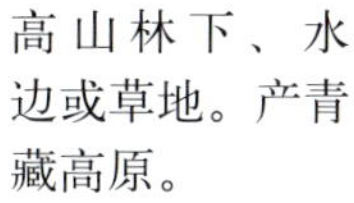

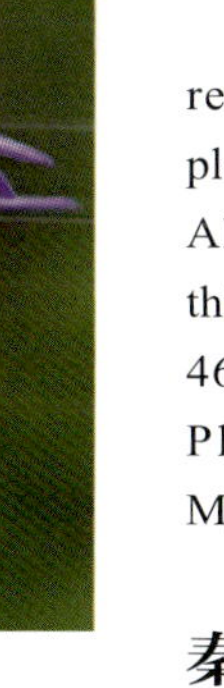

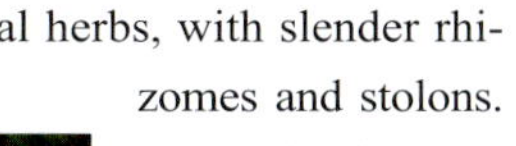

缘毛紫菀 *Aster souliei*

Perennial herbs, with slender rhizomes and stolons. Stems slender, long hairy, unbranched. Leaves spathulate, oblong or linear-lanceolate, long strigillose, lateral and triplicostate veins conspicuous. Capitula solitary, ca. 3-6.5 cm diam; phyllaries oblong-lanceolate, densely hairy; ray florets blue or reddish; pappus purple-brown. Fl. Jun-Aug. Fr. Jul-Sep. Alpine, subalpine forests, watersides or grasslands at 2500-4000 m. Distributed in Tibetan Plateau.

秦中紫菀 *Aster giraldii*

缘毛紫菀

Aster souliei Franch.

多年生草本。根状茎粗壮，木质。茎单生，纤细，不分枝，被长粗毛。莲座状叶和茎基部叶倒卵圆形、长圆状匙形或倒披针形。头状花序单生，直径3-6厘米；总苞半球形；总苞片长圆形；舌状花蓝紫色；冠毛紫褐色。花期5-7月，果期8月。生海拔2700-4600米的高山针林外缘、灌丛或山坡草地。产青藏高原。不丹和缅甸亦有。

Perennial herbs. Rhizomes stout, woody. Stems solitary, slender, unbranched, long strigillose. Rosette and basal cauline leaves obovate, oblong-spathulate or oblanceolate. Capitula solitary, 3-6 cm diam; involucres semiglobose; phyllaries oblong; ray florets bluish-purple; pappus purple-brown. Fl. May-Jul. Fr. Aug. Alpine coniferous forest edges, thickets or grassy slopes at 2700-4600 m. Distributed in Tibetan Plateau. Also in Bhutan and Myanmar.

秦中紫菀

Aster giraldii Diels

多年生草本。茎直立纤细。基部叶在花期枯萎；下部叶长圆状匙形到披针形；中部叶长圆状或倒卵圆状披针形。头状花序单生或2-4；总苞半球钟形；舌状花蓝紫色；冠毛红褐色或黄色。花果期8-12月。生海拔1800-2600米的山谷溪岸及疏林下。产陕西。

Perennial herbs. Stems erect, slender. Basal leaves withered at anthesis; lower cauline leaves oblong-spatulate to oblanceolate; middle leaves oblong or lanceolate. Capitula solitary or 2-4; involucres hemispheric-campanulate; ray florets blue-purple; pappus reddish brown or yellowish. Fl. and fr. Aug-Dec. Ravines, open forests at 1800-2600 m. Distributed in Shaanxi.

柄。头状花序在茎端单生；总苞片2-3层，带紫色；舌状花40-50(-70)朵，蓝紫色；管状花紫褐色或黄色。瘦果窄倒卵圆形，扁，褐色；冠毛白色至淡黄色。花期7-8月，果期9-10月。生海拔3600-4800米的高山草甸及针叶林下。产中国西南。东喜马拉雅亦有。

Perennial herbs. Basal leaves broadly winged petiolate. Capitula terminal, solitary; phyllaries 2-3-seriate, purplish; ray florets 40-50(-70), blue-purple; disk florets purple-brown or yellow. Achenes brown, narrowly obovoid, compressed; pappus whitish to straw-colored. Fl. Jul-Aug. Fr. Sep-Oct. Alpine meadows and *Pinus* forests at 3600-4800 m. Distributed in SW China. Also in E Himalaya.

怒江紫菀 *Aster salwinensis*

怒江紫菀

Aster salwinensis Onno

多年生草本。基部和下部茎叶具翅柄，倒卵形至倒披针形；茎部叶长卵形至倒卵形或卵圆披针形，半抱茎，无柄。头状花序单生；总苞半球形。舌状花蓝色或蓝紫色，稀白色，管状花黄色；冠毛淡黄色至褐色。花期7-9月，果期8-10月。生海拔3300-4600米的高山山坡草地或石砾地。产中国西南。印度东北部和缅甸亦有。

Perennial herbs. Basal and lower cauline leaves broadly to narrowly winged petiolate, blade obovate to oblanceolate; cauline leaves sessile, subclasping, obovate to ovate or ovate-lanceolate. Capitula solitary; involucres hemispheric. Ray florets blue to blue-purple, rarely white; disk florets yellow; pappus straw-colored to brownish. Fl. Jul-Sep. Fr. Aug-Oct. Alpine meadows on slopes or rocks, openings in Rhododendron-bamboo thickets at 3300-4600 m. Distributed in SW China. Also in NE India and Myanmar.

须弥紫菀

Aster himalaicus C. B. Clarke

多年生草本。莲座状叶具宽翅

须弥紫菀 *Aster himalaicus*

匍生紫菀

Aster stracheyi Hook. f.

多年生草本。基生叶匙形或椭圆形，边缘锯齿形；茎生叶无柄或具短柄，线形或椭圆形至倒披针形或倒卵形。头状花序

匍生紫菀 *Aster stracheyi*

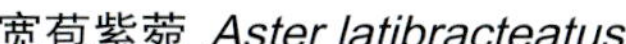
宽苞紫菀 *Aster latibracteatus*

线舌紫菀 *Aster bietii*

单生；总苞半球形；舌状花舌片蓝紫色，管状花黄色；冠毛污白色或带红色。花果期7-9月。生海拔3300-4800米的岩石溪岸及高山泥沼地。产西藏。喜马拉雅亦有。

Perennial herbs. Basal leaves spatulate or elliptic, margin serrate; cauline leaves sessile or shortly petiolate, blade linear or oblong to oblanceolate or obovate. Capitula solitary; involucres hemispheric; ray florets violet, disk florets yellow; pappus 3- or 4-seriate, white or whitish. Fl. and fr. Jul-Sep. Rocks near streamsides, among boulders in alpine tundra at 3300-4800 m. Distributed in Xizang. Also in Himalaya.

宽苞紫菀

Aster latibracteatus Franch.

多年生草本。茎密被黄褐色长柔毛。顶部叶较密而包围头状花序。头状花序单生于茎端；总苞片被黄褐色或深褐色长茸毛及长缘毛；舌状花25-30朵，舌片蓝色或紫色；管状花黄色。瘦果长圆形，褐色，被疏贴毛；冠毛3层，基部黄色，顶端褐色。花果期7-10月。生海拔2800-4000米的高山坡地或石砾地。产云南西北部。缅甸北部亦有。

Perennial herbs. Stems densely villous. Uppermost leaves dense, surrounding capitula. Capitula terminal, solitary; phyllaries 3-seriate, margin sparsely to densely villous-ciliate; ray florets 25-30, blue or purplish; disk florets yellow. Achenes pale brown, oblanceolate, compressed, sparsely strigose; pappus 3-seriate, yellowish at base, brown distally. Fl. and fr. Jul-Oct. Stony alpine pastures or slopes at 2800-4000 m. Distributed in NW Yunnan. Also in N Myanmar.

线舌紫菀

Aster bietii Franch.

多年生草本。叶基生和茎生。头状花序单生茎顶；舌状花50-70朵，舌片蓝紫色；管状花黄色。瘦果狭倒卵形；冠毛红褐色。花期7-8月，果期8-11月。生海拔3300-4600米的高山、亚高山山坡草地、沙地及石砾地。产西藏和云南。

Perennial herbs. Leaves basal and cauline. Capitula terminal, solitary; ray florets 50-70, purplish blue; disk florets yellow. Achenes obovoid; pappus reddish brown. Fl. Jul-Aug. Fr. Aug-Nov. Alpine and subalpine meadows and screes, ledges of cliffs at 3300-4600 m. Distributed in Xizang and Yunnan.

短毛紫菀

Aster brachytrichus Franch.

多年生草本。莲座状叶有长柄；下部叶和中部叶无柄。头状花序在茎端单生；舌状花30-60朵，舌片蓝色或紫色；管状花由橙变红。瘦果极扁，倒卵球形，边缘有翅；冠毛紫褐色到褐色。花期6-8月，果期8-10月。生海拔2500-4900米的高山草甸、山坡、高山针叶林、灌木丛或牧场。产云南、四川和贵州。缅甸亦有。

Perennial herbs. Basal long petiolate; lower and middle cauline leaves sessile. Capitula terminal, solitary; ray florets 30-60, blue or purplish; disk florets orange becoming reddish. Achenes obovoid, compressed, margin winged; pappus purple brown to brown. Fl. Jun-Aug. Fr. Aug-Oct. Alpine meadows, mountain slopes, alpine coniferous forests, thickets or pastures at 2500-4900 m. Distributed in Yunnan, Sichuan and Guizhou. Also in Myanmar.

短毛紫菀 *Aster brachytrichus*

滇西北紫菀

Aster jeffreyanus Diels

多年生草本。基部叶在花期生存，具短柄，半抱茎。头状花序单生于茎端；总苞半球形；总苞片2-3层，有毛；舌状花蓝色或紫色；管状花长3.5-4毫米。瘦果倒卵圆形，压扁，边缘有翅；冠毛基部黄色，上部紫褐色。花期6-7月。生海拔2800-3800米的高山或亚高山开阔坡地。产云南、四川和贵州。

Perennial herbs. Basal leaves usually present at anthesis, shortly petiolate; sessile, base subclasping. Capitula terminal, solitary; involucres hemispheric; phyllaries 2-3-seriate, villous; ray florets, blue to purple; disk florets 3.5-4 mm. Achenes broadly obovoid, compressed, margin winged; pappus purplish brown. Fl. Jun-Jul. Alpine or subalpine open slopes, meadows at 2800-3800 m. Distributed in Yunnan, Sichuan and Guizhou.

滇西北紫菀 *Aster jeffreyanus*

星舌紫菀 *Aster asteroides*

星舌紫菀

Aster asteroides (DC.) Kuntze

多年生草本。茎纤细，花葶状，紫色或下部绿色，中部以上或上部常无叶。舌状花35-60朵，舌片蓝紫色到深紫红色；管状花橙黄色。瘦果长圆形，被短糙毛；冠毛2层，淡黄色。花果期6-8月。生海拔3200-4600米的高山灌丛、湿润草地或冰碛物上。产中国西南和西北。喜马拉雅亦有。

Perennial herbs. Stems erect, solitary, scapiform, purplish or green below, base without marcescent leaf remains. Capitula terminal, solitary; Ray florets 35-60, bluish purple to deep mauve; disk florets orange-yellow. Achenes narrowly obovoid, strigillose; pappus 2-seriate, straw-colored. Fl. and fr. Jun-Aug. Alpine thickets, damp grasslands or marshy areas at 3200-4600 m. Distributed in SW and NW China. Also in Himalaya.

丽江紫菀 *Aster likiangensis*

丽江紫菀

Aster likiangensis Franch.

多年生草本，具数个块根。茎被紫色腺毛。基生叶密集，倒卵形、菱形或长圆形，下面被密毛，具离基三出脉。头状花序单生，直径2.5-5厘米；总苞半球形；总苞片披针形，革质，紫绿色；舌状花蓝紫色。瘦果长圆形；冠毛白色。花果期6-8月。生海拔3500-4500米的高山草甸、开阔坡地、河谷或泥炭沼泽地。产中国西南。不丹亦有。

Perennial herbs, with several tuberous roots. Stems purple glandular hairy. Basal leaves dense, obovate, rhombic or oblong, densely pubescent abaxially, triplicostate. Capitula solitary, 2.5-5 cm diam; involucres semiglobose; phyllaries lanceolate, coriaceous, purplish-green; ray florets bluish-purple. Achenes oblong; pappus white. Fl. and fr. Jun-Aug. Alpine meadows, open slopes, valleys or peat swamps at 3500-4500 m. Distributed in SW China. Also in Bhutan.

萎软紫菀

Aster flaccidus Bunge

多年生草本，不分枝，密被白色长毛。基部叶和莲座状叶多数，匙形，茎生叶3-5个，长圆

萎软紫菀 *Aster flaccidus*

形到长圆状披针形；半抱茎，全部叶质薄，具离基三出脉。头状花序在茎端单生，直径3.5-7厘米；总苞片条状披针形；舌状花紫色，稀浅红色；冠毛白色。瘦果长圆形。花果期6-11月。生海拔1800-5100米的高山草地、灌丛或石砾地。产中国西南、华北和西北。喜马拉雅、中亚、俄罗斯和蒙古亦有。

Perennial herbs, unbranched, long white hairy. Basal and rosette leaves numerous, spathulate, cauline ones 3-5, oblong or oblong-lanceolate, semi-amplexicaul, all leaves thin, triplicostate. Capitula terminal, solitary, 3.5-7 cm diam; phyllaries linear-lanceolate; ray florets purple, rarely reddish; pappus white. Achenes oblong. Fl. and fr. Jun-Nov. Alpine grasslands, thickets or rocky places at 1800-5100 m. Distributed in SW, N and NW China. Also in Himalaya, C Asia, Russia and Mongolia.

重冠紫菀

Aster diplostephioides (DC.) C. B. Clarke

多年生草本。叶长圆形至倒披针形。头状花序单个顶生；总苞半球形；总苞片2或3层，条状披针形，色深；舌状花2层，淡紫色至紫色或蓝紫色；管状花黄色至橘黄色；冠毛3层；白色。花期7-9月，果期9-12月。生海拔2700-4600米的草地、灌丛、沼泽地、溪岩、冲积平原或杂木林。产青藏高原。喜马拉雅亦有。

Perennial herbs. Leaves oblong to oblanceolate. Capitula terminal, solitary; involucres hemispheric; phyllaries 2- or 3-seriate, linear-lanceolate, dark colored; ray florets 2-seriate, mauve to purple or lilac-blue; disk florets yellow to orange; pappus 3-seriate, white. Fl. Jul-Sep. Fr. Sep-Dec. Grasslands, thickets, boggy areas, stream banks, floodplains or mixed forests at 2700-4600 m. Distributed in Tibetan Plateau. Also in Himalaya.

重冠紫菀 *Aster diplostephioides*

云南紫菀

Aster yunnanensis Franch.

多年生草本。叶长圆形、卵圆形或条形，上面被疏毛，有腺，中脉在下面凸起，离基三出脉和侧脉显明。头状花序直径4-8.5厘米，在茎和枝端单生；总苞片披针形，密被长柔毛；舌状花蓝色或浅蓝色；冠毛白色。花期7-9月，果期9-10月。生海拔2300-4500米的高山或亚高山草地或杂木林中。产青藏高原。

Perennial herbs. Leaves oblong, ovate or linear, sparsely hairy and glandular adaxially, midrib elevated abaxially, triplicostate and lateral veins conspicuous. Capitula 4-8.5 cm diam, solitary on tops of stems and branches; phyllaries lanceolate, densely villose; ray florets blue or bluish; pappus white. Fl. Jul-Sep. Fr. Sep-Oct. Alpine or subalpine grasslands or mixed forests at 2300-4500 m. Distributed in Tibetan Plateau.

云南紫菀 *Aster yunnanensis*

狭苞紫菀 *Aster farreri*

狭苞紫菀

Aster farreri W. W. Sm. et Jeffrery

多年生草本，被长毛。茎不分枝。叶狭匙形、条状披针形或条形，基部半抱茎，质薄。头状花序单生茎顶，直径5-8厘米；总苞片条形；舌状花紫蓝色；冠毛白色或污白色。瘦果长圆形。花期7-8月，果期8-9月。生海拔1300-4100米的高山草甸中。产四川、河北、山西、甘肃和青海。

Perennial herbs, long hairy. Stems simple. Leaves narrowly spathulate, linear-lanceolate or linear, base semi-amplexicaul, thin. Capitula solitary, terminal, 5-8 cm diam; invulucral bracts linear; ray florets purplish-blue. pappus white or dirty white. Achenes oblong. Fl. Jul-Aug. Fr. Aug-Sep. Alpine meadows at 1300-4100 m. Distributed in Sichuan, Hebei, Shanxi, Gansu and Qinghai.

短茎紫菀

Aster brevis Hand.-Mazz.

多年生丛生草本，高5-12厘米。直根粗壮。基生叶多数，匙形，茎生叶4-7个，长圆形，全部叶全缘，质厚。头状花序单生，直径3.5-4厘米；总苞2层；总苞片披针形，带紫色；舌状花深蓝色，舌片条状长圆形；管状花橙黄色；冠毛白色。瘦果被毛。花期7月。生海拔3400-3900米的高山山坡上。产云南西北部。

Perennial fasciate herbs, 5-12 cm tall. Taproots stout. Basal leaves numerous, spathulate, cauline leaves 4-7, oblong, all leaves entire, thick. Capitula solitary, 3.5-4 cm diam; phyllaries 2 rows, lanceolate, purple; ray florets dark blue, ligules linear-oblong; tubular florets orange-yellow; pappus white. Achenes pubescent. Fl. Jul. Alpine slopes at 3400-3900 m. Distributed in NW Yunnan.

巴塘紫菀

Aster batangensis Bur. et Franch.

亚灌木。基出条短或长达2厘米，有密集的叶和顶生的莲座状叶丛。叶匙形。花茎纤细，叶条形；全部叶厚，被柔毛和缘毛。头状花序单生，直径3-4.5厘米；总苞片条状披针形；舌状花紫色；冠毛2层，白色或稍红色。花期5-9月，果期9-10月。生海拔3400-4600米的林缘、灌丛、草地或石砾地。产云南、四川和西藏。

Subshrubs. Basal vegetative shoots very short or to 2 cm long, with dense spathulate leaves and terminal rosette leaves. Leaves spatulate. Flower stems slender, with linear leaves; all leaves thick, pubescent, ciliate. Capitula solitary, 3-4.5 cm diam; phyllaries linear-lanceolate; ray florets purple; pappus 2 layers, white or reddish. Fl. May-Sep. Fr. Sep-Oct. Forest edges, thickets, grasslands or rocky places at 3400-4600 m. Distributed in Yunnan, Sichuan and Xizang.

短茎紫菀 *Aster brevis*

巴塘紫菀 *Aster batangensis*

匙叶巴塘紫菀

Aster batangensis Bur. et Franch. var. **staticefolius** (Franch.) Ling

亚灌木。基出条短或长达2厘米，有密集的叶和顶生的莲座状叶丛。叶匙形；叶长2-5厘米，宽0.4-0.6厘米，叶两面无毛，仅沿脉有微柔毛。总苞片背面有腺，但无毛；舌状花紫色；冠毛2层，白色或稍红色。花期5-9月，果期9-10月。生海拔2500-4000米的开阔坡地、砾石质悬崖和林缘。产云南和四川。

Subshrubs. Basal vegetative shoots very short or to 2 cm long, with dense spathulate leaves and terminal rosette leaves. Leaves spatulate; leaves 2-5 cm long, 0.4-0.6 cm wide, glabrous except puberulous along midrib and nerves on both surfaces. Phyllaries glandular but glabrous abaxially; ray florets purple; pappus 2 layers, white or reddish. Fl. May-Sep. Fr. Sep-Oct. Open slopes, gravelly cliffs and forest edges at 2500-4000 m. Distributed in Yunnan and Sichuan.

匙叶巴塘紫菀 *Aster batangensis* var. *staticefolius*

灰枝紫菀

Aster poliothamnus Diels

丛生亚灌木，高15-100厘米。叶长圆形或线状长圆形。头状花序密集成伞房状或单生；总苞片4-6层，不等长；舌状花10-20朵，淡紫色；管状花黄色。瘦果2-3肋；冠毛4层，稻草色。花期6-9月，果期8-10月。生海拔800-4200米的干山坡、石崖、多石地区、蒿-针茅草丛、旱生灌丛、河岸或路旁。产中国西南和西北。

Subshrubs, 15-100 cm tall, caespitose. Leaves oblong or linear-oblong. Capitula in corymbiform or solitary. Phyllaries 4-6-seriate, unequal; ray florets 10-20, lavender or lilac to purple; disk florets yellow. Achenes 2-3-ribbed; pappus 4-seriate, straw-colored. Fl. Jun-Sep. Fr. Aug-Oct. Dry slopes, rocky cliffs, stony places, *Artemisia-Stipa* formations, xerophytic shrublands, riverbanks, roadsides at 800-4200 m. Distributed in SW and NW China.

灰枝紫菀 *Aster poliothamnus*

厚棉紫菀 *Aster prainii*

兴安一枝黄花 *Solidago dahurica*

厚棉紫菀

Aster prainii (J. R. Drumm.) Y. L. Chen

多年生草本，高4-10厘米。基生叶花期枯萎，匙形，有长柄；上部叶线形，常围裹总苞。头状花序单生于茎端；总苞宽钟状；总苞片3层，外层顶端紫色；舌状花38-40朵，深蓝色；管状花橙色渐变略带紫色。瘦果具4肋；冠毛4层，禾草色。花期8-9月。生海拔4200-5400米的流石滩。产西藏南部。不丹亦有。

Perennial herbs, 4-10 cm tall. Basal leaves present at anthesis, spatulate, long petiolate; upper cauline leaves linear, often surrounding involucre. Capitula solitary; involucres broadly campanulate; phyllaries 3-seriate; outer phyllaries tip purplish; ray florets 38-40, deep blue; disk florets orange becoming purplish. Achenes 4-ribbed; pappus 4-seriate, straw-colored. Fl. Aug-Sep. Stony screes at 4200-5400 m. Distributed in S Xizang. Also in Bhutan.

兴安一枝黄花

Solidago dahurica (Kitag.) Kitag. ex Juzep.

多年生草本，具根状茎。下部茎叶具柄，叶片椭圆形、长椭圆形或披针形，稀卵形；中部茎叶无柄，椭圆形、长椭圆形或披针形。头状花序排列成密圆锥花序或疏圆锥花序；总苞钟状；花冠黄色；冠毛白色。花果期7-9月。生海拔300-2100米的森林或林中草甸。产中国东北、华北和新疆。中亚、蒙古和俄罗斯亦有。

Perennial herbs, rhizomes. Leaves: lower cauline petioles, blade elliptic, long elliptic or lanceolate, rarely ovate; mid cauline sessile, blade elliptic, long elliptic or lanceolate. Capitula dense or lax paniculiform synflorescences; involucre campanulate; corolla yellow; pappus white. Fl. and fr. Jul-Sep. Forests, forest glade at 300-2100 m. Distributed in NE, N China and Xinjiang. Also in C Asia, Mongolia and Russia.

一枝黄花

Solidago decurrens Lour.

多年生草本。叶质厚，椭圆形、长圆形、卵形或宽披针形，边缘全缘或上部具细齿；叶柄具翅。头状花序小，多数在茎上部排成总状或伞房圆锥花序；总苞片披针形；舌状花黄色，椭圆形。花果期4-11月。生海拔100-2900米的林缘、林下、灌丛或山坡草地。广布华南。

Perennial herbs. Leaves thick, elliptic, oblong, ovate or broadly lanceolate, margin entire or serrulate on the upper part; petioles winged. Capitula small, numerous arranged in racemose or corymbose panicles on upper part of stem; phyllaries lanceolate; ray florets yellow, elliptic. Fl. and fr. Apr-Nov. Forest edges, forests, thickets or grassy slopes at 100-2900 m. Widely distributed in S China.

一枝黄花 *Solidago decurrens*

橙花飞蓬 *Erigeron aurantiacus*

短葶飞蓬 *Erigeron breviscapus*

橙花飞蓬

Erigeron aurantiacus Regel

多年生草本。莲座叶长圆状披针形或倒披针形，或有时倒卵形或椭圆形，花期宿存，有长柄，茎叶半抱茎，中部和上部叶无柄。头状花序单生于茎顶端；总苞片半球形，3层，近等长，线状披针形；舌状花3层，舌片橘红色，或黄色至红褐色；管状花黄色。瘦果被较密多少贴生的短毛；冠毛2层。花期7-9月。生海拔2100-3400米的高山草甸或林缘。产新疆北部。哈萨克斯坦亦有。

Perennial herbs. Basal leaves rosulate, present at anthesis, long petiolate, blade oblong-oblanceolate or oblanceolate, sometimes obovate or elliptic; cauline leaves subclasping. Capitula solitary; involucre hemispheric; phyllaries 3-seriate, subequal, linear-lanceolate; ray florets 3-seriate, orange, sometimes yellow to brick-red; disk florets yellow. Achenes compressed, strigose; pappus 2-seriate. Fl. Jul-Sep. Forest margins or alpine meadows at 2100-3400 m. Distributed in N Xinjiang. Also in Kazakhstan.

短葶飞蓬

Erigeron breviscapus (Vaniot) Hand.-Mazz.

多年生草本，被短硬毛。基生叶密集，莲座状；茎生叶2-4个，狭披针形或条形。头状花序单生茎或分枝顶端；总苞半球形；总苞片条状披针形；外围雌花舌状，蓝色或粉紫色；中央两性花管状，黄色。瘦果狭长圆形；冠毛淡褐色。花期3-10月。生海拔1200-3600米的山坡草地、林缘、溪边、弃荒地或松林下。产云南、四川、西藏、贵州、广西和湖南。

Perennial herbs, shortly strigillose. Basal leaves dense, rosette; cauline leaves 2-4, narrowly lanceolate or linear. Capitula solitary, terminal; involucres semiglobose; phyllaries linear-lanceolate; outer pistillate flowers ligulate, blue or pinkish-purple; central bisexual flowers tubular, yellow. Achenes narrowly oblong; pappus brownish. Fl. Mar-Oct. Grassy slopes, forest edges, streamsides, disturbed slopes or pine forests at 1200-3600 m. Distributed in Yunnan, Sichuan, Xizang, Guizhou, Guangxi and Hunan.

珠峰飞蓬

Erigeron himalajensis Vierh.

多年生草本。茎紫色，或上部绿色。基生叶莲座状，基生叶和底部茎叶倒披针形至长圆形；上部叶线形。头状花序径2至多数排列成伞房状，具1-2个线形的苞叶；总苞片3层。瘦果4棱；冠毛2层。花期7-9月。生海拔2000-3600米的多石地或林缘。产中国西南。阿富汗和喜马拉雅亦有。

Perennial herbs. Stems purple or green above. Leaves: basal rosulate, basal cauline oblanceolate to oblong; uppermost linear. Capitula 2 to many in corymbiform synflorescences; bracts 1-2, linear; phyllaries 3-seriate. Achenes 4-veined; pappus 2-seriate. Fl. Jul-Sep. Forest margins, stony slopes at 2000-3600 m. Distributed in SW China. Also in Afghanistan and Himalaya.

珠峰飞蓬 *Erigeron himalajensis*

棉毛飞蓬 *Erigeron lanuginosus*

多舌飞蓬 *Erigeron multiradiatus*

棉毛飞蓬

Erigeron lanuginosus Y. L. Chen

多年生草本，簇生，根状茎木质。茎绿色，具紫色条纹。基部叶莲座状，花期枯萎或生存，具3脉；茎叶长圆状倒披针形或长圆形。头状花序单生或数个排成伞房状；总苞半球形；总苞片3层，紫色；外围的舌花2-3层，鲜紫色；两性花黄色；冠毛2层。花期8月。生海拔3200-4200米的山坡草地。产西藏。

Perennial herbs, caespitose; rhizome woody. Stems green, purplish striate. Leaves: basal rosulate, withered or present at anthesis, 3-veined; cauline oblong-oblanceolate or oblong. Capitula solitary or few in corymbose synflorescences. Involucre hemispheric; phyllaries 3-seriate, purple. Ray florets 2- or 3-seriate, bright purple; disk florets yellow. Pappus 2-seriate. Fl. Aug. Alpine grasslands on slopes at 3200-4200 m. Distributed in Xizang.

多舌飞蓬

Erigeron multiradiatus (Lindl.) Benth.

多年生草本，被短硬毛。基生叶密集，莲座状，倒披针形；上部叶卵状披针形或条形。头状花序直径3-4厘米或更大，1至数个，顶生；总苞半球形；总苞片条状披针形；外围雌花舌状，紫色；冠毛污白色或淡褐色。花期5-9月。生海拔2300-4600米的草地、山坡或林缘。产云南、四川和西藏。印度北部、尼泊尔和阿富汗亦有。

Perennial herbs, shortly strigillose. Basal leaves dense, rosette; oblanceolate; upper leaves ovate-lanceolate or linear. Capitula 3-4 cm diam or larger, solitary to several, terminal; involucres semiglobose; phyllaries linear-lanceolate; outer pistillate flowers ligulate, purple; pappus dirty white or brownish. Fl. May-Sep. Grasslands, slopes or forest edges at 2300-4600 m. Distributed in Yunnan, Sichuan and Xizang. Also in N India, Nepal and Afghanistan.

展苞飞蓬

Erigeron patentisquama Jeffrey

多年生草本。基生叶密集，莲座状，匙形或倒披针状匙形，全缘；茎生叶披针形或条形。头状花序直径约2.5厘米，单生或2-4个排列成伞房状；总苞半球形；总苞片条状披针形；外围雌花舌状，紫红色；冠毛淡黄色。花期7-9月。生海拔2400-4100米的高山草地、草甸或林缘。产云南和四川。

展苞飞蓬 *Erigeron patentisquama*

一年蓬 *Erigeron annuus*

Perennial herbs. Basal leaves dense, rosette, spathulate or oblanceolate-spathualate, margin entire; cauline leaves lanceolate or linear. Capitula ca. 2.5 cm diam, solitary or 2-4 arranged in corymb; involucres semiglobose; phyllaries linear-lanceolate; outer pistillate flowers ligulate, purple; pappus yellowish. Fl. Jul-Sep. Alpine grasslands, meadows or forest edges at 2400-4100 m. Distributed in Yunnan and Sichuan.

一年蓬

Erigeron annuus (L.) Pers.

一年生或二年生草本，被硬毛。下部叶长圆形或宽卵形；上部叶披针形至条形。头状花序数个或多数，排成疏圆锥花序；总苞半球形；总苞片草质，披针形，密被腺毛；外围雌花舌状，白色或淡天蓝色，条形；冠毛白色。花期6-9月。生海拔约1100米的路边、旷野或山坡荒地。广布中国大部分地区。原产北美洲。

Annual or biennial herbs, strigose. Lower leaves oblong or broadly ovate; upper leaves lanceolate to linear. Capitula several or numerous, arranged in lax panicle; involucres semiglobose; phyllaries herbaceous, lanceolate, densely glandular hairy; outer pistillate flowers ligulate, white or sky-bluish, linear; pappus white. Fl. Jun-Sep. Roadsides, open fields or waste slopes at ca. 1100 m. Widely distributed throughout most parts of China. Native to North America.

飞蓬

Erigeron acris L.

二年生直立草本，被长硬毛。基生叶密集；茎生叶披针形或条形。头状花序较小，多数，排成密集狭圆锥花序；总苞半球形；总苞片条状披针形；外层雌花舌状，淡红紫色，内层细管状；中央两性花管状，黄色；冠毛白色。花期6-9月。生海拔700-3500米的山坡草地、牧场或林缘。产中国西南、华北、东北和西北。北亚和北美洲亦有。

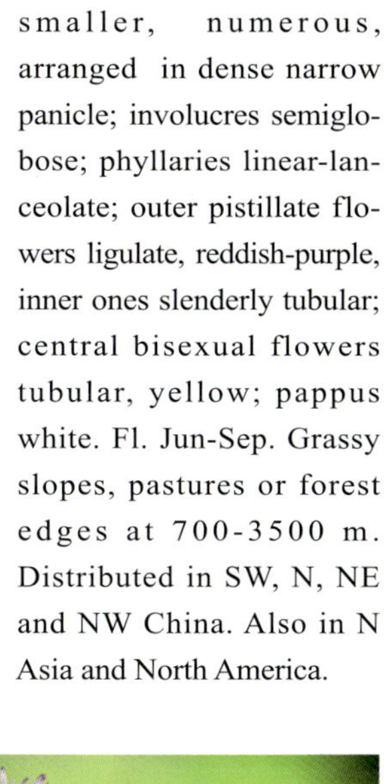

Biennial erect herbs, long strigillose. Basal leaves dense; cauline leaves lanceolate to linear. Capitula smaller, numerous, arranged in dense narrow panicle; involucres semiglobose; phyllaries linear-lanceolate; outer pistillate flowers ligulate, reddish-purple, inner ones slenderly tubular; central bisexual flowers tubular, yellow; pappus white. Fl. Jun-Sep. Grassy slopes, pastures or forest edges at 700-3500 m. Distributed in SW, N, NE and NW China. Also in N Asia and North America.

飞蓬 *Erigeron acris*

小蓬草 *Erigeron canadensis*

小蓬草
Erigeron canadensis L.

一年生草本。叶密集，披针形或条形，边缘具疏锯齿或全缘。头状花序多数，排成顶生多分枝的大型圆锥花序；总苞近圆柱状；总苞片条状披针形或条形；雌花白色，舌状；两性花淡黄色。瘦果条状披针形；冠毛污白色。花期5-9月。生海拔3000米以下的旷野、荒地、田边或路旁。中国大部分地区广泛入侵。世界各地广布。原产北美洲。

Annual herbs. Leaves dense, lanceolate or linear, margin sparsely serrate or entire. Capitula numerous, arranged in terminal and multi-branched panicles; involucres subcylindric; phyllaries linear-lanceolate or linear; pistillate flowers white, ligulate; bisexual flowers yellowish. Achenes linear-lanceolate; pappus dirty white. Fl. May-Sep. Open places, wastelands, fieldsides or roadsides below 3000 m. Widely invaded throughout most parts of China. Widely distributed around the world. Native to North America.

香丝草
Erigeron bonariensis L.

一年生或二年生草本，密被贴短毛，杂有开展疏长毛。叶密集，披针形或条形，边缘具粗齿、羽状浅裂或全缘。头状花序多数，排成总状或总状圆锥花序；总苞椭圆状卵形；总苞片条形；雌花白色，细管状；两性花淡黄色。瘦果条状披针形；冠毛污白色。花期5-10月。生海拔3100米以下的荒地，田边或路旁。中国西南、华南、华中和华东地区归化。世界热带和亚热带地区广泛分布。原产南美洲。

Annual or biennial herbs, densely and appressedly short hairy, and sparsely pilose. Leaves dense, lanceolate or linear, margin grossly serrate, pinnately lobed

香丝草 *Erigeron bonariensis*

or entire. Capitula numerous, arranged in raceme or racemose panicle; involucres ellipsoid-ovoid; phyllaries linear; pistillate flowers white, tubular; bisexual flowers yellowish. Achenes linear-lanceolate; pappus dirty white. Fl. May-Oct. Wastelands, fieldsides or roadsides below 3100 m. Naturalized in SW, S, C and E China. Widely distributed throughout tropical and subtropical regions in the world. Native to South America.

钻叶紫菀
Symphyotrichum subulatum

钻叶紫菀

Symphyotrichum subulatum (Michx.) G. L. Nesom

一年生草本，全株光滑无毛，肉质。基生叶倒披针形，花后凋落；茎生叶条状披针形，全缘。头状花序小，多数排成圆锥状；总苞狭钟状；总苞片3-4层，条状钻形；舌状花细狭，淡红色；冠毛污白色。瘦果长圆形或椭圆体形。花果期8-10月。生海拔2000米以下的荒地、路旁、河漫滩、沼泽地或山坡灌丛中。中国西南、华中、华东和华北等地归化。原产北美洲。

Annual herbs, wholly smooth and glabrous, succulent. Basal leaves oblanceolate, withered after anthesis; cauline leaves linear-lanceolate, margin entire. Capitula small, many arranged in panicle; involucres narrowly campanulate; phyllaries 3-4 rows, linear-subulate; ray florets minute, linear, reddish; pappus dirty white. Achenes oblong or ellipsoid. Fl. and fr. Aug-Oct. Wastelands, roadsides, flood plains, swamps or slope thickets below 2000 m. Naturalized in SW, C, E and N China. Native to North America.

石胡荽

Centipeda minima (L.) A. Br. et Asch.

一年生匍匐小草本，高5-20厘米。茎多分枝。叶互生，细小，楔状倒披针形，边缘有少数锯齿。头状花序小，单生叶腋，无梗，扁球形；总苞片椭圆状披针形；小花管状，淡绿黄色或淡紫红色。瘦果椭圆体形；无冠毛。花果期6-10月。生海拔1500(-2500)米以下的路边、荒地或阴湿地。产中国大部分地区。印度、马来西亚、朝鲜半岛、日本和大洋洲亦有。

Annual small creeping herbs, 5-20 cm tall. Stems multi-branched. Leaves alternate, minute, cuneate-oblanceolate, margin sparsely serrate. Capitula small, solitary, axillary, sessile, depressed globose; phyllaries elliptic-lanceolate; corolla tubular, greenish-yellow or purplish-red. Achenes ellipsoid; pappus absent. Fl. and fr. Jun-Oct. Roadsides, wastelands or damp places below 1500(-2500) m. Distributed in most parts of China. Also in India, Malaysia, Korean Peninsula, Japan and Oceania.

石胡荽 *Centipeda minima*

芫荽菊 *Cotula anthemoides*

分枝亚菊 *Ajania ramosa*

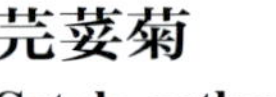

芫荽菊

Cotula anthemoides L.

一年生铺散小草本。叶互生，长圆形或椭圆形，二回羽状分裂。头状花序小，单个顶生或腋生，具纤细花序梗；总苞盘状；总苞片矩圆形，绿色，中脉红色；花冠管状，黄色。瘦果倒卵状矩圆形，具翅。花果期9月至翌年3月。生海拔1000-1100米的河边、湿地或稻田。产中国西南、华南、华中和东南。南亚、东南亚和非洲亦有。

Annual diffused small herbs. Leaves alternate, oblong or elliptic, bipinnatifid. Capitula small, solitary, terminal or axillary, slenderly pedunculate; involucres discoid; phyllaries oblong, green, midrib red; corolla tubular, yellow. Achenes obovate-oblong, winged. Fl. and fr. Sep to next Mar. Riversides, wetlands or rice paddies at 1000-1100 m. Distributed in SW, S, C and SE China. Also in S and SE Asia, and Africa.

裸柱菊

Soliva anthemifolia (Juss.) R. Br.

一年生矮小草本。叶莲座状，长5-10厘米，二至三回羽状分裂。头状花序近球形，无梗，生于茎基部；总苞片2-3层，矩圆形或披针形，边缘干膜质；边缘的雌花多数，无花冠；中央的两性花少数，花冠管状，黄色。瘦果倒披针形，扁平，有厚翅，花柱宿存，下部翅上有横皱纹；无冠毛。花果期全年。生荒地或栽培地。华南和华东地区有归化。原产南美洲。

Annual small herbs. Leaves in basal rosettes, 5-10 cm long; leaves 2-3-pinnatifid. Capitula subspherical at base of stem, sessile; involucres hemispheric; phyllaries in 2-3 rows, oblong or lanceolate, margin scarious; marginal female florets numerous corolla absent, styles persistent; disk florets corolla yellow. Achenes oblanceolate, with thick corky lateral wings; style persistent, spinescent; pappus absent. Fl. and fr. year-round. Waste ground and cultivated areas. Naturalized in S and E China. Native to South America.

裸柱菊 *Soliva anthemifolia*

分枝亚菊

Ajania ramosa (C. C. Chang) C. Shih in C. Shih et G. X. Fu

灌木，高80-150厘米。叶椭圆形、倒披针形或倒长卵形，羽状深裂，裂片3-4对，长椭圆形、披针形或镰刀形；两面异色，背面白色或灰白色。头状花序在枝端排成复伞房花序；总苞钟状，径5-6毫米；总苞片4层，边缘黄褐色；花冠黄色，外面有腺点。花果期8-9月。生海拔2900-4600米的山坡及河谷阶地。产四川、西藏、湖北和陕西。

Shrubs, 80-150 cm tall. Leaves elliptic, oblanceolate or long ovate, abaxially white or gray-white, pinnatipartite; lobes 3-4-paired, narrowly elliptic, lanceolate or falcate. Capitula compound-corymbose at apices of branches; involucres campanulate, 5-6 mm diam; phyllaries in 4 rows, scarious margin yellow-brown; corolla yellow, exterior with sessile glands. Fl. and fr. Aug-Sep. Mountain slopes and river valleys at 2900-4600 m. Distributed in Sichuan, Xizang, Hubei and Shaanxi.

灌木亚菊

Ajania fruticulosa (Ledeb.) Poljakov

小半灌木。叶两面同色，中部茎叶二回掌状或掌式羽状3-5分裂；上部和下部的叶掌状3-5全裂。头状花序少数或多数排成伞房花序或复伞房花序；总苞钟状；小花黄色，边缘雌花5-8朵。花果期6-10月。生海拔500-4400米的荒漠及荒漠草原。产内蒙古、华东和西南。

灌木亚菊 *Ajania fruticulosa*

中亚、蒙古和俄罗斯亦有。

Subshrubs. Leaves concolorous; middle stem leaves bipalmately 3-5-sect or bipalmate, pinnately 3-5-sect; lower and upper leaves palmately 3-5-sect. Capitula few to many around in flat-topped panicle; involucres campanulate; florets yellow, marginal female florets 5-8. Fl. and fr. Jun-Oct. Deserts and desert steppes at 500-4400 m. Distributed in Neimenggu, E and SW China. Also in C Asia, Mongolia and Russia.

短裂亚菊

Ajania breviloba (Franch. ex Hand.-Mazz.) Y. Ling et C. Shih

多年生草本。叶两面异色；中部茎叶卵形、半圆形或扇形；二回掌状或规则或不规则掌式羽状3-5分裂。头状花序多数或少数排成伞房或圆锥花序；总苞钟状；小花黄色。花果期9-10月。生海拔2800-4100米的林间空地或山坡石砾或沙质土壤上。产四川、云南、湖北、陕西和吉林。

Perennial herbs. Leaves discolorous; middle stem leaves ovate, suborbicular or flabelliform, bipinnatisect or irregularly 3-5-bipalmate-pinnatisect. Capitula many or few around in flattopped panicle, or panicle; involucres campanulate; florets yellow. Fl. and fr. Sep-Oct. Open places in forests, gravelly places on mountain slopes at 2800-4100 m. Distributed in Sichuan, Yunnan, Hubei, Shaanxi and Jilin.

铺散亚菊

Ajania khartensis (Dunn) C. Shih

多年生草本。中部叶全形圆形、半圆形、扇形或宽楔形，二回掌状或掌状3-5全裂；顶部叶和下部叶3裂。头状花序少数或多数排成伞房花序；总苞钟状，小花黄色；边缘雌花6-8朵。花果期7-9月。生海拔2500-5300米的山坡。产内蒙古、西南和西北。印度北部亦有。

Perennial herbs. Middle stem leaves orbicular, suborbicular, flabelliform or broadly cuneate, bipalmatisect or 3-5-palmatisect; lower and uppermost stem leaves trisect. Capitula few or many around in flat-topped cyme; involucres campanulate; florets yellow; marginal female florets 6-8. Fl. and fr. Jul-Sep. Mountain slopes at 2500-5300 m. Distributed in Neimenggu, SW and NW China. Also in N India.

短裂亚菊 *Ajania breviloba*

铺散亚菊 *Ajania khartensis*

多花亚菊 *Ajania myriantha*

多花亚菊
Ajania myriantha (Franch.) Ling ex C. Shih

多年生草本或半灌木，高25-100厘米。中部叶二回羽状分裂。头状花序多数在茎枝顶端排成复伞房花序；总苞钟状，直径2.5-3毫米；全部花冠顶端有腺点。花果期7-10月。生海拔2250-3600米的山坡及河谷。产青藏高原。

Perennial herbs or subshrubs, 25-100 cm tall. Median stem leaves bipinnatifid. Capitula numerous, termianal, arranged in compound corymbs; involucres campanulate, 2.5-3 mm diam; all corolla apex glandular punctuate. Fl and fr. Jul-Oct. Slopes and river valleys at 2250-3600 m. Distributed in Tibetan Plateau.

黄花亚菊
Ajania nubigena (Wall.) C. Shih

多年生草本。叶两面同色，灰白色，密被短柔毛；中部茎叶圆形或宽卵形，三回羽状分裂或不显著三回羽状分裂。头状花序多数排成复伞房花序，花序梗及花梗细，被白色稠密短柔毛；总苞钟状；小花黄色。花期8月。生海拔3400-4100米的山坡。产青藏高原。喜马拉雅亦有。

Perennial herbs. Leaves concolorous, gray-white, densely pubescent; middle stem leaves orbicular or broadly ovate, tripinnatisect or inconspicuously tripinnatisect. Capitula many around in compound flat-topped panicle, synflorescence branches and peduncles white, densely pubescent; involucres campanulate; florets yellow. Fl. Aug. Mountain slopes at 3400-4100 m. Distributed in Tibetan Plateau. Also in Himalaya.

亚菊
Ajania pallasiana (Fisch. ex Besser) Poljakov

多年生草本。叶两面异色；基生叶和下部茎叶花期枯萎脱落；中部茎叶卵形、长椭圆形或菱形。头状花序排成复伞房花序；总苞宽钟状；小花黄色。花果期8-9月。生海拔200-2900米的山坡或灌丛中。产中国东北和西北。朝鲜半岛、蒙古和俄罗斯亦有。

Perennial herbs. Leaves discolorous; basal and lower stem leaves withered at anthesis; middle stem leaves with petioles leaves ovate, narrowly elliptic or rhomboid. Synflorescence a terminal compound flat-topped panicle; involucres broadly campanulate; florets yellow. Fl. and fr. Aug-Sep. Thickets, mountain slopes at 200-2900 m. Distributed in NE and NW China. Also in Korean Peninsula, Mongolia and Russia.

紫花亚菊
Ajania purpurea C. Shih

小半灌木，高4-25厘米。老枝淡褐色；当年枝被稠密的短绒毛。叶有柄，椭圆形或偏斜椭圆形。头状花序排成伞房花序，或复头状花序式；全部苞片外面被短绒毛，边缘紫黑色；边缘雌花约6朵；全部花冠自中部以上紫红色。花果期

黄花亚菊 *Ajania nubigena*

亚菊 *Ajania pallasiana*

紫花亚菊 *Ajania purpurea*

疏齿亚菊 *Ajania remotipinna*

8-10月。生海拔4800-5300米的高山砾石堆和高山草甸及灌丛中。产西藏。

Subshrubs, 4-25 cm tall. Old branches pale brown; young branches densely tomentose. Leaves petiolate; leaves elliptic or obliquely elliptic. Synflorescence a terminal flat-topped cyme; phyllaries tomentose, scarious margin dark purple; florets purple above middle; marginal female florets 6. Fl. and fr. Aug-Oct. Alpine meadows, thickets, gravel mounds at 4800-5300 m. Distributed in Xizang.

栎叶亚菊

Ajania quercifolia (W. W. Sm.) Y. Ling et C. Shih

亚灌木，高60-150厘米。叶长椭圆形、披针形、倒卵状长圆形，少有宽线形的，边缘粗齿或缺刻状浅裂或半裂。头状花序多数，排成紧密的伞房花序；总苞钟状，直径5-6毫米。花果期8-10月。生海拔3200-3900米的林下及林缘灌丛中。产云南西北部及四川西南部。

Subshurbs, 60-150 cm tall. Leaves narrowly elliptic, lanceolate, obovate-oblong, broadly filiform, margin gross-dentate, incised-lobed or parted. Capitula numerous, arranged in dense corymb; involucres campanulate, 5-6 mm diam. Fl and fr. Aug-Oct. Forests, forest edges, thickets at 3200-3900 m. Distributed in NW Yunnan and SW Sichuan.

疏齿亚菊

Ajania remotipinna (Hand.-Mazz.) Y. Ling et C. Shih

多年生草本，被短柔毛。中部茎叶全形椭圆形、卵形或倒卵形，二回羽状分裂，末回裂片长椭圆形或镰刀形，上部和下部茎叶渐小。头状花序小，多数在茎顶排成直径3-11厘米的复伞房花序；总苞钟状；花冠黄色。花果期8-10月。生海拔200-3800米的山坡。产四川、西藏、陕西和甘肃。

Perennial herbs, puberulent. Middle cauline leaves elliptic, ovate or obovate in outline, bipinnatifid, final lobes oblong or falcate, upper and lower cauline leaves become smaller gradually. Capitula small, arranged in a compound corymb with 3-11 cm diam on top of stem; involucres campanulate; corolla yellow. Fl. and fr. Aug-Oct. Montane slopes at 200-3800 m. Distributed in Sichuan, Xizang, Shaanxi and Gansu.

栎叶亚菊 *Ajania quercifolia*

柳叶亚菊 *Ajania salicifolia*

柳叶亚菊

Ajania salicifolia (Mattf.) Poljakov

小半灌木，高30-60厘米。有长20-30厘米的不育短枝。叶线形、狭线形或披针形，全缘。头状花序多数在枝端排成密集的伞房花序；总苞钟状，直径4-6毫米。花果期6-9月。生海拔2600-4600米的山坡。产四川西北部、甘肃东部、陕西西南部和青海东部。

Subshrubs, 30-60 cm tall. Sterile short branch, 20-30 cm long. Leaves filiform, narrowly filiform or lanceolate, margin entire. Capitula numerous, arranged in terminal clustered corymb; phyllaries campanulate, 4-6 mm diam. Fl. and fr. Jun-Sep. Mountain slopes at 2600-4600 m. Distributed in NW Sichuan, E Gansu, SW Shaanxi and E Qinghai.

单头亚菊

Ajania scharnhorstii (Regel et Schmalh.) Tzvelev

小半灌木，高4-10厘米。根木质。叶两面同色，灰白色，被等量稠密的短柔毛；全形近圆形或扇形。头状花序单生枝端；总苞宽钟状；总苞片边缘黄褐色干膜质；花黄色，边缘雌花5。花果期7-8月。生海拔3900-5100米的山坡石缝或石灰岩碎石山坡或山坡灌丛中。产新疆、甘肃、青海和西藏。

Subshrubs, 4-10 cm tall, with thick woody rootstock. Leaves suborbicular or flabelliform, gray-white, densely pubescens. Synflorescence a solitary terminal capitula; involucres campanulate; phyllaries scarious margin brown; florets yellow; marginal female florets 5. Fl. and fr. Jul-Aug. Fissures of rocks on mountain slopes, calcareous talus slopes or thickets at 3900-5100 m. Distributed in Xinjiang, Gansu, Qinghai and Xizang.

细叶亚菊

Ajania tenuifolia Tzvelev

多年生草本。茎匍匐或直立，匍茎上生稀疏的宽卵形浅褐色的苞鳞。中部茎叶二回羽状分裂，近圆形或三角状卵形或扇形。头状花序少数排成伞房花序；苞片边缘宽膜质；小花黑色或橘黄色，边缘雌花7-13朵。花果期8-10月。生海拔2200-4600米的山坡草地。产甘肃、青海、西藏、云南、四川和江苏。

Perennial herbs. Stems procumbent and erect; procumbent stems with many brown ovate cataphylls. Middle stem leaves suborbicular, triangular-ovate, or flabelliform, 2-pinnatisect. Synflorescence a terminal flat-topped cyme; capitula few; phyllaries scarious margin; florets dark or orange yellow; marginal female florets 7-13. Fl. and fr. Aug-Oct. Grasslands on mountain slopes at 2200-4600 m. Distributed in Gansu, Qinghai, Xizang, Yunnan, Sichuan and Jiangsu.

单头亚菊 *Ajania scharnhorstii*

细叶亚菊 *Ajania tenuifolia*

西藏亚菊 *Ajania tibetica*

异叶亚菊 *Ajania variifolia*

西藏亚菊

Ajania tibetica (Hook. f. et Thoms.) Tzvelev

亚灌木，高4-20厘米。老枝黑褐色。叶有柄，椭圆形或倒披针形，二回羽状分裂。头状花序排成伞房花序；总苞钟状；小花黄色。花果期8-9月。生海拔3900-4700米的山坡。产西藏和四川。印度、巴基斯坦和哈萨克斯坦亦有。

Subshrubs, 4-20 cm tall. Old branches dark brown. Leaves petiolate, elliptic or oblanceolate, 2-pinnatisect. Synflorescence a terminal flat-topped panicle; involucres campanulate; florets yellow. Fl. and fr. Aug-Sep. Mountain slopes at 3900-4700 m. Distributed in Xizang and Sichuan. Also in India, Pakistan and Kazakhstan.

异叶亚菊

Ajania variifolia (C. C. Chang) Tzvelev

亚灌木，高约30厘米。中部叶有柄全形卵形。头状花序多数，排成复伞房花序；总苞钟状；总苞片边缘黄褐色膜质；小花黄色，边缘雌花约6朵。花果期8-9月。生海拔1200-3500米的岩坡或高山草甸。产湖北、陕西和黑龙江。朝鲜半岛和俄罗斯亦有。

Subshrubs, ca. 30 cm tall. Middle leaves of flowering branches with petioles; leaves ovate. Synflorescence a terminal compound flat-topped panicle; capitula many; involucres campanulate; phyllaries scarious margin yellow-brown; florets yellow; marginal female florets ca. 6. Fl. and fr. Aug-Sep. Rocky slopes or alpine meadows at 1200-3500 m. Distributed in Hubei, Shaanxi and Heilongjiang. Also in Korean Peninsula and Russia.

野菊

Chrysanthemum indicum L.

多年生草本，被疏毛。根状茎横走。叶卵形或椭圆状卵形，羽状分裂，基部截形至楔形。头状花序多数，直径1.5-2.5厘米，排成伞房圆锥花序或伞房花序；总苞片卵形至长圆形；舌状花黄色。瘦果小。花果期6-11月。生海拔100-2900米的草地、山坡、灌丛、湿地、海滨、田边或路旁。产中国西南、华南、华中、华东、华北和东北。南亚、中亚和东北亚亦有。

Perennial herbs, sparsely pubescent. Rhizomes horizontal. Leaves ovate or elliptic-ovate, pinnately lobed, base truncate to cuneate. Capitula numerous, 1.5-2.5 cm diam, arranged in corymbose-panicule or corymb; phyllaries ovate to oblong; ray florets yellow. Achenes small. Fl. and fr. Jun-Nov. Grasslands, slopes, thickets, wetlands, seashores, fieldsides or roadsides at 100-2900 m. Distributed in SW, S, C, E, N and NE China. Also in S, C and NE Asia.

野菊 *Chrysanthemum indicum*

小红菊 *Chrysanthemum chanetii*

小红菊

Chrysanthemum chanetii Lévl.

多年生草本，被疏毛。根状茎横走。叶具柄，肾形、半圆形、近圆形或宽卵形，掌状分裂或羽状分裂，基部稍心形或截形。头状花序顶生，直径2.5-5厘米；总苞片条形至长圆形；舌状花白色、粉红色或紫色。瘦果小。花果期7-10月。生海拔300-2700米的草原、山坡、林缘、灌丛、岩缝或沟边。产华北、东北和西北。俄罗斯、蒙古和朝鲜半岛亦有。

Perennial herbs, sparsely pubescent. Rhizomes horizontal. Leaves petiolate, reniform, semi-orbicular, suborbicular or broadly ovate, palmately or pinnately lobed, base subcordate or cuneate. Capitula terminal, 2.5-5 cm diam; phyllaries linear to oblong; ray florets white, pink or purple. Achenes small. Fl. and fr. Jul-Oct. Grasslands, slopes, forest edges, thickets, rocky cracks or ditchsides at 300-2700 m. Distributed in N, NE and NW China. Also in Russia, Mongolia and Korean Peninsula.

菊花

Chrysanthemum morifolium Ramat.

多年生草本，被柔毛。叶具短柄，卵形至披针形，羽状浅裂或半裂。头状花序径2.5-20厘米；总苞被柔毛；舌状花颜色形状多变；管状花黄色。花期9-11月。著名观赏植物。

Perennial herbs, pubescent. Leaves shortly

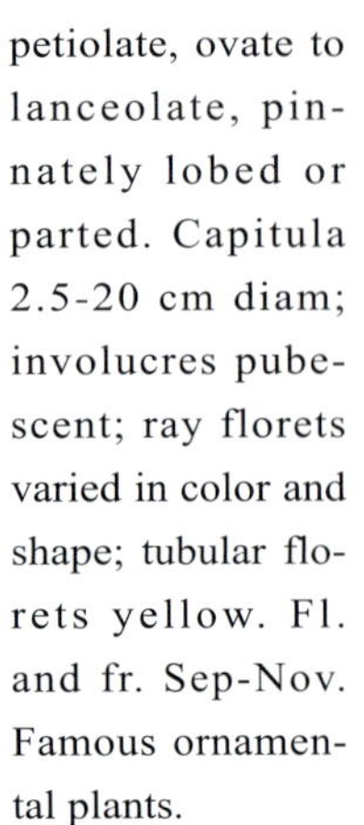

petiolate, ovate to lanceolate, pinnately lobed or parted. Capitula 2.5-20 cm diam; involucres pubescent; ray florets varied in color and shape; tubular florets yellow. Fl. and fr. Sep-Nov. Famous ornamental plants.

小山菊

Chrysanthemum oreastrum Hance

多年生草本，具匍匐根茎。叶二回三出羽状深裂；末端裂片条形或宽条形，下面密被柔毛。头状花序单个顶生；总苞浅碟状；苞片4层，边缘深棕色，膜质；辐射花白色或粉色；顶端3齿裂。花果期6-8

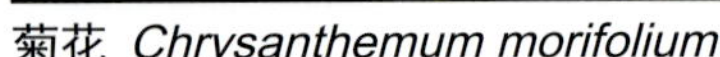

菊花 *Chrysanthemum morifolium*

小山菊
Chrysanthemum oreastrum

月。生海拔1800-3000米的草甸。产山西、河北和吉林。

Perennial herbs, with procumbent rhizomes. Leaves biternate-pinnatisect; ultimate segments linear or broadly linear, abaxially densely villous. Capitula apically solitary; involucres coryliform; phyllaries in 4 rows, margin dark brown, scarious; ray florets white or pink; rays 3-denticulate at apex. Fl. and fr. Jun-Aug. Meadows at 1800-3000 m. Distributed in Shanxi, Hebei and Jilin.

甘菊

Chrysanthemum lavandulifolium (Fisch. ex Trautv.) Makino

多年生草本，被疏柔毛。根状茎横走。叶卵形，二回羽状分裂，两面同色。头状花序直径1-2厘米，多数，排成顶生复伞房花序；总苞片条形至披针形，先端圆形，边缘膜质；舌状花黄色，椭圆形。花果期5-11月。生海拔600-2800米的山坡、岩石缝、河谷、河岸、荒地或黄土丘陵地。产华南以外的大部分地区。日本亦有。

Perennial herbs, sparsely pubescent. Rhizomes horizontal. Leaves ovate, bipinnatiparted, concolor. Capitula 1-2 cm diam, numerous, arranged in terminal compound corymb; phyllaries linear to lanceolate, apex rotundate, margin membranous; ray florets yellow, oblong. Fl. and fr. May-Nov. Slopes, rocky crevices, valleys, river banks, wastelands or loess hills at 600-2800 m. Distributed in most parts of China except S China. Also in Japan.

紫花野菊 *Chrysanthemum zawadskii*

甘菊 *Chrysanthemum lavandulifolium*

紫花野菊

Chrysanthemum zawadskii Herbich

多年生草本，疏被短柔毛。根状茎横走；茎中下部紫红色。叶具柄，卵形、菱形或长椭圆形，羽状分裂。头状花序直径1.5-4.5厘米，2-5个排成顶生疏松伞房花序；总苞片条形或椭圆形，边缘膜质；舌状花白色或紫红色。花果期7-9月。生海拔800-1800米的山坡、草原、草甸、林下或溪边。产华北、东北和西北。俄罗斯、蒙古和欧洲亦有。

Perennial herbs, sparsely and shortly pubescent. Rhizomes horizontal; stems purplish-red at medium and lower part. Leaves petiolate, ovate, rhombic or narrowly elliptic, pinnatiparted. Capitula 1.5-4.5 cm diam, 2-5 arranged in terminal lax corymb; phyllaries linear or elliptic, margin membranous; ray florets white or purple. Fl. and fr. Jul-Sep. Mountain slopes, grasslands, meadows, forests or streamsides at 800-1800 m. Distributed in N, NE and NW China. Also in Russia, Mongolia and Europe.

纤秆蒿 *Artemisia demissa*

叉枝蒿 *Artemisia divaricata*

纤秆蒿
Artemisia demissa Krasch.

一年生或两年生草本。底部茎叶长圆形或卵形，二回羽状分裂；中上部茎叶羽状分裂；苞叶线形，全缘。伞房花序排成圆锥花序；总苞卵形。花果期7-9月。生海拔2600-4800米的山谷、山坡、路边、草地和石丘。产华西。印度、阿富汗和塔吉克斯坦亦有。

Annual or biennial herbs. Lower stem oblong or ovate, 2-pinnatisect; middle and upper stem leaves pinnatisect; leaflike bracts linear, entire. Synflorescence a narrow, spikelike panicle; involucre ovoid. Fl. and fr. Jul-Sep. Valleys, slopes, roadsides, grasslands and rocky hills at 2600-4800 m. Distributed in W China. Also in India, Afghanistan and Tajikistan.

叉枝蒿
Artemisia divaricata (Pamp.) Pamp.

多年生草本。下部叶有柄，二回羽状全裂；中部叶卵形或长卵形，二回羽状分裂；上部叶一至二回羽状全裂。头状花序长圆形或长卵形，成圆锥花序。瘦果长圆形或长椭圆形。花果期8-10月。生海拔2000-3400米的荒山、草坡、路旁等地。产四川、云南西部和湖北西部。

Perennial herbs. Lowermost leaves long petiolate, 2-pinnatisect; middle stem leaves ovate or elliptic-ovate, 2-pinnatisect; uppermost leaves 1- or 2-pinnatisect. Capitula in broad, much-branched panicles, oblong or oblong-ovoid. Achenes oblong or ellipsoid. Fl. and fr. Aug-Oct. Hills, slopes, roadsides at 2000-3400 m. Distributed in Sichuan, W Yunnan and W Hubei.

牛尾蒿
Artemisia dubia L. ex B. D. Jacks.

半灌木。下部叶卵形或长圆形，5深裂；中部叶卵形；上部叶掌状3深裂；苞片全缘，叶椭圆状披针形或披针形。头状花序多数；小花8-20，略带紫色；边缘花6-8；盘花2-12，两性花。花果期8-10月。生低海拔至3500米的干山坡、草原、河岸、路边、山谷、峡谷及林缘。产中国大部分地区。不丹、印度、尼泊尔、泰国和日本亦有。

Subshrubs. Lowermost leaves ovate or oblong, 5-partite; middle stem leaves ovate; uppermost leaves ternate, 3-partite; leaflike bracts entire, elliptic-lanceolate or lanceolate. Capitula many; florets 8-20, purplish; marginal florets 6-8; disk florets 2-12, bisexual florets. Fl. and fr. Aug-Oct. Slopes, steppes, riverbanks, roadsides, valleys, canyons, forest margins at low elevations to 3500 m. Distributed in most parts of China. Also in Bhutan, India, Nepal, Thailand and Japan.

牛尾蒿 *Artemisia dubia*

盐蒿
Artemisia halodendron Turcz. ex Besser

小灌木。中下部叶宽卵形或近圆形，一至二回羽状全裂；上部叶与苞片叶3-5全裂或不分裂。聚伞花序排成复总状花序，并在茎上组成大型、开展的圆锥花序；雌花4-8朵；盘花8-15朵，雄性。花果期7-10月。生荒漠、沙丘、草原、森

盐蒿 *Artemisia halodendron*

林和砾石坡。产华北、东北和西北。蒙古和俄罗斯亦有。

Shrubs. Lower and middle stem leave broadly ovate or suborbicular, 1- or 2-pinnatisect; uppermost leaves and leaflike bracts 3-5-sect or entire. Synflorescence a broad, much-branched panicle; involucre ovoid; marginal female florets 4-8; disk florets 8-15, male. Fl. and fr. Jul-Oct. Dunes of desert areas, desert steppes, steppes, forest steppes and rocky slopes. Distributed in N, NE and NW China. Also in Mongolia and Russia.

臭蒿

Artemisia hedinii Ostenf.

一年生草本，紫色，有臭味。中下部叶莲座状或椭圆形，二回羽状分裂；上部叶与苞片叶羽状分裂。头状花序排成狭窄的圆锥花序；边缘雌花3-8；两性花15-30。花果期7-10月。生海拔1000-4800(-5000)米的沼泽地、河滩、砾质坡地、路边、林缘。产内蒙古、西南和西北。印度、尼泊尔、巴基斯坦和塔吉克斯坦亦有。

Annual herbs, purple, fetid. Basal and middle leaves many rosulate or elliptic, 2-pinnatisect; uppermost leaves and leaflike bracts pinnately divided. Capitula in dense narrow panicles; marginal female florets 3-8; disk florets 15-30, bisexual. Fl. and fr. Jul-Oct. Grassy marshlands, floodlands, rocky slopes, waysides, outer forest margins at 1000-4800(-5000) m. Distributed in Neimenggu, SW and NW China. Also in India, Nepal, Pakistan and Tajikistan.

臭蒿 *Artemisia hedinii*

野艾蒿

Artemisia lavandulifolia DC.

多年生草本或灌木。基生叶与下部叶卵形或近圆形，二回羽状全裂；中部叶卵形、长圆形或近圆形；上部叶和苞叶羽状全裂或3全裂或不分裂。雌花4-9；两性花10-20。花果期7-10月。生海拔400-3000米的路旁、林缘、山坡、草地、山谷、灌丛及河湖滨草地等。产中国大部分地区。日本、朝鲜半岛、蒙古和俄罗斯亦有。

Perennial herbs or shrubs. Basal and lowermost leaves ovate or suborbicular, 2-pinnatisect; middle stem leaves ovate, ovate-elliptic or suborbicula. uppermost leaves and leaflike bracts pinnatisect or 3-lobed or entire. Marginal female florets 4-9; disk florets 10-20, bisexual. Fl. and fr. Jul-Oct. Roadsides, forest margins, slopes, steppes, canyons, riverbanks or lakesides, brushlands at 400-3000 m. Distributed in most parts of China. Also in Japan, Korean Peninsula, Mongolia and Russia.

野艾蒿 *Artemisia lavandulifolia*

中亚旱蒿 *Artemisia marschalliana*

灰苞蒿 *Artemisia roxburghiana*

中亚旱蒿
Artemisia marschalliana Spreng.

半灌木或小灌木。下部叶长圆状卵形或卵形，二(或三)回羽状全裂；中上部叶一或二回羽状全裂；苞片叶3-5全裂或不分裂。聚伞花序组成圆锥花序；边缘雌花3-8；盘花5-15。花果期7-10月。生海拔500-2200米的干草原、森林草原、荒坡、山丘、砾质坡地。产新疆北部。哈萨克斯坦、俄罗斯和欧洲亦有。

Subshrubs or small shrubs. Lower stem leaves oblong-ovate or ovate, 2(or 3)-pinnatisect; middle and upper stem leaves 1- or 2-pinnatisect; leaflike bracts 3-5-sect or entire. Synflorescence a conical panicle; marginal female florets 3-8; disk florets 5-15, male. Fl. and fr. Jul-Oct. Steppes, forest steppes, wastelands, rocky slopes, hills at 500-2200 m. Distributed in N Xinjiang. Also in Kazakhstan, Russia and Europe.

黏毛蒿
Artemisia mattfeldii Pamp.

多年生草本，具浓香。中下部叶长圆形或长圆状卵形，二或三回羽状深裂；上部叶和苞叶一或二回羽状深裂。聚伞花序排成狭圆锥花序；总苞长圆形或宽卵形。花果期7-10月。生海拔2600-4800米的林缘、草地、山坡和路边。产中国西南和华中。

Perennial herbs, strongly aromatic. Lowermost and middle stem leaves oblong or oblong-ovate, (2 or)3-pinnatisect; uppermost leaves and leaflike bracts 1- or 2-pinnatisect. Synflorescence a narrow panicle; involucre oblong or broadly ovoid. Fl. and fr. Jul-Oct. Forest margins, grasslands, slopes and roadsides at 2600-4800 m. Distributed in SW and C China.

灰苞蒿
Artemisia roxburghiana Beeser

半灌木。中下部叶卵形或长圆状椭圆形，二回羽状深裂；上部叶和苞叶羽状全裂或3-5全裂或不分裂。小花黄色或紫褐色；雌花5-7；两性花10-20(-30)。花果期8-10月。生海拔700-3900米的荒地、干河谷、阶地、路旁、草地。产中国西南和华中。阿富汗、印度、尼泊尔、巴基斯坦和泰国亦有。

Subshrubs. Lowermost and middle stem leaves ovate or oblong-elliptic, 2-pinnatisect; uppermost leaves and leaflike bracts pinnatisect or 3-5-lobed or entire. Florets yellow or purple-brown tinged. female florets 5-7; disk-florets 10-20(-30). Fl. and fr. Aug-Oct. Roadsides, slopes, dry canyons, grasslands, waste areas, terraces at 700-3900 m. Distributed in SW and C China. Also in Afghanistan, India, Nepal, Pakistan and Thailand.

蒙古蒿
Artemisia mongolica (Fischer ex Besser) Nakai

多年生草本。叶片卵形、近圆形或椭圆状卵形，背面密被灰白色蛛丝状绒毛，二回羽状全裂或深裂，第一回全裂，每侧有裂片2-3枚，裂片椭圆形或长圆形。上部叶与苞片叶卵形或

黏毛蒿 *Artemisia mattfeldii*

蒙古蒿 *Artemisia mongolica*

冻原白蒿 *Artemisia stracheyi*

线叶蒿 *Artemisia subulata*

长卵形，羽状全裂或3或5全裂，裂片披针形或线形。头状花序无梗，在分枝上排成密集的圆锥花序；总苞片背面密被灰白色蛛丝状毛。花果期8-10月。生海拔2000米以下的山坡、灌丛、河边、湖边、路边、草原、林缘、草地或干山谷。产中国西南、东南、华中、华北、东北和西北。朝鲜半岛、俄罗斯和中亚亦有。

Perennial herbs. Leaves ovate, suborbicular or elliptic-ovate, densely arachnoid tomentose abaxially, 2-pinnatisect; segments 2 or 3 pairs. Lobules lanceolate linear. Uppermost leaves and leaflike bracts ovate or elliptic-ovate, pinnatisect or 3- or 5-sect. Capitula in narrow panicles, ellipsoid; phyllaries gray arachnoid pubescent. Fl. and fr. Aug-Oct. Slopes, shrublands, riverbanks, lakeshores, roadsides, steppes, forest steppes, dry valleys below 2000 m. Distributed in SW, SE, C, N, NE and NW China. Also in Korean Peninsula, Russia and C Asia.

冻原白蒿

Artemisia stracheyi Hook. f. et Thoms. ex C. B. Clarke

多年生草本，植株有臭味。基生叶二至三回羽状全裂；中上部叶一至二回羽状全裂；苞片叶羽状全裂或不分裂。头状花序下垂，排成总状或穗状花序；总苞和花序托半球形。花果期7-11月。生海拔4300-5200米的山坡、河滩、湖边、砾石地、草甸和灌丛。产西藏。喜马拉雅西北部亦有。

Perennial herbs, caespitose, fetid. Basal leaves 2-3-pinnatisect; middle and uppermost leaves 1-2-pinnatisect; leaflike bracts pinnatisect or entire. Synflorescence racemelike, sometimes spikelike; capitula nodding; involucre and receptacle hemispheric. Fl. and fr. Jul-Nov. Hills, floodlands, lakesides, rocky slopes, meadows, shrublands at 4300-5200 m. Distributed in Xizang. Also in NW Himalaya.

线叶蒿

Artemisia subulata Nakai

多年生草本。叶无柄；基生叶与茎下部叶倒披针形或倒披针状线形；中部叶线形或线状披针形或钩状；上部叶与苞片叶线形。头状花序排成狭穗状花序式的总状花序边缘雌花10或11，盘花10-15，两性。花果期8-10月。生低海拔的湿润、半湿润山坡、林缘、河岸、草甸。产华北和东北。东北亚亦有。

Perennial herbs. Leaves sessile; lowermost leaves oblanceolate or oblanceolate-linear; middle stem leaves linear, linear-lanceolate, or falcate; uppermost leaves and leaflike bracts linear. Synflorescence a narrow, racemelike panicle. Marginal female florets 10 or 11. Disk florets 10-15, bisexual. Fl. and fr. Aug-Oct. Humid and semihumid slopes, forest margins, riverbanks, meadows at low elevations. Distributed in N and NE China. Also in NE Asia.

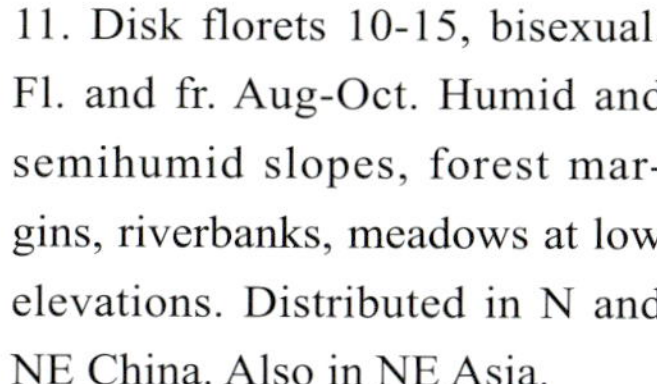

阴地蒿

Artemisia sylvatica Maxim.

多年生草本。中下部叶卵形或宽卵形；上部叶羽状深裂。总苞半球形、卵形或椭圆形；雌花4-7朵；两性花8-14朵。花果期8-10月。生海拔100-1300米的林缘、灌丛、潮湿地、斜坡和路边。产中国大部分地区。朝鲜半岛、蒙古和俄罗斯东部亦有。

Perennial herbs. Lower and middle stem leaves ovate, broadly ovate, or oblong; uppermost leaves pinnatipartite. Involucre subglobose, ovoid or elliptic; marginal female florets 4-7; disk florets 8-14, bisexual. Fl. and fr. Aug-Oct. Forest margins, shrublands, humid areas, slopes and roadsides at 100-1300 m. Distributed in most parts of China. Also in Korean Peninsula, Mongolia and E Russia.

阴地蒿 *Artemisia sylvatica*

甘青蒿 *Artemisia tangutica*

亚东蒿 *Artemisia yadongensis*

甘青蒿

Artemisia tangutica Pamp.

多年生草本。中下部叶椭圆状卵形或椭圆形，二回羽状全裂；上部叶羽状全裂；苞叶3裂或5裂或全缘。聚散花序排成圆锥花序；头状花序下垂；总苞长圆形或宽卵形；边缘雌花3-8朵，两性花5-15朵。花果期7-10月。生海拔2700-3800米的斜坡、河堤沙丘、草地和路边。产青藏高原和湖北西部。

Perennial herbs. Lower and middle stem leaves elliptic-ovate or elliptic, 2-pinnatisect; uppermost leaves pinnatipartite; leaflike bracts 3- or 5-partite or entire. Synflorescence in a panicle; capitula nodding; involucre oblong or broadly ovoid. Fl. and fr. Jul-Oct. Slopes, dunes along riverbanks, grasslands and roadsides at 2700-3800 m. Distributed in Tibetan Plateau and W Hubei.

藏沙蒿 *Artemisia wellbyi*

藏沙蒿

Artemisia wellbyi Hemsl. et H. Pearson

半灌木状草本。下部叶卵形或长卵形，二回羽状全裂；中部叶长卵形，一(或二)回羽状全裂；上部叶3或5全裂或不分裂，线形。头状花序组成狭窄、疏松的圆锥花序；总苞卵形或近球形。花果期7-11月。生海拔3600-5300米的河堤、湖岸、丘陵、砾石地、高山草原和草甸。产西藏。印度北部亦有。

Subshrubs. Lower leaves ovate or elliptic-ovate, 2-pinnatisect; middle stem leaves elliptic-ovate, 1(or 2)-pinnatisect; uppermost leaves and leaflike bracts 3- or 5-sect or entire, linear. Synflorescence a lax, narrow panicle; involucre ovoid or subglobose. Fl. and fr. Jul-Nov. Riverbanks, lakeshores, hills, rocky slopes, alpine steppes and alpine meadows at 3600-5300 m. Distributed in Xizang. Also in N India.

亚东蒿

Artemisia yadongensis Y. Ling et Y. R. Ling

多年生草本或半灌木状。中部叶长圆形或长卵形，二回羽状分裂；上部叶椭圆形或卵形，一(或二)回羽状深裂。头状花序多数在分枝上排成复穗状花序，并在茎上组成中圆锥花序；总苞近球形或宽卵形。花果期8-10月。生海拔约2900米的草坡。产西藏南部。

Perennial herbs or nearly subshrubs. Middle stem leaves oblong or oblong-ovate, 2-pinnatisect; uppermost leaves elliptic or ovate, 1 (or 2)-pinnatipartite. Synflorescence a moderately broad panicle; involucre subglobose or broadly ovoid. Fl. and fr. Aug-Oct. Grassy slopes at ca. 2900 m. Distributed in S Xizang.

藏白蒿

Artemisia younghusbandii J. R. Drumm. ex Pamp.

半灌木或草本，簇生。中下部叶宽卵形、卵形或近肾形，一(或二)回羽状全裂；上部叶与苞片叶羽状全裂、3裂或不分裂。头状花序下垂，排成开展的圆锥花序。花果期7-10月。生海拔4000-4700米的峡谷、丘陵、砾石坡和路边。产西藏。

Subshrubs or herbs, caespitose. Basal and middle stem leaves ovate, broadly ovate or subreniform, 1(or 2)-pinnatisect; uppermost leaves and leaflike bracts pinnatisect, 3-lobed or entire. Synflorescence a broad panicle; capitula nodding. Fl. and fr. Jul-Oct. Canyons, hills, rocky slopes and waysides at 4000-4700 m. Distributed in Xizang.

大花蒿

Artemisia macrocephala Jacq. ex Bess.

一年生草本，被灰白色微柔

藏白蒿 *Artemisia younghusbandii*

毛。叶草质，卵形，二回羽状全裂，小裂片狭条形，基部具假托叶。头状花序近球形，直径5-15毫米，下垂；花冠外面紫色，里面黄色。瘦果卵球形或椭圆体形。花果期8-10月。生海拔1500-5500米的草原、荒漠、山谷、洪积扇、湖边、沙砾地、盐碱地、草坡或路边。产西藏、甘肃、宁夏、青海和新疆。印度、巴基斯坦、阿富汗、伊朗、俄罗斯和蒙古亦有。

Annual herbs, offwhite puberulent. Leaves herbaceous, ovate, bipinnatisected, lobelets narrowly linear, base pseudo-stipulate. Capitula subglobose, 5-15 mm diam, pendulous; corolla purple outside, yellow inside. Achenes ovoid or ellipsoid. Fl. and fr. Aug-Oct. Grasslands, deserts, valleys, proluvial fans, lakesides, sandy places, saline and kaline places, grassy slopes or roadsides at 1500-5500 m. Distributed in Xizang, Gansu, Ningxia, Qinghai and Xinjiang. Also in India, Pakistan, Afghanistan, Iran, Russia and Mongolia.

大籽蒿

Artemisia sieversiana Ehrh. ex Willd.

一年生或二年生草本。叶草质，宽卵形，羽状全裂，小裂片条形。复合花序长狭圆锥状；头状花序半球形或近球形；花冠黄色；边缘雌花20-30；盘心两性花80-120。瘦果长圆形。花果期6-10月。生海拔4200米以下的路旁、荒地、河漫滩、草原、山坡或林缘。产中国西南、华北、东北和西北。印度、巴基斯坦、阿富汗、俄罗斯、蒙古、朝鲜半岛和日本亦有。

Annual or biennial herbs. Leaves herbaceous, broadly ovate, pinnatisected, lobelets linear. Synflorescences a long, narrow panicle; capitula semiglobose or subglobose; corolla yellow; marginal female florets 20-30; disk florets 80-120, bisexual. Achenes oblong. Fl. and fr. Jun-Oct. Roadsides, wastelands, flood plains, grasslands, slopes or forest edges below 4200 m. Distributed in SW, N, NE and NW China. Also in India, Pakistan, Afghanistan, Russia, Mongolia, Korean Peninsula and Japan.

大花蒿
Artemisia macrocephala

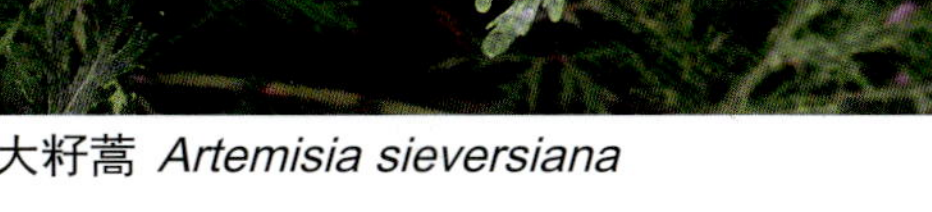
大籽蒿 *Artemisia sieversiana*

香叶蒿 *Artemisia rutifolia*

香叶蒿

Artemisia rutifolia Steph. ex Spreng.

半灌木状草本，具香气，被灰白色柔毛。茎多数，成丛。叶近半圆形或肾形，二回三出全裂，小裂片披针形。头状花序半球形或近球形，排成总状花序；花冠紫色；盘心两性花(10-)12-15(-28)。瘦果椭圆体状倒卵球形。花果期7-10月。生海拔1300-5000米干燥地区的山坡、河谷、盆地、草原或半荒漠。产西藏、青海和新疆。巴基斯坦、阿富汗、伊朗、俄罗斯和蒙古亦有。

Suffruticose herbs, fragrant, offwhite pubescent. Stems numerous, clustered. Leaves subsemiorbicular or reniform, doubly trisected, lobelets lanceolate. Capitula semiglobose or subglobose, arranged in raceme; corolla purple; disk florets (10-)12-15(-28), bisexual. Achenes ellipsoid-obovoid. Fl. and fr. Jul-Oct. Slopes, valleys, basins, grasslands or semideserts of dry regions at 1300-5000 m. Distributed in Xizang, Qinghai and Xinjiang. Also in Pakistan, Afghanistan, Iran, Russia and Mongolia.

碱蒿

Artemisia anethifolia Weber ex Stechm.

一年生或两年生草本。基生叶椭圆形或卵状椭圆形，二或三回羽状全裂；上部叶与苞片叶3-5裂或全缘。头状花序作宽圆锥状花序式排列，半球形或宽卵球形；雌花3-6朵；两性花18-28朵；冠檐红色或黄色。瘦果椭圆体形或倒卵球形。花果期8-10月。生海拔800-2500米的山坡、盐生草地、弃荒地或固定沙丘。产华北、华西、西北和东北。俄罗斯和蒙古亦有。

Annual or biennial herbs. Basal leaves elliptic or ovate-elliptic, 2- or 3-pinnatisect; uppermost leaves and leafy bracts 3-5-lobed or entire. Capitula in broad paniculate inflorescences, semiglobose or broadly ovoid; pistillate florets 3-6; bisexual florets 18-28; corolla limbs yellow or red. Achenes ellipsoid or obovoid. Fl. and fr. Aug-Oct. Slopes, saline steppe, wastelands or stable dunes at 800-2500 m. Distributed in N, W, NW and NE China. Also in Russia and Mongolia.

细裂叶莲蒿

Artemisia gmelinii Weber ex Stechm.

半灌木状草本。茎丛生，褐色。叶纸质，上面绿色，背面被灰白色平贴短柔毛，2-3回栉齿状羽状分裂，小裂片披针形。头状花序多数，近球形；总苞片密被灰白色短柔毛；花冠黄色。瘦果狭椭圆状卵球形或狭圆锥形。花果期8-10月。生海拔1000-4900米的山坡、路旁、灌丛或草原。广泛分布除高寒地区外的中国各地。印度、尼泊尔、巴基斯坦、阿富汗、俄罗斯、蒙古、朝鲜半岛和日本亦有。

Suffruticose herbs. Stems clustered, brown. Leaves chartaceous, green adaxially, shortly appressed offwhite-pubescent abaxially, bi- to tri-pectinate-pinnatifid, lobelets lanceolate. Capitula numerous, subglobose;

碱蒿 *Artemisia anethifolia*

细裂叶莲蒿 *Artemisia gmelinii*

phyllaries densely short offwhite-pubescent outside; corolla yellow. Achenes narrowly elliptic-ovoid or conical. Fl. and fr. Aug-Oct. Slopes, roadsides, thickets or grasslands at 1000-4900 m. Widely distributed throughout most parts of China, except extremely frigid zones. Also in India, Nepal, Pakistan, Afghanistan, Russia, Mongolia, Korean Peninsula and Japan.

宽叶蒿

Artemisia latifolia Ledeb.

多年生草本。茎单生，被柔毛。叶无毛或渐无毛，具密小凹点，长圆形，一至二回羽状深裂。头状花序近球形或半球形，直径3-4毫米；总苞片无毛，黄褐色；花冠黄色。瘦果倒卵球形。花果期7-10月。生草原、林缘、灌丛、草甸或盐碱地。产内蒙古、甘肃、黑龙江、吉林和辽宁。俄罗斯、蒙古和朝鲜半岛亦有。

Perennial herbs. Stems solitary, pubescent. Leaves glabrous or glabrescent, densely with minute pits, oblong, pinnatifid or bipinnatifid. Capitula subglobose or semiglobose, 3-4 mm diam; phyllaries glabrous, yellowish-brown; corolla yellow. Achenes obovoid. Fl. and fr. Jul-Oct. Grasslands, forest edges, thickets, meadows or saline and kaline places. Distributed in Neimenggu, Gansu, Heilongjiang, Jilin and Liaoning. Also in Russia, Mongolia and Korean Peninsula.

褐苞蒿

Artemisia phaeolepis Krasch.

多年生草本，具浓烈香气。叶具腺点，一至三回羽状全裂；叶苞全缘。少数头状花序作狭圆锥形总状花序式排列，半球形，稍下垂；总苞具褐色宽边缘；雌花12-18朵；两性花40-80朵。瘦果长圆形或长圆状倒卵球形。花果期7-10月。生海拔2500-3600米的山丘、路边、沼泽草地、干洪泛地、森林和灌丛外缘、多石山丘或半荒漠草原。产中国西南、华北、华西和西北。哈萨克斯坦、俄罗斯和蒙古亦有。

Perennial herbs, strongly aromatic. Leaves glandular punctuate, 1-3-pinnatisect; leafy bracts entire. Capitula in narrow racemose-paniculate inflorescences, few, semiglobose, nodding; phyllaries with broad brown margins; pistillate florets 12-18; bisexual florets 40-80. Achenes oblong or oblong-obovoid. Fl. and fr. Jul-Oct. Hills, waysides, grassy marshlands, dry floodlands, outer margins of forests and shrublands, rocky hills or semidesert steppe at 2500-3600 m. Distributed in SW, N, W and NW China. Also in Kazakstan, Russia and Mongolia.

宽叶蒿 *Artemisia latifolia*

褐苞蒿 *Artemisia phaeolepis*

裂叶蒿 *Artemisia tanacetifolia*

裂叶蒿
Artemisia tanacetifolia L.

多年生草本。叶椭圆状长圆形或椭圆状卵形，疏被腺毛，二或三回羽状全裂。头状花序作狭长的圆锥花序式排列，球形或半球形，稍下垂；雌花8-15朵；两性花30-40朵。瘦果椭圆体状倒卵球形。花果期7-10月。生低海拔至2400米的草原、林缘、盐土、山丘或灌丛。产华北、华西和东北。哈萨克斯坦、俄罗斯、蒙古、朝鲜半岛、欧洲中部与东部和北美洲亦有。

Perennial herbs. Leaves elliptic-oblong or elliptic-ovate, sparsely glandular punctate, 2- or 3-pinnatisect. Capitula in narrow elongated paniculate inflorescences, globose or semiglobose, nodding; pistillate florets 8-15; bisexual florets 30-40. Achenes ellipsoid-obovoid. Fl. and fr. Jul-Oct. Steppe, outer forest edges, saline soils, hills or shrublands at low elevations to 2400 m. Distributed in N, W and NE China. Also in Kazakstan, Russia, Mongolia and Korean Peninsula, C and E Europe, and North America.

青蒿
Artemisia caruifolia Buch.-Ham. ex Roxb.

一年生草本，具香气。茎单生，无毛。叶绿色，无毛，栉齿状羽状分裂，基部具假托叶。头状花序半球形，直径3.5-4.5毫米；总苞片无毛；花冠淡黄色。瘦果长圆形至椭圆体形。花果期6-9月。生海拔2500米以下的湿润山谷、河边、沙地、林缘、路旁或海滨。广布中国各地。印度、尼泊尔、缅甸、越南、朝鲜半岛和日本亦有。

Annual herbs, fragrant. Stems solitary, glabrous. Leaves green, glabrous, pectinately parted, base pseudo-stipulate. Capitula semiglobose, 3.5-4.5 mm diam; phyllaries glabrous; corolla yellowish. Achenes oblong or ellipsoid. Fl. and fr. Jun-Sep. Damp valleys, riversides, sandy grounds, forest edges, roadsides or coastal regions below 2500 m. Widely distributed throughout most parts of China. Also in India, Nepal, Myanmar, Vietnam, Korean Peninsula and Japan.

山蒿
Artemisia brachyloba Franch.

半灌木状草本。茎多数，基生，成丛。叶硬纸质，背面被白色绒毛，卵形或宽卵形，二回羽状全裂，小裂片狭条形。头状花序卵球形或卵状钟形；花冠黄色。瘦果卵球形。花果期7-10月。生低海拔至中海拔的阳坡草地、砾质坡地、半荒漠草原、戈壁或岩石缝中。产河北、山西、内蒙古、陕西、甘肃和宁夏。蒙古亦有。

Suffruticose herbs. Stems basal, many, clustered. Leaves firmly chartaceous, white tomentose abaxially, ovate or broadly ovate, bipinnatisected, lobelets narrowly linear. Capitula ovoid or ovoid-campanulate; corolla yellow. Achenes ovoid. Fl. and fr. Jul-Oct. Sunny grasslands, pebbly slopes, semi-desert grasslands, gobi or stone cracks at low to medium altitudes. Distributed in Hebei, Shanxi, Neimenggu, Shaanxi, Gansu and Ningxia. Also in Mongolia.

艾
Artemisia argyi Lévl. et Vaniot

多年生草本或亚灌木，具浓烈香气和灰色蛛丝状柔毛。叶卵形、三角状卵形或近菱形，一(至二)回羽状深裂或全裂；上部叶和叶苞羽状半裂。头状花

青蒿 *Artemisia caruifolia*

山蒿 *Artemisia brachyloba*

序作狭圆锥花序状排列，椭圆形；雌花6-10朵；两性花8-12朵。瘦果卵球状长圆形或长圆形。花果期7-10月。生低海拔至1500米的弃荒地、路边、山坡或森林草原。产中国西南、华南、东南、华中、华北、华西、华东和东北。俄罗斯、蒙古和朝鲜半岛亦有。

Perennial herbs or subshrubs, strongly aromatic, incanous arachnoid-pubescent. Leaves ovate, triangle-ovate or subrhombic, 1(or 2)-pinnatipartite or cleft; uppermost leaves and leafy bracts pinnatifid. Capitula in narrow paniculate inflorescences, ellipsoid; pistillate florets 6-10; bisexual florets 8-12. Achenes ovoid-oblong or oblong. Fl. and fr. Jul-Oct. Waste places, roadsides, slopes or forest steppe at low elevations to 1500 m. Distributed in SW, S, SE, C, N, W, E and NE China. Also in Russia, Mongolia and Korean Peninsula.

艾 *Artemisia argyi*

矮蒿

Artemisia lancea Vaniot

多年生草本。叶下面密被灰色或黄色蛛丝状柔毛，上面疏被蛛丝状柔毛及腺点或渐无毛；叶卵形，二回羽状全裂；上部叶及叶苞3-5裂或全缘。头状花序多数，密集作圆锥花序式排列，卵球形或椭圆体状卵形；雌花1-3朵；两性花2-5朵。瘦果长圆形。花果期8-10月。生海拔300-1700米的林缘、路边、山坡或弃荒地。产中国西南、华南、东南、华中、华北、华西、华东和东北。印度、俄罗斯东部、朝鲜半岛和日本亦有。

Perennial herbs. Leaves densely incanous or yellowish arachnoid-pubescent abaxially, glandular punctate and sparsely arachnoid-puberulous, or glabrescent adaxially; leaves ovate, 2-pinnatisect; uppermost and leafy bracts 3-5-lobed or entire. Capitula in dense paniculate inflorescences, numerous, ovoid or ellipsoid-ovoid; pistillate florets 1-3; bisexual florets 2-5. Achenes oblong. Fl. and fr. Aug-Oct. Forest edges, roadsides, slopes or waste areas at 300-1700 m. Distributed in SW, S, SE, C, N, W, E and NE China. Also in India, E Russia, Korean Peninsula and Japan.

矮蒿 *Artemisia lancea*

五月艾 *Artemisia indica*

秦岭蒿 *Artemisia qinlingensis*

秦岭蒿

Artemisia qinlingensis Ling ex Y. R. Ling

多年生草本。茎具纵棱。叶纸质，上面具腺点，下面密被灰白色蛛丝状绵毛，二回羽状分裂，基部具假托叶。头状花序长圆形或近卵圆形；总苞片绿色；花冠红色或紫色；雌花10-15朵，盘心两性花15-25朵。瘦果小，倒卵球形。花果期7-10月。生海拔1300-1500米的山坡、路旁或林缘。产河南、陕西和甘肃。

Perennial herbs. Stems ridged. Leaves chartaceous, glandular punctuate adaxially, densely off-white-arachnoid-tomentose abaxially, bipinnatifid, base pseudo-stipulate. Capitula oblong or subovoid; phyllaries green; corolla red or purple; female florets 10-15; disk florets 15-25, bisexual. Achenes small, obovoid. Fl. and fr. Jul-Oct. Slopes, roadsides or forest edges at 1300-1500 m. Distributed in Henan, Shaanxi and Gansu.

五月艾

Artemisia indica Willd.

半灌木状草本，具香气。茎褐色。叶上面渐无毛，背面密被灰白色蛛丝状绒毛，卵形至椭圆形，羽状分裂，具小型假托叶。头状花序卵形；总苞片渐无毛；花冠紫红色。瘦果长圆形或倒卵球形。花果期8-10月。生低海拔至2000米的湿润地区的路旁、林缘、坡地、灌丛或草地。广泛分布中国各地。南亚、东南亚、东北亚、北美洲和中美洲亦有。

Suffruticose herbs, fragrant. Stems brown. Leaves glabrescent adaxially, densely gray arachnoid-tomentose abaxially, ovate to elliptic, pinnatifid, base small pseudo-stipulate. Capitula ovoid; phyllaries glabrescent; corolla purple. Achenes oblong or obovoid. Fl. and fr. Aug-Oct. Roadsides, forest edges, slopes, thickets or grasslands of damp regions at low elevations to 2000 m. Widely distributed throughout most parts of China. Also in S, SE and NE Asia, N and S America.

魁蒿

Artemisia princeps Pamp.

多年生草本。叶纸质，上面无毛，下面密被灰白色蛛丝状绒毛，卵形或长卵形，羽状深裂，基部具假托叶。头状花序长圆形或长卵形，无梗，密集；总苞片绿色；花冠黄色，檐部紫红色；雌花5-7朵，盘心两性花4-9朵。瘦果椭圆体形。花果期7-11月。生海拔100-1400米的路旁、山坡、灌丛、林缘或沟边。广泛分布中国各地。朝鲜半岛和日本亦有。

Perennial herbs. Leaves chartaceous, glabrous adaxially, densely offwhite arachnoid-tomentose abaxially, ovate or narrowly ovate, pinnatifid, base pseudo-stipulate. Capitula oblong or narrowly ovoid, sessile, dense; phyllaries green; corolla yellow, limb purple; female florets 5-7; disk florets 4-9, bisexual. Achenes ellipsoid. Fl. and fr. Jul-Nov. Roadsides, slopes, thickets, forest edges or steamsides at 100-1400 m. Widely distributed throughout most parts of China. Also in Korean Peninsula and Japan.

宽叶山蒿

Artemisia stolonifera Maxim.

多年生草本。叶倒卵状椭圆形、卵状椭圆形或卵形。复合花序为狭圆锥状、长达40厘米；边缘雌花小花10-12朵；中央两性管状花12-15朵。瘦果长圆形或椭圆体形。花果期7-11月。生低海拔的林缘、路边、山坡或峡谷。产中国东南、华中、华北、华东、西北和东北。东北亚亦有。

魁蒿 *Artemisia princeps*

Perennial herbs. Leaves obovate-elliptic, ovate-elliptic or ovate. Synflorescences a narrow panicle, to 40 cm long; marginal female florets 10-12; disk florets 12-15, bisexual. Achenes oblong or ellipsoid. Fl. and fr. Jul-Nov. Forest edges, roadsides, slopes or canyons at low elevations. Distributed in SE, C, N, E, NW and NE China. Also in NE Asia.

宽叶山蒿 *Artemisia stolonifera*

柳叶蒿

Artemisia integrifolia L.

多年生草本，被蛛丝状柔毛。下部叶边缘具少量锯齿；茎中部叶椭圆形、椭圆状披针形或条状披针形，边缘具1-3锯齿；上部叶全缘。头状花序作狭圆锥花序式排列，椭圆形或长圆形；雌花10-15朵；两性花20-30朵。瘦果倒卵球形或长圆形。花果期8-10月。生中低海拔的林缘、路边、河岸、草甸或灌丛。产河北、内蒙古和东北。俄罗斯、蒙古和朝鲜半岛亦有。

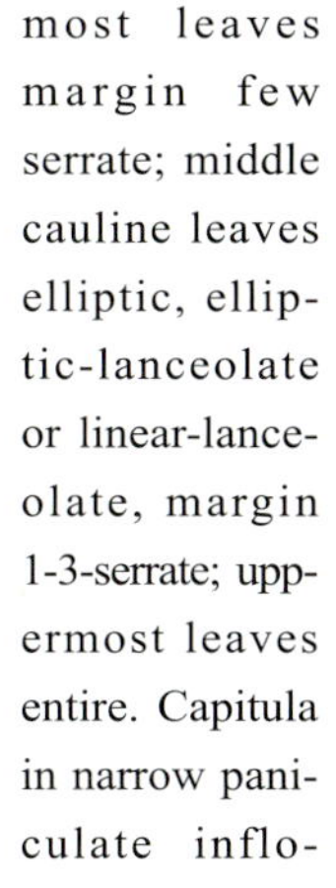

Perennial herbs, arachnoid-puberulous. Lowermost leaves margin few serrate; middle cauline leaves elliptic, elliptic-lanceolate or linear-lanceolate, margin 1-3-serrate; uppermost leaves entire. Capitula in narrow paniculate inflorescences, elliptic or oblong; pistillate florets 10-15; bisexual florets 20-30. Achenes obovoid or oblong. Fl. and fr. Aug-Oct. Forest edges, roadsides, river banks, meadows or shrublands at low to middle elevations. Distributed in Hebei, Neimenggu and NE China. Also in Russia, Mongolia and Korean Peninsula.

柳叶蒿 *Artemisia integrifolia*

蒌蒿 *Artemisia selengensis*

蒌蒿

Artemisia selengensis Turcz. ex Bess.

多年生草本，具清香。叶薄纸质，背面密被平贴灰白色蛛丝状绵毛，掌状或指状分裂，小裂片条形或披针形，边缘具锯齿。头状花序长圆形或宽卵形，排成密穗状花序；总苞片黄褐色；花冠黄色；边缘雌花8-12朵；盘心两性花10-15朵。花果期7-10月。生海拔2500米以下的河边、湖岸、沼泽、疏林、山坡、路旁或荒地。广泛分布中国各地。俄罗斯、蒙古和朝鲜半岛亦有。

Perennial herbs, sweet scented. Leaves thinly chartaceous, densely and adpressedly gray-arachnoid-lanate abaxially, palmately or ternately parted, lobes linear or lanceolate, margin serrate. Capitula oblong or broadly ovoid, arranged in dense spikes on branches; phyllaries yellowish-brown; corolla yellow; marginal female florets 8-12; disk florets 10-15, bisexual. Fl. and fr. Jul-Oct. Riversides, lake banks, swamps, sparse forests, slopes, roadsides or wastelands below 2500 m. Widely distributed throughout most parts of China. Also in Russia, Mongolia and Korean Peninsula.

白苞蒿

Artemisia lactiflora Wall. ex DC.

多年生草本。茎单生。叶纸质，渐无毛，卵形，羽状全裂，基部具细小的假托叶。头状花序长圆形，无梗，数个在小枝上排成密穗状花序；总苞片白色，无毛；花冠白色。花果期8-11月。生海拔3000米以下湿润或略为干燥地区的林下、林缘、灌丛或山谷。广泛分布秦岭以南。南亚和东南亚亦有。

Perennial herbs. Stems solitary. Leaves chartaceous, glabrescent, ovate, pinnatisected, base with pseudo-stipulate. Capitula oblong, sessile, numerous arranged in compact spikes; phyllaries white, grabrous; corolla white. Fl. and fr. Aug-Nov. Forests, forest edges, thickets or valleys of damp or somewhat dry regions below 3000 m. Widely distributed in the south of Qinling Mountain. Also in S and SE Asia.

奇蒿

Artemisia anomala S. Moore

多年生草本。叶卵形、卵状椭圆形或卵状披针形，厚纸质，具锯齿；上部叶与苞叶具锯齿。头状花序作圆锥花序式排

白苞蒿 *Artemisia lactiflora*

奇蒿 *Artemisia anomala*

光沙蒿 *Artemisia oxycephala*

列，长圆形或卵形；雌花4-6朵；两性花6-8朵。瘦果倒卵球形或倒卵球状长圆形。花果期6-11月。生海拔200-1200米的林缘、路边、峡谷、河岸、灌丛或山坡。产中国西南、华南、东南、华中、华北和华东。

Perennial herbs. Leaves ovate, ovate-elliptic or ovate-lanceolate thickly papery, serrate; uppermost leaves and leafy bracts, serrate. Capitula paniculate inflorescences, oblong or ovoid; pistillate florets 4-6; bisexual florets 6-8. Achenes obovoid or obovoid-oblong. Fl. and fr. Jun-Nov. Forest edges, roadsides, canyons, river banks, shrublands or slopes at 200-1200 m. Distributed in SW, S, SE, C, N and E China.

光沙蒿

Artemisia oxycephala Kitag.

亚灌木。叶宽卵形或近圆形，具柔毛或后变无毛，二回羽状全裂；上部叶及叶苞3-5深裂或全缘。头状花序作圆锥花序式排列，长圆形；雌花8-14朵；两性花3-10朵。瘦果长圆形。花果期8-10月。生低海拔草原、旱山丘、沙丘、盐碱土、湖岸或森林草原。产华北和东北。

Subshrubs. Leaves broadly ovate or suborbicular, puberulous or glabrescent, 2-pinnatisect; uppermost leaves and leafy bracts 3-5-sect or entire. Capitula in lax paniculate inflorescences, oblong; pistillate florets 8-14; bisexual florets 3-10. Achenes oblong. Fl. and fr. Aug-Oct. Steppe, dry hills, dunes, saline-alkali soils, lakeshores or forest steppe at low elevations. Distributed in N and NE China.

柔毛蒿

Artemisia pubescens Ledeb.

多年生草本。茎褐红色，渐无毛。叶纸质，被短柔毛，卵形或长卵形，羽状全裂，基部具假托叶。头状花序多数，长圆形、近球形或卵球形；总苞片无毛；花冠黄色；雌花8-15朵；盘心雄花10-15朵。花果期8-10月。生草原、草甸、林缘、荒坡、丘陵、砾石地或路旁。产华北、东北和西北。俄罗斯、蒙古和日本亦有。

Perennial herbs. Stems purplish-brown, glabrescent. Leaves chartaceous, shortly pubescent, ovate or narrowly ovate, pinnatisected, base pseudo-stipulate. Capitula numerous, oblong, subglobose or ovoid; phyllaries glabrous; corolla yellow; female florets 8-15; disk florets 10-15, male. Fl. and fr. Aug-Oct. Grasslands, meadows, forest edges, waste slopes, hills, rocky places or roadsides. Distributed in N, NE and NW China. Also in Russia, Mongolia and Japan.

柔毛蒿 *Artemisia pubescens*

南牡蒿 *Artemisia eriopoda*

南牡蒿

Artemisia eriopoda Bunge

多年生草本。叶近圆形，宽卵形或倒卵形，一或二回羽状全裂。头状花序多数作宽的多分枝的圆锥花序排列，宽卵形或近球形；雌花4-8朵；两性花6-10朵。瘦果长圆形。花果期6-11月。生海拔2100米以下的林缘、路边、草地、海边沙丘、灌丛、弃荒地或山坡。产中国西南、华北、华西、华东和东北。蒙古东部、朝鲜半岛和日本亦有。

Perennial herbs. Leaves suborbicular, broadly ovate or obovate, 1- or 2-pinnatisect. Capitula in broad, much branched paniculate inflorescences, numerous, broadly ovoid or subglobose; pistillate florets 4-8; bisexual florets 6-10. Achenes oblong. Fl. and fr. Jun-Nov. Forest edges, roadsides, grasslands, seashore dunes, shrublands, waste areas or slopes below 2100 m. Distributed in SW, N, W, E and NE China. Also in E Mongolia, Korean Peninsula and Japan.

牡蒿

Artemisia japonica Thunb.

多年生草本，具强烈芳香。叶簇生于茎顶；叶匙形，羽状裂。狭圆锥状复合花序；小头状花序多数；下垂；总苞卵形或近球形；叶状苞无毛；小花12-15(-20)朵，黄色。瘦果深褐色，倒卵球形。花果期7-11月。生海拔3300米以下的林缘、弃荒地、灌丛、山坡或路边。产中国西南、华南、东南、华中、华北、华西、华东和东北。南亚、东南亚和东北亚亦有。

Perennial herbs, strongly aromatic. Leaves clustered at apex; leaves spatulate, pinnately lobed. Synflorescences a ± narrow panicle, panicle; capitula many, nodding; involucres ovoid or subglobose; phyllaries glabrous; florets 12-15(-20), yellow. Achenes dark brown, obovoid. Fl. and fr. Jul-Nov. Forest edges, waste areas, shrublands, slopes, roadsides below 3300 m. Distributed in SW, S, SE, C, N, W, E and NE China. Also in S, SE and NE Asia.

牡蒿 *Artemisia japonica*

昆仑蒿

Artemisia nanschanica Krasch.

多年生草本，有臭味。茎下部叶与营养枝叶匙形、倒卵形或宽卵形，羽状或近于掌状深裂或浅裂，稀少近全裂，耳状小裂片无柄，椭圆形或披针形；中部叶无柄匙形或倒卵状楔形，自叶上端斜向基部的(2-)3(-4)深裂，稀少近全裂，耳状裂片椭圆形或线形；上部叶匙形，自叶上端斜向基部的2-3深裂或浅裂或不分裂。头状花序组成窄圆锥花序。花果期7-10月。生海拔2100-5300米的干山坡、草原、滩地、砾质坡地。产西藏、甘肃、青海和新疆南部。

Perennial herbs, fetid. Lower stem leaves: leaves spatulate,

昆仑蒿 *Artemisia nanschanica*

obovate or ovate, obliquely pinnati-subpalmatipartite or -cleft; lobes elliptic, oblong, or elliptic-lanceolate. Middle stem leaves sessile; leaves spatulate or obovate-spatulate, obliquely (2-)3(-4)-partite, rarely -sect; lobes elliptic or linear; uppermost leaves and leaflike bracts spatulate, obliquely 2- or 3-partite, -cleft or entire. Synflorescence a narrow panicle. Fl. and fr. Jul-Oct. Dry slopes, steppes, rocky terraces or slopes at 2100-5300 m. Distributed in Xizang, Gansu, Qinghai and S Xinjiang.

龙蒿

Artemisia dracunculus L.

亚灌木。叶具柔毛或渐无毛；上部叶和叶苞小，条形或披针形。头状花序作圆锥花序式排列，近球形、卵形或半球形；雌花6-10朵；两性花4-14朵。瘦果倒卵球形或倒卵球状椭圆体形。花果期7-10月。生海拔500-3800米的草原、林缘、弃荒地、梯田、路边或盐碱土中。产华北、华西、西北和东北。中亚，西亚，东北亚，欧洲中部、东部和西部，北美洲亦有。

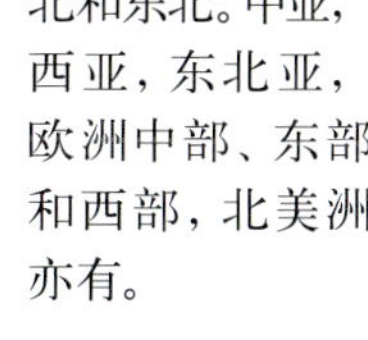

Subshrubs. Leaves puberulous or glabrescent; uppermost leaves and leafy bracts minute, linear or lanceolate. Capitula paniculate inflorescences, subglobose, ovoid or semiglobose; pistillate florets 6-10; bisexual florets 4-14. Achenes obovoid or obovoid-ellipsoid. Fl. and fr. Jul-Oct. Dry places in steppe, forest edges, waste areas, terraces, roadsides or saline-alkali soils at 500-3800 m. Distributed in N, W, NW and NE China. Also in C, W and NE Asia, C, E and W Europe, and North America.

龙蒿 *Artemisia dracunculus*

线叶菊 *Filifolium sibiricum*

线叶菊

Filifolium sibiricum (L.) Kitam.

多年生草本。茎密丛生，基部密被撕裂纤维状叶鞘。叶二或三回羽状全裂；末端裂片线形。头状花序伞房状；总苞半球形；苞片3层；小花黄色；外轮雌花6朵；中央两性花多数。瘦果倒卵球形或椭圆体形。花果期6-9月。生海拔1500-2600米的山坡草地。产华北和东北。东北亚亦有。

Perennial herbs. Stems densely fascicled, basally with densely fibrous-lacerate leaf sheath. Leaves bi- or tri-pinnatisect; ultimate segments filiform. Capitula corymbose; involucres hemispheric; phyllaries in 3 rows; florets yellow; outer female florets 6; central hermaphroditic florets numerous. Achenes obovoid or ellipsoid. Fl. and fr. Jun-Sep. Grasslands on mountain slopes at 1500-2600 m. Distributed in N and NE China. Also in NE Asia.

栉叶蒿 *Neopallasia pectinata*

栉叶蒿

Neopallasia pectinata (Pall.) Poljakov

多年生草本。叶无毛，篦齿状羽状全裂；裂片条状钻形。头状花序排成狭穗状圆锥花序式；苞片3或4裂；外部可育雄花3或4朵，狭管状；中央两性花管状，5裂。瘦果棕色，椭圆形。花果期6-9月。生海拔1300-3400米的沙地、河边砾石地、沟谷或弃荒地。产中国西南、华北、华西、西北和东北。哈萨克斯坦、俄罗斯和蒙古亦有。

Perennial herbs. Leaves glabrous, pectinate-pinnatisect; segments linear-subulate. Capitula in a narrowly spiciform panicle; phyllaries in 3 or 4 rows; outer female florets 3 or 4, fertile, narrowly tubular; central florets hermaphroditic, tubular, 5-lobed. Achenes brown, elliptic. Fl. and fr. Jun-Sep. Deserts, gravelly places of river valleys or wastelands at 1300-3400 m. Distributed in SW, N, W, NW and NE China. Also in Kazakstan, Russia and Mongolia.

扁毛菊

Allardia glabra Decne.

多年生草本，根状茎多分枝。叶莲座状，无柄，匙形，顶端3-5裂。头状花序单生茎端或枝端；总苞半球形；总苞片约5层；舌状花粉红色至白色；冠毛紫红色或淡棕色。花果期7-8月。生海拔(3500-)4900-5500米高山石坡石缝中。产四川西部和西藏。喜马拉雅和中亚亦有。

Perennial herbs, with much-branched rhizomes. Leaves in dense rosettes, sessile; leaves cuneate, apex 3(-5)-lobed or -parted. Capitula solitary, sessile; involucres hemispheric,phyllaries in 5 rows; ray florets deep pink to almost white; corona scales tinged pinkish violet or brownish. Fl. Jul-Sep. Scree slopes, rock crevices at (3500-)4900-5500 m. Distributed in W Sichuan and Xizang. Also in Himalaya and C Asia.

三指扁毛菊

Allardia tridactylites (Kar. et Kir.) Sch.Bip.

多年生草本，根状茎多分枝。叶叶莲座状，无柄，匙形，顶端3-5裂。头状花序单生茎端或枝端，直径约2厘米；总苞半球形；总苞片3-4层；舌状花白色、紫红色或粉红色；冠毛刚毛状，红色或棕色。花期7-9月。生海拔3000-4000米的高山冲积地、石坡石缝中。产西藏和新疆。哈萨克斯坦、蒙古和俄罗斯亦有。

Perennial herbs, with procumbent, much-branched rhizome. Leaves in dense rosettes, sessile; leaves cuneate, apex 3(-5)-lobed or -parted. Capitula solitary, sessile, ca. 2 cm diam; involucres hemispheric; phyllaries in 3-4 rows; ray florets pinkish white, pink or purple-red; corona scale bristlelike, usually tinged reddish or brownish. Fl. Jul-Sep. Floodlands, talus on mountain slopes at 3000-4000 m. Distributed in Xizang and Xinjiang. Also in Kazakhstan, Mongolia and Russia.

灌木小甘菊

Cancrinia maximowiczii C. Winkl.

亚灌木，高40-50厘米，多分枝。叶具柄，长圆状线形，羽状深裂；最上部叶线形，全缘或有齿。头状花序2-5个在枝端排成伞房状；总苞钟状或宽钟状，直径5-7毫米；总苞片3-4层；花冠黄色，5裂。花果期

扁毛菊 *Allardia glabra*

三指扁毛菊 *Allardia tridactylites*

灌木小甘菊 *Cancrinia maximowiczii*

7-10月。生海拔2100-3600米多砾石的山坡及河岸冲积扇上。产云南、甘肃、内蒙古、青海和新疆。蒙古亦有。

Subshrubs, 40-50 cm tall, much branched. Leaves petiolate, leavesoblong-linear, pinnatipartite; uppermost leaves linear, entire or denticulate. Synflorescence a flat-topped cyme; capitula 2-5; involucres campanulate or broadly campanulate, 5-7 mm diam; phyllaries in 3-4 rows; florets yellow, 5-lobed. Fl. and fr. Jul-Oct. Gravelly mountain slopes, alluvial fans by riverbanks at 2100-3600 m. Distributed in Yunnan, Gansu, Neimenggu, Qinghai and Xinjiang. Also in Mongolia.

小甘菊

Cancrinia discoidea (Ledeb.) Poljakov ex Tzvelev

二年生草本，高5-20厘米。茎葶状，被白色棉毛。叶灰绿色，长圆形或卵形，二回羽状深裂，每个裂片有2-5深裂或浅裂。头状花序单生茎端；总苞直径7-12毫米，被疏棉毛至几无毛；总苞片3-4层；花托明显凸起，锥状球形；花黄色，管状，檐部5齿裂。花果期4-9月。生海拔500-2000米的山坡、荒地和戈壁。产西藏、甘肃、内蒙古和新疆。哈萨克斯坦、蒙古和俄罗斯亦有。

Biennial herbs, 5-20 cm tall. Stems scapose, conspicuously lanate. Leaves grey-greenleaves, oblong or ovate, 2-pinnatipartite; primary lateral lobes 2-5-paired. Capitula solitary, terminal; involucres, 7-12 mm diam; phyllaries 3-4-seritate; receptacle conspicuously convex, conical-globose; florets yellow, all tubular, 5-lobed. Fl. and fr. Apr-Sep. Mountain slopes, Gobi Desert at 500-2000 m. Distributed in Xizang, Gansu, Neimenggu and Xinjiang. Also in Kazakhstan, Mongolia and Russia.

天山小甘菊

Cancrinia tianschanica (Krasch.) Tzvelev

多年生草本。基生叶有柄，长圆形或卵形，二回羽状分裂；

小甘菊 *Cancrinia discoidea*

茎叶全缘。头状花序单生；总苞直径4-6毫米，长10-17毫米。总苞3层。花黄色，5裂。花果期7-8月。生海拔3200-3900米的山坡多石地、草甸和河滩。产新疆。哈萨克斯坦亦有。

Perennial herbs. Basal leaves petiolate, oblong or ovate, 2-pinnatipartite; stem leaves margin entire. Capitula solitary; involucres coryliform 4-6 mm diam. Phyllaries in 3 rows. Florets yellow, 5-lobed. Fl. and fr. Jul-Aug. Rocky places on mountain slopes, meadows, floodlands at 3200-3900 m. Distributed in Xinjiang. Also in Kazakhstan.

天山小甘菊 *Cancrinia tianschanica*

灰叶匹菊 *Richteria pyrethroides*

川滇女蒿 *Hippolytia delavayi*

灰叶匹菊

Richteria pyrethroides Kar. et Kir.

多年生草本。基生叶长椭圆形，二回羽状分裂。头状花序单生茎顶，极少2-3个；总苞直径10-14毫米；总苞片约4层，边缘黑褐色、膜质；舌花白色或淡红色；冠状冠毛分裂至基部。花果期6-8月。生海拔2300-3700米的草甸、草原、岩坡及河滩。产新疆。克什米尔地区、中亚和俄罗斯亦有。

Perennial herbs. Basal leaves narrowly elliptic, 2-pinnatisect. Capitula solitary, rarely 2-3, terminal; involucre 10-14 mm diam; phyllaries ca. 4 rows, margin blackish brown, membranous; ray florets white or pale red; corona divided to base. Fl. and fr. Jun-Aug. Meadows, rocky mountain slopes, floodlands at 2300-3700 m. Distributed in Xinjiang. Also in Kashmir, C Asia and Russia.

川滇女蒿

Hippolytia delavayi (Franch. ex W. W. Sm.) C. Shih

多年生草本。基生叶椭圆形，二回羽状分裂，小裂片条形，下面被长柔毛。头状花序6-12个，排成顶生束状伞房花序；总苞钟状；总苞片披针形，黄白色，具光泽，硬草质；小花管状，花冠黄色。瘦果近纺锤形；无冠毛。花果期8-10月。生海拔3300-4000米的高山草甸。产云南和四川。

Perennial herbs. Basal leaves elliptic, bipinnatifid, lobelets linear, villose abaxially. Capitula 6-12, arranged in terminal fasciate corymb; involucres campanulate; phyllaries lanceolate, yellowish-white, lucidous, hard herbaceous; florets all tubular, corolla yellow. Achenes subfusiform; pappus absent. Fl. and fr. Aug-Oct. Alpine meadows at 3300-4000 m. Distributed in Yunnan and Sichuan.

束伞女蒿

Hippolytia desmantha C. Shih

亚灌木，高10-15厘米。叶卵形、椭圆形或偏斜椭圆形或长扇形，二回羽状全裂；接花序下部的叶有时羽状全裂。头状花序3-5个在枝端排成束状伞房花序；总苞钟状；总苞片4层；全部苞片硬草质，有光泽，黄白色或麦秆黄色，边缘浅褐色。花果期8-9月。生海拔3800-3900米的草甸或沟谷岩石裸露处。产青海西南部。

Subshrubs, 10-15 cm tall. Leaves ovate, elliptic, obliquely elliptic or narrowly flabelliform, 2-pinnatisect; leaves below synflorescence sometimes pinnatisect. Capitula 3-5, fascicled-corymbose at apices of branches; involucres campanulate; phyllaries 4-seritate, rigidly herbaceous,

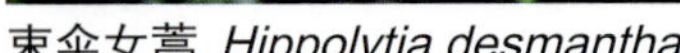
束伞女蒿 *Hippolytia desmantha*

棉毛女蒿 *Hippolytia gossypina*

glossy, yellow-white or straw-colored, margin brownish. Fl. and fr. Aug-Sep. Meadows, rock outcrops on valley slopes at 3800-3900 m. Distributed in SW Qinghai.

棉毛女蒿

Hippolytia gossypina (C. B. Clarke) C. Shih

垫状植物，高约7厘米。茎枝被密厚的残叶。叶片匙形，白色或灰白色，两面密被长绵毛，先端圆形或截形，3-6裂。头状花序约10个，排列密集，伞房状；总苞片3层；小花黄色，长约3毫米。花期8-10月。生海拔4500-5400米的荒漠、高山砾石堆或山顶裸岩上。产西藏南部。喜马拉雅亦有。

Cushion plants, ca. 7 cm tall. Leafy shoots densely compact, with dense overlapping leaves. Leaves spatulate, white or gray-white, both surfaces densely long lanate, apex truncate or rounded, 3-6-divided. Capitula ca. 10, in a dense corymb; involucres phyllaries in 3 rows; florets yellow, ca. 3 mm. Fl. Aug-Oct. Deserts, alpine gravel heaps, exposed rocks at mountain summits at 4500-5400 m. Distributed in S Xizang. Also in Himalaya.

垫状女蒿

Hippolytia kennedyi (Dunn) Ling

垫状植物，密生，高2-4厘米。茎多次分枝。靠近地面的叶莲座状，叶片圆形或扇形，二回三出掌状全裂或者羽状全裂，两面密被污黄色，灰白色长柔毛或绵毛。头状花序多数在茎顶排成紧密的伞房花序，或花叶复台体紧密形成直径达10厘米的团伞花序。花果期8-9月。生海拔4700-5200米的高山荒漠石砾处。产西藏南部。印度北部亦有。

Cushion plants, dense, 2-4 cm tall. Stems much branched. Leaves in dense rosettes at or near ground level; leaves orbicular or flabelliform, biternately palmatisect or 2-pinnatisect, both surfaces densely dirty yellow, gray-white villous or lanate. Synflorescence a large, dense, hemispheric glomerule, ca. 10 cm diam. Fl. and fr. Aug-Sep. Gravelly places of alpine deserts at 4700-5200 m. Distributed in S Xizang. Also in N India.

合头女蒿

Hippolytia syncalathiformis C. Shih

多年生草本，无茎。叶集中在团伞花序下部，全长卵形、长椭圆形，二回羽状分裂。头状花序多数，在顶端排列成半球状团伞花序；总苞楔钟状；总苞片3层，被长柔毛，边缘黑褐色膜质；花黄色，长约3毫米。花期9月。生海拔4500-5500米的高山草甸或岩石坡。产西藏。

Perennial herbs, stemless. Leaves in a single dense basal rosette around synflorescence; leaves narrowly ovate, narrowly elliptic, bipinnatiparted. Synflorescence a dense hemispheric glomerule; capitula many; involucres cuneate-campanulate; phyllaries in 3 rows, abaxially villous, scarious margin dark brown; florets yellow, ca. 3 mm. Fl. Sep. Alpine meadows or rocky slopes at 4500-5500 m. Distributed in Xizang.

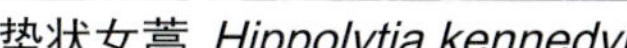
垫状女蒿 *Hippolytia kennedyi*

合头女蒿 *Hippolytia syncalathiformis*

喀什菊 *Kaschgaria komarovii*

喀什菊

Kaschgaria komarovii (Krasch. et Rubtzov) Poljakov

亚灌木，高30-55厘米。茎中部叶条形、匙形或倒披针形，大多顶端3-5裂或少有羽状分裂，两面疏生星状毛；上部叶全缘。头状花序卵形，3-5(-8)个在枝端排成束状伞房花序或紧密伞房花序；总苞狭杯状；总苞片2-4层。花果期7-8月。生海拔800-2000的荒漠山坡。产新疆。蒙古西部和哈萨克斯坦亦有。

Subshrubs, 30-55 cm tall. Middle stem leaves spatulate or oblanceolate, apex 3-5-partite, rarely pinnatisect; Sparsely stellate Pubescent upper stem leaves entire. Synflorescence a dense flattopped fascicle sometimes with additional axillary capitula below; capitula 3-5(-8); involucres copular; phyllaries in 2-4 rows. Fl. and fr. Jul-Aug. Barren rocky slopes, rocky floors of gorges at 800-2000 m. Distributed in Xinjiang. Also in W Mongolia and Kazakhstan.

太行菊

Opisthopappus taihangensis (Ling) C. Shih

多年生草本。基生叶卵形、宽卵形或椭圆形，一、二回全部全裂，一回侧裂片2-3对；最上部的叶常羽裂。头状花序单生枝端，或枝生1-3个头状花序；总苞浅盘状；总苞片约4层；舌状花粉红色或白色；冠毛芒片状，4-6个。花果期6-9月。生海拔800-1200米的山坡岩石上。产山西、河南和河北南部。

Perennial herbs. Basal and stem leaves ovate, broadly ovate or elliptic, 1- or 2-pinnatisect; primary lateral segments 2- or 3-paired; uppermost leaves pinnatifid. Capitula 1-3 at apices of branches; involucre coryliform; phyllaries in 4 rows; ray florets white or pink; corona scales 4-6. Fl. and fr. Jun-Sep. Rocks of mountain slopes, cliffs at 800-1200 m. Distributed in Shanxi, Henan and S Hebei.

蓍

Achillea millefolium L.

多年生草本。根状茎匍匐。叶披针形，二至三回羽状全裂。头状花序多数，密集成复伞房状；总苞矩圆形或近卵形；舌状花5朵；舌片白色、粉红色或淡紫红色，先端2或3齿裂；盘心花黄色，管状。瘦果矩圆形；无冠毛。花果期7-9月。中国各地庭园常有栽培，中国东北和西北有野生。亚洲、欧洲和非洲亦有。

Perennial herbs. Rhizomes creeping. Leaves laceolate, bipinnatisect to tripinnatisect. Capitula numerous, clustered to compound corymb; phyllaries oblong or subovoid; ray florets 5; laminas white, pink or violet-red, apex 2- or 3-denticulate; disk florets yellow, tubular. Achenes oblong; pappus absent. Fl. and fr. Jul-Sep. Commonly cultivated in gardens, wild in NE and NW China. Also in Asia, Europe and Africa.

齿叶蓍

Achillea acuminata (Ledeb.) Sch.-Bip.

多年生草本。叶披针形，边缘具重锯齿。头状花序密集成疏伞房状；总苞半球形；舌状花10-23朵；舌片白色，先端3齿裂；盘心花白色，管状。瘦果倒披针形；无冠毛。花果期7-8月。生海拔500-2900米的山坡潮湿处、草甸或林缘。产华北、东北和西北。东北亚和蒙古亦有。

Perennial herbs. Leaves lanceolate, margin biserrate. Capitula clustered to lax corymb; involucres semiglobose; ray florets

太行菊 *Opisthopappus taihangensis*

蓍 *Achillea millefolium*

10-23; laminas white, apex 3-crenate; disk florets white, tubular. Achenes oblanceolate; pappus absent. Fl. and fr. Jul-Aug. Slope damp sites, meadows or forest edges at 500-2900 m. Distributed in N, NE and NW China. Also in NE Asia and Mongolia.

高山蓍
Achillea alpina L.

多年生草本。叶条状披针形，篦齿状羽状浅裂至深裂，叶基部抱茎。头状花序密集排成顶生伞房花序；边缘舌状花6-8朵；舌片白色，顶部具3浅齿；无冠毛。花果期7-9月。生海拔800-2400米的山坡草地、林缘、灌丛下、草甸或河谷。产中国西南、东南、华中、华北、华西、华东和东北。东北亚和蒙古亦有。

Perennial herbs. Leaves linear-lanceolate, pectinate-pinnatilobed or pinnatipartite, basally amplexicaul. Capitula in a dense apical corymb; ray florets 6-8, lamina white, apex 3-denticulate; pappus absent. Fl. and fr. Jul-Sep. Grasslands on mountain slopes, forest edges, under thickets, meadows or river valleys at 800-2400 m. Distributed in SW, SE, C, N, W, E and NE China. Also in NE Asia and Mongolia.

齿叶蓍 *Achillea acuminata*

高山蓍 *Achillea alpina*

丝叶蓍 *Achillea setacea*

云南蓍 *Achillea wilsoniana*

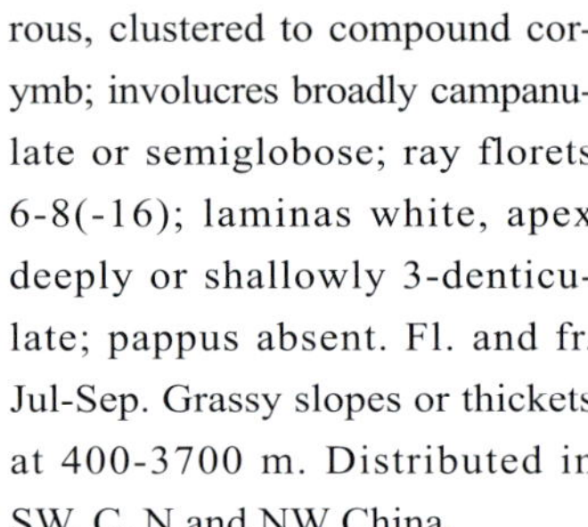

丝叶蓍

Achillea setacea Waldst. et Kit.

多年生草本。叶线状批针形，二或三回羽状分裂。头状花序多数；总苞狭长圆形或卵状长圆形；舌状花5朵，舌片黄白色，先端截形或3齿裂；无冠毛。花果期7-8月。生海拔500-2400米的荒地、林缘或山坡草地。产黑龙江和新疆。俄罗斯、蒙古、中亚、西南亚、北非和欧洲亦有。

Perennial herbs. Leaves linear-lanceolate, (2 or) 3-pinnatisect. Capitula many; involucres narrowly oblong or ovoid-oblong; ray florets 5; lamina yellowish white, apex subtruncate or 3-dentate; pappus absent. Fl. and fr. Jul-Aug. Wastelands, forest margins or grasslands slopes at 500-2400 m. Distributed in Heilongjiang and Xinjiang. Also in Russia, Mongolia, C and SW Asia, N Africa and Europe.

云南蓍

Achillea wilsoniana Heimerl ex Hand.-Mazz.

多年生草本。根状茎短。叶矩圆形，二回羽状全裂。头状花序多数，集成复伞房花序；总苞宽钟形或半球形；舌状花6-8(-16)朵；舌片白色，先端深或浅3齿裂；无冠毛。花果期7-9月。生海拔400-3700米的草坡或灌丛中。产中国西南、华中、华北和西北。

Perennial herbs. Rhizomes short. Leaves oblong, bipinnatisect. Capitula numerous, clustered to compound corymb; involucres broadly campanulate or semiglobose; ray florets 6-8(-16); laminas white, apex deeply or shallowly 3-denticulate; pappus absent. Fl. and fr. Jul-Sep. Grassy slopes or thickets at 400-3700 m. Distributed in SW, C, N and NW China.

臭春黄菊

Anthemis cotula L.

多年生草本，有臭味。茎聚伞状分枝。叶卵状长圆形，二回羽状全裂。头状花序单生于茎顶；总苞片3层，边缘狭膜质；舌状花白色。瘦果长圆状陀螺形，具结节及8条纵棱；无冠毛。花果期6-7月。中国东北和内蒙古有栽培。原产欧洲。

Annual herbs, fetid. Stems corymbose-branched. Leaves ovate-oblong, bipinnatisect. Capitula solitary at apices of branches; phyllaries in 3 rows, margin narrowly scarious; ray florets white. Achenes oblong-turbinate, tuberculate, 8-ribbed; pappus absent. Fl and fr. Jun-Jul. Cultivated in NE China and Neimenggu. Native to Europe.

臭春黄菊 *Anthemis cotula*

岩匹菊 *Tanacetum petraeum*

藏匹菊 *Pyrethrum atkinsonii*

岩匹菊

Tanacetum petraeum (C. Shih) K. Bremer et Humphries

小亚灌木。叶卵形、倒卵形或倒披针形，两面无毛，二回羽状深裂；远轴叶羽状全裂，小头状花序单个顶生，或2或3个形成不规则的伞房状；总苞浅碟形；总苞片4层；舌状花粉色，长椭圆形，顶部2齿裂。花果期8月。生海拔1800-2600米的山坡岩石上。产新疆北部。

Small subshrubs. Leaves ovate, obovate or oblanceolate, both surfaces glabrous, bipinnatipartite; distal leaves pinnatifid, small. Capitula termianal, solitary, or 2 or 3 in an irregular corymb; involucres coryliform; phyllaries in 4 rows; ray florets pink, long-elliptic, apex 2-denticulate. Fl. and fr. Aug. Rocks of mountain slopes at 1800-2600 m. Distributed in N Xinjiang.

藏匹菊

Pyrethrum atkinsonii (C. B. Clarke) Ling ex Ling et C. Shih

多年生草本。茎单生，直立。基生叶长4-10厘米，倒披针形或长椭圆形，三回羽状分裂。头状花序单生茎顶；总苞宽1-2.5厘米；总苞片4层；舌状花和管状花黄色；冠毛长不足0.1毫米。花期7月。产西藏南部。喜马拉雅亦有。

Perennial herbs. Stems solitary, erect. Basal le aves 4-10 cm long, oblanceolate or narrowly elliptic, tripinnatifid. Capitula termianal, solitary; involucres 1-2.5 cm wide; phyllaries 4-seritate; ray florets and disc florets yellow; pappus shorter than 0.1 mm. Fl. Jul. Distributed in S Xizang. Also in Himalaya.

川西小黄菊

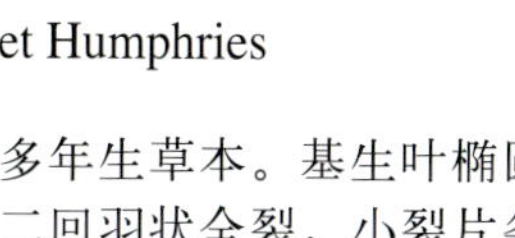

Tanacetum tatsienense (Bureau et Franch.) K. Bremer et Humphries

多年生草本。基生叶椭圆形，二回羽状全裂，小裂片条形，绿色。头状花序单生茎顶；总苞直径1-2厘米；总苞片披针形至条形，边缘褐色，膜质；舌状花橘黄色，条形。瘦果具5-8条纵肋；冠毛冠状。花果期7-9月。生海拔3500-5200米的高山草甸、灌丛或山坡砾石地。产云南、四川、西藏和青海。

Perennial herbs. Basal leaves elliptic, bipinnatisected, lobelets linear, green. Capitula solitary, terminal; involucres 1-2 cm diam; phyllaries lanceolate to linear, margin brown, membranous; ray florets orange-yellow, linear. Achenes 5-8 ridged; pappus crowned. Fl. and fr. Jul-Sep. Alpine meadows, thickets or gravelly places on slopes at 3500-5200 m. Distributed in Yunnan, Sichuan, Xizang and Qinghai.

川西小黄菊 *Tanacetum tatsienense*

无舌川西小黄菊

Tanacetum tatsienense var. **tanacetopsis** (W. W. Sm.) Grierson

头状花序同性。所有的小花两性，管状。生海拔3500-5000米的高山草甸或灌丛。产西藏和云南。

Capitula homogamous. All florets tubular, bisexual. Alpine meadows or thickets at 3500-5000 m. Distributed in Xizang and Yunnan.

无舌川西小黄菊 *Tanacetum tatsienense* var. *tanacetopsis*

黑苞匹菊 *Tanacetum krylovianum*

黑苞匹菊

Tanacetum krylovianum (Krasch.) K. Bremer et Humphries

多年生草本。叶长椭圆形，二回羽状全裂。头状花序1-3生于顶端，不明显伞房状；总苞浅碟状；总苞片4层；舌状花白色；舌状花长椭圆形，顶部2或3齿裂。花果期8-9月。生海拔2500-3200米的林下或碱土。产新疆。哈萨克斯坦、俄罗斯和蒙古亦有。

Perennial herbs. Leaves long-elliptic, bipinnatisect. Capitula 1-3 at apex, inconspicuously corymbose; involucres coryliform; phyllaries in 4 rows; ray florets white; rays long-elliptic, apex 2- or 3-denticulate. Fl. and fr. Aug-Sep. Forests or alkaline soils at 2500-3200 m. Distributed in Xinjiang. Also in Kazakstan, Russia and Mongolia.

密头菊蒿

Tanacetum crassipes (Stschgel.) Tzvel.

多年生草本。叶长椭圆形，两面密被“丁”字毛和单毛，二回羽状全裂。头状花序3-7个生顶部，密伞房状；总苞钟形；总苞片3或4层；舌状花黄色，具3齿。花果期6-8月。生海拔约2100米的多石山坡、林下或草地。产新疆(阿尔泰山)。哈萨克斯坦和俄罗斯亦有。

Perennial herbs. Leaves long-elliptic, both surfaces appressed pilose with T-shaped and simple hairs, bipinnatisect. Capitula 3-7 at apex, densely corymbose; involucres campanulate; phyllaries in 3 or 4 rows; ray florets yellow; rays 3-denticulate. Fl and fr. Jun-Aug. Rocky slopes, forests or grasslands at ca. 2100 m. Distributed in Xinjiang (Aertai Mountain). Also in Kazakstan and Russia.

密头菊蒿 *Tanacetum crassipes*

东北三肋果

Tripleurospermum tetragonospermum (F. Schmidt) Pobed.

一年生草本。叶两面无毛，二或三回羽状全裂。头状花序数个或单生于枝和茎顶；总苞半

东北三肋果
Tripleurospermum tetragonospermum

球形；苞片4层；舌状花白色；两性花黄色，5裂。瘦果棕色，三角形，具结节，厚纵棱下面1条，侧面2条。花果期6-8月。生海拔约300米的河边路边或河边沙地。产中国东北。俄罗斯和日本亦有。

Annual herbs. Leaves both surfaces glabrous, bi- or tri-pinnatisect. Capitula several, or solitary at apices of branches and stems; involucres hemispheric; phyllaries in 4 rows; ray florets white; disk florets yellow, 5-lobed. Achenes brownish, triquetrous, tuberculate, with 1 adaxial and 2 lateral thick ribs. Fl. and fr. Jun-Aug. Roadsides, sandy places by rivers at ca. 300 m. Distributed in NE China. Also in Russia and Japan.

三肋果

Tripleurospermum limosum (Maxim.) Pobed.

一年生或两年生草本。叶倒披针状长圆形或长圆形，两面无毛，羽状三深裂。头状花序排成顶生伞房状；总苞半球形；苞片2或3层；舌状花白色；两性花黄色，5裂；每个裂片具1个脂囊。瘦果棕色，三棱形，具皱，白色纵棱上面1条，侧面2条。花果期6-8月。生河边沙地、草甸或沙质旱山坡。产河北、内蒙古和辽宁。中亚和东北亚亦有。

Annual or biennial herbs. Leaves oblanceolate-oblong or oblong, both surfaces glabrous, tripinnatisect. Capitula in an apical corymb; involucres hemispheric; phyllaries in 2 or 3 rows; ray florets white; disk florets yellow, 5-lobed; lobes each with a resin sac. Achenes brown, triquetrous, wrinkled, with 1 adaxial and 2 lateral whitish ribs. Fl. and fr. Jun-Aug. Sandy places by rivers, meadows or dry sandy mountain slopes. Distributed in Hebei, Neimenggu and Liaoning. Also in C and NE Asia.

母菊

Matricaria chamomilla L.

一年生草本。叶矩圆形、倒披针形或卵形，二回羽状全裂，无柄，裂片条形。头状花序异型，排成伞房状，具花序梗；总苞片苍绿色，边缘膜质；花托长圆锥状，中空；舌状花白色，反折；管状花黄色。瘦果小，侧扁；无冠毛。花果期5-7月。生海拔1800-3300米的河谷、旷野或田边。产新疆。亚洲(北部和西部)和欧洲亦有。

Annual herbs. Leaves oblong, oblanceolate or ovate, bipinnatisected, sessile, lobes linear. Capitula heterotypic, arranged in corymb, pedunculate; phyllaries glaucous, margin membranous; receptacles narrowly conical, canulate; ray florets white, revolate; tubular florets yellow. Achenes small, laterally compressed. Corona absent. Fl. and fr. May-Jul. Valleys, open places or fieldsides at 1800-3300 m. Distributed in Xinjiang. Also in N and W Asia and Europe.

三肋果 *Tripleurospermum limosum*

母菊 *Matricaria chamomilla*

茼蒿 *Glebionis coronaria*

茼蒿

Glebionis coronaria (L.) Cass. ex Spach

一年生或二年生草本，光滑无毛。叶无柄，长圆形或长圆状倒卵形，二回羽状分裂，侧裂片4-10对，小裂片卵形或条形。头状花序单生茎顶，花梗长15-20厘米；总苞直径1.5-3厘米；总苞片4层，顶端膜质；舌状花黄色或白色；管状花黄色。瘦果具棱。花果期9月。中国各地广泛栽培，河北和山东等地有野生。

Annual or biennial herbs, glabrous. Leaves sessile, oblong or oblong-obovate, bipinnatifid, lateral lobes 4-10 pairs, lobelets ovate or linear. Capitula solitary, terminal, peduncles 15-20 cm long; involucres 1.5-3 cm diam; phyllaries 4 rows, apex membranous; ray florets yellow or white; disc florets yellow. Achenes ridged. Fl. and fr. Sep. Widely cultivated in China, wild in Hebei and Shandong.

匙叶合冠鼠麴草

Gamochaeta pensylvanica (Wiild.) Cabrera

一年生草本。基生叶花期枯萎；茎生叶无柄，倒披针形至匙形。头状花序多数，在叶腋簇生，成圆锥花序；外层总苞片卵状披针形或披针形；中央花2-3；冠毛白色，基部联合成环。花期1-5月。生海拔1500米以下的废弃田野、路边。产中国西南、华南、华中和华东。世界范围亦有。

Annual herbs. Basal leaves withering at anthesis; cauline leaves sessile, oblanceolate to spatulate. Capitula numerous in axillary clusters, forming panicles; outer phyllaries ovate-lanceolate or lanceolate; central florets 2 or 3; pappus white, connate at base into a ring. Fl. Jan-May. Waste fields, roadsides below 1500 m. Distributed in SW, S, C and E China. Also worldwide.

单头火绒草

Leontopodium monocephalum Edgew.

多年生草本，具匍匐茎。叶匙形至长圆状倒披针形，淡黄色，稀浅灰色，被毛。头状花序1-10个，雌雄异体；苞叶倒披针形或匙形；总苞片倒披针形，稀卵状倒披针形。生海拔4000-5000米的高山砾石地和草甸。产西藏。喜马拉雅亦有。

Perennial stoloniferous herbs. Leaves spatulate to oblong-oblanceolate, yellowish, rarely grayish, tomentose. Capitula 1-10, dioecious; bracteal leaves oblanceolate or spatulate; phyllaries oblanceolate, rarely ovate-

匙叶合冠鼠麴草 *Gamochaeta pensylvanica*

单头火绒草 *Leontopodium monocephalum*

藓状火绒草 *Leontopodium muscoides*

oblanceolate. Alpine gravelly slopes, meadows at 4000-5000 m. Distributed in Xizang. Also in Himalaya.

藓状火绒草

Leontopodium muscoides Hand.-Mazz.

多年生草本，丛生。地下茎多分枝。叶线状披针形，被浅灰色或淡黄色毛。头状花序(1-)3-6，密集；苞叶6-10，星状排列；总苞半球形，被棉毛；总苞片褐色。瘦果椭圆形；冠毛白色。花期7-8月，果期9月。生海拔4000-4200米的高山湿润草甸、湿润或干燥石砾地、岩石上。产云南西北部和西藏东南部。

Perennial herbs, subpulvinate. Rhizome many branched. Leaves linear-lanceolate, grayish or yellowish lanuginous. Capitula (1-)3-6, densely congested; bracteal leaves 6-10, forming a star; involucre subglobose, abaxially lanate; phyllaries apex dark brown. Achenes oblong; pappus white. Fl. Jul-Aug. Fr. Sep. Alpine meadows, thickets, Abies forests, rocky places at 4000-4200 m. Distributed in NW Yunnan and SE Xizang.

黄白火绒草

Leontopodium ochroleucum Beauv.

多年生草本。根状茎分枝，有多数莲座状叶丛和花茎。头状花序世代交替，雌雄异体，密集；苞叶椭圆形或长圆披针形，规则的星状排列；总苞片披针形；冠毛污白色。花期5-8月，果期7-9月。生海拔2200-5000米的高山湿润或干燥草地、苔原、砾石坡。产西藏、青海和新疆。印度、哈萨克斯坦、蒙古和俄罗斯亦有。

Perennial herbs. Rhizome often branching, forming tufts of numerous stems and sterile leaf rosettes. Capitula heterogamous and unisexual-dioecious, densely aggregated; bracteal leaves oblong-elliptic or lanceolate, forming regular multiradiate star; phyllaries lanceolate; pappus dirty white. Fl. May-Aug. Fr. Jul-Sep. Mountain tundra, humid or dry meadows, stony fields of slopes at 2200-5000 m. Distributed in Xizang, Qinghai and Xinjiang. Also in India, Kazakhstan, Mongolia and Russia.

短星火绒草

Leontopodium brachyactis Gand.

多年生草本，具匍匐枝，被毛。花茎多个。基生叶莲座状，匙形；茎叶密被羊毛状毛，1脉，顶端稍尖。头状花序3-5；苞叶星状排列，包围头状花序；总苞片深褐色。花期6-8月。生海拔2200-4100米的岩石、开阔地。产西藏西部和新疆。西喜马拉雅亦有。

Perennial herbs, stoloniferous, tomentose. Stems many. Basal leaves in dense rosettes, spatulate; cauline leaves densely lanate-tomentose, 1-veined, apex shortly mucronate. Capitula 3-5, subtended by bracteal leaves forming a distinct star, obtusely lanceolate; phyllaries dark brownish. Fl. Jun-Aug. Among rocks, open stony ground at 2200-4100 m. Distributed in W Xizang and Xinjiang. Also in W Himalaya.

黄白火绒草 *Leontopodium ochroleucum*

短星火绒草 *Leontopodium brachyactis*

山野火绒草 *Leontopodium campestre*

山野火绒草
Leontopodium campestre (Ledeb.) Hand.-Mazz.

多年生草本。基部莲座叶狭披针形。苞叶线形或披针状线形，形成不规则的辐射星状，微黄色或近白色。头状花序雌雄异株，紧密排列。花期5-9月，果期6-9月。生海拔(700-)1400-4500米的干旱草原、干燥坡地、河谷阶地沙地或石砾地、较湿润的林间草地。产青海和新疆。哈萨克斯坦、蒙古和俄罗斯亦有。

Perennial herbs. Radical rosette leaves narrowly lanceolate. Bracteal leaves linear or linear-lanceolate, forming irregular multiradiate star, yellowish or almost white capitula dioecious, closely aggregated. Fl. May-Sep. Fr. Jun-Sep. Steppes, dry or marshy meadows, dry pebbly and stony hills, herb communities on sands, forests at (700-)1400-4500 m. Distributed in Qinghai and Xinjiang. Also in Kazakhstan, Mongolia and Russia.

弱小火绒草
Leontopodium pusillum (Beauv.) Hand.-Mazz.

多年生草本，簇生。根状茎分枝，顶端有莲座状不育叶丛和多数花茎。叶匙形或长圆状匙形。头状花序径3-7个密集，稀1个；苞叶密集成星状；总苞被白色长柔毛状茸毛；总苞片深褐色。花期7-8月，果期8-9月。生海拔3500-5600米的高山草地、石砾地。产青藏高原。印度北部和克什米尔地区亦有。

Perennial herbs, subpulvinate. Rhizome many branched, with dense radical rosettes of sterile leaves and numerous flowering stems. Leaves spatulate to oblong-spatulate. Capitula (1-) 3-7, densely congested; bracteal leaves congested, forming star; involucre abaxially white lanate; phyllaries dark brown. Fl. Jul-Aug. Fr. Aug-Sep. Alpine grasslands, rocky screes, gravelly slopes, salt lake banks and shores at 3500-5600 m. Distributed in Tibetan Plateau. Also in N India and Kashmir.

匍枝火绒草
Leontopodium stoloniferum Hand.-Mazz.

多年生草本，簇生。根状茎有分枝，不育的莲座状叶丛和花茎较疏生。叶匙状披针形，被白色密茸毛。头状花序3-6个密集；苞叶6-10，星状排列；总苞被白色蛛丝状毛；总苞片倒卵状披针形，顶端褐色；冠毛白色。花期7-9月，果期10月。生海拔2900-3600米的较湿润的河边。产四川西部。

Perennial herbs, caespitose. Rhizome branched, with rosette suckers and flowering stems. Leaves spatulate-lanceolate, dense white tomentum. Capitula 3-6, densely congested; bracteal leaves 6-10, forming a star; involucre densely white arachnoid tomentose; phyllaries obovate-lanceolate, apex dark brown; pappus white. Fl. Jul-Sep. Fr. Oct. Moist streamsides at 2900-3600 m. Distributed in W Sichuan.

亚灌木火绒草
Leontopodium suffruticosum Y. L. Chen

亚灌木。叶线状匙形或线形，被白色蛛丝状毛和腺毛。头状花序3-5(-7)，雌雄异体，密集；苞叶少数，线形，星状排列；总苞半球形；苞片顶端棕色；冠毛白色，比花冠略长。花期6-8月，果期8-9月。生海拔约3200米的干枯河床。产西藏东部。

Subshrubs. Leaves linear-spatulate or linear, white arachnoid tomentose and glandula. Capitula

弱小火绒草 *Leontopodium pusillum*

匍枝火绒草 *Leontopodium stoloniferum*

3-5(-7), heterogamous or dioecious, densely congested; bracteal leaves few, linear, forming a star; involucre hemispheric; phyllaries apex brown; corolla yellow; pappus white, slightly longer than corollas. Fl. Jun-Aug. Fr. Aug-Sep. Dry riverbeds at ca. 3200 m. Distributed in E Xizang.

亚灌木火绒草 *Leontopodium suffruticosum*

松毛火绒草 *Leontopodium andersonii*

松毛火绒草

Leontopodium andersonii C. B. Clarke

多年生草本，被白色茸毛。根出条木质，有顶生密集的缨状叶丛。花茎多数，长5-30厘米。叶条形或条状钻形；苞叶多数，披针形，被长柔毛状厚茸毛。头状花序直径3-4毫米，10-40个密集成复伞房状或团伞状；总苞被白色厚茸毛；冠毛白色。花期8-9月，果期9-10月。生海拔1000-3600米的荒原、草甸、砾石坡或林缘。产贵州、云南和四川。

Perennial herbs, white tomentose. Stolons woody, apex with dense fringed leaves. Floral stems numerous, 5-30 cm long. Leaves linear or linear-subulate; bracteal leaves numerous, lanceolate, thickly villose-tomentose. Capitula 3-4 mm diam, 10-40 congested in corymb or glomerate; involucres thickly white tomentose; pappus white. Fl. Aug-Sep. Fr. Sep-Oct. Deserts, meadows, gravelly slopes or forest edges at 1000-3600 m. Distributed in Guizhou, Yunnan and Sichuan.

薄雪火绒草 *Leontopodium japonicum*

薄雪火绒草

Leontopodium japonicum Miq.

多年生草本，被白色薄绒毛。茎数个，簇生，直立。叶狭披针形，基部急狭，基出3-5脉；苞叶多数，卵圆形或长圆形，密被灰白色茸毛。头状花序小，多数，疏散；总苞钟形或半球形，密被白色茸毛；总苞片顶端钝，无毛；冠毛白色。花期6-9月，果期9-10月。生海拔700-2300米的山地、灌丛、草坡和林下。产华东、华中和秦岭。日本亦有。

Perennial herbs, thinly white tomentose. Stems several, clustered, erect. Leaves narrowly lanceolate, base acute-attenuate, basal 3-5 veined; bracteal leaves numerous, ovate or oblong, densely offwhite tomentose. Capitula small, numerous, lax; involucres campanulate or semiglobose, densely white tomentose; phyllaries obtuse and glabrous at apex; pappus white. Fl. Jun-Sep. Fr. Sep-Oct. Montanes, thickets, grassy slopes and forests at 700-2300 m. Distributed in E and C China, and Qingling Mountain. Also in Japan.

华火绒草 *Leontopodium sinense*

华火绒草
Leontopodium sinense Hemsl.

多年生草本，被白色或黄白色茸毛。茎数个，簇生，较粗壮，具腋芽或密叶小枝。叶矩圆状条形，基部狭耳形，无柄，上面黄绿色至黑绿色，下面被白色或黄白色厚茸毛；苞叶多数，卵圆状长圆形。头状花序较大，不等形；总苞被白色茸毛。花期7-9月，果期9-11月。生海拔(700-)1300-3600米的干旱草地、草甸、灌丛或林中。产云南、四川和贵州。

Perennial herbs, white or yellowish-white tomentose. Stems numerous, clustered, somewhat stout, with axillary buds or densely foliate twigs. Leaves oblong-linear, base narrowly auriculate, sessile, yellowish-green to dark green adaxially, densely white or yellowish-white tomentose abaxially; bracteal leaves numerous, ovate-oblong. Capitula larger, varied in shape; involucres white tomentose. Fl. Jul-Sep. Fr. Sep-Nov. Dry grasslands, meadows, thickets or forests at (700-)1300-3600 m. Distributed in Yunnan, Sichuan and Guizhou.

香芸火绒草
Leontopodium haplophylloides Hand.-Mazz.

多年生草本。茎多数，簇生。叶草质，黑绿色，披针形，两面被短茸毛，下面杂有黑色长腺毛；苞叶多数，披针形。头状花序直径约5毫米，5-7个密集；总苞被白色柔毛状茸毛；总苞片顶端无毛，宽，尖，褐色；花冠白色；冠毛白色。花期8-9月，果期9-10月。生海拔2400-4000米的高山草地、石砾地、灌丛或林缘。产四川、甘肃和青海。

Perennial herbs. Stems numerous, clustered. Leaves herbaceous, dark green, lanceolate, both surfaces shortly tomentose, mixed with black long glandular hairs abaxially; bracteal leaves numerous, lanceolate. Capitula ca. 5 mm diam, 5-7 congested; involucres white villose-tomentose; apex of phyllaries glabrous, broad, acute, and brown; corolla white; pappus white. Fl. Aug-Sep. Fr. Sep-Oct. Alpine grasslands, stony places, thickets or forest edges at 2400-4000 m. Distributed in Sichuan, Gansu and Qinghai.

香芸火绒草 *Leontopodium haplophylloides*

坚杆火绒草
Leontopodium franchetii Beauv.

多年生草本，密被黄色腺毛。茎簇生，不分枝，挺直。叶密集，直立，条形，边缘极反卷，中脉显著，下面密被白色棉毛；苞叶多数，条形，密被白色茸毛。头状花序径3-5毫米，多数；总苞被疏棉毛；总苞片顶端尖，无毛；花冠浅黄色；冠毛白色。花期7-9月，果期9-10月。生海拔3000-4000米的干燥草地、石砾坡地和河滩湿地。产云南和四川。

Perennial herbs, densely yellow glandular hairy. Stems clustered, unbranched, straight. Leaves dense, erect, linear, margin strongly revolute, midrib distinct, densely white villose abaxially; bracteal leaves numerous, linear, densely white tomentose. Capitula 3-5 mm diam, numerous; involucres sparsely villose; phyllaries acute and glabrous at apex; corolla yellowish; pappus white. Fl. Jul-Sep. Fr. Sep-Oct. Dry grasslands, gravelly slopes or flood wetlands at 3000-4000 m. Distributed in Yunnan and Sichuan.

毛香火绒草
Leontopodium stracheyi (Hook. f.) C. B. Clarke ex Hemsl.

多年生草本，被黄褐色短腺毛。叶较薄，卵状披针形或卵状条形，基部宽圆形或扩大，近心形抱茎，上面绿色，被密腺毛，下面被灰白色茸毛；苞叶多数，卵形或卵状披针形。头状花序直径4-5毫米；总苞被长柔毛；冠毛基部褐色。花期7-9月，果期9-10月。生海拔

坚杆火绒草 *Leontopodium franchetii*

2000-4700米的山谷、溪岸、草地、砾石坡、灌丛或外缘。产云南、四川和西藏。喜马拉雅亦有。

Perennial herbs, shortly yellowish-brown glandular hairy. Leaves thin, ovate-lanceolate or ovate-linear, base broadly rotundate or dilated, subcordately amplexicaul, green adaxially, densely glandular hairy, offwhite tomentose abaxially; bracteal leaves numerous, ovate or ovate-lanceolate. Capitula 4-5 mm diam; involucres villose; pappus brown at base. Fl. Jul-Sep. Fr. Sep-Oct. Valleys, riversides, grasslands, gravelly slopes, thickets or forest edges at 2000-4700 m. Distributed in Yunnan, Sichuan and Xizang. Also in Himalaya.

毛香火绒草 *Leontopodium stracheyi*

艾叶火绒草 *Leontopodium artemisiifolium*

艾叶火绒草

Leontopodium artemisiifolium (Lévl.) Beauv.

多年生草本。茎多数，木质，直立，不分枝。茎生叶19-35，长圆状披针形。头状花序在茎端组成疏松的复合伞房花序；苞叶11-13，披针形，密被白色茸毛；总苞半球形，直径4-6毫米；总苞片3层；冠毛白色。花期8-9月，果期9-10月。生海拔2100-3200米的草坡、牧场、林缘、山谷或溪边。产云南北部和四川西部。

Perennial herbs. Stems several, woody, erect, not branched. Cauline leaves 19-35, oblong-lanceolate. Capitula in dense corymbs at ends of stems or in sparsely compound corymbs; bracteal leaves 11-13, lanceolate, densely white tomentose; involucre subglobose, 4-6 mm diam; phyllaries 3-seriate; pappus white. Fl. Aug-Sep. Fr. Sep-Oct. Grasslands, forest margins, riverbanks at 2100-3200 m. Distributed in N Yunnan and W Sichuan.

矮火绒草 *Leontopodium nanum*

戟叶火绒草
Leontopodium dedekensii

戟叶火绒草
Leontopodium dedekensii (Bur. et Franch.) Beauv.

多年生草本，密被蛛丝状毛或灰白色茸毛。叶条形，基部宽大，心形或箭形，抱茎，边缘波状；苞叶多数，披针形或条形。头状花序直径4-5毫米，5-30个密集；总苞密被白色长柔毛状茸毛；总苞片顶端无毛，干膜质；冠毛白色。花期6-7月，果期7-9月。生海拔1400-4100米的林中、灌丛或草地。产云南、四川、西藏、贵州、湖南、陕西和甘肃。缅甸北部亦有。

Perennial herbs, densely arachnoid hairy or offwhite tomentose. Leaves linear, base dilated, cordate or sagittate, amplexicaul, margin undulate; bracteal leaves numerous, lanceolate or linear. Capitula 4-5 mm diam, 5-30 congested; involucres densely white villose-tomentose; phyllaries glabrous at apex, dry membranous; pappus white. Fl. Jun-Jul. Fr. Jul-Sep. Forests, thickets or grasslands at 1400-4100 m. Distributed in Yunnan, Sichuan, Xizang, Guizhou, Hunan, Shaanxi and Gansu. Also in N Myanmar.

矮火绒草
Leontopodium nanum (Hook. f. et Thoms.) Hand.-Mazz.

多年生草本，被白色绵毛状厚茸毛，垫状丛生。莲座状叶丛存在；叶匙形或条状匙形，被长柔毛状密茸毛；苞叶少数。头状花序直径6-13毫米，单生或3个密集；总苞被绵毛；冠毛亮白色。花期5-6月，果期6-8月。生海拔2100-5000米的湿地、泥炭地或砾石坡。产青藏高原。印度北部、巴基斯坦和中亚亦有。

Perennial herbs, densely white villose-tomentose, pulvinate-clustered. Rosette leaves present; leaves spathulate or linear-spathalate, densely villose-tomentose; bracteal leaves several. Capitula 6-13 mm diam, solitary or 3 congested; involucres villose; pappus lightly white. Fl. May-Jun. Fr. Jun-Aug. Wetlands, peatlands or gravelly slopes at 2100-5000 m. Distributed in Tibetan Plateau. Also in N India, Pakistan and C Asia.

长叶火绒草
Leontopodium junpeianum Kitam.

多年生草本，被柔毛或密茸毛。根状茎短。莲座状叶丛存在；莲座叶或茎基部叶狭匙形，基部宽大成紫红色无柄的长鞘部；中部叶条形；苞叶多数，披针形，被长柔毛状茸毛；冠毛白色。花期7-8月，果期9-10月。生海拔1100-4800米的草地、洼地、灌丛或岩石上。产中国西南、华北和西北。克什米尔地区亦有。

Perennial herbs, villose or densely tomentose. Rhizomes short. Rosette leaves present; rosette and lower cauline leaves narrowly spathulate, base dilated, purple, sessile, sheathlike; middle leaves linear; bracteal leaves numerous, lanceolate, villose-tomentose; pappus white. Fl. Jul-Aug. Fr. Sep-Oct. Grasslands, lowlands, thickets or rocks at 1100-4800 m. Distributed in SW, N and NW China. Also in Kashmir.

长叶火绒草 *Leontopodium junpeianum*

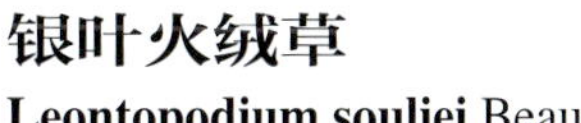

银叶火绒草 *Leontopodium souliei*

银叶火绒草

Leontopodium souliei Beauv.

多年生草本，被白色蛛丝状长柔毛。莲座状叶丛存在；叶狭条形或舌状条形，被银白色绢状茸毛，上部叶基部扩大，半抱茎，被长柔毛；苞叶多数，密集。头状花序直径5-7毫米，少数密集；总苞被长柔毛状密茸毛；总苞片先端无毛，褐色；冠毛白色。花期7-8月，果期9月。生海拔2700-4500米的林地、灌丛、湿润草地、沼泽地、山谷或溪岸。产中国西南。

Perennial herbs, white arachnoid villose. Rosette leaves present; leaves narrowly linear or ligulate-linear, argent sericeous-tomentose, upper leaves dilated, semi-amplexicaul and villose at base; bracteal leaves numerous, dense. Capitula 5-7 mm diam, several congested; involucres densely villose-tomentose; phyllaries glabrous and brown at apex; pappus white. Fl. Jul-Aug. Fr. Sep. Forests, thickets, damp grasslands, swamps, valleys or riversides at 2700-4500 m. Distributed in SW China.

绢茸火绒草

Leontopodium smithianum Hand.-Mazz.

多年生草本，被白色绵毛或密绢毛。地下茎粗壮；茎多数，簇生。叶直立，条形或条状披针形，无柄；苞叶多数，矩圆形或条形，组成稀疏不整齐的苞叶群。头状花序直径7-10毫米，3-7个密集；总苞半球形，被白色绵毛；冠毛基部稍黄色。花期6-8月，果期8-10月。生海拔1600-2900米的草地或干燥山坡。产河北、山西、内蒙古、陕西和甘肃。

Perennial herbs, white villose or densely sericeous. Rhizomes stout; stems numerous, clustered. Leaves erect, linear or linear-lanceolate, sessile; bracteal leaves numerous, oblong or linear, arranged in sparse, irregular group. Capitula 7-10 mm diam, 3-7 congested; involucres semiglobse, white villose; pappus yellow at base. Fl. Jun- Aug. Fr. Aug-Oct. Grasslands or dry slopes at 1600-2900 m. Distributed in Hebei, Shanxi, Neimenggu, Shaanxi and Gansu.

绢茸火绒草 *Leontopodium smithianum*

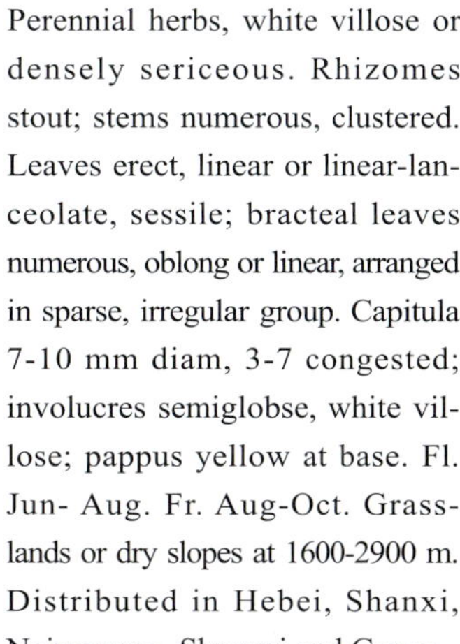

火绒草 *Leontopodium leontopodioides*

火绒草

Leontopodium leontopodioides (Willd.) Beauv.

多年生草本，被长柔毛或绢状毛。叶直立，条形或条状披针形，无柄，上面灰绿色，被柔毛，下面被白色密绵毛；苞叶少数，矩圆形或条形，被白色厚茸毛。头状花序3-7个密集；总苞半球形，被白色绵毛；冠毛基部黄色。花期6-9月，果期8-10月。生海拔100-3800米的草原、黄土坡地或草地。产华北、东北和西北。东北亚亦有。

Perennial herbs, villose or sericeous. Leaves erect, linear or linear-lanceolate, sessile, glaucous and pubescent adaxially, densely white villose abaxially; bracteal leaves several, oblong or linear, thickly white tomentose. Capitula 3-7 congested; involucres semiglobse, white villose; pappus yellow at base. Fl. Jun-Sep. Fr. Aug-Oct. Grasslands, loess slopes or grasslands at 100-3800 m. Distributed in N, NE and NW China. Also in NE Asia.

美头火绒草

Leontopodium calocephalum Beauv.

多年生草本。茎不分枝，高10-50厘米。基生叶披针形或线状披针形，基部成为疏松抱茎的鞘。头状花序5-20，聚生；苞叶10-18，线形，形成直径5-8厘米的星状苞叶群，苞叶两面密被白色或黄色厚绒毛，基部较宽，先端渐尖。花果期7-10月。生海拔2600-4200米的草甸、石砾坡地、湖岸、沼泽地、灌丛、冷杉和其他针叶林下或林缘。产青海东部、甘肃西南部、四川和云南西北部。

Perennial herbs. Stems unbranched, 10-50 cm tall. Basal Leaves lanceolate or linear-lanceolate, base forming sparse sheath around stem. Capitula 5-20, closely aggregated; bracteal leaves 10-18, linear, forming a multiradiate star of 5-8 cm diam, both surfaces densely white or yellowish tomentose, base broader, apex acuminate. Fl. and fr. Jul-Oct. Alpine meadows, grasslands, thickets, marshes, conifer forests, gravelly slopes, lake banks at 2600-4200 m. Distribuated in E Qinghai, SW Gansu, Sichuan and NW Yunnan.

美头火绒草 *Leontopodium calocephalum*

云岭火绒草

Leontopodium delavayanum Hand.-Mazz.

多年生草本，高6-13厘米。茎直立，不分枝。叶开展，披针形或披针状长圆形；苞叶多数，披针形，开展成径4-6.5厘米的星状苞叶群。头状花序半球形，多数密集；总苞长约4毫米；总苞片约3层。花期7-8月。生海拔3800-4000米的高山山顶石砾草地或岩石上。产云南西北部。缅甸东北部亦有。

云岭火绒草 *Leontopodium delavayanum*

Perennial herbs, 6-13 cm tall. Stems erect, unbranched. Leaves lanceolate or oblong-lanceolate, bracteal leaves numerous, lanceolate, forming a multiradiate star of 4-6.5 cm diam. Capitula hemispherical, numerous, dense; involucre ca. 4 mm; phyllaries 3-seriate. Fl. Jul-Aug. Alpine gravelly slopes, rocky places at 3800-4000 m. Distributed in NW Yunnan. Also in NE Myanmar.

鼠麴火绒草 *Leontopodium forrestianum*

鼠麴火绒草

Leontopodium forrestianum Hand.-Mazz.

多年生草本。花茎直立，高2-10厘米，不分枝或基部有分枝。叶线形、长圆状或倒披针状线形。头状花序径密集成伞房花序或单生；苞叶多数，与上部叶等大或略宽。花期7-9月，果期9-10月。生海拔3400-4000米的高山亚高山干燥草地、石砾地及灌丛的边缘。产云南西北部和西藏东南部。缅甸北部亦有。

Perennial herbs. Stems slender, erect, 2-10 cm tall, simple or branching at base. Leaves linear or oblong to oblanceolate. Capitula arranged in dense corymbs or solitary; bracteal leaves numerous, as large as upper leaves or broader. Fl. Jul-Sep. Fr. Sep-Oct. Alpine grasslands, gravelly slopes and thickets at 3400-4000 m. Distributed in NW Yunnan and SE Xizang. Also in N Myanmar.

珠峰火绒草 *Leontopodium himalayanum*

珠峰火绒草

Leontopodium himalayanum DC.

多年生垫状草本。茎生叶线形或线状披针形；苞叶披针状线形，形成径2-7厘米苞叶群。头状花序3-5个，密集。花期7-8月。生海拔3000-5100米石砾坡地、岩石缝隙或湿润草地。产云南西北部和西藏南部。喜马拉雅亦有。

Perennial herbs, densely tufted. Cauline leaves linear to linear-lanceolate; bracteal leaves lanceolate-filiform, forming a multiradiate star of 2-7 cm diam. Capitula 3-5, crowded. Fl. Jul-Aug. Alpine gravelly slopes, meadows at 3000-5100 m. Distributed in NW Yunnan and S Xizang. Also in Himalaya.

雅谷火绒草

Leontopodium jacotianum Beauv.

多年生草本，高6-28厘米。叶片线形、披针形或长圆状批针形。苞叶批针形至长圆状椭圆形，密被白色绒毛，开展成径2-3厘米的星状苞叶群。头状花序4-9(-18)个密集；总苞长3-4毫米，被长柔毛；冠毛白色，羽毛状。花期7-8月，果期8-10月。生海拔2200-4400米的高山草甸和石砾地。产西藏南部。喜马拉雅亦有。

Perennial herbs, 6-28 cm tall. Leaves linear, lanceolate or oblong-lanceolate. Bracteal leaves lanceolate to oblong-elliptic, more densely whitish tomentose than cauline leaves. Capitula 4-9(-18), usually densely crowded; pappus white, plumose. Fl. Jul-Aug. Fr. Aug-Oct. Alpine meadows, gravelly slopes at 2200-4400 m. Distributed in S Xizang. Also in Himalaya.

雅谷火绒草 *Leontopodium jacotianum*

川西火绒草

Leontopodium wilsonii Beauv.

多年生草本。茎木质，无分枝。叶线状披针形；苞叶多数，密集，形成直径4-5.5 厘米的苞叶群，两面密被灰白色绒毛。头状花序7-11个，疏散。总苞长约4毫米，被白色长柔毛。花期6-9月。生海拔2000-2500米的草坡、灌丛或岩石上。产四川西部和甘肃南部。

Perennial herbs. Stems woody, not branched. Leaves linear-lanceolate. Capitula 7-11; bracteal leaves 15-20, densely arranged, oblong, larger than upper leaves, forming a star of 4-5.5 cm diam, both surfaces densely grayish white tomentose, apex acute. Fl. Jun-Aug. Fr. Sep-Oct. Grasslands, thickets or rocks at 2000-2500 m. Distributed in W Sichuan and S Gansu.

川西火绒草 *Leontopodium wilsonii*

蝶须 *Antennaria dioica*

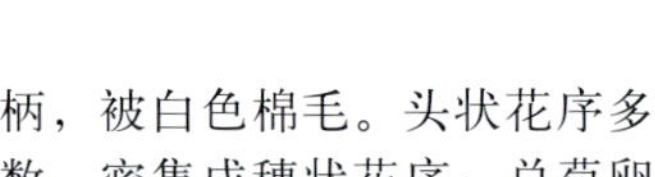
多茎鼠麴草 *Gnaphalium polycaulon*

蝶须

Antennaria dioica (L.) Gaertn.

多年生草本。茎直立，高6-25厘米，不分枝，被密棉毛。茎基部叶匙形，上部叶线形。头状花序通常3-5个，排列成伞房花序；雌株的头状花序总苞片约5层；雄株总苞片仅3层；雄花花冠管状，有5裂片。花期5-8月。生海拔600-2700米的草甸、干燥多沙石山坡或针叶林下。产甘肃、新疆和黑龙江。日本、俄罗斯、蒙古、哈萨克斯坦、欧洲和美国(阿拉斯加)亦有。

Perennial herbs. Stems 6-25 cm tall, simple, densely villose. Basal leaves spatulate or rhombic-spatulate; cauline leaves linear. Capitula 3-5, arranged in dense corymb; female capitula phyllaries ca. 5-seriate; male capitula 3-seriate only; staminate flowers tubular, lobes 5. Fl. May-Aug. Dry slopes on tundra, meadows or conifer forests at 600-2700 m. Distributed in Gansu, Xinjiang and Heilongjiang. Also in Japan, Russia, Mongolia, Kazakhstan, Europe and USA (Alaska).

多茎鼠麴草

Gnaphalium polycaulon Pers.

一年生草本，高10-25厘米。茎基部多分枝，下部匍匐或斜升。叶倒披针形或长圆形，无柄，被白色棉毛。头状花序多数，密集成穗状花序；总苞卵形；总苞片麦秆黄色或污黄色，膜质。瘦果圆柱形；冠毛污白色。花期1-4月。生耕地、草地或湿润山地上。产中国西南、华南和华东。印度、泰国、澳大利亚、埃及和热带非洲亦有。

Annual herbs, 10-25 cm tall. Stems branched at base, lower part creeping or obliquely ascending. Leaves oblanceolate or oblong, sessile, white sericeous. Capitula numerous, congested in spike; involucres ovoid; phyllaries straw yellow or dirty yellow, membranous. Achenes cylindric; pappus dirty white. Fl. Jan-Apr. Fields, grasslands or damp montanes. Distributed in SW, S and E China. Also in India, Thailand, Australia, Egypt and tropical Africa.

矮鼠麴草

Gnaphalium stewartii C. B. Clarke ex Hook. f.

草本。基部簇生，具绵毛。叶条形，两面具白色绵毛。头状花序少数排成总状；苞片2或3层；雌花冠线形，顶部3齿裂；两性花少量；花冠管状，冠檐齿裂。瘦果圆柱形；冠毛白色。花期6-9月。生海拔2500-

矮鼠麴草 *Gnaphalium stewartii*

尖叶香青 *Anaphalis acutifolia*

4000米的高山草甸。产西藏和新疆。南亚和西南亚亦有。

Herbs. Base tufted and lanate. Leaves linear, both surfaces white lanate; Heads few, in racemose; phyllaries in 2 or 3 series; female florets corolla filiform, apex 3-toothed; bisexual florets few; corolla tubular, limbs toothed. Achenes cylindrical; pappus white. Fl. Jun-Sep. Alpine meadows at 2500-4000 m. Distributed in Xizang and Xinjiang. Also in S and SW Asia.

尖叶香青

Anaphalis acutifolia Hand.-Mazz.

多年生草本。茎高12-23厘米，基部稍木质，被灰白色蛛丝状密毛。中部叶多少开展、线形或线状披针形，基部渐狭，不沿茎下延，边缘平，顶端渐尖，草质，两面被灰白色或幼时淡黄色的棉毛及腺毛，有1中脉。头状花序5-10个，密集成5-10毫米的团球状伞房花序；总苞球状；总苞片7-8层，放射状开展，近等长。花果期7-9月。生海拔3900-4200米的砾坡地和沙地。产西藏东南部。

Perennial herbs. Stem 12-23 cm tall, base minutely woody, white arachnoid-villose. Middle leaves spreading, linear or filiform, base narrowly, margin entire, apex acuminate, herbaceous, both surfaces appressed yellowish white cottony and glandular punctuate, midrib 1. Capitula 5-10, arranged in terminal clustered-glomerate corymb; involucres globose; phyllaries 7-8 rows, nearly equal. Fl. Jul-Sep. Gavelly slopes and sandy grounds at 3900-4200 m. Distributed in SE Xizang.

黄腺香青

Anaphalis aureopunctata Lingelsh. et Borza

多年生草本。茎草质或基部木质，被白色或灰白色蛛丝状棉毛。叶离基三或五出脉，或有单脉。头状花序多数成复伞房状；总苞钟状或狭钟状；花托有缝状凸起。花期7-9月，果期9-10月。生海拔1000-4200米的林下、林缘、草坡、岩石坡。产中国西南、华中和华南。

Perennial herbs. Stems herbaceous or woody at base, white or pallid arachnoid tomentose, or lower parts somewhat glabrous. Leaves 3- or 5-veined, or 1-veined. Capitula numerous, compound corymbiform; involucre campanulate or narrowly campanulate; receptacle with fimbrillate appendage. Fl. Jul-Sep. Fr. Sep-Oct. Forests, forest margins, hills, grasslands, bamboo or grassy slopes, rocky places, valleys, wetlands at 1000-4200 m. Distributed in SW, C and S China.

蛛毛香青

Anaphalis busua DC.

二年生草本。茎直立，高50-130厘米。中部叶线形或线状披针形，基部近等宽，下延成翼柄，顶端尖。头状花序极多数，在枝端密集成复伞房状；总苞片4-5层。花期9-10月，果期10月。生海拔1500-2800米的山谷、坡地、林地或草地。产中国西南。喜马拉雅亦有。

Biennial herbs. Stems erect, 50-130 cm tall. Middle leaves linear or linear-filiform, basal equal wide, decurrent as winged petioles, apex acuminate. Capitulal numerous, arranged in terminal clustered corymb; phyllaries 4-5 rows. Fl. Sep-Oct. Fr. Oct. Valleys of damp, slopes, forests or grasslands at 1500-2800 m. Distributed in SW China. Also in Himalaya.

黄腺香青 *Anaphalis aureopunctata*

蛛毛香青 *Anaphalis busua*

苞衣香青 *Anaphalis chlamydophylla*

伞房香青 *Anaphalis corymbifera*

苞衣香青

Anaphalis chlamydophylla Diels

半灌木状草本，高3-25厘米，全株被白色或灰白色棉毛。基部叶卵圆形、长圆形或匙状长圆形；中部叶线状披针形，基部沿茎下延成狭翅；上部叶小，线形。头状花序7-20个，排成复伞房状；总苞宽钟状；总苞片4-5层。花期7-8月，果期9月。生海拔2700-3700米的亚高山草甸、针叶林或稀疏杂木林中和石灰岩层上。产云南西北部。

Suffruticose herbs, 3-25 cm tall, white or whit cottony. Basal leaves ovate, narrowly ovate or spathualte-oblong; middle leaves lanceolate, and decurrent as narrow wings on stems; upper leaves small, linear. Capitula 7-20, arranged in compound corymb; involucre broadly campanulate; phyllaries 4-5 rows. Fl. Jul-Aug. Fr. Sep. Subalpine meadows, copses or calcareous ground at 2700-3700 m. Distributed in NW Yunnan.

伞房香青

Anaphalis corymbifera C. C. Chang

多年生草本。根状茎匍枝，有披针形膜质鳞片及顶生的莲座状叶丛。叶两面异色，倒卵形、匙形、倒披针状长圆形或椭圆形，有明显的离基三出脉。头状花序2-5个密集于枝端，排列成复伞房状；总苞钟状。花期6-8月。生海拔3000-3200米的高山草地和荒地。产云南西北部。缅甸东北亦有。

Perennial herbs. Rhizome repent, with lanceolate membranous scales and terminal rosette leaves. Leaves discolorous, obovate, spatulate, oblanceolate-oblong or elliptic, conspicuously 3-veined. Capitula 2-5 congested and terminal on branches, compound corymbiform; involucre campanulate. Fl. Jun-Aug. Alpine grasslands and deserts at 3000-3200 m. Distributed in NW Yunnan. Also in NE Myanmar.

江孜香青 *Anaphalis deserti*

江孜香青

Anaphalis deserti J. R. Drumm.

多年生草本，高30-40厘米。中部叶长圆状线形，基部等宽或稍狭，稍沿茎下延成狭翅，边缘平，顶端尖或近圆形，有长尖头；上部叶较小，上部暗绿色，中脉不明显。头状花序密集成团球状的伞房花序；总苞宽钟状；总苞片4-5层。花期6-8月。生海拔3900-4000米的草坡或林下。产西藏。

Perennial herbs, 30-40 cm tall. Middle leaves oblong-linear, base equilateral or slightly narrow, decurrent on stems and becoming a narrow wing, margin flat, apex acute or subrounded, with long cusp; upper leaves considerably small. Capitula congested to globose-corymbiform synflorescences; involucres campanulate; phyllaries 4-5-seriate. Fl. Jun-Aug. Grassy slopes or forests at 3900-4000 m. Distributed in Xizang.

乳白香青

Anaphalis lactea Maxim.

多年生草本，高10-40厘米。莲座叶披针状或匙状长圆形；茎下部叶稍小；中部及上部叶长椭圆形到线形。头状花序多数，密集成复伞房状；总苞钟状；总苞片4-5层。花果期6-9月。生海拔2000-3400米的亚高山及低山草地及针叶林下。产中国西北。

Perennial herbs, 10-40 cm tall. Rosette leaves lanceolate or spatulate-oblonge; lower leaves smaller than rosette leave; middle and upper leaves oblong linear-lanceolate to linear. Capitula numerous, densely compound corymbiform; involucres campanulate; phyllaries 4-5-seriate. Fl. and fr. Jun-Sep. Subalpine or low mountain grasslands or taiga at 2000-3400 m. Distributed in NW China.

乳白香青 *Anaphalis lactea*

丽江香青 *Anaphalis likiangensis*

德钦香青

Anaphalis larium Hand.-Mazz.

多年生草本。根状茎木质，被膜质卵圆形鳞片状叶，有顶生的莲座状叶丛。全部叶质稍厚，两面被浅黄色密棉毛，有向叶上部渐消失的离基三出脉。头状花序达10或多数，密集成复伞房状；总苞宽钟状；雄株头状花序只有雄花。花期8-9月。生海拔3000-4300米的高山或亚高山干燥坡地。产云南西北部。

Perennial herbs. Rhizome woody, with membranous oval squamiform leaves, rosette leaves terminal. Leaves thickish, densely light yellow lanate, 3-veined and veins gradually disappearing toward apex. Capitula 10 or numerous, dense, compound corymbiform; involucre broadly campanulate; predominantly male capitula with male florets only. Fl. Aug-Sep. Alpine or subalpine dry slopes at 3000-4300 m. Distributed in NW Yunnan.

丽江香青

Anaphalis likiangensis (Franch.) Y. Ling

多年生草本。根状茎木质。根出条或匍枝具鳞片状枯叶和顶生的莲座状叶丛。叶两面异色，顶部叶有1脉。头状花序10-30个，密集成复伞房状；花托有缝状短毛；雌株头状花序有多层雌花，中央有1-3个雄花。花期8-9月。生海拔3100-3400米的山谷、沟旁、草地或云杉林下。产云南西北部。

Perennial herbs. Rhizome woody. Sobols and stolons with dense scalelike withered leaves and terminal rosette leaves. Leaves discolorous, apical leaves 1-veined. Capitula 10-30, dense, corymbiform; receptacle with fimbrillate short hairs; center of predominantly female capitula with 1-3 male florets. Fl. Aug-Sep. Valleys, ditchsides, grasslands, under *Picea* forests at 3100-3400 m. Distributed in NW Yunnan.

木里香青

Anaphalis muliensis (Hand.-Mazz.) Hand.-Mazz.

多枝亚灌木。花枝生于不育枝的顶端，被白色厚棉毛。叶上面被蛛丝状毛或有时脱毛。头状花序5-25个，密集成复伞房状；总苞宽钟状，直径5-7毫米；花托有膜质凸起；雌株头状花序中央有1-2雄花。花期6-9月，果期9-10月。生海拔3400-4000米的高山针叶林下、草地或岩石上。产四川西南部和云南西北部。尼泊尔亦有。

Subshrubs, multibranched. Floriferous branches growing at tip of sterile branch, white tomentose. Leaves white tomentose abaxially. Capitula 5-25, densely compound corymbiform; involucre broadly campanulate, 5-7 mm diam; receptacle with membranous appendages; center of predominantly female capitula with 1-2 male florets. Fl. Jun-Sep. Fr. Sep-Oct. Alpine taiga, rocky places or grasslands at 3400-4000 m. Distributed in SW Sichuan and NW Yunnan. Also in Nepal.

德钦香青 *Anaphalis larium*

木里香青 *Anaphalis muliensis*

污毛香青 *Anaphalis pannosa*

污毛香青
Anaphalis pannosa Hand.-Mazz.

多年生草本。根状茎稍粗壮，灌木状。多分枝叶两面被黄褐色或褐色厚棉毛，有不明显的三出脉。头状花序7-10个，密集成团球伞房状；总苞宽钟状，直径约10毫米；花托有膜片状凸起；雌株头状花序中央有3-6个雄花。花期7-8月，果期9月。生海拔3800-4300米的高山干燥石砾坡地。产云南和四川西南部。

Perennial herbs. Rhizome thickish, much branched. Leaves fulvous brown tomentose, indistinctly 3-veined. Capitula 7-10, densely glomerulate; involucre broadly campanulate, ca. 10 mm diam; receptacle with chaffy process; predominantly female capitula center with 3-6 male florets. Fl. Jul-Aug. Fr. Sep. Alpine dry rocky slopes at 3800-4300 m. Distributed in Yunnan and SW Sichuan.

紫苞香青
Anaphalis porphyrolepis Y. Ling et Y. L. Chen

多年生草本。茎直立，高23-30厘米，被白色或灰白色密棉毛。下部叶鳞片状，长圆形，花期枯萎；中部以上叶向茎端渐大，长圆状披针形；顶部叶等大或稍小，密集于花序外围且排列成放射状。头状花序3-5个密集于茎端成径1.5-2厘米的伞房状，常无花序梗；总苞钟状，直径约5毫米；总苞片4-5层，紫红色或边缘稍白色。花期8月。生海拔约4000米的高山草甸。产西藏东南部。

Perennial herbs. Stems erect, 23-30 cm tall, white pallid tomentose. Lower leaves squamate, oblong, withered in efflorescence; leaves gradually larger from middle part to apex of stem, slightly expanding, oblong-lanceolate; uppermost leaves as large as or smaller than middle leaves, densely arranged in synflorescence, radiate. Capitula 3-5, densely arranged at top of stems, compound corymbiform, without peduncle; involucres campanulate, ca. 5 mm diam; phyllaries 4-5-seriate, mauve or margin white. Fl. Aug. Alpine meadows at ca. 4000 m. Distributed in SE Xizang.

须弥香青
Anaphalis royleana DC.

草本或亚灌木。叶无柄，长圆状线形或长圆状批针形，基部稍下延，背面杯白色蛛丝状密绒毛，正面近无毛。头状花序7-9个，排列成伞房状；总苞片无毛，雪白色，外层卵圆形，较短，中层卵状长圆形，较花长，内层线状长圆形。总苞直径5-7毫米，苞片雪白色。花期8-9月。生海拔3600-4200米的高山灌丛或草甸。产西藏。喜马拉雅亦有。

Herbs or subshrubs. Leaves sessile, somewhat adnate to stem, linear-oblong or oblong-lanceolate, abaxially white arachnoid tomentose or gray or reddish brown lanuginous, adaxially subglabrous, base weakly decurrent. Capitula 7-9, densely arranged in corymbiform synflorescences; involucres 5-7 mm diam; phyllaries snow white. Fl. Aug-Sep. Alpine meadows or thickets at 3600-4200 m. Distributed in

紫苞香青 *Anaphalis porphyrolepis*

须弥香青 *Anaphalis royleana*

灰叶香青 *Anaphalis spodiophylla*

蜀西香青 *Anaphalis souliei*

Xizang. Also in Himalaya.

蜀西香青

Anaphalis souliei Diels

多年生灌木状草本。茎直立，高5-30厘米，不分枝。莲座状叶披针形或倒卵状椭圆形；茎下部叶约与莲座状叶同形，但较小；中部和上部叶倒披针长圆形或线形。头状花序多数，密集成复伞房状；总苞宽钟状，径5-6毫米；总苞片5-6层。花果期6-9月。生海拔3000-4200米的高山及亚高山山坡山脊、草地或林下。产四川西部。

Perennial shrubby herbs. Stems erect, 5-30 cm tall, unbrabched. Rosette leaves lanceolate or obovate-elliptic; lower leaves homomorphic with rosette leaves, smaller; middle and upper leaves somewhat expanding, or erect and adherent to stem, oblanceolate-oblong or linear. Capitula numerous, densely compound corymbiform; involucres broadly campanulate, 5-6 mm diam; phyllaries 5-6-seriate. Fl. and Fr. Jun-Sep. Alpine or subalpine ridges, grasslands or forests at 3000-4200 m. Distributed in W Sichuan.

灰叶香青

Anaphalis spodiophylla Y. Ling et Y. L. Chen

多年生灌木状草本，茎高20-45厘米，被灰白色密棉毛。基部叶倒卵圆形；中部叶倒卵形或倒披针状匙形；上部叶苞状，披针状线形或线形，细尖；叶质稍厚。头状花序极多数，成复伞房状；总苞狭钟状，直径约4毫米；总苞片约5层，外层浅褐色，内层白色或污白色。花果期8-9月。生海拔3000-3600米的山谷林缘或山坡灌丛。产西藏东南部。

Perennial shrubby herbs, stems erect, 20-45 cm, densely hairy. Rosette leaves obovate; lower leaves caducous and lost by anthesis; middle leaves obovate, oblanceolate-spatulate; upper leaves bractlike, lanceolate-linear or linear, apiculate; all leaves slightly thickish. Capitula considerably numerous and dense, compound corymbiform; involucres narrowly campanulate, ca. 4 mm diam; phyllaries 5-seriate, outer ones light brownish, inner ones white or dirty white. Fl. and fr. Aug-Sep. Forests margin or thicket slopes at 3000-3600 m. Distributed in SE Xizang.

狭苞香青

Anaphalis stenocephala Ling et C. Shih

多枝亚灌木，高达60厘米。叶开展，椭圆状或长圆状线形，沿茎下延成狭长的翅，有不明显的离基三出脉或单脉。头状花序9个至多数密集成伞房花序；总苞圆筒状，直径2-3毫米。花果期7-8月。生海拔2600-3200米的亚高山干燥山地及松林下。产西藏和云南。

Subshrubs, multibranched, ca. 60 cm tall. Leaves expanding, elliptic or oblong-linear, indistinctly 1- or 3-veined, decurrent on stem into a long and narrow wing. Capitula 9 to numerous capitula densely forming corymbiform synflorescence; involucre cylindric, 2-3 mm diam. Fl. and fr. Jul-Aug. Mountains grassy slopes or *Pinus* forests at 2600-3200 m. Distributed in Xizang and Yunnan.

狭苞香青 *Anaphalis stenocephala*

四川香青

Anaphalis szechuanensis Ling et Y. L. Chen

多枝亚灌木。根状茎粗壮，扭曲，有顶生的莲座状叶丛及花茎。叶两面被白色或灰白色密棉毛，有不明显的三出脉或单脉。头状花序5-15个，密集成伞房状或复伞房状；总苞狭钟状，直径4-5毫米；雌株头状花序中央有5朵雄花。花期7-8月，果期9月。生海拔3500-4500米的高山石质草坡和石灰岩上。产四川西南部。

Subshrubs, multibranched. Rhizome thickish, contorted, withered terminal rosette leaves, and flowering stems. Leaves white-canescent tomentose, indistinctly 1- or 3-veined. Capitula 5-15, densely corymbiform or compound corymbiform; involucre narrowly campanulate, 4-5 mm diam; predominantly female capitula center with 5 male floretsy. Fl. Jul-Aug. Fr. Sep. Alpine rocky grassy slopes and calcareous ground at 3500-4500 m. Distributed in SW Sichuan.

四川香青 *Anaphalis szechuanensis*

黄绿香青 *Anaphalis virens*

黄绿香青

Anaphalis virens C. C. Chang

多枝亚灌木。根状茎粗壮，木质，有莲座状叶丛和密集丛生的花茎和不育茎。叶黄绿色，质薄，两面被具柄腺毛，有离基三出脉。头状花序极多数，成复伞房状；总苞宽钟形，直径5-8毫米。花期7-9月。生海拔1800-3600米的草坡或岩石间。产云南西北部和四川西南部。

Subshrubs. Rhizome thickish, woody, with rosette leaves and densely caespitose flowering stems and sterile stems. Leaves yellow-green, thin, with stalked glandular hairs, 3-veined. Capitula numerous, dense, compound corymbiform; involucre broadly campanulate, 5-8 mm diam; receptacle with fimbrillate short hairs. Fl. Jul-Sep. Hill grasslands or between rocks at 1800-3600 m. Distributed in NW Yunnan and SW Sichuan.

木根香青 *Anaphalis xylorhiza*

木根香青

Anaphalis xylorhiza Sch.Bip. ex Hook. f.

多年生灌木状草本。茎直立或匍匐，高3-17厘米，草质。莲座状叶与茎下部叶匙形；上部叶渐小，倒披针状或线状长圆形；全部叶两面被毛。头状花序5-10个密集成复伞房状；总苞宽钟状或倒锥状，直径约6毫米；总苞片约5层。花期6-9月，果期8-10月。生海拔3800-4000米的高山草甸和苔藓中。产西藏。不丹、印度北部和尼泊尔亦有。

Perennial shrubby herbs. Stems erect or creeping, 3-17 cm tall, herbaceous. Rosette leaves and lower leaves spatulate; upper leaves gradually smaller, oblanceolate or linear-oblong; all leaves sparsely canescent taupe tomentose on both surfaces. Capitula 5-10, densely compound corymbiform; involucres broadly campanulate or obconical ca. 6 mm diam; phyllaries 5-seriate. Fl. Jun-Sep. Fr. Aug-Oct. Alpine stonely meadows and lichen-covered areas at 3800-4000 m. Distributed in Xizang. Also in Bhutan, N India and Nepal.

云南香青

Anaphalis yunnanensis (Franch.) Diels

多枝亚灌木。根状茎粗壮，扭曲；花茎直立。叶两面被灰白色或黄白色密棉毛，有1脉或明显的离基三出脉。头状花序多数密集成复伞房状；总苞宽钟状，直径4-6毫米。花期6-9月，果期8-9月。生海拔2800-4000米的草地、林缘、湖岸及岩石上。产四川西南部和云南西北部。

Subshrubs, multibranched. Rhizome thickish, contorted; floriferous stems erect. Leaves pallid yellowish white tomentose, 1-veined or distinctly 3-veined. Capitula numerous, densely compound corymbiform; involucre broadly campanulate, 4-6 mm diam; receptacle with fimbrillate short hairs. Fl. Jun-Sep. Fr. Aug-Sep. Grassy s lopes, forest margins, banks or rocky slopes at 2800-4000 m. Distributed in SW Sichuan and NW Yunnan.

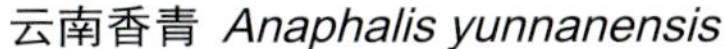
云南香青 *Anaphalis yunnanensis*

黏毛香青 *Anaphalis bulleyana*

黏毛香青

Anaphalis bulleyana (Jeffrey) C. C. Chang

一年生或二年生草本，具强烈芳香，被蛛丝状长棉毛和锈褐色的多节黏毛。叶倒卵形、倒披针形或倒卵状匙形，沿茎下延成宽翅，具3脉；总苞倒卵圆状，直径4-7毫米；总苞片4-5层，直立，浅褐色，膜质，透明，下部浅黄色，不开展。花期8-9月，果期9-10月。生海拔1100-3300米的亚高山阴湿坡地或低山草地。产云南、四川和贵州。

Annual or biennial herbs, strongly fragrant, covered with long arachnoid hairs and rusty septate-and viscid-hairs. Leaves obovate, oblanceolate or obovate-spathulate, decurrent along stems as broad wings, 3-veined; involucre obovid, 4-7 mm diam; phyllaries 4-5 rows, erect, brownish, membranous, transparent, base yellowish, not spreading. Fl. Aug-Sep. Fr. Sep-Oct. Subalpine damp slopes or montane grasslands at 1100-3300 m. Distributed in Yunnan, Sichuan and Guizhou.

旋叶香青

Anaphalis contorta (D. Don) Hook. f.

多年生草本，多分枝，被白色密棉毛和蛛丝状毛。叶密集，条形，基部宽大而抱茎，顶端渐尖。头状花序极多数；总苞钟形，直径4-6毫米；总苞片5-6层，浅黄褐色至白色，或带紫红色。花果期8-10月。生海拔1700-3500米的山坡和草地。产中国西南。喜马拉雅亦有。

Perennial herbs, much branched, densely white villose and arachnoid. Leaves clustered, linear, base widened and amplexicaul, apex acuminate. Capitula many; involucres campanulate, 4-6 mm diam; phyllaries 5-6 rows, pale brown to white, or tanged with purple. Fl. and fr. Aug-Oct. Slopes and grasslands at 1700-3500 m. Distributed in SW China. Also in Himalaya.

旋叶香青 *Anaphalis contorta*

苍山香青 *Anaphalis delavayi*

香青 *Anaphalis sinica*

苍山香青

Anaphalis delavayi (Franch.) Diels

多年生草本。叶下面密被蛛丝毛，上面疏被蛛丝毛，后除中脉及边缘外无毛。头状花序多，密集排成伞房状；总苞钟状，直径约5毫米；总苞5层；中央头状花序具雌花及少数两性不育花；雄性头状花序仅具雄花。花期5-8月，果期8-9月。生海拔3000-4000米的高山草地或林缘。产云南西北部。

Perennial herbs. Leaves densely arachnoid abaxially, sparsely arachnoid hairy adaxially, later glabrous except for midvein and margin. Capitula numerous, corymbiform, dense; involucre campanulate, ca. 5 mm diam; phyllaries arranged in 5 series; center of heads with pistillate florets and a few bisexual sterile florets; staminate heads with staminate florets only. Fl. May-Aug. Fr. Aug-Sep. Alpine grasslands or forest edges at 3000-4000 m. Distributed in NW Yunnan.

二色香青

Anaphalis bicolor (Franch.) Diels

多年生草本，密被白色厚茸毛。莲座状营养叶丛发达；花茎中部和上部叶条形或长圆状条形，基部沿茎下延成狭长的翅，具1脉。头状花序多数，在茎和枝端密集成复伞房花序；总苞钟状，直径6-8毫米；总苞片5-6层。瘦果长圆形。花期7-10月，果期9-11月。生海拔2000-3800米的高山草地、荒地、灌丛或针叶林下。产青藏高原。

Perennial herbs, densely white lanate. Vegetative rosette leaves developed; middle and upper cauline leaves linear or oblong-linear, base decurrent along stems as narrow wings, 1-veined. Capitula many, compound corymbiform, dense, terminal on stems and branches; involucres campanulate, 6-8 mm diam; phyllaries 5-6 rows. Fl. Jul-Oct. Fr. Sep-Nov. Alpine grasslands, wastelands, desert, thickets or coniferous forests at 2000-3800 m. Distributed in Tibetan Plateau.

香青

Anaphalis sinica Hance

多年生草本，被灰白色棉毛。匍匐枝长达8厘米。叶密生，长圆形、倒披针状长圆形或条形，基部沿茎下延成翅。头状花序多数；总苞钟状或近倒圆锥状，直径4-6毫米；总苞片6-7层，浅褐色至乳白色。花期7-9月，果期8-10月。生海拔400-2100米的灌丛、草地、山坡或溪边。产华南、华中、华东和华北。尼泊尔、朝鲜半岛和日本亦有。

Perennial herbs, white villose. Stolons to 8 cm long. Leaves dense, oblong, oblanceolate-oblong or linear, base decurrent as wings. Capitula many; involucre campanulate or obconical, 4-6 mm diam; phyllaries 6-7 rows, pale brown to milky white. Fl. Jul-Sep. Fr. Aug-Oct. Thickets, grasslands, slopes or streamsides at 400-2100 m. Distributed in S, C, E and N China. Also in Nepal, Korean Peninsula and Japan.

珠光香青

Anaphalis margaritacea (L.) Benth. et Hook. f.

多年生草本。叶稍革质，下面密被淡灰色至棕红色绵毛，上面被蛛丝毛。头状花序多，排成复伞房花序；总苞半球形，直径8-13毫米；总苞片5-7层。花果期7-11月。生海拔300-3400米的针叶林和白桦林中、旱土地、亚高山、低山草地、灌丛、多石沟谷或路边。产中

二色香青 *Anaphalis bicolor*

国西南、华中、华北和西北。南亚、东北亚和北美洲亦有。

Perennial herbs. Leaves slightly leathery, abaxially densely ash gray to reddish brown cottony tomentose, adaxially arachnoid. Capitula numerous, arranged in compound corymbiform; involucres semiglobose, 8-13 mm diam; phyllaries arranged in 5-7 series. Fl. and fr. Jul-Nov. *Conifer* and *Betula* forests, dry soils, subalpine, low mountain grasslands, shrubs, rocky valleys or roadsides at 300-3400 m. Distributed in SW, C, N and NW China. Also in S and NE Asia and North America.

纤枝香青

Anaphalis gracilis Hand.-Mazz.

半灌木，强烈分枝。腋芽和顶芽显著不同。茎纤细，不分枝，被蛛丝状毛。叶条形至披针形，基部下延成翅。头状花序5-50个；总苞狭钟状，直径4-5(-6)毫米；总苞片约6层，褐色至白色。花期7-8月，果期8-9月。生海拔2000-4000米的高山干旱坡地山谷或石砾地。产云南西北部和四川西部。

Subshrubs, strongly branched. Axillary buds and terminal buds distinct. Stems slender, simple, arachnoid. Leaves linear to lanceolate, base decurrent as wings. Capitula 5-50; involucres narrowly campanulate, 4-5(-6) mm diam; phyllaries ca. 6 rows, brown to white. Fl. Jul-Aug. Fr. Aug-Sep. Alpine dry slopes, valleys or stony places at 2000-4000 m. Distributed in NW Yunnan and W Sichuan.

绿香青

Anaphalis viridis Cumm.

垫状草本。根状茎多分枝。茎被灰白色棉毛。基生叶倒卵形或匙状椭圆形，茎中部和上部叶少数，披针形或倒披针状长圆形，基部下延为翅；所有叶两面黄绿色，被灰色或黄色棉毛。头状花序2-15个排列成团球伞房状或单生；总苞宽钟形；苞片4-5层；外层苞片褐色，被棉毛；中层和内层苞片白色；冠毛与管状花花冠等长。花期7-8月，果期8-9月。生海拔3000-4100米的山顶、高山冰蚀谷的岩石上或草坡。产四川西部。

珠光香青 *Anaphalis margaritacea*

Pulvinate herbs. Rhizome numerously branched. Stems canescent tomentose. Basal leaves obovate or spatulate-elliptic; middle and upper leaves few, lanceolate or oblanceolate-oblong, base decurrent on stem into a short cuneate wing; all leaves keeled on both surfaces, canescent yellow tomentose. Capitula 2-15, glomerulate or simple; involucre broadly campanulate; phyllaries 4- or 5-seriate; outer phyllaries brown, tomentose; middle and inner ones white; pappus equaling corolla. Fl. Jul-Aug. Fr. Aug-Sep. Mountain tops, granite of alpine glacial valleys, rocky or grassy slopes at 3000-4100 m. Distributed in W Sichuan.

纤枝香青 *Anaphalis gracilis*

绿香青 *Anaphalis viridis*

淡黄香青 *Anaphalis flavescens*

尼泊尔香青 *Anaphalis nepalensis*

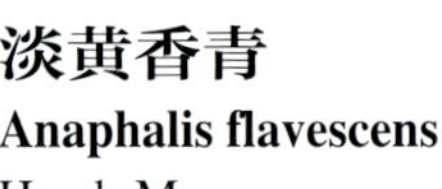

淡黄香青

Anaphalis flavescens Hand.-Mazz.

多年生草本，被灰白色或黄白蛛丝状棉毛或白色厚棉毛。叶披针形，直立，基部沿茎下延成狭翅，具3脉。头状花序6-16个；总苞宽钟状，直径约1厘米；总苞片4-5层，黄褐色。瘦果长圆形，密具乳突。花期8-9月，果期9-10月。生海拔2800-4700米的林下、山坡或草地。产青藏高原。

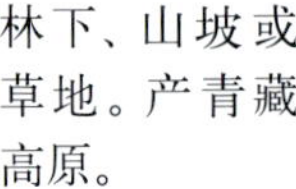

Perennial herbs, gray or yellowish-white arachnoid, or densely white tomentose. Leaves lanceolate, erect, base decurrent as narrow wings on stem, 3-veined. Capitula 6-16, clustered in corymb or compound corymb; involucres broadly campanulate, ca. 1 cm diam; phyllaries 4-5 rows, yellowish-brown. Achenes oblong, densely papillose. Fl. Aug-Sep. Fr. Sep-Oct. Forests, montane slopes or grasslands at 2800-4700 m. Distributed in Tibetan Plateau.

铃铃香青

Anaphalis hancockii Maxim.

多年生草本，被黄白色蛛丝状毛和腺毛。叶直立，条形或条状披针形，具3脉。头状花序9-15个，为紧密的复合伞房状；总苞宽钟状，直径8-10毫米；总苞片4-5层，褐色至白色。花期6-8月，果期8-9月。生海拔2000-3700米的亚高山山顶或草坡。产中国西南、华北和西北。

Perennial herbs, yellowish-white arachnoid and glandular pubescent. Leaves erect, linear or linear-lanceolate, 3-veined. Capitula 9-15, densely compound corymbiform; involucres broadly campanulate, 8-10 mm diam; phyllaries 4-5 rows, brown to white. Fl. Jun-Aug. Fr. Aug-Sep. Subalpine summits or grassy slopes at 2000-3700 m. Distributed in SW, N and NW China.

尼泊尔香青

Anaphalis nepalensis (Spreng.) Hand.-Mazz.

多年生草本，被白色密棉毛。匍匐枝长达20厘米。叶匙形、倒披针形或长圆状披针形，基部稍抱茎，不下延。头状花序1或少数；总苞多少球状，直径15-20毫米；总苞片8-9层，深褐色至白色。花期6-9月，果期8-10月。生海拔2400-4500米的草地、灌丛、松林、山谷、岩石区域、地衣坡地或河滩地。产青藏高原。喜马拉雅亦有。

铃铃香青 *Anaphalis hancockii*

Perennial herbs, densely white villose. Stolons to 20 cm long. Leaves spathulate, oblanceolate or oblong-lanceolate, base somewhat amplexicaul, no decurrent.

三脉香青 *Anaphalis triplinervis*

Capitula solitary or few; involucres somewhat globose, 15-20 mm diam; phyllaries 8-9 rows, dark brown to white. Fl. Jun-Sep. Fr. Aug-Oct. Grasslands, thickets, pine forests, valleys, rocks, among lichens on slopes or flood lands at 2400-4500 m. Distributed in Tibetan Plateau. Also in Himalaya.

三脉香青
Anaphalis triplinervis (Sims) Sims ex C. B. Clarke

多年生草本。茎叶长圆形或椭圆形，具三脉，无柄，有时抱茎。基部和中部叶披针形。头状花序大；总苞片白色，顶端尖。花期8-10月，果期10-11月。生海拔约2300米的草坡。产西藏。喜马拉雅亦有。

Perennial herbs. Cauline leaves oblong or elliptic, 3-veined or more, base sessile, sometimes amplexicaul, apex acute; lower and middle leaves oblong-lanceolate to broadly lanceolate. Capitula large; phyllaries white, apex acute. Fl. Aug-Oct. Fr. Oct-Nov. Grassy slopes at ca. 2300 m. Distributed in Xizang. Also in Himalaya.

单头香青
Anaphalis monocephala DC.

多年生草本。几无茎，高6厘米，莲座叶密生。叶匙状至倒披针状长圆形，背面密被白绵毛，正面被蛛丝状毛，1脉或不显著的基出3脉。头状花序单生茎顶，少有2或3个；总苞片白色。生海拔4100-4500米的高山潮湿斜坡地衣、石缝或河岸。产四川西部、西藏和云南。喜马拉雅亦有。

Perennial herbs. Plants acaulescent, or stems low, up to 6 cm tall, congested with rosette leaves. Leaves spatulate to oblanceolate-oblong, abaxially densely white lanate, adaxially arachnoid tomentose, 1-veined or inconspicuously 3-veined. Capitula solitary, terminal, rarely 2 or 3 capitula among rosette leaves; phyllaries white. Among lichens on alpine dank slopes, rock crevices, riverbanks at 4100-4500 m. Distributed in W Sichuan, Xizang and Yunnan. Also in Himalaya.

单头香青 *Anaphalis monocephala*

美丽香青
Anaphalis nubigena DC.

多年生草本，高5-15厘米，无匍匐茎。根状茎多分枝。茎多数，不分枝，纤细。叶披针形、长圆形或卵形，被白色绵毛。头状花序单生，少数2-3；总苞片白色、显著。花果期8-10月。生海拔3500-4800米的高山流石滩、石砾地、草甸。产西藏和四川西部。喜马拉雅亦有。

Perennial herbs, 5-15 cm tall, without stolons. Rhizome usually numerously branched. Stems numerous, unbranched, slender. Leaves lanceolate, oblong or ovate, densely white lanate. Capitula solitary, terminal, rarely 2-3; phyllaries white, striking. Fl. and fr. Aug-Oct. Alpine scree, gravelly slope, meadows at 3500-4800 m. Distributed in Xizang and W Sichuan. Also in Himalaya.

美丽香青 *Anaphalis nubigena*

宽叶拟鼠麹草 *Pseudognaphalium adnatum*

宽叶拟鼠麹草

Pseudognaphalium adnatum (DC.) Y. S. Chen

一年生粗壮草本，密被白色棉毛。叶长圆形至条形，近革质，全缘。头状花序径5-6毫米，密集成球状，再排成伞房花序；总苞近球形；总苞片干膜质，淡黄色；雌花多数，结实；花冠线形；两性花5-7朵；花冠管状。瘦果圆柱形；冠毛白色。花期8-10月。生海拔500-3000米的路旁、林缘、灌丛或山坡草地。产中国西南、华南和华东。南亚和东南亚亦有。

Annual stout herbs, densely white villose. Leaves oblong to linear, subcoriaceous, entire. Capitula 5-6 mm diam, glomerate and arranged in corymb; involucres subglobose; phyllaries dry membranous, yellowish; pistillate flowers numerous, fruited; corolla filiform; bisexual flowers 5-7; corolla tubular. Achenes cylindric; pappus white. Fl. Aug-Oct. Roadsides, forest edges, thickets or grassy slopes at 500-3000 m. Distributed in SW, S and E China. Also in S and SE Asia.

拟鼠麹草

Pseudognaphalium affine (D. Don) Anderb.

一年生草本，高10-40厘米，被白色厚棉毛。茎基部分枝，斜升。叶无柄，匙状倒披针形或倒卵状匙形。头状花序小，近无柄，密集成伞房花序；总苞钟形；总苞片金黄色或柠檬黄色，膜质，有光泽；小花多数，黄色至淡黄色。瘦果倒卵球形；冠毛污白色。花期1-4月，果期8-11月。生海拔2000米以下的干地或湿润草地上。产中国大部分地区。南亚、东南亚、东亚亦有。

Annual herbs, 10-40 cm tall, thickly white sericeous. Stems branched at base, obliquely ascending. Leaves sessile, spathulate-oblanceolate or obovate-spathulate. Capitula small, subsessile, congested in corymb; involucres campanulate; phyllaries golden-yellow or limon-yellow, membranous, lucidus; florets numerous, yellow to yellowish. Achenes obovoid; pappus dirty white. Fl. Jan-Apr. Fr. Aug-Nov. Dry places or damp grasslands below 2000 m. Wide distributed throughout most parts of China. Also in S, SE and E Asia.

金头拟鼠麹草

Pseudognaphalium chrysocephalum (Franch.) Hilliard et B. L. Burtt

多年生草本。叶长披针形或线形披针形，正面被灰色绵状毛，基部狭楔形，先端极尖。密伞房花序，钟状；外苞片具绵状毛；中苞片金黄色，光滑，倒卵形椭圆形，先端圆形；内苞片发白，匙形，基部皮质，先端膜质；花两性；冠毛白色。花期6-10月。生海拔2600-2800米的草坡。产四川西南部和云南西北部。

Perennial herbs. Leaves narrowly lanceolate or linear-lanceolate, adaxially gray lanate, base narrowly cuneate, apex acute. Capitula densely corymbose, campanulate; outer phyllaries lanate; middle ones golden yellow and polished, obovate-oblong, apex rounded; innermost ones pale, spatulate, base leathery, apex membranous; florets bisexual; pappus white. Fl. Jun-Oct. Grassy slopes at 2600-2800 m. Distributed in SW Sichuan and NW Yunnan.

拟鼠麹草 *Pseudognaphalium affine*

金头拟鼠麹草 *Pseudognaphalium chrysocephalum*

心叶天名精 *Carpesium cordatum*

秋拟鼠麹草
Pseudognaphalium hypoleucum (DC.) Hilliard et B. L. Burtt

粗壮草本，高达70厘米。茎直立，基部木质。下部叶条形，无柄，上面被腺毛，下面被白色棉毛。头状花序多数，密集成伞房花序；总苞球形；总苞片金黄色或黄色，有光泽，膜质；花冠黄色；冠毛污黄色。花期4-10月。生海拔2700米以下的空旷沙地、山地路旁或山坡上。产中国西南、华南、华中、华东和西北。南亚、东南亚和东亚亦有。

Stout herbs, to 70 cm tall. Stems erect, woody at base. Lower cauline leaves linear, sessile, glandular hairy adaxially, white sericeous abaxially. Capitula numerous, congested in corymb; involucres globose; phyllaries golden-yellow or yellow, lucidus, membranous; corolla yellow; pappus dirty yellow. Fl. Apr-Oct. Open sandy places, montane roadsides or slopes below 2700 m. Distributed in SW, S, C, E and NW China. Also in S, SE and E Asia.

心叶天名精
Carpesium cordatum F. H. Chen et C. M. Hu

多年生草本，高达60厘米。茎有纵条纹。基叶花期枯萎；茎叶卵形或长圆状卵形。头状花序单生茎端及枝端；苞叶3-5，卵形或卵状披针形；内层总苞片干膜质。小花全为管状，黄色。花期7-8月，果期9-10月。生海拔2300-3500米的山坡草地和针叶林。产中国西南部。喜马拉雅亦有。

Perennial herbs, up to 60 cm tall. Stems vertically striped. Basal leaves withered at anthesis; cauline leaves ovate or oblong. Capitula terminal on stems and branches; bracteal leaves 3-5, ovate or ovate-oblong; inner phyllaries dry membranous. Florets all tubular, yellow. Fl. Jul-Aug. Fr. Sep-Oct. Grassy slopes and conifer forests at 2300-3500 m. Distributed in SW China. Also in Himalaya.

秋拟鼠麹草 *Pseudognaphalium hypoleucum*

尼泊尔天名精 *Carpesium nepalense*

莛茎天名精
Carpesium scapiforme

莛茎天名精

Carpesium scapiforme F. H. Chen et C. M. Hu

多年生草本。叶卵状披针形至倒披针形；底部叶密集，有长柄。头状花序1-3，被匙状、草质的苞叶包围；总苞片长圆形，干膜质；小花全为管状，黄色。花期7-9月，果期9-10月。生海拔3000-4100米的高山草甸、林缘或河边。产中国西南部。喜马拉雅亦有。

Perennial herbs. Leaves ovate-lanceolate to oblanceolate; lower leaves close together, long petiolate. Capitula 1-3, surrounded by spatulate, herbaceous bracts; phyllaries oblong, scarious; florets all tubular, yellow. Fl. Jul-Sep. Fr. Sep-Oct. Alpine meadows, forest margins, streamsides at 3000-4100 m. Distributed in SW China. Also in Himalaya.

大花金挖耳

Carpesium macrocephalum Franch. et Savat.

多年生草本。叶具翼柄或无柄，叶片广卵形、椭圆形至披针形，边缘具重牙齿。头状花序单生茎端，下垂；苞叶多枚，椭圆形至披针形，叶状；总苞盘状，直径2.5-3.5厘米；总苞片叶状，披针形；小花全为筒状，黄色；无冠毛。花期6-9月，果期9-10月。生海拔700-2300米的山坡、灌丛、林缘或草地。产中国西南、华中、华北、东北和西北。东北亚亦有。

Perennial herbs. Leaves winged-petiolate or sessile, leaves broadly ovate, elliptic to lanceolate, margin doubly dentate. Capitula solitary, terminal, drooping; bracteal leaves numerous, elliptic to lanceolate, foliaceous; involucres discoid, 2.5-3.5 cm diam; phyllaries lanceolate, foliaceous; florets all tubular, yellow; pappus absent. Fl. Jun-Sep. Fr. Sep-Oct. Slopes, thickets, forest edges or grasslands at 700-2300 m. Distributed in SW, C, N, NE and NW China. Also in NE Asia.

尼泊尔天名精

Carpesium nepalense Less.

多年生草本。茎下部叶具长柄，卵形。头状花序直径9-11毫米，单生于叶腋，具长花梗，具苞叶；总苞杯状，苞片4层；外轮花花冠长约1.5毫米，冠檐5裂。花期6-9月，果期9-11月。生海拔1100-3200米的山坡或林中。产中国西南、华南、华中和华西。喜马拉雅亦有。

Perennial herbs. Lower cauline leaves long petiolate, ovate. Capitula 9-11 mm diam, solitary, long pedunculate, subtended by leafy bracts; involucres cupuliform, phyllaries 4-seriate; corolla of outer florets ca. 1.5 mm long, limbs 5-lobed. Fl. Jun-Sep. Fr. Sep-Nov. Mountain slopes or forests at 1100-3200 m. Distributed in SW, S, C and W China. Also in Himalaya.

大花金挖耳 *Carpesium macrocephalum*

棉毛尼泊尔天名精
Carpesium nepalense var. lanatum

区、澳大利亚和欧洲亦有。

Perennial herbs, white villose. Leaves s elliptic, oblong or spathulate-elliptic, glandular spotted, white villose abaxially, petioles narrowly winged. Capitula solitary, terminal, drooping; bracteal leaves several, unequal in size, foliaceous; involucres cupuliform, 1-2 cm diam; phyllaries foliaceous; florets tulular, yellow. Achenes narrowly cylindric, ridged, pappus absent. Fl. Jun-Aug. Fr. Sep-Oct. Roadsides, wastelands, ditchsides, slopes or forest edges below 2900(-3400) m. Distributed throughout most parts of China. Also in most parts of Asia, Australia and Europe.

高原天名精
Carpesium lipskyi Winkl.

多年生草本。茎常带紫色，被长柔毛。叶具柄，匙形、椭圆形至披针形，下面被白色疏长柔毛。头状花序单生，下垂；苞叶5-7枚，披针形，大小近相等；总苞盘状；总苞片披针形，叶状；小花漏斗状，黄色，被柔毛。瘦果狭圆柱形，有棱；无冠毛。花期7-9月，果期8-10月。生海拔2000-3700米的林缘、山坡或灌丛中。产云南、四川、甘肃、陕西和青海。

Perennial herbs. Stems generally purple, villose. Leaves petiolate, spathualate, elliptic to lanceolate, sparsely white villose abaxially. Capitula solitary, drooping; bracteal leaves 5-7, subequal in size; involucres discoid; phyllaries lanceolate, foliaceous; florets infundibular, yellow, pubescent. Achenes narrowly cylindric, ridged; pappus absent. Fl. Jul-Sep. Fr. Aug-Oct. Forest edges, slopes or thickets at 2000-3700 m. Distributed in Yunnan, Sichuan, Gansu, Shaanxi and Qinghai.

棉毛尼泊尔天名精
Carpesium nepalense var. **lanatum** (C. B. Clarke) Kitam.

全株被白色棉毛，茎上尤密。头状花序稍大，直径12-20毫米；苞片锐尖；花冠有时被稀疏柔毛。花期7-9月，果期9-11月。生海拔1100-2700米的山地斜坡。产中国西南、华中和陕西。东喜马拉雅亦有。

Entire plant white lanate, especially dense on stems. Capitula 12-20 mm diam; phyllaries acute; corolla sometimes sparsely pubescent. Fl. Jul-Sep. Fr. Sep-Nov. Montane slopes at 1100-2700 m. Distributed in SW and C China, and Shaanxi. Also in E Himalaya.

烟管头草
Carpesium cernuum L.

多年生草本，被白色长柔毛。叶片椭圆形、长圆形或匙状椭圆形，具腺点，下面被白色长柔毛；叶柄具狭翅。头状花序单生茎枝端，下垂；苞叶多枚，大小不等，叶状；总苞壳斗状，直径1-2厘米；总苞片叶状；小花筒状，黄色。瘦果狭圆柱形，有棱，无冠毛。花期6-8月，果期9-10月。生海拔2900(-3400)米以下的路边、荒地、沟边、山坡或林缘。产中国大部分地区。亚洲大部分地

烟管头草 *Carpesium cernuum*

高原天名精 *Carpesium lipskyi*

绒毛天名精 *Carpesium velutinum*

绒毛天名精

Carpesium velutinum Winkl.

多年生草本，被污黄色绒毛状长柔毛。叶椭圆形或椭圆状披针形，边缘具锯齿，下面密被绒毛状长柔毛。头状花序单生茎端和叶腋，排成总状花序状；苞叶披针形，1-2枚稍大；总苞扁球形；苞片被长柔毛；小花筒状，黄色。瘦果狭圆柱形，有棱；无冠毛。花期7-9月，果期9-10月。生海拔2000-3200米的山坡、路边或林下。产四川西部。

Perennial herbs, dirty yellow tomentose-villose. Leaves elliptic or elliptic-lanceolate, margin serrate, densely tomentose-villose abaxially. Capitula solitary, terminal and axillary, arranged in raceme; bracteal leaves lanceolate, 1-2 larger; involucres depressed globose; phyllaries villose; florets tubular, yellow. Achenes narrowly cylindric, ridged; pappus absent. Fl. Jul-Sep. Fr. Sep-Oct. Slopes, roadsides or forests at 2000-3200 m. Distributed in W Sichuan.

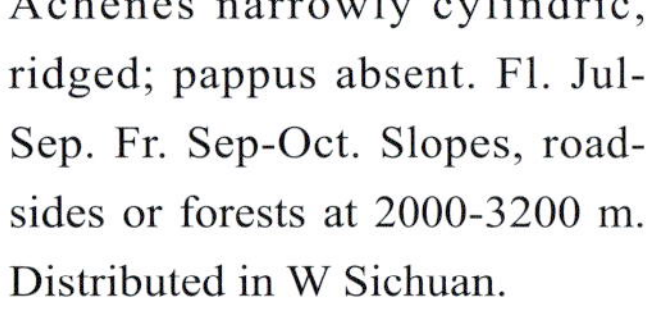

暗花金挖耳

Carpesium triste Maxim.

多年生草本。茎下部叶卵状长圆形，具不规则尖齿，叶柄长，具翅。头状花序数个至多数生于枝上；总苞钟形；苞片3层，外层膜质；边花雌性；花冠狭管状；筒状花两性。花期7-9月，果期9-10月。生海拔700-3700米的林中或溪边。产中国大部分地区。东北亚亦有。

Perennial herbs. Lower cauline leaves ovate-oblong, irregularly mucronate-toothed, petioles long, winged. Heads several to many on branches; involucres campanulate; phyllaries in 3 series, outer one scarious; marginal flowers female; corolla narrow tubular; disc flowers bisexual. Fl. Jul-Sep. Fr. Sep-Oct. Forests or streamsides at 700-3700 m. Distributed in most parts of China. Also in NE Asia.

暗花金挖耳 *Carpesium triste*

小花金挖耳

Carpesium minus Hemsl.

多年生草本。茎下部叶薄，卵状长圆形。头状花序多数，单生于枝上叶腋处，花期稍下垂，具多数苞叶；总苞钟状球形；总苞片4层；小花约80朵；外轮花冠圆柱形，中央花的花冠管状，冠檐4裂。花期7-9月，果期9-10月。生海拔700-1000米的草质山坡或林中。产中国西南、华南、华中和华东。日本亦有。

Perennial herbs. Lower cauline leaves thin, ovate-oblong. Heads numerous, solitary on branches, nodding at anthesis, subtended by many leafy bracts; involucres campanulate-globose; phyllaries in 4 series; florets ca. 80; corolla

小花金挖耳 *Carpesium minus*

金挖耳 *Carpesium divaricatum*

of outer florets cylindric; corolla of central florets tubular, limbs 4-lobed. Fl. Jul-Sep. Fr. Sep-Oct. Grassy slopes or forests at 700-1000 m. Distributed in SW, S, C and E China. Also in Japan.

金挖耳

Carpesium divaricatum Siebdd et Zucc.

多年生草本。叶具柄，卵形、长圆形或披针形，边缘具粗大牙齿，上面被柔毛，下面被白色短柔毛。头状花序单生枝端；苞叶3-5枚，2枚较大；总苞卵球形，直径6-10毫米；总苞片被柔毛；小花筒状，黄色。花期7-10月，果期10-11月。生海拔600-1600米的路旁、山坡或灌丛中。产中国西南、华南、华中、华东和东北。朝鲜半岛和日本亦有。

Perennial herbs. Leaves petiolate, ovate, oblong or lanceolate, margin gross-dentate, pubescent adaxially, white puberulent abaxially. Capitula solitary, terminal; bracteal leaves 3-5, 2 large; involucres ovoid, 6-10 mm diam; phyllaries villose; florets tulular, yellow. Fl. Jul-Oct. Fr. Oct-Nov. Roadsides, slopes or thickets at 600-1600 m. Distributed in SW, S, C, E and NE China. Also in Korean Peninsula and Japan.

粗齿天名精

Carpesium tracheliifolium Less.

多年生草本。叶卵形，近全缘至具粗齿。头状花序1-8个于枝顶叶腋，作总状或穗状花序式排列，近直立或下垂，具苞叶；总苞4或5层；苞片多数长圆形，膜质。花期7-8月，果期9-10月。生海拔2000-3500米的沟谷或林中。产云南、四川、西藏和台湾。尼泊尔亦有。

Perennial herbs. Leaves ovate, subentire to coarsely serrate. Capitula 1-8 in racemes or spikes at branch ends, suberect to pendulous, surrounded by leafy bracts; involucres 4- or 5-seriate; phyllaries mostly oblong, scarious. Fl. Jul-Aug. Fr. Sep-Oct. Valleys or forests at 2000-3500 m. Distributed in Yunnan, Sichuan, Xizang and Taiwan. Also in Nepal.

粗齿天名精 *Carpesium tracheliifolium*

中日金挖耳 *Carpesium faberi*

天名精 *Carpesium abrotanoides*

中日金挖耳
Carpesium faberi Winkl.

多年生草本。茎常紫色。茎下部叶具无翅长柄，卵状长圆形。头状花序多数，单生于枝上，花期下垂，由多数苞叶所托；总苞钟状球形；苞片4层；小花约80朵；冠檐4或5裂。花期8-9月，果期10-11月。生海拔700-2000米的草坡或灌丛。产四川、贵州、广西、台湾和湖北。日本亦有。

Perennial herbs. Stems often purplish. Lower cauline leaves long petiolate with petioles wingless, ovate-oblong. Capitula numerous, solitary on branches, nodding at anthesis, subtended by many leaflike bracts; involucres campanulate-globose; phyllaries 4-seriate; florets ca. 80; limbs 4- or 5-lobed. Fl. Aug-Sep. Fr. Oct-Nov. Grassy slopes or thickets at 700-2000 m. Distributed in Sichuan, Guizhou, Guangxi, Taiwan and Hubei. Also in Japan.

长叶天名精 *Carpesium longifolium*

长叶天名精
Carpesium longifolium F. H. Chen et C. M. Hu

多年生草本。叶面近无毛或被极疏短柔毛，椭圆状披针形至披针形。头状花序作穗状花序式排列；总苞半球形；苞片4层；雌花3或4层；花冠管状，5齿裂；两性花管状，冠檐5齿裂。花期7-9月，果期9-10月。生海拔600-2300米的潮湿林中、河边或草地。产四川、贵州、湖北、陕西和甘肃。

Perennial herbs. Leaves surfaces nearly glabrous or very sparsely pilose, elliptic-lanceolate to lanceolate. Heads in spicate; involucres hemispherical; phyllaries in 4 series; female florets in 3 or 4 series; corolla tubular, 5-dentate; bisexual florets tubular, limbs 5-dentate. Fl. Jul-Sep. Fr. Sep-Oct. Moist forests, riversides or grasslands at 600-2300 m. Distributed in Sichuan, Guizhou, Hubei, Shaanxi and Gansu.

天名精
Carpesium abrotanoides L.

多年生草本。叶椭圆形或披针形，被短柔毛。头状花序多数，顶生或腋生，近无梗，排成穗状花序；苞叶无或2-4枚，椭圆形或披针形；总苞卵球形或扁球形，直径6-8毫米；苞片被短柔毛；小花筒状；花冠黄色。瘦果狭圆柱形，有棱；无冠毛。花期8-10月，果期10-12月。生海拔2800(-3400)米以下的路边、灌丛、荒地、溪边或林缘。产中国西南、华南、华中、华东和华北。南亚、西南亚、东南亚、东北亚和欧洲亦有。

Perennial herbs. Leaves elliptic or lanceolate, puberulent. Capitula numerous, terminal or axillary, subsessile, arranged in spike; bracteal leaves absent or 2-4, elliptic or lanceolate; involucres ovoid or depressedly globose, 6-8 mm diam; phyllaries puberulent; florets tubular; corolla yellow. Achenes narrowly cylindric, ridged; pappus absent. Fl. Aug-Oct. Fr. Oct-Dec. Roadsides, thickets, wastelands, streamsides or forest edges below 2800 (-3400) m. Distributed in SW, S, C, E, and N China. Also in S, SW, SE and NE Asia, and Europe.

金仙草
Pulicaria chrysantha (Diels) Ling

多年生草本。茎直立，高30-60厘米。叶线状披针形至长圆状披针形。头状花序在茎端和枝端单生；总苞宽钟状；总苞片多层；花黄色；冠毛白色。花期7-8月，果期8-9月。生海拔

臭蚤草 *Pulicaria insignis*

金仙草 *Pulicaria chrysantha*

2500-3000米的草地或林缘。产四川西部。

Perennial herbs. Stem 30-60 cm tall. Leaves linear-lanceolate to oblong-lanseolate. Capitula solitary, terminal, 15-35 mm diam; involucres broadly campanulate, phyllaries several seriates; flowers yellow; pappus white. Fl. Jul-Aug. Fr. Aug-Sep. Grasslands or forest margin at 2500-3000 m. Distributed in W Sichuan.

臭蚤草

Pulicaria insignis Drumm. ex Dunn

多年生草本。茎直立或斜升，分枝或不分枝。基部叶倒披针形，茎部叶长圆形或卵圆状长圆形。头状花序茎端单生，少侧生；总苞宽钟状；总苞片多层；舌状花黄色，舌片狭长；冠毛白色。花期7-9月。生海拔3400-4600米的岩石上、石砾坡地或草丛中。产西藏和青海西南部。印度北部亦有。

Perennial herbs. Stems erect or ascending, brabched or not. Basal leaves oblanceolate, stem leaves oblong or ovate-oblong. Capitula solitary, few lateral; involucres broadly campanulate; phyllaries several seriates; ray forests yellow, laminas long and narrow; pappus white. Fl. Jul-Sep. Rock, gravelly slopes and grasslands at 3400-4600 m. Distributed in Xizang and SW Qinghai. Also in N India.

蚤草

Pulicaria vulgaris Gaertn.

一年生草本，高10-30厘米。茎从下部或中部起多分枝。叶长圆形、针形或倒披针形，全缘。头状花序小，单生，顶生；总苞半球形，长4-4.5毫米；总苞片约4层；两性花花冠黄色，管状；冠毛白色。花期6-9月。生海拔600-2800米的草地、沙地、沟渠沿岸和路旁。产新疆。蒙古、巴基斯坦、俄罗斯、中亚、西南亚、欧洲和北非亦有。

Annual herbs, 10-30 cm tall. Stem branched at base or middle of stem. Leaves oblong, lanceolate or oblanceolate, margin entire. Capitula small, solitary, terminal; involucres hemispheric, 4-4.5 mm long; phyllaries ac. 4-seritate; forest bisexual, yellow, tubular; pappus white. Fl. Jun-Sep. Dry grasslands, sandy places, riverbeds and roadsides at 600-2800 m. Distributed in Xinjiang. Also in Mongolia, Pakistan, Russia, C and SW Asia, Europe and N Africa.

蚤草 *Pulicaria vulgaris*

东风草 *Blumea megacephala*

东风草
Blumea megacephala (Randeria) C. T. Chang et C. H. Yu ex Y. Ling

多年生攀援小灌木或灌木。茎生叶厚，长圆形，长9-11厘米，无毛或表面稍被短柔毛。头状花序半球形，作疏松圆锥花序式排列；总苞3或4层；花冠黄色，管状；两性花5裂；外部花3或4裂；冠毛白色。花期8月至翌年4月。生海拔100-1900米的灌丛或林缘草坡。产中国西南、华南、东南、华中和华东。泰国、越南和琉球群岛亦有。

Perennial undershrubs or shrubs, scandent. Cauline leaves thick, oblong, 9-11 cm long, glabrate or minutely pilose on surfaces. Heads hemispheric in lax panicles; phyllaries in 3 or 4 series; corolla yellow, tubular; bisexual florets 5 lobes; outer florets 3- or 4-lobed; pappus white. Fl. Aug to next Apr. Thickets or grassy slopes at 100-1900 m. Distributed in SW, S, SE, C and E China. Also in Thailand, Vietnam and Ryukyu Islands.

艾纳香
Blumea balsamifera (L.) DC.

多年生草本或亚灌木。茎粗壮，稍木质，被黄褐色密柔毛。叶宽椭圆形或长圆状披针形，上面被柔毛，下面密被绢毛。头状花序多数，直径5-8毫米，排成开展大型圆锥花序；总苞钟形；总苞片草质，背面被柔毛；花冠黄色；冠毛红褐色。花期全年。生海拔1200米以下的林下、河边、林缘、山谷或草地。产云南、贵州、广西、广东、台湾和福建。南亚和东南亚亦有。

Perennial herbs or suffruticoses. Stems stout, somewhat woody, yellowish-brown villose. Leaves broad elliptic or oblong-lanceolate, villose adaxially, densely sericeous abaxially. Capitula numerous, 5-8 mm diam, arranged in large and lax panicle; involucres campanulate; phyllaries herbaceous, villose outside; corolla yellow; pappus brown. Fl. year-around. Forests, riversides, forest edges, valleys or grasslands below 1200 m. Distributed in Yunnan, Guizhou, Guangxi, Guangdong, Taiwan and Fujian. Also in S and SE Asia.

千头艾纳香
Blumea lanceolaria (Roxb.) Druce

多年生草本或亚灌木，高1-2.5米。茎基部木质，中空。叶狭长圆状或倒披针形。头状花序多数，排列成顶生、塔形的大圆锥花序；总苞圆柱形或近钟形；中央花黄色。瘦果长圆形，有10条棱；冠毛浅红色。花期1-4月。生海拔400-1500米的林缘、山坡、路旁、草地或溪边。产中国西南和华南。南亚和东南亚亦有。

Perennial herbs or subshrubs, 1-2.5 m tall. Stems woody at base, hollow in center. Leaves narrowly oblong to oblanceolate. Capitula numerous, terminal, in a pyramidal panicle; involucre campanulate-globose; central florets yellowish. Achenes oblong, 10-ribbed. Pappus pale reddish. Fl. Jan-Apr. Forests, grassy slopes, riversides at 400-1500 m. Distributed in SW and S China. Also in S and SE Asia.

长圆叶艾纳香
Blumea oblongifolia Kitam.

多年生草本。叶长圆形，被柔毛，近无柄，顶端短尖或钝，边缘具重锯齿，稍反卷。头状花序直径8-12毫米，多数排成顶生开展疏圆锥花序，花序柄长达2厘米；总苞球状钟形；总苞片绿色，背面被密长柔毛；花冠黄色。瘦果圆柱形；冠毛

艾纳香 *Blumea balsamifera*

千头艾纳香 *Blumea lanceolaria*

白色。花期8月至翌年4月。生路旁、田边、草地、山谷或溪边。产广东、台湾、福建、浙江和江西。

Perennial herbs. Leaves oblong, villose, subsessile, apex acute or obtuse, margin doubly serrate, slightly revolute. Capitula 8-12 mm diam, many arranged in terminal spread and lax panicle, peduncles to 2 cm long; involucres globose-campanulate; phyllaries green, densely villose outside; corolla yellow. Achenes cylindric; pappus white. Fl. Aug to next Apr. Roadsides, fieldsides, grasslands, valleys or streamsides. Distributed in Guangdong, Taiwan, Fujian, Zhejiang and Jiangxi.

节节红
Blumea fistulosa (Roxb.) Kurz

多年生草本。茎纤细，紫红色，被短柔毛或短绒毛。叶倒卵形、倒披针形或倒卵状长圆形，被长柔毛，边缘有疏粗齿或细尖齿。头状花序通常2-5个球状簇生，再排成穗状圆锥花序；总苞圆柱形或近钟形；总苞片紫红色；花冠黄色。瘦果圆柱形；冠毛白色。花期10月至翌年4月。生海拔300-1900米的杂木林、林缘、草地或溪边。产云南、贵州、广西和广东。喜马拉雅和中南半岛亦有。

Perennial herbs. Stems slender, purple, shortly pubescent or tomentose. Leaves obovate, oblanceolate or obovate-oblong, villose, margin sparsely gross-serrate or serrulate. Capitula generally 2-5 ones glomerate and arranged in spicate-panicle again; involucres cylindric or sub-campanulate; phyllaries purple; corolla yellow. Achenes cylindric; pappus white. Fl. Oct to next Apr. Mixed forests, forest edges, grasslands or streamsides at 300-1900 m. Distributed in Yunnan, Guizhou, Guangxi and Guangdong. Also in Himalaya and Indo-China Peninsula.

长圆叶艾纳香 *Blumea oblongifolia*

节节红 *Blumea fistulosa*

土木香 *Inula helenium*

土木香
Inula helenium L.

多年生高大草本。根状茎块状。下部叶大型，椭圆状披针形，长达60厘米，宽达25厘米，基部翅柄长达20厘米。头状花序少数，直径6-8厘米，排成伞房状花序；总苞片宽大，草质，被茸毛；舌状花条形；冠毛污白色。花期6-9月。生海拔2000米以下的路边、湿地及河边地带。产新疆，其他地区有栽培。亚洲、欧洲和北美洲广布。

Perennial large herbs. Rhizomes tuberous. Lower leaves large, elliptic-lanceolate, to 60 cm long, 25 cm wide, base with winged petioles to 20 cm long. Capitula several, 6-8 cm diam, arranged in corymb; phyllaries broad, herbaceous, tomentose; ray florets linear; pappus dirty white. Fl. Jun-Sep. Roadsides, wastelands, streamsides below 2000 m. Distributed in Xinjiang, cultivated in other parts of China. Widely distributed in Asia, Europe and North America.

总状土木香
Inula racemosa Hook. f.

多年生草本。叶革质，莲座状叶椭圆状披针形，渐狭成叶柄。头状花序直径4-8厘米，呈总状排列；苞片5或6层。瘦果无毛；冠毛白色。花期8-9月。生海拔1500-3100米的河岸沙地或草地。产新疆。克什米尔地区亦有。

Perennial herbs. Leaves leathery, radical leaves elliptic-lanceolate, narrowed into petioles. Capitula 4-8 cm diam, racemose; phyllaries in 5 or 6 series. Achenes glabrous; pappus white. Fl. Aug-Sep. River beaches or grasslands at 1500-3100 m. Distributed in Xinjiang. Also in Kashmir.

锈毛旋覆花
Inula hookeri C. B. Clarke

多年生草本。叶椭圆状披针形，具柔毛和短腺毛。头状花序单生于小枝顶端；总苞半球形；苞片3-4层，具棕色柔毛；舌状花多数，可育，雌性；花冠黄色。花期7-10月，果期10月。生海拔2400-3600米的山坡、灌丛、疏林中或草地。产云南西北部和西藏东南部。东喜马拉雅亦有。

Perennial herbs. Leaves elliptic-lanceolate, pubescent and short glandular on surfaces. Heads single on top of stem of branchlet; involucres hemispheric; phyllaries in 3-4 series, brown villous; ray florets many, pistillate, fertile; corolla yellow. Fl. Jul-Oct. Fr. Oct. Mountain slopes, shrublands, sparse forests

总状土木香 *Inula racemosa*

锈毛旋覆花 *Inula hookeri*

柳叶旋覆花 *Inula salicina*

or grasslands at 2400-3600 m. Distributed in NW Yunnan and SE Xizang. Also in E Himalaya.

柳叶旋覆花

Inula salicina L.

多年生草本。叶披针形，基部心形或有圆形小耳，半抱茎，边缘具细齿和密糙毛。头状花序直径2.5-4厘米，单生茎枝端，常为密集苞状叶所围绕；总苞半球形；总苞片披针形或长圆形；花黄色。瘦果无毛；冠毛白色。花期6-9月，果期9-10月。生海拔200-1000米的山顶、草坡或湿草地。产华北、东北和新疆。东北亚、中亚和欧洲亦有。

Perennial herbs. Leaves lanceolate, base cordate or rotundate-auriculate, semi-amplexicaul, margin serrulate or densely strigillose. Capitula 2.5-4 cm diam, solitary, terminal, surrounded by dense bracteal leaves; involucres semiglobose; phyllaries lanceolate or oblong; flowers yellow. Achenes glabrous; pappus white. Fl. Jun-Sep. Fr. Sep-Oct. Mountain summits, grassy slopes or damp grasslands at 200-1000 m. Distributed in N and NE China, and Xinjiang. Also in NE and C Asia, and Europe.

蓼子朴

Inula salsoloides (Turcz.) Ostenf.

多年生亚灌木。茎高达35厘米，有棱。叶披针状或长圆状线形，无柄，全缘。头状花序多数，茎尖单生；小花黄色。瘦果褐色，圆柱形，有纵棱。花期5-8月，果期7-9月。生海拔500-2000米的干草原、沙滩、冲积土。产中国东北、西北和华北。阿富汗、蒙古和俄罗斯亦有。

Perennials. Stems up to 35 cm tall, thinly angulate. Leaves sessile, lanceolate or lanceolate-linear, margin entire. Capitula numerous, solitary at ends of sprigs; florets yellow. Achenes brown, cylindric, longitudinally finely ribbed. Fl. May-Aug. Fr. Jul-Sep. Dry grasslands, sand banks, alluvium at 500-2000 m. Distributed in NE, NW and N China. Also in Afghanistan, Mongolia and Russia.

蓼子朴 *Inula salsoloides*

水朝阳旋覆花 *Inula helianthus-aquatilis*

水朝阳旋覆花

Inula helianthus-aquatilis C. Y. Wu ex Y. Ling

多年生草本，被薄柔毛。叶卵形或披针形，基部半抱茎，边缘具细密尖锯齿，下面具黄色腺点，脉上有短柔毛。头状花序单个顶生，直径2.5-4.5厘米；总苞半球形；总苞片条形至条状披针形；小花黄色。瘦果圆柱形；冠毛污白色。花期6-10月，果期9-10月。生海拔1200-3000米的湿润坡地、溪边、稻田或灌丛。产云南、四川、贵州和甘肃。药用植物。

Perennial herbs, thinly pubescent. Leaves ovate or lanceolate, bases semiamplexicaul, margin densely acute-serrulate, abaxially yellow-glandular, shortly pubescent on nerves. Capitula solitary, terminal, 2.5-4.5 cm diam; involucres semi-globose; phyllaries linear to linear-lanceolate; flowers yellow. Achenes cylindric; pappus dirty white. Fl. Jun-Oct. Fr. Sep-Oct. Damp slopes, streamsides, rice paddies or thickets at 1200-3000 m. Distributed in Yunnan, Sichuan, Guizhou and Gansu. Medicinal plant.

旋覆花 *Inula japonica*

旋覆花

Inula japonica Thunb.

多年生草本，被长伏毛。叶长圆形或披针形，基部渐狭或急狭，无柄，半抱茎，边缘具疏齿或全缘，下面被疏伏毛。头状花序顶生，直径3-4厘米，排成伞房花序；总苞半球形；总苞片草质，条状披针形；花黄色；冠毛白色。花期6-10月。生海拔100-2400米的山坡、路旁、河岸、湿润草地或荒地。产中国大部分地区。东北亚亦有。

Perennial herbs, appressed villose. Leaves oblong or lanceolate, base attenuate or acute, sessile, semi-amplexicaul, margin sparsely serrate or entire, sparsely appressed pubescent abaxially. Capitula terminal, 3-4 cm diam, arranged in corymb; involucres semi-globose; phyllaries herbaceous, linear-lanceolate; flowers yellow; pappus white. Fl. Jun-Oct. Slopes, roadsides, river banks, damp grasslands or wastelands at 100-2400 m. Distributed in most parts of China. Also in NE Asia.

欧亚旋覆花 *Inula britannica*

欧亚旋覆花

Inula britannica L.

多年生草本，被长柔毛。叶长椭圆形或披针形，半抱茎，下面密被伏柔毛。头状花序1-5个，顶生，直径2.5-5厘米；总苞半球形；总苞片草质，披针形，被长柔毛；舌状花条形，黄色。花期6-9月。生海拔300-1700米的河岸、路旁、田埂或湿润坡地。产河北、内蒙古、黑龙江和新疆。俄罗斯、中亚和欧洲亦有。

里海旋覆花 *Inula caspica*

线叶旋覆花 *Inula linariifolia*

Perennial herbs, villose. Leaves narrowly elliptic or lanceolate, semi-amplexicaul, densely appressed pubescent abaxially. Capitula 1-5, terminal, 2.5-5 cm diam; involucres semi-globose; phyllaries herbaceous, lanceolate, villose; ray florets linear, yellow. Fl. Jun-Sep. River banks, roadsides, field ridges or damp slopes at 300-1700 m. Distributed in Hebei, Neimenggu, Heilongjiang and Xinjiang. Also in Russia, C Asia and Europe.

里海旋覆花

Inula caspica F. K. Blum ex Ledeb.

二年生草本。下部叶长圆状线形或狭披针形；上部叶线状披针形至线形，无柄。头状花序多数，排列成伞房花序；小花黄色。瘦果褐色，有棱。花期7-9月。生海拔200-2400米的荒漠河边。产西藏和新疆。南亚、中亚、西南亚和俄罗斯亦有。

Biennial herbs. Lower leaves narrowly oblong or lanceolate; upper leaves sessile, linear-lanceolate to linear. Capitula numerous, in loose corymbiform synflorescences; florets yellow. Achenes brown, ribbed. Fl. Jul-Sep. Streamsides or riverbanks in deserts at 200-2400 m. Distributed in Xizang and Xinjiang. Also in S, C and SW Asia, and Russia.

线叶旋覆花

Inula linariifolia Turcz.

多年生草本，被短柔毛。叶条状披针形至条形，边缘反卷，具不明显小锯齿，下面被短柔毛或长伏毛。头状花序顶生，直径1.5-2.5厘米，(1-)3-5个，排成伞房花序；总苞半球形；总苞片条状披针形；舌状花长圆状条形。花期7-8月，果期8-10月。生海拔1800米以下的山坡、荒地、路旁或河岸。产华中、华东、华北和东北。东北亚亦有。

Perennial herbs, shortly pubescent. Leaves linear-lanceolate to linear, margin revolute, inconspicuously serrulate, shortly pubescent or appressed villose abaxially. Capitula terminal, 1.5-2.5 cm diam, (1-)3-5, arranged in corymb; involucres semi-globose; phyllaries linear-lanceolate; ray florets oblong-linear. Fl. Jul-Aug. Fr. Aug-Oct. Slopes, wastelands, roadsides or river banks below 1800 m. Distributed in C, E, N and NE China. Also in NE Asia.

羊眼花

Inula rhizocephala Schrenk

多年生草本，无茎，莲座状。叶片长圆形或长圆状卵形，全缘。头状花序8-20个密集成半球状团伞花序，无总花序梗。总苞半球形；总苞片多层；花黄色。花期6-8月。生海拔1700-3800米的针叶林下、石砾坡地及草坡上。产新疆。南亚、西南亚和中亚亦有。

Perennial herbs, acaulescent, stemless. Leaves many, arranged in rosette, blade oblong or oblong-ovate, entire. Capitula numerous, arranged in dense semispherical sessile subcapitate synflorescences; flowers yellow. Fl. Jun-Aug. Conifer forests, meadows, or gravelly sloples at 1700-3800 m. Distribuated in Xinjiang. Also in S, SW and C Asia.

羊眼花 *Inula rhizocephala*

羊耳菊 *Duhaldea cappa*

羊耳菊

Duhaldea cappa (Buch.-Ham. ex DC.) Anderb.

灌木，高70-200厘米。茎被绒毛，有分枝。叶椭圆形、披针形或狭长圆形，背面被白色绒毛，边缘有稀疏的细齿。头状花序辐射状或盘状，密伞房状排列；总苞片6层，披针形，被毛；花冠黄色。花期6-10月，果期8-12月。生海拔200-3200米的干山坡、荒地、灌丛或草丛中。产中国西南、华南和华东。南亚和东南亚亦有。

Shrubs, 70-200 cm tall. Stems lanate-tomentose, branched. Leaves elliptic, lanceolate or narrowly oblong, whitish lanate abaxially, margin remotely serrulate. Capitula radiate or disciform, in dense corymbs; phyllaries 6-seriate, lanceolate, tomentose; corollas yellow. Fl. Jun-Oct. Fr. Aug-Dec. Dry hills, waste fields, thickets or grasslands at 200-3200 m. Distributed in SW, S and E China. Also in S and SE Asia.

绒毛戴星草

Sphaeranthus indicus L.

芳香草本，多分枝，被长柔毛或绒毛。叶倒卵状长圆形，基部渐狭，沿茎下延成狭翅，顶端钝，边缘有细尖重锯齿。复头状花序球形或近椭圆形，红紫色，单生于枝顶。瘦果圆柱形。花期12月至翌年4月。生海拔700-1000米的河边沙滩、草地或灌丛中。产云南和广东。亚洲热带地区、澳大利亚北部和非洲亦有。

Fragrant herbs, multi-branched, villose or tomentose. Leaves obovate-oblong, bases attenuate and decurrent as narrow wings on stems, apex obtuse, margin acutely double-serrulate. Compound heads globose or nearly ellipsoid, reddish-purple, terminal, solitary. Achenes cylindric. Fl. Dec to next Apr. Beaches, grasslands or thickets by rivers at 700-1000 m. Distributed in Yunnan and Guangdong. Also in tropical Asia, N Australia and Africa.

光梗阔苞菊

Pluchea pteropoda Hemsl.

草本或亚灌木。叶无柄，倒卵形或倒披针形，基部长渐狭，顶端钝或浑圆，边缘有疏锯齿或有时浅裂。头状花序多数，径6-8毫米，在茎枝顶端排列成伞房花序；总苞卵状球形或阔钟形，总苞片5-6层，外层阔卵形。花期5-12月。生海滨沙地、石缝或潮水能到达之地。产广东、广西、海南和台湾。越南亦有。

Herbs or subshrubs. Leaves sessile, obovate to oblanceolate, base gradually narrowed, margin irregularly and sparsely dentate, apex obtuse to rounded. Capitula 6-8 mm diam, densely congested in terminal corymbs; involucre ovoid to broadly campanulate; phyllaries 5-6-seriate, glabrous,

绒毛戴星草 *Sphaeranthus indicus*

光梗阔苞菊 *Pluchea pteropoda*

outer ones broadly ovate. Fl. May-Dec. Coastal brackish areas and tidal flats, often associated with mangroves and near graveyards on seashores. Distributed in Guangdong, Guangxi, Hainan and Taiwan. Also in Vietnam.

花花柴

Karelinia caspia (Pall.) Less.

多年生草本。单叶轮生，耳状半抱茎。2-9个多花的异形头状花序作顶生的伞房花序式排列；总苞圆柱状钟形，6或7层；边缘雌花多层，红色；中部可育两性花红色，具10-20花。瘦果三角形，基部具环。花期7-9月，果期9-10月。生海拔900-1300米的戈壁滩、沙丘或盐生草甸。产内蒙古、甘肃、青海和新疆。西南亚、中亚、东北亚和欧洲亦有。

Perennial herbs. Leaves simple, alternate, semiamplexicaul-auriculate. Heads heterogamous, many-flowered, arranged in terminal corymbiform inflorescences, 2-9; involucres cylindrical-campanulate, 6 or 7 rows; marginal florets pistillate, multiseriate, reddish; central florets bisexual, sterile, 10-20, reddish. Achenes triquetrous, with a ring at base. Fl. Jul-Sep. Fr. Sep-Oct. Gobi, dunes or saline meadows at 900-1300 m. Distributed in Neimenggu, Gansu, Qinghai and Xinjiang. Also in SW, C and NE Asia, and Europe.

花花柴 *Karelinia caspia*

翼齿六棱菊

Laggera crispata (Vahl) Hepper et J. R. I. Wood

多年生草本。茎翅具不整齐粗齿或细齿。叶倒卵形、椭圆形或长圆形，边缘具密重齿。头状花序多数，排成大型圆锥花序；总苞近钟形；总苞片长圆形、披针形至条形，被柔毛；花多数，黄色，杂性。瘦果近纺锤形；冠毛白色。花期1-7月。生海拔2000米以下的空旷草地或山谷疏林。产中国西南、华南和华中。印度、中南半岛和热带非洲亦有。

Perennial herbs. Stems irregularly serrate or serrulate. Leaves obovate, elliptic or oblong, margin densely double-serrate. Capitula numerous, arranged in large panicle; involucres subcampanulate; phyllaries oblong, lanceolate to linear, pubescent; flowers numerous, yellow, polygamous. Achenes subfusiform; pappus white. Fl. Jan-Jul. Open grasslands or sparse forests in valleys below 2000 m. Distributed in SW, S and C China. Also in India, Indo-China Peninsula and tropical Africa.

翼齿六棱菊 *Laggera crispata*

六棱菊（四棱锋）*Laggera alata*

六棱菊（四棱锋）

Laggera alata (D. Don) Schulz-Bip. ex Oliv.

多年生草本。叶互生，披针形或长圆形，边缘有疏细齿，基部渐狭，无柄，下延于茎成全缘的翅，密被短疣毛。头状花序直径1-1.5厘米，多数，排成圆锥状；总苞片条状披针形，坚硬，被短腺毛；花多数，黄色，杂性。瘦果圆柱状；冠毛白色。花期3-10月。生海拔2300米以下的旷野、路旁或山坡阳处。产中国西南、华南和华东。南亚、东南亚和非洲亦有。

Perennial herbs. Leaves alternate, lanceolate or oblong, margin remotely serrulate, base attenuate, sessile, decurrent along stems as entire wings, densely and shortly warty hairy. Capitula 1-1.5 cm diam, numerous, arranged in panicle; phyllaries linear-lanceolate, rigid, shortly glandular hairy; flowers numerous, yellow, polygamous. Achenes cylindric; pappus white. Fl. Mar-Oct. Open places, roadsides or sunny slopes below 2300 m. Dstributed in SW, S and E China. Also in S and SE Asia, and Africa.

山黄菊 *Anisopappus chinensis*

山黄菊

Anisopappus chinensis Hook. et Arn.

一年生草本。中部茎叶卵状披针形或狭长圆形，基部截形或宽楔形，边缘有钝锯齿。头状花序单生或数个排列成顶生的伞房状花序，总苞半球形；总苞片4或5层；托片龙骨状，膜质；小花黄色；冠毛污白色，膜片状。花期8-11月。生海拔2400米以下的干燥山坡、沙地、荒地、草地上或潮湿山坡及林缘附近，或为杂草生路旁或宅旁。产云南、四川、广东、广西、江西和福建。印度、缅甸、泰国、非洲和马达加斯加亦有。

Annual herbs. Cauline leaves ovate-lanceolate or narrowly oblong, base truncate or broadly cuneate, margin obtusely serrate. Capitula solitary or several in terminal corymbs; involucre hemispheric; phyllaries in 3 series; paleae keeled, membranous; florets yellow; pappus grayish white, chaffy, of 4 or 5 bristles. Fl. Aug-Nov. Grassy slopes, waste fields, forest margins below 2400 m. Distributed in Yunnan, Sichuan, Guangdong, Guangxi, Jiangxi and Fujian. Also in India, Myanmar, Thailand, Africa and Madagascar.

万寿菊

Tagetes erecta L.

一年生草本。叶羽状分裂，裂片长圆形或披针形，边缘具锐锯齿。头状花序大，单生，直径5-8厘米；花序梗顶端棍棒状

万寿菊 *Tagetes erecta*

秋英（波斯菊）
Cosmos bipinnatus

膨大；总苞杯状；舌状花黄色或暗橙色；管状花黄色。瘦果条形；冠毛二型，长芒状或短鳞片状。花期6-10月。中国常见栽培，归化于云南和广东。原产墨西哥。

Annual herbs. Leaves pinnatisected, lobes oblong or lanceolate, margin sharply serrate. Capitula large, solitary, 5-8 cm diam; peduncles clavate-dilated at top; involucres cotyloid; ray florets yellow or dark orange; tulular flowers yellow. Achenes linear; pappus dichotypic, long aristate or short squamiform. Fl. Jun-Oct. Commonly cultivated in China, and naturalized in Yunnan and Guangdong. Native to Mexico.

秋英（波斯菊）
Cosmos bipinnatus Cav.

一年生或多年生草本。叶对生，二回羽状深裂。头状花序单生，径3-6厘米；花序梗长6-18厘米；外层总苞片披针形，淡绿色，具深紫色条纹；舌状花红色、紫红色、粉红色或白色，顶端有3-5钝齿；管状花黄色。瘦果黑紫色，无毛，顶端具2-3尖刺。花期6-8月。生海拔2700米以下的花圃、路边、田埂或溪岸。中国各地广泛栽培或归化。原产墨西哥。

Annual or perennial herbs. Leaves opposite, bipinnately parted. Capitula solitary, 3-6 cm diam; peduncles 6-18 cm long; outer phyllaries lanceolate, greenish, dark purple striate; ray florets red, purple, pink or white, apex 3-5-crenulate; tubular florets yellow. Achenes black-purple, glabrous, apex 2-3-aristate. Fl. Jun-Aug. Flower gardens, roadsides, field ridges or streamsides below 2700 m. Cultivated or naturalized all over China. Native to Mexico.

黄秋英（硫黄菊）
Cosmos sulfureus Cav.

一年生或多年生草本。叶对生，二至三回羽状深裂，裂片较宽，披针形至椭圆形。头状花序单生，具长花序梗；舌状花金黄色或橘黄色。瘦果有粗毛，长18-25毫米。花期7-8月。生海拔1500米以下的花圃或山坡。中国偶有栽培，归化于云南。原产墨西哥和巴西。

Annual or perennial herbs. Leaves opposite, bi- to tri-pinnately parted, lobes broader, lanceolate to elliptic. Capitula solitary, long pedunculate; ray florets golden or orange. Achenes strigillose, 18-25 mm long. Fl. Jul-Aug. Flower gardens or slopes below 1500 m. Occasionally cultivated in China, naturalized in Yunnan. Native to Mexico and Brazil.

黄秋英（硫黄菊） *Cosmos sulfureus*

柳叶鬼针草 *Bidens cernua*

柳叶鬼针草
Bidens cernua L.

一年生草本。叶对生，披针形，无毛，基部半抱茎，边缘具疏锯齿。头状花序单生，直径达4厘米，下垂；总苞盘状；外层总苞片条状披针形，叶状；舌状花多数，黄色，卵状椭圆形。瘦果狭楔形，压扁，顶端截形，顶端具4枚芒刺。花期8-10月。生海拔3600米以下的草甸或沼泽，有时生水中。产中国西南、华北和东北。亚洲、欧洲和北美洲亦有。

Annual herbs. Leaves opposite, lanceolate, glabrous, base semi-amplexicaul, margin sparsely serrate. Capitula solitary, to 4 cm diam, pendulous; involucres discoid; outer phyllarieslinear-lanceolate, foliaceous; ray florets numerous, yellow, ovate-elliptic. Achenes narrowly cuneate, compressed, truncate at apex, apex with 4 aristae. Fl. Aug-Oct. Meadows or swamps below 3600 m, sometimes submerged in water. Distributed in SW, N and NE China. Also in Asia, Europe and North America.

大狼耙草
Bidens frondosa L.

一年生草本。茎常带紫色。一回羽状复叶对生，具柄，小叶3-5枚，披针形，先端渐尖，边缘有粗锯齿。头状花序单生茎枝顶端，直径1.2-2.5厘米；总苞钟状或半球形；外层苞片披针形，叶状；无舌状花。瘦果扁平，狭楔形，顶端具2枚芒刺。花期8-9月。生沼泽地、水边、河漫滩、荒地或田野中。华中、华东、华北和东北地区有归化。原产北美洲。

Annual herbs. Stems generally purplish; pinnate compound leaves opposite, petiolate, leaflets 3-5, lanceolate, apex acuminate, margin gross-serrate. Capitula solitary on tops of stems and branches, ca. 1.5-2.5 cm diam; involucres campanulate or semiglobose; outer phyllaries lanceolate, foliaceous; ray florets absent. Achenes compressed, narrowly cuneate, apex with 2 aristae. Fl. Aug-Sep. Swamps, watersides, flood plains, wastelands or fields. Naturalized in C, E, N and NE China. Native to North America.

狼杷草
Bidens tripartita L.

一年生草本。茎圆柱形。茎中部叶对生，具柄，长圆状披针形，通常3-5深裂，裂片披针形，边缘具疏锯齿。头状花序单生茎枝顶端，直径1-3厘米；总苞盘状；外层总苞片条形或匙状倒披针形，叶状；无舌状花。瘦果扁平，楔形，顶端具2枚芒刺。花期7-10月。生路边、荒地、水边或沼泽地。广布中国各地。亚洲、欧洲、非洲北部和大洋洲东南部亦有。

Annual herbs. Stems terete. Middle cauline leaves opposite, petiolate, oblong-lanceolate, generally 3-5-partite, lobes lanceolate, margin sparsely serrate. Capitula solitary on tops of stems and branches, 1-3 cm diam; involucres discoid; outer phyllaries linear or spathulate-lanceolate, foliaceous; ray florets absent. Achenes flat, cuneate, apex with 2 aristae. Fl. Jul-Oct. Roadsides, wastelands, watersides or swamps. Distributed throughout most parts of China. Also in Asia, Europe, N Africa and SE Oceania.

小花鬼针草
Bidens parviflora Willd.

一年生草本。叶二或三回羽状全裂，裂片狭披针形或条状披针形至条形，常疏被毛或近无毛。头状花序单生(或2-3个)于茎顶，近圆柱状；小花全管状，黄色。瘦果条状四棱形，顶部具2枚芒刺。花期7-9月。生沼泽地或溪边。产中国西南、华北、华西、华东、西北和东北。东北亚亦有。

Annual herbs. Leaves bipinnatisect or tripinnatisect with narrow lanceolate or linear-lanceolate to

大狼耙草 *Bidens frondosa*

狼杷草 *Bidens tripartita*

linear segments, usually sparsely fine hairy to subglabrous. Heads solitary (or 2-3) at apices of the stems, subcylindrical; florets all tubular, yellow. Achenes linear-tetragonate, 2 upright awns, at the top of achenes. Fl. Jul-Sep. Marshes or streamsides. Distributed in SW, N, W, E, NW and NE China. Also in NE Asia.

鬼针草

Bidens pilosa L.

一年生草本。茎钝四棱形。茎中部叶具柄，三出复叶，小叶椭圆形或长圆形，先端锐尖，基部偏斜，边缘有锯齿，下部和上部叶较小，3裂或不分裂。头状花序直径8-9毫米；总苞基部被短柔毛；总苞片条状匙形；无舌状花。瘦果条形，黑色，顶端具3-4枚芒刺。花期全年。生海拔2500米以下的荒地、路边和草地。中国西南、华南、华中和华东地区归化。亚洲和美洲热带亚热带地区广布。

Annual herbs. Stems subtetragonal. Middle cauline leaves petiolate, trifoliolate, leaflets elliptic or oblong, apex acute, base oblique, margin serrate, lower and upper cauline leaves smaller, trifid or undivided. Capitula 8-9 mm diam; involucres shortly pilose at base; phyllaries linear-spathulate; ray florets absent. Achenes linear, black, apex with 3-4 aristae. Fl. year-round. Wastelands, roadsides or grasslands below 2500 m. Naturalized in SW, S, C and E China. Widely distributed in tropical and subtropical regions of Asia and America.

小花鬼针草 *Bidens parviflora*

鬼针草 *Bidens pilosa*

金盏银盘 *Bidens biternata*

金盏银盘

Bidens biternata (Lour.) Merr. et Sherff.

一年生草本。茎略四棱。羽状复叶对生，无柄或具短柄，小叶卵形或披针形，被柔毛，边缘具锯齿，下部1对三出复叶状分裂或仅一侧具1裂片。头状花序直径7-10毫米；外层总苞片条形；舌状花3-5朵，淡黄色，长圆形。瘦果条形，黑色，顶端具3-4枚芒刺。花期9-11月。生海拔1300米以下的路边、荒地或村旁。产中国大部分地区。亚洲其他地区、非洲和大洋洲亦有。

Annual herbs. Stems somewhat tetragonus. Pinnate compound leaves opposite, sessile or shortly petiolate, leaflets ovate or lanceolate, pubescent, margin serrate, lower pair of leaflets ternate or only with 1 lobe on one side. Capitula 7-10 mm diam; outer phyllaries linear; ray florets 3-5, yellowish, oblong. Achenes linear, black, apex crowned with 3-4 aristae. Fl. Sep-Nov. Roadsides, wastelands or by villages below 1300 m. Distributed in most parts of China. Also in most parts of Asia, Africa and Oceania.

婆婆针

Bidens bipinnata L.

一年生草本。茎略四棱形。叶对生，具柄，二回羽状分裂。头状花序直径6-10毫米；总苞杯形，8-12；外层总苞片条形；舌状花1-3朵，黄色。瘦果条形，顶端具3-4枚芒刺。花果期8-10月。生海拔1800(-3000)米以下的路边、荒地、田间或山坡。产中国大部分地区。美洲、亚洲、欧洲和非洲东部广布。

Annual herbs. Stems somewhat tetragonal. Leaves opposite, petiolate, bipinnatisected. Capitula 6-10 mm diam; involucres 8-12, cotyloid; outer phyllaries linear; ray florets 1-3, yellow. Achenes linear, apex crowned with 3-4 aristae. Fl. and fr. Aug-Oct. Roadsides, wastelands, fields and slopes below 1800(-3000) m. Distributed throughout most parts of China. Also in America, Asia, Europe and E Africa.

两色金鸡菊

Coreopsis tinctoria Nutt.

一年生草本，无毛。叶对生，具长柄，二回羽状全裂，裂片条形或条状披针形，全缘。头状花序多数，有细长花序梗；总苞半球形；总苞片卵状长圆形；舌状花黄色，下部红褐色；管状花红褐色，狭钟形。瘦果长圆形或纺锤形，顶端有2细芒。花期6-8月。中国各地栽培。原产北美洲。

Annual herbs, glabrous. Leaves opposite, long petiolate, bipinnatisected, lobes linear or linear-lanceolate, margin entire. Capitula numerous, with fine and long petioles; involucres semiglobose; phyllaries ovate-oblong; ray florets yellow, lower part reddish-brown; tubular florets reddish-brown, narrowly campanulate. Achenes oblong or fusiform, apex 2-aristate. Fl. Jun-Aug. Cultivated all over China. Native to North America.

沼菊

Enydra fluctuans Lour.

多年生草本。茎稍肉质，下部匍匐，长50-80厘米。叶近无柄，长圆形或线状长圆形，基部抱茎，边缘有疏齿。头状花序直径8-10毫米；总苞具4个苞片；舌状花长约3毫米；管状花5裂。花期11月至翌年4月。生沼泽地或溪边。产云南和海南。南亚、东南亚和澳大利亚亦有。

Perennial herbs. Stems slightly fleshy, prostrate in lower part, 50-80 cm. Leaves subsessile, oblong or linear-oblong, base amplexicaul, margin sparsely ser-

婆婆针 *Bidens bipinnata*

两色金鸡菊 *Coreopsis tinctoria*

rate. Capitula 8-10 mm diam; involucre of 4 phyllaries; ray florets ca. 3 mm; disk florets 5-lobed. Fl. Nov to next Apr. Marshes, streamsides. Distributed in Yunnan and Hainan. Also in S and SE Asia, and Australia.

金纽扣

Acmella paniculata (Wall. ex DC.) R. K. Jansen

一年生草本。茎多分枝，带紫红色。叶对生，卵形至椭圆形，全缘、波状或具波状钝锯齿。头状花序卵圆形，具花序梗；花托锥形；花冠黄色。瘦果长圆形；冠毛1-2个，细芒状。花果期4-11月。生海拔800-1900米的田边、沟边、溪边、荒地、路旁或林缘。产云南、广西、广东、海南和台湾。南亚、东南亚和日本亦有。

Annual herbs. Stems multi-branched, purplish. Leaves opposite, ovate to elliptic, margin entire, undulate or undulate-crenate. Capitula ovoid, pedunculate; receptacles conical; corolla yellow. Achenes oblong; pappus 1-2, minutely aristate. Fl. and fr. Apr-Nov. Fields, ditchsides, streamsides, wastelands, roadsides or forest edges at 800-1900 m. Distributed in Yunnan, Guangxi, Guangdong, Hainan and Taiwan. Also in S and SE Asia, and Japan.

沼菊 *Enydra fluctuans*

金纽扣 *Acmella paniculata*

美形金钮扣 *Acmella calva*

美形金钮扣

Acmella calva (DC.) R. K. Jansen

多年生疏散草本。茎匍匐或平卧，节上常生须根。叶对生，宽披针形或披针形，边缘有尖锯齿或近缺刻。头状花序卵状圆锥形，具细长花序梗；花托圆柱状锥形；花冠黄色。瘦果长圆形；冠毛2个，细芒状。花果期5-12月。生海拔1000-1900米的水边、林缘或荒地。产云南。南亚和东南亚亦有。

Perennial diffuse herbs. Stems creeping or procumbent, nodes with fibrous roots. Leaves opposite, broadly lanceolate or lanceolate, margin acutely serrate or nearly incised. Capitula ovoid-conical, slenderly pedunculate; receptacles cylindric-conical; corolla yellow. Achenes oblong; pappus 2, minutely aristate. Fl. and fr. May-Dec. Watersides, forest edges or wastlands at 1000-1900 m. Distributed in Yunnan. Also in S and SE Asia.

羽芒菊 *Tridax procumbens*

百日菊

Zinnia elegans Jacq.

一年生草本，被糙毛。叶对生，宽卵圆形或长圆状椭圆形，基部稍心形抱茎，基出三脉。头状花序大，单生枝端；舌状花深红色、玫瑰色、紫堇色或白色。瘦果倒卵球形或倒卵状楔形。花期6-9月，果期7-10月。中国各地广泛栽培，有时逸生。原产墨西哥。

Annual herbs, strigillose. Leaves opposite, broadly ovate or oblong-elliptic, base cordately axplexicaul, basally trinervate. Capitula large, solitary; ray florets dark red, roseous, violet or white. Achenes obovoid or obovoid-cuneate. Fl. Jun-Sep. Fr. Jul-Oct. Widely cultivated in China, sometimes naturalized. Native to Mexico.

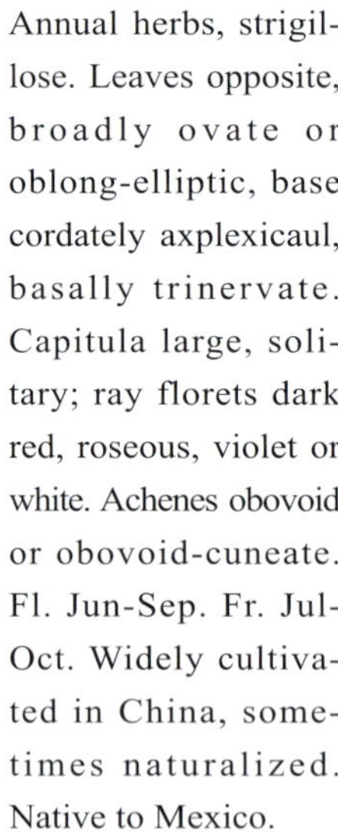

百日菊 *Zinnia elegans*

羽芒菊

Tridax procumbens L.

多年生铺地草本。茎略四方形，节处常生多数不定根。叶披针形或卵状披针形，边缘具粗齿或浅裂，基生三出脉。头状花序单生枝顶，具长花序梗；总苞钟形；总苞片绿色，被密毛；舌状花白色，顶端2-3浅裂。瘦果陀螺形；冠毛羽毛状。花期11月至翌年3月。生低海拔的旷野、荒地、坡地或路旁阳处。中国南部地区有归化。南亚和东南亚亦有。原产美洲热带地区。

Perennial procumbent herbs. Stems slightly tetragonous, nodes with numerous adventitious roots. Leaves lanceolate or ovate-lanceolate, margin gross-dentate or shallowly lobed, basally trinervate. Capitula terminal, solitary, long pedunculate; involucres

牛膝菊 *Galinsoga parviflora*

campanulate; phyllaries green, densely pilose; ray florets white, apex 2-3-lobed. Achenes top-shaped; pappus featherlike. Fl. Nov to next Mar. Open places, wastelands, slopes or sunny roadsides. Naturalized in south of China. Also in S and SE Asia. Native to tropical America.

牛膝菊
Galinsoga parviflora Cav.

一年生草本。叶对生，卵形或披针形，边缘具锯齿，基出三脉。头状花序半球形；总苞半球形或宽钟状；总苞片卵形，白色，膜质；舌状花白色，顶端3齿裂；管状花多数，黄色。瘦果小，黑色；冠毛白色。花期7-10月。生山谷、荒野、路边、田间或溪边。中国大部分地区归化。俄罗斯和欧洲亦有。原产南美洲。

Annual herbs. Leaves opposite, ovate or lanceolate, margin serrate, basal trinerved. Capitula semiglobose; involucres semiglobose or broadly campanulate; phyllaries ovate, white, membranous; ray florets white, apex 3-dentate; tubular florets numerous, yellow. Achenes small, black; pappus white. Fl. Jul-Oct. Valleys, wastelands, roadsides, fields or streamsides. Widely distributed throughout most parts of China. Also in Russia and Europe. Native to South America.

粗毛牛膝菊
Galinsoga quadriradiata Ruiz et Pav.

多年生草本。茎枝密被开展长柔毛。叶边缘有粗锯齿或犬齿。总苞半球形至钟形；总苞片早落；舌状花常白色，有时粉色；盘花15-35。花期7-10月。生林下路旁。云南、江西、陕西、甘肃、新疆、辽宁和吉林等地有归化。原产美洲。

Perennial herbs. Stems densely and spreadly villose. Margin of leaves gross-serrate or caniniform. Involucres hemispheric to campanulate; phyllaries deciduous; ray florets usually white, sometimes pink; disk florets 15-35. Fl. Jul-Oct. Roadsides in forests. Naturalized in Yunnan, Jiangxi, Shaanxi, Gansu, Xinjiang, Liaoning and Jilin. Native to America.

刺苞果
Acanthospermum hispidum DC.

一年生草本，节明显。叶中部以上有锯齿，基部多少抱茎。头状花序小；内层总苞片果期增厚变硬，顶端具2直刺。成熟瘦果倒卵球状三角形，顶端截形，有2个不等长的开展硬刺，周围有钩状的刺。花期6-7月，果期8-10月。生海拔1900米以下的山坡、河边或沟旁。云南有归化。原产南美洲。

Annual herbs, nodes distinct. Leaves serrate above the middle, somewhat amplexicaul at base. Capitula small; inner phyllaries thickened and hardened at fruits, with 2 spines at apex. Mature Achenes obovoid-deltoid, apex truncate, with 2 patent unequal spines, surface with hooked spines. Fl. Jun-Jul. Fr. Aug-Oct. Slopes, river banks or valleys below 1900 m. Naturalized in Yunnan. Native to South America.

粗毛牛膝菊 *Galinsoga quadriradiata*

刺苞果 *Acanthospermum hispidum*

豨莶 *Sigesbeckia orientalis*

豨莶

Sigesbeckia orientalis L.

一年生草本，被灰白色短柔毛。叶对生，纸质，三角状卵圆形或卵状披针形，下延成翼柄，边缘浅裂，具粗齿至全缘，基出三脉。头状花序小，顶生，排成具叶圆锥花序；总苞阔钟状；总苞片叶质，背面被紫褐色头状具柄腺毛；花冠黄色。花期4-9月，果期6-11月。生海拔100-2800米的山地、荒地、灌丛、林缘、林下或田间。中国南部地区广布。日本、南亚、东南亚、非洲、大洋洲和热带美洲亦有。

Annual herbs, offwhite puberulent. Leaves opposite, chartaceous, triangular-ovate or ovate-lanceolate, decurrent as winged petioles, margin lobed, gross-dentate to entire, basal triveined. Capitula small, terminal, arranged in foliate panicle; involucres broadly campanulate; phyllaries foliaceous, outsides with dense purplish-brown spiky glandular hairs; corolla yellow. Fl. Apr-Sep. Fr. Jun-Nov. Montanes, wastelands, thickets, forest edges, forests or fields at 100-2800 m. Widely distributed throughout southern parts of China. Also in Japan, S and SE Asia, Africa, Oceania and tropical America.

毛梗豨莶

Siegesbeckia glabrescens Makino

一年生草本，被平伏短柔毛。叶对生，卵形或卵状披针形，顶端渐尖，边缘具规则锯齿、疏齿或全缘，基出三脉。头状花序顶生，多数，排成疏散圆锥花序；总苞钟状；总苞片叶质，背面密被紫褐色头状有柄的腺毛；花冠黄色。瘦果倒卵球形。花期4-9月，果期6-11月。生海拔300-2600米的路边、旷野、荒地或灌丛。产中国西南、华南、华中和华东。朝鲜半岛和日本亦有。

腺梗豨莶
Sigesbeckia pubescens

Annual herbs, appressed puberulent. Leaves opposite, ovate or ovate-lanceolate, apex acuminate, margin regularly serrate, sparsely serrate, or entire, basal triveined. Capitula terminal, numerous, arranged in lax panicle; involucres campanulate; phyllaries foliaceous, outsides with dense purplish-brown spiky glandular hairs; corolla yellow. Achenes obovoid. Fl. Apr-Sep. Fr. Jun-Nov. Roadsides, open places, wastelands or thickets at 300-2600 m. Distributed in SW, S, C and E China. Also in Korean Peninsula and Japan.

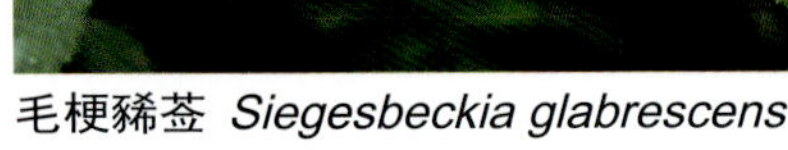

毛梗豨莶 *Siegesbeckia glabrescens*

腺梗豨莶

Sigesbeckia pubescens (Makino) Makino

一年生草本，被灰白色长柔毛和糙毛。叶对生，卵形至卵状披针形，边缘具粗齿，基出三

菊薯(雪莲果) *Smallanthus sonchifolius*

脉。头状花序多数，顶生，排成松散圆锥花序；总苞宽钟状；总苞片叶质，背面密生紫褐色头状具柄腺毛；花冠黄色。瘦果倒卵球形。花期5-8月，果期6-10月。生海拔3400米以下的山地、草地、河边、湿地、旷野或田间。中国大部分地区广布。印度、日本和朝鲜半岛亦有。

Annual herbs, offwhite villose and strigillose. Leaves opposite, ovate to ovate-lanceolate, margin gross-dentate, basal triveined. Capitula numerous, terminal, arranged in lax panicle; involucres broadly campanulate; phyllaries foliaceous, outsides with dense purplish-brown spiky glandular hairs; corolla yellow. Achenes obovoid. Fl. May-Aug. Fr. Jun-Oct. Montanes, grasslands, riversides, wetlands, open places or fields below 3400 m. Widely distributed throughout most parts of China. Also in India, Japan and Korean Peninsula.

菊薯(雪莲果)

Smallanthus sonchifolius (Poepp. et Endl.) H. Rob.

多年生草本，高1-3米。茎圆柱形，中空；地下部分分枝，有块茎。下部叶宽卵形，基部截形，叶柄有翅，基部有耳；上部叶卵状披针形，通常无裂片。花序顶生，1-5；总苞片5，1层，卵圆形；花冠黄色。花期6-9月。中国西南、华南、华东、东南、华中和华北有栽培。原产南美洲。

Perennial herbs, 1-3 m tall. Stems cylindric and hollow; underground part irregularly branched, often producing spindle-shaped tubers. Lower leaves broadly ovate, base hastate, petioles winged, connate and auriculate at base; upper leaves ovate-lanceolate, without lobes and hastate base. Synflorescence terminal, composed of 1-5 branches, each one with 3 capitula; phyllaries 5, 1-seriate, ovate; corollas yellow. Fl. Jun-Sep. Cultivated in SW, S, E, SE, C and N China. Native to South America.

百能葳

Blainvillea acmella (L.) Philipson

多年生草本。下部叶对生，上部叶常轮生。头状花序顶生或腋生；苞片2层；舌状花1层，黄色或黄白色，顶部具2-4齿；盘心花冠檐5齿裂。瘦果三棱形或压扁；冠毛短，不等长，2-5条，针状。花期4-6月。生海拔2600米以下的稀疏阔叶林中或草坡。产云南和海南。南亚、东南亚、澳大利亚、非洲和美洲亦有。

Perennial herbs. Lower leaves opposite, upper leaves usually alternate. Heads terminal or axillary; phyllaries in 2 series; ray florets 1 series, yellow or yellowish-white, apex 2-4-toothed; disc florets limbs 5-toothed. Achenes trigonous or compressed; pappus short, unequal, 2-5, spine like. Fl. Apr-Jun. Sparsely broadleaf forests or grassy slopes below 2600 m. Distributed in Yunnan and Hainan. Also in S and SE Asia, Australia, Africa and America.

百能葳 *Blainvillea acmella*

金腰箭 *Synedrella nodiflora*

金腰箭

Synedrella nodiflora (L.) Gaertn.

一年生草本，被贴生粗毛。叶对生，阔卵形至卵状披针形，边缘具不规则小齿，近基三出主脉。头状花序小，数个簇生于叶腋；总苞卵球形或长圆形；外层总苞片叶状；小花黄色。瘦果扁平，边缘具宽翅，翅缘具长硬尖刺；冠毛刚刺状。花期全年。生旷野、田地、路旁或荒地。中国西南、华南和华东地区归化。原产美洲。

Annual herbs, appressed hirsute. Leaves opposite, broadly ovate to ovate-lanceolate, margin irregularly serratulate, subbasal trinerved. Capitula small, several clustered in leaf-axil; involucres ovoid or oblong; outer phyllaries foliaceous; florets yellow. Achenes flat, margin widely winged, wings long and rigidly aculeate; pappus rigidly aculeate. Fl. year-round. Open places, fields, roadsides or wastelands. Distributed in SW, S and E China. Native from America.

鳢肠

Eclipta prostrata (L.) L.

一年生草本。叶对生，披针形，两面密被硬糙毛。头状花序小，成对单生叶腋，具细长花序梗；总苞球状钟形；总苞片绿色，草质，长圆形，被白色短伏毛；外围雌花舌状，白色；中央两性花多数，管状，白色。瘦果暗褐色；无冠毛。花期6-9月。生海拔1600米以下的河边、湿地、田边或路边。产中国大部分地区。世界热带和亚热带地区广布。

Annual herbs. Leaves opposite, lanceolate, both surfaces densely hirsute. Capitula small, solitary, axillary, opposite, slenderly pedunculate; involucres globose-campanulate; phyllaries green, herbaceous, oblong, shortly white strigose; outer pistillate flowers ligulate, white; central bisexual flowers numerous, tubular, white. Achenes dark brown; pappus absent. Fl. Jun-Sep. Riversides, wetlands, fieldsides or roadsides below 1600 m. Distributed in most parts of China. Also in tropical and subtropical regions of the world.

蟛蜞菊

Sphagneticola calendulacea (L.) Pruski

多年生匍匐草本。茎基部节处生不定根。叶对生，无柄，椭圆形、长圆形或条形，全缘或有1-3对疏粗齿，被短糙毛，侧脉1-2对。头状花序少数，单生枝顶或叶腋，具长花序梗；外层总苞片叶质；舌状花黄色。瘦果倒卵球形；无冠毛。花期3-9月。生路旁、田边、沟边或湿润草地上。产华南、华东和东北。南亚、东南亚和日本亦有。

Perennial creeping herbs. Basal nodes of stems with adventitious roots. Leaves opposite, sessile, elliptic, oblong or linear, margin entire or with 1-3 paris of gross-teeth, shortly strigillose, lateral veins 1-2 pairs. Capitula several, solitary, terminal or axillary, long pedunculate; outer phyllaries foliaceous; ray florets yellow.

鳢肠 *Eclipta prostrata*

蟛蜞菊 *Sphagneticola calendulacea*

Achenes obovid; pappus absent. Fl. Mar-Sep. Roadsides, fieldsides, ditchsides or damp grasslands. Distributed in S, E and NE China. Also in S and SE Asia, and Japan.

卤地菊
Melanthera prostrata (Hemsl.) W.L.Wagner et H.Rob.

多年生草本。茎长匍匐，节生生根。茎生叶长圆形，有时卵形或披针形，厚革质，3脉，边缘有疏齿。头状花序直径16-22毫米，通常顶端单生；总苞半球形，直径5-6毫米；舌花黄色；冠毛1-2，或缺失。生海滨沙滩。产广东和台湾。泰国、越南、日本和朝鲜半岛亦有。

Perennial herbs. Stems long creeping, rooting at node. Cauline leaves oblong, sometimes ovate or lanceolate, thickly leathery, 3-veined, margin loosely dentate. Capitula 16-22 mm diam, usually solitary, terminal; involucre hemispheric, 5-6 mm diam; ray florets yellow; pappus bristles 1-2, or obsolete. Littoral sand dunes, sandy seashores. Distribured in Guangdong and Taiwan. Also in Thailand, Vietnam, Japan and Korean Peninsula.

孪花菊 *Wollastonia biflora*

孪花菊
Wollastonia biflora (L.) DC.

亚灌木。茎生叶厚纸质，表面被平伏毛。头状花序3-6，顶生；舌状花黄色，1层，8-12朵；花冠2或3齿裂；两性花黄色，每个花序具20-35花；花冠顶部5齿裂。瘦果常三棱形；冠毛刚毛状。花期全年。产中国西南、华南、华中和华东。南亚、东南亚、东北亚和太平洋岛屿亦有。

Subshrubs. Cauline leaves thick chartaceous, appressed-strigose on surfaces. Heads 3-6, terminal; ray florets yellow, in one row, 8-12; corolla 2- or 3-dentate; disc florets yellow, 20-35 per head; corolla apex 5-dentate. Achenes often 3-angled; pappus bristles. Fl. year-round. Distributed in SW, S, C and E China. Also in S, SE and NE Asia, and Pacific Islands.

卤地菊 *Melanthera prostrata*

山蟛蜞菊 *Wollastonia montana*

黑心金光菊 *Rudbeckia hirta*

山蟛蜞菊

Wollastonia montana (Blume) DC.

多年生草本。叶两面粗糙。头状花序单生于延长的花梗上；总苞钟形；苞片2层；托苞长圆形，疏被柔毛；舌状花1层，黄色，顶部2-3齿裂；两性花管状，黄色，萼檐5齿裂。瘦果倒卵球状三角形，稍压扁，棕红色；冠毛2-3根。花期4-10月。生海拔500-1200(-3000)米的溪边或路边。产云南、贵州、广西、广东和海南。印度、尼泊尔、不丹、缅甸和泰国亦有。

Perennial herbs. Leaves both surfaces scabrous. Heads solitary on elongate peduncles; involucres campanulate; phyllaries in 2 series; paleaes oblong, sparsely pubescent; ray florets 1 series, yellow, apex 2-3-toothed; disc florets tubular, yellow, limbs 5-toothed. Achenes obovoid-triangonous, slightly compressed, red brown; pappus 2-3. Fl. Apr-Oct. Streamsides or roadsides at 500-1200(-3000) m. Distributed in Yunnan, Guizhou, Guangxi, Guangdong and Hainan. Also in India, Nepal, Bhutan, Myanmar and Thailand.

黑心金光菊

Rudbeckia hirta L.

一年生或二年生草本，全株被粗刺毛。叶长卵圆形、长圆形、匙形或长圆状披针形。头状花序顶生，直径5-7厘米，有长花序梗；总苞片长圆形至披针状条形，顶端钝；花托圆锥形；舌状花鲜黄色，长圆形；管状花暗褐色或暗紫色。瘦果四棱形。花果期6-9月。中国各地常见栽培。原产北美洲。

Annual or biennial herbs, wholly hirtose. Leaves narrowly ovate, oblong, spathulate or oblong-lanceolate. Capitula terminal, 5-7 cm diam, long pedunculate; phyllaries oblong to lanceolate-linear, apex obtuse; receptacles conical; ray florets daffadilly, oblong; tubular florets dark brown or dark purple. Achenes tetragonous. Fl. Jun-Sep. Commonly cultivated in China. Native to North America.

肿柄菊

Tithonia diversifolia A. Gray

一年生至多年生草本，高2-5米。茎粗壮，密被短柔毛。叶互生，具长柄，卵形、卵状三角形或近圆形，全缘或3-5深裂。头状花序大，直径5-15厘米，单个顶生，具粗壮长棒状的花序梗；舌状花黄色，长卵形。瘦果长圆形。花期9月至翌年1月。生海拔1500米以下的山坡、荒地、路边或灌丛中。中国西南、华南和华东地区栽培或归化。原产墨西哥。

肿柄菊 *Tithonia diversifolia*

肿柄菊 *Tithonia diversifolia*

Annual to perennial herbs, 2-5 m tall. Stems stout, densely pubescent. Leaves alternate, long petiolate, ovate, ovate-triangular or suborbicular, entire or 3-5-parted. Capitula large, 5-15 cm diam, solitary, terminal, with a thick long clavate peduncle; ray florets yellow, long ovate. Achenes oblong. Fl. Sep to next Jan. Slopes, wastelands, roadsides or thickets below 1500 m. Cultivated or naturalized in SW, S and E China. Native to Mexico.

向日葵

Helianthus annuus L.

一年生高大草本，密被粗硬刚毛。叶互生，宽卵形，顶端渐尖或急尖，基部心形或截形，边缘具粗锯齿，基出3脉，具长叶柄。头状花序单生茎端，盘状，直径可达35厘米；雌花舌状，金黄色；两性花筒状。瘦果矩卵形或椭圆体形，稍扁，灰色或黑色。花期7-9月，果期8-10月。中国各地广泛栽培。原产北美洲。

Annual large herbs, densely hirsute. Leaves alternate, broadly ovate, apex acuminate or acute, base cordate or truncate, margin gross-serrate, basal 3-veined, long petiolate. Capitula solitary, terminal, discoid, to 35 cm diam; pistillate flowers ligulate, golden-yellow; bisexual flowers tubular. Achenes oblong or ellipsoid, compressed, gray or black. Fl. Jul-Sep. Fr. Aug-Oct. Widely cultivated in China. Native to North America.

向日葵 *Helianthus annuus*

菊芋 *Helianthus tuberosus*

菊芋
Helianthus tuberosus L.

多年生草本。地下茎块状。叶通常对生，上部叶互生；叶片卵圆形、椭圆形至披针形，边缘有粗锯齿，具离基三出脉。头状花序直径2-5厘米，单生枝端；苞叶1-2个，条状披针形；总苞片披针形；舌状花12-20朵，黄色；管状花多数，黄色。瘦果小，楔形。花期8-9月。中国常见栽培。原产北美洲。

Perennial herbs. Rhizomes tuberous. Leaves usually opposite, upper cauline leaves alternate; leaves ovate, elliptic to lanceolate, margin gross-serrate, tripli-costate. Capitula 2-5 cm diam, solitary, terminal; bracteal leaves 1-2, linear-lanceolate; phyllaries lanceolate; ray florets 12-20, yellow; tubular florets numerous, yellow. Achenes small, cuneate. Fl. Aug-Sep. Commonly cultivated in China. Native to North America.

大丽花
Dahlia pinnata Cav.

多年生草本。块根巨大，棒状；茎粗壮，分枝。叶对生，羽状全裂，裂片卵形，无毛。头状花序大，具长花序梗；外层总苞片叶质，卵状椭圆形，内层膜质，椭圆状披针形；小花多数，形状颜色多变。瘦果长圆形，黑色，扁平。花果期6-10月。中国各地栽培，云南有归化。原产墨西哥。

Perennial herbs. Tubers large, clavate; stems stout, branched. Leaves opposite, pinnately sected, lobes ovate, glabrous. Capitula large, long pedunculate; phyllaries in outer layer foliaceous, ovate-elliptic, inner ones membranous, elliptic-lanceolate; florets numerous, varied in shape and color. Achenes oblong, black, flat. Fl. and fr. Jun-Oct. Cultivated throughout China, naturalized in Yunnan Native to Mexico.

苍耳
Xanthium strumarium L.

一年生草本，被糙毛。叶具长柄，三角状卵形或心形，具三基出脉。雄性头状花序球形；雌性头状花序椭圆形，结合成

大丽花 *Dahlia pinnata*

苍耳 *Xanthium strumarium*

囊状，宽卵球形或椭圆体形，成熟时坚硬，上端有1-2个坚硬的喙，外面具钩状直刺。瘦果2，倒卵形；无冠毛。花期7-8月，果期9-10月。生海拔2600米以下的平原、丘陵、荒地、路边或田野。产中国大部分地区。印度、俄罗斯、朝鲜半岛、日本和伊朗亦有。

Annual herbs, strigose. Leaves long petiolate, triangular-ovate or cordate, basally trinervate. Male capitula globose; female capitula ellipsoid, connate into a sac, sac broadly ovoid or ellipsoid, rigid at maturity, apex with 1-2 rigid beaks, outside with hooked spines. Achenes 2, obovoid; pappus absent. Fl. Jul-Aug. Fr. Sep-Oct. Plains, hills, wastelands, roadsides or fields below 2600 m. Distributed in most parts of China. Also in India, Russia, Korean Peninsula, Japan and Iran.

意大利苍耳
Xanthium italicum Moretti

一年生草本。茎直立，基部稍木质化。叶互生，三角状卵形至宽卵形，3-5浅裂，有3条主脉。总苞结果时长圆形，长1.9-3厘米，直径1.2-1.8厘米，总苞密生4-7毫米的倒钩刺，刺上被刚毛和短腺毛。花果期7-9月。生荒地或农田中。入侵植物，产华南、华北、东北和新疆等地。原产北美洲。

Annual herbs. Stem erect, woody in basal part. Leaves alternate, triangular-ovate to borad-ovate, 3-5-lobed, with 3 main veins. Involucres oblong, 1.9-3 cm long, 1.2-1.8 cm diam in fruit, outside with hooked spines 4-7 mm long, spine with bristles and short piloglandulose. Fl. and fr. Jul-Sep. Waste grassland or farmland. Invasive species, distributed in S, N and NE China, and Xinjiang. Native to North America.

三裂叶豚草
Ambrosia trifida L.

一年生粗壮草本。叶对生，有时互生，具叶柄和三基出脉，3-5裂，有时不裂，裂片卵状披针形或披针形，粗糙，边缘具锐锯齿。雄性头状花序多数，球形，总苞片有三肋；雌性头状花序在雄头状花序下面的叶腋内聚成团伞状；刺果金字塔形，具刺4或5。花期7-8月，果期9-10月。生海拔1600米以下的田野、路旁或河边湿地。归化于华北和东北。原产北美洲。

Annual robust herbs. Leaves opposite, sometimes alternate, with petioles and 3 basal veins, 3-5-lobed, sometimes entire, lobes ovate-lanceolate or lanceolate, corse, margin serrate. Male capitula many, globose, phyllaries with 3 ribs; female capitula glometate at axils of leaves below inflorescnese of male capitula; burs pyramidal, spines 4 or 5. Fl. Jul-Aug. Fr. Sep-Oct. Fields, roadsides or wetlands by river below 1600 m. Naturalized in N and NE China. Native to North America.

意大利苍耳 *Xanthium italicum*

三裂叶豚草 *Ambrosia trifida*

豚草 *Ambrosia artemisiifolia*

豚草
Ambrosia artemisiifolia L.

一年生草本。下部茎叶对生，二回羽状分裂；上部茎叶互生，羽状分裂。雄头状花序半球形或卵形，具短梗，下垂，在枝端密集成总状花序，总苞片全部合生；花冠淡黄色；雌头状花序无花序梗，腋生，1-3个密集成团伞状。瘦果倒卵球形，藏于坚硬的总苞中。花期7-10月，果期9-10月。生海拔1000米以下的田野、路旁、干燥土壤或河边湿地。归化于华中、华东、华北和东北。原产北美洲。

Annual herbs. Lower cauline leaves opposite, bipinnatifid; upper cauline leaves alternate, pinnatifid. Male capitula semiglobose or ovoid, shortly pedunculate, drooping, arranged in raceme on top of branches, all phyllaries connate; corolla yellowish; female capitula sessile, axillary, 1-3 glomerate. Achenes obovoid, included in hard involucral bract. Fl. Jul-Oct. Fr. Sep-Oct. Fields, roadsides, dry soils or wetlands by river below 1000 m. Naturalized in C, E, N and NE China. Native to North America.

银胶菊
Parthenium hysterophorus L.

一年生草本。下部叶具柄，卵形或椭圆形，二回羽状深裂，上面疏被糙毛；上部叶无柄，羽裂或指状三裂。头状花序盘状，直径3-4毫米，多数，排成伞房花序；总苞片卵形至圆形，先端钝，被短柔毛；花白色。瘦果倒卵球形；冠毛2，鳞片状。花期4-8月。生海拔1500米以下的荒地、田间、路旁、河边或坡地。产云南、贵州、广西、广东、海南、香港和福建。广布全球热带地区。原产美国和墨西哥。

Annual herbs. Lower cauline leaves petiolate, ovate or elliptic, bipinnatiparted, sparsely strigillose adaxially; upper cauline leaves sessile, pinnatifid or digitately 3-lobed. Capitula discoid, 3-4 mm diam, numerous, arranged in corymb; phyllaries ovate to orbicular, apex obtuse, puberulent; flowers white. Achenes obovoid; pappus 2, squamiform. Fl. Apr-Aug. Wastelands, fields, roadsides, riversides or slopes below 1500 m. Distributed in Yunnan, Guizhou, Guangxi, Guangdong, Hainan, Hong Kong and Fujian. Widely Distributed in tropical regions of the world. Native to USA and Mexico.

银胶菊 *Parthenium hysterophorus*

破坏草
Ageratina adenophora (Spreng.) R. M. King et H. Rob.

多年生草本。茎紫色，被柔毛。叶对生，卵状三角形或卵状菱形，边缘具粗大圆齿，基出脉3。头状花序40-50朵在茎顶排成伞房状花序；花冠紫色。瘦果无毛；冠毛10，基部贴生，白色。花果期4-10月。生海拔1000-2000米的灌丛、林缘或路边。归化于中国西南、广东和湖北。原产墨西哥。

Perennial herbs. Stems purple, pubescent. Leaves opposite, ovate-triangular or ovate-rhombic, margin crassly crenate, basal

破坏草 *Ageratina adenophora*

破坏草 *Ageratina adenophora*

veins 3. Capitula terminal, 40-50-floweredin corymbose inflorescences; corolla purplish. Achenes glabrous; pappus setae 10, basally connate, white. Fl. and fr. Apr-Oct. Thickets, forest edges or roadsides at 1000-2000 m. Naturalized in SW China, Guangdong and Hubei. Native to Mexico.

微甘菊

Mikania micrantha Kunth

草质分枝藤本。叶对生，叶片卵形，基部心形或深心形，边缘全缘或有粗锯齿。头状花序组成伞房状圆锥花序；总苞片4，长圆形，近等长；小花4，白色，管状。花果期全年。中国南方有归化。原产美洲。

Vines, branched. Leaves opposite; blade ovate, base cordate to deeply so, margin entire to coarsely dentate. Synflorescence a corymbose panicle; phyllaries 4, oblong, subequal; florets 4, white, tubular. Fl. and fr. year-round. Naturalized in south parts of China. Native to America.

微甘菊 *Mikania micrantha*

下田菊

Adenostemma lavenia (L.) O. Kuntze

一年生草本。叶对生，卵形至长椭圆状披针形，边缘有圆锯齿，具三出脉。头状花序小，顶生，少数，排成伞房圆锥状花序；总苞半球形；花冠白色。瘦果倒披针形，冠毛4，棒状。花果期8-10月。生海拔400-2300米的林下、灌丛、沼泽地或路边。产中国西南、华南、华中和华东。南亚、东南亚、东亚和澳大利亚亦有。

Annual herbs. Leaves opposite, ovate to long elliptic-lanceolate, margin crenate, basal veins 3. Capitula small, terminal, few in corymbose-paniculate inflorescences; involucres semiglobose; corolla white. Achenes oblanceolate, pappus 4, clavate. Fl. and fr. Aug-Oct. Forests, thickets, swamps or roadsides at 400-2300 m. Distributed in SW, S, C and E China. Also in S, SE and E Asia, and Australia.

下田菊 *Adenostemma lavenia*

野生或栽培。原产中美洲和南美洲，广泛归化。

Annual herbs, puberscent or villose. Leaves opposite, petiolate, ovate, elliptic or oblong, margin crenate, basal veins 3. Capitula many in compact corymb; involucres campanulate or semiglobose; corolla purplish. Achenes dark brown; pappus scales 5 or awned. Fl. and fr. year-round. Valleys, slope forests, forest edges, river banks, grassy slopes or wastelands at 100-1500 m. Cultivated or naturalized in SW, S and E China. Native to Central and South America, widely naturalized.

熊耳草 *Ageratum houstonianum*

熊耳草

Ageratum houstonianum Mill.

一年生草本，被绒毛。叶对生，具柄，卵形或三角状卵形，基部心形或楔形。头状花序多数排成伞房状花序；总苞钟状；花冠淡紫色。瘦果黑色；冠毛膜片状，5个，短，离生，先端芒刺状渐尖。花果期全年。生海拔100-1500米的草地、路旁或山谷坡地。云南、四川、广西、广东、江苏、山东和黑龙江有栽培或归化。原产墨西哥和毗邻地区，栽培或归化于亚洲、欧洲和非洲。

Annual herbs, tomentose. Leaves opposite, petiolate, ovate or triangular-ovate, base cordate or cuneate. Capitula many in corymbs; involucres campanulate; corolla purplish. Achenes black; pappus of 5 short free scales, apex aristate-acuminate. Fl. and fr. year-round. Grasslands, roadsides or slopes in valleys at 100-1500 m. Cultivated or naturalized in Yunnan, Sichuan, Guangxi, Guangdong, Jiangsu, Shandong and Heilongjiang. Native to Mexico and its vicinity, cultivated or naturalized in Asia, Europe and Africa.

藿香蓟

Ageratum conyzoides L.

一年生草本，被柔毛或长绒毛。叶对生，具柄，卵形、椭圆形或长圆形，边缘圆锯齿，基出脉3。头状花序多数组成紧密的伞房状花序；总苞钟状或半球形；花冠淡紫色。瘦果黑褐色，冠毛存在；冠毛膜片5或具芒。花果期全年。生海拔100-1500米的山谷、山坡林下、林缘、河边、草坡或荒地。中国西南、华南和华东有

林泽兰

Eupatorium lindleyanum DC.

多年生草本，密被白色柔毛。茎红色或淡紫红色。叶披针形，不分裂或三全裂，基出三脉，两面粗糙，被短粗毛和黄色腺点。头状花序多数，排成紧密伞房花序；总苞钟状，含5朵小花；总苞片披针形；花白色、粉红色或淡紫红色。瘦果椭圆状；冠毛白色。花果期5-12月。生海拔200-2600米的山谷、水湿地、林下或草原。产中国大部分地区(新疆除外)。东北亚亦有。

藿香蓟 *Ageratum conyzoides*

林泽兰 *Eupatorium lindleyanum*

Perennial herbs, densely white pubescent. Stems red or purplish-red. Leaves lanceolate, undivided or trisected, basal trinerved, both surfaces coarse, shortly strigillose and yellow punctate. Capitula numerous, arranged in dense corymb; involucres campanulate, with 5 florets; phyllaries lanceolate; florets white, pink or purplish. Achenes elliptic; pappus white. Fl. and fr. May-Dec. Valleys, wetlands, forests or grasslands at 200-2600 m. Distributed in most parts of China (except Xinjiang). Also in NE Asia.

大麻叶泽兰

Eupatorium cannabinum L.

多年生草本。茎淡紫红色。叶对生，有短柄，不规则分裂。头状花序排成复伞房花序；总苞钟状，含3-7朵小花；总苞片2-3层，覆瓦状排列；花紫红色、粉红色或淡白色。瘦果5棱；冠毛白色。生山顶、草地或竹林。产江苏和浙江。欧洲亦有。

Perennial herbs. Stems purplish red. Leaves opposite, shortly petiolate, irregularly lobed. Synflorescences compound corymbs; capitula numerous, 3-7-flowered; involucre campanulate; phyllaries 2-3-seriate, imbricate; corollas purple-red, pink or whitish. Achenes 5-ribbed; pappus white. Summits of small hills, grasslands or among bamboos. Distributed in Jiangsu and Zhejiang. Also in Europe.

佩兰

Eupatorium fortunei Turcz.

多年生草本。中部茎叶大，3全裂或3深裂，下部茎叶减小。头状花序多数

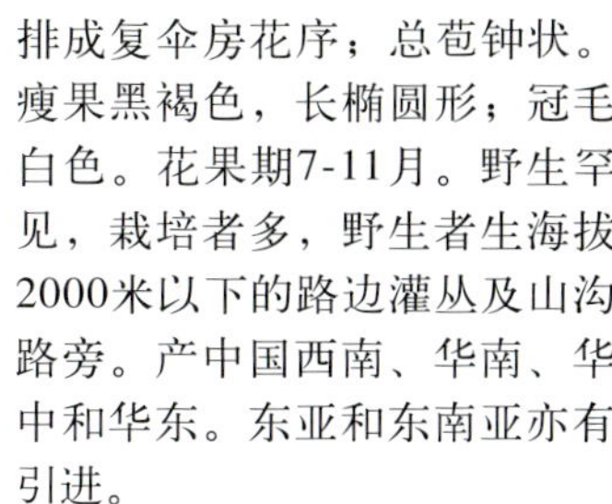
排成复伞房花序；总苞钟状。瘦果黑褐色，长椭圆形；冠毛白色。花果期7-11月。野生罕见，栽培者多，野生者生海拔2000米以下的路边灌丛及山沟路旁。产中国西南、华南、华中和华东。东亚和东南亚亦有引进。

Perennial herbs. Median stem leaves large, 3-sect or 3-partite; lower stem leaves gradually smaller. Capitula numerous in apical compound corymbs; involucre campanulate. Achenes black-brown, elliptic; pappus white. Fl. and fr. Jul-Nov. Rare as a wild plant, but commonly cultivated, usually in thickets or roadside ditches below 2000 m. Distributed in SW, S, C and E China. Introduced in E and SE Asia.

大麻叶泽兰 *Eupatorium cannabinum*

佩兰 *Eupatorium fortunei*

南川泽兰 *Eupatorium nanchuanense*

白头婆 *Eupatorium japonicum*

南川泽兰

Eupatorium nanchuanense Y. Ling et C. Shih

多年生草本。茎淡褐色、紫红色或暗紫红色。叶不规则对生；中部茎叶三全裂；上部叶三全裂或不规则三深裂；茎基部叶花期枯萎。头状花序排成复伞房花序；总苞钟状；总苞片3层，覆瓦状排列；花白色或淡红色。瘦果6棱。花果期6-7月。生海拔1200-1700米的山坡。产云南和重庆。

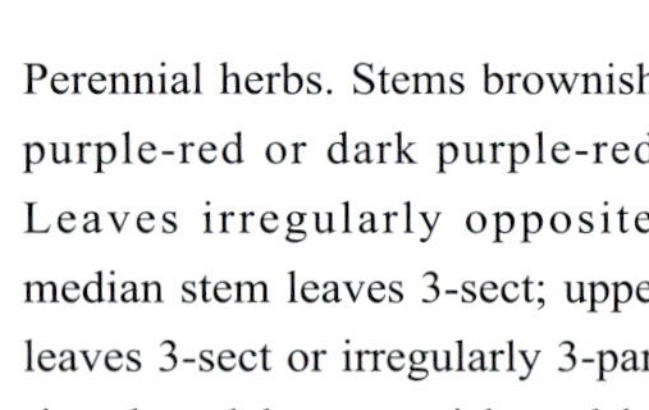

Perennial herbs. Stems brownish, purple-red or dark purple-red. Leaves irregularly opposite; median stem leaves 3-sect; upper leaves 3-sect or irregularly 3-partite; basal leaves withered by anthesis. Synflorescence terminal or compound corymbs; involucre campanulate; phyllaries 3-seriate, imbricate; corollas white or reddish. Achenes 6-angled. Fl. and fr. Jun-Jul.

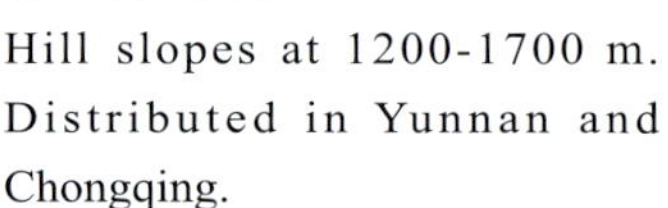

Hill slopes at 1200-1700 m. Distributed in Yunnan and Chongqing.

毛果泽兰 *Eupatorium shimadai*

毛果泽兰

Eupatorium shimadai Kitam.

多年生草本。中部茎叶卵状披针形、卵状长圆形或卵形，长8-10厘米，基部圆形或截形，顶端近尾状渐尖；全部叶基出三脉。头状花序多数在茎顶排成复伞房花序；头状花序有5朵小花；总苞钟状；总苞片2-3层，覆瓦状排列。花白色。瘦果椭圆形，被稀疏的白色长柔毛，上部的毛较密，无腺点。花果期5-6月。生草坡红岩石坡。产福建和台湾。

Perennial herbs. Middle cauline leaves ovate-lanceolate, ovate-oblong or ovate, 8-10 cm long, base rounded or truncate, apex caudate-acuminate, margin coarsely or shallowly serrate, abaxially with yellow glands, basally 3-veined. Capitula arranged in compound corymb; heads 5-flowered; involucre campanulate; bracts 2-3-seriate; corolla white. Achenes elliptic, sparsely villous, eglandular. Fl. and fr. May-Jun. Grasslands on slopes or rocky places. Distributed in Fujian and Taiwan.

白头婆

Eupatorium japonicum Thunb.

多年生草本，被白色短柔毛。茎直立，淡紫红色。叶对生，椭圆形或披针形，两面粗涩，被黄色腺点。头状花序多数，排成紧密伞房花序；总苞钟

田代氏泽兰 *Eupatorium clematideum*

状；总苞片披针形至长椭圆形，绿色或带紫红色；花冠白色或粉红色。瘦果椭圆体状；冠毛白色。花果期6-11月。生海拔100-3000米的山坡草地、林下、灌丛、湿地或河边。产中国大部分地区(中国西北除外)。朝鲜半岛和日本亦有。

Perennial herbs, shortly white pubescent. Stems erect, purplish. Leaves opposite, elliptic or lanceolate, both surfaces coarse, yellow punctuate. Capitula numerous, arranged in compound corymb; involucres campanulate; phyllaries lanceolate to narrowly elliptic, green or purple; corolla white or pink. Achenes ellipsoid; pappus white. Fl. and fr. Jun-Nov. Grassy slopes, forests, thickets, damp places or riversides at 100-3000 m. Distributed in most parts of China (except NW China). Also in Korean Peninsula and Japan.

田代氏泽兰

Eupatorium clematideum (Wall. ex DC.) Sch. Bip.

亚灌木。茎纤细，分枝，顶端具微柔毛。中部茎叶较大，长圆形至长圆状卵形，长6-9厘米，先端长渐尖，基部圆形或近截形，边缘具不明显地小尖头状锯齿，上面无毛，下面变无毛或疏被长柔毛，无腺点，具3脉。松散伞房花序具5花；苞片10-12，2轮，先端近急尖。瘦果圆柱形，长2.5毫米，无毛。生海拔0-1000米的山坡。产台湾。尼泊尔亦有。

Subshrubs. Stems slender, branched, puberulous at tip. Middle cauline leaves larger, oblong to oblong-ovate, 6-9 cm long, apex long-acuminate, base base rounded or suntruncate, margin obscurely mucronulate-serrate, adaxially glabrous, abaxially glabrate or sparsely pilose, not glandular, 3-veined. Inflorescences of lax corymbs 5-flowered; bracts 10-12, 2-seriate, apex subacute. Achenes cylindric, 2.5 mm long, glabrous. Mountain slopes at 0-1000 m. Distributed in Taiwan. Also in Nepal.

多须公

Eupatorium chinense L.

多年生草本或灌木。叶对生；茎中部叶具白色柔毛和腺点。头状花序组成大型疏松的复伞房花序；头状花序具5花；总苞钟形；苞片3层；花冠具黄色腺点。瘦果浅棕黑色，椭圆体形，具5条纵棱及黄色腺点；冠毛具白色硬毛。花果期6-11月。生海拔200-1900米的林中、灌丛或山坡草地。产中国西南、华南、东南、华中、华北、华西和华东。印度、尼泊尔、朝鲜半岛和日本亦有。

Perennial herbs or shrubs. Leaves opposite; median stem leaves white puberulent and glandular. Inflorescences terminal, of large laxly compound corymbs, 20-30 cm diam; capitula numerous, 5-flowered; involucres campanulate; phyllaries 3-seriate; corolla with yellow glands. Achenes pale black-brown, ellipsoid, 5-ribbed and yellow glandular; pappus setae white. Fl. and fr. Jun-Nov. Forests, thickets or grasslands on slopes at 200-1900 m. Distributed in SW, S, SE, C, N, W and E China. Also in India, Nepal, Korean Peninsula and Japan.

多须公 *Eupatorium chinense*

异叶泽兰 *Eupatorium heterophyllum*

异叶泽兰

Eupatorium heterophyllum DC.

多年生草本或半灌木，被白色短柔毛。茎淡褐色或紫红色。叶对生，三裂至不分裂，两面被黄色腺点，下面密被绒毛。头状花序多数，排成复伞房花序；总苞钟状；总苞片卵形至长椭圆形，紫红色；花冠白色。瘦果长椭圆体形；冠毛白色。花果期4-10月。生海拔1700-3000米的山坡、林下、林缘、草地或河谷。产中国西南。

Perennial herbs or subshrubs, shortly white pubescent. Stems brownish or purple. Leaves opposite, 3-sected to entire, both surfaces yellow punctuate, densely tomentose abaxially. Capitula numerous, arranged in compound corymb; involucres campanulate; phyllaries ovate to narrowly elliptic, purple; corolla white. Achenes narrowly ellipsoid; pappus white. Fl. and fr. Apr-Oct. Slopes, forests, forest edges, grasslands or valleys at 1700-3000 m. Distributed in SW China.

假臭草

Praxelis clematidea (Griseb.) R. M. King et H. Rob.

一年生草本，密被长柔毛。茎直立，多分枝。叶对生，具柄，卵圆形至菱形，具腺点，边缘具锯齿，先端急尖，基部圆楔形，具3脉。头状花序多数，顶生；总苞钟形；小花25-30朵，管状，蓝紫色。瘦果黑色；冠毛白色。花果期全年。生低海拔的路边、荒地、荒坡、滩涂、林地或果园。广东、广西、香港、海南、福建和台湾归化。东半球热带地区亦有。原产南美洲。

假臭草 *Praxelis clematidea*

Annual herbs, densely villose. Stems erect, multi-branched. Leaves opposite, petiolate, ovate to rhombic, glandular punctate, margin serrate, apex acute, base rotundate-cuneate, trinerved. Capitula numerous, terminal; involucres campanulate; florets 25-30, tubular, bluish-purple. Achenes black; pappus white. Fl. and fr. year-round. Roadsides, wastelands, waste slopes, beaches, woods or orchards at low altitude. Naturalized in Guangdong, Guangxi, Hong Kong, Hainan, Fujian and Taiwan. Also in tropical regions of Old World. Native to South America.

中文名索引

D

J

Q

R

S

Y

Z

拉丁学名索引

B

D

E

T

U

V

W

X

Y

Z